STOCHASTIC EQUATIONS THROUGH THE EYE OF THE PHYSICIST

Basic Concepts, Exact Results and Asymptotic Approximations

Translated from Russian
by A. Vinogradov

STOCHASTIC EQUATIONS THROUGH THE EYE OF THE PHYSICIST

Basic Concepts, Exact Results and Asymptotic Approximations

V.I. Klyatskin
Institute of Atmospheric Physics
Russian Academy of Sciences
Moscow 119017
Russia

2005

ELSEVIER

Amsterdam - Boston - Heidelberg - London - New York - Oxford - Paris
San Diego - San Francisco - Singapore - Sydney - Tokyo

ELSEVIER B.V.	ELSEVIER Inc.	ELSEVIER Ltd	ELSEVIER Ltd
Radarweg 29	525 B Street, Suite 1900	The Boulevard, Langford Lane	84 Theobalds Road
P.O. Box 211, 1000 AE	San Diego, CA 92101-4495	Kidlington, Oxford OX5 1GB	London WC1X 8RR
Amsterdam, The Netherlands	USA	UK	UK

First edition 2005

Library of Congress Cataloging in Publication Data
A catalog record is available from the Library of Congress.

British Library Cataloguing in Publication Data
A catalogue record is available from the British Library.

ISBN: 0-444-51797-9

♾ The paper used in this publication meets the requirements of ANSI/NISO Z39.48-1992 (Permanence of Paper).
Printed in The Netherlands.

To Sonya Klyatskina

Preface

The book gives the theory of stochastic equations (including ordinary differential equations, partial differential equations, boundary-value problems, and integral equations) in terms of the functional analysis. The developed approach yields exact solutions to stochastic problems for a number of models of fluctuating parameters among which are telegrapher's and generalized telegrapher's processes, Markovian processes with a finite number of states, Gaussian Markovian processes, and functions of the above processes. Asymptotic methods of analyzing stochastic dynamic systems, such as delta-correlated random process (field) approximation and diffusion approximation are also considered. These methods are used to describe the coherent phenomena in stochastic systems (particle and passive tracer clustering in random velocity field, dynamic localization of plane waves in randomly layered media, and caustic structure formation in multidimensional random media).

The book is destined for scientists dealing with stochastic dynamic systems in different areas, such as hydrodynamics, acoustics, radio wave physics, theoretical and mathematical physics, and applied mathematics, and can be useful for senior and postgraduate students.

Now, a few words are due on the structure of the text. The book is in five parts.

The first part may be viewed as an introductory text. It takes up a few typical physical problems to discuss their solutions obtained under random perturbations of parameters affecting the system behavior. More detailed formulations of these problems and relevant statistical analysis may be found in other parts of the book.

The second part is devoted to the general theory of statistical analysis of dynamic systems with fluctuating parameters described by differential and integral equations. This theory is illustrated by analyzing specific dynamic systems.

The third part treats asymptotic methods of statistical analysis such as the delta-correlated random process (field) approximation and diffusion approximation.

The fourth part deals with analysis of specific physical problems associated with coherent phenomena. These are clustering and diffusion of particles and passive ingredients in a random velocity field, dynamic localization of plane waves propagating in layered random media, and formation of caustics by waves propagating in random multidimensional media. These phenomena are described by ordinary differential equations and partial differential equations. Each of these formulations splits into many separate problems of individual physical interest.

In order to avoid crowding the book by mathematical niceties, it is appended by the fifth part that consists of three appendixes presenting detailed derivations of some mathematical expressions used in the text. Specifically, they give a definition and some rules to calculate variational derivatives; they discuss the properties of wavefield factorization in a homogeneous space and in layered media which drastically simplify analysis of statistical problems. In these appendixes, we also discuss a derivation of the method of imbedding that offers a possibility of reformulating boundary-value wave problems into initial value problems with respect to auxiliary variables.

It is worth noting that purely mathematical and physical papers devoted to considered issues run into thousands. It would be physically impossible to give an exhaustive bibliography. Therefore, in this book we confine ourselves to referencing those papers which are used or discussed in this book and also recent review papers and with extensive bibliography on the subject.

V. I. Klyatskin

Moscow

Contents

Mephistopheles (about algebra)

. . .
I want to warn you:
    careful you should
Be with this science
    which is very tricky.
It can involve your brain into the chaos
    of vain unnecessary transformations.
If you don't manage
    to embrace its basis,
You won't be able indexes to differ.
. . .
What's important -
    deepen into symbols
And, having mastered them,
    you bravely step
Onto the path, which leads into the Kingdom
of Formulas, eternal and exact.

Kurd Lasswitz, Prost!,
www. gutenberg2000.de/lasswitz/prost/prost.htm
Translated by Alla Parollo.

Introduction

Different areas of physics pose statistical problems in ever-greater numbers. Apart from issues traditionally obtained in statistical physics, many applications call for including fluctuation effects into consideration. While fluctuations may stem from different sources (such as thermal noise, instability, and turbulence), methods used to treat them are very similar. In many cases, the statistical nature of fluctuations may be deemed known (either from physical considerations or from problem formulation) and the physical processes may be modeled by differential, integro-differential or integral equations.

Today the most powerful tools used to tackle complicated statistical problems are the *Markov theory of random processes* and the theory of *diffusion type processes* evolved from *Brownian motion theory*. Mathematical aspects underlying these theories and their applications have been treated extensively in academic literature and textbooks ([63]), and therefore we will not dwell on these issues in this treatise.

We will consider a statistical theory of dynamic and wave systems with fluctuating parameters. These systems can be described by ordinary differential equations, partial differential equations, integro-differential equations and integral equations. A popular way to solve such systems is by obtaining a closed system of equations for statistical characteristics of such systems to study their solutions as comprehensively as possible.

We note that often wave problems are boundary-value problems. When this is the case, one may resort to the imbedding method to reformulate the equations at hand to initial-value problems, thus considerably simplifying the statistical analysis [136].

We shall dwell in depth on dynamic systems whose fluctuating parameters are Gaussian random processes (fields), although what we present in this book is a general theory valid for fluctuating parameters of any nature.

The purpose of this book is to demonstrate how different physical problems described by stochastic equations may be solved on the base of a general approach. This treatment reveals interesting similarities between different physical problems.

Examples of specific physical systems outlined below are mainly borrowed from statistical hydrodynamics, statistical radio wave physics and acoustics because of author's research in these fields. However, similar problems and solution techniques occur in such areas as plasma physics, solid-state physics, magnetofluid dynamics to name a few.

In stochastic problems with fluctuating parameters, the variables are functions. It would be natural therefore to resort to functional methods for their analysis. We will use a *functional method* devised by Novikov [255] for Gaussian fluctuations of parameters in a turbulence theory and developed by the author of this book [132], [134]–[136] for the general case of dynamic systems and fluctuating parameters of arbitrary nature.

However, only a few dynamic systems lend themselves to analysis yielding solutions in a general form. It proved to be more efficient to use an asymptotic method where the statistical characteristics of dynamic problem solutions are expanded in powers of a small parameter which is essentially a ratio of the random impact's correlation time to the time of observation or to other characteristic time scale of the problem (in some cases, these may be spatial rather than temporal scales). This method is essentially a generalization of the theory of Brownian motion. It is termed the *delta-correlated random process (field) approximation*. In Brownian motion theory, this approximation is consistent with a model obtained by neglecting the time between random collisions as compared to all other time scales.

For dynamic systems described by ordinary differential stochastic equations with Gaussian fluctuations of parameters, this method leads to a Markovian problem solving model, and the respective equation for transition probability density has the form of the *Fokker–Planck equation*. In this book, we will consider in depth the methods of analysis available for this equation and its boundary conditions. We will analyze solutions and validity conditions by way of integral transformations. In more complicated problems described by partial differential equations, this method leads to a generalized equation of Fokker–Planck type in which variables are the derivatives of the solution's characteristic functional. For dynamic problems with non-Gaussian fluctuations of parameters, this method also yields Markovian type solutions. Under the circumstances, the probability density of respective dynamic stochastic equations satisfies a closed operator equation. For example, systems with parameters fluctuating in a Poisson profile are converted into the *Kolmogorov–Feller* type of integro-differential equations.

In physical investigations, Fokker–Planck and similar equations are usually set up from rule of thumb considerations, and dynamic equations are invoked only to calculate the coefficients of these equations. This approach is inconsistent, generally speaking. Indeed, the statistical problem is completely defined by dynamic equations and assumptions on the statistics of random impacts. For example, the Fokker–Planck equation must be a logical sequence of the dynamic equations and some assumptions on the character of random impacts. It is clear that not all problems lend themselves for reducing to a Fokker–Planck equation. The functional approach allows one to derive a Fokker–Planck equation from the problem's dynamic equation along with its applicability conditions.

For a certain class of random processes (Markovian telegrapher's processes, Gaussian Markovian process and the like), the developed functional approach also yields closed equations for the solution probability density with allowance for a finite correlation time of random interactions.

For processes with Gaussian fluctuations of parameters, one may construct a better physical approximation than the delta-correlated random process (field) approximation, — the *diffusion approximation* that allows for finiteness of correlation time radius. In this approximation, the solution is Markovian and its applicability condition has transparent physical meaning, namely, the statistical effects should be small within the correlation time of fluctuating parameters. This book treats these issues in depth from a general standpoint and for some specific physical applications.

In recent time, the interest of both theoreticians and experimenters has been attracted to relation of the behavior of average statistical characteristics of a problem solution with the behavior of the solution in certain happenings (realizations). This is especially important for geophysical problems related to the atmosphere and ocean where, generally speaking, a respective averaging ensemble is absent and experimenters, as a rule, have to do with individual observations.

Seeking solutions to dynamic problems for these specific realizations of medium parameters is almost hopeless due to extreme mathematical complexity of these problems.

At the same time, researchers are interested in main characteristics of these phenomena without much need to know specific details. Therefore, the idea to use a well developed approach to random processes and fields based on ensemble averages rather than separate observations proved to be very fruitful. By way of example, almost all physical problems of atmosphere and ocean to some extent are treated by statistical analysis.

Randomness in medium parameters gives rise to a stochastic behavior of physical fields. Individual samples of scalar two-dimensional fields $\rho(\mathbf{R}, t)$, $\mathbf{R} = (x, y)$, say, recall a rough mountainous terrain with randomly scattered peaks, troughs, ridges and saddles. Common methods of statistical averaging (computing mean-type averages — $\langle\rho(\mathbf{R}, t)\rangle$, space-time correlation function — $\langle\rho(\mathbf{R}, t)\,\rho(\mathbf{R}', t')\rangle$ etc., where $\langle...\rangle$ implies averaging over an ensemble of random parameter samples) smooth the qualitative features of specific samples. Frequently, these statistical characteristics have nothing in common with the behavior of specific samples, and at first glance may even seem to be at variance with them. For example, the statistical averaging over all observations makes the field of average concentration of a passive tracer in a random velocity field ever more smooth, whereas each its realization sample tends to be more irregular in space due to mixture of areas with substantially different concentrations.

Thus, these types of statistical average usually characterize 'global' space-time dimensions of the area with stochastic processes but tell no details about the process behavior inside the area. For this case, details heavily depend on the velocity field pattern, specifically, on whether it is divergent or solenoidal. Thus, the first case will show with the total probability that *clusters* will be formed, i.e. compact areas of enhanced concentration of tracer surrounded by vast areas of low-concentration tracer. In the circumstances, all statistical moments of the distance between the particles will grow with time exponentially; that is, on average, a statistical recession of particles will take place.

In a similar way, in case of waves propagating in random media, an exponential spread of the rays will take place on average; but simultaneously, with the total probability, *caustics* will form at finite distances. One more example to illustrate this point is the *dynamic localization* of plane waves in layered randomly inhomogeneous media. In this phenomenon, the wavefield intensity exponentially decays inward the medium with the probability equal to unity when the wave is incident on the half-space of such a medium, while all statistical moments increase exponentially with distance from the boundary of the medium.

These physical processes and phenomena occurring with the probability equal to unity will be referred to as *coherent* processes and phenomena [157]. This type of *statistical coherence* may be viewed as some organization of the complex dynamic system, and retrieval of its *statistically stable characteristics* is similar to the concept of *coherence* as *self-organization* of multicomponent systems that evolve from the random interactions of their elements [254]. In the general case, it is rather difficult to say whether or not the phenomenon occurs with the probability equal to unity. However, for a number of applications amenable to treatment with the simple models of fluctuating parameters, this may be handled by analytical means. In other cases, one may verify this by performing numerical modeling experiments or analyzing experimental findings.

The complete statistic (say, the whole body of all n-point space-time moment functions), would undoubtedly contain all the information about the investigated dynamic system. In practice, however, one may succeed only in studying the simplest statistical characteristics associated mainly with one-time and one-point probability distributions. It would be reasonable to ask how with these statistics on hand one would look into the quantitative and qualitative behavior of some system happenings?

This question is answered by *methods of statistical topography.* These methods were highlighted by [319], who seems to had coined this term. Statistical topography yields a different philosophy of statistical analysis of dynamic stochastic systems, which may prove useful for experimenters planning a statistical processing of experimental data. These issues are treated in depths in this book.

Part I

Dynamical description of stochastic systems

Chapter 1

Examples, basic problems, peculiar features of solutions

In this chapter, we consider several dynamic systems described by differential equations of different types and discuss the features in the behaviors of solutions to these equations under random disturbances of parameters. Here, we content ourselves with the problems in the simplest formulation. More complete formulations will be discussed below, in the sections dealing with statistical analysis of corresponding systems.

1.1 Ordinary differential equations: initial value problems

1.1.1 Particle under random velocity field

In the simplest case, a particle under random velocity field is described by the system of ordinary differential equations of the first order

$$\frac{d}{dt}\mathbf{r}(t) = \mathbf{U}(\mathbf{r}, t), \quad \mathbf{r}(t_0) = \mathbf{r}_0, \tag{1.1}$$

where $\mathbf{U}(\mathbf{r}, t) = \mathbf{u}_0(\mathbf{r}, t) + \mathbf{u}(\mathbf{r}, t)$, $\mathbf{u}_0(\mathbf{r}, t)$ is the deterministic component of the velocity field (mean flow), and $\mathbf{u}(\mathbf{r}, t)$ is the random component. In the general case, field $\mathbf{u}(\mathbf{r}, t)$ can have both divergence-free (solenoidal, for which $\operatorname{div}\mathbf{u}(\mathbf{r}, t) = 0$) and divergent (for which $\operatorname{div}\mathbf{u}(\mathbf{r}, t) \neq 0$) components.

1.1.2 Particles under random velocity field

We dwell on stochastic features of the solution to problem (1.1) for a system of particles in the absence of mean flow ($\mathbf{u}_0(\mathbf{r}, t) = 0$). From Eq. (1.1) formally follows that every particle moves independently of other particles. However, if random field $\mathbf{u}(\mathbf{r}, t)$ has a finite spatial correlation radius l_{cor}, particles spaced by a distance shorter than l_{cor} appear in the common zone of infection of random field $\mathbf{u}(\mathbf{r}, t)$ and the behavior of such a system can show new collective features.

For steady velocity field $\mathbf{u}(\mathbf{r}, t) \equiv \mathbf{u}(\mathbf{r})$, Eq. (1.1) reduces to

$$\frac{d}{dt}\mathbf{r}(t) = \mathbf{u}(\mathbf{r}), \quad \mathbf{r}(0) = \mathbf{r}_0. \tag{1.2}$$

This equation clearly shows that steady points $\tilde{\mathbf{r}}$ (at which $\mathbf{u}(\tilde{\mathbf{r}}) = 0$) remain the fixed points. Depending on whether these points are stable or unstable, they will attract or repel nearby particles. In view of randomness of function $\mathbf{u}(\mathbf{r})$, points $\tilde{\mathbf{r}}$ are random too.

It is expected that the similar behavior will be also characteristic of the general case of the space-time random velocity field of $\mathbf{u}(\mathbf{r},t)$.

If some points $\tilde{\mathbf{r}}$ remain stable during sufficiently long time, then clusters of particles (i.e., compact regions with elevated particle concentration, which occur merely in rarefied zones) must arise around these points in separate realizations of random field $\mathbf{u}(\mathbf{r}, t)$. On the contrary, if the stability of these points alternates with instability sufficiently rapidly and particles have no time for significant rearrangement, no clusters of particles will occur.

Simulations (see [198, 271, 320]) show that the behavior of a system of particles essentially depends on whether the random field of velocities is divergence-free or divergent. By way of example, Fig.1.1*a* shows a schematic of the evolution of the two-dimensional system of particles uniformly distributed within the circle for a particular realization of the divergence-free steady field $\mathbf{u}(\mathbf{r})$.

Here, we use the dimensionless time related to statistical parameters of field $\mathbf{u}(\mathbf{r})$. In this case, the area of surface patch within the contour remains intact and particles relatively uniformly fill the region within the deformed contour. The only feature consists in the fractal-type irregularity of the deformed contour. On the contrary, in the case of the divergent field of velocities $\mathbf{u}(\mathbf{r})$, particles uniformly distributed in the square at the initial instance will form clusters during the temporal evolution. Results simulated for this case are shown in Fig. 1.1*b*. We emphasize that the formation of clusters is purely the kinematic effect. This feature of particle dynamics disappears on averaging over an ensemble of realizations of the random velocity field .

To demonstrate the process of particle clustering, we consider the simplest problem [161], in which the random velocity field $\mathbf{u}(\mathbf{r}, t)$ has the form

$$\mathbf{u}(\mathbf{r}, t) = \mathbf{v}(t) f(\mathbf{r}), \tag{1.3}$$

where $\mathbf{v}(t)$ is the random vector process and

$$f(\mathbf{r}) = \sin 2(\mathbf{kr}) \tag{1.4}$$

is the deterministic function of one variable. Note that this form of function $f(\mathbf{r})$ corresponds to the first term of the expansion in harmonic components and is commonly used in numerical simulations [198, 320].

In this case, Eq. (1.1) can be written in the form

$$\frac{d}{dt}\mathbf{r}(t) = \mathbf{v}(t) \sin 2(\mathbf{kr}), \quad \mathbf{r}(0) = \mathbf{r}_0.$$

In the context of this model, motions of a particle along vector $\mathbf{k}$ and in the plane perpendicular to vector $\mathbf{k}$ are independent and can be separated. If we direct the x-axis along vector $\mathbf{k}$, then the equations assume the form

$$\begin{aligned} \frac{d}{dt}x(t) &= v_x(t) \sin(2kx), \quad x(0) = x_0, \\ \frac{d}{dt}\mathbf{R}(t) &= \mathbf{v}_{\mathbf{R}}(t) \sin(2kx), \quad \mathbf{R}(0) = \mathbf{R}_0. \end{aligned} \tag{1.5}$$

The solution of the first equation in (1.5) is

$$x(t) = \frac{1}{k} \arctan \left[e^{T(t)} \tan(kx_0) \right], \tag{1.6}$$

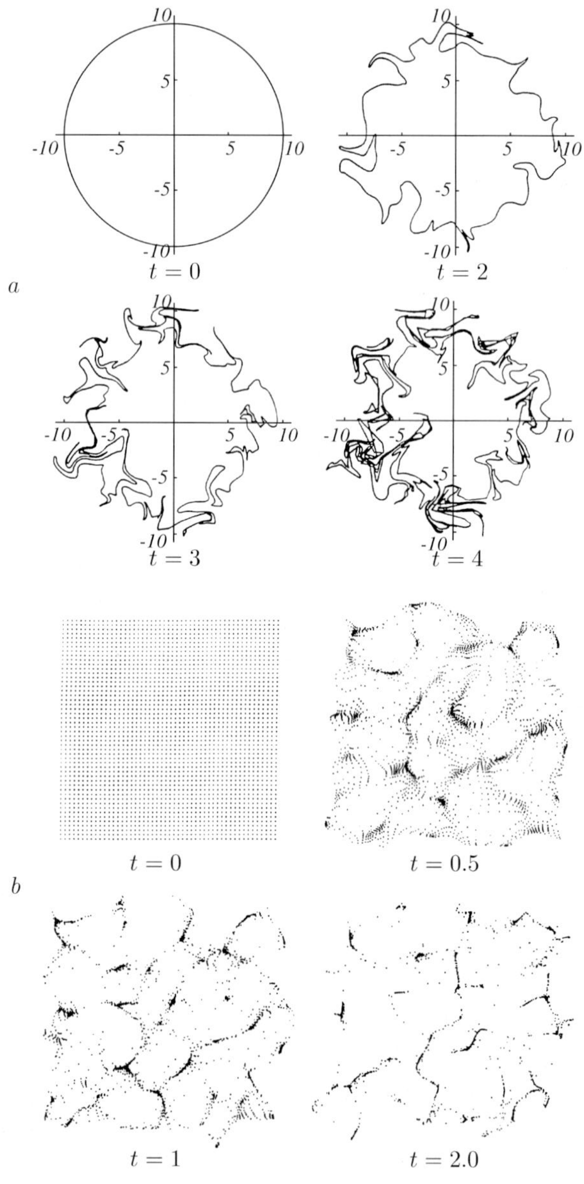

Figure 1.1: Diffusion of a system of particles described by Eqs. (1.2) numerically simulated for (*a*) solenoidal and (*b*) divergence-free random steady velocity field $\mathbf{u}(\mathbf{r})$.

where

$$T(t) = 2k \int_0^t d\tau v_x(\tau). \tag{1.7}$$

Taking into account the equality following from (1.6)

$$\sin(2kx) = \sin(2kx_0)\frac{1}{e^{-T(t)}\cos^2(kx_0) + e^{T(t)}\sin^2(kx_0)},$$

we can write the second equation in (1.5) in the form

$$\frac{d}{dt}\mathbf{R}(t|\mathbf{r}_0) = \sin(2kx_0)\frac{\mathbf{v_R}(t)}{e^{-T(t)}\cos^2(kx_0) + e^{T(t)}\sin^2(kx_0)}.$$

As a result, we have

$$\mathbf{R}(t|\mathbf{r}_0) = \mathbf{R}_0 + \sin(2kx_0)\int_0^t d\tau \frac{\mathbf{v_R}(\tau)}{e^{-T(\tau)}\cos^2(kx_0) + e^{T(\tau)}\sin^2(kx_0)}. \tag{1.8}$$

Consequently, if the initial particle position x_0 is such that

$$kx_0 = n\frac{\pi}{2}, \tag{1.9}$$

where $n = 0, \pm 1, ...,$ then the particle will be the fixed particle and $\mathbf{r}(t) \equiv \mathbf{r}_0$.

Equalities (1.9) define planes in the general case and points in the one-dimensional case. They correspond to zeros of the field of velocities. Stability of these points depends on the sign of function $\mathbf{v}(t)$, and this sign changes during the evolution process. As a result, we can expect that particles will be concentrated around these points if $v_x(t) \neq 0$, which just corresponds to clustering of particles.

In the case of a divergence-free velocity field, $v_x(t) = 0$ and, consequently, $T(t) \equiv 0$; as a result, we have

$$x(t|x_0) \equiv x_0, \quad \mathbf{R}(t|\mathbf{r}_0) = \mathbf{R}_0 + \sin 2(kx_0)\int_0^t d\tau \mathbf{v_R}(\tau),$$

which means that no clustering occurs.

Figure 1.2*a* shows a fragment of the realization of random process $T(t)$ obtained by numerical integration of Eq. (1.7) for a realization of random process $v_x(t)$; we used this fragment for simulating the temporal evolution of coordinates of four particles $x(t)$, $x \in (0, \pi/2)$ initially located at coordinates $x_0(i) = \frac{\pi}{2}\frac{i}{5}$ $(i = 1, 2, 3, 4)$ (see Fig. 1.2*b*). Figure 1.2*b* shows that particles form a cluster in the vicinity of point $x = 0$ at the dimensionless time $t \approx 4$. Further, at time $t \approx 16$ the initial cluster disappears and new one appears in the vicinity of point $x = \pi/2$. At moment $t \approx 40$, the cluster appears again in the vicinity of point $x = 0$, and so on. In this process, particles in clusters remember their past history and significantly diverge during intermediate temporal segments (see Fig. 1.2*c*).

Thus, we see that, in this example, the cluster does not move from one region to another; instead, it first collapses and then a new cluster is formed. Moreover, the lifetime of clusters significantly exceeds the duration of intermediate segments. It seems that this feature is characteristic of the particular model of the velocity field and follows from stationarity of points (1.9).

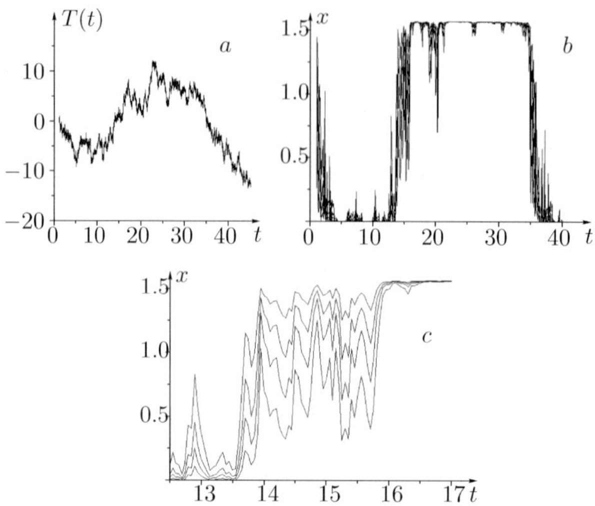

Figure 1.2: (a) Segment of a realization of random process $T(t)$ obtained by numerically integrating Eq. (1.7) for a realization of random process $v_x(t)$; (b), (c) x-coordinates simulated with this segment for four particles versus time.

As regards the particle diffusion along the y-direction, no cluster occurs in this direction.

Note that such clustering in a system of particles was found, to all appearance for the first time, in papers [243] – [246] as a result of simulating the so-called Eole experiment with the use of the simplest equations of atmospheric dynamics. In the context of this global experiment, 500 constant-density balloons were launched in Argentina in 1970-1971; these balloons traveled at a height of about 12 km and spread along the whole of the southern hemisphere.

Figure 1.3 shows the balloon distribution over the southern hemisphere for day 105 from the beginning of this process simulation [245]; this distribution clearly shows that balloons are concentrated in groups, which just corresponds to clustering. Results of statistical processing of balloon arrangement can be found, for example, in papers [61, 249].

Now, we dwell on another stochastic aspect related to dynamic equations of type (1.1); namely, we consider the *phenomenon of transfer* caused by random fluctuations.

Consider the one-dimensional nonlinear equation

$$\frac{d}{dt}x(t) = x\left(1 - x^2\right) + f(t), \quad x(0) = x_0; \ \lambda > 0, \tag{1.10}$$

where $f(t)$ is the random function of time. In the absence of randomness ($f(t) \equiv 0$), the solution of Eq. (1.10) has two stable steady states $x = \pm 1$ and one instable state $x = 0$. Depending on the initial value, solution of Eq. (1.10) arrives at one of the stable states. However, in the presence of small random disturbances $f(t)$, dynamic system (1.10) will first approach the vicinity of one of the stable states and then, after the lapse of certain time, it will be transferred into the vicinity of another stable state.

It is clear that the similar behavior can occur in more complicated situations.

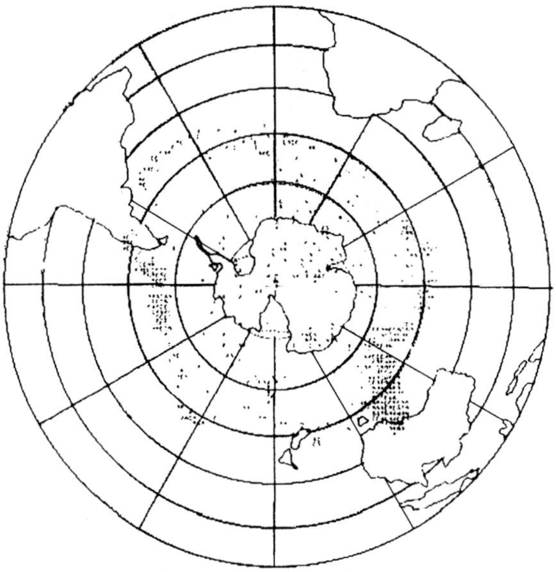

Figure 1.3: Balloon distribution in the atmosphere for day 105 from the beginning of process simulation.

As an example, consider the simplest hydrodynamic system described by the stochastic system of equations (see, e.g., [58])

$$\begin{aligned} \frac{d}{dt}v_0(t) &= v_2^2(t) - v_1^2(t) - v_0(t) + R + f_0(t), \\ \frac{d}{dt}v_1(t) &= v_0(t)v_1(t) - v_1(t) + f_1(t), \\ \frac{d}{dt}v_2(t) &= -v_0(t)v_2(t) - v_2(t) + f_2(t). \end{aligned} \tag{1.11}$$

This system describes the motion of a triplet (gyroscope) with the linear isotropic friction, which is driven by the force acting on the instable mode and having both regular (R) and random ($\mathbf{f}(t)$) components. Such a situation occurs, for example, for a liquid moving in the ellipsoidal cavity.

In the absence of random components ($\mathbf{f}(t) = 0$), dynamic system (1.11) has steady-state solutions depending on parameter R (an analog to the Reynolds number). In this problem, the critical value is $R_{\text{cr}} = 1$.

For $R < 1$, the system has the stable steady-state solution

$$v_1 = v_2 = 0, \quad v_0 = R.$$

For $R > 1$, this solution becomes instable with respect to small disturbances of parameters, and new steady mode

$$v_0 = 1, \quad v_2 = 0, \quad v_1 = \pm\sqrt{R-1}$$

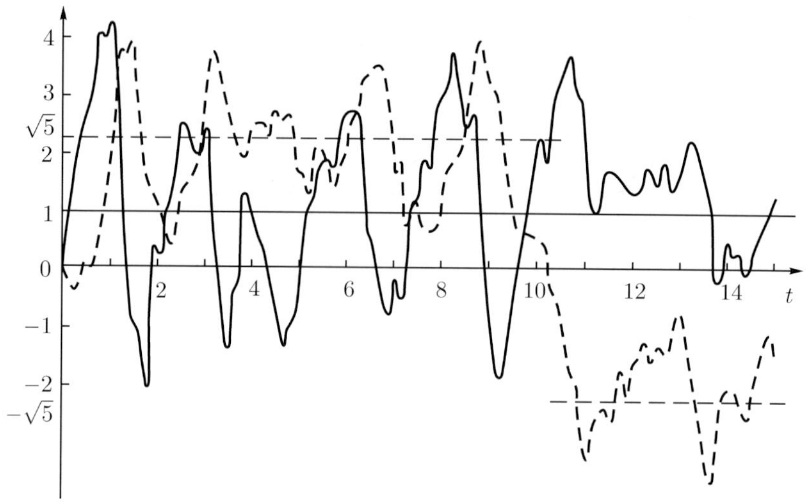

Figure 1.4: Phenomenon of transfer illustrated by simulating the system of equations (1.11) for $R = 6$ and $\sigma^2 = 0.1$ (the solid and dashed lines show components $v_0(t)$ and $v_1(t)$, respectively).

sets in. Some element of chance appears here, because parameter v_1 can be either positive or negative, depending on the amplitude of small disturbances.

In the presence of random actions, dynamic system (1.11) for $R > 1$ will first approach one of the stable states and then, after the lapse of certain time, it will be transferred to the vicinity of another stable state. Figure 1.4 shows the results simulated for this phenomenon with $R = 6$ for different realizations of random force $\mathbf{f}(t)$ whose components were modeled as Gaussian random processes.

1.1.3 Particles under random forces

The system of equations (1.1) describes also the behavior of a particle under the field of random external forces $\mathbf{f}(\mathbf{r}, t)$. In the simplest case, the behavior of a particle in the presence of linear friction is described by the system of the first-order differential equations

$$\begin{aligned} \frac{d}{dt}\mathbf{r}(t) &= \mathbf{v}(t), \quad \frac{d}{dt}\mathbf{v}(t) = -\lambda\mathbf{v}(t) + \mathbf{f}(\mathbf{r}, t), \\ \mathbf{r}(0) &= \mathbf{r}_0, \quad \mathbf{v}(0) = \mathbf{v}_0. \end{aligned} \tag{1.12}$$

The behavior of a particle under the deterministic potential field in the presence of linear friction and random forces is described by the system of equations

$$\begin{aligned} \frac{d}{dt}\mathbf{r}(t) &= \mathbf{v}(t), \quad \frac{d}{dt}\mathbf{v}(t) = -\lambda\mathbf{v}(t) - \frac{\partial U(\mathbf{r}, t)}{\partial \mathbf{r}} + \mathbf{f}(\mathbf{r}, t), \\ \mathbf{r}(0) &= \mathbf{r}_0, \quad \mathbf{v}(0) = \mathbf{v}_0, \end{aligned} \tag{1.13}$$

which is the simplest example of *Hamiltonian systems*. In statistical problems, equations of type (1.12), (1.13) are widely used for describing the *Brownian motion* of particles.

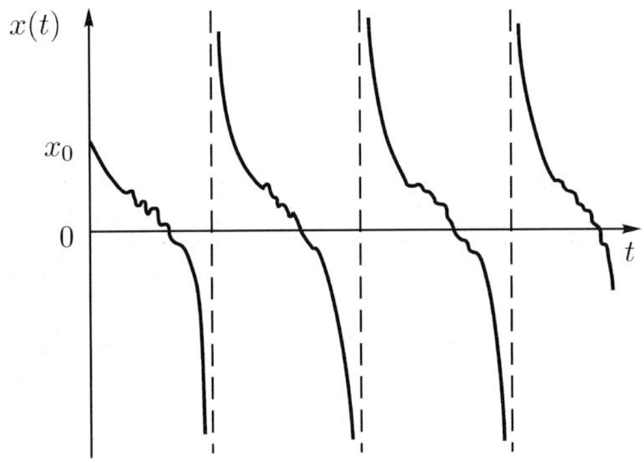

Figure 1.5: Typical realization of the solution to Eq. (1.14).

1.1.4 Systems with blow-up singularities

The simplest stochastic system showing singular behavior in time is described by the following equation commonly used in the statistical theory of waves

$$\frac{d}{dt}x(t) = -\lambda x^2(t) + f(t), \quad x(0) = x_0, \quad \lambda > 0, \tag{1.14}$$

where $f(t)$ is the random function of time.

In the absence of randomness ($f(t) = 0$), the solution to Eq.(1.14) has the form

$$x(t) = \frac{1}{\lambda (t - t_0)}, \quad t_0 = -\frac{1}{\lambda x_0}.$$

For $x_0 > 0$, we have $t_0 < 0$, and solution $x(t)$ monotonically tends to zero with increasing time. On the contrary, for $x_0 < 0$, solution $x(t)$ reaches $-\infty$ within a finite time $t_0 = -1/\lambda x_0$, which means that the solution becomes *singular* and shows the *blow-up behavior*. In this case, random force $f(t)$ has insignificant effect on the behavior of the system. The effect becomes significant only for positive parameter x_0. Here, the solution, slightly fluctuating, decreases with time as long as it remains positive. On reaching sufficiently small value $x(t)$, the force $f(t)$ *transfers* the solution into the region of negative values of x, where it will reach the value of $-\infty$ within a certain finite time.

Thus, in the stochastic case, the solution to problem (1.14) shows the blow-up behavior for arbitrary values of parameter x_0 and always reaches $-\infty$ within the finite time t_0. Figure 1.5 schematically shows the temporal realization of the solution $x(t)$ to problem(1.14) for $t > t_0$; its behavior resembles the *quasi-periodic* structure.

1.1.5 Oscillator with randomly varying frequency (stochastic parametric resonance)

In the above stochastic examples, we considered the effect of additive random impacts (forces) on the behavior of systems. The simplest nontrivial system with multiplicative (parametric) impact can be illustrated using the *stochastic parametric resonance* as an

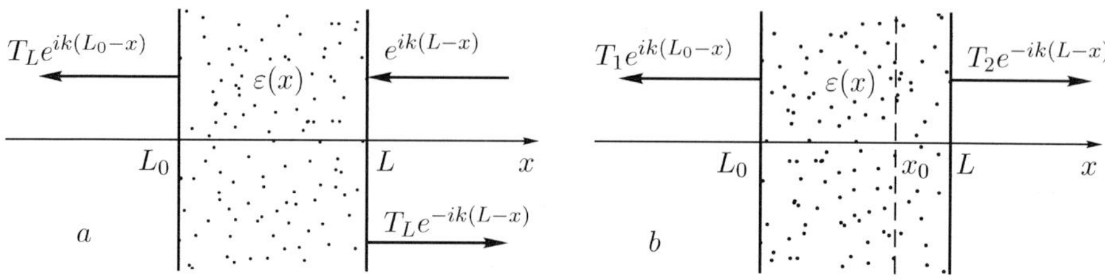

Figure 1.6: (*a*) Plane wave incident on the medium layer and (*b*) source inside the medium layer.

example. Such a system is described by the second-order equation

$$\frac{d^2}{dt^2}x(t) + \omega_0^2[1 + z(t)]x(t) = 0,$$
$$x(0) = x_0, \quad \frac{d}{dt}x(0) = v_0, \tag{1.15}$$

where $z(t)$ is the random function of time. This equation is characteristic of almost all fields of physics. Physically, it is obvious that dynamic system (1.15) is capable of parametric excitation, because random process $z(t)$ has harmonic components of all frequencies, including frequencies $2\omega_0/n\,(n = 1, 2, ...)$ that exactly correspond to the frequencies of parametric resonance in the system with periodic function $z(t)$, as it is the case, for example, in the *Mathieu equation*.

1.2 Linear ordinary differential equations: boundary-value problems

In the previous section, we considered several dynamic systems described by a system of ordinary differential equations with given initial values. Now, we consider the simplest linear boundary-value problem, namely, the stationary one-dimensional wave problem.

1.2.1 Plane waves in layered media: a wave incident on a medium layer

Let the layer of inhomogeneous medium occupies the segment of space $L_0 < x < L$ and let the unit-amplitude plane wave $u_0(x) = e^{-ik(x-L)}$ is incident on this layer from the region $x > L$ (Fig. 1.6*a*). The wavefield satisfies the *Helmholtz equation*,

$$\frac{d^2}{dx^2}u(x) + k^2(x)u(x) = 0, \tag{1.16}$$

where

$$k^2(x) = k^2[1 + \varepsilon(x)]$$

and function $\varepsilon(x)$ describes medium inhomogeneities. We assume that $\varepsilon(x) = 0$, i.e., $k(x) = k$ outside the layer; inside the layer, we set $\varepsilon(x) = \varepsilon_1(x) + i\gamma$, where $\varepsilon_1(x)$ is the real part responsible for wave scattering in the medium and the imaginary part $\gamma \ll 1$ describes the absorption of the wave in the medium.

In region $x > L$, the wavefield has the following structure

$$u(x) = e^{-ik(x-L)} + R_L e^{ik(x-L)},$$

where R_L is the complex reflection coefficient. In region $x < L_0$, the structure of the wavefield is

$$u(x) = T_L e^{ik(L_0-x)},$$

where T_L is the complex transmission coefficient. Boundary conditions for Eq. (1.16) are the continuity conditions for the field and the field derivative at the layer boundaries; they can be written as follows

$$u(L) + \frac{i}{k}\frac{du(x)}{dx}\bigg|_{x=L} = 2, \qquad u(L_0) - \frac{i}{k}\frac{du(x)}{dx}\bigg|_{x=L_0} = 0. \tag{1.17}$$

Thus, the wavefield in the layer of an inhomogeneous medium is described by the boundary-value problem (1.16), (1.17). Dynamic equation (1.16) coincides in form with Eq. (1.15). Note that the problem under consideration assumes that function $\varepsilon(x)$ is discontinuous at the layer boundaries. We will call the boundary-value problem (1.16), (1.17) the unmatched boundary-value problem. In such problems, wave scattering is caused not only by medium inhomogeneities, but also by discontinuities of function $\varepsilon(x)$ at layer boundaries.

If medium parameters (function $\varepsilon_1(x)$) are specified in the statistical form, then solving the stochastic problem (1.16), (1.17) consists in obtaining statistical characteristics of the reflection and transmission coefficients, which are related to the wavefield values at the layer boundaries by the relationships

$$R_L = u(L) - 1, \quad T_L = u(L_0),$$

and the wavefield intensity

$$I(x) = |u(x)|^2$$

inside the inhomogeneous medium. Determination of these characteristics constitutes the subject of the *statistical theory of radiative transfer.*

Note that, for $x < L$, from Eq. (1.16) follows the equality

$$k\gamma I(x) = \frac{d}{dx}S(x),$$

where energy-flux density $S(x)$ is determined by the relationship

$$S(x) = \frac{i}{2k}\left[u(x)\frac{d}{dx}u^*(x) - u^*(x)\frac{d}{dx}u(x)\right].$$

By virtue of boundary conditions, we have $S(L) = 1 - |R_L|^2$ and $S(L_0) = |T_L|^2$.

For non-absorptive media ($\gamma = 0$), conservation of energy-flux density is expressed by the equality

$$|R_L|^2 + |T_L|^2 = 1. \tag{1.18}$$

Consider some features characteristic of solutions to the stochastic boundary-value problem (1.16), (1.17). On the assumption that medium inhomogeneities are absent ($\varepsilon_1(x) = 0$) and absorption γ is sufficiently small, the intensity of the wavefield in the medium slowly decays with distance according to the exponential law

$$I(x) = |u(x)|^2 = e^{-k\gamma(L-x)}. \tag{1.19}$$

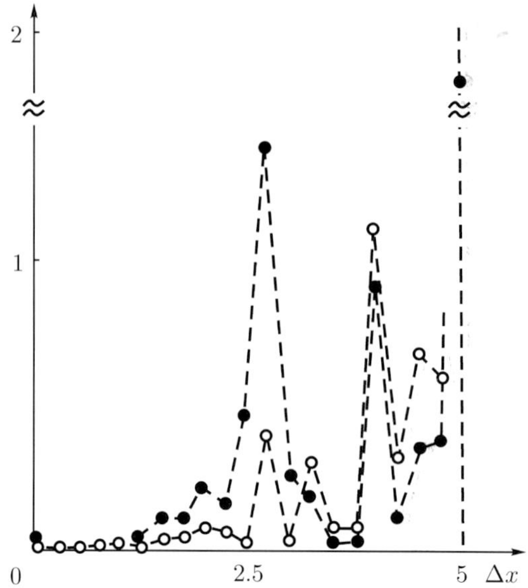

Figure 1.7: Dynamic localization phenomenon simulated for two realizations of medium inhomogeneities.

Figure 1.7 shows two realizations of the intensity of a wave in a sufficiently thick layer of medium. These realizations were simulated for two realizations of medium inhomogeneities [312]. Omitting the detailed description of problem parameters, we mention only that this figure clearly shows the prominent tendency of a sharp exponential decay (accompanied by significant spikes toward both higher and nearly zero-valued intensity values), which is caused by multiple reflections of the wave in the chaotically inhomogeneous random medium (the phenomenon of *dynamic localization*). Recall that absorption is small ($\gamma \ll 1$), so that it cannot significantly affect the dynamic localization.

It is well known that the introduction of the new function

$$\psi(x) = \frac{i}{k}\frac{d}{dx}\ln u(x)$$

reduces the second-order equation (1.16) to two first-order equations, and this function satisfies the closed equation following from Eq. (1.16):

$$\frac{d}{dx}\psi(x) = ik\left[\psi^2(x) - 1 - \varepsilon(x)\right], \quad \psi(L_0) = 1. \tag{1.20}$$

From the condition at boundary $x = L$ follows that

$$u(L) = \frac{2}{1 + \psi(L)}$$

and, consequently, the reflection coefficient is determined from the solution to Eq. (1.20) by the formula

$$R_L = \frac{1 - \psi(L)}{1 + \psi(L)}.$$

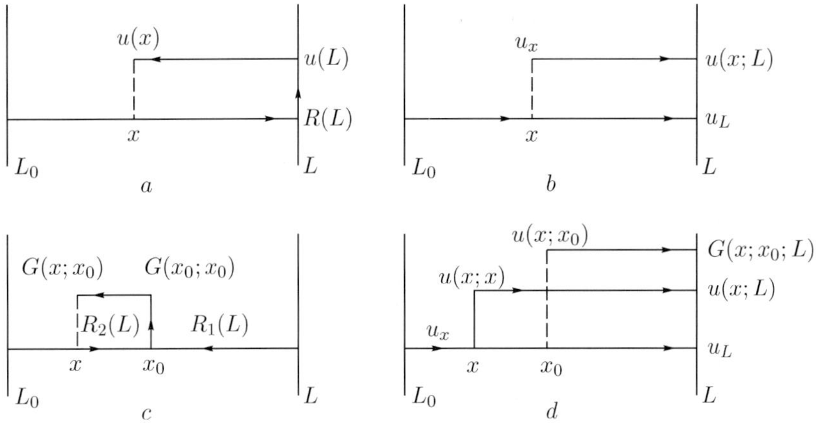

Figure 1.8: Schematic of solving boundary problem (1.16), (1.17) by (*a*) the sweep method and (*b*) the imbedding method and boundary-value problem (1.32) by (*c*) the sweep method and (*d*) the imbedding method.

Introducing the new function

$$R(x) = \frac{1-\psi(x)}{1+\psi(x)}, \qquad \psi(x) = \frac{1-R(x)}{1+R(x)},$$

we can rewrite Eq. (1.20) in the form of the equation

$$\frac{d}{dx}R(x) = 2ikR(x) + \frac{i}{2k}\varepsilon(x)\left(1+R(x)\right)^2, \qquad R(L_0) = 0 \tag{1.21}$$

whose solution at $x = L$ coincides with the reflection coefficient, i.e.,

$$R_L = R(L).$$

In terms of function $R(x)$, the wavefield $u(x)$ inside the medium is now expressed by the following equality

$$u(x) = [1 + R(L)] \exp\left[ik\int_x^L d\xi \frac{1-R(\xi)}{1+R(\xi)}\right]. \tag{1.22}$$

Figure 1.8*a* shows the traditional procedure of solving the problem. One solves Eq. (1.21) first and then reconstructs the wavefield by the formula (1.22). This is the well known approach called the sweep method. However, it is inappropriate for analyzing statistical problems.

Alternatively, the wavefield inside the medium can be represented in the form

$$\begin{aligned} u(x) &= u_1(x) + u_2(x), \\ \frac{d}{dx}u(x) &= -ik[u_1(x) - u_2(x)], \end{aligned}$$

where $u_1(x)$ and $u_2(x)$ are the complex *contradirectional modes*. Because these modes are related to the wavefield by the expressions

$$\begin{aligned} u_1(x) &= \frac{1}{2}\left[1+\frac{i}{k}\frac{d}{dx}\right]u(x), \quad u_1(L)=1, \\ u_2(x) &= \frac{1}{2}\left[1-\frac{i}{k}\frac{d}{dx}\right]u(x), \quad u_2(L_0)=0, \end{aligned} \tag{1.23}$$

we can rewrite the boundary-value problem (1.16), (1.17) in the form

$$\begin{aligned} \left(\frac{d}{dx}+ik\right)u_1(x) &= -\frac{ik}{2}\varepsilon(x)\left[u_1(x)+u_2(x)\right], \quad u_1(L)=1, \\ \left(\frac{d}{dx}-ik\right)u_2(x) &= -\frac{ik}{2}\varepsilon(x)\left[u_1(x)+u_2(x)\right], \quad u_2(L_0)=0. \end{aligned} \tag{1.24}$$

Note that function $R(x)$ introduced earlier is expressed in terms of modes $u_1(x)$ and $u_2(x)$ simply as the ratio

$$R(x)=\frac{u_2(x)}{u_1(x)}.$$

The *imbedding method* offers a possibility of reformulating the boundary-value problem (1.16), (1.17) to the dynamic problem with the initial values for parameter L (this parameter corresponds to the geometrical position of the layer right-hand boundary) by considering the solution to the boundary-value problem as a function of parameter L [135, 136, 142]. On such reformulation, the reflection coefficient R_L satisfies the *Riccati equation*

$$\frac{d}{dL}R_L=2ikR_L+\frac{ik}{2}\varepsilon(L)\left(1+R_L\right)^2, \quad R_{L_0}=0 \tag{1.25}$$

that coincides, naturally, with Eq. (1.24), and the wavefield in the medium layer $u(x)\equiv u(x;L)$ satisfies the linear equation

$$\begin{aligned} \frac{\partial}{\partial L}u(x;L) &= iku(x;L)+\frac{ik}{2}\varepsilon(L)\left(1+R_L\right)u(x;L), \\ u(x;x) &= 1+R_x \end{aligned} \tag{1.26}$$

that can be derived, for example, by differentiating Eq. (1.22) with respect to parameter L. Figure 1.8*b* shows the procedure of solving the problem in this formulation. Comparing this procedure with that of the sweep method (Fig. 1.8*a*), we see that solving procedure has changed the direction, and namely this fact will offer a possibility of constructing the statistical description of the solution to the problem in the stochastic formulation.

The equation for the squared modulus of the reflection coefficient $W_L=|R_L|^2$ follows from Eq. (1.25):

$$\frac{d}{dL}W_L=-2k\gamma W_L-\frac{ik}{2}\varepsilon_1(L)\left(R_L-R_L^*\right)\left(1-W_L\right), \quad W_{L_0}=0. \tag{1.27}$$

Note that condition $W_{L_0}=1$ will be the initial value to Eq. (1.27) in the case of totally reflecting boundary at L_0. In this case, the wave incident on the layer of a non-absorptive medium ($\gamma=0$) is totally reflected from the layer, i.e., $W_L=1$, so that the reflection coefficient can be written in the form $R_L=e^{i\phi_L}$. For the phase of the reflection coefficient, we have the dynamic equation following from Eq. (1.25)

$$\frac{d}{dL}\phi_L=2k+k\varepsilon_1(L)\left(1+\cos\phi_L\right). \tag{1.28}$$

It governs the phase varying in the whole range of values $(-\infty, +\infty)$. At the same time, the equation for the wavefield (1.26) depends only on trigonometric functions of the reflection coefficient phase. For this reason, it would be desirable to deal with the phase varying in interval $(-\pi, \pi)$. We can do this by introducing new function $z_L = \tan(\phi_L/2)$. This function satisfies the dynamic equation of type (1.14)

$$\frac{d}{dL} z_L = k\left(1 + z_L^2\right) + k\varepsilon_1(L),$$

whose solutions show singular behavior.

In the general case of arbitrarily reflecting boundary L_0, the steady-state (independent of L) solution $W_L = 1$ corresponding to the total reflection of incident wave formally exists for a half-space ($L_0 \to -\infty$) filled with non-absorptive random medium, too. This solution is actually realized in the statistical problem with probability equal to unity [135, 136, 142].

It is obvious that the division of the field into contradirectional modes (1.23) is of arbitrary nature; this is nothing more than the mathematical technique that reduces the second-order equation (1.16) to two first-order equations with the simplest boundary conditions.

If, in contrast to the above problem, we assume that function $k(x)$ is continuous at boundary $x = L$, i.e., if we assume that the wave number in the free half-space $x > L$ is equal to $k(L)$, then boundary conditions (1.17) of problem (1.16) will be replaced with the conditions

$$u(L) + \frac{i}{k(L)} \frac{du(x)}{dx}\bigg|_{x=L} = 2, \qquad u(L_0) - \frac{i}{k(L_0)} \frac{du(x)}{dx}\bigg|_{x=L_0} = 0. \tag{1.29}$$

We will call the boundary-value problem (1.16), (1.29) the matched boundary-value problem. In this case, it is convenient to represent the wavefield in the form

$$\begin{aligned} u(x) &= u_1(x) + u_2(x), \\ \frac{du(x)}{dx} &= -ik(x)\left[u_1(x) - u_2(x)\right], \end{aligned}$$

where the complex *contradirectional modes* $u_1(x)$ and $u_2(x)$ are now related to the wavefield by the expressions

$$\begin{aligned} u_1(x) &= \frac{1}{2}\left[1 + \frac{i}{k(x)} \frac{d}{dx}\right] u(x), \quad u_1(L) = 1, \\ u_2(x) &= \frac{1}{2}\left[1 - \frac{i}{k(x)} \frac{d}{dx}\right] u(x), \quad u_2(L_0) = 0 \end{aligned}$$

and satisfy the boundary-value problem

$$\begin{aligned} \left(\frac{d}{dx} + ik(x)\right) u_1(x) &= -\frac{k'(x)}{k(x)}\left[u_1(x) - u_2(x)\right], \quad u_1(L) = 1, \\ \left(\frac{d}{dx} - ik(x)\right) u_2(x) &= \frac{k'(x)}{k(x)}\left[u_1(x) - u_2(x)\right], \quad u_2(L_0) = 0, \end{aligned}$$

where $k'(x) = \frac{dk(x)}{dx}$. Function $R(x) = u_2(x)/u_1(x)$ is now described by the Riccati equation

$$\frac{d}{dx} R(x) = 2ikR(x) + \frac{k'(x)}{2k}\left[1 - R^2(x)\right], \quad R(L_0) = 0, \tag{1.30}$$

and the reflection coefficient is determined in terms of the solution to Eq. (1.30) from the relationship

$$R_L = R(L).$$

In the case of sufficiently small function $\varepsilon(x)$, we can rewrite Eq. (1.30) in the form

$$\frac{dR(x)}{dx} = 2ikR(x) + \frac{1}{4}\varepsilon'(x)\left(1 - R^2(x)\right),$$

where the derivative of function $\varepsilon(x)$ appears as distinct from Eq. (1.24).

Note that, for the matched boundary-value problem (1.16), (1.29), the equations of the imbedding method have the form

$$\begin{aligned}
&\frac{d}{dL}R_L = 2ikR_L + \frac{1}{4}\varepsilon'(L)\left(1 - R_L^2\right), \quad R_{L_0} = 0,\\
&\frac{\partial}{\partial L}u(x, L) = 2iku(x, L) + \frac{1}{4}\varepsilon'(L)(1 - R_L)u(x, L), \quad u(x, x) = 1 + R_x. \quad (1.31)
\end{aligned}$$

1.2.2 Plane waves in layered media: source inside the medium

The field of a point source located in the layer of random medium is described by the similar boundary-value problem for Green's function of the Helmholtz equation:

$$\begin{gathered}
\frac{d^2}{dx^2}G(x;x_0) + k^2[1+\varepsilon(x)]G(x;x_0) = 2ik\delta(x - x_0),\\
G(L;x_0) + \frac{i}{k}\frac{dG(x;x_0)}{dx}\bigg|_{x=L} = 0, \quad G(L_0;x_0) - \frac{i}{k}\frac{dG(x;x_0)}{dx}\bigg|_{x=L_0} = 0. \qquad (1.32)
\end{gathered}$$

Outside the layer, the solution has here the form of outgoing waves (Fig. 1.6*b*)

$$G(x;x_0) = T_1e^{ik(x-L)} \quad (x \ge L), \qquad G(x;x_0) = T_2e^{-ik(x-L_0)} \quad (x \le L_0).$$

Note that, for the source located at layer boundary $x_0 = L$, this problem coincides with the boundary-value problem (1.16), (1.17) on the wave incident on the layer, which yields

$$G(x;L) = u(x;L).$$

The solution to the boundary-value problem (1.32) has the structure

$$G(x;x_0) = G(x_0;x_0)\begin{cases}\exp\left[ik\int\limits_x^{x_0}\psi_1(\xi)\,d\xi\right], & x_0 \ge x,\\ \exp\left[ik\int\limits_{x_0}^{x}\psi_2(\xi)\,d\xi\right], & x_0 \le x,\end{cases} \qquad (1.33)$$

where the field at the source location, by virtue of the derivative gap condition

$$\frac{dG(x;x_0)}{dx}\bigg|_{x=x_0+0} - \frac{dG(x;x_0)}{dx}\bigg|_{x=x_0-0} = 2ik,$$

is determined by the formula

$$G(x_0;x_0) = \frac{2}{\psi_1(x_0) + \psi_2(x_0)}$$

and functions $\psi_i(x)$ satisfy the Riccati equations

$$\begin{aligned}\frac{d}{dx}\psi_1 &= ik\left[\psi_1^2 - 1 - \varepsilon(x)\right], \quad \psi_1(L_0) = 1,\\ \frac{d}{dx}\psi_2 &= -ik\left[\psi_2^2 - 1 - \varepsilon(x)\right], \quad \psi_2(L) = 1.\end{aligned} \tag{1.34}$$

Figure 1.8*c* shows the procedure of solving this problem by the sweep method. One solves two equations (1.34) first and then reconstructs the wavefield using Eq. (1.33).

Introduce new functions $R_i(x)$ related to functions $\psi_i(x)$ by the formula

$$\psi_i(x) = \frac{1 - R_i(x)}{1 + R_i(x)}, \qquad i = 1, 2.$$

With these functions, the wavefield in region $x < x_0$ can be written in the form

$$G(x; x_0) = \frac{[1 + R_1(x_0)][1 + R_2(x_0)]}{1 - R_1(x_0) R_2(x_0)} \exp\left[ik \int_x^{x_0} d\xi \frac{1 - R_1(\xi)}{1 + R_1(\xi)}\right], \tag{1.35}$$

where function $R_1(x)$ satisfies the Riccati equation (1.21).

For $x_0 = L$, expression (1.35) becomes

$$G(x; L) = u(x; L) = [1 + R_1(L)] \exp\left[ik \int_x^{L} d\xi \frac{1 - R_1(\xi)}{1 + R_1(\xi)}\right], \tag{1.36}$$

so that parameter $R_1(L) = R_L$ is the reflection coefficient of the plane wave incident on the layer from region $x > L$. In a similar way, quantity $R_2(x_0)$ is the reflection coefficient of the wave incident on the medium layer (x_0, L) from the homogeneous half-space $x < x_0$ (i.e., from region with $\varepsilon = 0$).

Using Eq. (1.36), we can rewrite Eq. (1.35) in the form

$$G(x; x_0) = \frac{1 + R_2(x_0)}{1 - R_1(x_0) R_2(x_0)} u(x; x_0), \qquad x \le x_0,$$

where $u(x; x_0)$ is the wavefield inside the inhomogeneous layer (L_0, x_0) in the case of the incident wave coming from the free half-space $x > x_0$.

Thus, for $x < x_0$, the field of the point source is proportional to the wavefield generated by the plane wave incident on layer (L_0, x_0) from the free half-space $x > x_0$. The layer segment (x_0, L) affects only parameter $R_2(x_0)$.

Note that, considering the wavefield as a function of parameter L (i.e., setting $G(x; x_0) \equiv G(x; x_0; L)$) , we can use the imbedding method to obtain the following system of equations with initial values:

$$\begin{aligned}&\frac{\partial}{\partial L} G(x; x_0; L) = i\frac{k}{2}\varepsilon(L)\, u(x_0; L)\, u(x; L),\\ &G(x; x_0; L)_{L=\max(x,x_0)} = \begin{cases} u(x; x_0), & x \ge x_0,\\ u(x_0; x), & x \le x_0,\end{cases}\\ &\frac{\partial}{\partial L} u(x; L) = ik\{1 + \varepsilon(L)\, u(L; L)\}\, u(x; L), \quad u(x; L)|_{L=x} = u(x; x),\\ &\frac{d}{dL} u(L; L) = 2ik[u(L; L) - 1] + i\frac{k}{2}\varepsilon(L)\, u^2(L; L), \quad u(L_0; L_0) = 1.\end{aligned} \tag{1.37}$$

Here, two last equations describe the wavefield appearing in the problem on the wave incident on the medium layer. Figure 1.8*d* shows the procedure of solving this problem.

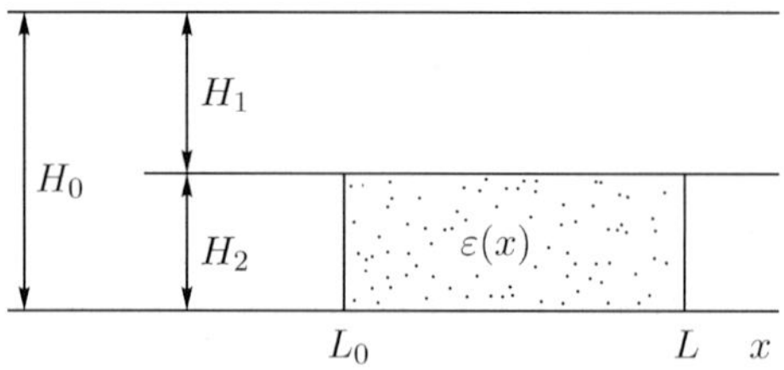

Figure 1.9: Two-layer model of medium.

1.2.3 Plane waves in layered media: two-layer model

Investigators often faces with multidimensional situations in which one wave modes can originate other wave mode due to dependence of problem parameters on spatial coordinates. Sometimes, such problems allow a parametrization by selecting certain direction and dividing the medium in this direction into the layers characterized by discrete values of certain parameters, whereas other parameters may vary continuously in these layers. As an example, we mention the large-scale and low-frequency motions in Earth's atmosphere and ocean, such as the Rossby waves. These waves can be described within the framework of the quasi-geostrophic model that describes the atmosphere and ocean as thin multilayer films characterized in the vertical direction by thicknesses and densities of layers [260]. At the same time, other parameters vary continuously in these layers. It is quite possible that the reason of the local property of the Rossby waves consists in the spatial variation of bottom topography inhomogeneities in the horizontal plane. The simplest one-layer model is equivalent to the one-dimensional Helmholtz equation and describes barotropic motions of the medium; the two-layer model (Fig. 1.9) includes additionally the baroclinic effects [91, 145, 175].

In the context of two-layer media, the simplest model describing the propagation of interacting waves is the system of equations [90]

$$\begin{aligned} &\frac{d^2}{dx^2}\psi_1 + k^2\psi_1 - \alpha_1 F(\psi_1 - \psi_2) = 0, \\ &\frac{d^2}{dx^2}\psi_2 + k^2[1+\varepsilon(x)]\psi_2 + \alpha_2 F(\psi_1 - \psi_2) = 0, \end{aligned} \tag{1.38}$$

where parameters $\alpha_1 = 1/H_1$, $\alpha_2 = 1/H_2$ (H_1 and H_2 are the thicknesses of the top and bottom layers), parameter F characterizes wave interaction, and function $\varepsilon(x)$ describes medium inhomogeneities in the bottom layer. Boundary conditions for system (1.38) are the radiation conditions at infinity.

Note that parameter F characterizing the medium parametrization in the vertical direction appears in system (1.38) as some sort of the horizontal scale responsible for generation of an additional wave. System (1.38) describes wave interaction (and, in particular, dependence of parameters α_i on layer thicknesses) in conformity with problems of geophysical hydrodynamics. For other problem types, the form of these relationships can change, which only slightly concerns the essence of the problem. The only essential point is the fact that wave interaction is the linear interaction.

Transition to the one-layer model is performed by setting $F = 0$, $\psi_1 = 0$ which transforms the corresponding wave equation to the Helmholtz equation (1.16). Proceeding to limit $H_1 \to 0$ also results in the transition to the one-layer model; in this case, $\psi_1 = \psi_2$. However, one can bear in mind that limit processes $L_0 \to -\infty$ (transition to the half-space) and $H_i \to 0$ do not commute in statistical problems. In this case, layer thicknesses H_i must be finite though arbitrarily small.

1.3 First-order partial differential equations

Consider now several dynamic systems (dynamic fields) described by partial differential equations.

1.3.1 Linear first-order partial differential equations: passive tracer in random velocity field

In the context of linear first-order partial differential equations, the simplest problems concern the equation of continuity for the density (concentration) of a conservative tracer and the transfer of a nonconservative passive tracer by the random velocity field $\mathbf{U}(\mathbf{r}, t)$:

$$\begin{aligned}\left(\frac{\partial}{\partial t} + \frac{\partial}{\partial \mathbf{r}}\mathbf{U}(\mathbf{r}, t)\right)\rho(\mathbf{r}, t) &= 0, \quad \rho(\mathbf{r}, 0) = \rho_0(\mathbf{r}), \qquad (1.39)\\ \left(\frac{\partial}{\partial t} + \mathbf{U}(\mathbf{r}, t)\frac{\partial}{\partial \mathbf{r}}\right) q(\mathbf{r}, t) &= 0, \quad q(\mathbf{r}, 0) = q_0(\mathbf{r}). \qquad (1.40)\end{aligned}$$

We can use the method of characteristics to solve the linear first-order partial differential equations (1.39), (1.40). Introducing *characteristic curves* (particles)

$$\frac{d}{dt}\mathbf{r}(t) = \mathbf{U}(\mathbf{r}, t), \qquad \mathbf{r}(0) = \mathbf{r}_0, \tag{1.41}$$

we can write these equations in the form

$$\begin{aligned}\frac{d}{dt}\rho(t) &= -\frac{\partial \mathbf{U}(\mathbf{r}, t)}{\partial \mathbf{r}}\rho(t), \quad \rho(0) = \rho_0(\mathbf{r}_0),\\ \frac{d}{dt}q(t) &= 0, \quad q(0) = q_0(\mathbf{r}_0). \qquad (1.42)\end{aligned}$$

This formulation of the problem corresponds to the *Lagrangian description*, while the initial dynamic equations (1.39), (1.40) correspond to the *Eulerian description*.

Here, we introduced the characteristic vector parameter $\mathbf{r}_0$ in the system of equations (1.41), (1.42). With this parameter, Eq. (1.41) coincides with Eq. (1.1) that describes particle dynamics in the random velocity field.

The solution of the system of equations (1.41), (1.42) depends on the initial value $\mathbf{r}_0$,

$$\mathbf{r}(t) = \mathbf{r}(t|\mathbf{r}_0), \quad \rho(t) = \rho(t|\mathbf{r}_0), \tag{1.43}$$

which we will isolate by the vertical bar symbol.

The first equality in Eq. (1.43) can be considered as the algebraic equation in characteristic parameter; the solution of this equation $\mathbf{r}_0 = \mathbf{r}_0(\mathbf{r}, t)$ exists because *divergence* $j(t|\mathbf{r}_0) = \det \|\partial r_i(t|\mathbf{r}_0)/\partial r_{0k}\|$ is different from zero. Consequently, we can write the solution of the initial equation (1.39) in the form

$$\rho(\mathbf{r}, t) = \rho(t|\mathbf{r}_0(\mathbf{r}, t)) = \int d\mathbf{r}_0 \rho(t|\mathbf{r}_0) j(t|\mathbf{r}_0) \delta\left(\mathbf{r}(t|\mathbf{r}_0) - \mathbf{r}\right).$$

From Eq. (1.41) follows the equation for divergence $j(t|\mathbf{r}_0)$

$$\frac{d}{dt} j(t|\mathbf{r}_0) = \frac{\partial \mathbf{U}(\mathbf{r},t)}{\partial \mathbf{r}} j(t|\mathbf{r}_0), \quad j(0) = 1. \tag{1.44}$$

Correlating it with Eq. (1.42), we see that

$$\rho(t|\mathbf{r}_0) = \frac{\rho_0(\mathbf{r}_0)}{j(t|\mathbf{r}_0)} \tag{1.45}$$

and, consequently, the density field can be rewritten in the form of the equality

$$\rho(\mathbf{r},t) = \int d\mathbf{r}_0 \rho(t|\mathbf{r}_0) j(t|\mathbf{r}_0) \delta\left(\mathbf{r}(t|\mathbf{r}_0) - \mathbf{r}\right) = \int d\mathbf{r}_0 \rho_0(\mathbf{r}_0) \delta\left(\mathbf{r}(t|\mathbf{r}_0) - \mathbf{r}\right) \tag{1.46}$$

that states the relationship between the Lagrangian and Eulerian characteristics. For the position of the Lagrangian particle, the delta-function appeared in the right-hand side of this equality is the *indicator function* (see the next chapter). Consequently, averaging this equality over an ensemble of realizations of the random velocity field, we obtain the well-known relationship between the average density in the Eulerian description and the one-time probability density $P(t,\mathbf{r}|\mathbf{r}_0) = \langle \delta\left(\mathbf{r}(t|\mathbf{r}_0) - \mathbf{r}\right) \rangle$ of the Lagrangian particle (see, e.g., [251]):

$$\langle \rho(\mathbf{r},t) \rangle = \int d\mathbf{r}_0 \rho_0(\mathbf{r}_0) P\left(t,\mathbf{r}|\mathbf{r}_0\right). \tag{1.47}$$

For a divergence-free velocity field ($\operatorname{div} \mathbf{U}(\mathbf{r},t) = 0$), both particle divergence and particle density are conserved, i.e.,

$$j(t|\mathbf{r}_0) = 1, \qquad \rho(t|\mathbf{r}_0) = \rho_0(\mathbf{r}_0), \quad q(t|\mathbf{r}_0) = q_0(\mathbf{r}_0).$$

Consider now stochastic features of the solutions to problem (1.39). A convenient way of analyzing random field dynamics consists in using *topographic* concepts. Indeed, in the case of the divergence-free velocity field, temporal evolution of the contour of constant concentration $\rho = \text{const}$ coincides with the dynamics of particles in this velocity field and, consequently, coincides with the dynamics shown in Fig. 1.1a, page 4. In this case, the area within the contour remains constant and, as it is seen from Fig. 1.1a, the pattern becomes highly indented, which is manifested in gradient sharpening and the appearance of contour dynamics for progressively shorter scales. In the other limiting case (the divergent velocity field), the area within the contour tends to zero, and the field of density condenses forming clusters. One can find examples simulated for this case in papers [198, 320]. These features of particle dynamics disappear on averaging over an ensemble of realizations. Cluster formation in the Eulerian description can be traced using the random velocity field in the form (1.3), (1.4). In this case, the density field $\rho(\mathbf{r},t)$ is described by the expression [161]

$$\rho(\mathbf{r},t) = \rho_0\left(\mathbf{r}_0\right) \frac{1}{e^{T(t)}\cos^2(kx) + e^{-T(t)}\sin^2(kx)}, \tag{1.48}$$

where function $T(t)$ is given by Eq. (1.7).

For the divergence-free velocity field $v_x(t) = 0$, $T(t) \equiv 0$, and we have

$$\rho(\mathbf{r},t) = \rho_0 \left(\mathbf{r} - \sin(2kx) \int\limits_0^t d\tau \mathbf{v}(\tau) \right).$$

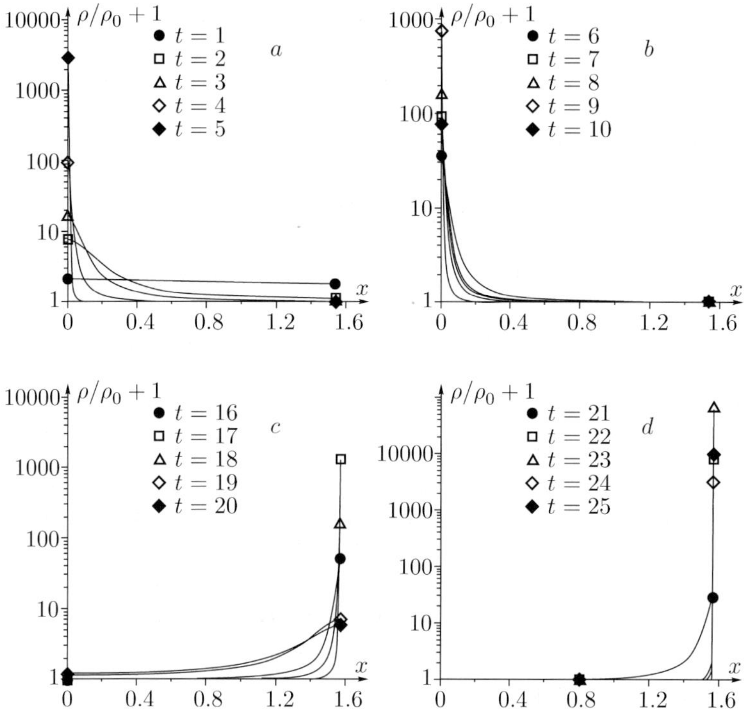

Figure 1.10: Space-time evolution of the Eulerian density field given by Eq. (1.49).

In the particular case of the initial density distribution independent of $\mathbf{r}$, i.e., if $\rho_0(\mathbf{r}) = \rho_0$, equality (1.48) is simplified and assumes the form

$$\rho(\mathbf{r}, t)/\rho_0 = \frac{1}{e^{T(t)} \cos^2(kx) + e^{-T(t)} \sin^2(kx)}. \tag{1.49}$$

Figure 1.10 shows the Eulerian density field $1 + \rho(\mathbf{r}, t)/\rho_0$ and its space-time evolution calculated by Eq. (1.49) in the dimensionless space-time variables (the density field is added with a unity to avoid the difficulties of dealing with nearly zero-valued densities in the logarithmic scale).

This figure shows successive patterns of density field rearrangement toward narrow neighborhoods of points $x \approx 0$ and $x \approx \pi/2$, i.e., the cluster formation. Figures 1.10*a* and 1.10*b* show the temporal pattern ($t = 1 \div 10$) of cluster formation around point $x \approx 0$. Figures 1.10*c* and 1.10*d* show the temporal pattern ($t = 16 \div 25$) of rearranging the density field from the neighborhood of point $x \approx 0$ toward the neighborhood of point $x \approx \pi/2$, i.e., they show the removal of the cluster near $x \approx 0$ and the birth of the new cluster near $x \approx \pi/2$. This process is then repeated in time. As is seen from figures, the lifetimes of such clusters coincide, on the order of magnitude, with the time of cluster formation.

Thus, we considered the simplest model for the diffusion of a tracer (particles and the Eulerian density field) in random velocity field, which clearly shows the process of cluster structure formation. A feature of the model considered consists in the fact that the points at which clusters are formed are the fixed points, which decreases the usefulness of the model.

However, this model provides an insight into the difference between the diffusion processes in divergent and divergence-free velocity fields. In divergence-free (incompressible) velocity fields, particles (and, consequently, density field) have no time for attracting to stable centers of attraction during the lifetime of these centers, and particles slightly fluctuate relative their initial location. On the contrary, in the divergent (compressible) velocity field, lifetime of stable centers of attraction is sufficient for particles to attract to them, because the speed of attraction increases exponentially, which is clearly seen from Eq. (1.49).

From the above description, it becomes obvious that dynamic equation (1.39) considered as the model equation describing actual physical phenomena can be used only on finite temporal intervals. A more complete analysis assumes the consideration of the field of tracer concentration gradient $\mathbf{p}(\mathbf{r},t) = \boldsymbol{\nabla}\rho(\mathbf{r},t)$ that satisfies the equation

$$\begin{aligned}\left(\frac{\partial}{\partial t}+\frac{\partial}{\partial \mathbf{r}}\mathbf{U}(\mathbf{r},t)\right)p_i(\mathbf{r},t) &= -p_k(\mathbf{r},t)\frac{\partial U_k(\mathbf{r},t)}{\partial r_i}-\rho(\mathbf{r},t)\frac{\partial^2\mathbf{U}(\mathbf{r},t)}{\partial r_i\partial\mathbf{r}},\\ \mathbf{p}(\mathbf{r},0) &= \mathbf{p}_0(\mathbf{r}) = \boldsymbol{\nabla}\rho_0(\mathbf{r}).\end{aligned} \tag{1.50}$$

In addition, one should also include the effect of the molecular diffusion (with the molecular diffusion coefficient μ) that smooths the mentioned gradient sharpen; this effect is described by the linear second-order partial differential equation

$$\left(\frac{\partial}{\partial t}+\frac{\partial}{\partial \mathbf{r}}\mathbf{U}(\mathbf{r},t)\right)\rho(\mathbf{r},t) = \mu\Delta\rho(\mathbf{r},t), \quad \rho(\mathbf{r},0)=\rho_0(\mathbf{r}). \tag{1.51}$$

1.3.2 Quasilinear equations

Consider now the simplest quasilinear equation for scalar quantity $q(\mathbf{r},t)$, which we write in the form

$$\left(\frac{\partial}{\partial t}+\mathbf{U}(t,q)\frac{\partial}{\partial \mathbf{r}}\right)q(\mathbf{r},t) = Q(t,q), \quad q(\mathbf{r},0)=q_0(\mathbf{r}), \tag{1.52}$$

where we assume for simplicity that functions $\mathbf{U}(t,q)$ and $Q(t,q)$ are explicitly independent of spatial variable $\mathbf{r}$.

Supplement Eq. (1.52) with the equation for the gradient $\mathbf{p}(\mathbf{r},t) = \boldsymbol{\nabla}q(\mathbf{r},t)$, which follows from Eq. (1.52), and the equation of continuity for conserved quantity $I(\mathbf{r},t)$:

$$\begin{aligned}&\left(\frac{\partial}{\partial t}+\mathbf{U}(t,q)\frac{\partial}{\partial \mathbf{r}}\right)\mathbf{p}(\mathbf{r},t)+\frac{\partial\left\{\mathbf{U}(t,q)\mathbf{p}(\mathbf{r},t)\right\}}{\partial q}\mathbf{p}(\mathbf{r},t) = \frac{\partial Q(t,q)}{\partial q}\mathbf{p}(\mathbf{r},t),\\ &\frac{\partial}{\partial t}I(\mathbf{r},t)+\frac{\partial}{\partial \mathbf{r}}\left\{\mathbf{U}(t,q)I(\mathbf{r},t)\right\}=0.\end{aligned} \tag{1.53}$$

From Eqs. (1.53) follows that

$$\int d\mathbf{r}I(\mathbf{r},t) = \int d\mathbf{r}I_0(\mathbf{r}). \tag{1.54}$$

In terms of characteristic curves determined from the system of ordinary differential equations, Eqs. (1.52) and (1.53) can be written in the form

$$\begin{aligned}
\frac{d}{dt}\mathbf{r}(t) &= \mathbf{U}(t,q), \quad \frac{d}{dt}q(t) = Q(t,q), \quad \mathbf{r}(0) = \mathbf{r}_0, \quad q(0) = q_0(\mathbf{r}_0), \\
\frac{d}{dt}\mathbf{p}(t) &= -\frac{\partial\{\mathbf{U}(t,q)\,\mathbf{p}(t)\}}{\partial q}\mathbf{p}(t) + \frac{\partial Q(t,q)}{\partial q}\mathbf{p}(t), \quad \mathbf{p}(0) = \frac{\partial q_0(\mathbf{r}_0)}{\partial \mathbf{r}_0}, \\
\frac{d}{dt}I(t) &= -\frac{\partial\{\mathbf{U}(t,q)\,\mathbf{p}(t)\}}{\partial q}I(t), \quad I(0) = I_0(\mathbf{r}_0).
\end{aligned} \tag{1.55}$$

Thus, the Lagrangian description considers the system (1.55) as the initial value problem. In this description, the two first equations form the closed system that defines characteristic curves.

Expressing now characteristic parameter $\mathbf{r}_0$ in terms of t and $\mathbf{r}$, one can write the solution to Eqs. (1.52) and (1.53) in the Eulerian description as

$$\begin{aligned}
q(\mathbf{r},t) &= \int d\mathbf{r}_0 q(t|\mathbf{r}_0)\, j(t|\mathbf{r}_0)\,\delta(\mathbf{r}(t|\mathbf{r}_0) - \mathbf{r}), \\
I(\mathbf{r},t) &= \int d\mathbf{r}_0 I(t|\mathbf{r}_0)\, j(t|\mathbf{r}_0)\,\delta(\mathbf{r}(t|\mathbf{r}_0) - \mathbf{r}).
\end{aligned} \tag{1.56}$$

The feature of the transition from the Lagrangian description (1.55) to the Eulerian description (1.56) consists in the general appearance of ambiguities, which yields discontinuous solutions. These ambiguities are related to the fact that the divergence — Jacobian $j(t|\mathbf{r}_0) = \det \left\| \frac{\partial}{\partial r_{0k}} r_i(t|\mathbf{r}_0) \right\|$ — can now vanish at certain moments.

Quantities $I(t|\mathbf{r}_0)$ and $j(t|\mathbf{r}_0)$ are not independent. Indeed, integrating $I(\mathbf{r},t)$ in Eq. (1.56) over $\mathbf{r}$ and taking into account Eq. (1.54), we see that there exists the evolution integral

$$j(t|\mathbf{r}_0) = \frac{I_0(\mathbf{r}_0)}{I(t|\mathbf{r}_0)}, \tag{1.57}$$

from which follows that zero-valued divergence $j(t|\mathbf{r}_0)$ is accompanied by the infinite value of conservative quantity $I(t|\mathbf{r}_0)$.

Example. Consider the one-dimensional Riemann equation

$$\frac{\partial}{\partial t}q(x,t) + q(x,t)\frac{\partial}{\partial x}q(x,t) = 0, \quad q(x,0) = q_0(x) \tag{1.58}$$

as the simplest example. This equation corresponds to Eq. (1.52) with $G(t,q) = 0$, $U(t,q) = q(x,t)$ and.describes free propagation of the *nonlinear Riemann wave.*

The method of characteristics applied to Eq. (1.58) gives

$$q(t|x_0) = q_0(x_0), \quad x(t|x_0) = x_0 + tq_0(x_0),$$

so that the solution of Eq. (1.58) can be written in the form of the transcendental equation

$$q(x,t) = q_0(x - q(x,t))$$

from which follows

$$\frac{\partial}{\partial x}q(x,t) = \frac{q_0'(x_0)}{1 + tq_0'(x_0)},$$

where

$$x_0 = x - tq(x,t) \quad \text{and} \quad q_0'(x_0) = \frac{d}{dx_0}q_0(x_0).$$

If $q_0'(x_0) < 0$, then derivative $\frac{\partial}{\partial x}q(x,t)$ becomes infinite during a finite time t_0 and the solution of Eq. (1.58) becomes discontinuous. For times prior to t_0, the solution is unique and representable in the form of a quadrature. To show this fact, we calculate the variational derivative (for variational derivative definitions and the corresponding operation rules, see Appendix A)

$$\frac{\delta q\left(x,t\right)}{\delta q_0\left(x_0\right)} = \frac{1}{1+tq_0'\left(x_0\right)}\delta\left(x - tq\left(x,t\right) - x_0\right).$$

Because $q(x,t) = q_0(x_0)$ and $x - tq_0(x_0) = x_0$, the argument of delta function vanishes at $x = F(x_0,t) = x_0 + tq_0(x_0)$. Consequently, we have

$$\frac{\delta q\left(x,t\right)}{\delta q_0\left(x_0\right)} = \delta(x - F(x_0,t)) = \delta(x - x_0 - tq_0(x_0)) = \frac{1}{2\pi}\int\limits_{-\infty}^{\infty} dk\, e^{ik(x-x_0)+iktq_0(x_0)}.$$

We can consider this equality as the functional equation in variable $q_0(x_0)$. Then, integrating this equation with the initial value

$$q\left(x,t\right)\big|_{q_0(x_0)=0} = 0,$$

in the functional space, we obtain the solution of the Riemann equation in the form of the quadrature

$$q\left(x,t\right) = \frac{1}{2\pi i t}\int\limits_{-\infty}^{\infty}\frac{dk}{k}\int\limits_{-\infty}^{\infty} dx_0 e^{ik(x-x_0)}\left[e^{iktq_0(x_0)} - 1\right].$$

The mentioned ambiguity can be eliminated by considering the *Burgers equation*

$$\frac{\partial}{\partial t}q(x,t) + q(x,t)\frac{\partial}{\partial x}q(x,t) = \mu\frac{\partial^2}{\partial x^2}q(x,t), \quad q(x,0) = q_0(x)$$

(it includes the molecular viscosity and also can be solved in quadratures) followed by the limit process $\mu \to 0$ (see, e.g., [101]). ♦

It is obvious that all these results can be easily extended to the case in which functions $\mathbf{U}(\mathbf{r},t,q)$ and $Q(\mathbf{r},t,q)$ explicitly depend on spatial variable $\mathbf{r}$ and Eq. (1.52) itself is the vector equation. As a particular physical example, we consider the equation for the velocity field $\mathbf{V}(\mathbf{r},t)$ of low-inertia particles moving in the hydrodynamic flow whose velocity field is $\mathbf{u}(\mathbf{r},t)$ (see, e.g., [240])

$$\left(\frac{\partial}{\partial t} + \mathbf{V}(\mathbf{r},t)\frac{\partial}{\partial \mathbf{r}}\right)\mathbf{V}(\mathbf{r},t) = -\lambda\left[\mathbf{V}(\mathbf{r},t) - \mathbf{u}(\mathbf{r},t)\right]. \qquad (1.59)$$

We will assume this equation the phenomenological equation.

In the general case, the solution to Eq. (1.59) can be nonunique, it can have discontinuities, etc. However, in the case of asymptotically small inertia property of particles (parameter $\lambda \to \infty$), which is of our concern here, the solution will be unique during reasonable temporal intervals. Note that, in the right-hand side of Eq. (1.59), term $\mathbf{F}(\mathbf{r},t) = \lambda\mathbf{V}(\mathbf{r},t)$ linear in the velocity field $\mathbf{V}(\mathbf{r},t)$ is, according to the known *Stokes formula*, the resistance force acting on a slowly moving particle. If we approximate the particle by the sphere of radius a, parameter λ will be $\lambda = 6\pi a\eta/m_{\mathrm{p}}$, where η is the coefficient of dynamic viscosity and m_{p} is the mass of the particle (see, e.g., [217, 218]).

From Eq. (1.59) follows that velocity field $\mathbf{V}(\mathbf{r},t)$ is the divergent field (div $\mathbf{V}(\mathbf{r},t) \neq 0$) even if hydrodynamic flow $\mathbf{u}(\mathbf{r},t)$ is divergence-free (div $\mathbf{u}(\mathbf{r},t) = 0$). As a consequence, particle number density $n(\mathbf{r},t)$ in divergence-free hydrodynamic flows, which satisfies the linear equation of continuity

$$\left(\frac{\partial}{\partial t} + \frac{\partial}{\partial \mathbf{r}}\mathbf{V}(\mathbf{r},t)\right) n(\mathbf{r},t) = 0, \quad n(\mathbf{r},0) = n_0(\mathbf{r}) \tag{1.60}$$

similar to Eq. (1.39), shows the cluster behavior.

For large parameters $\lambda \to \infty$ (inertialess particles), we have

$$\mathbf{V}(\mathbf{r},t) \approx \mathbf{u}(\mathbf{r},t), \tag{1.61}$$

which means that particle number density $n(\mathbf{r},t)$ shows no cluster behavior in divergence-free hydrodynamic flows.

The first-order partial differential equation (1.59) (Eulerian description) is equivalent to the system of ordinary differential characteristic equations (Lagrangian description)

$$\begin{aligned}\frac{d}{dt}\mathbf{r}(t) &= \mathbf{V}\left(\mathbf{r}(t),t\right), \quad \mathbf{r}(0) = \mathbf{r}_0,\\ \frac{d}{dt}\mathbf{V}(t) &= -\lambda\left[\mathbf{V}(t) - \mathbf{u}\left(\mathbf{r}(t),t\right)\right], \quad \mathbf{V}(0) = \mathbf{V}_0(\mathbf{r}_0)\end{aligned} \tag{1.62}$$

that describes the diffusion of a particle under the random external force and linear friction and coincide with Eq. (1.12). In the simplest case of the random force independent of spatial coordinates, we have the system

$$\begin{aligned}\frac{d}{dt}\mathbf{r}(t) &= \mathbf{v}(t), \quad \frac{d}{dt}\mathbf{v}(t) = -\lambda\left[\mathbf{v}(t) - \mathbf{f}(t)\right],\\ \mathbf{r}(0) &= \mathbf{r}_0, \quad \mathbf{v}(0) = \mathbf{v}_0,\end{aligned} \tag{1.63}$$

the stochastic solution to which has the form

$$\mathbf{v}(t) = \lambda\int\limits_0^t d\tau e^{-\lambda(t-\tau)}\mathbf{f}(\tau), \quad \mathbf{r}(t) = \int\limits_0^t d\tau\left[1 - e^{-\lambda(t-\tau)}\right]\mathbf{f}(\tau).$$

1.3.3 Boundary-value problems for nonlinear ordinary differential equations

Note that, using the imbedding method (see Appendix C), the boundary-value problems for nonlinear ordinary differential equations also can be reduced to quasilinear equations. This is the case, for example, for the nonlinear vector boundary-value problem

$$\frac{d}{dt}\mathbf{x}(t) = \mathbf{U}\left(t, \mathbf{x}(t)\right),$$

defined on segment $t \in [0,T]$ with boundary conditions

$$G\mathbf{x}(0) + H\mathbf{x}(T) = \mathbf{v},$$

where G and H are constant matrices. Consider the solution of this problem as a function of parameters T and $\mathbf{v}$, i.e., $\mathbf{x}(t) = \mathbf{x}(t;T,\mathbf{v})$. Then, function $\mathbf{R}(T,\mathbf{v}) = \mathbf{x}(T;T,\mathbf{v})$ as a function of parameters T and $\mathbf{v}$ is described by the quasilinear vector equation [83, 136]

$$\left(\frac{\partial}{\partial T} + \left[H\mathbf{U}\left(T, \mathbf{R}\left(t,\mathbf{v}\right)\right)\right]\frac{\partial}{\partial \mathbf{v}}\right)\mathbf{R}\left(T,\mathbf{v}\right) = \mathbf{U}\left(T, \mathbf{R}\left(t,\mathbf{v}\right)\right)$$

with the boundary condition for $T \to 0$

$$\mathbf{R}(T,\mathbf{v})|_{T=0} = (G+H)^{-1}\mathbf{v},$$

and function $\mathbf{x}(t;T,\mathbf{v})$ itself satisfies the linear equation

$$\frac{\partial x_i(t;T,\mathbf{v})}{\partial T} = -H_{kl}U_l(T,\mathbf{R}(T,\mathbf{v}))\frac{\partial x_i(t;T,\mathbf{v})}{\partial v_k}$$

with the boundary condition

$$\mathbf{x}(t;T,\mathbf{v})|_{T=t} = \mathbf{R}(t,\mathbf{v}).$$

1.3.4 Nonlinear first-order partial differential equations

In the general case, a nonlinear scalar first-order partial differential equation can be written in the form

$$\frac{\partial}{\partial t}q(\mathbf{r},t) + H(\mathbf{r},t,q,\mathbf{p}) = 0, \quad q(\mathbf{r},0) = q_0(\mathbf{r}), \tag{1.64}$$

where $\mathbf{p}(\mathbf{r},t) = \boldsymbol{\nabla} q(\mathbf{r},t)$.

In terms of the Lagrangian description, this equation can be rewritten in the form of the system of characteristic equations (see, e.g., [310]):

$$\begin{aligned}
\frac{d}{dt}\mathbf{r}(t|\mathbf{r}_0) &= \frac{\partial}{\partial \mathbf{p}}H(\mathbf{r},t,q,\mathbf{p}), \quad \mathbf{r}(0|\mathbf{r}_0) = \mathbf{r}_0;\\
\frac{d}{dt}\mathbf{p}(t|\mathbf{r}_0) &= -\left(\frac{\partial}{\partial \mathbf{r}} + \mathbf{p}\frac{\partial}{\partial q}\right)H(\mathbf{r},t,q,\mathbf{p}), \quad \mathbf{p}(0|\mathbf{r}_0) = \mathbf{p}_0(\mathbf{r}_0);\\
\frac{d}{dt}q(t|\mathbf{r}_0) &= \left(\mathbf{p}\frac{\partial}{\partial \mathbf{p}} - 1\right)H(\mathbf{r},t,q,\mathbf{p}), \quad q(0|\mathbf{r}_0) = \mathbf{q}_0(\mathbf{r}_0).
\end{aligned} \tag{1.65}$$

Now, we supplement Eq. (1.64) with the equation for the conservative quantity $I(\mathbf{r},t)$

$$\frac{\partial}{\partial t}I(\mathbf{r},t) + \frac{\partial}{\partial \mathbf{r}}\left\{\frac{\partial H(\mathbf{r},t,q,\mathbf{p})}{\partial \mathbf{p}}I(\mathbf{r},t)\right\} = 0, \quad I(\mathbf{r},0) = I_0(\mathbf{r}). \tag{1.66}$$

From Eq. (1.66) follows that

$$\int d\mathbf{r}I(\mathbf{r},t) = \int d\mathbf{r}I_0(\mathbf{r}). \tag{1.67}$$

In the Lagrangian description, the corresponding quantity satisfies the equation

$$\frac{d}{dt}I(t|\mathbf{r}_0) = -\frac{\partial^2 H(\mathbf{r},t,q,\mathbf{p})}{\partial \mathbf{r}\partial \mathbf{p}}I(\mathbf{r},t), \quad I(0|\mathbf{r}_0) = I_0(\mathbf{r}_0),$$

so that the solution to Eq. (1.66) has the form

$$I(\mathbf{r},t) = I(t|\mathbf{r}_0(t,\mathbf{r})) = \int d\mathbf{r}_0 I(t|\mathbf{r}_0)\, j(t|\mathbf{r}_0)\, \delta(\mathbf{r}(t|\mathbf{r}_0) - \mathbf{r}), \tag{1.68}$$

where $j(t|\mathbf{r}_0) = \det\|\partial r_i(t|\mathbf{r}_0)/\partial r_{0j}\|$ is the divergence (Jacobian).

Quantities $I(t|\mathbf{r}_0)$ and $j(t|\mathbf{r}_0)$ are related to each other. Indeed, substituting Eq. (1.68) for $I(\mathbf{r},t)$ in Eq. (1.67), we see that there exists the evolution integral

$$j\left(t|\mathbf{r}_0\right)=\frac{I_0\left(\mathbf{r}_0\right)}{I\left(t|\mathbf{r}_0\right)},$$

and Eq. (1.68) assumes the form

$$I\left(\mathbf{r},t\right)=\int d\mathbf{r}_0 I_0\left(\mathbf{r}_0\right)\delta\left(\mathbf{r}\left(t|\mathbf{r}_0\right)-\mathbf{r}\right).$$

Example. In the case of function $H(\mathbf{r},t,q,\mathbf{p})$ specified as

$$H\left(\mathbf{r},t,q,\mathbf{p}\right)=\frac{1}{2}\mathbf{p}^2\left(\mathbf{r},t\right)+U\left(\mathbf{r},t\right),$$

Eqs. (1.65) correspond to the Hamilton equations

$$\frac{d}{dt}\mathbf{r}(t)=\mathbf{p}(t),\quad \frac{d}{dt}\mathbf{p}(t)=-\frac{\partial}{\partial\mathbf{r}}U\left(\mathbf{r},t\right),\quad \frac{d}{dt}q(t)=-U\left(\mathbf{r},t\right),$$

whereas Eq. (1.64) becomes the *Hamilton–Jacobi equation*

$$\frac{\partial}{\partial t}q(\mathbf{r},t)+U\left(\mathbf{r},t\right)+\mathbf{p}^2\left(\mathbf{r},t\right)=0,\qquad q\left(\mathbf{r},0\right)=q_0\left(\mathbf{r}\right)$$

and function $\mathbf{p}(\mathbf{r},t)=\boldsymbol{\nabla}q(\mathbf{r},t)$ satisfies the quasilinear equation

$$\left(\frac{\partial}{\partial t}+\mathbf{p}\left(\mathbf{r},t\right)\frac{\partial}{\partial\mathbf{r}}\right)\mathbf{p}\left(\mathbf{r},t\right)+\frac{\partial}{\partial\mathbf{r}}U\left(\mathbf{r},t\right)=0,\quad \mathbf{p}\left(\mathbf{r},0\right)=\boldsymbol{\nabla}q_0\left(\mathbf{r}\right).\ \blacklozenge$$

1.4 Partial differential equations of higher orders

1.4.1 Stationary problems for Maxwell's equations

In the steady inhomogeneous medium, propagation of a monochromatic electromagnetic wave of frequency ω is described by *Maxwell's equations* (see, e.g., [294])

$$\mathrm{rot}\mathbf{E}(\mathbf{r})=ik\mathbf{H}(\mathbf{r}),\quad \mathrm{rot}\mathbf{H}(\mathbf{r})=-ik\varepsilon(\mathbf{r})\mathbf{E}(\mathbf{r}),\quad \mathrm{div}\varepsilon(\mathbf{r})\mathbf{E}(\mathbf{r})=0,\tag{1.69}$$

where $\mathbf{E}(\mathbf{r})$ and $\mathbf{H}(\mathbf{r})$ are the electric and magnetic strengths, $\varepsilon(\mathbf{r})$ is the dielectric permittivity of the medium, and $k=\omega/c=2\pi/\lambda$ is the wave number (λ is the wavelength and c is the velocity of wave propagation). Here, we assumed that magnetic permeability $\mu=1$, medium conductivity $\sigma=0$, and specified temporal factor $e^{-i\omega t}$ for all fields.

Equations (1.69) can be rewritten in the form of the equation closed in the electric field $\mathbf{E}(\mathbf{r})$

$$\left[\Delta+k^2\varepsilon(\mathbf{r})\right]\mathbf{E}(\mathbf{r})=-\boldsymbol{\nabla}\left(\mathbf{E}(\mathbf{r})\boldsymbol{\nabla}\ln\varepsilon(\mathbf{r})\right).\tag{1.70}$$

In this case, the magnetic field $\mathbf{H}(\mathbf{r})$ is calculated by the equality

$$\mathbf{H}(\mathbf{r})=\frac{1}{ik}\mathrm{rot}\mathbf{E}(\mathbf{r}).\tag{1.71}$$

We restrict ourselves with electromagnetic wave propagation in media with weakly fluctuating dielectric permittivity. We set

$$\varepsilon(\mathbf{r}) = 1 + \varepsilon_1(\mathbf{r}),$$

where $\varepsilon_1(\mathbf{r})$ stands for small fluctuations of the dielectric permittivity ($\langle\varepsilon_1(\mathbf{r})\rangle = 0$). Smallness of fluctuations $\varepsilon_1(\mathbf{r})$ assumes that $\langle|\varepsilon_1(\mathbf{r})|\rangle \ll 1$. With this assumption, Eq. (1.70) can be rewritten in the simplified form

$$\left[\Delta + k^2\right] \mathbf{E}(\mathbf{r}) = -k^2\varepsilon_1(\mathbf{r})\mathbf{E}(\mathbf{r}) - \boldsymbol{\nabla}\left(\mathbf{E}(\mathbf{r})\boldsymbol{\nabla}\varepsilon_1(\mathbf{r})\right). \tag{1.72}$$

Using the theory of perturbations, Tatarskii [296] and Kravtsov [204] estimated the light wave depolarization at propagation paths of about 1 *km* in the conditions of the actual atmosphere and showed that the depolarization is very small. In these conditions, we can neglect the last term in the right-hand side of Eq. (1.72). As a result, the problem reduces in fact to the scalar *Helmholtz equation*

$$\left[\Delta + k^2\right] U(\mathbf{r}) = -k^2\varepsilon_1(\mathbf{r})U(\mathbf{r}). \tag{1.73}$$

For Eq. (1.73) be meaningful, one must formulate boundary conditions and specify the source of radiation.

1.4.2 The Helmholtz equation (boundary-value problem) and the parabolic equation of quasi-optics (waves in randomly inhomogeneous media)

Let the layer of inhomogeneous medium occupies spatial segment $L_0 < x < L$ and let the point source is located at point $(x_0, \mathbf{R}_0)$, where $\mathbf{R}_0$ stands for the coordinates in the plane perpendicular to the x-axis. In this case, the field inside the layer $G(x, \mathbf{R}; x_0, \mathbf{R}_0)$ satisfies the equation for Green's function

$$\left\{\frac{\partial^2}{\partial x^2} + \Delta_{\mathbf{R}} + k^2\left[1 + \varepsilon\left(x, \mathbf{R}\right)\right]\right\} G\left(x, \mathbf{R}; x_0, \mathbf{R}_0\right) = \delta\left(x - x_0\right)\delta\left(\mathbf{R} - \mathbf{R}_0\right), \tag{1.74}$$

where k is the wave number, $\Delta_{\mathbf{R}} = \partial^2/\partial\mathbf{R}^2$, and $\varepsilon_1(\mathbf{r}) = \varepsilon(x, \mathbf{R})$ is the deviation of the refractive index (or dielectric permittivity) from unity. Let $\varepsilon(x, \mathbf{R}) = 0$ outside the layer. Then, the wavefield outside the layer satisfies the Helmholtz equation

$$\left\{\frac{\partial^2}{\partial x^2} + \Delta_{\mathbf{R}} + k^2\right\} G\left(x, \mathbf{R}; x_0, \mathbf{R}_0\right) = 0,$$

and continuity conditions for functions G and $\partial G/\partial x$ at the layer boundaries. Furthermore, the solution to Eq. (1.74) must satisfy the radiation conditions for $x \to \pm\infty$.

The wavefield outside the layer can obviously be represented in the form

$$G\left(x, \mathbf{R}; x_0, \mathbf{R}_0\right) = \begin{cases} \int d\mathbf{q}\, T_1\left(\mathbf{q}\right) \exp\left[-i\sqrt{k^2 - q^2}\left(x - L_0\right) + i\mathbf{q}\mathbf{R}\right], & x \le L_0; \\ \int d\mathbf{q}\, T_2\left(\mathbf{q}\right) \exp\left[i\sqrt{k^2 - q^2}\left(x - L\right) + i\mathbf{q}\mathbf{R}\right], & x \ge L. \end{cases}$$

Consequently, the boundary condition for Eq. (1.74) at $x = L_0$ can be written as

$$\left(\frac{\partial}{\partial x} + i\sqrt{k^2 + \Delta_{\mathbf{R}}}\right) G\left(x, \mathbf{R}; x_0, \mathbf{R}_0\right)\Big|_{x=L_0} = 0. \tag{1.75}$$

Similarly, the boundary condition at $x = L$ has the form

$$\left(\frac{\partial}{\partial x} - i\sqrt{k^2 + \Delta_{\mathbf{R}}}\right) G(x, \mathbf{R}; x_0, \mathbf{R}_0)\Big|_{x=L} = 0. \tag{1.76}$$

In the case of space infinite in coordinates $\mathbf{R}$, operator $\sqrt{k^2 + \Delta_{\mathbf{R}}}$ appeared in Eqs. (1.75), (1.76) can be defined in terms of the *Fourier transform*. Alternatively, this operator can be also treated as the linear integral operator whose kernel is expressed in terms of Green's function for free space (see Appendix B).

Thus, the field of the point source in the inhomogeneous medium is described by the boundary-value problem (1.74) – (1.76). This problem is equivalent to the integral equation

$$\begin{aligned} G(x, \mathbf{R}; x_0, \mathbf{R}_0) &= g(x - x_0, \mathbf{R} - \mathbf{R}_0) \\ &+ \int_{L_0}^{L} dx' \int d\mathbf{R}' g(x - x', \mathbf{R} - \mathbf{R}') \varepsilon(x', \mathbf{R}') G(x', \mathbf{R}'; x_0, \mathbf{R}_0), \end{aligned} \tag{1.77}$$

where $g(x, \mathbf{R})$ is Green's function in free space. In the three-dimensional case, we have

$$g(x, \mathbf{R}) = -\frac{1}{4\pi r} e^{ikr}, \quad r = \sqrt{x^2 + \mathbf{R}^2}.$$

The integral representation of this Green's function is as follows

$$g(x, \mathbf{R}) = \int d\mathbf{q}\, g(\mathbf{q}) \exp\left\{i\sqrt{k^2 - q^2}\,|x| + i\mathbf{q}\mathbf{R}\right\}, \quad g(\mathbf{q}) = \frac{1}{8i\pi^2\sqrt{k^2 - q^2}}. \tag{1.78}$$

It can be shown (see Appendix B) that operator $\sqrt{k^2 + \Delta_{\mathbf{R}}}$ applied to arbitrary function $F(\mathbf{R})$ acts as the integral operator

$$\sqrt{k^2 + \Delta_{\mathbf{R}}} F(\mathbf{R}) = \int d\mathbf{R}' K(\mathbf{R} - \mathbf{R}') F(\mathbf{R}') \tag{1.79}$$

whose kernel is

$$K(\mathbf{R} - \mathbf{R}') = \sqrt{k^2 + \Delta_{\mathbf{R}}} \delta(\mathbf{R} - \mathbf{R}') = 2i\left(k^2 + \Delta_{\mathbf{R}}\right) g(0, \mathbf{R} - \mathbf{R}'). \tag{1.80}$$

The corresponding kernel of the inverse operator is

$$L(\mathbf{R} - \mathbf{R}') = \left(k^2 + \Delta_{\mathbf{R}}\right)^{-1/2} \delta(\mathbf{R} - \mathbf{R}') = 2ig(0, \mathbf{R} - \mathbf{R}'). \tag{1.81}$$

If the point source resides at the layer boundary $x_0 = L$, then the wavefield inside the layer $L_0 < x < L$ satisfies the equation

$$\left\{\frac{\partial^2}{\partial x^2} + \Delta_{\mathbf{R}} + k^2 [1 + \varepsilon(x, \mathbf{R})]\right\} G(x, \mathbf{R}; L, \mathbf{R}_0) = 0 \tag{1.82}$$

with the boundary conditions following from conditions (1.75), (1.76)

$$\begin{aligned} \left(\frac{\partial}{\partial x} + i\sqrt{k^2 + \Delta_{\mathbf{R}}}\right) G(x, \mathbf{R}; L, \mathbf{R}_0)\Big|_{x=L_0} &= 0, \\ \left(\frac{\partial}{\partial x} - i\sqrt{k^2 + \Delta_{\mathbf{R}}}\right) G(x, \mathbf{R}; L, \mathbf{R}_0)\Big|_{x=L} &= -\delta(\mathbf{R} - \mathbf{R}_0). \end{aligned} \tag{1.83}$$

Boundary-value problem (1.82), (1.83) can be reduced to an equivalent integral equation

$$
\begin{aligned}
&G(x,\mathbf{R};L,\mathbf{R}_0)\,g(x-L,\mathbf{R}-\mathbf{R}_0)\\
&+\int\limits_{L_0}^{L}dx'\int d\mathbf{R}'g(x-x',\mathbf{R}-\mathbf{R}')\,\varepsilon(x',\mathbf{R}')\,G(x',\mathbf{R}';L,\mathbf{R}_0)
\end{aligned}
\tag{1.84}
$$

coinciding with Eq. (1.77) for $x_0 = L$.

If the wave $u_0(x,\mathbf{R})$ is incident on the layer from region $x > L$ (in the negative direction of the x-axis), then the wavefield $U(x,\mathbf{R})$ inside the layer satisfies the Helmholtz equation

$$
\left\{\frac{\partial^2}{\partial x^2}+\Delta_{\mathbf{R}}+k^2\left[1+\varepsilon(x,\mathbf{R})\right]\right\}U(x,\mathbf{R})=0,
\tag{1.85}
$$

with the boundary conditions

$$
\begin{aligned}
\left(\frac{\partial}{\partial x}+i\sqrt{k^2+\Delta_{\mathbf{R}}}\right)U(x,\mathbf{R})\Big|_{x=L_0} &= 0,\\
\left(\frac{\partial}{\partial x}-i\sqrt{k^2+\Delta_{\mathbf{R}}}\right)U(x,\mathbf{R})\Big|_{x=L} &= -2i\sqrt{k^2+\Delta_{\mathbf{R}}}u_0(L,\mathbf{R}).
\end{aligned}
\tag{1.86}
$$

Similarly to the one-dimensional case, we can represent field $U(x,\mathbf{R})$ in the form

$$
\begin{aligned}
U(x,\mathbf{R}) &= u_1(x,\mathbf{R})+u_2(x,\mathbf{R}),\\
\frac{\partial}{\partial x}U(x,\mathbf{R}) &= -ik\sqrt{k^2+\Delta_{\mathbf{R}}}\left\{u_1(x,\mathbf{R})+u_2(x,\mathbf{R})\right\},
\end{aligned}
\tag{1.87}
$$

where we replaced function $U(x,\mathbf{R})$ with the sum of two functions $u_1(x,\mathbf{R})$ and $u_2(x,\mathbf{R})$ corresponding to the waves propagating in the negative and positive directions of the x-axis, respectively. These functions are related to field $U(x,\mathbf{R})$ through the expressions

$$
\begin{aligned}
u_1(x,\mathbf{R}) &= \frac{i}{2\sqrt{k^2+\Delta_{\mathbf{R}}}}\left(\frac{\partial}{\partial x}-i\sqrt{k^2+\Delta_{\mathbf{R}}}\right)U(x,\mathbf{R}),\\
u_2(x,\mathbf{R}) &= -\frac{i}{2\sqrt{k^2+\Delta_{\mathbf{R}}}}\left(\frac{\partial}{\partial x}+i\sqrt{k^2+\Delta_{\mathbf{R}}}\right)U(x,\mathbf{R})
\end{aligned}
\tag{1.88}
$$

following from Eq. (1.87)

Differentiating Eq. (1.88) with respect to x and using Eq. (1.85), we obtain the system of equations for functions $u_1(x,\mathbf{R})$ and $u_2(x,\mathbf{R})$ and derive the corresponding boundary conditions from (1.86) [236]

$$
\begin{aligned}
&\left(\frac{\partial}{\partial x}+i\sqrt{k^2+\Delta_{\mathbf{R}}}\right)u_1(x,\mathbf{R}) = -\frac{ik^2}{2\sqrt{k^2+\Delta_{\mathbf{R}}}}\left\{\varepsilon(x,\mathbf{R})U(x,\mathbf{R})\right\},\\
&\left(\frac{\partial}{\partial x}-i\sqrt{k^2+\Delta_{\mathbf{R}}}\right)u_2(x,\mathbf{R}) = \frac{ik^2}{2\sqrt{k^2+\Delta_{\mathbf{R}}}}\left\{\varepsilon(x,\mathbf{R})U(x,\mathbf{R})\right\},\\
&u_1(L,\mathbf{R}) = u_0(L,\mathbf{R}), \qquad u_2(L_0,\mathbf{R}) = 0.
\end{aligned}
\tag{1.89}
$$

Function $u_2(x,\mathbf{R})$ describes the wave propagating in the direction inverse to the direction of the incident wave, i.e., it describes the *backscattered field*.

Neglecting the backscattering effects, i.e., setting $u_2(x, \mathbf{R}) = 0$, we obtain the *generalized parabolic equation*

$$\begin{aligned}\left(\frac{\partial}{\partial x} + i\sqrt{k^2 + \Delta_{\mathbf{R}}}\right) U(x, \mathbf{R}) &= -\frac{ik^2}{2\sqrt{k^2 + \Delta_{\mathbf{R}}}} \{\varepsilon(x, \mathbf{R}) U(x, \mathbf{R})\}, \\ U(L, \mathbf{R}) &= u_0(L, \mathbf{R}),\end{aligned} \tag{1.90}$$

valid for waves scattered by arbitrary angles (less than $\pi/2$). In the case of small-angle scattering ($\Delta_{\mathbf{R}} \ll k^2$), we represent field $U(x, \mathbf{R})$ in the form

$$U(x, \mathbf{R}) = e^{-ik(x-L)} u(x, \mathbf{R}).$$

If we assume that the wave is incident on the inhomogeneous medium from half-space $x < 0$ (i.e., if we replace $L - x$ with x), then Eq. (1.90) reduces to the *parabolic equation of quasi-optics*,

$$\frac{\partial}{\partial x} u(x, \mathbf{R}) = \frac{i}{2k} \Delta_{\mathbf{R}} u(x, \mathbf{R}) + \frac{ik}{2} \varepsilon(x, \mathbf{R}) u(x, \mathbf{R}), \qquad u(0, \mathbf{R}) = u_0(\mathbf{R}), \tag{1.91}$$

which concerns the wave propagation in media with large-scale three-dimensional inhomogeneities responsible for small-angle scattering. It was successfully used in many problems on wave propagation in Earth's atmosphere and ocean.

There is the waste literature on derivation and basing of both parabolic and generalized parabolic equations. Appendix C, page 491 gives such a derivation in terms of the imbedding method.

Introducing the amplitude-phase representation of the wavefield in Eq. (1.91) by the formula

$$u(x, \mathbf{R}) = A(x, \mathbf{R}) e^{iS(x, \mathbf{R})},$$

we can write the equation for the wavefield intensity $I(x, \mathbf{R}) = u(x, \mathbf{R}) u^*(x, \mathbf{R})$ in the form

$$\frac{\partial}{\partial x} I(x, \mathbf{R}) + \frac{1}{k} \nabla_{\mathbf{R}} \{\nabla_{\mathbf{R}} S(x, \mathbf{R}) I(x, \mathbf{R})\} = 0, \qquad I(0, \mathbf{R}) = I_0(\mathbf{R}). \tag{1.92}$$

From this equation follows that the power of a wave in plane $x = \text{const}$ is conserved in the general case of arbitrary incident wave beam:

$$E_0 = \int I(x, \mathbf{R}) d\mathbf{R} = \int I_0(\mathbf{R}) d\mathbf{R}.$$

Equation (1.92) coincides in form with Eq. (1.39). Consequently, we can treat it as the transport equation for the conservative tracer in the potential velocity field. However, this tracer can be considered the passive tracer only in the *geometrical optics approximation*, in which case the phase of the wave, the transverse gradient of the phase $\mathbf{p}(x, \mathbf{R}) = \frac{1}{k} \nabla_{\mathbf{R}} S(x, \mathbf{R})$, and the matrix of the phase second derivatives $u_{ij}(x, \mathbf{R}) = \frac{1}{k} \frac{\partial^2}{\partial R_i \partial R_j} S(x, \mathbf{R})$ characterizing the *curvature of the phase front* $S(x, \mathbf{R}) = \text{const}$ satisfy the closed system of equations [134, 135]

$$\begin{aligned}&\frac{\partial}{\partial x} S(x, \mathbf{R}) + \frac{k}{2} \mathbf{p}^2(x, \mathbf{R}) = \frac{k}{2} \varepsilon(x, \mathbf{R}), \\ &\left(\frac{\partial}{\partial x} + \mathbf{p}(x, \mathbf{R}) \nabla_{\mathbf{R}}\right) \mathbf{p}(x, \mathbf{R}) = \frac{1}{2} \nabla_{\mathbf{R}} \varepsilon(x, \mathbf{R}), \\ &\left(\frac{\partial}{\partial x} + \mathbf{p}(x, \mathbf{R}) \nabla_{\mathbf{R}}\right) u_{ij}(x, \mathbf{R}) + u_{ik}(x, \mathbf{R}) u_{kj}(x, \mathbf{R}) = \frac{1}{2} \frac{\partial^2}{\partial R_i \partial R_j} \varepsilon(x, \mathbf{R}),\end{aligned} \tag{1.93}$$

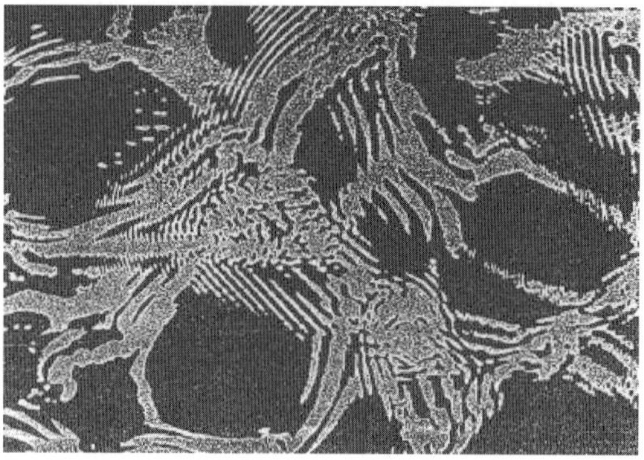

Figure 1.11: Transverse section of a laser beam in turbulent medium.

Figure 1.12: Caustics in a pool.

In the general case, i.e., with the inclusion of diffraction effects, this tracer is the active tracer.

According to the material of the previous section, realizations of intensity must show the cluster behavior, which manifests itself in the appearance of *caustic* structures. An example demonstrating the appearance of the wavefield caustic structures is given in Fig. 1.11, which is a fragment of the photo on the back of the cover — the flyleaf — of book [268] that shows the transverse section of the laser beam propagating in the turbulent atmosphere (see also papers [64, 65, 103] for the results of laboratory investigations and simulations).

A photo of the pool in Fig. 1.12 also shows the prominent caustic structure of the wavefield on the pool bottom. Such structures appear due to light refraction and reflection by the water rough surface, which corresponds to scattering by the so-called *phase screen*.

Consider now the geometrical optics approximation (1.93) for parabolic equation (1.91). In this approximation, the equation for the phase of the wave is the Hamilton–Jacobi equation and the equation for the transverse gradient of the phase (1.93) is the closed quasilinear first-order partial differential equation, and we can solve it by the method of characteristics (see, e.g., [310]). Equations for the *characteristic curves* (*rays*) have the form

$$\frac{d}{dx}\mathbf{R}(x) = \mathbf{p}(x), \quad \frac{d}{dx}\mathbf{p}(x) = \frac{1}{2}\nabla_{\mathbf{R}}\varepsilon(x, \mathbf{R}), \tag{1.94}$$

and the wavefield intensity and matrix of the phase second derivatives along the characteristic curves will satisfy the equations

$$\begin{aligned} &\frac{d}{dx}I(x) = -I(x)u_{ii}(x), \\ &\frac{d}{dx}u_{ij}(x) + u_{ik}(x)u_{kj}(x) = \frac{1}{2}\frac{\partial^2}{\partial R_i \partial R_j}\varepsilon(x, \mathbf{R}). \end{aligned} \tag{1.95}$$

Equations (1.94) coincide in appearance with the equations for a particle under random external forces in the absence of friction (1.12) and form the system of the Hamilton equations.

In the two-dimensional case ($R = y$), Eqs. (1.94), (1.95) become significantly simpler and assume the form

$$\begin{aligned} &\frac{d}{dx}y(x) = p(x), \quad \frac{d}{dx}p(x) = \frac{1}{2}\frac{\partial}{\partial y}\varepsilon(x, y), \\ &\frac{d}{dx}I(x) = -I(x)u(x), \quad \frac{d}{dx}u(x) + u^2(x) = \frac{1}{2}\frac{\partial^2}{\partial y^2}\varepsilon(x, y). \end{aligned} \tag{1.96}$$

The last equation for $u(x)$ in (1.96) is similar to Eq. (1.14) whose solution shows the singular behavior. The only difference between these equations consists in the random term that has now a more complicated structure. Nevertheless, it is quite clear that solutions to stochastic problem (1.96) will show the blow-up behavior; namely, function $u(x)$ will reach minus infinity and intensity will reach plus infinity at a finite distance. Such a behavior of a wavefield in randomly inhomogeneous media corresponds to *random focusing*, i.e., to the formation of caustics, which means the appearance of points of multivaluedness (and discontinuity) in the solutions to quasilinear equation (1.93) for the transverse gradient of the wavefield phase.

1.4.3 The Navier–Stokes equation: random forces in hydrodynamic theory of turbulence

Consider now the turbulent motion model that assumes the presence of external forces $\mathbf{f}(\mathbf{r}, t)$ acting on the liquid. Such a model is evidently only imaginary, because there is no actual analogues for these forces. However, assuming that forces $\mathbf{f}(\mathbf{r}, t)$ on average ensure an appreciable energy income only to large-scale velocity components, we can expect that, within the concepts of the theory of local isotropic turbulence, the imaginary nature of field $\mathbf{f}(\mathbf{r}, t)$ will only slightly affect statistical properties of small-scale turbulent components [251]. Consequently, this model is quite appropriate for describing small-scale properties of turbulence.

Motion of an incompressible liquid under external forces is governed by the *Navier–Stokes equation*

$$\begin{aligned} &\left(\frac{\partial}{\partial t} + \mathbf{u}(\mathbf{r}, t)\frac{\partial}{\partial \mathbf{r}}\right)\mathbf{u}(\mathbf{r}, t) = -\frac{1}{\rho_0}\frac{\partial}{\partial \mathbf{r}}p(\mathbf{r}, t) + \nu\Delta\mathbf{u}(\mathbf{r}, t) + \mathbf{f}(\mathbf{r}, t), \\ &\frac{\partial}{\partial \mathbf{r}}\mathbf{u}(\mathbf{r}, t) = 0, \quad \frac{\partial}{\partial \mathbf{r}}\mathbf{f}(\mathbf{r}, t) = 0. \end{aligned} \tag{1.97}$$

Here, ρ_0 is the density of the liquid, ν is the kinematic viscosity, and pressure field $p(\mathbf{x}, t)$ is expressed in terms of the velocity field at the same instant by the relationship

$$p(\mathbf{r}, t) = -\rho_0 \int \Delta^{-1}(\mathbf{r}, \mathbf{r}') \frac{\partial^2 (u_i(\mathbf{r}', t) u_j(\mathbf{r}', t))}{\partial r'_i \partial r'_j} d\mathbf{r}', \tag{1.98}$$

where $\Delta^{-1}(\mathbf{r}, \mathbf{r}')$ is the integral operator inverse to the Laplace operator (repeated indexes assume summation).

If we substitute Eq. (1.98) in Eq. (1.97) to exclude the pressure field, then we obtain in the three-dimensional case that the Fourier transform of the velocity field with respect to spatial coordinates

$$\hat{u}_i(\mathbf{k}, t) = \int d\mathbf{r} u_i(\mathbf{r}, t) e^{-i\mathbf{k}\mathbf{r}}, \quad u_i(\mathbf{r}, t) = \frac{1}{(2\pi)^3} \int d\mathbf{k} \hat{u}_i(\mathbf{k}, t) e^{i\mathbf{k}\mathbf{r}},$$

($\hat{u}_i^*(\mathbf{k}, t) = \hat{u}_i(-\mathbf{k}, t)$) satisfies the nonlinear integro-differential equation

$$\begin{aligned} &\frac{\partial}{\partial t} \hat{u}_i(\mathbf{k}, t) + \frac{i}{2} \int d\mathbf{k}_1 \int d\mathbf{k}_2 \Lambda_i^{\alpha\beta}(\mathbf{k}_1, \mathbf{k}_2, \mathbf{k}) \hat{u}_\alpha(\mathbf{k}_1, t) \hat{u}_\beta(\mathbf{k}_2, t) - \nu k^2 \hat{u}_i(\mathbf{k}, t) \\ &\quad = \hat{f}_i(\mathbf{k}, t), \end{aligned} \tag{1.99}$$

where

$$\Lambda_i^{\alpha\beta}(\mathbf{k}_1, \mathbf{k}_2, \mathbf{k}) = \frac{1}{(2\pi)^3} \{k_\alpha \Delta_{i\beta}(\mathbf{k}) + k_\beta \Delta_{i\alpha}(\mathbf{k})\} \delta(\mathbf{k}_1 + \mathbf{k}_2 - \mathbf{k}),$$

$$\Delta_{ij}(\mathbf{k}) = \delta_{ij} - \frac{k_i k_j}{\mathbf{k}^2} \quad (i, \alpha, \beta = 1, 2, 3),$$

and $\widehat{\mathbf{f}}(\mathbf{k}, t)$ is the spatial Fourier harmonics of external forces,

$$\widehat{\mathbf{f}}(\mathbf{k}, t) = \int d\mathbf{r} \mathbf{f}(\mathbf{r}, t) e^{-i\mathbf{k}\mathbf{r}}, \quad \mathbf{f}(\mathbf{r}, t) = \frac{1}{(2\pi)^3} \int d\mathbf{k} \widehat{\mathbf{f}}(\mathbf{k}, t) e^{i\mathbf{k}\mathbf{r}}.$$

A specific feature of the three-dimensional hydrodynamic motions consists in the fact that the absence of external forces and viscosity-driven effects is sufficient for energy conservation.

It appears convenient to describe the stationary turbulence in terms of the space–time Fourier harmonics of the velocity field

$$\hat{u}_i(\mathbf{K}) = \int d\mathbf{x} \int_{-\infty}^{\infty} dt u_i(\mathbf{x}, t) e^{-i(\mathbf{k}\mathbf{x} + \omega t)}, \quad u_i(\mathbf{x}, t) = \frac{1}{(2\pi)^4} \int d\mathbf{k} \int_{-\infty}^{\infty} d\omega \hat{u}_i(\mathbf{K}) e^{i(\mathbf{k}\mathbf{x} + \omega t)},$$

where $\mathbf{K}$ is the four-dimensional wave vector $\{\mathbf{k}, \omega\}$ and field $\hat{u}_i^*(\mathbf{K}) = \hat{u}_i(-\mathbf{K})$ because field $u_i(\mathbf{r}, t)$ is real. In this case, we obtain the equation for component $\hat{u}_i(\mathbf{K})$ by accomplishing the Fourier transformation of Eq. (1.99) with respect to time:

$$\begin{aligned} &(i\omega + \nu \mathbf{k}^2) \hat{u}_i(\mathbf{K}) \\ &\quad + \frac{i}{2} \int d^4\mathbf{K}_1 \int d^4\mathbf{K}_2 \Lambda_i^{\alpha\beta}(\mathbf{K}_1, \mathbf{K}_2, \mathbf{K}) \hat{u}_\alpha(\mathbf{K}_1) \hat{u}_\beta(\mathbf{K}) = \hat{f}_i(\mathbf{K}), \end{aligned} \tag{1.100}$$

where

$$\Lambda_i^{\alpha\beta}(\mathbf{K}_1, \mathbf{K}_2, \mathbf{K}) = \frac{1}{2\pi} \Lambda_i^{\alpha\beta}(\mathbf{k}_1, \mathbf{k}_2, \mathbf{k}) \delta(\omega_1 + \omega_2 - \omega),$$

and $\hat{f}_i(\mathbf{K})$ are the space–time Fourier harmonics of external forces. The obtained Eq. (1.100) is now the integral (not integro-differential) nonlinear equation.

1.4.4 Equations of geophysical hydrodynamics

Consider now the description of hydrodynamic flows on the rotating Earth in the so-called *quasi-geostrophic approximation* [260]. In the simplest case of the one-layer model, the incompressible liquid flow in the two-dimensional plane $\mathbf{R} = (x, y)$ is described by the *stream function* that satisfies the equation

$$\frac{\partial}{\partial t}\Delta\psi(\mathbf{R},t)+\beta_0\frac{\partial}{\partial x}\psi(\mathbf{R},t) = J\left\{\Delta\psi(\mathbf{R},t) + h(\mathbf{R}); \psi(\mathbf{R},t)\right\}, \quad \psi(\mathbf{R},0) = \psi_0(\mathbf{R}), \tag{1.101}$$

where parameter β_0 is the derivative of the local *Coriolis parameter* f_0 with respect to latitude, $J\{\psi, \varphi\}$ is the Jacobian of two functions $\psi(\mathbf{R}, t)$ and $\varphi(\mathbf{R}, t)$

$$J\left\{\psi(\mathbf{R},t);\varphi(\mathbf{R},t)\right\} = \frac{\partial\psi(\mathbf{R},t)}{\partial x}\frac{\partial\varphi(\mathbf{R},t)}{\partial y} - \frac{\partial\varphi(\mathbf{R},t)}{\partial x}\frac{\partial\psi(\mathbf{R},t)}{\partial y},$$

and function $h(\mathbf{R}) = f_0\tilde{h}(\mathbf{R})/H_0$ is the deviation of bottom topography $\tilde{h}(\mathbf{R})$ relative its average thickness H_0. The velocity field is expressed in terms of the stream function by the relationship

$$\mathbf{v}(\mathbf{R},t) = \left(-\frac{\partial\psi(\mathbf{R},t)}{\partial y}, \frac{\partial\psi(\mathbf{R},t)}{\partial x}\right).$$

Note that, under the neglect of Earth's rotation and effects of underlying surface topography, Eq. (1.101) reduces to the standard equation of two-dimensional hydrodynamics (see, e.g., [217]).

Equation (1.101) describes the *barotropic* motion of a liquid. In the more general case of *baroclinic* motions, investigation is usually carried out within the framework of the two-layer model of hydrodynamic flows described by the system of equations [260]

$$\begin{aligned}
\frac{\partial}{\partial t}\left[\Delta\psi_1 - \alpha_1 F\left(\psi_1 - \psi_2\right)\right] + \beta_0\frac{\partial}{\partial x}\psi_1 &= J\left\{\Delta\psi_1 - \alpha_1 F\left(\psi_1 - \psi_2\right); \psi_1\right\}, \\
\frac{\partial}{\partial t}\left[\Delta\psi_2 - \alpha_2 F\left(\psi_2 - \psi_1\right)\right] + \beta_0\frac{\partial}{\partial x}\psi_2 &= J\left\{\Delta\psi_2 - \alpha_2 F\left(\psi_2 - \psi_1\right) + f_0\alpha_2 h; \psi_2\right\},
\end{aligned} \tag{1.102}$$

where additional parameters $F = f_0^2\rho/g(\Delta\rho)$ and $\Delta\rho/\rho = (\rho_2 - \rho_1)/\rho_0 > 0$ are introduced and $\alpha_1 = 1/H_1$ and $\alpha_2 = 1/H_2$ are the inverse thicknesses of layers.

Among the particular cases of Eqs. (1.101), (1.102) are the equations obtained by neglecting Earth's rotation (two-dimensional hydrodynamics) but with allowance for bottom topography and the linearized quasi-geostrophic equations similar to Eq. (1.38) that describe the effect of topography on propagation of the Rossby waves.

1.5 Solution dependence on medium parameters and initial value

Below, we considered a number of dynamic systems described by both ordinary and partial differential equations. Many applications concerning research of statistical characteristics of the solutions to these equations require the knowledge of the solution dependence (generally, in the functional form) on the medium parameters appeared in the equation as coefficients and the initial values. Some properties appear common of all such dependencies, and two of them are of special interest in the context of statistical descriptions. We illustrate these dependencies using the simplest problem, namely, the system of ordinary differential equations (1.1) that describes particle dynamics in random velocity field and which we reformulate in the form of the nonlinear integral equation

$$\mathbf{r}(t) = \mathbf{r}_0 + \int_{t_0}^{t} \mathbf{U}(\mathbf{r}(\tau), \tau) d\tau, \tag{1.103}$$

as an example.

The solution to Eq. (1.103) functionally depends on vector field $\mathbf{U}(\mathbf{r}', \tau)$ and initial values $\mathbf{r}_0$, t_0.

1.5.1 Principle of dynamic causality

Vary Eq. (1.103) with respect to field $\mathbf{U}(\mathbf{r}, t)$. Assuming that the initial position $\mathbf{r}_0$ is independent of field $\mathbf{U}$, we obtain the equation linear in variational derivative (the linear variational differential equation)

$$\frac{\delta r_i(t)}{\delta U_j(\mathbf{r}, t')} = \delta_{ij}\delta(\mathbf{r} - \mathbf{r}(t'))\,\theta(t' - t_0)\,\theta(t - t') + \int_{t_0}^{t} \frac{\partial U_i(\mathbf{r}(\tau), \tau)}{\partial r_k} \frac{\delta r_k(\tau)}{\delta U_j(\mathbf{r}, t')} d\tau, \tag{1.104}$$

where $\delta(\mathbf{r} - \mathbf{r}')$ is the Dirac delta function, and $\theta(z)$ is the Heaviside step function. From Eq. (1.104) follows that

$$\frac{\delta r_i(t)}{\delta U_j(\mathbf{r}, t')} = 0 \quad \text{for } t' > t \text{ or } t' < t_0, \tag{1.105}$$

which means that solution to the dynamic problem (1.103) $\mathbf{r}(t)$ as a functional of field $\mathbf{U}(\mathbf{r}, t')$ depends only on $\mathbf{U}(\mathbf{r}, t')$ for $t_0 < t' < t$. Consequently, function $\mathbf{r}(t)$ will remain unchanged if field $\mathbf{U}(\mathbf{r}, t')$ varies outside the interval (t_0, t), i.e., for $t' < t_0$ or $t' > t$. We will call condition (1.105) the *dynamic causality condition*.

Taking this condition into account, we can rewrite Eq. (1.104) in the form

$$\frac{\delta r_i(t)}{\delta U_j(\mathbf{r}, t')} = \delta_{ij}\delta(\mathbf{r} - \mathbf{r}(t'))\,\theta(t' - t_0)\,\theta(t - t') + \int_{t'}^{t} \frac{\partial U_i(\mathbf{r}(\tau), \tau)}{\partial r_k} \frac{\delta r_k(\tau)}{\delta U_j(\mathbf{r}, t')} d\tau. \tag{1.106}$$

As a consequence, proceeding to limit $t \to t' + 0$, we obtain the equality

$$\left.\frac{\delta r_i(t)}{\delta U_j(\mathbf{r}, t')}\right|_{t=t'+0} = \delta_{ij}\delta(\mathbf{r} - \mathbf{r}(t')). \tag{1.107}$$

Integral equation (1.106) in variational derivative is obviously equivalent to the linear differential equation with the initial value

$$\frac{\partial}{\partial t}\left(\frac{\delta r_i(t)}{\delta U_j(\mathbf{r}, t')}\right) = \frac{\partial U_i(\mathbf{r}(t), t)}{\partial r_k}\left(\frac{\delta r_k(t)}{\delta U_j(\mathbf{r}, t')}\right), \quad \left.\frac{\delta r_i(t)}{\delta U_j(\mathbf{r}, t')}\right|_{t=t'} = \delta_{ij}\delta(\mathbf{r} - \mathbf{r}(t')). \tag{1.108}$$

The dynamic causality condition is the general property of problems described by differential equations with initial values. The boundary-value problems possess no such property. Indeed, in the case of problem (1.16), (1.17) that describes propagation of a plane wave in a layer of inhomogeneous medium, wavefield $u(x)$ at point x and reflection and transmission coefficients depend functionally on function $\varepsilon(x)$ for all x of layer (L_0, L). However, using the imbedding method, we can convert this problem into the initial value problem with respect to an auxiliary parameter L and make use the causality property in terms of the equations of the imbedding method.

1.5.2 Solution dependence on initial value

We will use now the vertical bar symbol to isolate the dependence of the solution to Eq. (1.103) $\mathbf{r}(t)$ on the initial parameters $\mathbf{r}_0$ and t_0:

$$\mathbf{r}(t) = \mathbf{r}(t|\mathbf{r}_0, t_0), \quad \mathbf{r}_0 = \mathbf{r}(t_0|\mathbf{r}_0, t_0).$$

Let us differentiate Eq. (1.103) with respect to parameters r_{0k} and t_0. As a result, we obtain linear equations for Jacobi's matrix $\frac{\partial}{\partial r_{0k}} r_i(t|\mathbf{r}_0, t_0)$ and quantity $\frac{\partial}{\partial t_0} r_i(t|\mathbf{r}_0, t_0)$

$$\begin{aligned}
\frac{\partial r_i(t|\mathbf{r}_0, t_0)}{\partial r_{0k}} &= \delta_{ik} + \int_{t_0}^{t} d\tau \frac{\partial U_i(\mathbf{r}(\tau), \tau)}{\partial r_j} \frac{\partial r_j(\tau|\mathbf{r}_0, t_0)}{\partial r_{0k}}, \\
\frac{\partial r_i(t|\mathbf{r}_0, t_0)}{\partial t_0} &= -U_i(\mathbf{r}_0(t_0), t_0) + \int_{t_0}^{t} d\tau \frac{\partial U_i(\mathbf{r}(\tau), \tau)}{\partial r_j} \frac{\partial r_j(\tau|\mathbf{r}_0, t_0)}{\partial t_0}.
\end{aligned} \tag{1.109}$$

Multiplying now the first of these equations by $U_k(\mathbf{r}_0(t), t)$, summing over index k, adding the result to the second equation, and introducing the vector function

$$F_i(t|\mathbf{r}_0, t_0) = \left(\frac{\partial}{\partial t_0} + \mathbf{U}(\mathbf{r}_0, t_0) \frac{\partial}{\partial \mathbf{r}_0}\right) r_i(t|\mathbf{r}_0, t_0),$$

we obtain that this function satisfies the linear homogeneous equation

$$F_i(t|\mathbf{r}_0, t_0) = \int_{t_0}^{t} d\tau \frac{\partial U_i(\mathbf{r}(\tau), \tau)}{\partial r_k} F_k(\tau|\mathbf{r}_0, t_0),$$

whose solution is obviously $F_i(t|\mathbf{r}_0, t_0) = 0$. Therefore, we obtain the equality

$$\left(\frac{\partial}{\partial t_0} + \mathbf{U}(\mathbf{r}_0, t_0) \frac{\partial}{\partial \mathbf{r}_0}\right) r_i(t|\mathbf{r}_0, t_0) = 0, \tag{1.110}$$

which can be considered as the linear partial differential equation with the derivatives with respect to variables $\mathbf{r}_0$, t_0 and the initial value at $t_0 = t$

$$\mathbf{r}(t|\mathbf{r}_0, t) = \mathbf{r}_0. \tag{1.111}$$

The variable t appears now in problem (1.110), (1.111) as a parameter.

Equation (1.110) is solved using the time direction inverse to that used in solving problem (1.1); for this reason, we will call it *the backward equation*.

Equation (1.110) with initial value (1.111) obviously possess the property of dynamic causality with respect to parameter t_0. This means that

$$\frac{\delta \mathbf{r}(t|\mathbf{r}_0, t_0)}{\delta U_j(\mathbf{y}, t')} = 0, \quad \text{if } t' > t \quad \text{or} \quad t' < t_0$$

and, as follows from Eq. (1.110),

$$\left.\frac{\delta \mathbf{r}(t|\mathbf{r}_0, t_0)}{\delta U_j(\mathbf{r}, t')}\right|_{t'=t_0+0} = \delta(\mathbf{r}_0 - \mathbf{r}) \frac{\partial \mathbf{r}(t|\mathbf{r}_0, t_0)}{\partial r_{0j}}. \tag{1.112}$$

Chapter 2

Indicator function and Liouville equation

Modern apparatus of the theory of random processes is able of constructing closed descriptions of dynamic systems if these systems meet the condition of dynamic causality and are formulated in terms of linear partial differential equations or certain types of integral equations (see Chapter 5). One can use indicator functions to perform the transition from the initial, generally nonlinear system to the equivalent description in terms of the linear partial differential equations. However, this approach results in increasing the dimension of the space of variables. Consider such a transition using the dynamic systems described in the previous chapter.

2.1 Ordinary differential equations

Assume that a stochastic problem is described by the system of equations (1.1), page 2

$$\frac{d}{dt}\mathbf{r}(t) = \mathbf{U}\left(\mathbf{r}(t), t\right), \quad \mathbf{r}(t_0) = \mathbf{r}_0. \tag{2.1}$$

We introduce the scalar function

$$\varphi(t; \mathbf{r}) = \delta(\mathbf{r}(t) - \mathbf{r}), \tag{2.2}$$

which is concentrated on the section of the random process $\mathbf{r}(t)$ by a given plane $\mathbf{r}(t) =$ const and is usually called the *indicator function.*

Differentiating Eq. (2.2) with respect to time t, we obtain, using Eq. (2.1), the equality

$$\frac{\partial}{\partial t}\varphi(t; \mathbf{r}) = -\frac{\partial}{\partial \mathbf{r}}\delta(\mathbf{r}(t) - \mathbf{r})\frac{d\mathbf{r}(t)}{dt} = -\frac{\partial}{\partial \mathbf{r}}\delta(\mathbf{r}(t) - \mathbf{r})\mathbf{U}\left(\mathbf{r}(t), t\right).$$

Using then the *probing property* of the delta-function

$$\delta(\mathbf{r}(t) - \mathbf{r})\mathbf{U}\left(\mathbf{r}(t), t\right) = \delta(\mathbf{r}(t) - \mathbf{r})\mathbf{U}\left(\mathbf{r}, t\right),$$

we obtain the linear partial differential equation

$$\left(\frac{\partial}{\partial t} + \frac{\partial}{\partial \mathbf{r}}\mathbf{U}(\mathbf{r}, t)\right)\varphi(t; \mathbf{r}) = 0, \qquad \varphi(t_0; \mathbf{r}) = \delta(\mathbf{r}_0 - \mathbf{r}) \tag{2.3}$$

equivalent to the initial system. This equation is called the *Liouville equation.*

The transition from system (2.1) to Liouville equation (2.3) entails enlargement of the phase space $(t, \mathbf{r})$, which, however, has the finite dimension. Note that Eq. (2.3) coincides in form with the equation of tracer transfer by the velocity field $\mathbf{U}(\mathbf{r}, t)$ (1.39); the only difference consists in the initial values.

The solution to Eq. (2.1) and, consequently, function (2.2) depends on the initial values t_0, $\mathbf{r}_0$. Indeed, function $\mathbf{r}(t) = \mathbf{r}(t|\mathbf{r}_0, t_0)$ as a function of variables $\mathbf{r}_0$ and t_0 satisfies the linear first-order partial differential equation (1.110). The equations of such type also allow the transition to the equations in the indicator function $\varphi(t; \mathbf{r}|t_0, \mathbf{r}_0)$ (see the next section); in the case under consideration, this equation is again the linear first-order partial differential equation including the derivatives with respect to variables $\mathbf{r}_0$ and t_0

$$\left(\frac{\partial}{\partial t_0} + \mathbf{U}(\mathbf{r}_0, t_0)\frac{\partial}{\partial \mathbf{r}_0}\right)\varphi(t; \mathbf{r}|t_0, \mathbf{r}_0) = 0, \quad \varphi(t; \mathbf{r}|t, \mathbf{r}_0) = \delta(\mathbf{r}_0 - \mathbf{r}). \tag{2.4}$$

Equation (2.4) can be called the *backward Liouville equation*

2.2 First-order partial differential equations

If the initial value problem is formulated in terms of partial differential equations, we always can convert it to the equivalent formulation in terms of the linear variational differential equation in the infinite-dimensional space (the *Hopf equation*) [116]–[118] (see also [134, 135, 251]). For some particular types of problems, this approach is simplified. Indeed, if the initial dynamic system satisfies the first-order partial differential equation (either *linear* as Eq. (1.39), page 19, or *quasilinear* as Eq. (1.52), page 22, or in the general case *nonlinear* as Eq. (1.64), page 26), then the phase space of the corresponding indicator function will be the finite-dimension space [134, 135], which follows from the fact that first-order partial differential equations are equivalent to systems of ordinary (characteristic) differential equations. Consider these cases in more details.

2.2.1 Linear equations

Consider the problem on tracer transfer by random velocity field in more details. The problem is formulated in terms of Eq. (1.39), page 19 that we rewrite in the form

$$\left(\frac{\partial}{\partial t} + \mathbf{U}(\mathbf{r}, t)\frac{\partial}{\partial \mathbf{r}}\right)\rho(\mathbf{r}, t) + \frac{\partial \mathbf{U}(\mathbf{r}, t)}{\partial \mathbf{r}}\rho(\mathbf{r}, t) = 0, \quad \rho(\mathbf{r}, 0) = \rho_0(\mathbf{r}). \tag{2.5}$$

To describe the density field in the Eulerian description, we introduce the indicator function

$$\varphi(t, \mathbf{r}; \rho) = \delta(\rho(t, \mathbf{r}) - \rho), \tag{2.6}$$

which is similar to function (2.2) and is localized on surface $\rho(\mathbf{r}, t) = \rho = \text{const}$ in the three-dimensional case or on a contour in the two-dimensional case. An equation for this function can be easily obtained either immediately from Eq. (2.5), or from the Liouville equation in the Lagrangian description. Indeed, differentiating Eq. (2.6) with respect to time and using dynamic equation (2.5) and probing property of the delta-function, we obtain the equation

$$\frac{\partial}{\partial t}\varphi(t, \mathbf{r}; \rho) = \frac{\partial \mathbf{U}(\mathbf{r}, t)}{\partial \mathbf{r}}\frac{\partial}{\partial \rho}\rho\varphi(t, \mathbf{r}; \rho) + \mathbf{U}(\mathbf{r}, t)\frac{\partial \rho(\mathbf{r}, t)}{\partial \mathbf{r}}\frac{\partial}{\partial \rho}\varphi(t, \mathbf{r}; \rho). \tag{2.7}$$

However, this equation is not closed because the right-hand side includes term $\partial\rho(\mathbf{r},t)/\partial\mathbf{r}$ that cannot be explicitly expressed through $\rho(\mathbf{r},t)$.

On the other hand, differentiating function (2.6) with respect to $\mathbf{r}$, we obtain the equality

$$\frac{\partial}{\partial\mathbf{r}}\varphi(t,\mathbf{r};\rho) = -\frac{\partial\rho(\mathbf{r},t)}{\partial\mathbf{r}}\frac{\partial}{\partial\rho}\varphi(t,\mathbf{r};\rho). \tag{2.8}$$

Eliminating now the last term in Eq. (2.7) with the use of (2.8), we obtain the closed Liouville equation in the Eulerian description

$$\left(\frac{\partial}{\partial t}+\mathbf{U}(\mathbf{r},t)\frac{\partial}{\partial\mathbf{r}}\right)\varphi(t,\mathbf{r};\rho) = \frac{\partial\mathbf{U}(\mathbf{r},t)}{\partial\mathbf{r}}\frac{\partial}{\partial\rho}\left[\rho\varphi(t,\mathbf{r};\rho)\right], \quad \varphi(0,\mathbf{r};\rho)=\delta(\rho_0(\mathbf{r})-\rho). \tag{2.9}$$

To obtain a more complete description, we consider the extended indicator function including both density field $\rho(\mathbf{r},t)$ and its spatial gradient $\mathbf{p}(\mathbf{r},t)=\boldsymbol{\nabla}\rho(\mathbf{r},t)$

$$\varphi(t,\mathbf{r};\rho,\mathbf{p}) = \delta\left(\rho(\mathbf{r},t)-\rho\right)\delta\left(\mathbf{p}(\mathbf{r},t)-\mathbf{p}\right). \tag{2.10}$$

Differentiating Eq. (2.10) with respect to time, we obtain the equality

$$\frac{\partial}{\partial t}\varphi(t,\mathbf{r};\rho,\mathbf{p}) = -\left[\frac{\partial}{\partial\rho}\frac{\partial\rho(\mathbf{r},t)}{\partial t}+\frac{\partial}{\partial p_i}\frac{\partial p_i(\mathbf{r},t)}{\partial t}\right]\varphi(t,\mathbf{r};\rho,\mathbf{p}). \tag{2.11}$$

Using now dynamic equations (2.5) for density and Eq. (1.50), page 22 for the density spatial gradient, we can rewrite Eq. (2.11) as the equation

$$\begin{aligned}\frac{\partial}{\partial t}\varphi(t,\mathbf{r};\rho,\mathbf{p}) &= \frac{\partial}{\partial\rho}\left[\frac{\partial\mathbf{U}(\mathbf{r},t)}{\partial\mathbf{r}}\rho+\mathbf{U}(\mathbf{r},t)\mathbf{p}\right]\varphi(t,\mathbf{r};\rho,\mathbf{p})\\ &+\frac{\partial}{\partial p_i}\left[\mathbf{U}(\mathbf{r},t)\frac{\partial p_i(\mathbf{r},t)}{\partial\mathbf{r}}+\frac{\partial\mathbf{U}(\mathbf{r},t)}{\partial\mathbf{r}}p_i+p_k\frac{\partial U_k(\mathbf{r},t)}{\partial r_i}+\rho\frac{\partial^2\mathbf{U}(\mathbf{r},t)}{\partial r_i\partial\mathbf{r}}\right]\varphi(t,\mathbf{r};\rho,\mathbf{p}),\end{aligned} \tag{2.12}$$

which is not closed because of the term $\partial p_i(\mathbf{r},t)/\partial\mathbf{r}$ in the right-hand side.

Differentiating function (2.10) with respect to $\mathbf{r}$, we obtain the equality

$$\frac{\partial}{\partial\mathbf{r}}\varphi(t,\mathbf{r};\rho,\mathbf{p}) = -\left[\mathbf{p}\frac{\partial}{\partial\rho}+\frac{\partial p_i(\mathbf{r},t)}{\partial\mathbf{r}}\frac{\partial}{\partial p_i}\right]\varphi(t,\mathbf{r};\rho,\mathbf{p}). \tag{2.13}$$

Now, multiplying Eq. (2.13) by $\mathbf{U}(\mathbf{r},t)$ and adding the result to Eq. (2.12), we obtain the closed Liouville equation for the extended indicator function

$$\begin{aligned}&\left(\frac{\partial}{\partial t}+\mathbf{U}(\mathbf{r},t)\frac{\partial}{\partial\mathbf{r}}\right)\varphi(t,\mathbf{r};\rho,\mathbf{p})\\ &=\left[\frac{\partial U_k(\mathbf{r},t)}{\partial r_i}\frac{\partial}{\partial p_i}p_k+\frac{\partial\mathbf{U}(\mathbf{r},t)}{\partial\mathbf{r}}\left(\frac{\partial}{\partial\rho}\rho+\frac{\partial}{\partial\mathbf{p}}\mathbf{p}\right)+\frac{\partial^2\mathbf{U}(\mathbf{r},t)}{\partial r_i\partial\mathbf{r}}\frac{\partial}{\partial p_i}\rho\right]\varphi(t,\mathbf{r};\rho,\mathbf{p}),\\ &\quad\varphi(0,\mathbf{r};\rho,\mathbf{p})=\delta\left(\rho_0(\mathbf{r})-\rho\right)\delta\left(\mathbf{p}_0(\mathbf{r})-\mathbf{p}\right).\end{aligned} \tag{2.14}$$

Derive now Eq. (2.9) starting from the Lagrangian description of the dynamic system. In the Lagrangian representation, the behavior of passive tracer is described in terms of ordinary differential equations (1.41), (1.42), and (1.44), page 20. Using these equations, one can easily derive the linear Liouville equation in the corresponding phase space for the function

$$\varphi_{\rm Lag}(t;\mathbf{r},\rho,j|\mathbf{r}_0)=\delta(\mathbf{r}(t|\mathbf{r}_0)-\mathbf{r})\delta(\rho(t|\mathbf{r}_0)-\rho)\delta(j(t|\mathbf{r}_0)-j),\tag{2.15}$$

which explicitly assumes that the solution to the initial dynamic equations is a function of the Lagrangian coordinates $\mathbf{r}_0$. This equation has the form

$$\begin{aligned}&\left(\frac{\partial}{\partial t}+\frac{\partial}{\partial\mathbf{r}}\mathbf{U}(\mathbf{r},t)\right)\varphi_{\rm Lag}(t;\mathbf{r},\rho,j|\mathbf{r}_0)=\frac{\partial\mathbf{U}(\mathbf{r},t)}{\partial\mathbf{r}}\left(\frac{\partial}{\partial\rho}\rho-\frac{\partial}{\partial j}j\right)\varphi_{\rm Lag}(t;\mathbf{r},\rho,j|\mathbf{r}_0),\\&\varphi_{\rm Lag}(0;\mathbf{r},\rho,j|\mathbf{r}_0)=\delta(\mathbf{r}_0-\mathbf{r})\delta(\rho_0(\mathbf{r}_0)-\rho)\delta(j-1).\end{aligned}\tag{2.16}$$

Taking into account the equality

$$\delta(\mathbf{r}(t|\mathbf{r}_0)-\mathbf{r})=\frac{1}{\|\partial r_\alpha/\partial r_{0\beta}\|}\delta(\mathbf{r}_0-\mathbf{r}_0(t,\mathbf{r}))=\frac{1}{j(t|\mathbf{r}_0)}\delta(\mathbf{r}_0-\mathbf{r}_0(t,\mathbf{r})),$$

we can rewrite function (2.15) in the form

$$\varphi_{\rm Lag}(t;\mathbf{r},\rho,j|\mathbf{r}_0)=\frac{1}{j}\delta(\mathbf{r}_0-\mathbf{r}_0(t,\mathbf{r}))\delta(j(t|\mathbf{r}_0)-j)\varphi(t,\mathbf{r};\rho),$$

where $\varphi(t,\mathbf{r};\rho)$ is the indicator function (2.6). Consequently,

$$\varphi(t,\mathbf{r};\rho)=\int d\mathbf{r}_0\int\limits_0^\infty jdj\varphi_{\rm Lag}(t;\mathbf{r},\rho,j|\mathbf{r}_0).\tag{2.17}$$

Multiplying Eq. (2.16) by j and integrating the result over j and $\mathbf{r}_0$, we obtain the corresponding Liouville equation in the Eulerian representation (2.9).

In the case of divergence-free velocity field, Eqs. (2.5), (2.9), and (2.16) coincide. Fundamental differences appear only if the potential component is available in the velocity field.

Note that solutions to the dynamic problems have the one-time and one-point probability densities that coincide with the corresponding indicator functions averaged over an ensemble of realizations

$$\begin{aligned}&P(t;\mathbf{r},\rho,j|\mathbf{r}_0)=\left\langle\varphi_{\rm Lag}\left(t;\mathbf{r},\rho,j|\mathbf{r}_0\right)\right\rangle,\\&P(t,\mathbf{r};\rho)=\left\langle\varphi\left(t;\mathbf{r};\rho\right)\right\rangle,\quad P(t,\mathbf{r};\rho,\mathbf{p})=\left\langle\varphi\left(t;\mathbf{r};\rho,\mathbf{p}\right)\right\rangle.\end{aligned}$$

This point explains the special interest to indicator functions in the statistical dynamics of systems. In addition, indicator functions provide a good deal of data on geometric structure of random fields, which can be obtained using statistical topography of random fields (see Sect. 3.2.2).

2.2.2 Quasilinear equations

Consider now the simplest quasilinear equation for scalar quantity $q(\mathbf{r},t)$ (1.52), page 22

$$\left(\frac{\partial}{\partial t}+\mathbf{U}(t,q)\frac{\partial}{\partial\mathbf{r}}\right)q(\mathbf{r},t)=Q\left(t,q\right),\qquad q(\mathbf{r},0)=q_0(\mathbf{r}).\tag{2.18}$$

In this case, an attempt of deriving a closed equation for the indicator function $\varphi(t, \mathbf{r}; q) = \delta(q(\mathbf{r}, t) - q)$ on the analogy of the linear problem will fail. Here, we must supplement Eq. (2.18) with Eq. (1.53) for the gradient field $\mathbf{p}(\mathbf{r}, t) = \boldsymbol{\nabla} q(\mathbf{r}, t)$

$$\left(\frac{\partial}{\partial t} + \mathbf{U}(t, q)\frac{\partial}{\partial \mathbf{r}}\right)\mathbf{p}(\mathbf{r}, t) + \frac{\partial\left\{\mathbf{U}(t, q)\mathbf{p}(\mathbf{r}, t)\right\}}{\partial q}\mathbf{p}(\mathbf{r}, t) = \frac{\partial Q(t, q)}{\partial q}\mathbf{p}(\mathbf{r}, t) \tag{2.19}$$

and consider the extended indicator function

$$\varphi(t, \mathbf{r}; q, \mathbf{p}) = \delta(q(\mathbf{r}, t) - q)\delta(\mathbf{p}(\mathbf{r}, t) - \mathbf{p}). \tag{2.20}$$

Differentiating Eq. (2.20) with respect to time and using Eqs. (2.18) and (2.19), we obtain the equation

$$\frac{\partial}{\partial t}\varphi(t, \mathbf{r}; q, \mathbf{p}) = \left\{\frac{\partial}{\partial q}\left[\mathbf{p}\mathbf{U}(t, q) - Q(t, q)\right] + \frac{\partial}{\partial p_k}\mathbf{U}(t, q)\frac{\partial p_k(\mathbf{r}, t)}{\partial \mathbf{r}}\right\}\varphi(t, \mathbf{r}; q, \mathbf{p})$$
$$+\frac{\partial}{\partial p_k}\left[p_k\left(\mathbf{p}\frac{\partial \mathbf{U}(t, q)}{\partial q} - \frac{\partial Q(t, q)}{\partial q}\right)\right]\varphi(t, \mathbf{r}; q, \mathbf{p}), \tag{2.21}$$

which is not closed, however, because of the term $\partial p_k(\mathbf{r}, t)/\partial \mathbf{r}$ in the right-hand side.

Differentiating function (2.20) with respect to $\mathbf{r}$, we obtain the equality

$$\frac{\partial}{\partial \mathbf{r}}\varphi(t, \mathbf{r}; q, \mathbf{p}) = -\left[\mathbf{p}\frac{\partial}{\partial q} + \frac{\partial p_k(\mathbf{r}, t)}{\partial \mathbf{r}}\frac{\partial}{\partial p_k}\right]\varphi(t, \mathbf{r}; q, \mathbf{p}), \tag{2.22}$$

from which follows that

$$\frac{\partial}{\partial p_k}\frac{\partial p_k(\mathbf{r}, t)}{\partial \mathbf{r}}\varphi(t, \mathbf{r}; q, \mathbf{p}) = -\left[\frac{\partial}{\partial \mathbf{r}} + \mathbf{p}\frac{\partial}{\partial q}\right]\varphi(t, \mathbf{r}; q, \mathbf{p}).$$

Consequently, Eq. (2.21) can be rewritten in the closed form

$$\left(\frac{\partial}{\partial t} + \mathbf{U}(t, q)\frac{\partial}{\partial \mathbf{r}}\right)\varphi(t, \mathbf{r}; q, \mathbf{p}) = \frac{\partial}{\partial q}\left\{\left[\mathbf{p}\mathbf{U}(t, q) - Q(t, q)\right]\varphi(t, \mathbf{r}; q, \mathbf{p})\right\}$$
$$+\frac{\partial}{\partial \mathbf{p}}\left\{\mathbf{p}\left(\frac{\partial \mathbf{p}\mathbf{U}(t, q)}{\partial q} - \frac{\partial Q(t, q)}{\partial q}\right)\varphi(t, \mathbf{r}; q, \mathbf{p})\right\},$$

which is just the desired Liouville equation for the quasilinear equation (2.18) in the extended phase space $\{q, \mathbf{p}\}$ with the initial value

$$\varphi(0, \mathbf{r}; q, \mathbf{p}) = \delta(q_0(\mathbf{r}) - q)\delta(\mathbf{p}_0(\mathbf{r}) - \mathbf{p}). \tag{2.23}$$

Note that the equation of continuity for conserved quantity $I(\mathbf{r}, t)$

$$\frac{\partial}{\partial t}I(\mathbf{r}, t) + \frac{\partial}{\partial \mathbf{r}}\left\{\mathbf{U}(t, q)I(\mathbf{r}, t)\right\} = 0, \qquad I(\mathbf{r}, 0) = I_0(\mathbf{r}) \tag{2.24}$$

can be combined with Eqs. (2.18), (2.19). In this case, the indicator function has the form

$$\varphi(t, \mathbf{r}; q, \mathbf{p}, I) = \delta(q(\mathbf{r}, t) - q)\delta(\mathbf{p}(\mathbf{r}, t) - \mathbf{p})\delta(I(\mathbf{r}, t) - I) \tag{2.25}$$

and derivation of the closed equation for this function appears possible in space $\{q, \mathbf{p}, I\}$, which follows from the fact that, in the Lagrangian description, quantity inverse to $I(\mathbf{r}, t)$ coincides with the divergence.

2.2.3 General-form nonlinear equations

Consider now the scalar nonlinear first-order partial differential equation in the general form (1.64), page 26

$$\frac{\partial}{\partial t}q(\mathbf{r},t) + H\left(\mathbf{r},t,q,\mathbf{p}\right) = 0, \qquad q\left(\mathbf{r},0\right) = q_0\left(\mathbf{r}\right), \tag{2.26}$$

where $\mathbf{p}(\mathbf{r},t) = \partial q(\mathbf{r},t)/\partial\mathbf{r}$. In order to derive the closed Liouville equation in this case, we must supplement Eq. (2.26) with equations for vector $\mathbf{p}(\mathbf{r},t)$ and second derivative matrix $U_{ik}(\mathbf{r},t) = \partial^2 q(\mathbf{r},t)/\partial r_i \partial r_k$.

Introduce now the extended indicator function

$$\varphi(t,\mathbf{r};q,\mathbf{p},U,I) = \delta(q(\mathbf{r},t)-q)\delta(\mathbf{p}(\mathbf{r},t)-\mathbf{p})\delta(U(\mathbf{r},t)-U)\delta(I(\mathbf{r},t)-I), \tag{2.27}$$

where we included for generality an additional conserved variable $I(\mathbf{r},t)$ satisfying the equation of continuity (1.66)

$$\frac{\partial}{\partial t}I(\mathbf{r},t) + \frac{\partial}{\partial\mathbf{r}}\left\{\frac{\partial H\left(\mathbf{r},t,q,\mathbf{p}\right)}{\partial\mathbf{p}}I(\mathbf{r},t)\right\} = 0, \qquad I\left(\mathbf{r},0\right) = I_0\left(\mathbf{r}\right). \tag{2.28}$$

Equations (2.26), (2.28) describe, for example, wave propagation in inhomogeneous media within the frames of the geometrical optics approximation of the parabolic equation of quasi-optics. Differentiating function (2.27) with respect to time and using dynamic equations for functions $q(\mathbf{r},t)$, $\mathbf{p}(\mathbf{r},t)$, $U(\mathbf{r},t)$ and $I(\mathbf{r},t)$, we generally obtain an unclosed equation containing third-order derivatives of function $q(\mathbf{r},t)$ with respect to spatial variable $\mathbf{r}$. However, the combination

$$\left(\frac{\partial}{\partial t} + \frac{\partial H\left(\mathbf{r},t,q,\mathbf{p}\right)}{\partial\mathbf{p}}\frac{\partial}{\partial\mathbf{r}}\right)\varphi(t,\mathbf{r};q,\mathbf{p},U,I)$$

will not include the third-order derivatives; as a result, we obtain the closed Liouville equation in space $\{q,\mathbf{r},U,I\}$ [134, 135].

2.3 Higher-order partial differential equations

If the initial dynamic system includes higher-order derivatives (e.g., Laplace operator), derivation of a closed equation for the corresponding indicator function becomes impossible. In this case, only the variational differential equation (the *Hopf equation*) can be derived in the closed form for the functional whose average over an ensemble of realizations coincides with the *characteristic functional* of the solution to the corresponding dynamic equation. Consider such a transition using the partial differential equations considered in Chapter 1 as examples.

2.3.1 Parabolic equation of quasi-optics

The first example concerns wave propagation in a random medium within the frames of the linear parabolic equation (1.91), page 31

$$\frac{\partial}{\partial x}u(x,\mathbf{R}) = \frac{i}{2k}\Delta_{\mathbf{R}}u(x,\mathbf{R}) + \frac{ik}{2}\varepsilon(x,\mathbf{R})u(x,\mathbf{R}), \qquad u(0,\mathbf{R}) = u_0(\mathbf{R}). \tag{2.29}$$

Consider the functional

$$\varphi[x;v(\mathbf{R}'),v^*(\mathbf{R}')] = \exp\left\{i\int d\mathbf{R}'\left[u(x,\mathbf{R}')v(\mathbf{R}') + u^*(x,\mathbf{R}')v^*(\mathbf{R}')\right]\right\}, \tag{2.30}$$

where wavefield $u(x,\mathbf{R})$ satisfies Eq. (2.29) and $u^*(x,\mathbf{R})$ is the complex conjugated function. Differentiating (2.30) with respect to x and using dynamic equation (2.29) and its complex conjugate, we obtain the equality

$$\begin{aligned}
&\frac{\partial}{\partial x}\varphi[x;v(\mathbf{R}'),v^*(\mathbf{R}')] \\
&= -\frac{1}{2k}\int d\mathbf{R}\left[v(\mathbf{R})\Delta_{\mathbf{R}}u(x,\mathbf{R}) - v^*(\mathbf{R})\Delta_{\mathbf{R}}u^*(x,\mathbf{R})\right]\varphi[x;v(\mathbf{R}'),v^*(\mathbf{R}')] \\
&-\frac{k}{2}\int d\mathbf{R}\varepsilon(x,\mathbf{R})\left[v(\mathbf{R})u(x,\mathbf{R}) - v^*(\mathbf{R})u^*(x,\mathbf{R})\right]\varphi[x;v(\mathbf{R}'),v^*(\mathbf{R}')],
\end{aligned}$$

which can be written as the variational differential equation

$$\begin{aligned}
\frac{\partial}{\partial x}\varphi[x;v(\mathbf{R}'),v^*(\mathbf{R}')] &= \frac{ik}{2}\int d\mathbf{R}\varepsilon(x,\mathbf{R})\widehat{M}(\mathbf{R})\varphi[x;v(\mathbf{R}'),v^*(\mathbf{R}')] \\
&+\frac{i}{2k}\int d\mathbf{R}\left[v(\mathbf{R})\Delta_{\mathbf{R}}\frac{\delta}{\delta v(\mathbf{R})} - v^*(\mathbf{R})\Delta_{\mathbf{R}}\frac{\delta}{\delta v^*(\mathbf{R})}\right]\varphi[x;v(\mathbf{R}'),v^*(\mathbf{R}')]
\end{aligned} \tag{2.31}$$

with the Hermitian operator

$$\widehat{M}(\mathbf{R}) = v(\mathbf{R})\frac{\delta}{\delta v(\mathbf{R})} - v^*(\mathbf{R})\frac{\delta}{\delta v^*(\mathbf{R})},$$

and Eq. (2.31) is equivalent to the input Eq. (2.29). The equality

$$\frac{\delta}{\delta\varepsilon(x-0,\mathbf{R})}\varphi[x;v(\mathbf{R}'),v^*(\mathbf{R}')] = \frac{ik}{2}\widehat{M}(\mathbf{R})\varphi[x;v(\mathbf{R}'),v^*(\mathbf{R}')] \tag{2.32}$$

is a consequence of Eq (2.31).

2.3.2 Random forces in hydrodynamic theory of turbulence

Consider now integro-differential equation (1.99), page 34 for the Fourier harmonics $\widehat{\mathbf{u}}(\mathbf{k},t)$ of the solution to the Navier–Stokes equation (1.97)

$$\begin{aligned}
&\frac{\partial}{\partial t}\hat{u}_i(\mathbf{k},t) + \frac{i}{2}\int d\mathbf{k}_1\int d\mathbf{k}_2\Lambda_i^{\alpha\beta}(\mathbf{k}_1,\mathbf{k}_2,\mathbf{k})\,\hat{u}_\alpha(\mathbf{k}_1,t)\,\hat{u}_\beta(\mathbf{k}_2,t) - \nu k^2\hat{u}_i(\mathbf{k},t) = \hat{f}_i(\mathbf{k},t), \\
&\Lambda_i^{\alpha\beta}(\mathbf{k}_1,\mathbf{k}_2,\mathbf{k}) = \frac{1}{(2\pi)^3}\left\{k_\alpha\Delta_{i\beta}(\mathbf{k}) + k_\beta\Delta_{i\alpha}(\mathbf{k})\right\}\delta(\mathbf{k}_1+\mathbf{k}_2-\mathbf{k}), \quad \Delta_{ij}(\mathbf{k}) = \delta_{ij} - \frac{k_ik_j}{\mathbf{k}^2},
\end{aligned} \tag{2.33}$$

where $\widehat{\mathbf{f}}(\mathbf{k},t)$ is the spatial Fourier harmonics of external forces.

We introduce the functional

$$\varphi[t;\mathbf{z}] = \varphi[t;\mathbf{z}(\mathbf{k}')] = \exp\left\{i\int d\mathbf{k}'\widehat{\mathbf{u}}(\mathbf{k}',t)\mathbf{z}(\mathbf{k}')\right\}. \tag{2.34}$$

Differentiating this functional with respect to time t and using dynamic equation (2.33), we obtain the equality

$$\frac{\partial}{\partial t}\varphi[t;\mathbf{z}] = \frac{1}{2}\int d\mathbf{k}z_i(\mathbf{k})\int d\mathbf{k}_1\int d\mathbf{k}_2\Lambda_i^{\alpha\beta}(\mathbf{k}_1,\mathbf{k}_2,\mathbf{k})\,\hat{u}_\alpha(\mathbf{k}_1,t)\,\hat{u}_\beta(\mathbf{k}_2,t)\,\varphi[t;\mathbf{z}]$$
$$-i\int d\mathbf{k}z_i(\mathbf{k})\left\{\nu k^2\hat{u}_i(\mathbf{k},t) - \hat{f}_i(\mathbf{k},t)\right\}\varphi[t;\mathbf{z}],$$

which can be rewritten in the functional space as the linear *Hopf equation* containing variational derivatives

$$\frac{\partial}{\partial t}\varphi[t;\mathbf{z}] = -\frac{1}{2}\int d\mathbf{k}z_i(\mathbf{k})\int d\mathbf{k}_1\int d\mathbf{k}_2\Lambda_i^{\alpha\beta}(\mathbf{k}_1,\mathbf{k}_2,\mathbf{k})\,\frac{\delta^2\varphi[t;\mathbf{z}]}{\delta z_\alpha(\mathbf{k}_1)\,\delta z_\beta(\mathbf{k}_2)}$$
$$-\int d\mathbf{k}z_i(\mathbf{k})\left\{\nu k^2\frac{\delta}{\delta z_i(\mathbf{k})} - i\hat{f}_i(\mathbf{k},t)\right\}\varphi[t;\mathbf{z}]. \tag{2.35}$$

A consequence of Eq. (2.35) is the equality

$$\frac{\delta}{\delta\hat{\mathbf{f}}(\mathbf{k},t-0)}\varphi[t;\mathbf{z}] = i\mathbf{z}(\mathbf{k})\varphi[t;\mathbf{z}]. \tag{2.36}$$

In a similar way, considering the space-time harmonics of the velocity field $\hat{u}_i(\mathbf{K})$, where $\mathbf{K}$ is the four-dimensional wave vector $\{\mathbf{k},\omega\}$ and $\hat{u}_i^*(\mathbf{K}) = \hat{u}_i(-\mathbf{K})$ because field $u_i(\mathbf{r},t)$ is real, we obtain the nonlinear integral equation (1.100), page 34

$$(i\omega+\nu\mathbf{k}^2)\hat{u}_i(\mathbf{K}) + \frac{i}{2}\int d^4\mathbf{K}_1\int d^4\mathbf{K}_2\Lambda_i^{\alpha\beta}(\mathbf{K}_1,\mathbf{K}_2,\mathbf{K})\,\hat{u}_\alpha(\mathbf{K}_1)\,\hat{u}_\beta(\mathbf{K}) = \hat{f}_i(\mathbf{K}),$$
$$\Lambda_i^{\alpha\beta}(\mathbf{K}_1,\mathbf{K}_2,\mathbf{K}) = \frac{1}{2\pi}\Lambda_i^{\alpha\beta}(\mathbf{k}_1,\mathbf{k}_2,\mathbf{k})\,\delta(\omega_1+\omega_2-\omega), \tag{2.37}$$

where $\hat{f}_i(\mathbf{K})$ are the space-time Fourier harmonics of external forces. In this case, dealing with the functional

$$\varphi[\mathbf{z}] = \varphi[\mathbf{z}(\mathbf{K}')] = \exp\left\{i\int d^4\mathbf{K}'\hat{\mathbf{u}}(\mathbf{K}')\mathbf{z}(\mathbf{K}')\right\}, \tag{2.38}$$

we derive the linear variational integro-differential equation of the form

$$(i\omega+\nu\mathbf{k}^2)\frac{\delta\varphi[\mathbf{z}]}{\delta z_i(\mathbf{K})} = -\frac{1}{2}\int d^4\mathbf{K}_1\int d^4\mathbf{K}_2\Lambda_i^{\alpha\beta}(\mathbf{K}_1,\mathbf{K}_2,\mathbf{K})\,\frac{\delta\varphi[\mathbf{z}]}{\delta z_\alpha(\mathbf{K}_1)\,\delta z_\beta(\mathbf{K}_2)}$$
$$+i\hat{f}_i(\mathbf{K})\,\varphi[\mathbf{z}]. \tag{2.39}$$

Part II

Stochastic equations

Chapter 3

Random quantities, processes, and fields

Prior to consider statistical descriptions of the problems mentioned in Part 1, we discuss basic concepts of the theory of random quantities, processes, and fields.

This chapter concerns those basic properties of random quantities, processes, and fields that are widely used in analyzing dynamic systems with fluctuating parameters, but only slightly elucidated in textbooks. Here, we will follow monographs [132, 134, 135], in which one can also found the fundamental references concerning the problem.

3.1 Random quantities and their characteristics

The probability for a random quantity ξ to fall in interval $-\infty < \xi < z$ is the monotonous function

$$F(z) = P(-\infty < \xi < z) = \langle \theta(z - \xi) \rangle_{\xi}, \quad F(\infty) = 1, \tag{3.1}$$

where

$$\theta(z) = \begin{cases} 1, & \text{if } z > 0, \\ 0, & \text{if } z < 0 \end{cases}$$

is the Heaviside step function and $\langle \ldots \rangle_{\xi}$ denotes averaging over an ensemble of realizations of random quantity ξ. This function is called the *probability distribution function* or the *integral distribution function*. Definition (3.1) reflects the real-world procedure of finding the probability according to the rule

$$P(-\infty < \xi < z) = \lim_{N \to \infty} \frac{n}{N},$$

where n is the integer equal to the number of realizations of event $\xi < z$ in N independent trials. Consequently, the probability for a random quantity ξ to fall into interval $z < \xi < z + dz$, where dz is the infinitesimal increment, can be written in the form

$$P(z < \xi < z + dz) = p(z)dz,$$

where function $p(z)$ called the *probability density* is represented by the formula

$$p(z) = \frac{d}{dz} P(-\infty < \xi < z) = \langle \delta(z - \xi) \rangle_{\xi}, \tag{3.2}$$

where $\delta(z)$ is the Dirac delta function. In terms of probability density $p(z)$, the integral distribution function is expressed by the formula

$$F(z) = P(-\infty < \xi < z) = \int_{-\infty}^{z} d\xi p(\xi), \tag{3.3}$$

so that

$$p(z) > 0, \qquad \int_{-\infty}^{\infty} dz p(z) = 1.$$

Multiplying Eq. (3.2) by arbitrary function $f(z)$ and integrating over the whole domain of variable z, we express the mean value of the arbitrary function of random quantity in the following form

$$\langle f(\xi)\rangle_{\xi} = \int_{-\infty}^{\infty} dz p(z) f(z). \tag{3.4}$$

The *characteristic function* defined by the equality

$$\Phi(v) = \left\langle e^{iv\xi}\right\rangle_{\xi} = \int_{-\infty}^{\infty} dz e^{ivz} p(z)$$

is a very important quantity that exhaustively describes all characteristics of random quantity ξ. The characteristic function being known, we can obtain the probability density (via the Fourier transform)

$$p(x) = \frac{1}{2\pi} \int_{-\infty}^{\infty} dv \Phi(v) e^{-ivx},$$

moments

$$M_n = \langle \xi^n \rangle = \int_{-\infty}^{\infty} dz p(z) z^n = \left(\frac{d}{idv}\right)^n \Phi(v)\bigg|_{v=0},$$

cumulants (or semi-invariants)

$$K_n = \left(\frac{d}{idv}\right)^n \Theta(v)\bigg|_{v=0},$$

where $\Theta(v) = \ln \Phi(v)$, and other statistical characteristics. In terms of moments and cumulants of random quantity ξ, functions $\Theta(v)$ and $\Phi(v)$ are the Taylor series

$$\Phi(v) = \sum_{n=0}^{\infty} \frac{i^n}{n!} M_n v^n, \qquad \Theta(v) = \sum_{n=1}^{\infty} \frac{i^n}{n!} K_n v^n. \tag{3.5}$$

In the case of multidimensional random quantity $\boldsymbol{\xi} = \{z_1, ..., z_n\}$, the exhaustive statistical description assumes the multidimensional characteristic function

$$\Phi(\mathbf{v}) = \left\langle e^{i\mathbf{v}\boldsymbol{\xi}}\right\rangle_{\boldsymbol{\xi}}, \qquad \mathbf{v} = \{v_1, ..., v_n\}. \tag{3.6}$$

The corresponding joined probability density for quantities $\xi_1, ..., \xi_n$ is the Fourier transform of characteristic function $\Phi(\mathbf{v})$, i.e.,

$$P(\mathbf{x}) = \frac{1}{(2\pi)^n} \int d\mathbf{v} \Phi(\mathbf{v}) e^{-i\mathbf{v}\mathbf{x}}, \qquad \mathbf{x} = \{x_1, ..., x_n\}. \tag{3.7}$$

Substituting function $\Phi(\mathbf{v})$ defined by Eq. (3.6) in Eq. (3.7) and integrating the result over $\mathbf{v}$, we obtain the obvious equality

$$P(\mathbf{x}) = \langle\delta(\boldsymbol{\xi} - \mathbf{x})\rangle_{\boldsymbol{\xi}} = \langle\delta(\xi_1 - x_1)...\delta(\xi_n - x_n)\rangle \tag{3.8}$$

that can serve the definition of the probability density of random vector quantity $\boldsymbol{\xi}$.

In this case, the moments and cumulants of random quantity $\boldsymbol{\xi}$ are defined by the expressions

$$M_{i_1,...,i_n} = \left.\frac{\partial^n}{i^n\partial v_{i_1}...\partial v_{i_n}}\Phi(\mathbf{v})\right|_{\mathbf{v}=0}, \quad K_{i_1,...,i_n} = \left.\frac{\partial^n}{i^n\partial v_{i_1}...\partial v_{i_n}}\Theta(\mathbf{v})\right|_{\mathbf{v}=0},$$

where $\Theta(\mathbf{v}) = \ln\Phi(\mathbf{v})$, and functions $\Theta(\mathbf{v})$ and $\Phi(\mathbf{v})$ are expressed in terms of moments $M_{i_1,...,i_n}$ and cumulants $K_{i_1,...,i_n}$ via the Taylor series

$$\Phi(\mathbf{v}) = \sum_{n=0}^{\infty}\frac{i^n}{n!}M_{i_1,...,i_n}v_{i_1}...v_{i_n}, \quad \Theta(\mathbf{v}) = \sum_{n=1}^{\infty}\frac{i^n}{n!}K_{i_1,...,i_n}v_{i_1}...v_{i_n}. \tag{3.9}$$

Note that, for quantities ξ assuming only discrete values ξ_i $(i = 1, 2, ...)$ with probabilities p_i, formula (3.8) is replaced with its discrete analog

$$p_k = \left\langle\delta_{z,\xi_k}\right\rangle,$$

where $\delta_{i,k}$ is the Kronecker delta ($\delta_{i,k} = 1$ for $i = k$ and 0 otherwise).

Consider now statistical average $\langle\xi f(\xi)\rangle_{\xi}$, where $f(z)$ is arbitrary deterministic function such that the above average exists. We calculate this average using the procedure that will be widely used in what follows. Instead of $f(\xi)$, we consider function $f(\xi + \eta)$, where η is arbitrary deterministic quantity. Expand function $f(\xi + \eta)$ in the Taylor series in powers of ξ, , i.e., represent it the form

$$f(\xi + \eta) = \sum_{n=0}^{\infty}\frac{1}{n!}f^{(n)}(\eta)\xi^n = e^{\xi\frac{d}{d\eta}}f(\eta),$$

where we introduced the shift operator with respect to η. Then we can write the equality

$$\begin{aligned}\langle\xi f(\xi + \eta)\rangle_{\xi} &= \left\langle\xi e^{\xi\frac{d}{d\eta}}\right\rangle_{\xi} f(\eta) \\ &= \frac{\left\langle\xi e^{\xi\frac{d}{d\eta}}\right\rangle_{\xi}}{\left\langle e^{\xi\frac{d}{d\eta}}\right\rangle_{\xi}}\left\langle e^{\xi\frac{d}{d\eta}}\right\rangle_{\xi} f(\eta) = \Omega\left(\frac{d}{id\eta}\right)\langle f(\xi + \eta)\rangle_{\xi},\end{aligned} \tag{3.10}$$

where function

$$\Omega(v) = \frac{\left\langle\xi e^{i\xi v}\right\rangle_{\xi}}{\left\langle e^{i\xi v}\right\rangle_{\xi}} = \frac{d}{idv}\ln\Phi(v) = \frac{d}{idv}\Theta(v),$$

and $\Phi(v)$ is the characteristic function of random quantity ξ. Using now the Taylor series (3.5) for function $\Theta(v)$, we obtain function $\Omega(v)$ in the form of the series

$$\Omega(v) = \sum_{n=0}^{\infty}\frac{i^n}{n!}K_{n+1}v^n. \tag{3.11}$$

Because variable η appears in the right-hand side of Eq. (3.10) only as the term of sum $\xi + \eta$, we can replace differentiation with respect to η with differentiation with respect to ξ (in so doing, operator $\Omega(d/id\xi)$ should be introduced into averaging brackets) and set $\eta = 0$. As a result, we obtain the equality

$$\langle \xi f(\xi) \rangle_\xi = \left\langle \Omega\left(\frac{d}{id\xi}\right) f(\xi) \right\rangle_\xi,$$

which can be rewritten, using expansion (3.11) for $\Omega(v)$, as the series in cumulants K_n

$$\langle \xi f(\xi) \rangle_\xi = \sum_{n=0}^{\infty} \frac{1}{n!} K_{n+1} \left\langle \frac{d^n f(\xi)}{d\xi^n} \right\rangle_\xi. \tag{3.12}$$

Note that, setting $f(\xi) = \xi^{n-1}$ in Eq. (3.12), we obtain the recurrence formula that relates moments and cumulants of random quantity ξ in the form

$$M_n = \sum_{k=1}^{n} \frac{(n-1)!}{(k-1)!(n-k)!} K_k M_{n-k} \quad (M_0 = 1, \quad n = 1, 2, ...). \tag{3.13}$$

In a similar way, we can obtain the following operator expression for statistical average $\langle g(\xi) f(\xi) \rangle_\xi$

$$\begin{aligned} &\langle g(\xi + \eta_1) f(\xi + \eta_2) \rangle_\xi \\ &= \exp\left\{\Theta\left(\frac{d}{id\eta_1} + \frac{d}{id\eta_2}\right) - \Theta\left(\frac{d}{id\eta_1}\right) - \Theta\left(\frac{d}{id\eta_2}\right)\right\} \langle g(\xi + \eta_1)\rangle_\xi \langle f(\xi + \eta_2)\rangle_\xi. \end{aligned} \tag{3.14}$$

In the particular case $g(z) = e^{\omega z}$, where parameter ω assumes complex values too, we obtain the expression

$$\left\langle e^{\omega\xi} f(\xi + \eta) \right\rangle_\xi = \exp\left\{\Theta\left(\frac{1}{i}\left(\omega + \frac{d}{d\eta}\right)\right) - \Theta\left(\frac{d}{id\eta}\right)\right\} \langle f(\xi + \eta)\rangle_\xi. \tag{3.15}$$

To illustrate practicability of the above formulas, we consider two types of random quantities ξ as examples.

1. Let ξ be the Gaussian random quantity with the probability density

$$p(z) = \frac{1}{\sqrt{2\pi}\sigma} \exp\left\{-\frac{z^2}{2\sigma^2}\right\}.$$

Then, we have

$$\Phi(v) = \exp\left\{-\frac{v^2\sigma^2}{2}\right\}, \quad \Theta(v) = -\frac{v^2\sigma^2}{2},$$

so that

$$M_1 = K_1 = \langle \xi \rangle = 0, \quad M_2 = K_2 = \sigma^2 = \left\langle \xi^2 \right\rangle, \quad K_{n>2} = 0.$$

In this case, the recurrence formula (3.13) assumes the form

$$M_n = (n-1)\sigma^2 M_{n-2}, \quad n = 2, ..., \tag{3.16}$$

from which follows that

$$M_{2n+1} = 0, \quad M_{2n} = (2n-1)!!\sigma^{2n}.$$

For averages (3.12) and (3.15), we obtain the expressions

$$\langle \xi f(\xi) \rangle_\xi = \sigma^2 \left\langle \frac{df(\xi)}{d\xi} \right\rangle_\xi, \quad \left\langle e^{\omega\xi} f(\xi) \right\rangle_\xi = \exp\left\{ \frac{\omega^2\sigma^2}{2} \right\} \left\langle f(\xi + \omega\sigma^2) \right\rangle_\xi. \tag{3.17}$$

Additional useful formulas can be derived from Eqs. (3.17); for example,

$$\left\langle e^{\omega\xi} \right\rangle_\xi = \exp\left\{ \frac{\omega^2\sigma^2}{2} \right\}, \quad \left\langle \xi e^{\omega\xi} \right\rangle_\xi = \omega\sigma^2 \exp\left\{ \frac{\omega^2\sigma^2}{2} \right\},$$

and so on.

If we deal with the random Gaussian vector whose components are ξ_i ($\langle \xi_i \rangle_\xi = 0$, $i = 1, ..., n$), then the characteristic function is given by the equality

$$\Phi(\mathbf{v}) = \exp\left\{ -\frac{1}{2} B_{ij} v_i v_j \right\}, \quad \Theta(\mathbf{v}) = -\frac{1}{2} B_{ij} v_i v_j,$$

where matrix $B_{ij} = \left\langle \xi_i \xi_j \right\rangle$ and repeated indexes assume summation. In this case, Eq. (3.17) is replaced with the equality

$$\langle \boldsymbol{\xi} f(\boldsymbol{\xi}) \rangle_{\boldsymbol{\xi}} = B \left\langle \frac{df(\boldsymbol{\xi})}{d\boldsymbol{\xi}} \right\rangle_{\boldsymbol{\xi}}. \tag{3.18}$$

2. Let $\xi \equiv n$ be the integer random quantity governed by Poisson distribution

$$p_n = \frac{\bar{n}^n}{n!} e^{-\bar{n}},$$

where $\bar{n}$ is the average value of quantity n. In this case, we have

$$\Phi(v) = \exp\left\{ \bar{n} \left(e^{iv} - 1 \right) \right\}, \quad \Theta(v) = \bar{n} \left(e^{iv} - 1 \right).$$

The recurrence formula (3.13) and Eq. (3.12) assume for this random quantity the forms

$$M_l = \bar{n} \sum_{k=0}^{l-1} \frac{(l-1)!}{k!(l-1-k)!} M_k \equiv \bar{n} \left\langle (n+1)^{l-1} \right\rangle, \quad \langle n f(n) \rangle = \bar{n} \langle f(n+1) \rangle. \tag{3.19}$$

3.2 Random processes, fields, and their characteristics

3.2.1 General remarks

If we deal with random function (random process) $z(t)$, then all this function statistical characteristics at any fixed instant t are exhaustively described in terms of the one-point (one-time) probability density

$$P(t; z) = \langle \delta(z(t) - z) \rangle \tag{3.20}$$

dependent parametrically on time t by the following relationship

$$\langle f(z(t))\rangle = \int\limits_{-\infty}^{\infty} dz f(z) P(t;z).$$

The integral distribution function for this process, i.e. the probability of the event that process $z(t) < Z$ at instant t, is calculated by the formula

$$F(t,Z) = P(z(t) < Z) = \int\limits_{-\infty}^{Z} dz P(t;z)$$

from which follows that

$$F(t,Z) = \langle \theta(Z - z(t))\rangle, \quad F(t,\infty) = 1, \tag{3.21}$$

where $\theta(z)$ is the Heaviside step function equal to zero for $z < 0$ and unity for $z > 0$.

Note that the singular Dirac delta function

$$\varphi(t;z) = \delta\left(z(t) - z\right)$$

appearing in Eq. (3.20) in angle brackets of averaging is called the *indicator function.*

Similar definitions hold for the two-point probability density

$$P(t_1, t_2; z_1, z_2) = \langle \varphi(t_1, t_2, ; z_1, z_2)\rangle$$

and for the general case of the n-point probability density

$$P(t_1, ..., t_n; z_1, ..., z_n) = \langle \varphi(t_1, ..., t_n; z_1, ..., z_n)\rangle,$$

where

$$\varphi(t_1, ..., t_n; z_1, ..., z_n) = \delta(z(t_1) - z_1)...\delta(z(t_n) - z_n)$$

is the n-point indicator function.

Process $z(t)$ is called *stationary* if all its statistical characteristics are invariant with respect to arbitrary temporal shift, i.e., if

$$P(t_1 + \tau, ..., t_n + \tau; z_1, ..., z_n) = P(t_1, ..., t_n; z_1, ..., z_n).$$

In particular, the one-point probability density of stationary process is at all independent of time, and the correlation function depends only on difference of times,

$$B_z(t_1, t_2) = \langle z(t_1) z(t_2)\rangle = B_z(t_1 - t_2).$$

Temporal scale τ_0 characteristic of correlation function $B_z(t)$ is called the temporal correlation radius of process $z(t)$. We can determine this scale, say, by the equality

$$\int\limits_{0}^{\infty} \langle z(t+\tau) z(t)\rangle \, d\tau = \tau_0 \left\langle z^2(t)\right\rangle. \tag{3.22}$$

Note that the Fourier transform of the stationary process correlation function

$$\Phi_z(\omega) = \int\limits_{-\infty}^{\infty} dt B_z(t) e^{i\omega t}$$

is called the *temporal spectral function* (or simply *temporal spectrum*).

For random field $f(\mathbf{x},t)$, the one- and n-point probability densities are defined similarly

$$P(t,\mathbf{x};f) = \langle \varphi(t,\mathbf{x};f) \rangle , \tag{3.23}$$

$$P(t_1,...,t_n,\mathbf{x}_1,...\mathbf{x}_n;f_1,...f_n) = \langle \varphi(t_1,...,t_n,\mathbf{x}_1,...\mathbf{x}_n;f_1,...f_n) \rangle , \tag{3.24}$$

where the indicator functions are defined as follows:

$$\begin{aligned} \varphi(t,\mathbf{x};f) &= \delta(f(\mathbf{x},t)-f), \\ \varphi(t_1,...,t_n,\mathbf{x}_1,...\mathbf{x}_n;f_1,...f_n) &= \delta\left(f(\mathbf{x}_1,t_1)-f_1\right)...\delta\left(f(\mathbf{x}_n,t_n)-f_n\right). \end{aligned} \tag{3.25}$$

For clarity, we use here variables $\mathbf{x}$ and t as spatial and temporal coordinates; however, in many physical problems, some preferred spatial coordinate can play the role of the temporal coordinate.

Random field $f(\mathbf{x},t)$ is called the spatially homogeneous field if all its statistical characteristics are invariant relative to spatial translations by arbitrary vector $\mathbf{a}$, i.e., if

$$P(t_1,...,t_n,\mathbf{x}_1+\mathbf{a},...\mathbf{x}_n+\mathbf{a};f_1,...f_n) = P(t_1,...,t_n,\mathbf{x}_1,...\mathbf{x}_n;f_1,...f_n).$$

In this case, the one-point probability density $P(t,\mathbf{x};f) = P(t;f)$ is independent of $\mathbf{x}$, and the spatial correlation function $B_f(\mathbf{x}_1,t_1;\mathbf{x}_2,t_2)$ depends on the difference $\mathbf{x}_1 - \mathbf{x}_2$

$$B_f(\mathbf{x}_1,t_1;\mathbf{x}_2,t_2) = \langle f(\mathbf{x}_1,t_1)f(\mathbf{x}_2,t_2)\rangle = B_f(\mathbf{x}_1-\mathbf{x}_2;t_1,t_2).$$

If random field $f(\mathbf{x},t)$ is additionally invariant with respect to rotation of all vectors $\mathbf{x}_i$ by arbitrary angle, i.e., with respect to rotations of the reference system, then field $f(\mathbf{x},t)$ is called the homogeneous isotropic random field. In this case, the correlation function depends on length $|\mathbf{x}_1-\mathbf{x}_2|$:

$$B_f(\mathbf{x}_1,t_1;\mathbf{x}_2,t_2) = \langle f(\mathbf{x}_1,t_1)f(\mathbf{x}_2,t_2)\rangle = B_f(|\mathbf{x}_1-\mathbf{x}_2|;t_1,t_2).$$

The corresponding Fourier transform of the correlation function with respect to the spatial variable defines the spatial spectral function (called also the angular spectrum)

$$\Phi_f(\mathbf{k},t) = \int d\mathbf{x} B_f(\mathbf{x},t)e^{i\mathbf{k}\mathbf{x}},$$

and the Fourier transform of the correlation function of random field $f(\mathbf{x},t)$ stationary in time and homogeneous in space defines the space-time spectrum

$$\Phi_f(\mathbf{k},\omega) = \int d\mathbf{x} \int_{-\infty}^{\infty} dt B_f(\mathbf{x},t)e^{i(\mathbf{k}\mathbf{x}+\omega t)}.$$

In the case of isotropic random field $f(\mathbf{x},t)$, the space-time spectrum appears isotropic in the $\mathbf{k}$-space:

$$\Phi_f(\mathbf{k},\omega) = \Phi_f(k,\omega).$$

An exhaustive description of random function $z(t)$ can be given in terms of the characteristic functional

$$\Phi[v(\tau)] = \left\langle \exp\left\{ i \int_{-\infty}^{\infty} d\tau v(\tau)z(\tau) \right\}\right\rangle ,$$

where $v(t)$ is arbitrary (but sufficiently smooth) function. Functional $\Phi[v(\tau)]$ being known, one can determine such characteristics of random function $z(t)$ as mean value $\langle z(t)\rangle$, correlation function $\langle z(t_1)z(t_2)\rangle$, n-point moment function $\langle z(t_1)...z(t_n)\rangle$, etc.

Indeed, expanding functional $\Phi[v(\tau)]$ in the functional Taylor series, we obtain the representation of characteristic functional in terms of the moment functions of process $z(t)$:

$$\Phi[v(\tau)] = \sum_{n=0}^{\infty} \frac{i^n}{n!} \int_{-\infty}^{\infty} dt_1 ... \int_{-\infty}^{\infty} dt_n M_n(t_1, ..., t_n) v(t_1)...v(t_n),$$

$$M_n(t_1, ..., t_n) = \langle z(t_1)...z(t_n)\rangle = \frac{1}{i^n} \frac{\delta^n}{\delta v(t_1)...\delta v(t_n)} \Phi[v(\tau)]\bigg|_{v=0}.$$

Consequently, the moment functions of random process $z(t)$ are expressed in terms of variational derivatives of the characteristic functional. See Appendix A for variational derivative definitions and the corresponding operation rules.

Represent now functional $\Phi[v(\tau)]$ in the form $\Phi[v(\tau)] = \exp\{\Theta[v(\tau)]\}$. Functional $\Theta[v(\tau)]$ also can be expanded in the functional Taylor series

$$\Theta[v(\tau)] = \sum_{n=1}^{\infty} \frac{i^n}{n!} \int_{-\infty}^{\infty} dt_1 ... \int_{-\infty}^{\infty} dt_n K_n(t_1, ..., t_n) v(t_1)...v(t_n), \tag{3.26}$$

where function

$$K_n(t_1, ..., t_n) = \frac{1}{i^n} \frac{\delta^n}{\delta v(t_1)...\delta v(t_n)} \Theta[v(\tau)]\bigg|_{v=0}$$

is called the nth-order *cumulant function* of random process $z(t)$.

The characteristic functional and the nth-order cumulant functions of scalar random field $f(\mathbf{x}, t)$ are defined similarly

$$\Phi[v(\mathbf{x}', \tau)] = \left\langle \exp\left\{ i \int d\mathbf{x} \int_{-\infty}^{\infty} dt v(\mathbf{x},t) f(\mathbf{x}, t) \right\} \right\rangle = \exp\left\{\Theta[v(\mathbf{x}', \tau)]\right\},$$

$$M_n(\mathbf{x}_1, t_1, ..., \mathbf{x}_n, t_n) = \frac{1}{i^n} \frac{\delta^n}{\delta v(\mathbf{x}_1, t_1)...\delta v(\mathbf{x}_n, t_n)} \Phi[v(\mathbf{x}', \tau)]\bigg|_{v=0},$$

$$K_n(\mathbf{x}_1, t_1, ..., \mathbf{x}_n, t_n) = \frac{1}{i^n} \frac{\delta^n}{\delta v(\mathbf{x}_1, t_1)...\delta v(\mathbf{x}_n, t_n)} \Theta[v(\mathbf{x}', \tau)]\bigg|_{v=0}.$$

In the case of vector random field $\mathbf{f}(\mathbf{x}, t)$, we must assume that $\mathbf{v}(\mathbf{x},t)$ is the vector function.

As we noted earlier, characteristic functionals ensure the exhaustive description of random processes and fields. However, even one-point probability densities provide certain information about temporal behavior and spatial structure of random processes for arbitrary long temporal intervals. The ideas of statistical topography of random processes and fields can assist in obtaining this information.

3.2.2 Statistical topography of random processes and fields

The term *statistical topography* was seemingly for the first time introduced in book [319], though the underlying ideas of this approach can be traced back to much earlier works (see, e.g., books [3, 47] and review [121] with detailed reference lists on the problem).

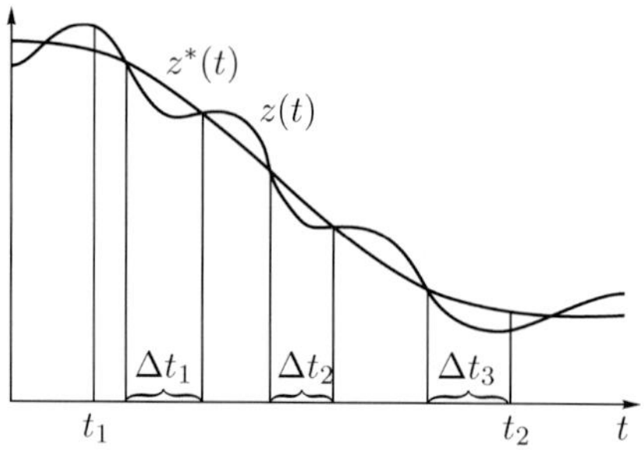

Figure 3.1: To the definition of the typical realization curve of a random process.

Random processes

Following works [143, 166], we discuss first the concept of *typical realization curve* of random process $z(t)$. This concept concerns the fundamental features of the behavior of a separate process realization as a whole for temporal intervals of arbitrary duration.

Random process typical realization curve We will call the typical realization curve of random process $z(t)$ the deterministic curve $z^*(t)$, which is the *median of the integral distribution function* (3.21) and is determined as the solution to the algebraic equation

$$F(t, z^*(t)) = 1/2. \tag{3.27}$$

The reason to this definition rests on the median property consisting in the fact that, for any temporal interval (t_1, t_2), random process $z(t)$ entwines about curve $z^*(t)$ in a way to force the identity of average times during which the inequalities $z(t) > z^*(t)$ and $z(t) < z^*(t)$ hold (Fig. 3.1):

$$\left\langle T_{z(t)>z^*(t)} \right\rangle = \left\langle T_{z(t)<z^*(t)} \right\rangle = \frac{1}{2}(t_2 - t_1). \tag{3.28}$$

Indeed, integrating Eq. (3.27) over temporal interval (t_1, t_2), we obtain

$$\int_{t_1}^{t_2} dt F(t, z^*(t)) = \frac{1}{2}(t_2 - t_1). \tag{3.29}$$

On the other hand, in view of definition of the integral distribution function (3.21), the integral in the right-hand side of Eq. (3.29) can be represented as

$$\int_{t_1}^{t_2} dt F(t, z^*(t)) = \langle T(t_1, t_2) \rangle, \tag{3.30}$$

where $T(t_1, t_2) = \sum_1^N \Delta t_k$ is the combined time during which the realization of process $z(t)$ appears above curve $z^*(t)$ in interval (t_1, t_2). Combining Eqs. (3.29) and (3.30), we obtain Eq. (3.28).

Curve $z^*(t)$ can significantly differ from any particular realization of process $z(t)$ and cannot describe possible magnitudes of spikes. Nevertheless, the definitional domain of typical realization curve $z^*(t)$ of random process $z(t)$ derived from the one-point probability density is the whole of temporal axis $t \in (0, \infty)$.

Consideration of specific random processes allow obtaining the additional information concerning the realization spikes relative to this curve.

Statistics of random process cross points with a line The one-point probability density (3.20) of random process $z(t)$ is a result of averaging the singular indicator function over an ensemble of realizations of this process. This function is concentrated at points at which process $z(t)$ crosses line $z =$ const. Because the cross points are determined as roots of algebraic equation

$$z(t_n) = z \quad (n = 0, 1, ..., \infty),$$

we can rewrite the indicator function in the following form

$$\varphi(t; z) = \sum_{k=1}^{n} \frac{1}{|p(t_k)|} \delta(t - t_k),$$

where $p(t) = \frac{d}{dt} z(t)$.

The number of cross points by itself is obviously a random quantity described by the formula

$$n(t, z) = \int_{-\infty}^{t} d\tau |p(\tau)| \varphi(\tau; z). \tag{3.31}$$

As a consequence, the average number of points where process $z(t)$ crosses line $z =$ const can be described in terms of the correlation between the process derivative with respect to time and the process indicator function, or in terms of joint one-point probability density of process $z(t)$ and its derivative with respect to time $\frac{d}{dt} z(t)$.

In a similar way, we can determine certain elements of statistics related to some other special points (such as points of maxima or minima) of random process $z(t)$.

Random fields

Similarly to common topography of mountain ranges, the statistical topography studies the systems of contours (level lines in the two-dimensional case and surfaces of constant values in the three-dimensional case) specified by the equality $f(\mathbf{r}, t) = f =$ const.

For analyzing a system of contours (in this section, we will deal for simplicity with the two-dimensional case and assume $\mathbf{r} = \mathbf{R}$), we introduce the singular indicator function (3.25) concentrated on these contours.

The convenience of function (3.25) consists, in particular, in the fact that it allows simple expressions for quantities such as the total area of regions where $f(\mathbf{R}, t) > f$ (i.e., within level lines $f(\mathbf{R}, t) = f$) and the total mass of the field within these regions [167]

$$\begin{aligned} S(t; f) &= \int \theta(f(\mathbf{R}, t) - f) d\mathbf{R} = \int_{f}^{\infty} df' \int d\mathbf{R} \varphi(t, \mathbf{R}; f'), \\ M(t; f) &= \int f(\mathbf{R}, t) \theta(f(\mathbf{R}, t) - f) d\mathbf{R} = \int_{f}^{\infty} f' df' \int d\mathbf{R} \varphi(t, \mathbf{R}; f'). \end{aligned}$$

As we mentioned earlier, the mean value of indicator function (3.25) over an ensemble of realizations determines the one-time (in time) and one-point (in space) probability density

$$P(t, \mathbf{R}; f) = \langle \varphi(t, \mathbf{R}; f) \rangle = \langle \delta\left(f(\mathbf{R}, t) - f\right) \rangle .$$

Consequently, this probability density immediately determines ensemble-averaged values of the above expressions.

If we include into consideration the spatial gradient $\mathbf{p}(\mathbf{R}, t) = \boldsymbol{\nabla} f(\mathbf{R}, t)$, we can obtain additional information on details of the structure of field $f(\mathbf{R}, t)$. For example, quantity

$$l(t; f) = \int d\mathbf{R}\, |\mathbf{p}(\mathbf{R}, t)|\, \delta(f(\mathbf{R}, t) - f) = \oint dl \tag{3.32}$$

is the total length of contours [227] – [231] and extends formula (3.31) to random fields.

The integrand in Eq. (3.32) is described in terms of the extended indicator function

$$\varphi(t, \mathbf{R}; f, \mathbf{p}) = \delta\left(f(\mathbf{R}, t) - f\right) \delta\left(\mathbf{p}(\mathbf{R}, t) - \mathbf{p}\right), \tag{3.33}$$

so that the average value of total length (3.32) is related to the joint one-time probability density of field $f(\mathbf{R}, t)$ and its gradient $\mathbf{p}(\mathbf{R}, t)$, which is defined as the ensemble average of indicator function (3.33), i.e., as the function

$$P(t, \mathbf{R}; f, \mathbf{p}) = \langle \delta\left(f(\mathbf{R}, t) - f\right) \delta\left(\mathbf{p}(\mathbf{R}, t) - \mathbf{p}\right) \rangle .$$

Inclusion of second-order spatial derivatives into consideration allows estimating the total number of contours $f(\mathbf{R}, t) = f = \text{const}$ by the approximate formula (neglecting unclosed lines) [293]

$$N(t; f) = N_{\text{in}}(t; f) - N_{\text{out}}(t; f) = \frac{1}{2\pi} \int d\mathbf{R} \kappa(t, \mathbf{R}; f)\, |\mathbf{p}(\mathbf{R}, t)|\, \delta\left(f(\mathbf{R}, t) - f\right),$$

where $N_{\text{in}}(t; f)$ and $N_{\text{out}}(t; f)$ are the numbers of contours for which vector $\mathbf{p}$ is directed along internal and external normals, respectively; and $\kappa(t, \mathbf{R}; f)$ is the curvature of the level line.

Recall that, in the case of the spatially homogeneous field $f(\mathbf{R}, t)$, the corresponding one-point probability densities $P(t, \mathbf{R}; f)$ and $P(t, \mathbf{R}; f, \mathbf{p})$ are independent of $\mathbf{R}$. In this case, if statistical averages of the above expressions (without integration over $\mathbf{R}$) exist, they will characterize the corresponding specific (per unit area) values of these quantities.

Consider now several examples of random processes.

3.2.3 Gaussian random process

We start the discussion with continuous processes; namely, we consider the Gaussian random process $z(t)$ with zero-valued mean ($\langle z(t) \rangle = 0$) and correlation function $B(t_1, t_2) = \langle z(t_1) z(t_2) \rangle$. The corresponding characteristic functional assumes the form

$$\Phi[v(\tau)] = \exp\left\{ -\frac{1}{2} \int\limits_{-\infty}^{\infty} \int\limits_{-\infty}^{\infty} dt_1 dt_2 B(\tau_1, \tau_2) v(\tau_1) v(\tau_2) \right\}.$$

Only one cumulant function (the correlation function $K_2(t_1, t_2) = B(t_1, t_2)$) is different from zero for this process, so that

$$\Theta[v(\tau)] = -\frac{1}{2} \int\limits_{-\infty}^{\infty} \int\limits_{-\infty}^{\infty} dt_1 dt_2 B(\tau_1, \tau_2) v(\tau_1) v(\tau_2). \tag{3.34}$$

Consider the nth-order variational derivative of functional $\Phi[v(\tau)]$. It satisfies the following line of equalities:

$$\frac{\delta^n}{\delta v(t_1)...\delta v(t_n)}\Phi[v(\tau)] = \frac{\delta^{n-1}}{\delta v(t_2)...\delta v(t_n)}\frac{\delta\Theta[v(\tau)]}{\delta v(t_1)}\Phi[v(\tau)]$$
$$= \frac{\delta^2\Theta[v(\tau)]}{\delta v(t_1)\delta v(t_2)}\frac{\delta^{n-2}}{\delta v(t_3)...\delta v(t_n)}\Phi[v(\tau)] + \frac{\delta^{n-2}}{\delta v(t_3)...\delta v(t_n)}\frac{\delta\Theta[v(\tau)]}{\delta v(t_1)}\frac{\delta\Phi[v(\tau)]}{\delta v(t_2)}.$$

Setting now $v = 0$, we obtain that moment functions of the Gaussian process $z(t)$ satisfy the recurrence formula

$$M_n(t_1, ..., t_n) = \sum_{k=2}^{n} B(t_1, t_2)M_{n-2}(t_2, ..., t_{k-1}, t_{k+1}, ..., t_n). \tag{3.35}$$

From this formula follows that, for the Gaussian process with zero-valued mean, all moment functions of odd orders are identically equal to zero and the moment functions of even orders are represented as sums of terms which are the products of averages of all possible pairs $z(t_i)z(t_k)$.

If we assume that function $v(\tau)$ in Eq. (3.34) is different from zero only in interval $0 < \tau < t$, the characteristic function

$$\Phi[t; v(\tau)] = \left\langle \exp\left(i\int_0^t d\tau z(\tau)v(\tau)\right)\right\rangle = \exp\left\{-\int_0^t d\tau_1\int_0^{\tau_1} d\tau_2 B(\tau_1,\tau_2)v(\tau_1)v(\tau_2)\right\} \tag{3.36}$$

becomes a function of time t and satisfies the ordinary differential equation

$$\frac{d}{dt}\Phi[t; v(\tau)] = -v(t)\int_0^t d\tau B(t,\tau)v(\tau)\Phi[t; v(\tau)], \quad \Phi[0; v(\tau)] = 1. \tag{3.37}$$

3.2.4 Discontinuous random processes

Consider now some examples of discontinuous processes. The discontinuous processes are the random functions that change their time-dependent behavior at discrete instants $t_1, t_2, ...$ given statistically. The description of discontinuous processes requires first of all either the knowledge of the statistics of these instants, or the knowledge of the statistics of number $n(0,t)$ of instants t_i falling in time interval $(0,t)$. In the latter case, we have the equality

$$n(0,t) = n(0,t') + n(t',t), \quad 0 \le t' \le t.$$

The quantity $n(0,t)$ by itself is a random process, and Fig. 3.2 shows its possible realization.

The set of points of discontinuity t_1, t_2, ... of process $z(t)$ is called the *stream of points*. In what follows, we will consider *Poisson stationary stream of points* in which the probability of falling n points in interval (t_1, t_2) is specified by the Poisson formula

$$P_{n(t_1,t_2)=n} = \frac{\left[\overline{n(t_1,t_2)}\right]^n}{n!}e^{-\overline{n(t_1,t_2)}} \tag{3.38}$$

with the mean number of points in interval (t_1, t_2) given by the formula

$$\overline{n(t_1,t_2)} = \nu|t_1 - t_2|,$$

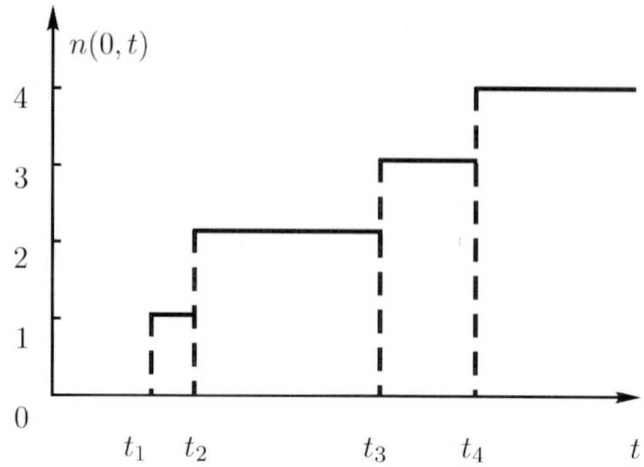

Figure 3.2: A possible realization of process $n(0, t)$.

where ν is the mean number of points per unit time. It is assumed here that the numbers of points falling in nonoverlapping intervals are statistically independent and the instants at which points were fallen in interval (t_1, t_2) under the condition that their total number was n are also statistically independent and uniformly distributed over the interval (t_1, t_2). The length of the interval between adjacent points of discontinuity satisfies the exponential distribution.

Poisson stream of points is an example of the Markovian processes (see Sect. 3.3).

Note that quantity (3.38)

$$P(t; n) = \langle \delta (n(0, t) - n) \rangle ,$$

which is the probability density of falling n points in time interval $(0, t)$, satisfies as a function of parameter t the recurrence equations

$$\begin{aligned} \frac{d}{dt} P(t; n) &= -\nu \left[P(t; n-1) - P(t; n) \right], \quad P(0; n) = 0 \quad (n = 1, 2, \ldots), \\ \frac{d}{dt} P(t; 0) &= -\nu P(t; 0), \quad P(0; 0) = 1. \end{aligned} \tag{3.39}$$

Equations (3.39) are the special case of the *Kolmogorov–Feller equations*.

Consider now random processes whose points of discontinuity form Poisson streams of points. Currently, three types of such processes — *Poisson process*, *telegrapher's process*, and *generalized telegrapher's process* — are mainly used in the model problems of physics. Below, we focus our attention on these processes.

Poisson (impulse) random process

Poisson (impulse) random process $z(t)$ is the process described by the formula

$$z(t) = \sum_{i=1}^{n} \xi_i g(t - t_i), \tag{3.40}$$

where random quantities ξ_i are statistically independent and distributed with probability density $p(\xi)$; random points t_k are uniformly distributed on interval $(0, T)$, so that their number n obeys the Poisson law with parameter $\bar{n} = \nu T$; and function $g(t)$ is the deterministic function that describes the pulse envelope ($g(t) = 0$ for $t < 0$).

The characteristic functional of the Poisson random process $z(t)$ assumes the form

$$\Phi[t; v(\tau)] = \exp\left\{\nu \int_0^t dt' \left[W\left(\int_{t'}^t d\tau v(\tau) g(t-t')\right) - 1\right]\right\},$$

where

$$W(v) = \int_{-\infty}^{\infty} d\xi p(\xi) e^{i\xi v}$$

is the characteristic function of random quantity ξ. Consequently, functional $\Theta[t; v(\tau)]$ and cumulant functions assume the forms

$$\Theta[t; v(\tau)] = \nu \int_0^t dt' \int_{-\infty}^{\infty} d\xi p(\xi) \left\{\exp\left[i\xi \int_{t'}^t d\tau v(\tau) g(t-t')\right] - 1\right\}, \tag{3.41}$$

$$K_n(t_1, ..., t_n) = \nu \langle \xi^n \rangle \int_0^{\min\{t_1,...,t_n\}} dt' g(t-t')...g(t_n - t').$$

We consider Poisson processes of two types important for applications.

1. Let $g(t) = \theta(t) = \begin{cases} 1, & t > 0, \\ 0, & t < 0, \end{cases}$, i.e., $z(t) = \sum_{i=1}^n \xi_i \theta(t - t_i)$. In this case,

$$K_n(t_1, ..., t_n) = \nu \langle \xi^n \rangle \min\{t_1, ..., t_n\}.$$

If additionally $\xi = 1$, then process $z(t) \equiv n(0, t)$, and we have

$$K_n(t_1, ..., t_n) = \nu \min\{t_1, ..., t_n\}, \quad \Theta[t; v(\tau)] = \nu \int_0^t dt' \left\{\exp\left[i \int_{t'}^t d\tau v(\tau)\right] - 1\right\}. \tag{3.42}$$

2. Let now $g(t) = \delta(t)$. In this case, process

$$z(t) = \sum_{i=1}^n \xi_i \delta(t - t_i)$$

is usually called the *shot noise process*. This process is a particular case of the *delta-correlated processes* (see Sect. 4.7, page 89). For such a process, functional $\Theta[t; v(\tau)]$ and cumulant functions assume the forms

$$\begin{aligned} \Theta[t; v(\tau)] &= \nu \int_0^t d\tau \int_{-\infty}^{\infty} d\xi p(\xi) \left\{e^{i\xi v(\tau)} - 1\right\}, \\ K_n(t_1, ..., t_n) &= \nu \langle \xi^n \rangle \delta(t_1 - t_2)\delta(t_2 - t_3)...\delta(t_{n-1} - t_n). \end{aligned} \tag{3.43}$$

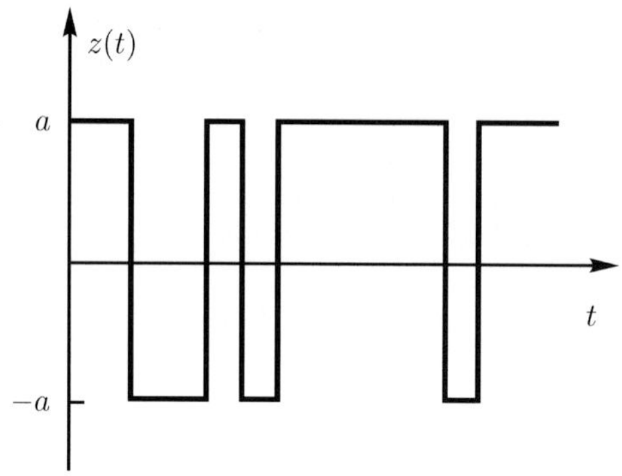

Figure 3.3: A possible realization of telegrapher's random process.

Telegrapher's random process

Consider now statistical characteristics of telegrapher's random process (Fig. 3.3) defined by the formula

$$z(t) = a(-1)^{n(0,t)} \quad (z(0) = a, \quad z^2(t) \equiv a^2), \tag{3.44}$$

where $n(t_1, t_2)$ is the random sequence of integers equal to the number of points of discontinuity in interval (t_1, t_2).

We consider two cases.

1. We will assume first that amplitude a is the deterministic quantity.

For the two first moment functions of process $z(t)$, we have the expressions

$$\langle z(t)\rangle = a \sum_{n(0,t)=0}^{\infty} (-1)^{n(0,t)} P_{n(0,t)} = ae^{-2\overline{n(0,t)}} = ae^{-2\nu t},$$

$$\langle z(t_1)z(t_2)\rangle = a^2 \left\langle (-1)^{n(0,t_1)+n(0,t_2)} \right\rangle = a^2 \left\langle (-1)^{n(t_2,t_1)} \right\rangle$$

$$= a^2 e^{-2\overline{n(t_2,t_1)}} = a^2 e^{-2\nu(t_1-t_2)} \quad (t_1 \geq t_2).$$

The higher moment functions for $t_1 \geqslant t_2 \geqslant ... \geqslant t_n$ satisfy the recurrence relationship

$$M_n(t_1, ..., t_n) = \langle z(t_1)...z(t_n)\rangle = a^2 \left\langle (-1)^{n(0,t_1)+n(0,t_2)+n(0,t_3)+...+n(0,t_n)} \right\rangle$$

$$= a^2 \left\langle (-1)^{n(t_2,t_1)} \right\rangle \left\langle (-1)^{n(0,t_3)+...+n(0,t_n)} \right\rangle = \langle z(t_1)z(t_2)\rangle \, M_{n-2}(t_3, ..., t_n). \tag{3.45}$$

This relationship is very similar to Eq. (3.35) for the Gaussian process with correlation function $B(t_1, t_2)$. The only difference is that the right-hand side of Eq. (3.45) coincides with only one term of the sum in Eq. (3.35), namely, with the term that corresponds to the above order of times.

Consider now the characteristic functional of this process

$$\Phi_a[t; v(\tau)] = \left\langle \exp\left\{ i \int_0^t d\tau z(\tau)v(\tau) \right\}\right\rangle,$$

where index a means that amplitude a is the deterministic quantity. Expanding the characteristic functional in the functional Taylor series and using recurrence formula (3.45), we obtain the expansion

$$
\begin{aligned}
\Phi_a[t; v(\tau)] &= \sum_{n=0}^{\infty} \frac{i^n}{n!} \int_0^t \ldots \int_0^t dt_1 \ldots dt_n M_n(t_1, \ldots, t_n) v(t_1) \ldots v(t_n) \\
&= 1 + ia \int_0^t dt_1 e^{-2\nu t_1} v(t_1) - a^2 \int_0^t dt_1 \int_0^{t_1} dt_2 e^{-2\nu(t_1 - t_2)} v(t_1) v(t_2) \\
&\times \sum_{n=2}^{\infty} i^n \int_0^{t_2} dt_3 \ldots \int_0^{t_{n-1}} dt_n M_n(t_3, \ldots, t_n) v(t_3) \ldots v(t_n). \qquad (3.46)
\end{aligned}
$$

The sum in the right-hand side of Eq. (3.46) can be expressed in terms of the characteristic functional; as a result, we obtain the integral equation

$$
\begin{aligned}
\Phi_a[t; v(\tau)] &= 1 + ia \int_0^t dt_1 e^{-2\nu t_1} v(t_1) \\
&- a^2 \int_0^t dt_1 \int_0^{t_1} dt_2 e^{-2\nu(t_1 - t_2)} v(t_1) v(t_2) \Phi_a[t_2; v(\tau)]. \qquad (3.47)
\end{aligned}
$$

Differentiating Eq. (3.47) with respect to t, we obtain the integro-differential equation

$$
\frac{d}{dt} \Phi_a[t; v(\tau)] = iae^{-2\nu t} v(t) - a^2 v(t) \int_0^t dt_1 e^{-2\nu(t - t_1)} v(t_1) \Phi_a[t_1; v(\tau)]. \qquad (3.48)
$$

No general solution to Eq. (3.48) is known. It can be shown that this equation is equivalent to the second-order differential equation

$$
\begin{gathered}
\left\{ \frac{d^2}{dt^2} + \left[2\nu + \frac{d \ln v(t)}{dt} \right] \frac{d}{dt} + a^2 v(t) \right\} \Phi_a[t; v(\tau)] = 0, \\
\Phi_a[0; v(\tau)] = 1, \quad \left. \frac{d}{dt} \Phi_a[t; v(\tau)] \right|_{t=0} = iav(0).
\end{gathered}
$$

2. Let now amplitude a be the random quantity with probability density $p(a)$. To obtain the characteristic functional of process $z(t)$ in this case, we should average Eq. (3.48) with respect to random amplitude a. In the general case, such averaging cannot be performed analytically. Analytical averaging of Eq. (3.48) appears possible only if probability density of random amplitude a has the form

$$
p(a) = \frac{1}{2} \left[\delta (a - a_0) + \delta (a + a_0) \right]
$$

with $\langle a \rangle = 0$ and $\langle a^2 \rangle = a_0^2$ (in fact, this very case is what is called usually telegrapher's process). As a result, we obtain the integro-differential equation

$$
\frac{d}{dt} \Phi[t; v(\tau)] = -a_0^2 v(t) \int_0^t dt_1 e^{-2\nu(t - t_1)} v(t_1) \Phi[t_1; v(\tau)] \qquad (3.49)
$$

equivalent to the second-order equation

$$\left\{\frac{d^2}{dt^2} + \left[2\nu + \frac{d\ln v(t)}{dt}\right]\frac{d}{dt} + a_0^2 v(t)\right\}\Phi[t;v(\tau)] = 0,$$

$$\Phi[0;v(\tau)] = 1, \quad \left.\frac{d}{dt}\Phi[t;v(\tau)]\right|_{t=0} = 0. \tag{3.50}$$

Note that in the special case of $v(t) \equiv v$, Eq. (3.50) can be solved analytically, and the solution has the form

$$\Phi[t;v] = \left\langle \exp\left\{iv\int_0^t d\tau z(\tau)\right\}\right\rangle$$

$$= e^{-\nu t}\left\{\cosh\sqrt{\nu^2 - a_0^2 v^2}t + \frac{\nu}{\sqrt{\nu^2 - a_0^2 v^2}}\sinh\sqrt{\nu^2 - a_0^2 v^2}t\right\}. \tag{3.51}$$

One can easily see that this expression is the one-point characteristic function of random process $\xi(t) = \int_0^t d\tau z(\tau)$.

Now, we dwell on an important limit theorem concerning telegrapher's random processes.

Consider the random process

$$\xi_N(t) = z_1(t) + \ldots + z_N(t),$$

where all $z_k(t)$ are statistically independent telegrapher's processes with zero-valued means and correlation functions

$$\langle z(t)z(t+\tau)\rangle = \frac{\sigma^2}{N}e^{-\alpha|\tau|}.$$

In this case, the characteristic functional of process $z_k(t)$ satisfies Eq. (3.49)

$$\frac{d}{dt}\Phi[t;v(\tau)] = -\frac{\sigma^2}{N}v(t)\int_0^t dt_1 e^{-\alpha(t-t_1)}v(t_1)\Phi[t_1;v(\tau)],$$

from which follows that $\Phi[t;v(\tau)] \to 1$ for $N \to \infty$.

For the characteristic functional of random process $\xi_N(t)$, we have the expression

$$\Phi_N[t;v(\tau)] = \left\langle \exp\left\{i\int_0^t d\tau \xi_N(\tau)v(\tau)\right\}\right\rangle = \{\Phi[t;v(\tau)]\}^N.$$

Consequently, it satisfies the equation

$$\frac{d}{dt}\ln\Phi_N[t;v(\tau)] = -\sigma^2 v(t)\int_0^t dt_1 e^{-\alpha(t-t_1)}v(t_1)\frac{\Phi[t_1;v(\tau)]}{\Phi[t;v(\tau)]}.$$

In the limit $N \to \infty$, we obtain the equation

$$\frac{d}{dt}\ln\Phi_\infty[t;v(\tau)] = -\sigma^2 v(t)\int_0^t dt_1 e^{-\alpha(t-t_1)}v(t_1),$$

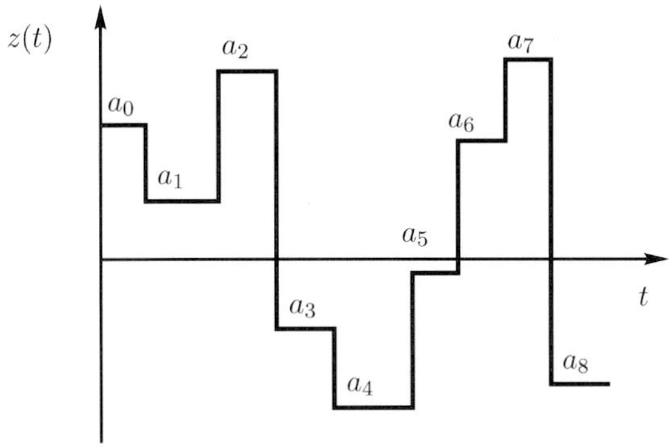

Figure 3.4: A possible realization of generalized telegrapher's random process.

which means that process $\xi(t) = \lim_{N\to\infty} \xi_N(t)$ is the Gaussian random process with the exponential correlation function

$$\langle \xi(t)\xi(t+\tau)\rangle = \sigma^2 e^{-\alpha|\tau|},$$

i.e., the Gaussian Markovian process (see Sect. 3.3). Thus, process $\xi_N(t)$ for finite N is the finite-number-of-states process approximating the Gaussian Markovian process. This approximation appears practicable for studying various functions of the Gaussian Markovian processes rather than only the Gaussian Markovian processes by themselves. As an example, for process

$$z(t) = x^2(t) - \left\langle x^2(t)\right\rangle,$$

where $x(t)$ is the Gaussian Markovian process with the exponential correlation function, the finite-series approximation assumes the form

$$z_N(t) = \sum_{i\neq j=1}^{N} z_i(t)z_j(t), \quad z(t) = \lim_{N\to\infty} z_N(t).$$

This representation is much more convenient for analyzing stochastic equations than the immediate use of processes $x(t)$ and $z(t)$.

Generalized telegrapher's random process

Consider now generalized telegrapher's process defined by the formula

$$z(t) = a_{n(0,t)}. \tag{3.52}$$

Here, $n(0,t)$ is the sequence of integers described above and quantities a_k are assumed statistically independent with distribution function $p(a)$. Figure 3.4 shows a possible realization of such a process.

For process $z(t)$, we have

$$\langle z(t)\rangle = \sum_{k=0}^{\infty} \left\langle a_k \delta_{k,n(0,t)} \right\rangle = \langle a \rangle ,$$

$$\langle z(t_1) z(t_2)\rangle = \sum_{k=0}^{\infty}\sum_{k=0}^{\infty} \langle a_k a_l \rangle \left\langle \delta_{k,n(0,t_1)} \delta_{l,n(0,t_2)} \right\rangle$$

$$= \left\langle a^2 \right\rangle \left\{ \sum_{k=0}^{\infty} \left\langle \delta_{k,n(0,t_1)} \right\rangle \left\langle \delta_{0,n(t_2,t_1)} \right\rangle + 1 - \sum_{k=0}^{\infty} \left\langle \delta_{k,n(0,t_2)} \right\rangle \left\langle \delta_{0,n(t_2,t_1)} \right\rangle \right\}$$

$$= \left\langle a^2 \right\rangle e^{-\nu(t_1-t_2)} + \left\langle a^2 \right\rangle \left(1 - e^{-\nu(t_1-t_2)}\right) \quad (t_1 \geqslant t_2),$$

and so on. In addition, the probability of absence of points of discontinuity in interval (t_2, t_1) is given by the formula

$$P_{n(t_2,t_1)=0} = \left\langle \delta_{0,n(t_2,t_1)} \right\rangle = e^{-\nu|t_1-t_2|}.$$

For such a process, no relationship similar to Eq. (3.46) can be obtained, and derivation of the equation for the characteristic functional is essentially based on the fact that this process is the Markovian process. The resulting equation is the integro-differential equation

$$\Phi[t, v(\tau)] = \left\langle \exp\left\{ ia \int_0^t d\tau v(\tau) \right\} \right\rangle e^{-\nu t}$$
$$+\nu \int_0^t dt_1 e^{-\nu(t-t_1)} \left\langle \exp\left\{ ia \int_{t_1}^t d\tau v(\tau) \right\} \right\rangle \Phi[t_1, v(\tau)]. \tag{3.53}$$

The first term in the right-hand side of Eq. (3.53) corresponds to the absence of points of discontinuity in interval $(0, t)$, and the second term corresponds to the situations in which the number of points of discontinuity in interval $(0, t)$ can vary from one to infinity. Here, time t_1 is the instance at which the last point of discontinuity appears.

Note that, for probability density of the form

$$p(a) = \frac{1}{2}[\delta(a - a_0) + \delta(a + a_0)],$$

Eq. (3.53) coincides (after replacing ν with $\nu/2$) with the equation for telegrapher's process. This fact is quite expectable, because, unlike telegrapher's process, the process $z(t)$ considered here can change the sign at a point of discontinuity with a probability of 1/2, which just results in doubling the mean time between the discontinuities.

Earlier, we noted that the Poisson stream of points and processes based on these streams are the Markovian processes. Below, we consider this important class of random processes in detail.

3.3 Markovian processes

3.3.1 General properties

In the foregoing section, we considered the characteristic functional that describes all statistical characteristics of random process $z(t)$. Specification of the argument of the functional in the form

$$v(t)=\sum_{k=1}^{n} v_k\delta(t-t_k)$$

transforms the characteristic functional into the joint characteristic function of random quantities $z_k = z(t_k)$

$$\Phi_n(v_1,...,v_n)=\left\langle \exp\left\{i\sum_{k=1}^{n} v_k z(t_k)\right\}\right\rangle,$$

whose Fourier transform is the joint probability density of process $z(t)$ at discrete instants

$$P_n(z_1,t_1;...;z_n,t_n)=\langle\delta(z(t_1)-z_1)...\delta(z(t_n)-z_n)\rangle\,. \tag{3.54}$$

Assume that the above instants are ordered according to the line of inequalities

$$t_1\geq t_2\geq ...\geq t_n.$$

Then, by definition of the conditional probability, we have

$$P_n(z_1,t_1;...;z_n,t_n)=p_n(z_1,t_1|z_2,t_2;...;z_n,t_n)P_{n-1}(z_2,t_2;...;z_n,t_n), \tag{3.55}$$

where p_n is the conditional probability density of the value of process $z(t)$ at instant t_1 under the condition that function $z(t)$ was equal to z_k at instants t_k for $k=2,...,n$ $(z(t_k)=z_k,\ k=2,...,n)$. If process $z(t)$ is such that the conditional probability density for all $t_1>t_2$ is unambiguously determined by the value z_2 of the process at instant t_2 and is independent of the previous history, i.e., if

$$p_n(z_1,t_1|z_2,t_2;...;z_n,t_n)=p(z_1,t_1|z_2,t_2), \tag{3.56}$$

then this process is called the Markovian process, or the memoryless process. In this case, function

$$p(z,t|z_0,t_0)=\langle\delta(z(t)-z)|z(t_0)=z_0\rangle\quad(t>t_0) \tag{3.57}$$

is called the *transition probability density*. Setting $t=t_0$ in Eq. (3.57), we obtain the equality

$$p(z,t_0|z_0,t_0)=\delta(z-z_0).$$

Substituting expression (3.55) in Eq. (3.54), we obtain the recurrence formula for the n-time probability density of process $z(t)$. Iterating this formula, we find the relationship of probability density P_n with the one-time probability density $(t_1\geq t_2\geq ...\geq t_n)$

$$P_n(z_1,t_1;...;z_n,t_n)=p(z_1,t_1|z_2,t_2)...p(z_{n-1},t_{n-1}|z_n,t_n)P(t_n,z_n). \tag{3.58}$$

Thus, only two functions — transition probability density $p(z,t|z_0,t_0)$ and one-time probability density $P(t,z)$ — are sufficient to exhaustively describe all statistical characteristics of the Markovian process $z(t)$. It appears that transition probability density as a function of its arguments satisfies the nonlinear integral equation called the *Smolukhovsky equation* (or the *Kolmogorov–Chapman equation*). In the context of this equation derivation, we note the following fact: if process $z(t)$ assumes values $z(t_0)=z_0$, $z(t_1)=z_1$, $z(t)=z$ at fixed instants $t_0<t_1<t$, then the coordination condition

$$\int_{-\infty}^{\infty} dz_1 P_3(z,t;z_1,t_1;z_0,t_0)=P_2(z,t;z_0,t_0). \tag{3.59}$$

holds. Substituting now P_3 and P_2 expressed in the form of Eq (3.58) in Eq. (3.59), we obtain the desired equation

$$p(z,t;z_0,t_0) = \int_{-\infty}^{\infty} dz_1 p(z,t|z_1,t_1)p(z_1,t_1;z_0,t_0). \tag{3.60}$$

Integrating Eq. (3.60) over z_0, we obtain the linear integral equation for the one-point (one-time) probability density $P(t,z)$

$$P(t,z) = \int_{-\infty}^{\infty} dz_1 p(z,t|z_1,t_1)P(t_1,z_1). \tag{3.61}$$

Integral equations (3.60) and (3.61) offer a possibility of deriving differential or integro-differential equations for simple Markovian processes. The simplest Markovian processes with continuous time can be classified as follows:

1) Discrete processes,

2) Continuous processes, and

3) Discrete-continuous processes that can undergo discontinuous variations at certain instants and behave as continuous processes between these instants.

Discrete Markovian process

Consider the discrete Markovian process $z(t)$. This assumes that the process can take on only discrete values $z_1, ..., z_n$ and switching between the values occurs at random time instants. We introduce the transition probability density

$$p_{ij}(t,t_0) = \langle \delta(z(t) - z_i)|z(t_0) = z_j \rangle, \quad \sum_i p_{ij}(t,t_0) = 1 \quad (t_0 < t), \tag{3.62}$$

which is the conditional probability of the event that process $z(t)$ assumes value z_i at instant t under the condition that its value at instant t_0 was z_j. It is obvious that

$$p_{ij}(t_0,t_0) = 1. \tag{3.63}$$

For short temporal intervals $\Delta t \to 0$, we have

$$p_{ij}(t+\Delta t,t) = \delta_{ij} + a_{ij}(t)\Delta t + o(\Delta t), \tag{3.64}$$

where $a_{ij}(t)\Delta t$ is the transition probability from state z_j at instant t to state z_i during time Δt. It is assumed that

$$a_{ij}(t) > 0 \quad (i \neq j), \quad a_{jj}(t) = -\sum_{i(i\neq j)} a_{ij}(t), \tag{3.65}$$

because normalization condition (3.62) must hold.

Using Eq. (3.64), one can easily show from the Smolukhovsky equation (3.60) that probability $p_{ij}(t,t_0)$ satisfies the system of linear differential equations

$$\frac{d}{dt} p_{ij}(t,t_0) = \sum_{k=1}^{n} a_{ik}(t) p_{kj}(t,t_0) \quad (i,j = 1,...,n). \tag{3.66}$$

Representing the one-point probability $P_i(t)$ in the form

$$P_i(t) = \sum_j p_{ij}(t, t_0) p_j^0, \tag{3.67}$$

where p_j^0 are the initial probabilities of states ($p_j^0 = P_j(t_0)$), we obtain that this one-point probability satisfies the system of equations

$$\frac{d}{dt} P_i(t) = \sum_{k=1}^{n} a_{ik}(t) P_k(t), \quad P_i(t_0) = p_i^0. \tag{3.68}$$

Consider three examples as illustrations of the above consideration.

1. Let random process $z(t) = n(0,t)$ represent the number of discontinuities occurred in interval $(0,t)$ at random instants (see Fig. 3.2 for a possible realization of this process). It is assumed that process $z(t)$ takes on only integer values $0, 1, 2, \ldots$, and it is obvious that

$$p_{ij}(t, t_0) = 0 \quad \text{for} \quad i < j, \quad t \geq t_0.$$

Assuming additionally that, in temporal interval $(t, t + \Delta t)$, the probability of one change of state is $\nu \Delta t + o(\Delta t)$ and the probability of the absence of discontinuities is $1 - \nu \Delta t + o(\Delta t)$ and neglecting the possibility of two and more changes of state in this interval (these assumptions are just the assumptions that govern the Poisson stream of instants at which the discontinuities appear), we can write the system of equations (3.68) for this process. In the case under consideration, this system assumes the form

$$\begin{aligned} \frac{d}{dt} P_0(t) &= -\nu P_0(t), \quad P_0(0) = 1, \\ \frac{d}{dt} P_i(t) &= -\nu \left[P_i(t) - P_{i-1}(t) \right], \quad P_{i \neq 0}(0) = 0. \end{aligned} \tag{3.69}$$

and coincides with the system of equations (3.39). Index i in system (3.69) corresponds to the value $n(0,t) = n$.

2. As the second example, we consider the simplest Markovian process with the finite number of states, namely, telegrapher's random process that can take on only two values $z(t) = \pm a$. In the foregoing section, we considered this process from another viewpoint. Here, we assume that the probabilities of transitions ($a \to -a$) and ($-a \to a$) during short interval Δt coincide and are $\nu \Delta t + o(\Delta t)$, the corresponding probabilities of state preservation during interval Δt are $1 - \nu \Delta t + o(\Delta t)$, and probabilities of initial states are p_a^0 and $p_{-a}^0 = 1 - p_a^0$. In this case, the transition probabilities satisfy the system of equations (3.66) with parameters

$$a_{11} = a_{22} = -\nu, \quad a_{12} = a_{21} = v.$$

The solution of this system is ($\tau = t - t_0$)

$$p_{11}(\tau) = p_{22}(\tau) = \frac{1}{2} \left[1 + e^{-2\nu\tau} \right], \quad p_{12}(\tau) = p_{21}(\tau) = \frac{1}{2} \left[1 - e^{-2\nu\tau} \right]. \tag{3.70}$$

Expressions for the one-point probabilities are obtained similarly:

$$P_1(\tau) = \frac{1}{2} + \left[p_a^0 - \frac{1}{2} \right] e^{-2\nu\tau}, \quad P_2(\tau) = \frac{1}{2} - \left[p_a^0 - \frac{1}{2} \right] e^{-2\nu\tau}. \tag{3.71}$$

If process $z(t)$ had at the initial instant the fixed value $z(t_0) = a$, then $p_a^0 = 1$ and Eqs. (3.71) assume the form

$$P_1(\tau) = \frac{1}{2}\left[1 + e^{-2\nu\tau}\right], \quad P_2(\tau) = \frac{1}{2}\left[1 - e^{-2\nu\tau}\right]. \tag{3.72}$$

For $t \to \infty$, these probability distributions tend to steady-state values $P_{1,2}(\infty) = 1/2$ and the process behavior tends to steady-state regime. If $p_a^0 = p_{-a}^0 = 1/2$ at the initial instant, the process $z(t)$ is always stationary.

Note that, in the case of telegrapher's process, formulas (3.70) can be combined in one formula; namely,

$$p(z, t|z_0, t_0) = \delta(z - z_0)P_1(\tau) + \delta(z + z_0)P_2(\tau), \tag{3.73}$$

where $P_1(\tau)$ and $P_2(\tau)$ are given by Eqs. (3.72) and $\tau = t - t_0$. Differentiating Eq. (3.73) with respect to time, we obtain the equation for the transition probability density $p(z, t|z_0, t_0)$

$$\frac{\partial}{\partial t}p(z, t|z_0, t_0) = -\nu\left\{p(z, t|z_0, t_0) - p(-z, t|z_0, t_0)\right\} \tag{3.74}$$

with the initial value

$$p(z, t_0|z_0, t_0) = \delta(z - z_0).$$

Thus, the transition probability density of telegrapher's process $p(z, t|z_0, t_0)$ satisfies the linear operator equation

$$\frac{\partial}{\partial t}p(z, t|z_0, t_0) = \hat{L}(z)p(z, t|z_0, t_0), \tag{3.75}$$

where operator $\hat{L}(z)$ is defined by the equality

$$\hat{L}(z)f(z) = -\nu\left\{f(z) - f(-z)\right\}. \tag{3.76}$$

Note that this is the property characteristic of all Markovian processes. However, the equation for the transition probability density not always allows the compact representation such as (3.75). In the general case of arbitrary Markovian process with a finite number of states, operator $\hat{L}(z)$ is matrix $||a_{ij}||$ appeared in Eq. (3.66) and probability density $p(z, t|z_0, t_0)$ itself is the matrix function. In this case, any realization of process $z(t)$ satisfies the identity

$$[z(t) - z_1][z(t) - z_2]...[z(t) - z_n] = 0. \tag{3.77}$$

Opening the brackets in Eq. (3.77), we see that different powers of process $z(t)$ satisfy the algebraic relationship

$$z^n(t) = (z_1 + ... + z_n)z^{n-1}(t) + ... + (-1)^{n+1}z_1z_2...z_n. \tag{3.78}$$

In the case of telegrapher's random process, i.e., the process with two possible states $z(t) = \pm a$, identity (3.78) reduces to

$$z^2(t) = a^2,$$

which appears very useful for analyzing stochastic equations whose parameters fluctuate by the law of telegrapher's process.

3. Consider now generalized telegrapher's process as an example of the spasmodic process. This process is defined by Eq. (3.52), and its transition probability density has the form

$$\begin{aligned} p(z,t|z_0,t_0) &= \langle \delta(z(t)-z)|z(t_0)=z_0\rangle \\ &= \delta(z-z_0)P_{n(0,t)=0} + \langle \delta(z-a)\rangle_a \left\{P_{n(0,t)=1} + P_{n(0,t)=2} + \ldots\right\}. \end{aligned} \tag{3.79}$$

Taking into account the normalization condition

$$\sum_{n=0}^{\infty} P_{n(0,t)=n} = 1,$$

we obtain the final expression in the form

$$p(z,t|z_0,t_0) = \delta(z-z_0)P_0(t,t_0) + p_a(z)\left\{1 - P_0(t,t_0)\right\}, \tag{3.80}$$

where $P_0(t,t_0) = e^{-\nu(t-t_0)}$ is the probability of the absence of jumps within temporal interval (t_0,t) and $p_a(z)$ is the probability of the event that random quantity a assumes value z.

The one-point probability distribution of process $z(t)$ is obviously the steady-state distribution

$$P(t,z) = p_a(z). \tag{3.81}$$

It is obvious that quantity (3.80) satisfies, as a function of variable t, the differential equation

$$\frac{\partial}{\partial t}p(z,t|z_0,t_0) = -\nu\left\{p(z,t|z_0,t_0) - p_a(z)\right\}, \tag{3.82}$$

which can be rewritten in the operator form

$$\frac{\partial}{\partial t}p(z,t|z_0,t_0) = \hat{L}(z)p(z,t|z_0,t_0), \tag{3.83}$$

where $\hat{L}(z)$ in this particular case is the integral operator

$$\hat{L}(z)f(z) = -\nu\left\{f(z) - p_a(z)\int_{-\infty}^{\infty} dz' f(z')\right\}. \tag{3.84}$$

Continuous Markovian processes

Consider now the continuous Markovian processes. In this case, the transition probability density $p(z,t|z_0,t_0)$ satisfies the operator equation (this equation is a consequence of the Smolukhovsky equation (3.60))

$$\frac{\partial}{\partial t}p(z,t|z_0,t_0) = \sum_{n=1}^{\infty}\frac{(-1)^n}{n!}\frac{\partial^n}{\partial z^n}\left[B_n(z,t)p(z,t|z_0,t_0)\right], \tag{3.85}$$

where functions $B_n(z,t)$ are determined by the equalities

$$\begin{aligned} B_n(z,t) &= \lim_{\Delta t\to 0}\frac{1}{\Delta t}\left\langle \{z(t+\Delta t)-z(t)\}^n \,|z(t)\right\rangle \\ &= \lim_{\Delta t\to 0}\frac{1}{\Delta t}\int_{-\infty}^{\infty} dz\,\{z(t+\Delta t)-z(t)\}^n p(z,t+\Delta t|z,t). \end{aligned} \tag{3.86}$$

A consequence of Eq. (3.85) is the similar equation for the one-point probability density $P(t,z) = \langle \delta(z(t)-z) \rangle$:

$$\frac{\partial}{\partial t} P(t,z) = \sum_{n=1}^{\infty} \frac{(-1)^n}{n!} \frac{\partial^n}{\partial z^n} \left[B_n(z,t) P(t,z) \right].$$

Continuous processes for which all coefficients B_n in Eq. (3.85) with $n \geq 3$ vanish form an important particular class. Markovian processes having this property are called the *diffusion processes*. In the context of such processes, Eq. (3.85) assumes the form

$$\frac{\partial}{\partial t} p(z,t|z_0,t_0) = -\frac{\partial}{\partial z} \left[B_1(z,t) p(z,t|z_0,t_0) \right] + \frac{1}{2} \frac{\partial^2}{\partial z^2} \left[B_2(z,t) p(z,t|z_0,t_0) \right]. \tag{3.87}$$

This equation is called the *Fokker–Planck equation*, and functions $B_1(z,t)$ and $B_2(z,t)$ are called the *drift* and *diffusion coefficients*, respectively.

In the particular case of $B_2(z,t) = \text{const}$ and $B_1(z,t) = -B_1 z$, Markovian process $z(t)$ is the Gaussian process with the exponential correlation function

$$\langle z(t) z(t+\tau) \rangle = \sigma^2(t) e^{-B_1|\tau|}.$$

In this case Eq. (3.87) is replaced with the equation

$$\frac{\partial}{\partial t} p(z,t|z_0,t_0) = \hat{L}(z) p(z,t|z_0,t_0),$$

where operator

$$\hat{L}(z) = B_1 \frac{\partial}{\partial z} z + \frac{1}{2} B_2 \frac{\partial^2}{\partial z^2}. \tag{3.88}$$

Note that the converse is also valid; namely, any Gaussian process with the exponential correlation function is the Markovian process.

Discrete-continuous Markovian processes

Consider now the one-dimensional discrete-continuous Markovian process. Two cases are possible here: the case of a purely discontinuous (spasmodic) process and the case of a process varying both continuously and discontinuously. In the first case, two functions — $q(z,t)$ and $u(z,z',t)$ — characterize random process $z(t)$. The meaning of these functions is as follows: within the short temporal interval $(t, t+\Delta t)$, the probability for the process to preserve its previous value is $1 - q(z,t)\Delta t$ and the probability for the process to change its value from z to z'' is $u(z,z',t)\Delta t \Delta z'$ (here, $z' < z'' < z' + \Delta z'$). Of cause, the normalization condition

$$\int_{-\infty}^{\infty} dz' u(z,z',t) = q(z,t). \tag{3.89}$$

is assumed additionally. For this process, a consequence of the Smolukhovsky equation (3.60) is the integro-differential equation

$$\frac{\partial}{\partial t} p(z,t|z_0,t_0) = -q(z,t) p(z,t|z_0,t_0) + \int_{-\infty}^{\infty} dz' u(z,z',t) p(z',t|z_0,t_0), \tag{3.90}$$

which is called the *Kolmogorov–Feller equation*. The equation for the one-point probability density has the similar form.

If, in addition to jumps, the process allows continuous variation, then the right-hand side of Eq. (3.90) is added with the right-hand side of Eq. (3.87). Note that random process $z(t) = n(0, t)$ (the number of jumps within temporal interval $(0, t)$) that we considered earlier is a special case of spasmodic processes; correspondingly, difference-differential equations (3.69) are special cases of the integro-differential equation (3.90).

It is obvious that Eq. (3.83) for generalized telegrapher's process is the Kolmogorov–Feller equation (3.90) with specially defined parameters $q(z, t) = \nu$ and $u(z, z', t) = \nu p_a(z)$.

Up to this point, we dealt with the one-dimensional processes; it is clear however that all results remain valid for multidimensional processes, i.e., for vector random functions $\mathbf{z}(t)$. In particular, the transition probability density

$$p(\mathbf{z}, t|\mathbf{z}_0, t_0) = \langle \delta(\mathbf{z}(t) - \mathbf{z})|\mathbf{z}(t_0) = \mathbf{z}_0) \rangle$$

will satisfy the linear operator equation

$$\frac{\partial}{\partial t} p(\mathbf{z}, t|\mathbf{z}_0, t_0) = \hat{L}(\mathbf{z}) p(\mathbf{z}, t|\mathbf{z}_0, t_0). \tag{3.91}$$

Note additionally that transition probability density $p(\mathbf{z}, t|\mathbf{z}_0, t_0)$ satisfies, as a function of its arguments, not only Eq. (3.91) (we will call this equation the *forward equation*), but also the equation with respect to variable t_0

$$\frac{\partial}{\partial t_0} p(\mathbf{z}, t|\mathbf{z}_0, t_0) = \hat{L}^{+}(\mathbf{z}_0) p(\mathbf{z}, t|\mathbf{z}_0, t_0), \tag{3.92}$$

which we will call the *backward equation*. Here, $\hat{L}^{+}(\mathbf{z}_0)$ is the operator conjugated to operator $\hat{L}(\mathbf{z})$. Equation (3.92) is convenient for analyzing the problems that deal with dependencies on initial locations of space-time points.

We mentioned earlier that two functions — transition probability density $p(z, t|z_0, t_0)$ and one-point probability density $P(t, z)$ — are sufficient to exhaustively describe all statistical characteristics of the Markovian process $z(t)$. Nevertheless, statistical analysis of stochastic equations requires additionally the knowledge of the characteristic functional of random process $z(t)$.

3.3.2 Characteristic functional of the Markovian process

For the Markovian process $z(t)$, no closed equation can be derived in the general case for the characteristic functional $\Phi[t; v(\tau)] = \langle \varphi[t; v(\tau)] \rangle$, where

$$\varphi[t; v(\tau)] = \exp\left\{ i \int_0^t d\tau z(\tau) v(\tau) \right\}.$$

Instead, we can derive the closed equation for the functional

$$\Psi[z, t; v(\tau)] = \langle \delta(z(t) - z) \varphi[t; v(\tau)] \rangle \tag{3.93}$$

describing correlations of process $z(t)$ with its prehistory. The characteristic functional $\Phi[t; v(\tau)]$ can be obtained from functional $\Psi[z, t; v(\tau)]$ by the formula

$$\Phi[t; v(\tau)] = \int_{-\infty}^{\infty} dz \Psi[z, t; v(\tau)]. \tag{3.94}$$

To derive the equation for functional $\Psi[z,t;v(\tau)]$, we note that the following equality

$$\varphi[t;v(\tau)] = 1 + i\int\limits_0^t dt_1 z(t_1)v(t_1)\varphi[t_1;v(\tau)] \tag{3.95}$$

holds. Substituting Eq. (3.95) in Eq. (3.93), we obtain the expression

$$\Psi[t,z;v(\tau)] = P(t,z) + i\int\limits_0^t dt_1 v(t_1)\left\langle \delta(z(t)-z)z(t_1)\varphi[t_1;v(\tau)]\right\rangle, \tag{3.96}$$

where $P(t,z) = \langle \delta(z(t)-z)\rangle$ is the one-point probability density of random quantity $z(t)$.

We rewrite Eq. (3.96) in the form

$$\begin{aligned}\Psi[t,z;v(\tau)] &= P(t,z)\\ &+i\int\limits_0^t dt_1 v(t_1)\int\limits_{-\infty}^{\infty} dz_1 z_1 \left\langle \delta(z(t)-z)\delta(z(t_1)-z_1)\varphi[t_1;v(\tau)]\right\rangle.\end{aligned} \tag{3.97}$$

Taking into account the fact that process $z(t)$ is the Markovian process, we can perform averaging in (3.97) to obtain the closed integral equation

$$\Psi[t,z;v(\tau)] = P(t,z) + i\int\limits_0^t dt_1 v(t_1)\int\limits_{-\infty}^{\infty} dz_1 z_1 p(z,t|z_1,t_1)\Psi[t_1,z_1;v(\tau)], \tag{3.98}$$

where $p(z,t;z_0,t_0)$ is the transition probability density.

We note that the integral equation similar to Eq. (3.98) can be derived also for the functional

$$\Psi[t',t,z;v(\tau)] = \left\langle \delta(z(t')-z)\varphi[t;v(\tau)]\right\rangle \quad (t' \geq t). \tag{3.99}$$

This equation has the form

$$\Psi[t',t,z;v(\tau)] = P(t',z) + i\int\limits_0^t dt_1 v(t_1)\int\limits_{-\infty}^{\infty} dz_1 z_1 p(z,t'|z_1,t_1)\Psi[t_1,z_1;v(\tau)]. \tag{3.100}$$

Integrating Eq. (3.98) with respect to z, we obtain an additional relationship between the characteristic functional $\Phi[t;v(\tau)]$ and functional $\Psi[z,t;v(\tau)]$. This relationship has the form

$$\frac{1}{iv(t)}\frac{d}{dt}\Phi[t;v(\tau)] = \int\limits_{-\infty}^{\infty} dz_1 z_1 \Psi[t,z_1;v(\tau)] = \Psi[t;v(\tau)]. \tag{3.101}$$

Multiplying Eq. (3.98) by z and integrating the result over z, we obtain the relationship between functionals $\Psi[t;v(\tau)]$ and $\Psi[t,z;v(\tau)]$

$$\Psi[t;v(\tau)] = \langle z(t)\rangle + i\int\limits_0^t dt_1 v(t_1)\int\limits_{-\infty}^{\infty} dz_1 \left\langle z(t)|z_1,t_1\right\rangle \Psi[t_1,z_1;v(\tau)]. \tag{3.102}$$

Equation (3.98) is in the general case a complicated integral equation whose explicit form depends on functions $P(t,z)$ and $p(z,t;z_0,t_0)$, i.e., on parameters of the Markovian process. Preliminarily differentiating this equation with respect to t and using Eq. (3.75), we can convert it into the integro-differential equation

$$\begin{aligned}\frac{d}{dt}\Psi[t,z;v(\tau)] &= izv(t)\Psi[t,z;v(\tau)] + \hat{L}(z)\Psi[t,z;v(\tau)],\\ \Psi[0,z;v(\tau)] &= P(0,z).\end{aligned} \quad (3.103)$$

In this case, functional $\Psi[t',t,z;v(\tau)]$ (3.99) as a function of variable t' satisfies the equation with the initial value at $t'=t$

$$\begin{aligned}\frac{d}{dt'}\Psi[t',t,z;v(\tau)] &= \hat{L}(z)\Psi[t',t,z;v(\tau)] \quad (t'>t),\\ \Psi[t,t,z;v(\tau)] &= \Psi[t,z;v(\tau)].\end{aligned} \quad (3.104)$$

Thus, Eq. (3.103) together with Eqs. (3.101) and (3.102) forms the starting point for the determination of the characteristic functional of the Markovian process.

We demonstrate this fact using the processes considered earlier as examples.

For telegrapher's process, Eq. (3.73) gives

$$\langle z(t)|z_1,t_1\rangle = z_1 e^{-2\nu(t-t_1)}, \quad \langle z(t)\rangle = 0,$$

and we obtain Eq. (3.18).

Consider now generalized telegrapher's process. By virtue of Eq. (3.84), Eq. (3.103) for functional $\Psi[t,z;v(\tau)]$ assumes in this case the form

$$\begin{aligned}\frac{d}{dt}\Psi[t,z;v(\tau)] &= \{izv(t)-\nu\}\,\Psi[t,z;v(\tau)] + \nu p_a(z)\Phi[t;v(\tau)],\\ \Psi[0,z;v(\tau)] &= p_a(z).\end{aligned} \quad (3.105)$$

Deriving Eq (3.105), we used equality (3.94). Solving Eq. (3.105) in functional $\Psi[t,z;v(\tau)]$, we relate it to the characteristic functional

$$\begin{aligned}\Psi[t,z;v(\tau)] &= p_a(z)\exp\left\{-\nu t + iz\int_0^t d\tau v(\tau)\right\}\\ &+\nu p_a(z)\int_0^t dt_1\Phi[t_1;v(\tau)]\exp\left\{-\nu(t-t_1)+iz\int_{t_1}^t d\tau v(\tau)\right\}.\end{aligned} \quad (3.106)$$

Integrating Eq. (3.106) over z, we obtain the closed integral equation for the characteristic functional $\Phi[t,v(\tau)]$

$$\begin{aligned}\Phi[t,v(\tau)] &= \left\langle \exp\left\{ia\int_0^t d\tau v(\tau)\right\}\right\rangle_a e^{-\nu t}\\ &+\nu\int_0^t dt_1 e^{-\nu(t-t_1)}\left\langle \exp\left\{ia\int_{t_1}^t d\tau v(\tau)\right\}\right\rangle_a \Phi[t_1,v(\tau)],\end{aligned} \quad (3.107)$$

which coincides with Eq. (3.53).

Multiplying Eq. (3.106) by arbitrary function $F(z)$ and integrating the result over z, we obtain the equality

$$\left\langle F\left(z(t)\right)\exp\left\{i\int_0^t d\tau z(\tau)v(\tau)\right\}\right\rangle = \left\langle F(a)\exp\left\{ia\int_0^t d\tau v(\tau)\right\}\right\rangle_a e^{-\nu t}$$
$$+\nu\int_0^t dt_1 e^{-\nu(t-t_1)}\left\langle F(a)\exp\left\{ia\int_{t_1}^t d\tau v(\tau)\right\}\right\rangle_a \Phi[t_1, v(\tau)]. \tag{3.108}$$

In the particular case of $F(z) = z$, Eq. (3.108) can be reduced to the integro-differential equation for the characteristic functional $\Phi[t, v(\tau)]$

$$\frac{d}{iv(t)dt}\Phi[t, v(\tau)] = \left\langle a\exp\left\{ia\int_0^t d\tau v(\tau)\right\}\right\rangle_a e^{-\nu t}$$
$$+\nu\int_0^t dt_1 e^{-\nu(t-t_1)}\left\langle a\exp\left\{ia\int_{t_1}^t d\tau v(\tau)\right\}\right\rangle_a \Phi[t_1, v(\tau)], \tag{3.109}$$

which is equivalent to Eq. (3.107).

In the case of generalized telegrapher's process, we can additionally establish the relationship between functionals $\Psi[t', t, z; v(\tau)]$ and $\Psi[t, z; v(\tau)]$. This relationship has the form $(t' \geq t)$

$$\Psi[t', t, z; v(\tau)] = \Psi[t, z; v(\tau)]e^{-\nu(t'-t)} + p_a(z)\Phi[t; v(\tau)][1 - e^{-\nu(t'-t)}]. \tag{3.110}$$

Chapter 4

Correlation splitting

4.1 General remarks

For simplicity, we content themselves here with the one-dimensional random processes (extensions to multidimensional cases are obvious). We need the ability of calculating correlation $\langle F[z(\tau)]R[z(\tau)]\rangle$, where $F[z(\tau)]$ is the functional explicitly dependent on process $z(t)$ and $R[z(\tau)]$ is the functional that can depend on process $z(t)$ both explicitly and implicitly.

To calculate this average, we consider auxiliary functionals $F\left[z(\tau)+\eta_1(\tau)\right]$ and $R\left[z(\tau)+\eta_2(\tau)\right]$, where $\eta_i(t)$ are arbitrary deterministic functions, and calculate the correlation

$$\langle F[z(\tau)+\eta_1(\tau)]R[z(\tau)+\eta_2(\tau)]\rangle .$$

The correlation of interest will be obtained by setting $\eta_i(\tau)=0$ in the final result.

We can expand the above auxiliary functionals in the functional Taylor series with respect to $z(\tau)$. The result can be represented in the form

$$F\left[z(\tau)+\eta_1(\tau)\right]=e^{\int\limits_{-\infty}^{\infty} d\tau z(\tau)\frac{\delta}{\delta\eta_1(\tau)}} F\left[\eta_1(\tau)\right], \quad R\left[z(\tau)+\eta_2(\tau)\right]=e^{\int\limits_{-\infty}^{\infty} d\tau z(\tau)\frac{\delta}{\delta\eta_2(\tau)}} R\left[\eta_2(\tau)\right],$$

where we introduced the functional shift operators. With this representation, we can obtain the following expression for the correlation

$$\begin{aligned}
&\langle F[z(\tau)+\eta_1(\tau)]R[z(\tau)+\eta_2(\tau)]\rangle \\
&= \left\langle \exp\left\{\Theta\left[\frac{1}{i}\left(\frac{\delta}{\delta\eta_1(\tau)}+\frac{\delta}{\delta\eta_2(\tau)}\right)\right]-\Theta\left[\frac{1}{i}\frac{\delta}{\delta\eta_1(\tau)}\right]-\Theta\left[\frac{1}{i}\frac{\delta}{\delta\eta_2(\tau)}\right]\right\}\right\rangle \\
&\times\langle F[z(\tau)+\eta_1(\tau)]\rangle\,\langle R[z(\tau)+\eta_2(\tau)]\rangle . \qquad (4.1)
\end{aligned}$$

This formula expresses the average of the product of functionals through the product of averages of the functionals themselves. The main problem here consists in calculating the action of the functional operator

$$\left\langle \exp\left\{\Theta\left[\frac{1}{i}\left(\frac{\delta}{\delta\eta_1(\tau)}+\frac{\delta}{\delta\eta_2(\tau)}\right)\right]-\Theta\left[\frac{1}{i}\frac{\delta}{\delta\eta_1(\tau)}\right]-\Theta\left[\frac{1}{i}\frac{\delta}{\delta\eta_2(\tau)}\right]\right\}\right\rangle$$

on the product of average functionals.

In a number of statistical problems, the intensity of parameter fluctuation can be considered small. In these situations, we can expand functional $F[z(\tau)]$ in the Taylor series with respect to process $z(\tau)$ and content themselves with the linear term of the expansion. In the case of the linear functional $F[z(\tau)] \equiv z(t')$, we obtain the following expression for the correlation

$$\langle z(t')R[z(\tau)+\eta(\tau)]\rangle = \Omega\left[t'; \frac{\delta}{i\delta\eta(\tau)}\right]\langle R[z(\tau)+\eta(\tau)]\rangle,$$

where functional

$$\Omega[t';v(\tau)] = \frac{\left\langle z(t')\exp\left\{i\int\limits_{-\infty}^{\infty} d\tau z(\tau)v(\tau)\right\}\right\rangle}{\left\langle \exp\left\{i\int\limits_{-\infty}^{\infty} d\tau z(\tau)v(\tau)\right\}\right\rangle} = \frac{\delta}{i\delta v(t')}\Theta[v(\tau)].$$

Setting now $\eta(\tau)=0$, we obtain the expression

$$\langle z(t')R[z(\tau)\rangle = \left\langle \Omega\left[t'; \frac{\delta}{i\delta z(\tau)}\right] R[z(\tau)]\right\rangle. \tag{4.2}$$

If we expand functional $\Theta[v(\tau)]$ in the functional Taylor series (3.26)

$$\Omega[t';v(\tau)] = \sum_{n=0}^{\infty}\frac{i^n}{n!}\int\limits_{-\infty}^{\infty} dt_1 \ldots \int\limits_{-\infty}^{\infty} dt_n K_{n+1}(t',t_1,\ldots,t_n)v(t_1)\ldots v(t_n),$$

then expression (4.2) assumes the form

$$\langle z(t')R[z(\tau)]\rangle = \sum_{n=0}^{\infty}\frac{1}{n!}\int\limits_{-\infty}^{\infty} dt_1 \ldots \int\limits_{-\infty}^{\infty} dt_n K_{n+1}(t',t_1,\ldots,t_n)\left\langle \frac{\delta^n R[z(\tau)]}{\delta z(t_1)\ldots\delta z(t_n)}\right\rangle. \tag{4.3}$$

Note that, if functional $R[z(\tau)]$ has the form of the power monomial

$$R[z(\tau)] = z(t_1)\ldots z(t_n),$$

then Eq. (4.3) recursively relates the n-point moment of process $z(t)$ to its cumulants.

If process $z(t)$ is simply random quantity z, operator $\int dt\delta/\delta z(t)$ reduces to the ordinary derivative d/dz and Eq. (4.3) grades into Eq. (3.12), page 51. Thus, Eq. (4.3) extends Eq. (3.12) to random processes.

In physical problems satisfying the condition of dynamical causality in time t, statistical characteristics of the solution at instant t depend on the statistical characteristics of process $z(\tau)$ for $0 \le \tau \le t$, which are completely described by the characteristic functional

$$\Phi[t;v(\tau)] = \exp\{\Theta[t;v(\tau)]\} = \left\langle \exp\left\{i\int\limits_0^t d\tau z(\tau)v(\tau)\right\}\right\rangle.$$

In this case, the obtained formulas hold also for calculating statistical averages $\langle z(t')R[t;z(\tau)]\rangle$ for $t'<t,\ \ \tau \le t$, i.e., we have the equality

$$\langle z(t')R[t;z(\tau)]\rangle = \left\langle \Omega\left[t',t; \frac{\delta}{i\delta z(\tau)}\right] R[t;z(\tau)]\right\rangle \quad (0<t'<t), \tag{4.4}$$

where

$$\Omega[t',t;v(\tau)] = \frac{\delta}{i\delta v(t')]}\Theta[t;v(\tau)]$$
$$+\sum_{n=0}^{\infty}\frac{i^n}{n!}\int_0^t dt_1...\int_0^t dt_n K_{n+1}(t',t_1,...,t_n)v(t_1)...v(t_n). \tag{4.5}$$

For $t' = t - 0$, formula (4.4) holds as before, i.e.

$$\langle z(t)R[t;z(\tau)]\rangle = \left\langle \Omega\left[t,t;\frac{\delta}{i\delta z(\tau)}\right] R[t;z(\tau)]\right\rangle. \tag{4.6}$$

However, expansion (4.5) not always gives the correct result in the limit $t' \to t - 0$ (which means that the limiting process and the procedure of expansion in the functional Taylor series can appear non-commutable). In this case,

$$\Omega[t,t;v(\tau)] = \frac{\left\langle z(t)\exp\left\{i\int_0^t d\tau z(\tau)v(\tau)\right\}\right\rangle}{\left\langle \exp\left\{i\int_0^t d\tau z(\tau)v(\tau)\right\}\right\rangle} = \frac{d}{iv(t)dt}\Theta[t;v(\tau)], \tag{4.7}$$

and statistical averages in Eqs. (4.4) and (4.6) can be discontinuous at $t' = t - 0$.

Consider several examples of random processes.

4.2 Gaussian process

In the case of the Gaussian random process $z(t)$, all formulas obtained in the previous section become significantly simpler. In this case, the logarithm of characteristic functional $\Phi[v(\tau)]$ is given by Eq. (3.34), page 58 (we assume that the mean value of process $z(t)$ is zero); as a consequence, Eq. (4.1) assumes the form

$$\langle F[z(\tau)+\eta_1(\tau)]R[z(\tau)+\eta_2(\tau)]\rangle$$
$$= e^{\int_{-\infty}^{\infty}\int_{-\infty}^{\infty} d\tau_1 d\tau_2 B(\tau_1,\tau_2)\frac{\delta^2}{\delta\eta_1(\tau_1)\delta\eta_2(\tau_2)}} \langle F[z(\tau)+\eta_1(\tau)]\rangle \langle R[z(\tau)+\eta_2(\tau)]\rangle. \tag{4.8}$$

We can easily calculate variational derivative of Eq. (4.8) with respect to $\eta_1(\tau)$ (this operation reduces to functional shift) and set $\eta_1(\tau) = 0$. As a result, we obtain the equality

$$\langle F[z(\tau)]R[z(\tau)+\eta(\tau)]\rangle$$
$$= \left\langle F\left[z(\tau)+\int_{-\infty}^{\infty} d\tau_1 B(\tau,\tau_1)\frac{\delta}{\delta\eta(\tau_1)}\right]\right\rangle \langle R[z(\tau)+\eta(\tau)]\rangle. \tag{4.9}$$

Let $F[z(\tau)] = z(t)$ for example. Then Eq. (4.9) assumes the form

$$\langle z(t)R[z(\tau)+\eta(\tau)]\rangle = \int_{-\infty}^{\infty} d\tau_1 B(t,\tau_1)\frac{\delta}{\delta\eta(\tau_1)} \langle R[z(\tau)+\eta(\tau)]\rangle. \tag{4.10}$$

Replacing now differentiation with respect to $\eta(\tau)$ by differentiation with respect to $z(\tau)$ and setting $\eta(\tau) = 0$, we obtain the equality

$$\langle z(t)R[z(\tau)]\rangle = \int_{-\infty}^{\infty} d\tau_1 B(t,\tau_1)\left\langle \frac{\delta}{\delta z(\tau_1)} R[z(\tau)]\right\rangle \tag{4.11}$$

commonly known in physics as the *Furutsu–Novikov formula* [69, 255]. Note that this formula can be obtained by partial integration in appropriate functional space [59].

One can easily obtain the multi-dimensional extension of Eq. (4.11); it can be written in the form

$$\langle z_{i_1,\ldots,i_n}(\mathbf{r})R[\mathbf{z}]\rangle = \int d\mathbf{r}' \langle z_{i_1,\ldots,i_n}(\mathbf{r}) z_{j_1,\ldots,j_n}(\mathbf{r}')\rangle \left\langle \frac{\delta R[\mathbf{z}]}{\delta z_{j_1,\ldots,j_n}(\mathbf{r}')}\right\rangle, \tag{4.12}$$

where $\mathbf{r}$ stands for all continuous arguments of random vector field $\mathbf{z}(\mathbf{r})$ and $i_1, \ldots, i_n$ are the discrete (index) arguments. Repeated index arguments in the right-hand side of Eq. (4.12) assume summation.

If we set in Eq. (4.9) $F[z(\tau)] = \exp\left\{i \int_{-\infty}^{\infty} d\tau\, z(\tau)v(\tau)\right\}$, then we obtain, at $\eta(\tau) = 0$, the equality

$$\left\langle e^{i\int_{-\infty}^{\infty} d\tau z(\tau)v(\tau)} R[z(\tau)]\right\rangle = \Phi[v(\tau)]\left\langle R\left[z(\tau) + i\int_{-\infty}^{\infty} d\tau_1 B(\tau,\tau_1)v(\tau_1)\right]\right\rangle, \tag{4.13}$$

in which random process $z(\tau)$ within the averaging brackets in the right-hand side is added with the deterministic imaginary term. Formulas (4.11), (4.12) and (4.13) extend formulas (3.17), (3.18), page 52 to the Gaussian random processes.

If random process $z(\tau)$ is defined only on time interval $[0,t]$, then functional $\Theta[t, v(\tau)]$ will assume the form

$$\Theta[t, v(\tau)] = -\frac{1}{2}\int_0^t\int_0^t d\tau_1 d\tau_2 B(\tau_1,\tau_2)v(\tau_1)v(\tau_2), \tag{4.14}$$

and functionals $\Omega[t', t; v(\tau)]$ and $\Omega[t, t; v(\tau)]$ will be the linear functionals

$$\begin{aligned}
\Omega[t', t; v(\tau)] &= \frac{\delta}{i\delta v(t')}\Theta[t, v(\tau)] = i\int_0^t d\tau B(t',\tau)v(\tau), \\
\Omega[t, t; v(\tau)] &= \frac{d}{iv(t)dt}\Theta[t, v(\tau)] = i\int_0^t d\tau B(t,\tau)v(\tau).
\end{aligned} \tag{4.15}$$

As a consequence, Eqs. (4.4), (4.6) will assume the form

$$\langle z(t')R[t, z(\tau)]\rangle = \int_0^t d\tau B(t',\tau)\left\langle \frac{\delta R[z(\tau)]}{\delta z(\tau)}\right\rangle \quad (t' \leqslant t) \tag{4.16}$$

that coincides with Eq. (4.11) if the condition

$$\frac{\delta R[t; z(\tau)]}{\delta z(\tau)} = 0 \quad \text{for} \quad \tau < 0, \quad \tau > t \tag{4.17}$$

holds. Note that Eq. (4.13) assumes in this case the form

$$\left\langle e^{i\int\limits_0^t d\tau z(\tau)v(\tau)} R[t;z(\tau)]\right\rangle = \Phi[t;v(\tau)]\left\langle R\left[t;z(\tau)+i\int\limits_0^t d\tau_1 B(\tau,\tau_1)v(\tau_1)\right]\right\rangle, \qquad (4.18)$$

where $\Phi[t;v(\tau)]$ is the characteristic functional of the Gaussian random process $z(t)$.

4.3 Poisson process

The Poisson process is defined by Eq. (3.40), page 60, and its characteristic functional logarithm is given by Eq. (3.41). In this case, formulas (4.5) and (4.7) for functionals $\Omega[t',t;v(\tau)]$ and $\Omega[t,t;v(\tau)]$ assume the forms

$$\begin{aligned}\Omega[t',t;v(\tau)] &= \frac{\delta}{i\delta v(t')}\Theta[t,v(\tau)] = -i\int\limits_0^{t'} d\tau g(t'-\tau)\dot{W}\left(\int\limits_\tau^t d\tau_1 v(\tau_1)g(\tau_1-\tau)\right),\\ \Omega[t,t;v(\tau)] &= \frac{d}{iv(t)dt}\Theta[t,v(\tau)] = -i\int\limits_0^{t} d\tau g(t-\tau)\dot{W}\left(\int\limits_\tau^t d\tau_1 v(\tau_1)g(\tau_1-\tau)\right),\end{aligned} \qquad (4.19)$$

where $\dot{W}(v) = \frac{dW(v)}{dv} = i\int\limits_{-\infty}^{\infty} d\xi\xi p(\xi)e^{i\xi v}$.

Changing the integration order, we can rewrite equalities (4.19) in the form

$$\Omega[t',t;v(\tau)] = i\int\limits_{-\infty}^{\infty} d\xi\xi p(\xi)\int\limits_0^{t'} d\tau g(t'-\tau)\exp\left\{i\xi\int\limits_\tau^t d\tau_1 v(\tau_1)g(\tau_1-\tau)\right\} \quad (t'\leqslant t). \qquad (4.20)$$

As a result, we obtain that correlations of the Poisson random process $z(t)$ with functionals of this process are described by the expression

$$\langle z(t')R[t;z(\tau)]\rangle = \nu\int\limits_{-\infty}^{\infty} d\xi\xi p(\xi)\int\limits_0^{t'} d\tau' g(t'-\tau')\langle R[t;z(\tau)+\xi g(\tau-\tau')]\rangle \quad (t'\leqslant t). \qquad (4.21)$$

As we mentioned earlier, random process $n(0,t)$ describing the number of jumps during temporal interval $(0,t)$ is the special case of the Poisson process. In this case, $p(\xi)=\delta(\xi-1)$ and $g(t)=\theta(t)$, so that Eq. (4.21) assumes the extra-simple form

$$\langle n(0,t)R[t;n(0,\tau)]\rangle = \nu\int\limits_0^{t'} d\tau\langle R[t;n(0,\tau)+\theta(t'-\tau)]\rangle \quad (t'\leqslant t). \qquad (4.22)$$

Equality (4.22) extends formula (3.19) for the Poisson random quantities to the Poisson random processes.

4.4 Telegrapher's random process

Now, we dwell on telegrapher's random process defined by formula (3.44), page 62

$$z(t) = a(-1)^{n(0,t)}, \tag{4.23}$$

where a is the deterministic quantity. The nth-order moment functions of this process satisfy recurrence equation (3.45) from which immediately follows the relationship

$$\langle z(t_1)z(t_2)R[z(\tau)]\rangle = \langle z(t_1)z(t_2)\rangle \langle R[z(\tau)]\rangle, \tag{4.24}$$

which holds for arbitrary functional $R[z(\tau)]$ under the condition that $\tau \le t_2 \le t_1$ [29]. The proof of Eq. (4.24) consists in expanding functional $R[z(\tau)]$ in the Taylor series in $z(\tau)$ and using formula (3.45).

Let now quantity a be the random quantity with probability distribution density

$$p(a) = \frac{1}{2}\left[\delta(a - a_0) + \delta(a + a_0)\right]. \tag{4.25}$$

In this case, $M_{2k+1} = 0$ and, in addition to Eq. (4.24), the following equality holds [29]

$$\begin{aligned} &\langle F[z(\tau_1)]z(t_1)z(t_2)R[z(\tau_2)]\rangle \\ &\quad = \langle F[z(\tau_1)\rangle] \langle z(t_1)z(t_2)\rangle \langle R[z(\tau_2)]\rangle + \langle F[z(\tau_1)]z(t_1)\rangle \langle z(t_2)R[z(\tau_2)]\rangle, \end{aligned} \tag{4.26}$$

which is valid for any $\tau_1 \le t_1 \le t_2 \le \tau_2$ and arbitrary functionals $F[z(\tau_1)]$ and $R[z(\tau_2)]$. Indeed, in terms of the Taylor expansion in $z(\tau)$, functional $R[z(\tau_2)]$ can be represented in the form

$$R[z(\tau_2)] = \sum_{2k} + \sum_{2k+1},$$

where the first sum consists of terms with the even number of co-factors $z(\tau)$ and the second sum consists of terms with the odd number of such co-factors. Then, taking into account Eq. (3.45) and equalities

$$\langle R[z(\tau_2)]\rangle = \left\langle \sum_{2k} \right\rangle, \quad \langle z(t_2)R[z(\tau_2)]\rangle = \left\langle z(t_2) \sum_{2k+1} \right\rangle,$$

we obtain Eq. (4.26).

Formula (4.26) allows another representation. Denote functional $F[z(\tau_1)]z(t_1)$ as $F[t_1; z(\tau_1)]$, where $\tau_1 \le t_1$, and functional $z(t_2)R[z(\tau_2)]$ as $R[t_2; z(\tau_2)]$, where $t_1 \le t_2 \le \tau_2$. Then, Eq. (4.26) can be written in the form

$$\begin{aligned} \langle F[t_1; z(\tau_1)]R[t_2; z(\tau_2)]\rangle &= \langle F[t_1; z(\tau_1)]\rangle \langle R[t_2; z(\tau_2)]\rangle \\ &+ \frac{1}{a_0^2} e^{-2\nu(t_2 - t_1)} \langle z(t_1)F[t_1; z(\tau_1)]\rangle \langle z(t_2)R[t_2; z(\tau_2)]\rangle. \end{aligned} \tag{4.27}$$

Because functionals $F[z(\tau_1)]$ and $R[z(\tau_2)]$ in Eq. (4.26) are arbitrary functionals, functionals $F[t_1; z(\tau_1)]$ and $R[t_2; z(\tau_2)]$ in (4.27) are also arbitrary functionals.

Formulas (4.24) and (4.26) include bilinear combinations of process $z(t)$, which is not always practicable.

As we mentioned earlier, calculation of correlator $\langle z(t)R[t; z(\tau)]\rangle$ for $\tau \le t$ assumes the knowledge of the characteristic functional of process $z(t)$; unfortunately, the characteristic functional is unavailable in this case. We know only Eqs. (3.48) and (3.49), page 63 that describe the relationship of the functional

$$\Psi[t;v(\tau)] = \frac{1}{iv(t)}\frac{d}{dt}\Phi[t;v(\tau)] = \left\langle z(t)\exp\left\{i\int_0^t d\tau z(\tau)v(\tau)\right\}\right\rangle$$

and the characteristic functional $\Phi[t;v(\tau)]$ by itself in the form

$$\left\langle z(t)\exp\left\{i\int_0^t d\tau z(\tau)v(\tau)\right\}\right\rangle$$
$$= ae^{-2\nu t} + ia^2\int_0^t dt_1 e^{-2\nu(t-t_1)}v(t_1)\left\langle \exp\left\{i\int_0^{t_1} d\tau z(\tau)v(\tau)\right\}\right\rangle.$$

This relationship is sufficient to split correlator $\langle z(t)R[t;z(\tau)]\rangle$, i.e., to express it immediately in terms of the average functional $\langle R[t;z(\tau)]\rangle$.

Indeed, the equality

$$\langle z(t)R[t;z(\tau)+\eta(\tau)]\rangle = \left\langle z(t)\exp\left\{\int_0^t d\tau z(\tau)\frac{\delta}{\delta\eta(\tau)}\right\}\right\rangle R[t;\eta(\tau)]$$
$$= \quad aR[t;\eta(\tau)]e^{-2\nu t} + a^2\int_0^t dt_1 e^{-2\nu(t-t_1)}\frac{\delta}{\delta\eta(t_1)}\left\langle R[t;z(\tau)\theta(t_1-\tau)+\eta(\tau)]\right\rangle, \tag{4.28}$$

where $\eta(t)$ is arbitrary deterministic function, holds for any functional $R[t;z(\tau)]$ for $\tau \le t$. The final result is obtained by the limit process $\eta \to 0$:

$$\langle z(t)R[t;z(\tau)]\rangle = aR[t;0]e^{-2\nu t} + a^2\int_0^t dt_1 e^{-2\nu(t-t_1)}\left\langle\frac{\delta}{\delta z(t_1)}\tilde{R}[t,t_1;z(\tau)]\right\rangle, \tag{4.29}$$

where functional $\tilde{R}[t,t_1;z(\tau)]$ is defined by the formula

$$\tilde{R}[t,t_1;z(\tau)] = R[t;z(\tau)\theta(t_1-\tau+0)]. \tag{4.30}$$

In the case of random quantity a distributed according to probability density distribution (4.25), additional averaging (4.29) over a results in the equality

$$\langle z(t)R[t;z(\tau)]\rangle = a_0^2\int_0^t dt_1 e^{-2\nu(t-t_1)}\left\langle\frac{\delta}{\delta z(t_1)}\tilde{R}[t,t_1;z(\tau)]\right\rangle. \tag{4.31}$$

Formula (4.31) is very similar to the formula for splitting the correlator of the Gaussian process $z(t)$ characterized by the exponential correlation function (i.e., the Gaussian Markovian process) with functional $R[t;z(\tau)]$. The difference consists in the fact that the right-hand side of Eq. (4.31) depends on the functional that is cut by the process $z(\tau)$ rather than on functional $R[t;z(\tau)]$ itself.

Note that correlation $\langle z(t')R[t;z(\tau)]\rangle$,where $t' \geq t$ and $\tau < t$, can be represented in the form

$$\langle z(t)R[t;z(\tau)]\rangle = \frac{1}{a_0^2}\langle z(t')z(t)z(t)R[t;z(\tau)]\rangle .$$

As a consequence, the equality

$$\langle z(t')R[t;z(\tau)]\rangle = e^{-2\nu(t'-t)}\langle z(t)R[t;z(\tau)]\rangle \quad (t' \geqslant t). \tag{4.32}$$

holds according to formula (4.24).

In a similar way, we can obtain the expression

$$\langle z(t')R[t_0,t;z(\tau)]\rangle = e^{-2\nu(t_0-t')}\langle z(t_0)R[t_0,t;z(\tau)]\rangle \quad (t' \geqslant t), \tag{4.33}$$

where $t' \leq t_0 \leq \tau \leq t$ and

$$\langle z(t_0)R[t_0,t;z(\tau)]\rangle = a_0^2 \int\limits_{t_0}^{t} dt_1 e^{-2\nu(t_1-t_0)}\left\langle \frac{\delta}{\delta z(t_1)}R[t_0,t,;z(\tau)\theta(\tau-t_1+0]\right\rangle . \tag{4.34}$$

In the case of the general-form correlator $\langle z(\xi)R[t_0,t;z(\tau)]\rangle$, where $t_0 \leqslant \xi \leqslant t$ and $t_0 \leqslant \tau \leqslant t$, one can show [134, 135] the validity of the following equality:

$$\begin{aligned}&\langle z(\xi)R[t_0,t;z(\tau)]\rangle\\ &= a_0^2 \int\limits_{t_0}^{\xi} dt_1 e^{-2\nu(\xi-t_1)}\left\langle \frac{\delta R[t_0,t,;z_1(\tau)\theta(\tau-\xi)+z_2(\tau)\theta(t_1-\tau+0]}{\delta z_2(t_1)}\right\rangle\\ &+a_0^2 \int\limits_{\xi}^{t} dt_1 e^{-2\nu(t_1-\xi)}\left\langle \frac{\delta R[t_0,t,;z_1(\tau)\theta(\xi-\tau)+z_2(\tau)\theta(\tau-t_1+0]}{\delta z_2(t_1)}\right\rangle ,\end{aligned} \tag{4.35}$$

where $z_1(t)$ and $z_2(t)$ are statistically independent telegrapher's processes characterized by the same correlation function of the form

$$\langle z(t)z(t')\rangle = a_0^2 e^{-2\nu|t-t'|}.$$

The limits of integration t_0 and t in Eq. (4.35) can assume arbitrary values from $-\infty$ to ∞. At $\xi = t$ or $\xi = t_0$, Eq. (4.35) grades into Eq. (4.31) or (4.34), respectively.

If we set $v(\tau) = v$ and $R[t_0,t;z(\tau)] = \exp\left\{iv\int\limits_{t_0}^{t} d\tau z(\tau)\right\}$ in Eq. (4.35) and take into account Eq. (3.51), page 64, we obtain the expression

$$\begin{aligned}&\left\langle z(\xi)\exp\left\{iv\int\limits_{t_0}^{t} d\tau z(\tau)\right\}\right\rangle\\ &= \frac{iva_0^2}{\lambda}e^{-\nu(t-t_0)}\left\{\sinh\lambda(t-t_0) + \frac{2\nu}{\lambda}\sinh\lambda(t-\xi)\sinh\lambda(\xi-t_0)\right\},\end{aligned} \tag{4.36}$$

where $\lambda = \sqrt{\nu^2 - a_0^2 v^2}$.

Let us differentiate Eq. (4.29) with respect to time t. Taking into account the fact that there is no need in differentiating with respect to the upper limit of the integral in the right-hand side of Eq. (4.29), we obtain the expression [276, 277]

$$\left(\frac{d}{dt}+2\nu\right)\langle z(t)R[t;z(\tau)]\rangle=\left\langle z(t)\frac{d}{dt}R[t;z(\tau)]\right\rangle \tag{4.37}$$

called usually the *differentiation formula.*

One essential point should be noticed. Functional $R[t;z(\tau)]$ in differentiation formula (4.37) is an arbitrary functional and can simply coincide with process $z(t-0)$. In the general case, the realization of telegrapher's process is the generalized function. The derivative of this process is also the generalized function (the sequence of delta-functions), so that

$$z(t)\frac{d}{dt}z(t)\neq\frac{1}{2}\frac{d}{dt}z^2(t)\equiv 0.$$

in the general case. These generalized functions, as any generalized functions, are defined only in terms of functionals constructed on them. In the case of our interest, such functionals are the average quantities denoted by angle brackets $\langle ...\rangle$, and the above differentiation formula describes the differential constraint between different functionals related to random process $z(t)$ and its one-sided derivatives for $t\to t-0$, such as dz/dt, d^2z/dt^2. For example, formula (4.37) allows derivation of equalities, such as

$$\left\langle z(t)\frac{d}{d\tau}z(\tau)\right\rangle\Bigg|_{\tau=t-0}=2\nu\left\langle z^2\right\rangle,\quad \left\langle z(t)\frac{d^2}{d\tau^2}z(\tau)\right\rangle\Bigg|_{\tau=t-0}=4\nu^2\left\langle z^2\right\rangle.$$

It is clear that these formulas can be obtained immediately by differentiating the correlation function $\langle z(t)z(t')\rangle$ with respect to t' ($t'<t$) for $t'\to t-0$.

Above, we considered the correlation of random process $z(t)$ with a functional of this process. If we deal with an arbitrary function of telegrapher's process $F(z(t))$, then, clearly, the equality

$$F(z(t))=\frac{F(a)+F(-a)}{2}+\frac{F(a)-F(-a)}{2a}z(t) \tag{4.38}$$

will hold, and all results valid for telegrapher's process $z(t)$ will be valid (with small variations) for process $F(z(t))$.

4.5 Generalized telegrapher's random process

Consider now generalized telegrapher's process described by Eq. (3.52), page 65. In this case, functional

$$\Psi[t;v(\tau)]=\frac{1}{iv(t)}\frac{d}{dt}\Phi[t;v(\tau)]=\left\langle z(t)\exp\left\{i\int_0^t d\tau z(\tau)v(\tau)\right\}\right\rangle$$

is related to the characteristic functional $\Phi[t;v(\tau)]$ of process $z(t)$ by Eq. (3.109)

$$\left\langle z(t)\exp\left\{i\int_0^t d\tau z(\tau)v(\tau)\right\}\right\rangle=\left\langle a\exp\left\{ia\int_0^t d\tau v(\tau)\right\}\right\rangle_a e^{-\nu t}$$
$$+\nu\int_0^t dt_1 e^{-\nu(t-t_1)}\left\langle a\exp\left\{ia\int_{t_1}^t d\tau v(\tau)\right\}\right\rangle_a \Phi[t_1,v(\tau)].$$

As in the case of telegrapher's process, this formula allows one to express correlator $\langle z(t)R[t;z(\tau)]\rangle$, where $R[t;z(\tau)]$ is arbitrary functional, in terms of the mean value of the functional. Indeed, if we replicate operations used for telegrapher's process, we obtain the equality

$$\langle z(t)R[t;z(\tau)+\eta(\tau)]\rangle = \Psi\left[t;\frac{\delta}{i\delta\eta(\tau)}\right]R[t;\eta(\tau)] = \langle \tilde{a}R[t;\eta(\tau)+\tilde{a}]\rangle_{\tilde{a}}\, e^{-\nu t}$$
$$+\nu\int\limits_0^t dt_1 e^{-\nu(t-t_1)}\left\langle \tilde{a}R\left[t;\eta(\tau)+\tilde{a}\theta(\tau-t_1)+z\left(\tau\right)\theta(t_1-\tau)\right]\right\rangle_{\tilde{a},z}, \tag{4.39}$$

where $\eta(\tau)$ is arbitrary function and random quantity $\tilde{a}$ is distributed with probability density $p(\tilde{a})$ and is statistically independent of process $z(t)$. Setting now $\eta(\tau) = 0$, we obtain the final expression

$$\langle z(t)R[t;z(\tau)]\rangle = \langle \tilde{a}R[t;\tilde{a}]\rangle_{\tilde{a}}\, e^{-\nu t}$$
$$+\nu\int\limits_0^t dt_1 e^{-\nu(t-t_1)}\left\langle \tilde{a}R\left[t;\tilde{a}\theta(\tau-t_1)+z\left(\tau\right)\theta(t_1-\tau)\right]\right\rangle_{\tilde{a},z}. \tag{4.40}$$

Note additionally that, in the case of generalized telegrapher's process, correlation $\langle F(z(t))R[t;z(\tau)]\rangle$, where $F(z)$ is arbitrary function, is described, in view of Eq.(3.108), by the formula similar to Eq. (4.40)

$$\langle F\left(z(t)\right)R[t;z(\tau)]\rangle = \langle F\left(\tilde{a}\right)R[t;\tilde{a}]\rangle_{\tilde{a}}\, e^{-\nu t}$$
$$+\nu\int\limits_0^t dt_1 e^{-\nu(t-t_1)}\left\langle F\left(\tilde{a}\right)R\left[t;\tilde{a}\theta(\tau-t_1)+z\left(\tau\right)\theta(t_1-\tau)\right]\right\rangle_{\tilde{a},z}. \tag{4.41}$$

4.6 General-form Markovian processes

Processes such as the above telegrapher's processes are the simplest examples of the Markovian processes. Here, we consider what consequences can be derived for the correlation of functionals from the sole assumption that process $z(t)$ is the Markovian process.

In the case of the general-form Markovian process $z(t)$, we have no equation for the characteristic functional. We have only integral equation (3.98), page 74 for the functional

$$\Psi[t,z;v(\tau)] = \left\langle \delta\left(z(t){-}z\right)\exp\left\{i\int\limits_0^t d\tau z(\tau)v(\tau)\right\}\right\rangle$$

that describes statistical relationship of process $z(t)$ at instant t with its prehistory

$$\Psi[t,z;v(\tau)] = P(t,z) + i\int\limits_0^t dt_1 v(t_1)\int\limits_{-\infty}^{\infty} dz_1 z_1 p(z,t|z_1,t_1)\Psi[t_1,z_1;v(\tau)], \tag{4.42}$$

where $P(t,z)$ is the one-time probability density and $p(z,t|z_0,t_0)$ is the transition probability density of process $z(t)$. In this case, we can again use the method described above to obtain the expression for the correlator

$$\langle \delta\left(z(t){-}z\right)R[t;z(\tau)+\eta(\tau)]\rangle \quad (\tau \leqslant t)$$

in the form of the integral equality with variational derivatives

$$\langle \delta\left(z(t)-z\right) R[t; z(\tau)+\eta(\tau)]\rangle = \Psi\left[t, z; \frac{\delta}{i\delta\eta(\tau)}\right] R[t; \eta(\tau)], \tag{4.43}$$

where $\eta(t)$ is arbitrary function.

For the Markovian processes $z(t)$, functions $P(t,z)$ and $p(z,t|z_0,t_0)$ satisfy linear operator equations (3.75), page 70

$$\frac{\partial}{\partial t}P(t,z) = \hat{L}(z)P(t,z), \quad \frac{\partial}{\partial t}p(z,t|z_0,t_0) = \hat{L}(z)p(z,t|z_0,t_0), \tag{4.44}$$

where $\hat{L}(z)$ is the integro-differential operator with respect to variable z.

Let us differentiate Eq. (4.43) with respect to time t and take into account that variational derivative

$$\frac{\delta}{\delta\eta(t_1)}R[t;\eta(\tau)] \sim \theta(t-t_1)$$

by the definition of functional $R[t; z(\tau)]$, so that we have no need in differentiating the integral in Eq. (4.42) with respect to the upper limit (we can set it to ∞). An additional point consists in the fact that the differentiation operation commutes with the variational differentiation operation (see Appendix A, Eq. (A.12), page 431):

$$\frac{\partial}{\partial t}\frac{\delta}{\delta\eta(t_1)}R[t;\eta(\tau)] = \frac{\delta}{\delta\eta(t_1)}\frac{\partial}{\partial t}R[t;\eta(\tau)].$$

Taking into account Eqs. (4.44), we obtain the formula for the derivative of the correlation of interest with respect to time (function $\eta(t)$ can be set to zero) [133]–[135]

$$\frac{\partial}{\partial t}\langle \delta\left(z(t)-z\right) R[t; z(\tau)]\rangle$$
$$= \left\langle \delta\left(z(t)-z\right) \frac{\partial}{\partial t}R[t; z(\tau)]\right\rangle + \hat{L}(z)\langle \delta\left(z(t)-z\right) R[t; z(\tau)]\rangle \tag{4.45}$$

Multiply now Eq. (4.45) by arbitrary function $f(z)$ and integrate the result over z. The result will be the differentiation formula

$$\frac{\partial}{\partial t}\langle f\left(z(t)\right) R[t; z(\tau)]\rangle = \left\langle f\left(z(t)\right) \frac{\partial}{\partial t}R[t; z(\tau)]\right\rangle$$
$$+ \int\limits_{-\infty}^{\infty} dz f(z)\hat{L}(z)\langle \delta\left(z(t)-z\right) R[t; z(\tau)]\rangle, \tag{4.46}$$

that can be rewritten in the form

$$\frac{\partial}{\partial t}\langle f\left(z(t)\right) R[t; z(\tau)]\rangle - \left\langle f\left(z(t)\right) \frac{\partial}{\partial t}R[t; z(\tau)]\right\rangle = \left\langle R[t; z(\tau)]\left[\hat{L}^{+}(z)f(z)\right]\right\rangle, \tag{4.47}$$

where we introduced operator $\hat{L}^{+}(z)$ conjugated to operator $\hat{L}(z)$.

Thus, Eqs. (4.45)–(4.47) govern the rules of differentiating with respect to time the correlators of functions of the Markovian process $z(t)$ with functionals of this process.

Note that, if the mean value of process $z(t)$ is equal to zero, the right-hand side of Eq. (4.47) can be expressed in terms of the desired correlation $\langle z(t)R[t; z(\tau)]\rangle$ for all Markovian processes considered earlier. This is most probably the evidence of low practicability of this formula. However, for telegrapher's and generalized telegrapher's processes, Eq. (4.47) appears practicable for analyzing linear stochastic equations. Indeed, for telegrapher's process $z(t)$, the right-hand side of Eq. (4.47) for correlation $\langle z(t)R[t;z(\tau)]\rangle$ has the form

$$-2\nu\left\langle z(t)R[t;z(\tau)]\right\rangle. \tag{4.48}$$

For generalized telegrapher's process $z(t)$, the right-hand side of Eq. (4.47) under the condition that $\langle f(a)\rangle = 0$ assumes the form

$$-\nu\left\langle f\left(z(t)\right)R[t;z(\tau)]\right\rangle. \tag{4.49}$$

In the special case $f(z) = z$, Eq. (4.49) reduces to

$$-\nu\left\langle z(t)R[t;z(\tau)]\right\rangle. \tag{4.50}$$

Now, we dwell on some extensions of the above formulas. First of all, we note that if we deal with the vector Markovian process $\mathbf{z}(t) = \{z_1(t), ..., z_N(t)\}$ described by operator $\hat{L}(\mathbf{z})$, then functional

$$\Psi[t,\mathbf{z};\mathbf{v}(\tau)] = \left\langle \delta\left(\mathbf{z}(t)-\mathbf{z}\right)\exp\left\{i\int\limits_0^t d\tau\mathbf{z}(\tau)\mathbf{v}(\tau)\right\}\right\rangle,$$

where $\mathbf{v}(t) = \{v_1(t), ..., v_N(t)\}$, satisfies the equation

$$\frac{\partial}{\partial t}\Psi[t,\mathbf{z};\mathbf{v}(\tau)] = \left\{\hat{L}(\mathbf{z})+i\mathbf{z}\mathbf{v}\right\}\Psi[t,\mathbf{z};\mathbf{v}(\tau)] \tag{4.51}$$

with the initial value

$$\Psi[0,\mathbf{z};\mathbf{v}(\tau)] = P_0(\mathbf{z}). \tag{4.52}$$

With this remark, we can easily derive the formula of differentiating the correlator $\langle F(\mathbf{z}(t))R[t;\mathbf{z}(\tau)]\rangle$ with respect to time; it assumes the form

$$\frac{\partial}{\partial t}\left\langle F\left(\mathbf{z}(t)\right)R[t;\mathbf{z}(\tau)]\right\rangle = \left\langle F\left(\mathbf{z}(t)\right)\frac{\partial}{\partial t}R[t;\mathbf{z}(\tau)]\right\rangle + \left\langle R[t;\mathbf{z}(\tau)]\left[\hat{L}^{+}(\mathbf{z})F(\mathbf{z})\right]\right\rangle, \tag{4.53}$$

where $\hat{L}^{+}(\mathbf{z})$ is the operator conjugated to operator $\hat{L}(\mathbf{z})$.

An important special case corresponds to the situation in which all components of vector $\mathbf{z}(t)$ are the statistically independent Markovian processes described by the same operator $\hat{L}(\mathbf{z})$; in this case, Eq. (4.53) reduces to the form

$$\frac{\partial}{\partial t}\left\langle F\left(\mathbf{z}(t)\right)R[t;\mathbf{z}(\tau)]\right\rangle = \left\langle F\left(\mathbf{z}(t)\right)\frac{\partial}{\partial t}R[t;\mathbf{z}(\tau)]\right\rangle + \sum_{k=1}^{N}\left\langle R[t;\mathbf{z}(\tau)]\left[\hat{L}^{+}(z_k)F(\mathbf{z})\right]\right\rangle, \tag{4.54}$$

For example, all above Markovian processes having the exponential correlation function

$$\langle z(t)z(t+\tau)\rangle = \left\langle z^2\right\rangle e^{-\alpha|\tau|} \tag{4.55}$$

satisfy the equality

$$\left(\frac{\partial}{\partial t}+\alpha k\right)\left\langle z_1(t)....z_k(t)R[t;\mathbf{z}(\tau)]\right\rangle = \left\langle z_1(t)....z_k(t)\frac{\partial}{\partial t}R[t;\mathbf{z}(\tau)]\right\rangle. \tag{4.56}$$

Formula (4.56) defines the rule of factoring the operation of differentiating with respect to time out of angle brackets of averaging; in particular, we have

$$\left\langle z_1(t)....z_k(t)\frac{\partial^n}{\partial t^n}R[t;\mathbf{z}(\tau)]\right\rangle = \left(\frac{\partial}{\partial t}+\alpha k\right)^n\left\langle z_1(t)....z_k(t)R[t;\mathbf{z}(\tau)]\right\rangle, \tag{4.57}$$

where $k = 1, ..., N$.

4.7 Delta-correlated random processes

Random processes $z(t)$ that can be treated as delta-correlated processes are of special interest in physics. The importance of this approximation follows first of all from the fact that it is physically obvious in the context of many physical problems, and the corresponding dynamic systems allow obtaining closed equations for probability densities of the solutions to these systems.

For the Gaussian delta-correlated (in time) process, correlation function has the form

$$B(t_1, t_2) = \langle z(t_1) z(t_2) \rangle = B(t_1)\delta(t_1 - t_2), \quad (\langle z(t) \rangle = 0).$$

In this case, functionals $\Theta[t; v(\tau)]$, $\Omega[t', t; v(\tau)]$, and $\Omega[t, t; v(\tau)]$ introduced above by Eqs. (4.14), (4.15), page 80 assume the forms

$$\begin{aligned} \Theta[t; v(\tau)] &= -\frac{1}{2}\int\limits_0^t d\tau B(\tau) v^2(\tau), \\ \Omega[t', t; v(\tau)] &= iB(t')v(t'), \quad \Omega[t, t; v(\tau)] = \frac{i}{2}B(t)v(t), \end{aligned}$$

and Eqs. (4.16) appear significantly simpler

$$\begin{aligned} \langle z(t') R[t; v(\tau)] \rangle &= B(t') \left\langle \frac{\delta}{\delta z(t')} R[t; v(\tau)] \right\rangle \quad (0 < t' < t), \\ \langle z(t) R[t; v(\tau)] \rangle &= \frac{1}{2} B(t) \left\langle \frac{\delta}{\delta z(t)} R[t; v(\tau)] \right\rangle. \end{aligned} \tag{4.58}$$

Formulas (4.58) show that statistical averages of the Gaussian delta-correlated process considered here are discontinuous at $t' = t$. This discontinuity is completely dictated by the fact that this process is delta-correlated; if process is not delta-correlated, no discontinuity occurs (see Eq. (4.16), page 80).

The Poisson delta-correlated random process corresponds to the limit process

$$g(t) \to \delta(t).$$

In this case, the logarithm of the characteristic functional has the simple form (see (3.43), page 61); as a consequence, Eqs. (4.20), page 81 for functionals $\Omega[t', t; v(\tau)]$ and $\Omega[t, t; v(\tau)]$ assume the forms

$$\begin{aligned} \Omega[t', t; v(\tau)] &= i\int\limits_{-\infty}^{\infty} d\xi \xi p(\xi) e^{i\xi v(t')} \quad (t' < t), \\ \Omega[t, t; v(\tau)] &= \frac{\nu}{iv(t)} \int\limits_{-\infty}^{\infty} d\xi \xi p(\xi) \left[e^{i\xi v(t)} - 1 \right] = \nu \int\limits_{-\infty}^{\infty} d\xi p(\xi) \int\limits_0^{\xi} d\eta e^{i\eta v(t)}, \end{aligned}$$

and we obtain the following expression for the correlation of the Poisson random process $z(t)$ with a functional of this process

$$\begin{aligned} \langle z(t') R[t; z(\tau)] \rangle &= \nu \int\limits_{-\infty}^{\infty} d\xi \xi p(\xi) \langle R[t; z(\tau) + \xi\delta(\tau - \tau')] \rangle \quad (t' \leqslant t), \\ \langle z(t) R[t; z(\tau)] \rangle &= \nu \int\limits_{-\infty}^{\infty} d\xi p(\xi) \int\limits_0^{\xi} d\eta \langle R[t; z(\tau) + \eta\delta(t - \tau)] \rangle. \end{aligned} \tag{4.59}$$

These expressions also show that statistical averages are discontinuous at $t' = t$. As in the case of the Gaussian process, this discontinuity is completely dictated by the fact that this process is delta-correlated.

In the general case of delta-correlated process $z(t)$, we can expand the logarithm of the characteristic functional in the functional Taylor series

$$\Theta[t; v(\tau)] = \sum_{n=1}^{\infty} \frac{i^n}{n!} \int_0^t d\tau K_n(\tau) v^n(\tau), \tag{4.60}$$

where cumulant functions assume the form

$$K_n(t_1, ..., t_n) = K_n(t_1)\delta(t_1 - t_2)...\delta(t_{n-1} - t_n).$$

As can be seen from Eq. (4.60), a characteristic feature of these processes consists in the validity of the equality

$$\dot{\Theta}[t; v(\tau)] = \dot{\Theta}[t; v(t)] \quad \left(\dot{\Theta}[t; v(\tau)] = \frac{d}{dt}\Theta[t; v(\tau)]\right), \tag{4.61}$$

which is of fundamental significance. This equality shows that, in the case of arbitrary delta-correlated process, quantity $\Theta[t; v(\tau)]$ appears not a functional, but simply a function of time t. In this case, functionals $\Omega[t', t; v(\tau)]$ and $\Omega[t, t; v(\tau)]$ are

$$\begin{aligned} \Omega[t', t; v(\tau)] &= \sum_{n=0}^{\infty} \frac{i^n}{n!} K_{n+1}(t') v^n(t') \quad (t' < t), \\ \Omega[t, t; v(\tau)] &= \sum_{n=0}^{\infty} \frac{i^n}{(n+1)!} K_{n+1}(t) v^n(t), \end{aligned}$$

and formulas for correlation splitting assume the forms

$$\begin{aligned} \langle z(t') R[t; z(\tau)] \rangle &= \sum_{n=0}^{\infty} \frac{i^n}{n!} K_{n+1}(t') \left\langle \frac{\delta^n R[t; z(\tau)]}{\delta z^n(t')} \right\rangle \quad (t' < t), \\ \langle z(t') R[t; z(\tau)] \rangle &= \sum_{n=0}^{\infty} \frac{i^n}{(n+1)!} K_{n+1}(t) \left\langle \frac{\delta^n R[t; z(\tau)]}{\delta z^n(t)} \right\rangle. \end{aligned} \tag{4.62}$$

These formulas describe the discontinuity of statistical averages at $t' = t$ in the general case of delta-correlated processes.

Note that, for $t' > t$, delta-correlated processes satisfy the obvious equality

$$\langle z(t') R[t; z(\tau)] \rangle = \langle z(t') \rangle \langle R[t; z(\tau)] \rangle. \tag{4.63}$$

Now, we dwell on the concept of random delta-correlated (in time) fields.

We will deal with vector field $\mathbf{f}(\mathbf{x}, t)$, where $\mathbf{x}$ describes the spatial coordinates and t is the temporal coordinate. In this case, the logarithm of the characteristic functional can be expanded in the Taylor series with coefficients expressed in terms of cumulant functions of random field $\mathbf{f}(\mathbf{x}, t)$ (see Sect. 3.2). In the special case of cumulant functions

$$K_n^{i_1,...,i_m}(\mathbf{x}_1, t_1; ..., \mathbf{x}_n, t_n) = K_n^{i_1,...,i_m}(\mathbf{x}_1, ..., \mathbf{x}_n; t_1)\delta(t_1 - t_2)...\delta(t_{n-1} - t_n), \tag{4.64}$$

we will call field $\mathbf{f}(\mathbf{x},t)$ the random field delta-correlated in time t. In this case, functional $\Theta[t;\boldsymbol{\psi}(\mathbf{x}',\tau)]$ assumes the form

$$\Theta[t;\boldsymbol{\psi}(\mathbf{x}',\tau)] = \sum_{n=1}^{\infty}\frac{i^n}{n!}\int_0^t d\tau\int\ldots\int d\mathbf{x}_1\ldots d\mathbf{x}_n K_n^{i_1,\ldots,i_m}(\mathbf{x}_1,\ldots,\mathbf{x}_n;\tau)\psi_{i_1}(\mathbf{x}_1,\tau)\ldots\psi_{im}(\mathbf{x}_n,\tau). \tag{4.65}$$

An important feature of this functional is the fact that it satisfies the equality similar to Eq. (4.61):

$$\dot{\Theta}[t;\boldsymbol{\psi}(\mathbf{x}',\tau)] = \dot{\Theta}[t;\boldsymbol{\psi}(\mathbf{x}',t)]. \tag{4.66}$$

4.7.1 Asymptotic meaning of delta-correlated processes and fields

The nature knows no delta-correlated processes. All actual processes and fields are characterized by a finite temporal correlation radius, and delta-correlated processes and fields result from asymptotic expansions in terms of their temporal correlation radii.

We illustrate the appearance of delta-correlated processes using the stationary Gaussian process with correlation radius τ_0 as an example. In this case, the logarithm of the characteristic functional is described by the expression

$$\Theta[t;v(\tau)] = -\int_0^t d\tau_1 v(\tau_1)\int_0^{\tau_1} d\tau_2 B\left(\frac{\tau_1-\tau_2}{\tau_0}\right)v(\tau_2). \tag{4.67}$$

Setting $\tau_1-\tau_2=\xi\tau_0$, we transform Eq. (4.67) to the form

$$\Theta[t;v(\tau)] = -\tau_0\int_0^t d\tau_1 v(\tau_1)\int_0^{\tau_1/\tau_0} d\xi B(\xi)\, v(\tau_1-\xi\tau_0).$$

Assume now that $\tau_0\to 0$. In this case, the leading term of the asymptotic expansion in parameter τ_0 is given by the formula

$$\Theta[t;v(\tau)] = -\tau_0\int_0^\infty d\xi B(\xi)\int_0^t d\tau_1 v^2(\tau_1)$$

that can be represented in the form

$$\Theta[t;v(\tau)] = -B^{\text{eff}}\int_0^t d\tau_1 v^2(\tau_1), \tag{4.68}$$

where

$$B^{\text{eff}} = \int_0^\infty d\tau B\left(\frac{\tau}{\tau_0}\right) = \frac{1}{2}\int_{-\infty}^{\infty} d\tau B\left(\frac{\tau}{\tau_0}\right). \tag{4.69}$$

Certainly, asymptotic expression (4.68) holds only for the functions $v(t)$ that vary during times about τ_0 only slightly, rather than for arbitrary functions $v(t)$. Indeed, if we specify this function as $v(t) = v\delta(t - t_0)$, asymptotic expression (4.68) appears invalid; in this case, we must replace Eq. (4.67) with the expression

$$\Theta[t; v(\tau)] = -\frac{1}{2}B(0)v^2 \quad (t > t_0)$$

corresponding to the characteristic function of process $z(t)$ at a fixed time $t = t_0$.

Consider now correlation $\langle z(t)R[t; z(\tau)]\rangle$ given, according to the Furutsu–Novikov formula (4.16), page 80, by the expression

$$\langle z(t)R[t; z(\tau)]\rangle = \int\limits_0^t dt_1 B\left(\frac{t - t_1}{\tau_0}\right)\left\langle \frac{\delta}{\delta z(t_1)} R[t; z(\tau)] \right\rangle.$$

The change of integration variable $t - t_1 \to \xi\tau_0$ transforms this expression to the form

$$\langle z(t)R[t; z(\tau)]\rangle = \tau_0 \int\limits_0^{t/\tau_0} d\xi B(\xi) \left\langle \frac{\delta}{\delta z(t - \xi\tau_0)} R[t; z(\tau)] \right\rangle, \tag{4.70}$$

which grades for $\tau_0 \to 0$ into the equality obtained earlier for the Gaussian delta-correlated process

$$\langle z(t)R[t; z(\tau)]\rangle = B^{\text{eff}} \left\langle \frac{\delta}{\delta z(t)} R[t; z(\tau)] \right\rangle$$

provided that the variational derivative in Eq. (4.70) varies only slightly during times about τ_0.

Thus, the approximation of process $z(t)$ by the delta-correlated one is conditioned by small variations of functionals of this process during times of about process' temporal correlation radius.

Consider now telegrapher's and generalized telegrapher's processes. In the case of telegrapher's process, the characteristic functional satisfies Eq. (3.49), page 63. The correlation radius of this process is $\tau_0 = 1/2\nu$, and, for $\nu \to \infty$ ($\tau_0 \to 0$), this equation grades for sufficiently smooth functions $v(t)$ into the equation

$$\frac{d}{dt}\Phi[t; v(\tau)] = -\frac{a_0^2}{2\nu} v^2(t)\Phi[t; v(\tau)], \tag{4.71}$$

which corresponds to the Gaussian delta-correlated process. If we additionally assume that $a_0^2 \to \infty$ and

$$\lim_{\nu\to\infty} a_0^2/2\nu = \sigma_0^2,$$

then Eq. (4.71) appears independent of parameter ν. Of cause, this fact does not mean that telegrapher's process looses its telegrapher's properties for $\nu \to \infty$. Indeed, for $\nu \to \infty$, the one-point probability distribution of process $z(t)$ will as before correspond to telegrapher's process, i.e., to the process with two possible states. As regards the correlation function and higher-order moment functions, they will possess for $\nu \to \infty$ all properties of delta-functions in view of the fact that

$$\lim_{\nu\to\infty} \left\{2\nu e^{-2\nu|\tau|}\right\} = \begin{cases} 0, & \text{if } \tau \neq 0, \\ \infty, & \text{if } \tau = 0. \end{cases}$$

Such functions should be considered the generalized functions; their delta-functional behavior will manifest itself in the integrals of them (see, e.g., [74]). As Eq. (4.71) shows, the limit process $\nu \to \infty$ is equivalent for these quantities to the replacement of process $z(t)$ by the Gaussian delta-correlated process. This situation is completely similar to the approximation of the Gaussian random process with a finite correlation radius τ_0 by the delta-correlated process for $\tau_0 \to 0$.

In a similar way, we obtain that generalized telegrapher's process whose characteristic functional satisfies integro-differential equation (3.109), page 76 is governed for $\nu \to \infty$ and sufficiently smooth functions $v(t)$ by the equation (here, we assume $\langle a \rangle = 0$ for simplicity)

$$\frac{d}{dt}\Phi[t;v(\tau)] = -\frac{\langle a^2 \rangle}{\nu} v^2(t)\Phi[t;v(\tau)],$$

which again corresponds to the Gaussian delta-correlated process.

Consider the square of the Gaussian stationary process, i.e., process $z(t) = \xi^2(t)$, where $\xi(t)$ is the Gaussian process with parameters

$$\langle \xi(t) \rangle = 0, \quad \langle \xi(t_1)\xi(t_2) \rangle = B(t_1 - t_2),$$

as a more complicated example.

Let as calculate the characteristic functional of this process

$$\Phi[t;v(\tau)] = \langle \varphi[t;\xi(\tau)] \rangle, \quad \varphi[t;\xi(\tau)] = \exp\left\{ i \int\limits_0^t d\tau v(\tau)\xi^2(\tau) \right\}. \tag{4.72}$$

The characteristic functional of process $z(t)$ satisfies the stochastic equation

$$\frac{d}{dt}\Phi[t;v(\tau)] = iv(t)\left\langle \xi^2(t)\varphi[t;\xi(\tau)] \right\rangle. \tag{4.73}$$

Consider quantity $\Psi(t_1,t) = \langle \xi(t_1)\xi(t)\varphi[t;\xi(\tau)] \rangle$. According to the Furutsu–Novikov formula (4.11), page 80,

$$\Psi(t_1,t) = \int\limits_0^t dt' B(t_1 - t') \left\langle \frac{\delta}{\delta\xi(t')} \xi(t)\varphi[t;\xi(\tau)] \right\rangle. \tag{4.74}$$

Calculating now the variational derivative in the right-hand side of Eq. (4.74) (using in this process the explicit expression for functional $\varphi[t;\xi(\tau)]$), we obtain the integral equation for function $\Psi(t_1,t)$

$$\Psi(t_1,t) = B(t_1 - t)\Phi[t;v(\tau)] + 2i \int\limits_0^t d\tau B(t_1 - \tau)v(\tau)\Psi(\tau,t). \tag{4.75}$$

Function $\Psi(t_1,t)$ is representable in the form

$$\Psi(t_1,t) = S(t_1,t)\Phi[t;v(\tau)], \tag{4.76}$$

where function $S(t_1,t)$ satisfies the linear integral equation

$$S(t_1,t) = B(t_1 - t) + 2i \int\limits_0^t d\tau B(t_1 - \tau)v(\tau)S(\tau,t). \tag{4.77}$$

As a consequence, characteristic functional $\Phi[t; v(\tau)]$ can be represented in the form

$$\Phi[t; v(\tau)] = \exp\left\{ i \int_0^t d\tau v(\tau) S(\tau, \tau) \right\}. \tag{4.78}$$

Thus, the expansion of quantity $S(t,t)$ in the functional Taylor series in $v(\tau)$ determines the cumulants of process $z(t) = \xi^2(t)$. Because Eq. (4.77) is the linear integral equation, we can represent its solution as the iterative series

$$S(t,t) = \sum_{n=0}^{\infty} S^{(n)}(t,t),$$

$$S^{(n)}(t,t) = (2i)^n \int_0^t \ldots \int_0^t d\tau_1 \ldots d\tau_n v(\tau_1) \ldots v(\tau_n) B(t-\tau_1) B(\tau_1 - \tau_2) \ldots B(\tau_n - t). \tag{4.79}$$

If function $v(t)$ varies slowly during correlation time τ_0 of process $\xi(t)$ (which means that we omit from consideration the one-time characteristic functions of process $z(t) = \xi^2(t)$), we can proceed to the limit $\tau_0 \to 0$. As a result, we obtain the expressions

$$\begin{aligned} S^{(0)}(t,t) &= B(0), \\ S^{(n)}(t,t) &= (2i)^n v^n(t) \int_0^\infty \ldots \int_0^\infty d\tau_1 \ldots d\tau_n B(\tau_1) B(\tau_1 - \tau_2) \ldots B(\tau_n), \end{aligned} \tag{4.80}$$

from which follows that process $z(t) = \xi^2(t)$ in this limit can be considered the delta-correlated (in time t) random process. The effective expansion parameter of quantity $S(t,t)$ in series (4.79) is in this case $\beta = \tau_0 B(0) v(t)$. If $\beta \ll 1$, we can content themselves with the first term of series (4.80), which corresponds to the standard perturbation theory. But if $\beta \sim 1$, one needs take into account the whole series for function $S(t,t)$.

The Gaussian Markovian process $\xi(t)$ with correlation function $B(\tau) = \sigma^2 e^{-\alpha|\tau|}$, where $\alpha = 1/\tau_0$, allows a more detailed analysis. In this case, integral equation (4.77) assumes for $t_1 < t$ the form

$$S(t_1,t) = \sigma^2 e^{-\alpha(t-t_1)} + 2i\sigma^2 \int_0^t d\tau e^{-\alpha|t_1-\tau|} v(\tau) S(\tau,t). \tag{4.81}$$

The solution to this equation as a function of parameter t can be described as the initial-value problem

$$\begin{aligned} &\frac{\partial}{\partial t} S(t_1,t) = \left\{ -\alpha + 2iv(t) S(t,t) \right\} S(t_1,t), \quad S(t_1,t)\big|_{t=t_1} S(t_1,t_1), \\ &\frac{d}{dt} S(t,t) = -2\alpha \left[S(t,t) - \sigma^2 \right] + 2iv(t) S^2(t,t), \quad S(t,t)\big|_{t=0} = \sigma^2, \end{aligned} \tag{4.82}$$

which corresponds to the imbedding method with respect to parameter t (see Appendix C, page 451). The asymptotic solution of the latter equation for $\alpha \to \infty$ $(\tau_0 \to 0)$ has the form

$$S(t,t) = \frac{1 - \sqrt{1 - 4i\sigma^2 \tau_0 v(t)}}{2i\tau_0 v(t)}, \quad S^{(n)}(t,t) = \left[2i\tau_0 v(t)\right]^n \sigma^{2(n+1)} \frac{(2n-1)!!}{(n+1)!},$$

which coincides with the solution of Eq. (4.82) under the assumption that function $v(t)$ is a constant. As a consequence, the logarithm of the characteristic functional in this asymptotic case is given by the formula

$$\Theta[t; v(\tau)] = \frac{1}{2\tau_0} \int_0^t d\tau \left(1 - \sqrt{1 - 4i\sigma^2 \tau_0 v(\tau)}\right),$$

and cumulants of process $z(t) = \xi^2(t)$ have the forms

$$K_1 = \sigma^2, \quad K_n(t_1, ..., t_n) = (2\tau_0)^{n-1} \sigma^{2n} (2n-3)!! \delta(t_1 - t_2) ... \delta(t_{n-1} - t_n),$$

which correspond to Eqs. (4.80); in this case, we have

$$I_n = \int_0^\infty d\tau_1 ... \int_0^\infty d\tau_n \exp\{-\tau_1 - |\tau_1 - \tau_2| - ... - |\tau_{n-1} - \tau_n| - \tau_n\} = \frac{(2n-1)!!}{(n+1)!}.$$

Process $z(t) = z\xi(t)$, where $\xi(t)$ is the Gaussian delta-correlated process with parameters $\langle \xi(t) \rangle = 0$, $\langle \xi(t)\xi(t') \rangle = 2\sigma^2 \tau_0 \delta(t - t')$, and quantity z is the random quantity with probability density $p(z)$, is an example of the process that cannot be considered the delta correlated process. In this case, the characteristic functional is given by the equality

$$\Phi[t; v(\tau)] = \left\langle \exp\left\{ iz \int_0^t d\tau v(\tau) \xi(\tau) \right\} \right\rangle_{z,\xi} = \int_{-\infty}^{\infty} dz p(z) \exp\left\{ -z^2 \sigma^2 \tau_0 \int_0^t d\tau v^2(\tau) \right\}, \tag{4.83}$$

and process $z(t)$ cannot be considered the delta-correlated process, because it does not satisfy Eq. (4.61) despite its second cumulant has the form

$$K_2(t_1, t_2) = \langle z(t_1) z(t_2) \rangle = 2 \left\langle z^2 \right\rangle \sigma^2 \tau_0 \delta(t_1 - t_2). \tag{4.84}$$

This follows from the fact that process $z(t)$ is formed as the product of two processes — process z with the infinite correlation radius and process $\xi(t)$ with the zero-valued correlation radius.

Now, we proceed to the direct statistical analysis of stochastic dynamic systems.

Chapter 5

General approaches to analyzing stochastic dynamic systems

In this chapter, we will consider basic methods of determining statistical characteristics of solutions to the stochastic equations.

Consider a linear (differential, integro-differential, or integral) stochastic equation. Averaging of such an equation over an ensemble of realizations of fluctuating parameters will not result generally in a closed equation for the corresponding average value. To obtain the closed equation, we must deal with an additional extended space whose dimension appears infinite in most cases. This approach makes it possible to derive the linear equation for average quantity of interest, but this equation will contain variational derivatives.

Consider some special types of dynamic systems.

5.1 Ordinary differential equations

Let dynamics of vector function $\mathbf{x}(t)$ is described by the ordinary differential equation

$$\frac{d}{dt}\mathbf{x}(t) = \mathbf{v}(\mathbf{x}, t) + \mathbf{f}(\mathbf{x}, t), \quad \mathbf{x}(t_0) = \mathbf{x}_0. \tag{5.1}$$

Here, function $\mathbf{v}(\mathbf{x}, t)$ is the deterministic function and $\mathbf{f}(\mathbf{x}, t)$ is the random function.

The solution to Eq. (5.1) is a functional of $\mathbf{f}(\mathbf{y}, \tau) + \mathbf{v}(\mathbf{y}, \tau)$ with $\tau \in (t_0, t)$, i.e.,

$$\mathbf{x}(t) = \mathbf{x}[t; \mathbf{f}(\mathbf{y}, \tau) + \mathbf{v}(\mathbf{y}, \tau)].$$

From this fact follows the equality

$$\frac{\delta}{\delta f_j(\mathbf{y}, \tau)} F(\mathbf{x}(t)) = \frac{\delta}{\delta v_j(\mathbf{y}, \tau)} F(\mathbf{x}(t)) = \frac{\partial F(\mathbf{x}(t))}{\partial x_i} \frac{\delta x_i(t)}{\delta f_j(\mathbf{y}, \tau)},$$

valid for arbitrary function $F(\mathbf{x})$. In addition, we have

$$\frac{\delta}{\delta f_j(\mathbf{y}, t-0)} x_i(t) = \frac{\delta}{\delta v_j(\mathbf{y}, t-0)} x_i(t) = \delta_{ij} \delta\left(\mathbf{x}(t) - \mathbf{y}\right).$$

The corresponding Liouville equation for the indicator function $\varphi(\mathbf{x}, t) = \delta(\mathbf{x}(t) - \mathbf{x})$ follows from Eq. (5.1) and has the form

$$\frac{\partial}{\partial t}\varphi(\mathbf{x}, t) = -\frac{\partial}{\partial \mathbf{x}}\left\{[\mathbf{v}(\mathbf{x}, t) + \mathbf{f}(\mathbf{x}, t)]\,\varphi(\mathbf{x}, t)\right\}, \quad \varphi(\mathbf{x}, t_0) = \delta(\mathbf{x} - \mathbf{x}_0), \tag{5.2}$$

from which follows the equality

$$\frac{\delta}{\delta \mathbf{f}(\mathbf{y}, t-0)}\varphi(\mathbf{x}, t) = \frac{\delta}{\delta \mathbf{v}(\mathbf{y}, t-0)}\varphi(\mathbf{x}, t) = -\frac{\partial}{\partial \mathbf{x}}\left\{\delta(\mathbf{x}-\mathbf{y})\varphi(\mathbf{x}, t)\right\}. \tag{5.3}$$

Using this equality, we can rewrite Eq. (5.2) in the form, which may look at first sight more complicated

$$\left(\frac{\partial}{\partial t} + \frac{\partial}{\partial \mathbf{x}}\mathbf{v}(\mathbf{x}, t)\right)\varphi(\mathbf{x}, t) = \int d\mathbf{y} \mathbf{f}(\mathbf{y}, t)\frac{\delta}{\delta \mathbf{v}(\mathbf{y}, t)}\varphi(\mathbf{x}, t). \tag{5.4}$$

Consider now the one-time probability density for solution $\mathbf{x}(t)$ of Eq. (5.1)

$$P(\mathbf{x}, t) = \langle \varphi(\mathbf{x}, t)\rangle = \langle \delta(\mathbf{x}(t) - \mathbf{x})\rangle .$$

Here, $\mathbf{x}(t)$ is the solution of Eq. (5.1) corresponding to the particular realization of random field $\mathbf{f}(\mathbf{x}, t)$, and angle brackets $\langle ... \rangle$ denote averaging over an ensemble of realizations of field $\mathbf{f}(\mathbf{x}, t)$.

Averaging Eq. (5.4) over an ensemble of realizations of field $\mathbf{f}(\mathbf{x}, t)$, we obtain the expression

$$\left(\frac{\partial}{\partial t} + \frac{\partial}{\partial \mathbf{x}}\mathbf{v}(\mathbf{x}, t)\right)P(\mathbf{x}, t) = \int d\mathbf{y}\frac{\delta}{\delta \mathbf{v}(\mathbf{y}, t)}\langle \mathbf{f}(\mathbf{y}, t)\varphi(\mathbf{x}, t)\rangle . \tag{5.5}$$

Quantity $\langle \mathbf{f}(\mathbf{y}, t)\varphi(\mathbf{x}, t)\rangle$ in the right-hand side of Eq. (5.5) is the correlator of random field $\mathbf{f}(\mathbf{y}, t)$ with function $\varphi(\mathbf{x}, t)$, which is a functional of random field $\mathbf{f}(\mathbf{y}, \tau)$ and is given either by Eq. (5.2), or by Eq. (5.4).

The characteristic functional

$$\Phi[t, t_0; \mathbf{u}(\mathbf{y}, \tau)] = \left\langle \exp\left\{ i \int\limits_{t_0}^{t} d\tau \int d\mathbf{y} \mathbf{f}(\mathbf{y}, \tau)\mathbf{u}(\mathbf{y}, \tau)\right\}\right\rangle = \exp\left\{\Theta[t, t_0; \mathbf{u}(\mathbf{y}, \tau)]\right\}$$

exhaustively describes all statistical characteristics of random field $\mathbf{f}(\mathbf{y}, \tau)$ for $\tau \in (t_0, t)$.

We split correlator $\langle \mathbf{f}(\mathbf{y}, t)\varphi(\mathbf{x}, t)\rangle$ using the technique of functionals. Introducing functional shift operator with respect to field $\mathbf{v}(\mathbf{y}, \tau)$, we represent functional $\varphi[t, \mathbf{x}; \mathbf{f}(\mathbf{y}, \tau) + \mathbf{v}(\mathbf{y}, \tau)]$ in the operator form

$$\varphi[t, \mathbf{x}; \mathbf{f}(\mathbf{y}, \tau) + \mathbf{v}(\mathbf{y}, \tau)] = \exp\left[\int\limits_{t_0}^{t} d\tau \int d\mathbf{y} \mathbf{f}(\mathbf{y}, \tau)\frac{\delta}{\delta \mathbf{v}(\mathbf{y}, \tau)}\right]\varphi[t, \mathbf{x}; \mathbf{v}(\mathbf{y}, \tau)].$$

With this representation, the term in the right-hand side of Eq. (5.5) assumes the form

$$\int d\mathbf{y}\frac{\delta}{\delta v_j(\mathbf{y}, t)}\frac{\left\langle f_j(\mathbf{y}, t)\exp\left\{\int\limits_{t_0}^{t} d\tau \int d\mathbf{y}'\mathbf{f}(\mathbf{y}', \tau)\frac{\delta}{\delta \mathbf{v}(\mathbf{y}', \tau)}\right\}\right\rangle}{\left\langle \exp\left\{\int\limits_{t_0}^{t} d\tau \int d\mathbf{y}'\mathbf{f}(\mathbf{y}', \tau)\frac{\delta}{\delta \mathbf{v}(\mathbf{y}', \tau)}\right\}\right\rangle}P(\mathbf{x}, t)$$

$$= \dot{\Theta}_t\left[t, t_0; \frac{\delta}{i\delta \mathbf{v}(\mathbf{y}, \tau)}\right]P(\mathbf{x}, t),$$

where we introduced the functional

$$\dot{\Theta}_t[t, t_0; \mathbf{u}(\mathbf{y}, \tau)] = \frac{d}{dt}\ln \Phi[t, t_0; \mathbf{u}(\mathbf{y}, \tau)].$$

Consequently, we can rewrite Eq. (5.5) in the form

$$\left(\frac{\partial}{\partial t}+\frac{\partial}{\partial \mathbf{x}}\mathbf{v}(\mathbf{x},t)\right)P(\mathbf{x},t)=\dot{\Theta}_t\left[t,t_0;\frac{\delta}{i\delta\mathbf{v}(\mathbf{y},\tau)}\right]P(\mathbf{x},t). \tag{5.6}$$

Equation (5.6) is the closed equation with variational derivatives in the functional space of all possible functions $\{\mathbf{v}(\mathbf{y},\tau)\}$. However, for a fixed function $\mathbf{v}(\mathbf{x},t)$, we arrive at unclosed equation [133]–[135]

$$\begin{aligned}\left(\frac{\partial}{\partial t}+\frac{\partial}{\partial \mathbf{x}}\mathbf{v}(\mathbf{x},t)\right)P(\mathbf{x},t) &= \left\langle\dot{\Theta}_t\left[t,t_0;\frac{\delta}{i\delta\mathbf{f}(\mathbf{y},t)}\right]\varphi[t,\mathbf{x};\mathbf{f}(\mathbf{y},\tau)]\right\rangle,\\ P(\mathbf{x},t_0) &= \delta(\mathbf{x}-\mathbf{x}_0).\end{aligned} \tag{5.7}$$

Equation (5.7) is the exact consequence of the initial dynamic equation (5.1). Statistical characteristics of random field $\mathbf{f}(\mathbf{x},t)$ appear in this equation only through functional $\dot{\Theta}_t[t,t_0;\mathbf{u}(\mathbf{y},\tau)]$ whose expansion in the functional Taylor series in powers of $\mathbf{u}(\mathbf{x},\tau)$ depends on all space-time cumulant functions of random field $\mathbf{f}(\mathbf{x},t)$.

Note that equation for the one-point probability density $P(\mathbf{x},t)$ preserve the form of Eq. (5.7) even for more general integro-differential equation

$$\frac{d}{dt}x_i(t)=v_i(\mathbf{x},t)+\int d\mathbf{y}D_{ij}(\mathbf{x},\mathbf{y},t)f_j(\mathbf{y},t),\quad \mathbf{x}(t_0)=\mathbf{x}_0,$$

in which case the variational derivative has the form

$$\frac{\delta}{\delta f_j(\mathbf{y},t-0)}x_i(t)=D_{ij}(\mathbf{x}(t),\mathbf{y},t).$$

As we mentioned earlier, Eq. (5.7) is not closed in the general case with respect to function $P(\mathbf{x},t)$, because quantity

$$\dot{\Theta}_t\left[t,t_0;\frac{\delta}{i\delta\mathbf{f}(\mathbf{y},\tau)}\right]\delta(\mathbf{x}(t)-\mathbf{x})$$

appearing in averaging brackets depends on the solution $\mathbf{x}(t)$ (which is a functional of random field $\mathbf{f}(\mathbf{y},\tau)$) for all times $t_0<\tau<t$. However, in some cases, the variational derivative in Eq. (5.7) can be expressed in terms of ordinary differential operators. In such conditions, equations like Eq. (5.7) will be the closed equations for the corresponding probability densities. The corresponding examples will be given below.

Note that Eq. (5.2) is the forward Liouville equation and describes the evolution of indicator function

$$\varphi(\mathbf{x},t)=\varphi(\mathbf{x},t|\mathbf{x}_0,t_0)=\delta\left(\mathbf{x}(t)-\mathbf{x}|\mathbf{x}(t_0)=\mathbf{x}_0\right)$$

in time t. For this reason, we can call Eq. (5.7) the *forward equation* for probability density.

In Chapter 2, we obtained the backward Liouville equation (2.4), page 39 for the indicator function, which describes the evolution of dynamic system (5.1) in terms of initial values t_0 and $\mathbf{x}_0$. In our case, this equation has the form,

$$\begin{aligned}\left(\frac{\partial}{\partial t_0}+\mathbf{v}(\mathbf{x}_0,t_0)\frac{\partial}{\partial \mathbf{x}_0}\right)\varphi(\mathbf{x},t|\mathbf{x}_0,t_0) &= -\mathbf{f}(\mathbf{x}_0,t_0)\frac{\partial}{\partial \mathbf{x}_0}\varphi(\mathbf{x},t|\mathbf{x}_0,t_0),\\ \varphi(\mathbf{x},t|\mathbf{x}_0,t) &= \delta(\mathbf{x}-\mathbf{x}_0).\end{aligned} \tag{5.8}$$

Averaging Eq. (5.8) over an ensemble of realizations of field $\mathbf{f}(\mathbf{x}_0, t_0)$ and performing the procedure similar to that used in derivation of Eq. (5.7), we obtain the equation for probability density $P(\mathbf{x}, t|\mathbf{x}_0, t_0) = \langle\delta(\mathbf{x}, t|\mathbf{x}_0, t_0)\rangle$ in the form

$$\begin{aligned}\left(\frac{\partial}{\partial t_0} + \mathbf{v}(\mathbf{x}_0, t_0)\frac{\partial}{\partial \mathbf{x}_0}\right) P(\mathbf{x}, t|\mathbf{x}_0, t_0) &= \left\langle \dot{\Theta}_{t_0}\left[t, t_0; \frac{\delta}{i\delta\mathbf{f}(\mathbf{y}, \tau)}\right] \delta\left(\mathbf{x}(t|\mathbf{x}_0, t_0) - \mathbf{x}\right)\right\rangle, \\ P(\mathbf{x}, t|\mathbf{x}_0, t) &= \delta(\mathbf{x} - \mathbf{x}_0), \end{aligned} \tag{5.9}$$

where

$$\dot{\Theta}_{t_0}[t, t_0; \mathbf{u}(\mathbf{y}, \tau)] = \frac{d}{dt_0} \ln \Phi[t, t_0; \mathbf{u}(\mathbf{y}, \tau)].$$

Equation (5.9) describes the evolution of the probability density as a function of initial parameters $\{\mathbf{x}_0, t_0\}$; for this reason, we can call it the *backward equation*.

The forward and backward equations are equivalent. The forward equation appears more convenient for studying the behavior of statistical characteristics of solutions to Eq. (5.1) in time domain. The backward equation is more convenient for studying the statistical characteristics that concern the residence of random process $\mathbf{x}(t)$ within certain region of space, such as residence duration within this region and time of arrival at region boundary. Indeed, the probability of residence of random process $\mathbf{x}(t)$ in spatial region V is given by the integral

$$G(t; \mathbf{x}_0, t_0) = \int\limits_V d\mathbf{x} P(x, t|\mathbf{x}_0, t_0),$$

which, according to Eq. (5.9), will satisfy the equation

$$\begin{aligned}&\left(\frac{\partial}{\partial t_0} + \mathbf{v}(\mathbf{x}_0, t_0)\frac{\partial}{\partial \mathbf{x}_0}\right) G(t|\mathbf{x}_0, t_0) \\ &= \left\langle \dot{\Theta}_{t_0}\left[t, t_0; \frac{\delta}{i\delta\mathbf{f}(\mathbf{y}, \tau)}\right] \int\limits_V d\mathbf{x}\delta\left(\mathbf{x}(t|\mathbf{x}_0, t_0) - \mathbf{x}\right)\right\rangle, \\ &G(t|\mathbf{x}_0, t_0) = \left\{\begin{array}{ll} 1 & (\mathbf{x}_0 \in V) \\ 0 & (\mathbf{x}_0 \notin V) \end{array}\right. . \end{aligned} \tag{5.10}$$

This equation must be supplemented with the boundary conditions following from the nature of each particular problem and depend on region V and region boundaries.

Proceeding in a similar way, one can obtain the equation similar to Eq. (5.7) for the m-time probability density that refers to m different instants $t_1 < t_2 < ... < t_m$

$$P_m(\mathbf{x}_1, t_1; ...; \mathbf{x}_m, t_m) = \langle\varphi_m(\mathbf{x}_1, t_1; ...; \mathbf{x}_m, t_m)\rangle, \tag{5.11}$$

where the indicator function is defined by the equality

$$\varphi_m(\mathbf{x}_1, t_1; ...; \mathbf{x}_m, t_m) = \delta(\mathbf{x}(t_1) - \mathbf{x}_1)...\delta(\mathbf{x}(t_m) - \mathbf{x}_m).$$

Differentiating Eq. (5.11) with respect to time t_m and using then dynamic equation (5.1), one can obtain the equation

$$\begin{aligned}&\left(\frac{\partial}{\partial t_m} + \frac{\partial}{\partial \mathbf{x}_m}\mathbf{v}(\mathbf{x}_m, t_m)\right) P_m(\mathbf{x}_1, t_1; ...; \mathbf{x}_m, t_m) \\ &= \left\langle \dot{\Theta}_{t_m}\left[t_m, t_0; \frac{\delta}{i\delta\mathbf{f}(\mathbf{y}, \tau)}\right] \varphi_m(\mathbf{x}_1, t_1; ...; \mathbf{x}_m, t_m)\right\rangle. \end{aligned} \tag{5.12}$$

No summation over index m is performed here. The initial value to Eq. (5.12) can be derived from Eq. (5.11). Setting $t_m = t_{m-1}$ in Eq. (5.11), we obtain the equality

$$P_m(\mathbf{x}_1, t_1; ...; \mathbf{x}_m, t_{m-1}) = \delta(\mathbf{x}_m - \mathbf{x}_{m-1}) P_{m-1}(\mathbf{x}_1, t_1; ...; \mathbf{x}_{m-1}, t_{m-1}),$$

which just determines the initial value for Eq. (5.12).

5.2 Partial differential equations

Above, we considered statistical description of dynamic systems starting from the Liouville equation (5.2) that matches the ordinary differential equation (5.1). It is quite obvious that the derivation procedure of Eqs. (5.6), (5.7), (5.12), and the like can be applied to other dynamic systems specified in terms of linear equations both in finite- and infinite-dimension spaces, i.e., in terms of partial differential equations of the first and higher orders. Below, we consider the use of these equations using specific examples, such as passive tracer transfer in random field of velocities Eqs. (1.39), page 19, (2.5), and (2.9), page 40, wave propagation in random media described within the framework of the parabolic equation of quasi-optics (1.91), page 31, and hydrodynamic turbulence evolution described by the integro-differential equation (1.99), page 34.

5.2.1 Passive tracer transfer in random field of velocities

The first example deals with Eq. (2.9), page 40

$$\left(\frac{\partial}{\partial t} + \mathbf{U}(\mathbf{r},t)\frac{\partial}{\partial \mathbf{r}}\right)\Phi(t,\mathbf{r};\rho) = \frac{\partial \mathbf{U}(\mathbf{r},t)}{\partial \mathbf{r}}\frac{\partial}{\partial \rho}\rho\Phi(t,\mathbf{r};\rho) \tag{5.13}$$

for indicator function

$$\Phi(t,\mathbf{r};\rho) = \delta\left(\rho(\mathbf{r},t) - \rho\right).$$

We assume that $\mathbf{U}(\mathbf{r},t) = \mathbf{u}_0(\mathbf{r},t) + \mathbf{u}(\mathbf{r},t)$, where $\mathbf{u}_0(\mathbf{r},t)$ is the deterministic component of the velocity field (mean flow) and $\mathbf{u}(\mathbf{r},t)$ is the random component.

A consequence of Eq. (5.13) is the equality

$$\begin{aligned}\frac{\delta}{\delta U_j(\mathbf{r}',t-0)}\Phi(t,\mathbf{r};\rho) &= \frac{\delta}{\delta u_j(\mathbf{r}',t-0)}\Phi(t,\mathbf{r};\rho) = \frac{\delta}{\delta u_{0j}(\mathbf{r}',t-0)}\Phi(t,\mathbf{r};\rho)\\ &= \left\{-\delta(\mathbf{r}-\mathbf{r}')\frac{\partial}{\partial r_j} + \frac{\partial\delta(\mathbf{r}-\mathbf{r}')}{\partial r_j}\frac{\partial}{\partial\rho}\rho\right\}\Phi(t,\mathbf{r};\rho).\end{aligned} \tag{5.14}$$

Statistical characteristics of random field $\mathbf{u}(\mathbf{r},t)$ can be exhaustively described in terms of the characteristic functional

$$\Phi[t;\psi(\mathbf{r}',\tau)] = \left\langle \exp\left\{ i\int_0^t d\tau \int d\mathbf{r}'\mathbf{u}(\mathbf{r}',\tau)\psi(\mathbf{r}',\tau)\right\}\right\rangle = \exp\left\{\Theta[t;\psi(\mathbf{r}',\tau)]\right\}.$$

Now, we average Eq. (5.13) over an ensemble of realizations of random field $\mathbf{u}(\mathbf{r},t)$. Then, replicating derivation of Eq.(5.7) and taking into account Eq. (5.14), we obtain that the one-point probability density

$$P(t,\mathbf{r};\rho) = \langle\Phi(t,\mathbf{r};\rho)\rangle = \langle\delta\left(\rho(\mathbf{r},t)-\rho\right)\rangle$$

satisfies the equation

$$\left(\frac{\partial}{\partial t}+\mathbf{u}_0(\mathbf{r},t)\frac{\partial}{\partial\mathbf{r}}\right)P(t,\mathbf{r};\rho)=\frac{\partial\mathbf{u}_0(\mathbf{r},t)}{\partial\mathbf{r}}\frac{\partial}{\partial\rho}\rho P(t,\mathbf{r};\rho)$$
$$+\left\langle\int d\mathbf{r}\mathbf{u}(\mathbf{r},t)\frac{\delta}{\delta\mathbf{u}_0(\mathbf{r},t-0)}\Phi(t,\mathbf{r};\rho)\right\rangle, \tag{5.15}$$

whose last term can be rewritten in the form

$$\left\langle\int d\mathbf{r}\mathbf{u}(\mathbf{r},t)\frac{\delta}{\delta\mathbf{u}_0(\mathbf{r},t-0)}\Phi[t,\mathbf{r};\rho;\mathbf{u}+\mathbf{u}_0]\right\rangle$$
$$=\left\langle\int d\mathbf{r}\mathbf{u}(\mathbf{r},t)\frac{\delta}{\delta\mathbf{u}_0(\mathbf{r},t-0)}\exp\left\{\int_0^t d\tau\int d\mathbf{r}'\mathbf{u}(\mathbf{r}',\tau)\frac{\delta}{\delta\mathbf{u}_0(\mathbf{r}',\tau)}\right\}\right\rangle\Phi[t,\mathbf{r};\rho;\mathbf{u}_0]$$
$$=\left\langle\frac{d}{dt}\exp\left\{\int_0^t d\tau\int d\mathbf{r}'\mathbf{u}(\mathbf{r}',\tau)\frac{\delta}{\delta\mathbf{u}_0(\mathbf{r}',\tau)}\right\}\right\rangle\Phi[t,\mathbf{r};\rho;\mathbf{u}_0]$$
$$=\dot{\Theta}_t\left[t;\frac{\delta}{i\delta\mathbf{u}_0(\mathbf{r}',\tau)}\right]\langle\Phi[t,\mathbf{r};\rho;\mathbf{u}+\mathbf{u}_0]\rangle=\dot{\Theta}_t\left[t;\frac{\delta}{i\delta\mathbf{u}_0(\mathbf{r}',\tau)}\right]P(t,\mathbf{r};\rho),$$

where

$$\dot{\Theta}_t\left[t;\boldsymbol{\psi}(\mathbf{r}',\tau)\right]=\frac{d}{dt}\ln\left\langle\exp\left\{i\int_0^t d\tau\int d\mathbf{r}'\mathbf{u}(\mathbf{r}',\tau)\boldsymbol{\psi}(\mathbf{r}',\tau)\right\}\right\rangle$$

is the derivative of the characteristic functional logarithm of random field $\mathbf{u}(\mathbf{r},t)$.

Thus, expression (5.15) can be represented as the functional linear variational differential equation in the functional space of functions $\{\mathbf{u}_0(\mathbf{r},t)\}$

$$\left(\frac{\partial}{\partial t}+\mathbf{u}_0(\mathbf{r},t)\frac{\partial}{\partial\mathbf{r}}\right)P(t,\mathbf{r};\rho)$$
$$=\frac{\partial\mathbf{u}_0(\mathbf{r},t)}{\partial\mathbf{r}}\frac{\partial}{\partial\rho}\rho P(t,\mathbf{r};\rho)+\dot{\Theta}_t\left[t;\frac{\delta}{i\delta\mathbf{u}_0(\mathbf{r}',\tau)}\right]P(t,\mathbf{r};\rho), \tag{5.16}$$

However, if we deal with a fixed mean flow $\mathbf{u}_0(\mathbf{r},t)$ (e.g., $\mathbf{u}_0(\mathbf{r},t)=0$), then Eq. (5.16) assumes the form of unclosed equality

$$\left(\frac{\partial}{\partial t}+\mathbf{u}_0(\mathbf{r},t)\frac{\partial}{\partial\mathbf{r}}\right)P(t,\mathbf{r};\rho)$$
$$=\frac{\partial\mathbf{u}_0(\mathbf{r},t)}{\partial\mathbf{r}}\frac{\partial}{\partial\rho}\rho P(t,\mathbf{r};\rho)+\left\langle\dot{\Theta}_t\left[t;\frac{\delta}{i\delta\mathbf{u}(\mathbf{r}',\tau)}\right]\Phi(t,\mathbf{r};\rho)\right\rangle. \tag{5.17}$$

5.2.2 Parabolic equation of quasi-optics

The second example deals with wave propagation in random medium within the framework of the linear parabolic equation (1.91), page 31

$$\frac{\partial}{\partial x}u(x,\mathbf{R})=\frac{i}{2k}\Delta_{\mathbf{R}}u(x,\mathbf{R})+i\frac{k}{2}\varepsilon(x,\mathbf{R})u(x,\mathbf{R}),\quad u(0,\mathbf{R})=u_0(\mathbf{R}). \tag{5.18}$$

In this case, functional (2.30)

$$\varphi[x;v(\mathbf{R}'),v^*(\mathbf{R}')]=\varphi[x;v,v^*]$$
$$=\exp\left\{i\int d\mathbf{R}'\left[u(x,\mathbf{R}')v(\mathbf{R}')+u^*(x,\mathbf{R}')v^*(\mathbf{R}')\right]\right\}$$

is described by the variational differential equation (the Hopf equation) (2.31), page 44

$$\frac{\partial}{\partial x}\varphi[x;v,v^*] = i\frac{k}{2}\int d\mathbf{R}'\varepsilon(x,\mathbf{R}')\hat{M}(\mathbf{R}')\varphi[x;v,v^*]$$
$$+\frac{i}{2k}\left\{\int d\mathbf{R}'\left[v(\mathbf{R}')\Delta_{\mathbf{R}'}\frac{\delta}{\delta v(\mathbf{R}')} - v^*(\mathbf{R}')\Delta_{\mathbf{R}'}\frac{\delta}{\delta v^*(\mathbf{R}')}\right]\right\}\varphi[x;v,v^*] \qquad (5.19)$$

equivalent to the initial equation (5.18). Here, $\hat{M}(\mathbf{R}')$ is the Hermitian operator

$$\hat{M}(\mathbf{R}') = v(\mathbf{R}')\frac{\delta}{\delta v(\mathbf{R}')} - v^*(\mathbf{R}')\frac{\delta}{\delta v^*(\mathbf{R}')}.$$

A consequence of Eq. (5.19) is the equality

$$\frac{\delta}{\delta\varepsilon(x-0,\mathbf{R}')}\varphi[x;v,v^*] = i\frac{k}{2}\hat{M}(\mathbf{R}')\varphi[x;v,v^*]. \qquad (5.20)$$

Averaging Eq. (5.19) over an ensemble of realizations of random field $\varepsilon(x,\mathbf{R})$ and replicating derivation of Eqs. (5.7) and (5.17), we obtain that the characteristic functional of the solution to problem (5.18)

$$\Phi[x;v(\mathbf{R}'),v^*(\mathbf{R}')] = \Phi[x;v,v^*] = \langle\varphi[x;v(\mathbf{R}'),v^*(\mathbf{R}')]\rangle$$

satisfies the variational derivative equation

$$\frac{\partial}{\partial x}\Phi[x;v,v^*] = \left\langle\dot{\Theta}_x\left[x;\frac{\delta}{i\delta\varepsilon(\xi,\mathbf{R}')}\right]\varphi[x;v,v^*]\right\rangle$$
$$+\frac{i}{2k}\left\{\int d\mathbf{R}'\left[v(\mathbf{R}')\Delta_{\mathbf{R}'}\frac{\delta}{\delta v(\mathbf{R}')} - v^*(\mathbf{R}')\Delta_{\mathbf{R}'}\frac{\delta}{\delta v^*(\mathbf{R}')}\right]\right\}\Phi[x;v,v^*], \qquad (5.21)$$

where

$$\dot{\Theta}_x\left[x;\psi(\xi,\mathbf{R}')\right] = \frac{d}{dx}\ln\left\langle\exp\left\{i\int\limits_0^x d\xi\int d\mathbf{R}'\varepsilon(\xi,\mathbf{R}')\psi(\xi,\mathbf{R}')\right\}\right\rangle$$

is the derivative of the characteristic functional logarithm of random field $\varepsilon(x,\mathbf{R})$.

5.2.3 Random forces in the theory of hydrodynamic turbulence

In Chapter 1, page 34 we obtained that stationary and homogeneous hydrodynamic turbulence can be described in terms of the Fourier transform of the velocity field ($\hat{u}_i^*(\mathbf{k},t) = \hat{u}_i(-\mathbf{k},t)$)

$$\hat{u}_i(\mathbf{k},t) = \int d\mathbf{r}u_i(\mathbf{r},t)e^{-i\mathbf{kr}}, \quad u_i(\mathbf{r},t) = \frac{1}{(2\pi)^3}\int d\mathbf{k}\hat{u}_i(\mathbf{k},t)e^{i\mathbf{kr}},$$

which satisfies the nonlinear integro-differential equation (1.99)

$$\frac{\partial}{\partial t}\hat{u}_i(\mathbf{k},t) + \frac{i}{2}\int d\mathbf{k}_1\int d\mathbf{k}_2\Lambda_i^{\alpha\beta}(\mathbf{k}_1,\mathbf{k}_2,\mathbf{k})\,\hat{u}_\alpha(\mathbf{k}_1,t)\,\hat{u}_\beta(\mathbf{k}_2,t)$$
$$-\nu k^2\hat{u}_i(\mathbf{k},t) = \hat{f}_i(\mathbf{k},t), \qquad (5.22)$$

where

$$\begin{aligned}\Lambda_i^{\alpha\beta}(\mathbf{k}_1,\mathbf{k}_2,\mathbf{k}) &= \frac{1}{(2\pi)^3}\left\{k_\alpha\Delta_{i\beta}(\mathbf{k})+k_\beta\Delta_{i\alpha}(\mathbf{k})\right\}\delta(\mathbf{k}_1+\mathbf{k}_2-\mathbf{k}),\\ \Delta_{ij}(\mathbf{k}) &= \delta_{ij}-\frac{k_ik_j}{\mathbf{k}^2}\quad (i,\,\alpha,\,\beta\,=\,1,2,3)\end{aligned}$$

and $\widehat{\mathbf{f}}(\mathbf{k},t)$ is the spatial Fourier harmonics of the external force.

A specific feature of the three-dimensional hydrodynamic motions consists in the existence of the integral of energy under the condition that external forces and effects related to the molecular viscosity are absent.

Furthermore, in Chapter 2, page 45 we obtained that functional

$$\varphi[t;\mathbf{z}(\mathbf{k}')]=\varphi[t;\mathbf{z}]=\exp\left\{i\int d\mathbf{k}'\widehat{\mathbf{u}}(\mathbf{k}',t)\mathbf{z}(\mathbf{k}')\right\}$$

satisfies the linear variational differential Hopf equation in functional space (2.35)

$$\begin{aligned}\frac{\partial}{\partial t}\varphi[t;\mathbf{z}] &= -\int d\mathbf{k}z_i(\mathbf{k})\left\{\nu k^2\frac{\delta}{\delta z_i(\mathbf{k})}-i\hat{f}_i(\mathbf{k},t)\right\}\varphi[t;\mathbf{z}]\\ &-\frac{1}{2}\int d\mathbf{k}z_i(\mathbf{k})\int d\mathbf{k}_1\int d\mathbf{k}_2\Lambda_i^{\alpha\beta}(\mathbf{k}_1,\mathbf{k}_2,\mathbf{k})\frac{\delta^2\varphi[t;\mathbf{z}]}{\delta z_\alpha(\mathbf{k}_1)\,\delta z_\beta(\mathbf{k}_2)}.\end{aligned}\tag{5.23}$$

A consequence of Eq. (5.23) is the equality

$$\frac{\delta}{\delta\widehat{\mathbf{f}}(\mathbf{k},t-0)}\varphi[t;\mathbf{z}]=i\mathbf{z}(\mathbf{k})\varphi[t;\mathbf{z}].\tag{5.24}$$

Average Eq. (5.23) over an ensemble of realizations of random force $\widehat{\mathbf{f}}(\mathbf{k},t)$. Then, replicating derivation of Eq. (5.7), we obtain that characteristic functional of the velocity field

$$\Phi[t;\mathbf{z}(\mathbf{k}')]=\Phi[t;\mathbf{z}]=\langle\varphi[t;\mathbf{z}(\mathbf{k}')]\rangle\,,$$

satisfies the unclosed variational differential equation

$$\begin{aligned}\frac{\partial}{\partial t}\Phi[t;\mathbf{z}] &= \left\langle\dot{\Theta}_t\left[t;\frac{\delta}{i\delta\mathbf{f}(\boldsymbol{\kappa},\tau)}\right]\varphi[t;\mathbf{z}]\right\rangle\\ &-\int d\mathbf{k}z_i(\mathbf{k})\left\{\frac{1}{2}\int d\mathbf{k}_1\int d\mathbf{k}_2\Lambda_{i,\alpha\beta}(\mathbf{k}_1,\mathbf{k}_2,\mathbf{k})\frac{\delta^2}{\delta z_\alpha(\mathbf{k}_1)\delta z_\beta(\mathbf{k}_2)}+\nu k^2\frac{\delta}{\delta z_i(\mathbf{k})}\right\}\Phi[t;\mathbf{z}],\end{aligned}\tag{5.25}$$

where

$$\dot{\Theta}_t\left[t;\boldsymbol{\psi}(\boldsymbol{\kappa},\tau)\right]=\frac{d}{dt}\ln\left\langle\exp\left\{i\int_0^t d\tau\int d\boldsymbol{\kappa}\mathbf{f}(\boldsymbol{\kappa},\tau)\boldsymbol{\psi}(\boldsymbol{\kappa},\tau)\right\}\right\rangle$$

is the derivative of the characteristic functional logarithm of the random field of external force $\widehat{\mathbf{f}}(\mathbf{k},t)$. The equation for the characteristic functional $\Phi[t;\mathbf{z}(\mathbf{k}')]$ describes all one-time statistical characteristics of the velocity field.

5.3 Stochastic integral equations (methods of quantum field theory)

Problems discussed in the above sections allow deriving the closed (or unclosed) statistical description in functional space due to the fact that every of these problems can be formulated in terms of some system of differential equations of the first-order with respect to time and initial values at $t = 0$. Such systems satisfy the causality condition formulated in Sect. 1.5.1, which reads as follows: problem solution at instant t depends only on fluctuations of system parameters for times preceding this instant and is independent of fluctuations for posterior times.

Problems formulated in terms of integral equations that cannot be generally reduced to the system of differential equations also can satisfy the causality condition.

However, prior to consider this class of stochastic problems, we dwell on general methods of statistical description of dynamic systems, which are borrowed from the quantum field theory. The essence of these methods consists in constructing a perturbation series for statistical characteristics of quantity of interest and analyzing the result with the use of the methods developed in the quantum field theory. It appears convenient to represent each term of these perturbation series diagrammatically (in the form of the so-called *Feynman diagrams*) and associate every diagram element with certain function or operator; as a result, each diagram corresponds to certain analytical expression. We will not consider the diagram technique as such (for its exhaustive description in the context of statistical problems, see, e.g., monographs [251, 294]); instead, we derive the basic results directly, using the functional methods described above [134, 135].

5.3.1 Linear integral equations

The input stochastic equation is the linear integral (or integro-differential) equation for Green's function

$$S(\mathbf{r},\mathbf{r}') = S_0(\mathbf{r},\mathbf{r}') + \int d\mathbf{r}_1 \int d\mathbf{r}_2 \int d\mathbf{r}_3 S_0(\mathbf{r},\mathbf{r}_1)\Lambda(\mathbf{r}_1,\mathbf{r}_2,\mathbf{r}_3)f(\mathbf{r}_2)S(\mathbf{r}_3,\mathbf{r}'), \tag{5.26}$$

where $\mathbf{r}$ denotes all arguments of functions S and f, including the index arguments that assume summation instead of integration. It is assumed that function $f(\mathbf{r})$ is the random field and function $S_0(\mathbf{r},\mathbf{r}')$ is Green's function of the problem without parameter fluctuations, i.e., at $f(\mathbf{r}) = 0$. In some problems, quantity $\Lambda(\mathbf{r}_1,\mathbf{r}_2,\mathbf{r}_3)$ can be an operator; the notation of Eq. (5.26) assumes that this operator acts on all factors appeared to the right from it. For example, the nonlinear system of ordinary differential equations can be reduced to the equation like Eq. (5.26) by constructing the equivalent linear stochastic partial differential equation (the Liouville equation) whose characteristics correspond to the solution of the nonlinear system. In this case, function S is Green's function of the stochastic Liouville equation and quantity Λ is the differential operator. For problems formulated in terms of systems of linear equations, quantity $\Lambda(\mathbf{r}_1,\mathbf{r}_2,\mathbf{r}_3)$ appears a function.

Below, we will assume for simplicity that quantity $\Lambda(\mathbf{r}_1,\mathbf{r}_2,\mathbf{r}_3)$ is not an operator, but a function. The consideration of operator quantity $\Lambda(\mathbf{r}_1,\mathbf{r}_2,\mathbf{r}_3)$ causes only insignificant differences. Indeed, if quantity $\Lambda(\mathbf{r}_1,\mathbf{r}_2,\mathbf{r}_3)$ is an operator, we can reduce the problem to the problem under consideration by introducing delta-functions with arguments coinciding with variables on which this operator is acting and adding the corresponding integrations.

Equation (5.26) can be represented in the symbolic form

$$S = S_0 + S_0\Lambda f S, \tag{5.27}$$

where integration is assumed with respect to all arguments of function $\Lambda(\{\mathbf{r}_i\})$.

We can solve Eq. (5.27) by the iteration method with function $S_0(\mathbf{r},\mathbf{r}')$ as the zero-order approximation. As a result, we obtain the solution in the form of a series that we again represent in the symbolic form

$$S = \{1 + S_0\Lambda f + S_0\Lambda f S_0\Lambda f + \ldots\} S_0 = \sum_{n=0}^{\infty} (S_0\Lambda f)^n S_0. \tag{5.28}$$

The same iteration series represents the solution of the integral equation

$$S = S_0 + S\Lambda f S_0.$$

Consequently. Eq. (5.27) is equivalent to the equation

$$S(\mathbf{r},\mathbf{r}') = S_0(\mathbf{r},\mathbf{r}') + \int d\mathbf{r}_1 d\mathbf{r}_2 d\mathbf{r}_3 S(\mathbf{r},\mathbf{r}_1)\Lambda(\mathbf{r}_1,\mathbf{r}_2,\mathbf{r}_3) f(\mathbf{r}_2) S_0(\mathbf{r}_3,\mathbf{r}'). \tag{5.29}$$

The solution to Eqs. (5.27) and (5.29) is the functional of field $f(\mathbf{r})$, i.e.,

$$S(\mathbf{r},\mathbf{r}') = S[\mathbf{r},\mathbf{r}'; f(\widetilde{\mathbf{r}})].$$

It is not difficult to show that Eqs. (5.27) and (5.29) are equivalent to the variational differential equation in functional space $\{f(\widetilde{\mathbf{r}})\}$

$$\frac{\delta}{\delta f(\mathbf{r}_0)} S[\mathbf{r},\mathbf{r}'; f(\widetilde{\mathbf{r}})] = \int d\mathbf{r}_1 d\mathbf{r}_2 S[\mathbf{r},\mathbf{r}_1; f(\widetilde{\mathbf{r}})]\Lambda(\mathbf{r}_1,\mathbf{r}_0,\mathbf{r}_2) S[\mathbf{r}_2,\mathbf{r}'; f(\widetilde{\mathbf{r}})] \tag{5.30}$$

with the initial value

$$S[\mathbf{r},\mathbf{r}'; f(\widetilde{\mathbf{r}})]_{f=0} = S_0(\mathbf{r},\mathbf{r}').$$

Indeed, varying Eq. (5.27) for $S(\mathbf{r},\mathbf{r}')$ with respect to function $f(\mathbf{r}_0)$, we obtain the equation

$$\frac{\delta}{\delta f} S = S_0\Lambda\delta S + S_0\Lambda f \frac{\delta}{\delta f} S,$$

where δ denotes the delta-function of the corresponding arguments. The solution to this equation can be represented as the iteration series

$$\frac{\delta}{\delta f} S = \left\{1 + S_0\Lambda f + (S_0\Lambda f)^2 + \ldots\right\} S_0\Lambda\delta S.$$

Taking into account the iteration series (5.28) for S, we obtain the desired formula (5.30).

Averaging now the obtained iteration series (5.28) over an ensemble of realizations of field $f(\mathbf{r})$, we obtain function $\langle S(\mathbf{r},\mathbf{r}')\rangle$ in the form of the iteration series dependent on all moment functions of field $f(\mathbf{r})$. Rearranging the terms of this series, we can then express the right-hand side of the expansion in terms of function $\langle S(\mathbf{r},\mathbf{r}')\rangle$ itself. This rearrangement produces new unknown functions specified by the corresponding iteration series and called, by analogy with the quantum field theory, the *mass* and *vertex functions*.

Consider instead of Eq. (5.26) the auxiliary equation

$$S[\mathbf{r},\mathbf{r}'; f+\eta] = S_0(\mathbf{r},\mathbf{r}') + \int d\mathbf{r}_1 \int d\mathbf{r}_2 \int d\mathbf{r}_3 S_0(\mathbf{r},\mathbf{r}_1)\Lambda(\mathbf{r}_1,\mathbf{r}_2,\mathbf{r}_3)[f(\mathbf{r}_2)+\eta(\mathbf{r}_2)] S[\mathbf{r}_3,\mathbf{r}'; f+\eta], \tag{5.31}$$

where $\eta(\mathbf{r})$ is arbitrary deterministic function. We can find the desired function $S(\mathbf{r},\mathbf{r}')$ by setting $\eta(\mathbf{r}) = 0$ in Eq. (5.31), i.e.,

$$S(\mathbf{r},\mathbf{r}') = S[\mathbf{r},\mathbf{r}';f(\mathbf{r})] = S[\mathbf{r},\mathbf{r}';f(\mathbf{r})+\eta(\mathbf{r})]_{\eta=0}.$$

Let us average Eq. (5.31). Splitting the correlator $\langle fS\rangle$ by formula (4.2), page 75, we obtain the equation

$$\begin{aligned} G[\mathbf{r},\mathbf{r}';\eta] &= S_0(\mathbf{r},\mathbf{r}') \\ &+ \int d\mathbf{r}_1 \int d\mathbf{r}_2 \int d\mathbf{r}_3 S_0(\mathbf{r},\mathbf{r}_1)\Lambda(\mathbf{r}_1,\mathbf{r}_2,\mathbf{r}_3)\eta(\mathbf{r}_2)G[\mathbf{r}_3,\mathbf{r}';\eta] \\ &+ \int d\mathbf{r}_1 \int d\mathbf{r}_2 \int d\mathbf{r}_3 S_0(\mathbf{r},\mathbf{r}_1)\Lambda(\mathbf{r}_1,\mathbf{r}_2,\mathbf{r}_3)\left\langle \Omega_{\mathbf{r}_2}\left[\frac{\delta}{i\delta f(\mathbf{r})}\right] S[\mathbf{r}_3,\mathbf{r}';f+\eta]\right\rangle. \end{aligned} \tag{5.32}$$

Here, functional $\Omega_{\mathbf{r}}\left[v(\mathbf{r})\right]$ is given by the formula

$$\Omega_{\mathbf{r}} = \frac{\delta}{i\delta v(\mathbf{r})}\Theta\left[v(\mathbf{r})\right],$$

functional

$$\Theta\left[v(\mathbf{r})\right] = \ln\left\langle \exp\left\{ i\int d\mathbf{r} f(\mathbf{r})v(\mathbf{r})\right\}\right\rangle$$

is the logarithm of the characteristic functional of random field $f(\mathbf{r})$, and

$$G[\mathbf{r},\mathbf{r}';\eta(\mathbf{r})] = \left\langle S[\mathbf{r},\mathbf{r}';f(\mathbf{r})+\eta(\mathbf{r})]\right\rangle.$$

Taking into account the fact that functional $S[\mathbf{r},\mathbf{r}';f(\mathbf{r})+\eta(\mathbf{r})]$ is the functional of argument $f(\mathbf{r})+\eta(\mathbf{r})$, we can replace variational differentiation with respect to $f(\mathbf{r})$ by differentiation with respect to $\eta(\mathbf{r})$ and rewrite Eq. (5.32) in the form of the closed variational differential equation similar to the *Schwinger equation* of the quantum field theory

$$\begin{aligned} G[\mathbf{r},\mathbf{r}';\eta] &= S_0(\mathbf{r},\mathbf{r}') \\ &+ \int d\mathbf{r}_1 \int d\mathbf{r}_2 \int d\mathbf{r}_3 S_0(\mathbf{r},\mathbf{r}_1)\Lambda(\mathbf{r}_1,\mathbf{r}_2,\mathbf{r}_3)\eta(\mathbf{r}_2)G[\mathbf{r}_3,\mathbf{r}';\eta] \\ &+ \int d\mathbf{r}_1 \int d\mathbf{r}_2 \int d\mathbf{r}_3 S_0(\mathbf{r},\mathbf{r}_1)\Lambda(\mathbf{r}_1,\mathbf{r}_2,\mathbf{r}_3)\Omega_{\mathbf{r}_2}\left[\frac{\delta}{i\delta\eta(\mathbf{r})}\right] G[\mathbf{r}_3,\mathbf{r}';\eta]. \end{aligned} \tag{5.33}$$

We can solve Eq. (5.33) for functional $G[\mathbf{r},\mathbf{r}';\eta(\mathbf{r})]$ by the iteration method with function $S_0(\mathbf{r},\mathbf{r}')$ as the zero-order approximation. Setting $\eta(\mathbf{r}) = 0$ in the resulting expansion, we obtain the iteration series for function $\langle S(\mathbf{r},\mathbf{r}')\rangle$.

To simplify further presentation, we rewrite Eq. (5.33) in the symbolic form (the corresponding complete expressions can be easily restored at every step)

$$G = S_0 + S_0\Lambda\left(\eta + \Omega\left[\frac{\delta}{i\delta\eta}\right]\right)G. \tag{5.34}$$

We introduce the inverse functional G^{-1}, such that

$$G^{-1}G = 1, \quad GG^{-1} = 1. \tag{5.35}$$

Here, the unity is understood as the corresponding delta-function. In addition, we introduce the functional

$$\Gamma = -\frac{\delta G^{-1}}{\delta \eta}, \tag{5.36}$$

which we call the *vertex functional.*

Varying Eq. (5.35) with respect to field η, we obtain the equality

$$\frac{\delta G}{\delta \eta} = G\Gamma G, \tag{5.37}$$

whose substitution in Eq. (5.34) results in the equation

$$G = S_0 + S_0 \Lambda \eta G + S_0 Q G. \tag{5.38}$$

We call the quantity

$$Q = \Lambda \left(\Omega \left[\frac{\delta}{i\delta \eta} \right] G \right) G^{-1} \tag{5.39}$$

the *mass functional.*

Multiplying now Eq. (5.38) by G^{-1} from the right and by S_0^{-1} from the left (and integrating over the corresponding arguments), we obtain the equation for functional G^{-1}

$$S_0^{-1} - G^{-1} = \Lambda \eta + Q. \tag{5.40}$$

Varying Eq. (5.40) with respect to field η, we obtain the equation for functional Γ

$$\Gamma = \Lambda + \frac{\delta}{\delta \eta} Q. \tag{5.41}$$

The system of functional equations (5.38) and (5.41) is closed in functionals G and Γ. An additional point is Eq. (5.37) that relates the solutions to these equations. We can solve Eq. (5.41) for Γ by the iteration method with quantity Λ as the zero-order approximation. If we use formula (5.37) to express variational derivatives of functional G with respect to η, we obtain the integral equations for Γ and G with infinite number of terms every of which includes no functionals other than G and Γ. Setting now $\eta(\mathbf{r}) = 0$, we can obtain the closed system of integral equations. In particular, Eq. (5.38) assumes the following form

$$\langle S \rangle = S_0 + S_0 Q \langle S \rangle, \quad \langle S \rangle = S_0 + \langle S \rangle Q S_0, \tag{5.42}$$

and is called the *Dyson equation.*

Now, we consider in more detail the case of the Gaussian random field $f(\mathbf{r})$ with correlation function $B(\mathbf{r},\mathbf{r}') = \langle f(\mathbf{r}) f(\mathbf{r}') \rangle$. In this case

$$\Omega_{\mathbf{r}} \left[v(\mathbf{r}) \right] = i \int d\mathbf{r}' B(\mathbf{r}, \mathbf{r}') v(\mathbf{r}'),$$

the mass functional assumes the form

$$Q = \Lambda B G \Gamma, \tag{5.43}$$

and Eqs. (5.38) and (5.41) assume the form

$$\begin{aligned} G &= S_0 + S_0 \Lambda \eta G + S_0 \Lambda B G \Gamma G, \\ \Gamma &= \Lambda + \Lambda B G \Gamma G \Gamma + \Lambda B G \frac{\delta \Gamma}{\delta \eta}. \end{aligned} \tag{5.44}$$

Setting now $\eta = 0$, we obtain the closed system of equations

$$\begin{aligned} \langle S \rangle &= S_0 + S_0 Q \langle S \rangle \quad \text{(the Dyson equation)}, \\ Q &= \Lambda B \langle S \rangle \tilde{\Gamma}, \\ \tilde{\Gamma} &= \Lambda + \Lambda B \langle S \rangle \tilde{\Gamma} \langle S \rangle \tilde{\Gamma} + \ldots \quad (\tilde{\Gamma} = \Gamma|_{\eta} = 0). \end{aligned} \tag{5.45}$$

System of equations (5.45) is very complicated and low understood. The simplest way of its simplification consists in cutting the infinite series in the equation for $\tilde{\Gamma}$. If we content ourselves with the first term, we obtain the closed nonlinear equation (the *Kraichnan approximation*)

$$\langle S \rangle = S_0 + S_0 Q_{\mathrm{Kr}} \langle S \rangle, \quad Q_{\mathrm{Kr}} = \Lambda B \langle S \rangle \Lambda. \tag{5.46}$$

Further, if we replace $\langle S \rangle$ in the expression for the mass function Q_{Kr} with S_0, we obtain the linear equation (the *Bourret approximation*)

$$\langle S \rangle = S_0 + S_0 \Lambda B S_0 \Lambda \langle S \rangle. \tag{5.47}$$

Functional Γ and, consequently, function $\tilde{\Gamma}$ are closely related to quantity $\langle SS \rangle$. Indeed, in view of formula (5.30), we can rewrite expression (5.37) at $\eta = 0$ in the form

$$\langle S \Lambda S \rangle = \langle S \rangle \tilde{\Gamma} \langle S \rangle. \tag{5.48}$$

Thus, different approximations for function $\tilde{\Gamma}$ are equivalent to certain hypotheses about splitting the correlation $\langle SS \rangle$. The Kraichnan approximation (5.46) corresponds to the equality

$$\langle S \Lambda S \rangle = \langle S \rangle \Lambda \langle S \rangle,$$

while the Bourret approximation (5.47) is equivalent to the requirement that

$$\langle S \Lambda S \rangle = S_0 \Lambda \langle S \rangle.$$

Splitting the correlation $\langle S \Lambda S \rangle$ by formula (4.9), page 79, we obtain the general operator expression

$$\langle S \Lambda S \rangle = G \left[i B \frac{\delta}{\delta \eta} \right] \Lambda G[\eta] |_{\eta = 0},$$

which is in essence equivalent to the introduction of the vertex function.

Note that the knowledge of functional $G[\mathbf{r},\mathbf{r}'; \eta(\mathbf{r})]$ is equivalent, in the case of the Gaussian field $f(\mathbf{r})$, to the knowledge of the functional

$$\Phi[\mathbf{r}, \mathbf{r}'; v(\mathbf{r})] = \left\langle S(\mathbf{r}, \mathbf{r}') e^{i \int d\mathbf{r} f(\mathbf{r}) v(\mathbf{r})} \right\rangle$$

describing all statistical correlations between function $S(\mathbf{r},\mathbf{r}')$ and field $f(\mathbf{r})$. Indeed, according to formula (4.13), page 80, we can rewrite functional $\Phi[\mathbf{r},\mathbf{r}'; v(\mathbf{r})]$ in the form

$$\Phi[\mathbf{r}, \mathbf{r}'; v(\mathbf{r})] = \left\langle e^{i \int d\mathbf{r} f(\mathbf{r}) v(\mathbf{r})} \right\rangle \left\langle S \left[\mathbf{r}, \mathbf{r}'; f(\mathbf{r}) + i \int d\mathbf{r}_1 B(\mathbf{r}, \mathbf{r}_1) \right] \right\rangle,$$

wherefrom we obtain the equality

$$\Phi[\mathbf{r}, \mathbf{r}'; v(\mathbf{r})] = \left\langle e^{i \int d\mathbf{r} f(\mathbf{r}) v(\mathbf{r})} \right\rangle G \left[\mathbf{r}, \mathbf{r}'; i \int d\mathbf{r}_1 B(\mathbf{r}, \mathbf{r}_1) \right].$$

To complete the picture, we dwell now on the so-called renormalization method. The point is that, even if the mass function is known, the Dyson equation (5.45) is the very complicated integral equation, which only rarely can be solved analytically. At the same time, the Dyson equation with the simplified mass function can be easily solved in a number of cases. The renormalization method lies in rearranging the Dyson equation in the integral equation in which function $S_0(\mathbf{r},\mathbf{r}')$ is replaced with the solution to the simplified problem.

Denote $\tilde{S}$ the solution of the Dyson equation (5.45) with the simplified mass function $\tilde{Q}$. Then, function $\tilde{S}$ will satisfy the equation

$$\tilde{S} = S_0 + S_0\tilde{Q}\tilde{S}. \tag{5.49}$$

In view of the fact that Eq. (5.49) is linear in $\tilde{S}$, it is obvious that it can be rewritten in the form

$$\tilde{S} = S_0 + \tilde{S}\tilde{Q}S_0 = (1 + \tilde{S}\tilde{Q})S_0, \tag{5.50}$$

where 1 is the unit operator.

To exclude function $S_0(\mathbf{r},\mathbf{r}')$ from the Dyson equation (5.45), we apply operator $(1+\tilde{S}\tilde{Q})$ to it. Then, we obtain, in view of Eq. (5.50), the equation

$$\langle S\rangle = \tilde{S} + \tilde{S}\{Q - \tilde{Q}\}\langle S\rangle\,. \tag{5.51}$$

Now, we can solve Eq. (5.51) by the iteration method with function $\tilde{S}$ as the zero-order approximation.

At $\tilde{Q} = 0$, function $\tilde{S} = S_0$, and we turn back to the Dyson equation (5.45). It is obvious that the above derivation of Eq. (5.51) holds not only for the Gaussian field $f(\mathbf{r})$, but for any arbitrary field $f(\mathbf{r})$, because the form of the Dyson equation is independent of the type of field $f(\mathbf{r})$.

Now, we dwell on the general-form Dyson equation (5.38). Note that we can represent functional $\Omega[v]$ in terms of the Taylor series in powers of v

$$\Omega[v] = \sum_{n=1}^{\infty} \frac{1}{n!} K_{n+1} v^n,$$

where K_n are the cumulant functions of random field $f(\mathbf{r})$. As a consequence, we can represent the mass functional (5.39) in the form

$$Q = \Lambda \sum_{n=0}^{\infty} \frac{1}{n!} K_{n+1} \left\{ \frac{\delta^n}{\delta\eta^n} G \right\} G^{-1},$$

where variational derivatives of functional G with respect to η are calculated by formula (5.37). In this case, the Dyson equation has very complicated structure. The standard ways of simplifying this equation are quite similar to those used in the case of the Gaussian parameter fluctuations.

If we set $\Gamma = \Lambda$, expression (5.37) assumes the form

$$\frac{\delta G}{\delta\eta} = G\Lambda G$$

and, consequently,

$$\frac{\delta^n G}{\delta\eta^n} = n!(G\Lambda)^n G = \int_0^{\infty} d\lambda e^{-\lambda} (G\lambda\Lambda)^n G.$$

In this case, we arrive at the generalized Kraichnan equation

$$\begin{aligned} \langle S \rangle &= S_0 + S_0 Q_{\mathrm{Kr}} \langle S \rangle , \\ Q_{\mathrm{Kr}} &= \Lambda \sum_{n=0}^{\infty} K_{n+1} \{ \langle S \rangle \Lambda \}^n = \Lambda \int_0^{\infty} d\lambda e^{-\lambda} \Omega[\langle S \rangle \lambda \Lambda]. \end{aligned} \tag{5.52}$$

If we replace $\langle S \rangle$ in operator Q_{Kr} in Eq. (5.52) with S_0, we obtain the generalized Bourret equation

$$Q_{\mathrm{B}} = \Lambda \int_0^{\infty} d\lambda e^{-\lambda} \Omega[S_0 \lambda \Lambda],$$

which coincides with the so-called *one-group approximation for the Dyson equation.*

Above, we considered the derivation of the equation for averaged Green's function (the Dyson equation). In a similar way, we can derive the equation for the correlation function

$$\Gamma(\mathbf{r}, \mathbf{r}'; \mathbf{r}_1, \mathbf{r}_1') = \langle S(\mathbf{r}, \mathbf{r}') S^*(\mathbf{r}_1, \mathbf{r}_1') \rangle .$$

With this goal in view, we multiply Eq. (5.26) by $S^*(\mathbf{r}_1,\mathbf{r}_1')$ and average the result over an ensemble of realizations of random field $f(\mathbf{r})$. In this way, we obtain the equation

$$\Gamma = S_0 \langle S^* \rangle + S_0 \Lambda \langle f S S^* \rangle . \tag{5.53}$$

Taking into account the Dyson equation (5.42)

$$\langle S \rangle = 1 + \langle S \rangle Q S_0,$$

we apply operator $\{1 + \langle S \rangle Q\}$ to Eq. (5.53). As a result, we obtain the equation

$$\Gamma = \langle S \rangle \langle S^* \rangle + \langle S \rangle \{ \Lambda \langle f S S^* \rangle - Q \Gamma \} . \tag{5.54}$$

In standard notation, Eq. (5.54) assumes the form

$$\begin{aligned} &\Gamma(\mathbf{r}, \mathbf{r}'; \mathbf{r}_1, \mathbf{r}_1') = \langle S(\mathbf{r}, \mathbf{r}') \rangle \langle S^*(\mathbf{r}_1, \mathbf{r}_1') \rangle \\ &+ \int d\mathbf{r}_2 d\mathbf{r}_2' d\mathbf{r}_3 d\mathbf{r}_3' \langle S(\mathbf{r}, \mathbf{r}_2) \rangle \langle S^*(\mathbf{r}_1, \mathbf{r}_2') \rangle K(\mathbf{r}_2, \mathbf{r}_2'; \mathbf{r}_3, \mathbf{r}_3') \Gamma(\mathbf{r}_3, \mathbf{r}'; \mathbf{r}_3', \mathbf{r}_1') \end{aligned} \tag{5.55}$$

and is called the *Bete–Salpeter equation.* Function $K(\mathbf{r}_2,\mathbf{r}_2';\mathbf{r}_3,\mathbf{r}_3')$ is called the *kernel of the intensity operator.*

The simplest approximation to this equation — the so-called *ladder approximation* — corresponds to function $K(\mathbf{r}_2,\mathbf{r}_2';\mathbf{r}_3,\mathbf{r}_3')$ of the form

$$K(\mathbf{r}_2, \mathbf{r}_2'; \mathbf{r}_3, \mathbf{r}_3') = \delta(\mathbf{r}_2 - \mathbf{r}_3) \delta(\mathbf{r}_2' - \mathbf{r}_3') B_f(\mathbf{r}_2, \mathbf{r}_2'), \tag{5.56}$$

where $B_f(\mathbf{r}_2,\mathbf{r}_2') = \langle f(\mathbf{r}_2) f(\mathbf{r}_2') \rangle$ is the correlation function of field $f(\mathbf{r})$.

5.3.2 Nonlinear integral equations

Consider now integral equation (5.26) extended to the case of an equation with quadratic nonlinearity. There are two possible cases. In the first — simplest — case, the solution can be expressed through the integral of the solution to the linear equation with respect to an auxiliary parameter; the second — more complex — case describes the space-time structure of hydrodynamic turbulence and is described by the integro-differential equation (1.100), page 34.

The simplest nonlinear integral equation

Consider the nonlinear equation

$$S(\mathbf{r},\mathbf{r}') = S_0(\mathbf{r},\mathbf{r}') + \int d\mathbf{r}_1 \int d\mathbf{r}_2 \int d\mathbf{r}_3 S(\mathbf{r},\mathbf{r}_1)\Lambda(\mathbf{r}_1,\mathbf{r}_2,\mathbf{r}_3) f(\mathbf{r}_2) S(\mathbf{r}_3,\mathbf{r}'). \tag{5.57}$$

Along with this equation we draw once again Eq. (5.26) and mark its solution by index Λ

$$S_\Lambda(\mathbf{r},\mathbf{r}') = S_0(\mathbf{r},\mathbf{r}') + \int d\mathbf{r}_1 \int d\mathbf{r}_2 \int d\mathbf{r}_3 S_0(\mathbf{r},\mathbf{r}_1)\Lambda(\mathbf{r}_1,\mathbf{r}_2,\mathbf{r}_3) f(\mathbf{r}_2) S_\Lambda(\mathbf{r}_3,\mathbf{r}'). \tag{5.58}$$

As was mentioned earlier, the solution to Eq. (5.58) can be represented as the iteration series

$$S_\Lambda = \sum_{n=0}^{\infty} (S_0 \Lambda f)^n S_0.$$

It is obvious that the solution to integral equation (5.57) has a similar iteration structure

$$S = \sum_{n=0}^{\infty} A_n (S_0 \Lambda f)^n S_0 \tag{5.59}$$

with an additional numeric parameter A_n. This parameter can be easily found from the quadratic equation

$$y = 1 + \lambda y^2$$

whose solution is

$$y = \frac{1-\sqrt{1-4\lambda}}{2\lambda} = \sum_{n=0}^{\infty} A_n \lambda^n.$$

Consequently,

$$A_n = 2^n \frac{(2n-1)!!}{(n+1)!} = \frac{2^{2n+1}}{\pi} \frac{\Gamma(n+1/2)\Gamma(3/2)}{\Gamma(n+2)} = \frac{2^{2n+1}}{\pi} B\left(n+\frac{1}{2}, \frac{3}{2}\right),$$

where $B(\gamma,\delta)$ is the beta-function whose integral representation is

$$B(\gamma,\delta) = \int_0^1 dp p^{\gamma-1}(1-p)^{\delta-1}.$$

As a result, we have

$$A_n = \frac{2}{\pi} \int_0^1 dp \, (4p)^n \sqrt{\frac{1-p}{p}}.$$

Substituting this expression in Eq. (5.59), we obtain

$$S = \frac{2}{\pi} \int_0^1 dp \sqrt{\frac{1-p}{p}} \sum_{n=0}^{\infty} (S_0 4p \Lambda f)^n S_0 = \frac{2}{\pi} \int_0^1 dp \sqrt{\frac{1-p}{p}} S_{4p\Lambda}. \tag{5.60}$$

Thus, the solution to the nonlinear equation (5.57) is expressed through the solution to the linear equation (5.58) as the integral with respect to an auxiliary parameter [253].

Space-time description of hydrodynamic turbulence

Consider now the nonlinear integral equation (1.100), page 34 for the space-time harmonics of the turbulent velocity field

$$(i\omega+\nu\mathbf{k}^2)\widehat{u}_i(\mathbf{K})+\frac{i}{2}\int d^4\mathbf{K}_1\int d^4\mathbf{K}_2\Lambda_i^{\alpha\beta}(\mathbf{K}_1,\mathbf{K}_2,\mathbf{K})\,\widehat{u}_\alpha(\mathbf{K}_1)\,\widehat{u}_\beta(\mathbf{K}_2)=\widehat{f}_i(\mathbf{K}),\quad(5.61)$$

where

$$\begin{aligned}\Lambda_i^{\alpha\beta}(\mathbf{K}_1,\mathbf{K}_2,\mathbf{K}) &= \frac{1}{(2\pi)^3}\left\{k_\alpha\Delta_{i\beta}(\mathbf{k})+k_\beta\Delta_{i\alpha}(\mathbf{k})\right\}\delta(\mathbf{k}_1+\mathbf{k}_2-\mathbf{k}),\delta(\omega_1+\omega_2-\omega),\\ \Delta_{ij}(\mathbf{k}) &= \delta_{ij}-\frac{k_ik_j}{\mathbf{k}^2}\quad(i,\,\alpha,\,\beta\,=\,1,2,3).\end{aligned}$$

Here, $\mathbf{K}$ is the four-dimensional wave vector with components $\{\mathbf{k},\omega\}$ and $\widehat{f}_i(\mathbf{K})$ are the space-time Fourier harmonics of external forces; in view of the fact that $u_i(\mathbf{x},t)$ is the real-valued field, we have $\widehat{u}_i^*(\mathbf{K})=\widehat{u}_i(-\mathbf{K})$.

Equation (5.61) can be juxtaposed with the equivalent linear variational differential Hopf equation

$$\begin{aligned}&(i\omega+\nu\mathbf{k}^2)\frac{\delta}{\delta z_i(\mathbf{K})}\varphi[\mathbf{z}]=i\widehat{f}_i(\mathbf{K})\,\varphi[\mathbf{z}]-\\ &-\frac{1}{2}\int d^4\mathbf{K}_1\int d^4\mathbf{K}_2\Lambda_i^{\alpha\beta}(\mathbf{K}_1,\mathbf{K}_2,\mathbf{K})\frac{\delta^2\varphi[\mathbf{z}]}{\delta z_\alpha(\mathbf{K}_1)\,\delta z_\beta(\mathbf{K}_2)}\end{aligned}\quad(5.62)$$

for the functional

$$\varphi[\mathbf{z}]=\exp\left\{i\int d^4\mathbf{K}'\widehat{\mathbf{u}}(\mathbf{K}')\,\mathbf{z}(\mathbf{K}')\right\}.\quad(5.63)$$

Averaging Eq. (5.62) over an ensemble of realizations of the external force field $\widehat{\mathbf{f}}(\mathbf{K})$, we obtain the equation

$$\begin{aligned}&(i\omega+\nu\mathbf{k}^2)\frac{\delta}{\delta z_i(\mathbf{K})}\Phi[\mathbf{z}]=i\left\langle\widehat{f}_i(\mathbf{K})\,\varphi[\mathbf{z}]\right\rangle\\ &-\frac{1}{2}\int d^4\mathbf{K}_1\int d^4\mathbf{K}_2\Lambda_i^{\alpha\beta}(\mathbf{K}_1,\mathbf{K}_2,\mathbf{K})\frac{\delta^2\Phi[\mathbf{z}]}{\delta z_\alpha(\mathbf{K}_1)\,\delta z_\beta(\mathbf{K}_2)}\end{aligned}\quad(5.64)$$

for the characteristic functional

$$\Phi[\mathbf{z}]=\langle\varphi[\mathbf{z}]\rangle\,.$$

We will now assume that external force random field $\widehat{\mathbf{f}}(\mathbf{K})$ is the Gaussian field homogeneous in space and stationary in time, whose different statistical characteristics are determined by the space-time spectral function

$$\left\langle\widehat{f}_i(\mathbf{K}_1)\widehat{f}_j(\mathbf{K}_2)\right\rangle=\frac{1}{2}\delta^4(\mathbf{K}_1+\mathbf{K}_2)\,F_{ij}(\mathbf{K}_1).$$

Because $\widehat{\mathbf{f}}(\mathbf{K})$ is the divergence-free (solenoidal) field, we have

$$F_{ij}(\mathbf{K})=\Delta_{ij}(\mathbf{k})F(\mathbf{K}),$$

where $F(\mathbf{K})$ is the space-time spectrum of external forces.

Splitting the correlator in the right-hand side of Eq. (5.64) by the Furutsu–Novikov formula, we can rewrite Eq. (5.64) in the form

$$(i\omega+\nu\mathbf{k}^2)\frac{\delta}{\delta z_i(\mathbf{K})}\Phi[\mathbf{z}] = -\frac{1}{2}F_{ij}(\mathbf{K})\int d^4\mathbf{K}_1 z_\alpha(\mathbf{K}_1)\,G_{\alpha j}[\mathbf{K}_1,-\mathbf{K};\mathbf{z}]$$
$$-\frac{1}{2}\int d^4\mathbf{K}_1\int d^4\mathbf{K}_2\Lambda_i^{\alpha\beta}(\mathbf{K}_1,\mathbf{K}_2,\mathbf{K})\frac{\delta^2\Phi[\mathbf{z}]}{\delta z_\alpha(\mathbf{K}_1)\,\delta z_\beta(\mathbf{K}_2)}, \tag{5.65}$$

where we introduced the new functional

$$G_{ij}[\mathbf{K},\mathbf{K}';\mathbf{z}] = \left\langle \frac{\delta u_i(\mathbf{K})}{\delta \widehat{f}_j(\mathbf{K}')}\varphi[\mathbf{z}]\right\rangle.$$

We can obtain the equation for quantity $\delta u_i(\mathbf{K})/\delta\widehat{f}_j(\mathbf{K}')$ by varying Eq. (5.61),

$$(i\omega+\nu\mathbf{k}^2)\frac{\delta\widehat{u}_i(\mathbf{K})}{\delta\widehat{f}_j(\mathbf{K}')} + i\int d^4\mathbf{K}_1\int d^4\mathbf{K}_2\Lambda_i^{\alpha\beta}(\mathbf{K}_1,\mathbf{K}_2,\mathbf{K})\,\widehat{u}_\alpha(\mathbf{K}_1)\frac{\delta\widehat{u}_\beta(\mathbf{K}_2)}{\delta\widehat{f}_j(\mathbf{K}')}$$
$$=\delta_{ij}\delta^4(\mathbf{K}-\mathbf{K}'). \tag{5.66}$$

As a consequence, functional $G_{ij}[\mathbf{K},\mathbf{K}';\mathbf{z}]$ satisfies the equation

$$(i\omega+\nu\mathbf{k}^2)G_{ij}[\mathbf{K},\mathbf{K}';\mathbf{z}] = \delta_{ij}\delta^4(\mathbf{K}-\mathbf{K}')\,\Phi[\mathbf{z}]$$
$$-\int d^4\mathbf{K}_1\int d^4\mathbf{K}_2\Lambda_i^{\alpha\beta}(\mathbf{K}_1,\mathbf{K}_2,\mathbf{K})\frac{\delta}{\delta z_\alpha(\mathbf{K}_1)}G_{\beta j}[\mathbf{K}_2,\mathbf{K}';\mathbf{z}] \tag{5.67}$$

The system of functional equations (5.65) and (5.67) for $\Phi[\mathbf{z}]$ and $G_{ij}[\mathbf{K},\mathbf{K}';\mathbf{z}]$ is closed and completely governs the statistical description of the velocity field [131] (see also [132, 135, 251]).

It is easy to show that average energy income of the velocity field harmonics due to work of the external force is given by the expression

$$\left\langle \widehat{u}_i(\mathbf{K})\,\widehat{f}_j(\mathbf{K}')\right\rangle = \frac{1}{2}F_{jl}(\mathbf{K}')\left\langle\frac{\delta\widehat{u}_i(\mathbf{K})}{\delta\widehat{f}_l(-\mathbf{K}')}\right\rangle = \frac{1}{2}F_{j\alpha}(\mathbf{K}')\,G_{i\alpha}[\mathbf{K},-\mathbf{K}';0], \tag{5.68}$$

which defines the physical meaning of quantity $G_{ij}[\mathbf{K},\mathbf{K}';\mathbf{z}]$ as the functional that describes correlations between the velocity field and the energy income rate (force power). Here, quantity $\delta\widehat{u}_i(\mathbf{K})/\delta\widehat{f}_l(\mathbf{K}')$ can be considered as some sort of Green's function for Eq. (5.61). Indeed, if we add some deterministic force $\boldsymbol{\eta}(\mathbf{K})$ to the right-hand side of Eq. (5.61), then the solution of the resulting equation will be a functional of argument $\widehat{\mathbf{f}}(\mathbf{K})+\boldsymbol{\eta}(\mathbf{K})$, i.e.,

$$\widehat{u}_i(\mathbf{K}) = \widehat{u}_i\left[\mathbf{K};\widehat{\mathbf{f}}(\mathbf{K}')+\boldsymbol{\eta}(\mathbf{K}')\right]. \tag{5.69}$$

Expand solution (5.69) in the functional series in $\boldsymbol{\eta}(\mathbf{K})$

$$\begin{aligned}\widehat{u}_i(\mathbf{K}) &= \widehat{u}_i\left[\mathbf{K};\widehat{\mathbf{f}}(\mathbf{K}')\right] + \int d^4\mathbf{K}'\left.\frac{\delta\widehat{u}_i(\mathbf{K})}{\delta\eta_j(\mathbf{K}')}\right|_{\boldsymbol{\eta}=0}\eta_j(\mathbf{K}')+\ldots\\ &= \widehat{u}_i(\mathbf{K}) + \int d^4\mathbf{K}'\frac{\delta\widehat{u}_i(\mathbf{K})}{\delta\widehat{f}_j(\mathbf{K}')}\eta_j(\mathbf{K}')+\ldots.\end{aligned} \tag{5.70}$$

The first term of the expansion is simply the solution of problem (5.61). The second term describes the dynamic system response on the infinitely small deterministic force, and quantity $\delta\widehat{u}_i(\mathbf{K})/\delta\widehat{f}_j(\mathbf{K}')$ appears the analog of Green's function for linear systems. Averaging Eq. (5.70) over an ensemble of realizations of random forces and taking into account the equality $\langle\widehat{u}_i(\mathbf{K})\rangle = 0$, we obtain the expression for the system average response

$$\langle\widehat{u}_i(\mathbf{K})\rangle = \int d^4\mathbf{K}' \left\langle \frac{\delta\widehat{u}_i(\mathbf{K})}{\delta\widehat{f}_j(\mathbf{K}')} \right\rangle \eta_j(\mathbf{K}') + \ldots .$$

Turn back to the system of equations (5.65) and (5.67) and represent functionals $\Phi[\mathbf{z}]$ and $G_{ij}[\mathbf{K},\mathbf{K}';\mathbf{z}]$ in the form

$$\Phi[\mathbf{z}] = e^{\phi[\mathbf{z}]}, \quad G_{ij}[\mathbf{K},\mathbf{K}';\mathbf{z}] = S_{ij}[\mathbf{K},\mathbf{K}';\mathbf{z}]\, e^{\phi[\mathbf{z}]}.$$

Then, equations for functionals $\phi[\mathbf{z}]$ and $S_{ij}[\mathbf{K},\mathbf{K}';\mathbf{z}]$ assume the form

$$\begin{aligned}
\frac{\delta}{\delta z_i(\mathbf{K})}\phi[\mathbf{z}] &= -\frac{1}{2}\int d^4\mathbf{K}'\int d^4\mathbf{K}_1 S^0_{i\gamma}(\mathbf{K},\mathbf{K}')\, F_{\gamma j}(\mathbf{K}') z_\alpha(\mathbf{K}_1)\, G_{\alpha j}[\mathbf{K}_1,-\mathbf{K};\mathbf{z}] \\
&-\frac{1}{2}\int d^4\mathbf{K}'\int d^4\mathbf{K}_1\int d^4\mathbf{K}_2 S^0_{i\gamma}(\mathbf{K},\mathbf{K}')\,\Lambda^{\alpha\beta}_\gamma(\mathbf{K}_1,\mathbf{K}_2,\mathbf{K}) \\
&\times\left\{\frac{\delta^2\phi[\mathbf{z}]}{\delta z_\alpha(\mathbf{K}_1)\,\delta z_\beta(\mathbf{K}_2)} + \frac{\delta^2\phi[\mathbf{z}]}{\delta z_\alpha(\mathbf{K}_1)}\frac{\delta\phi[\mathbf{z}]}{\delta z_\beta(\mathbf{K}_2)}\right\}, \qquad (5.71)\\
S_{ij}[\mathbf{K},\mathbf{K}';\mathbf{z}] &= S^0_{ij}(\mathbf{K},\mathbf{K}') \\
&-\int d^4\mathbf{K}''\int d^4\mathbf{K}_1\int d^4\mathbf{K}_2 S^0_{i\gamma}(\mathbf{K},\mathbf{K}'')\,\Lambda^{\alpha\beta}_\gamma(\mathbf{K}_1,\mathbf{K}_2,\mathbf{K}'') \\
&\times\left\{\frac{\delta\phi[\mathbf{z}]}{\delta z_\alpha(\mathbf{K}_1)}G_{\beta j}[\mathbf{K}_2,\mathbf{K}';\mathbf{z}] + \frac{\delta}{\delta z_\alpha(\mathbf{K}_1)}G_{\beta j}[\mathbf{K}_2,\mathbf{K}';\mathbf{z}]\right\}, \qquad (5.72)
\end{aligned}$$

where

$$S^0_{ij}(\mathbf{K},\mathbf{K}') = \left(i\omega + \nu\mathbf{k}^2\right)^{-1}\delta_{ij}\delta^4(\mathbf{K}-\mathbf{K}').$$

The last equation is analogous to the *Schwinger equation* of the quantum field theory.

Note that expansion of functional $\phi[\mathbf{z}]$ in the functional Taylor series in $\mathbf{z}(\mathbf{K})$ determines the velocity field cumulants, and expansion of functional $S_{ij}[\mathbf{K},\mathbf{K}';\mathbf{z}]$ determines the correlators between Green's function analog $G_{ij}(\mathbf{K},\mathbf{K}') = \delta u_i(\mathbf{K})/\delta f_j(\mathbf{K}')$ and the velocity field.

If only the behavior of the velocity correlation function is of interest, the system of functional equations (5.71), (5.72) appears redundant, and we can filter out useless information by representing the spectral function of velocity in terms of a specific perturbation series. To construct such a series, we introduce, as in the linear case, quantity $S^{-1}_{ij}[\mathbf{K},\mathbf{K}';\mathbf{z}]$ by the formula

$$\int d^4\mathbf{K}' S_{ij}[\mathbf{K},\mathbf{K}';\mathbf{z}]\, S^{-1}_{j\delta}[\mathbf{K}',\mathbf{K}'';\mathbf{z}] = \delta_{i\delta}\delta^4(\mathbf{K}-\mathbf{K}''). \qquad (5.73)$$

One can easily see that the relationship

$$\int d^4\mathbf{K}' S^{-1}_{ij}[\mathbf{K},\mathbf{K}';\mathbf{z}]\, S_{j\delta}[\mathbf{K}',\mathbf{K}'';\mathbf{z}] = \delta_{i\delta}\delta^4(\mathbf{K}-\mathbf{K}'') \qquad (5.74)$$

will also hold.

Introduce additionally the three-index functional

$$\Gamma^{j\delta}_\gamma[\mathbf{P},\mathbf{K}',\mathbf{K}'';\mathbf{z}] = \frac{\delta}{\delta z_\gamma(\mathbf{P})} S^{-1}_{j\delta}[\mathbf{K}',\mathbf{K}'';\mathbf{z}], \qquad (5.75)$$

which is similar to the *mass operator vertex portion* in the quantum field theory. Varying Eq. (5.73) with respect to $z_\gamma(\mathbf{P})$, we can express $\delta S/\delta z$ in terms of S and Γ

$$\begin{aligned}&\frac{\delta}{\delta z_\gamma(\mathbf{P})} S_{i\mu}[\mathbf{K},\mathbf{Q};\mathbf{z}] \\ &= -\int d^4\mathbf{K}' \int d^4\mathbf{K}'' S_{ij}[\mathbf{K},\mathbf{K}';\mathbf{z}]\, \Gamma_\gamma^{j\delta}[\mathbf{P},\mathbf{K}',\mathbf{K}'';\mathbf{z}]\, S_{\delta\mu}[\mathbf{K}'',\mathbf{Q};\mathbf{z}]\,. \end{aligned} \tag{5.76}$$

Using Eq. (5.76), we can rewrite Eq. (5.72) in the form

$$\begin{aligned} S_{ij}[\mathbf{K},\mathbf{K}';\mathbf{z}] &= S_{ij}^0(\mathbf{K},\mathbf{K}') \\ &- \int d^4\mathbf{P} \int d^4\mathbf{K}_1 \int d^4\mathbf{K}_2 S_{i\gamma}^0(\mathbf{K},\mathbf{P})\, \Lambda_\gamma^{\alpha\beta}(\mathbf{K}_1,\mathbf{K}_2,\mathbf{P}) \\ &\times \left\{ \frac{\delta\phi[\mathbf{z}]}{\delta z_\alpha(\mathbf{K}_1)} G_{\beta j}[\mathbf{K}_2,\mathbf{K}';\mathbf{z}] \right. \\ &\left. - \int d^4\mathbf{K}' \int d^4\mathbf{K}'' S_{\beta\sigma}[\mathbf{K}_2,\mathbf{K}';\mathbf{z}]\, \Gamma_\alpha^{\sigma\nu}[\mathbf{K}_1,\mathbf{K}',\mathbf{K}'';\mathbf{z}]\, S_{\nu j}[\mathbf{K}'',\mathbf{K}';\mathbf{z}] \right\}. \end{aligned} \tag{5.77}$$

Setting $\mathbf{z} = 0$ in Eq. (5.77), we obtain the equation that interconnects quantities $S|_{\mathbf{z}=0}$ and $\Gamma|_{\mathbf{z}=0}$ and is similar to the *Dyson equation* of the quantum field theory ($\delta\phi/\delta\mathbf{z} = 0$ at $\mathbf{z} = 0$).

Multiplying Eq. (5.77) by S^{-1} from the right and by S_0^{-1} from the left, integrating over the corresponding arguments, and varying the result with respect to $\mathbf{z}$, we obtain the following functional equation for Γ

$$\begin{aligned} \Gamma_\gamma^{\mu\rho}[\mathbf{P}_3,\mathbf{P}_2,\mathbf{P}_1;\mathbf{z}] &= \int d^4\mathbf{K}_1 \Lambda_\mu^{\alpha,\rho}(\mathbf{K}_1,\mathbf{P}_1,\mathbf{P}_2) \frac{\delta^2\phi[\mathbf{z}]}{\delta z_\alpha(\mathbf{K}_1)\,\delta z_\gamma(\mathbf{P}_3)} \\ &- \int d^4\mathbf{K}_1 \int d^4\mathbf{K}_2 \int d^4\mathbf{K}' \Lambda_\mu^{\alpha,\beta}(\mathbf{K}_1,\mathbf{P}_1,\mathbf{P}_2) \\ &\times \frac{\delta}{\delta z_\gamma(\mathbf{P}_3)} \left\{ S_{\beta\sigma}[\mathbf{K}_2,\mathbf{K}';\mathbf{z}]\, \Gamma_\alpha^{\sigma\rho}[\mathbf{K},\mathbf{K}',\mathbf{P}_2;\mathbf{z}] \right\}. \end{aligned} \tag{5.78}$$

The system of equations (5.71), (5.77), and (5.78) is closed; however, the solutions to this system are interconnected additionally by the relationship (5.76).

If we will now construct the perturbation series with absolute terms of Eqs. (5.71) and (5.78) as the zero-order approximations and will express appearing variations of S with respect to $\mathbf{z}$ using relationship (5.76), then we obtain the space-time velocity spectrum and function $\Gamma|_{\mathbf{z}=0}$ in the form of the infinite series, every term of which includes these very functions. These series will be integral equations with infinite number of terms and, being combined with Eq. (5.72) at $\mathbf{z} = 0$, they form the closed system of equations for quantities $\delta^2\phi/\delta\mathbf{z}\delta\mathbf{z}$, $S|_{\mathbf{z}=0}$, and $\Gamma|_{\mathbf{z}=0}$. However, explicit representation of even a few terms of these series is hardly possible because of cumbersome rearrangements and complicated structure of functional equations (5.71), (5.77), and (5.78). The reader can find an analysis of possible simplifications in Ref. [131] (see also [132, 135, 251]).

5.4 Completely solvable stochastic dynamic systems

Consider now several dynamic systems that allow sufficiently adequate statistical analysis for arbitrary random parameters.

5.4.1 Ordinary differential equations

Multiplicative action

As the first example, we consider the vector stochastic equation with initial value

$$\frac{d}{dt}\mathbf{x}(t) = z(t)g(t)\mathbf{F}(\mathbf{x}), \quad \mathbf{x}(0) = \mathbf{x}_0, \tag{5.79}$$

where $g(t)$ and $F_i(\mathbf{x})$, $i = 1, ..., N$, are the deterministic functions and $z(t)$ is the random process whose statistical characteristics are described by the characteristic functional

$$\Phi[t; v(\tau)] = \left\langle \exp\left\{ i \int_0^t d\tau z(\tau) v(\tau) \right\} \right\rangle = e^{\Theta[t; v(\tau)]}.$$

Equation (5.79) has a feature that offers a possibility of determining statistical characteristics of its solutions in the general case of arbitrarily distributed process $z(t)$. The point is that introduction of new 'random' time

$$T = \int_0^t d\tau z(\tau) g(\tau)$$

reduces Eq. (5.79) to the equation formally looking deterministic

$$\frac{d}{dT}\mathbf{x}(T) = \mathbf{F}(\mathbf{x}), \quad \mathbf{x}(0) = \mathbf{x}_0,$$

so that the solution to Eq. (5.79) has the following structure

$$\mathbf{x}(t) = \mathbf{x}(T) = \mathbf{x}\left(\int_0^t d\tau z(\tau) g(\tau) \right). \tag{5.80}$$

Varying Eq. (5.80) with respect to $z(\tau)$ and using Eq. (5.79), we obtain the equality

$$\frac{\delta}{\delta z(\tau)}\mathbf{x}(t) = g(\tau)\frac{d}{dT}\mathbf{x}(T) = g(\tau)\mathbf{F}(\mathbf{x}(t)). \tag{5.81}$$

Thus, variational derivatives of solution $\mathbf{x}(t)$ is expressed in terms of the same solution at the same time. This fact makes it possible to immediately write the closed equations for statistical characteristics of problem (5.79).

Let us derive the equation for the one-point probability density $P(\mathbf{x}, t) = \langle \delta(\mathbf{x}(t) - \mathbf{x}) \rangle$. It has the form

$$\frac{\partial}{\partial t} P(\mathbf{x}, t) = \left\langle \dot{\Theta}_t \left[t; \frac{\delta}{i \delta z(\tau)} \right] \delta\left(\mathbf{x}(t) - \mathbf{x}\right) \right\rangle. \tag{5.82}$$

Consider now the result of operator $\delta / \delta z(\tau)$ applied to the indicator function $\varphi(\mathbf{x}, t) = \delta(\mathbf{x}(t) - \mathbf{x})$. Using formula (5.81), we obtain the expression

$$\frac{\delta}{\delta z(\tau)} \delta\left(\mathbf{x}(t) - \mathbf{x}\right) = -g(\tau)\frac{\partial}{\partial \mathbf{x}}\left\{ \mathbf{F}(\mathbf{x})\varphi(\mathbf{x}, t) \right\}.$$

Consequently, we can rewrite Eq. (5.82) in the form of the closed operator equation

$$\frac{\partial}{\partial t} P(\mathbf{x}, t) = \dot{\Theta}_t \left[t; i g(\tau)\frac{\partial}{\partial \mathbf{x}} \mathbf{F}(\mathbf{x}) \right] P(\mathbf{x}, t), \quad P(\mathbf{x}, 0) = \delta(\mathbf{x} - \mathbf{x}_0), \tag{5.83}$$

whose particular form depends on the behavior of process $z(t)$.

For the two-time probability density

$$P(\mathbf{x},t;\mathbf{x}_1,t_1) = \langle\delta(\mathbf{x}(t)-\mathbf{x})\delta(\mathbf{x}(t_1)-\mathbf{x}_1)\rangle$$

we obtain similarly the equation (for $t > t_1$)

$$\frac{\partial}{\partial t}P(\mathbf{x},t;\mathbf{x}_1,t_1) = \dot{\Theta}_t\left[t; ig(\tau)\left\{\frac{\partial}{\partial \mathbf{x}}\mathbf{F}(\mathbf{x}) + \theta(t_1-\tau)\frac{\partial}{\partial \mathbf{x}_1}\mathbf{F}(\mathbf{x}_1)\right\}\right] P(\mathbf{x},t;\mathbf{x}_1,t_1) \quad (5.84)$$

with the initial value

$$P(\mathbf{x},t_1;\mathbf{x}_1,t_1) = \delta(\mathbf{x}-\mathbf{x}_1)P(\mathbf{x}_1,t_1),$$

where function $P(\mathbf{x}_1,t_1)$ satisfies Eq. (5.83).

One can see from Eq. (5.84) that multidimensional probability density cannot be factorized in terms of the transition probability (see Sect 3.3), so that process $\mathbf{x}(t)$ is not the Markovian process. The particular forms of Eqs. (5.83) and (5.84) is governed by the statistics of process $z(t)$.

If $z(t)$ is the Gaussian process whose mean value and correlation function are

$$\langle z(t)\rangle = 0, \quad B(t,t') = \langle z(t)z(t')\rangle\,,$$

then functional $\Theta[t;v(\tau)]$ has the form

$$\Theta[t;v(\tau)] = -\frac{1}{2}\int\limits_0^t dt_1 \int\limits_0^t dt_2 B(t_1,t_2)v(t_1)v(t_2)$$

and Eq. (5.83) assumes the following form

$$\frac{\partial}{\partial t}P(\mathbf{x},t) = g(t)\int\limits_0^t d\tau B(t,\tau)g(\tau)\frac{\partial}{\partial x_j}F_j(\mathbf{x})\frac{\partial}{\partial x_k}F_k(\mathbf{x})P(\mathbf{x},t) \quad (5.85)$$

and can be considered as the *extended Fokker–Planck equation*.

The class of problems formulated in terms of the system of equations

$$\frac{d}{dt}\mathbf{x}(t) = z(t)\mathbf{F}(\mathbf{x}) - \lambda\mathbf{x}(t), \quad \mathbf{x}(0) = \mathbf{x}_0, \quad (5.86)$$

where $\mathbf{F}(\mathbf{x})$ are the homogeneous polynomials of power k, can be reduced to problem (5.79). Indeed, introducing new functions

$$\mathbf{x}(t) = \tilde{\mathbf{x}}(t)e^{-\lambda t},$$

we arrive at problem (5.79) with function $g(t) = e^{-\lambda(k-1)t}$. In the important special case with $k = 2$ and functions $\mathbf{F}(\mathbf{x})$ such that $\mathbf{x}\mathbf{F}(\mathbf{x}) = 0$, the system of equations (5.79) describes hydrodynamic systems with the linear friction (see, e.g., [58, 75]). In this case, the interaction between the components appears random.

If $\lambda = 0$, energy conservation holds in hydrodynamic systems for any realization of process $z(t)$. For $t \to \infty$, there is the steady-state probability distribution $P(\mathbf{x})$, which is, under the assumption that no additional integrals of motion exist, the uniform distribution over sphere $x_i^2 = E_0$. If additional integrals of motion exist (as it is the case for finite-dimensional approximation of the two-dimensional motion of liquid), the domain of the steady-state probability distribution will coincide with the phase space region allowed by the integrals of motion.

Note that in the special case of the Gaussian process $z(t)$ appeared in the one-dimensional linear equation of type Eq. (5.86)

$$\frac{d}{dt}x(t) = -\lambda x(t) + z(t)x(t), \quad x(0) = 1,$$

which determines the simplest logarithmic-normal random process, we obtain, instead of Eq. (5.85), the extended Fokker–Planck equation

$$\left(\frac{\partial}{\partial t} + \lambda\frac{\partial}{\partial x}x\right) P(x,t) = \int_0^t d\tau B(t,\tau)\frac{\partial}{\partial x}x\frac{\partial}{\partial x}xP(x,t), \; P(x,0) = \delta(x-1). \tag{5.87}$$

Additive action

Consider now the class of linear equations

$$\frac{d}{dt}\mathbf{x}(t) = A(t)\mathbf{x}(t) + \mathbf{f}(t), \quad \mathbf{x}(0) = \mathbf{x}_0, \tag{5.88}$$

where $A(t)$ is the deterministic matrix and $\mathbf{f}(t)$ is the random vector function whose characteristic functional $\Phi[t;\mathbf{v}(\tau)]$ is known.

For the probability density of the solution to Eq. (5.88), we have

$$\frac{\partial}{\partial t}P(\mathbf{x},t) = -\frac{\partial}{\partial x_i}\left(A_{ik}(t)x_kP(\mathbf{x},t)\right) + \left\langle \dot{\Theta}_t\left[t;\frac{\delta}{i\delta\mathbf{f}(\tau)}\right]\delta\left(\mathbf{x}(t)-\mathbf{x}\right)\right\rangle. \tag{5.89}$$

In the problem under consideration, the variational derivative $\delta\mathbf{x}(t)/\delta\mathbf{f}(\tau)$ also satisfies (for $\tau < t$) the linear equation with the initial value

$$\frac{d}{dt}\frac{\delta}{\delta\mathbf{f}(\tau)}x_i(t) = A_{ik}(t)\frac{\delta}{\delta\mathbf{f}(\tau)}x_k(t), \quad \left.\frac{\delta}{\delta f_l(\tau)}x_i(t)\right|_{t=\tau} = \delta_{il}. \tag{5.90}$$

Equation (5.90) has no randomness and governs Green's function $G_{il}(t,\tau)$ of homogeneous system (5.88), which means that

$$\frac{\delta}{\delta f_l(\tau)}x_i(t) = G_{il}(t,\tau).$$

As a consequence, we have

$$\frac{\delta}{\delta f_l(\tau)}\delta(\mathbf{x}(t)-\mathbf{x}) = -\frac{\partial}{\partial x_k}G_{kl}(t,\tau)\delta(\mathbf{x}(t)-\mathbf{x}),$$

and Eq. (5.89) appears converted into the closed equation

$$\frac{\partial}{\partial t}P(\mathbf{x},t) = -\frac{\partial}{\partial \mathbf{x}_i}\left(A_{ik}(t)x_kP(\mathbf{x},t)\right) + \dot{\Theta}_t\left[t;iG_{kl}(t,\tau)\frac{\partial}{\partial x_k}\right]P(\mathbf{x},t). \tag{5.91}$$

From Eq. (5.91) follows that any moment of quantity $\mathbf{x}(t)$ will satisfy a closed linear equation that will include only a finite number of cumulant functions whose order will not exceed the order of the moment of interest.

For the two-time probability density

$$P(\mathbf{x},t;\mathbf{x}_1,t_1) = \langle \delta(\mathbf{x}(t)-\mathbf{x})\delta(\mathbf{x}(t_1)-\mathbf{x}_1)\rangle ,$$

we quite similarly obtain the equation

$$\begin{aligned}\frac{\partial}{\partial t}P(\mathbf{x},t;\mathbf{x}_1,t_1) &= -\frac{\partial}{\partial x_i}\left(A_{ik}(t)x_k P(\mathbf{x},t;\mathbf{x}_1,t_1)\right)\\ &+\dot{\Theta}_t\left[t;i\left\{G_{kl}(t,\tau)+G_{kl}(t_1,\tau)\right\}\frac{\partial}{\partial x_l}\right]P(\mathbf{x},t;\mathbf{x}_1,t_1) \quad (t>t_1)\end{aligned} \tag{5.92}$$

with the initial value

$$P(\mathbf{x},t_1;\mathbf{x}_1,t_1) = \delta(\mathbf{x}-\mathbf{x}_1)P(\mathbf{x}_1,t_1),$$

where $P(\mathbf{x}_1,t_1)$ is the one-point probability density satisfying Eq. (5.91). From Eq. (5.92) follows that $\mathbf{x}(t)$ is not the Markovian process. The particular form of Eqs. (5.91) and (5.92) depends on the structure of functional $\Phi[t;v(\tau)]$, i.e., on the random behavior of function $\mathbf{f}(t)$.

For the Gaussian vector process $\mathbf{f}(t)$ whose mean value and correlation function are as follows

$$\langle \mathbf{f}(t)\rangle = 0, \quad B_{ij}(t,t') = \langle f_i(t)f_j(t')\rangle ,$$

Eq. (5.91) assumes the form of the extended Fokker–Planck equation

$$\begin{aligned}\frac{\partial}{\partial t}P(\mathbf{x},t) &= -\frac{\partial}{\partial x_i}\left(A_{ik}(t)x_k P(\mathbf{x},t)\right)\\ &+\int\limits_0^t d\tau B_{jl}(t,\tau)G_{kj}(t,t)G_{ml}(t,\tau)\frac{\partial^2}{\partial x_k \partial x_m}P(\mathbf{x},t).\end{aligned} \tag{5.93}$$

Consider the dynamics of a particle under random forces in the presence of friction [154] as an example of such a problem.

Inertial particle under random forces The simplest example of particle diffusion under the action of random external force $\mathbf{f}(t)$ and linear friction is described by the linear system of equations (1.63), page 25

$$\begin{aligned}\frac{d}{dt}\mathbf{r}(t) &= \mathbf{v}(t), \quad \frac{d}{dt}\mathbf{v}(t) = -\lambda\left[\mathbf{v}(t)-\mathbf{f}(t)\right],\\ \mathbf{r}(0) &= 0, \quad \mathbf{v}(0) = 0.\end{aligned} \tag{5.94}$$

The stochastic solution to Eqs. (5.94) has the form

$$\mathbf{v}(t) = \lambda\int\limits_0^t d\tau e^{-\lambda(t-\tau)}\mathbf{f}(\tau), \quad \mathbf{r}(t) = \int\limits_0^t d\tau\left[1-e^{-\lambda(t-\tau)}\right]\mathbf{f}(\tau). \tag{5.95}$$

In the case of stationary random process $\mathbf{f}(t)$ with the correlation tensor $\langle f_i(t)f_j(t')\rangle = B_{ij}(t-t')$ and temporal correlation radius τ_0 determined from the relationship

$$\int\limits_0^\infty d\tau B_{ii}(\tau) = \tau_0 B_{ii}(0),$$

Eq. (5.94) allows obtaining analytical expressions for correlators between particle velocity components and coordinates

$$\langle v_i(t)v_j(t)\rangle = \lambda \int_0^t d\tau B_{ij}(\tau)\left[e^{-\lambda\tau} - e^{-\lambda(2t-\tau)}\right],$$

$$\frac{1}{2}\frac{d}{dt}\langle r_i(t)r_j(t)\rangle = \langle r_i(t)v_j(t)\rangle = \int_0^t d\tau B_{ij}(\tau)\left[1-e^{-\lambda t}\right]\left[1-e^{-\lambda(t-\tau)}\right]. \quad (5.96)$$

In the steady-state regime, when $\lambda t \gg 1$ and $t/\tau_0 \gg 1$, but parameter $\lambda\tau_0$ can be arbitrary, particle velocity is the stationary process with the correlation tensor

$$\langle v_i(t)v_j(t)\rangle = \lambda \int_0^\infty d\tau B_{ij}(\tau)e^{-\lambda\tau}, \quad (5.97)$$

and correlations $\langle r_i(t)v_j(t)\rangle$ and$\langle r_i(t)r_j(t)\rangle$ are as follows

$$\langle r_i(t)v_j(t)\rangle = \int_0^\infty d\tau B_{ij}(\tau), \quad \langle r_i(t)r_j(t)\rangle = 2t\int_0^\infty d\tau B_{ij}(\tau). \quad (5.98)$$

If we additionally assume that $\lambda\tau_0 \gg 1$, the correlation tensor grades into

$$\langle v_i(t)v_j(t)\rangle = B_{ij}(0), \quad (5.99)$$

which is consistent with (5.94), because $\mathbf{v}(t) \equiv \mathbf{f}(t)$ in this limit.

If the opposite condition $\lambda\tau_0 \ll 1$ holds, then

$$\langle v_i(t)v_j(t)\rangle = \lambda \int_0^\infty d\tau B_{ij}(\tau).$$

This result corresponds to random process $\mathbf{f}(t)$ in the delta-correlated approximation.

Probability distribution function Introduce now the indicator function of the solution to Eq. (5.94)

$$\varphi(\mathbf{r},\mathbf{v};t) = \delta\left(\mathbf{r}(t)-\mathbf{r}\right)\delta\left(\mathbf{v}(t)-\mathbf{v}\right),$$

which satisfies the Liouville equation

$$\left(\frac{\partial}{\partial t} + \mathbf{v}\frac{\partial}{\partial \mathbf{r}} - \lambda\frac{\partial}{\partial \mathbf{v}}\mathbf{v}\right)\varphi(\mathbf{r},\mathbf{v};t) = -\lambda\mathbf{f}(t)\frac{\partial}{\partial \mathbf{v}}\varphi(\mathbf{r},\mathbf{v};t),$$
$$\varphi(\mathbf{r},\mathbf{v};0) = \delta\left(\mathbf{r}\right)\delta\left(\mathbf{v}\right). \quad (5.100)$$

The mean value of the indicator function $\varphi(\mathbf{r},\mathbf{v};t)$ over an ensemble of realizations of random process $\mathbf{f}(t)$ is the joint one-time probability density of particle position and velocity

$$P(\mathbf{r},\mathbf{v};t) = \langle\varphi(\mathbf{r},\mathbf{v};t)\rangle = \left\langle\delta\left(\mathbf{r}(t)-\mathbf{r}\right)\delta\left(\mathbf{v}(t)-\mathbf{v}\right)\right\rangle_{\mathbf{f}}.$$

Averaging Eq. (5.100) over an ensemble of realizations of random process $\mathbf{f}(t)$, we obtain the unclosed equation

$$\left(\frac{\partial}{\partial t} + \mathbf{v}\frac{\partial}{\partial \mathbf{r}} - \lambda\frac{\partial}{\partial \mathbf{v}}\mathbf{v}\right)P(\mathbf{r},\mathbf{v};t) = -\lambda\frac{\partial}{\partial \mathbf{v}}\left\langle\mathbf{f}(t)\varphi(\mathbf{r},\mathbf{v};t)\right\rangle,$$
$$P(\mathbf{r},\mathbf{v};0) = \delta\left(\mathbf{r}\right)\delta\left(\mathbf{v}\right). \quad (5.101)$$

This equation contains correlation $\langle \mathbf{f}(t)\varphi(\mathbf{r},\mathbf{v};t)\rangle$ and is equivalent to the equality

$$\left(\frac{\partial}{\partial t}+\mathbf{v}\frac{\partial}{\partial \mathbf{r}}-\lambda\frac{\partial}{\partial \mathbf{v}}\mathbf{v}\right)P(\mathbf{r},\mathbf{v};t)=\left\langle \dot{\Theta}\left[t;\frac{\delta}{i\delta \mathbf{f}(\tau)}\right]\varphi(\mathbf{r},\mathbf{v};t)\right\rangle,$$
$$P(\mathbf{r},\mathbf{v};0)=\delta(\mathbf{r})\,\delta(\mathbf{v}), \tag{5.102}$$

where functional $\Theta\left[t;\mathbf{v}(\tau)\right]$ is related to the characteristic functional of random process $\mathbf{f}(t)$

$$\Phi\left[t;\boldsymbol{\psi}(\tau)\right]=\left\langle \exp\left\{i\int_0^t d\tau\,\boldsymbol{\psi}(\tau)\mathbf{f}(\tau)\right\}\right\rangle=e^{\Theta[t;\boldsymbol{\psi}(\tau)]}$$

by the formula

$$\dot{\Theta}\left[t;\boldsymbol{\psi}(\tau)\right]=\frac{d}{dt}\ln\Phi\left[t;\boldsymbol{\psi}(\tau)\right]=\frac{d}{dt}\Theta\left[t;\boldsymbol{\psi}(\tau)\right].$$

Functional $\Theta[t;\boldsymbol{\psi}(\tau)]$ can be expanded in the functional power series

$$\Theta[t;\boldsymbol{\psi}(\tau)]=\sum_{n=1}^{\infty}\frac{i^n}{n!}\int_0^t dt_1\ldots\int_0^t dt_n K^{(n)}_{i_1,\ldots,i_n}(t_1,\ldots,t_n)\psi_{i_1}(t_1)\ldots\psi_{i_n}(t_n),$$

where functions

$$K^{(n)}_{i_1,\ldots,i_n}(t_1,\ldots,t_n)=\frac{1}{i^n}\frac{\delta^n}{\delta\psi_{i_1}(t_1)\ldots\delta\psi_{i_n}(t_n)}\Theta[t;\boldsymbol{\psi}(\tau)]\bigg|_{\boldsymbol{\psi}=0}$$

are the n-th order cumulant functions of random process $\mathbf{f}(t)$.

Consider the variational derivative

$$\frac{\delta}{\delta f_j(t')}\varphi(\mathbf{r},\mathbf{v};t)=-\left[\frac{\partial}{\partial r_k}\frac{\delta r_k(t)}{\delta f_j(t')}+\frac{\partial}{\partial v_k}\frac{\delta v_k(t)}{\delta f_j(t')}\right]\varphi(\mathbf{r},\mathbf{v};t). \tag{5.103}$$

In the context of dynamic problem (5.94), the variational derivatives of functions $\mathbf{r}(t)$ and $\mathbf{v}(t)$ in Eq. (5.103) can be calculated from Eqs. (5.95) and have the forms

$$\frac{\delta v_k(t)}{\delta f_j(t')}=\lambda\delta_{kj}e^{-\lambda(t-t')},\quad \frac{\delta r_k(t)}{\delta f_j(t')}=\delta_{kj}\left[1-e^{-\lambda(t-t')}\right]. \tag{5.104}$$

Using Eq. (5.104), we can now rewrite Eq. (5.103) in the form

$$\frac{\delta}{\delta \mathbf{f}(t')}\varphi(\mathbf{r},\mathbf{v};t)=-\left\{\left[1-e^{-\lambda(t-t')}\right]\frac{\partial}{\partial \mathbf{r}}+\lambda e^{-\lambda(t-t')}\frac{\partial}{\partial \mathbf{v}}\right\}\varphi(\mathbf{r},\mathbf{v};t),$$

after which Eq. (5.102) assumes the closed form

$$\left(\frac{\partial}{\partial t}+\mathbf{v}\frac{\partial}{\partial \mathbf{r}}-\lambda\frac{\partial}{\partial \mathbf{v}}\mathbf{v}\right)P(\mathbf{r},\mathbf{v};t)$$
$$=\dot{\Theta}\left[t;i\left\{\left[1-e^{-\lambda(t-t')}\right]\frac{\partial}{\partial \mathbf{r}}+\lambda e^{-\lambda(t-t')}\frac{\partial}{\partial \mathbf{v}}\right\}\right]P(\mathbf{r},\mathbf{v};t),$$
$$P(\mathbf{r},\mathbf{v};0)=\delta(\mathbf{r})\,\delta(\mathbf{v}). \tag{5.105}$$

Note that from Eq. (5.105) follows that equations for the n-th order moment functions include cumulant functions of order not higher than n.

The Gaussian process $\mathbf{f}(t)$ Assume now that $\mathbf{f}(t)$ is the Gaussian stationary process with the zero mean value and correlation tensor

$$B_{ij}(t-t') = \langle f_i(t) f_j(t') \rangle .$$

In this case, the characteristic functional of process $\mathbf{f}(t)$ is

$$\Phi\left[t; \boldsymbol{\psi}(\tau)\right] = \exp\left\{ -\frac{1}{2} \int\limits_0^t \int\limits_0^t dt_1 dt_2 B_{ij}(t_1 - t_2) \psi_i(t_1) \psi_j(t_2) \right\},$$

functional $\dot{\Theta}[t; \boldsymbol{\psi}(\tau)]$ is given by the formula

$$\dot{\Theta}\left[t; \boldsymbol{\psi}(\tau)\right] = -\psi_i(t) \int\limits_0^t dt' B_{ij}(t-t') \psi_j(t'),$$

and Eq. (5.105) appears an extension of the Fokker–Planck equation

$$\left(\frac{\partial}{\partial t} + \mathbf{v}\frac{\partial}{\partial \mathbf{r}} - \lambda \frac{\partial}{\partial \mathbf{v}}\mathbf{v}\right) P(\mathbf{r}, \mathbf{v}; t) = \lambda^2 \int\limits_0^t d\tau B_{ij}(\tau) e^{-\lambda\tau} \frac{\partial^2}{\partial v_i \partial v_j} P(\mathbf{r}, \mathbf{v}; t)$$

$$+\lambda \int\limits_0^t d\tau B_{ij}(\tau)\left[1 - e^{-\lambda\tau}\right] \frac{\partial^2}{\partial v_i \partial r_j} P(\mathbf{r}, \mathbf{v}; t),$$

$$P(\mathbf{r}, \mathbf{v}; 0) = \delta(\mathbf{r})\, \delta(\mathbf{v}) . \tag{5.106}$$

Equation (5.106) is the exact equation and remains valid for arbitrary times t. From this equation follows that $\mathbf{r}(t)$ and $\mathbf{v}(t)$ are the Gaussian functions. For moment functions of processes $\mathbf{r}(t)$ and $\mathbf{v}(t)$, we obtain in the ordinary way the system of equations

$$\frac{d}{dt}\langle r_i(t) r_j(t)\rangle = 2\langle r_i(t) v_j(t)\rangle ,$$

$$\left(\frac{d}{dt} + \lambda\right)\langle r_i(t) v_j(t)\rangle = \langle v_i(t) v_j(t)\rangle + \lambda \int\limits_0^t d\tau B_{ij}(\tau)\left[1 - e^{-\lambda\tau}\right],$$

$$\left(\frac{d}{dt} + 2\lambda\right)\langle v_i(t) v_j(t)\rangle = 2\lambda^2 \int\limits_0^t d\tau B_{ij}(\tau) e^{-\lambda\tau}. \tag{5.107}$$

From system (5.107) follows that steady-state values of all one-time correlators for $\lambda t \gg 1$ and $t/\tau_0 \gg 1$ are given by the expressions

$$\langle v_i(t) v_j(t)\rangle = \lambda \int\limits_0^\infty d\tau B_{ij}(\tau) e^{-\lambda\tau}, \quad \langle r_i(t) v_j(t)\rangle = D_{ij},$$

$$\langle r_i(t) r_j(t)\rangle = 2t D_{ij}, \tag{5.108}$$

where

$$D_{ij} = \int\limits_0^\infty d\tau B_{ij}(\tau) \tag{5.109}$$

is the spatial diffusion tensor, which agrees with expressions (5.97) and (5.98).

Remark 1 ***Temporal correlation tensor and temporal correlation radius of process $\mathbf{v}(t)$.***

We can additionally calculate the temporal correlation radius of velocity $\mathbf{v}(t)$, i.e., the scale of correlator $\langle v_i(t)v_j(t_1)\rangle$. Using equalities (5.104), we obtain for $t_1 < t$ the equation

$$\left(\frac{d}{dt}+\lambda\right)\langle v_i(t)v_j(t_1)\rangle = \lambda^2 \int\limits_0^{t_1} dt' B_{ij}(t-t')e^{-\lambda(t_1-t')}$$
$$= \lambda^2 e^{\lambda(t-t_1)} \int\limits_{t-t_1}^{t} d\tau B_{ij}(\tau)e^{-\lambda\tau}, \tag{5.110}$$

with the initial value

$$\langle v_i(t)v_j(t_1)\rangle\,|_{t=t_1} = \langle v_i(t_1)v_j(t_1)\rangle\,. \tag{5.111}$$

In the steady-state regime, i.e., for $\lambda t \gg 1$ and $\lambda t_1 \gg 1$, but at fixed difference $(t-t_1)$, we obtain the equation with initial value $(\tau = t - t_1)$

$$\left(\frac{d}{d\tau}+\lambda\right)\langle v_i(t+\tau)v_j(t)\rangle = \lambda^2 e^{\lambda\tau}\int\limits_{\tau}^{\infty} d\tau_1 B_{ij}(\tau_1)e^{-\lambda\tau_1},$$
$$\langle v_i(t+\tau)v_j(t)\rangle_{\tau=0} = \langle v_i(t)v_j(t)\rangle\,. \tag{5.112}$$

One can easily write the solution to Eq. (5.112); however, our interest here concerns only the temporal correlation radius $\tau_{\mathbf{v}}$ of random process $\mathbf{v}(t)$. To obtain this quantity, we integrate Eq. (5.112) with respect to parameter τ over the interval $(0,\infty)$. The result is

$$\lambda\int\limits_0^{\infty} d\tau\,\langle v_i(t+\tau)v_j(t)\rangle = \langle v_i(t)v_j(t)\rangle + \lambda\int\limits_0^{\infty} d\tau_1 B_{ij}(\tau_1)\left[1-e^{-\lambda\tau_1}\right],$$

and we, using Eq. (5.108), arrive at the expression

$$\tau_{\mathbf{v}}\left\langle \mathbf{v}^2(t)\right\rangle = D_{ii} = \tau_0 B_{ii}(0), \tag{5.113}$$

i.e.,

$$\tau_{\mathbf{v}} = \frac{\tau_0 B_{ii}(0)}{\langle \mathbf{v}^2(t)\rangle} = \frac{\tau_0 B_{ii}(0)}{\lambda\int\limits_0^{\infty} d\tau B_{ii}(\tau)e^{-\lambda\tau}} = \begin{cases} \tau_0, & \text{for } \lambda\tau_0 \gg 1, \\ 1/\lambda, & \text{for } \lambda\tau_0 \ll 1. \end{cases} \quad \blacklozenge \tag{5.114}$$

Integrating Eq.(5.106) over $\mathbf{r}$, we obtain the closed equation for the probability density of particle velocity

$$\left(\frac{\partial}{\partial t}-\lambda\frac{\partial}{\partial \mathbf{v}}\mathbf{v}\right)P(\mathbf{v};t) = \lambda^2\int\limits_0^{t} d\tau B_{ij}(\tau)e^{-\lambda\tau}\frac{\partial^2}{\partial v_i\partial v_j}P(\mathbf{r},\mathbf{v};t),$$
$$P(\mathbf{r},\mathbf{v};0) = \delta\left(\mathbf{v}\right).$$

The solution to this equation corresponds to the Gaussian process $\mathbf{v}(t)$ with correlation tensor (5.96), which follows from the fact that the second equation of system (5.94) is closed. It can be shown that, if the steady-state probability density exists under the condition $\lambda t \gg 1$, then this probability density satisfies the equation

$$-\frac{\partial}{\partial \mathbf{v}}\mathbf{v}P(\mathbf{v};t) = \lambda \int\limits_0^\infty d\tau B_{ij}(\tau) e^{-\lambda\tau} \frac{\partial^2}{\partial v_i \partial v_j} P(\mathbf{v};t),$$

and the rate of establishing this distribution depends on parameter λ.

The equation for the probability density of particle coordinate $P(\mathbf{r};t)$ cannot be derived immediately from Eq. (5.106). Indeed, integrating Eq. (5.106) over $\mathbf{v}$, we obtain the equality

$$\frac{\partial}{\partial t}P(\mathbf{r},t) = -\frac{\partial}{\partial \mathbf{r}} \int \mathbf{v}P(\mathbf{r},\mathbf{v};t)d\mathbf{v}, \quad P(\mathbf{r},0) = \delta(\mathbf{r}). \tag{5.115}$$

For function $\int v_k P(\mathbf{r},\mathbf{v};t)d\mathbf{v}$, we have the equality

$$\begin{aligned}\left(\frac{\partial}{\partial t} + \lambda\right) \int v_k P(\mathbf{r},\mathbf{v};t)d\mathbf{v} &= -\frac{\partial}{\partial \mathbf{r}} \int v_k \mathbf{v} P(\mathbf{r},\mathbf{v};t)d\mathbf{v} \\ &-\lambda \int\limits_0^t d\tau B_{kj}(\tau)\left[1 - e^{-\lambda\tau}\right] \frac{\partial}{\partial r_j} P(\mathbf{r},t),\end{aligned} \tag{5.116}$$

and so on, i.e., this approach results in an infinite system of equations.

Random function $\mathbf{r}(t)$ satisfies the first equation of system (5.94) and, if we would know the complete statistics of function $\mathbf{v}(t)$ (i.e., the multi-time statistics), we could calculate all statistical characteristics of function $\mathbf{r}(t)$. Unfortunately, Eq. (5.106) describes only one-time statistical quantities, and only the infinite system of equations similar to Eqs. (5.115), (5.116), and so on appears equivalent to the multi-time statistics of function $\mathbf{v}(t)$. Indeed, function $\mathbf{r}(t)$ can be represented in the form

$$\mathbf{r}(t) = \int\limits_0^t dt_1 \mathbf{v}(t_1),$$

so that the spatial diffusion coefficient in the steady-state regime assumes, in view of Eq. (5.113), the form

$$\begin{aligned}\frac{1}{2}\frac{d}{dt}\left\langle \mathbf{r}^2(t)\right\rangle &= \int\limits_0^\infty d\tau \left\langle \mathbf{v}(t+\tau)\mathbf{v}(t)\right\rangle \\ &= \tau_{\mathbf{v}} \left\langle \mathbf{v}^2(t)\right\rangle = D_{ii} = \tau_0 B_{ii}(0),\end{aligned} \tag{5.117}$$

from which follows that it depends on the temporal correlation radius $\tau_{\mathbf{v}}$ and the variance of random function $\mathbf{v}(t)$.

However, in the case of this simplest problem, we know immediately variances and all correlations of functions $\mathbf{v}(t)$ and $\mathbf{r}(t)$ (see Eqs. (5.96)) and, consequently, can draw the equation for the probability density of particle coordinate $P(\mathbf{r};t)$. This equation is the diffusion equation

$$\frac{\partial}{\partial t}P(\mathbf{r};t) = D_{ij}(t)\frac{\partial^2}{\partial r_i \partial r_j}P(\mathbf{r},t), \quad P(\mathbf{r},0) = \delta(\mathbf{r}),$$

where

$$
\begin{aligned}
D_{ij}(t) &= \frac{1}{2}\frac{d}{dt}\left\langle r_i(t) r_j(t)\right\rangle = \frac{1}{2}\left\{\left\langle r_i(t) v_j(t)\right\rangle + \left\langle r_j(t) v_i(t)\right\rangle\right\} \\
&= \int_0^t d\tau B_{ij}(\tau)\left[1 - e^{-\lambda t}\right]\left[1 - e^{-\lambda(t-\tau)}\right]
\end{aligned}
$$

is the diffusion tensor (5.96). Under the condition $\lambda t \gg 1$, we obtain the equation

$$
\frac{\partial}{\partial t} P(\mathbf{r}; t) = D_{ij}\frac{\partial^2}{\partial r_i \partial r_j} P(\mathbf{r}, t), \quad P(\mathbf{r}, 0) = \delta(\mathbf{r}) \tag{5.118}
$$

with the diffusion tensor

$$
D_{ij} = \int_0^\infty d\tau B_{ij}(\tau). \tag{5.119}
$$

Note that conversion from Eq. (5.106) to the equation for the probability density of particle coordinate (5.118) with the diffusion coefficient (5.117) corresponds to the so-called *Kramers problem* (see, e.g., [303]).

Delta-correlated approximation ($\lambda\tau_0 \ll 1$). Under the assumption that $\lambda\tau_0 \ll 1$, where τ_0 is the temporal correlation radius of process $\mathbf{f}(t)$, Eq. (5.106) becomes simpler

$$
\begin{aligned}
&\left(\frac{\partial}{\partial t} + \mathbf{v}\frac{\partial}{\partial \mathbf{r}} - \lambda\frac{\partial}{\partial \mathbf{v}}\mathbf{v}\right) P(\mathbf{r}, \mathbf{v}; t) = \lambda^2 \int_0^t d\tau B_{ij}(\tau)\frac{\partial^2}{\partial v_i \partial v_j} P(\mathbf{r}, \mathbf{v}; t), \\
&P(\mathbf{r}, \mathbf{v}; 0) = \delta(\mathbf{r})\,\delta(\mathbf{v}),
\end{aligned}
$$

and corresponds to the approximation of random function $\mathbf{f}(t)$ by the delta-correlated process. If $t \gg \tau_0$, we can replace the upper limit of the integral with the infinity and proceed to the standard diffusion Fokker–Planck equation

$$
\begin{aligned}
&\left(\frac{\partial}{\partial t} + \mathbf{v}\frac{\partial}{\partial \mathbf{r}} - \lambda\frac{\partial}{\partial \mathbf{v}}\mathbf{v}\right) P(\mathbf{r}, \mathbf{v}; t) = \lambda^2 D_{ij}\frac{\partial^2}{\partial v_i \partial v_j} P(\mathbf{r}, \mathbf{v}; t), \\
&P(\mathbf{r}, \mathbf{v}; 0) = \delta(\mathbf{r})\,\delta(\mathbf{v}),
\end{aligned} \tag{5.120}
$$

with diffusion tensor (5.117). In this approximation, the combined random process $\{\mathbf{r}(t), \mathbf{v}(t)\}$ is the Markovian process.

Under the condition $\lambda t \gg 1$, there are the steady-state equation for the probability density of particle velocity

$$
-\lambda\frac{\partial}{\partial \mathbf{v}}\mathbf{v}P(\mathbf{v}) = \lambda^2 D_{ij}\frac{\partial^2}{\partial v_i \partial v_j} P(\mathbf{v}),
$$

and the nonsteady-state equation for the probability density of particle coordinate

$$
\frac{\partial}{\partial t} P(\mathbf{r}; t) = D_{ij}\frac{\partial^2}{\partial r_i \partial r_j} P(\mathbf{r}, t), \quad P(\mathbf{r}, 0) = \delta(\mathbf{r}).
$$

Another asymptotic limit ($\lambda\tau_0 \gg 1$). Consider now the limit $\lambda\tau_0 \gg 1$. In this case, we can rewrite Eq. (5.106) in the form

$$\left(\frac{\partial}{\partial t} + \mathbf{v}\frac{\partial}{\partial \mathbf{r}} - \lambda\frac{\partial}{\partial \mathbf{v}}\mathbf{v}\right) P(\mathbf{r},\mathbf{v};t)$$
$$= \lambda B_{ij}(0)\left[1 - e^{-\lambda t}\right]\frac{\partial^2}{\partial v_i \partial v_j}P(\mathbf{r},\mathbf{v};t)$$
$$-B_{ij}(0)\left[1 - e^{-\lambda t}\right]\frac{\partial^2}{\partial v_i \partial r_j}P(\mathbf{r},\mathbf{v};t) + \lambda\int\limits_0^t d\tau B_{ij}(\tau)\frac{\partial^2}{\partial v_i \partial r_j}P(\mathbf{r},\mathbf{v};t),$$
$$P(\mathbf{r},\mathbf{v};0) = \delta(\mathbf{r})\,\delta(\mathbf{v}).$$

Integrating this equation over $\mathbf{r}$, we obtain the equation for the probability density of particle velocity

$$\left(\frac{\partial}{\partial t} - \lambda\frac{\partial}{\partial \mathbf{v}}\mathbf{v}\right) P(\mathbf{v};t) = \lambda B_{ij}(0)\left[1 - e^{-\lambda t}\right]\frac{\partial^2}{\partial v_i \partial v_j}P(\mathbf{v};t),$$
$$P(\mathbf{v};0) = \delta(\mathbf{v}),$$

and, under the condition $\lambda t \gg 1$, we arrive at the steady-state Gaussian probability density with variance

$$\langle v_i(t)v_j(t)\rangle = B_{ij}(0).$$

As regards the probability density of particle position, it satisfies, under the condition $\lambda t \gg 1$, the equation

$$\frac{\partial}{\partial t}P(\mathbf{r};t) = D_{ij}\frac{\partial^2}{\partial r_i \partial r_j}P(\mathbf{r},t), \quad P(\mathbf{r},0) = \delta(\mathbf{r}) \tag{5.121}$$

with the same diffusion coefficient as previously. This is a consequence of the fact that Eq. (5.113) is independent of parameter λ. Note that this equation corresponds to the limit process $\lambda \to \infty$ in Eq. (5.94)

$$\frac{d}{dt}\mathbf{r}(t) = \mathbf{v}(t), \quad \mathbf{v}(t) = \mathbf{f}(t), \quad \mathbf{r}(0) = 0.$$

In the limit $\lambda \to \infty$ (or $\lambda\tau_0 \gg 1$), we have the equality

$$\mathbf{v}(t) \approx \mathbf{f}(t), \tag{5.122}$$

and all multi-time statistics of random functions $\mathbf{v}(t)$ and $\mathbf{r}(t)$ will be described in terms of statistical characteristics of process $\mathbf{f}(t)$. In particular, the one-time probability density of particle velocity $\mathbf{v}(t)$ is the Gaussian probability density with variance $\langle v_i(t)v_j(t)\rangle = B_{ij}(0)$, and the spatial diffusion coefficient is

$$D = \frac{1}{2}\frac{d}{dt}\left\langle \mathbf{r}^2(t)\right\rangle = \int\limits_0^\infty d\tau B_{ii}(\tau) = \tau_0 B_{ii}(0).$$

As we have seen earlier, in the case of process $\mathbf{f}(t)$ such that it can be correctly described in the delta-correlated approximation (i.e., if $\lambda\tau_0 \ll 1$), the approximate equality (5.122) appears inappropriate to determine statistical characteristics of process $\mathbf{v}(t)$. Nevertheless, Eq. (5.121) with the same diffusion tensor remains as before valid for the one-time statistical characteristics of process $\mathbf{r}(t)$, which follows from the fact that Eq. (5.117) is valid for any parameter λ and arbitrary probability density of random process $\mathbf{f}(t)$.

Above, we considered several types of stochastic ordinary differential equations that allow obtaining closed statistical description in the general form. It is clear that similar situations can appear in dynamic systems formulated in terms of partial differential equations.

5.4.2 Partial differential equations

First of all, we note that the first-order partial differential equation

$$\left(\frac{\partial}{\partial t} + z(t)g(t)\frac{\partial}{\partial \mathbf{x}}\mathbf{F}(\mathbf{x})\right)\rho(\mathbf{r},t) = 0$$

is equivalent to the system of ordinary differential equations (5.79) and, consequently, also allows the complete statistical description for arbitrary given random process $z(t)$.

Consider now the class of nonlinear partial differential equations whose parameters are independent of spatial variable $\mathbf{x}$,

$$\frac{\partial}{\partial t}q(t,\mathbf{x}) + \mathbf{z}(t)\frac{\partial}{\partial \mathbf{x}}q(t,\mathbf{x}) = F\left(t, q, \frac{\partial q}{\partial \mathbf{x}}, \frac{\partial}{\partial \mathbf{x}} \otimes \frac{\partial}{\partial \mathbf{x}}q, \ldots\right),$$

where $\mathbf{z}(t)$ is the vector random process and F is the deterministic function. Solution to this equation is representable in the form

$$q(t,\mathbf{x}) = Q\left(t, \mathbf{x} - \int_0^t d\tau \mathbf{z}(\tau)\right),$$

where function $Q(t,\mathbf{x})$ satisfies the deterministic equation

$$\frac{\partial}{\partial t}Q(t,\mathbf{x}) = F\left(t, Q, \frac{\partial Q}{\partial \mathbf{x}}, \frac{\partial}{\partial \mathbf{x}} \otimes \frac{\partial}{\partial \mathbf{x}}Q, \ldots\right),$$

and, consequently,

$$\frac{\delta}{\delta z_i(\tau)}q(t,\mathbf{x}) = -\theta(t-\tau)\frac{\partial}{\partial z_i(\tau)}q(t,\mathbf{x}).$$

In the case of such problems, statistical characteristics of the solution can be determined immediately by averaging the corresponding expressions constructed from the solution to the last equation. Proceeding in this way, one obtains that the desired function, say, function $\langle q(t,\mathbf{x})\rangle$, satisfies a closed equation containing derivatives of all orders with respect to $\mathbf{x}$.

Consider two examples.

One-dimensional diffusion of passive tracers

Consider the one-dimensional diffusion of passive tracers in random velocity field. This problem is formulated in terms of the equation

$$\frac{\partial}{\partial t}\rho(t,x) + v(t)\frac{\partial}{\partial x}f(x)\rho(t,x) = 0, \tag{5.123}$$

where we will assume that $v(t)$ is the stationary random Gaussian process with parameters

$$\langle v(t)\rangle = 0, \quad B(t-t') = \langle v(t)v(t')\rangle \quad \left(B_v(0) = \left\langle v^2(t)\right\rangle\right)$$

and $f(x)$ is the deterministic function. The indicator function $\varphi(t,x;\rho) = \delta\left(\rho(t,x) - \rho\right)$ for Eq. (5.123) satisfies the equation

$$\begin{aligned}\left(\frac{\partial}{\partial t} + v(t)f(x)\frac{\partial}{\partial x}\right)\varphi(t,x;\rho) &= v(t)\frac{df(x)}{dx}\frac{\partial}{\partial\rho}\rho\varphi(t,x;\rho),\\ \varphi(0,x;\rho) &= \delta(\rho_0(x) - \rho).\end{aligned}$$

We rewrite this equation in the form

$$\begin{aligned}\frac{\partial}{\partial t}\varphi(t,x;\rho) &= -v(t)\left\{\frac{\partial}{\partial x}f(x) - \frac{df(x)}{dx}\left(1 + \frac{\partial}{\partial\rho}\rho\right)\right\}\varphi(t,x;\rho),\\ \varphi(0,x;\rho) &= \delta(\rho_0(x) - \rho).\end{aligned} \tag{5.124}$$

Averaging Eq. (5.124) over an ensemble of realizations of random process $v(t)$, we obtain the expression

$$\begin{aligned}&\frac{\partial}{\partial t}P(t,x;\rho)\\ &= -\int\limits_0^t dt' B(t-t')\left\{\frac{\partial}{\partial x}f(x) - \frac{df(x)}{dx}\left(1 + \frac{\partial}{\partial\rho}\rho\right)\right\}\left\langle\frac{\delta}{\delta v(t')}\varphi(t,x;\rho)\right\rangle,\\ &P(0,x;\rho) = \delta(\rho_0(x) - \rho).\end{aligned} \tag{5.125}$$

The solution to Eq. (5.124) has the form

$$\varphi(t,x,\rho) = \varphi\left(T(t),x,\rho\right),$$

where $T(t) = \int\limits_0^t d\tau v(\tau)$ is the new (random) time and function $\varphi\left(T,x,\rho\right)$ as a function of its arguments satisfies the deterministic equation

$$\begin{aligned}\frac{\partial}{\partial T}\varphi(T,x;\rho) &= -\left\{\frac{\partial}{\partial x}f(x) - \frac{df(x)}{dx}\left(1 + \frac{\partial}{\partial\rho}\rho\right)\right\}\varphi(T,x;\rho),\\ \varphi(0,x;\rho) &= \delta(\rho_0(x) - \rho).\end{aligned} \tag{5.126}$$

Consequently,

$$\begin{aligned}\frac{\delta}{\delta v(t')}\varphi(t,x;\rho) &= \frac{\partial\varphi(T,x;\rho)}{\partial T}\frac{\delta}{\delta v(t')}T(t) = \frac{\partial\varphi(T,x;\rho)}{\partial T}\theta(t-t')\\ &= -\theta(t-t')\left\{\frac{\partial}{\partial x}f(x) - \frac{df(x)}{dx}\left(1 + \frac{\partial}{\partial\rho}\rho\right)\right\}\varphi(t,x;\rho),\end{aligned}$$

where $\theta(t)$ is the Heaviside step function (1 for $t > 0$ and 0 for $t < 0$), and Eq. (5.125) assumes the form of a closed equation

$$\begin{aligned}\frac{\partial}{\partial t}P(t,x;\rho) &= \int\limits_0^t dt' B(t-t')\left\{\frac{\partial}{\partial x}f(x) - \frac{df(x)}{dx}\left(1 + \frac{\partial}{\partial\rho}\rho\right)\right\}\\ &\times\left\{\frac{\partial}{\partial x}f(x) - \frac{df(x)}{dx}\left(1 + \frac{\partial}{\partial\rho}\rho\right)\right\}P(t,x;\rho),\\ P(0,x;\rho) &= \delta(\rho_0(x) - \rho).\end{aligned}$$

Burgers equation with random drift.

Consider the one-dimensional Burgers equation with random drift

$$\frac{\partial}{\partial t}q(t,x)+(q+z(t))\frac{\partial}{\partial x}q(t,x)=\nu\frac{\partial^2}{\partial x^2}q(t,x). \tag{5.127}$$

In this case, we have for the variational derivative of $q(t,x)$ with respect to $z(\tau)$

$$\frac{\delta}{\delta z(\tau)}q(t,x)=\frac{\delta}{\delta z(\tau)}Q\left(t,x-\int_0^t d\tau z(\tau)\right)=-\theta(t-\tau)\frac{\partial}{\partial x}q(t,x). \tag{5.128}$$

Assume now that random process $z(t)$ is the Gaussian process stationary in time and described by correlation function $B(t-t')=\langle z(t)z(t')\rangle$. Let us average Eq. (5.127) over an ensemble of realizations of process $z(t)$ to obtain

$$\frac{\partial}{\partial t}\langle q(t,x)\rangle+\frac{1}{2}\frac{\partial}{\partial x}\left\langle q^2(t,x)\right\rangle+\frac{\partial}{\partial x}\langle z(t)q(t,x)\rangle=\nu\frac{\partial^2}{\partial x^2}\langle q(t,x)\rangle. \tag{5.129}$$

We split the correlators in the left-hand side of this equation using formulas (see Sect. 4.2), page 79

$$\langle z(t)q(t,x)\rangle=\int_0^t d\tau B(t-\tau)\left\langle\frac{\delta}{\delta z(\tau)}q(t,x)\right\rangle,$$

$$\langle q[z(\tau)+\eta_1(\tau)]q[z(\tau)+\eta_2(\tau)]\rangle$$
$$=e^{\int_0^t d\tau_1\int_0^t d\tau_2 B(\tau_1-\tau_2)\frac{\delta^2}{\delta\eta_1(\tau_1)\delta\eta_2(\tau_2)}}\langle q[z(\tau)+\eta_1(\tau)]q[z(\tau)+\eta_2(\tau)]\rangle.$$

In view of Eq. (5.128) we can represent these formulas in the form

$$\langle z(t)q(t,x)\rangle=-\int_0^t d\tau B(\tau)\frac{\partial}{\partial x}\langle q(t,x)\rangle,$$

$$\left\langle q^2(t,x)\right\rangle=e^{2\int_0^t d\tau(t-\tau)B(\tau)\frac{\partial^2}{\partial\eta_1\partial\eta_2}}\langle q(t,x+\eta_1)\rangle\langle q(t,x+\eta_2)\rangle|_{\eta=0}$$
$$=\sum_{n=0}^{\infty}\frac{2^n}{n!}\left[\int_0^t d\tau(t-\tau)B(\tau)\right]^n\left[\frac{\partial^n}{\partial x^n}\langle q(t,x)\rangle\right]^2.$$

As a result, Eq. (5.129) becomes the closed equation

$$\frac{\partial}{\partial t}\langle q(t,x)\rangle+\frac{1}{2}\frac{\partial}{\partial x}\sum_{n=0}^{\infty}\frac{2^n}{n!}\left[\int_0^t d\tau(t-\tau)B(\tau)\right]^n\left[\frac{\partial^n}{\partial x^n}\langle q(t,x)\rangle\right]^2$$
$$=\left(\nu+\int_0^t d\tau B(\tau)\right)\frac{\partial^2}{\partial x^2}\langle q(t,x)\rangle. \tag{5.130}$$

However, in contrast to (5.127), this equation depends on all derivatives with respect to spatial variable x [134, 135].

Unfortunately, there is only limited number of equations that allow sufficiently complete analysis. In the general case, the analysis of dynamic systems appears possible only on the basis of various asymptotic and approximate techniques. In physics, techniques based on approximating actual random processes and fields by the fields delta-correlated in time are often and successfully used.

5.5 Delta-correlated fields and processes

In the case of random field $\mathbf{f}(\mathbf{x}, t)$ delta-correlated in time, the following equality holds (see Sect. 4.7)

$$\dot{\Theta}_t[t, t_0; \mathbf{v}(\mathbf{y}, \tau)] \equiv \dot{\Theta}_t[t, t_0; \mathbf{v}(\mathbf{y}, t)],$$

and situation becomes significantly simpler. The fact that field $\mathbf{f}(\mathbf{x}, t)$ is delta-correlated means that

$$\begin{aligned}&\Theta[t, t_0; \mathbf{v}(\mathbf{y}, \tau)]\\ &= \sum_{i=1}^{\infty} \frac{i^n}{n!} \int_{t_0}^{t} d\tau \int d\mathbf{y}_1 ... \int d\mathbf{y}_n K_n^{i_1,...,i_n}(\mathbf{y}_1, ..., \mathbf{y}_n; \tau) v_{i_1}(\mathbf{y}_1, \tau) ... v_{i_n}(\mathbf{y}_n, \tau),\end{aligned}$$

which, in turn, means that field $\mathbf{f}(\mathbf{x}, t)$ is characterized by cumulant functions of the form

$$K_n^{i_1,...,i_n}(\mathbf{y}_1, t_1; ...; \mathbf{y}_n, t_n) = K_n^{i_1,...,i_n}(\mathbf{y}_1, ..., \mathbf{y}_n; t_1)\delta(t_1 - t_2)...\delta(t_{n-1} - t_n).$$

In this case, Eqs. (5.7), (5.9), and (5.12) appear, in view of Eq. (5.3), the closed operator equations in functions $P(\mathbf{x}, t)$, $p(\mathbf{x}, t|\mathbf{x}_0, t_0)$, and $P_m(\mathbf{x}_1, t_1; ...; \mathbf{x}_m, t_m)$. Indeed, Eq. (5.7) is reduced to the equation

$$\left(\frac{\partial}{\partial t} + \frac{\partial}{\partial \mathbf{x}} \mathbf{v}(\mathbf{x}, t)\right) P(\mathbf{x}, t) = \dot{\Theta}_t \left[t, t_0; i\frac{\partial}{\partial \mathbf{x}} \delta(\mathbf{y} - \mathbf{x})\right] P(\mathbf{x}, t), \quad P(\mathbf{x}, 0) = \delta(\mathbf{x} - \mathbf{x}_0), \tag{5.131}$$

whose concrete form is governed by functional $\Theta[t, t_0; \mathbf{v}(\mathbf{y}, \tau)]$, i.e., by statistical behavior of random field $\mathbf{f}(\mathbf{x}, t)$. Correspondingly, Eq. (5.12) for the m-time probability density is reduced to the operator equation ($t_1 \le t_2 \le ... \le t_m$)

$$\begin{aligned}&\left(\frac{\partial}{\partial t} + \frac{\partial}{\partial \mathbf{x}_m} \mathbf{v}(\mathbf{x}_m, t_m)\right) P_m(\mathbf{x}_1, t_1; ...; \mathbf{x}_m, t_m)\\ &\quad = \dot{\Theta}_{t_m} \left[t_m, t_0; i\frac{\partial}{\partial \mathbf{x}_m} \delta(\mathbf{y} - \mathbf{x}_m)\right] P_m(\mathbf{x}_1, t_1; ...; \mathbf{x}_m, t_m),\\ &P_m(\mathbf{x}_1, t_1; ...; \mathbf{x}_m, t_{m-1}) = \delta(\mathbf{x}_m - \mathbf{x}_{m-1}) P_{m-1}(\mathbf{x}_1, t_1; ...; \mathbf{x}_{m-1}, t_{m-1}),\end{aligned} \tag{5.132}$$

We can seek the solution to Eq. (5.132) in the form

$$P_m(\mathbf{x}_1, t_1; ...; \mathbf{x}_m, t_m) = p(\mathbf{x}_m, t_m|\mathbf{x}_{m-1}, t_{m-1}) P_{m-1}(\mathbf{x}_1, t_1; ...; \mathbf{x}_{m-1}, t_{m-1}). \tag{5.133}$$

Because all differential operations in Eq. (5.132) concern only t_m and $\mathbf{x}_m$, we can substitute Eq. (5.133) in Eq. (5.132) to obtain the following equation for the transition probability density

$$\begin{aligned}&\left(\frac{\partial}{\partial t} + \frac{\partial}{\partial \mathbf{x}} \mathbf{v}(\mathbf{x}, t)\right) p(\mathbf{x}, t|\mathbf{x}_0, t_0) = \dot{\Theta}_t \left[t, t_0; i\frac{\partial}{\partial \mathbf{x}} \delta(\mathbf{y} - \mathbf{x})\right] p(\mathbf{x}, t|\mathbf{x}_0, t_0),\\ &p(\mathbf{x}, t|\mathbf{x}_0, t_0)|_{t \to t_0} = \delta(\mathbf{x} - \mathbf{x}_0).\end{aligned} \tag{5.134}$$

Here, we denoted variables $\mathbf{x}_m$ and t_m as $\mathbf{x}$ and t and variables $\mathbf{x}_{m-1}$ and t_{m-1} as $\mathbf{x}_0$ and t_0.

Using formula (5.133) $(m-1)$ times, we obtain the relationship

$$P_m(\mathbf{x}_1, t_1; ...; \mathbf{x}_m, t_m) = p(\mathbf{x}_m, t_m|\mathbf{x}_{m-1}, t_{m-1})...p(\mathbf{x}_2, t_2|\mathbf{x}_1, t_1)P(\mathbf{x}_1, t_1), \qquad (5.135)$$

where $P(\mathbf{x}_1, t_1)$ is the one-time probability density governed by Eq. (5.131). Equality (5.135) expresses the many-time probability density in terms of the product of transition probability densities, which means that random process $\mathbf{x}(t)$ is the Markovian process. The transition probability density is defined in this case as follows:

$$p(\mathbf{x}, t|\mathbf{x}_0, t_0) = \langle \delta(\mathbf{x}(t) - \mathbf{x})|\mathbf{x}_0, t_0\rangle .$$

Special models of parameter fluctuations can significantly simplify the obtained equations.

For example, in the case of the Gaussian delta-correlated field $\mathbf{f}(\mathbf{x}, t)$, the correlation tensor has the form ($\langle \mathbf{f}(\mathbf{x}, t)\rangle = 0$)

$$B_{ij}(\mathbf{x}, t; \mathbf{x}', t') = 2\delta(t - t')F_{ij}(\mathbf{x}, \mathbf{x}'; t).$$

Then, functional $\Theta[t, t_0; \mathbf{v}(\mathbf{y}, \tau)]$ assumes the form

$$\Theta[t, t_0; \mathbf{v}(\mathbf{y}, \tau)] = -\int\limits_{t_0}^{t} d\tau \int d\mathbf{y}_1 \int d\mathbf{y}_2 F_{ij}(\mathbf{y}_1, \mathbf{y}_2; \tau)v_i(\mathbf{y}_1, \tau)v_j(\mathbf{y}_2, \tau),$$

and Eq. (5.131) reduces to the Fokker–Planck equation

$$\left(\frac{\partial}{\partial t} + \frac{\partial}{\partial x_k}\left[v_k(\mathbf{x},t) + A_k(\mathbf{x},t)\right]\right) P(\mathbf{x}, t) = \frac{\partial^2}{\partial x_k \partial x_l}\left[F_{kl}(\mathbf{x}, \mathbf{x},t)P(\mathbf{x},t)\right], \qquad (5.136)$$

where

$$A_k(\mathbf{x}, t) = \left.\frac{\partial}{\partial x'_l} F_{kl}(\mathbf{x}, \mathbf{x}';t)\right|_{\mathbf{x}'=\mathbf{x}} .$$

Note that Eq. (5.9) in this case assumes the form of the backward Fokker–Planck equation (see, e.g., [72])

$$\begin{aligned} &\left(\frac{\partial}{\partial t_0} + \left[v_k(\mathbf{x}_0,t_0)+A_k(\mathbf{x}_0,t_0)\right]\frac{\partial}{\partial x_{0k}}\right) P(\mathbf{x}, t|\mathbf{x}_0, t_0) \\ = \; &-F_{kl}(\mathbf{x}_0, \mathbf{x}_0; t_0)\frac{\partial^2}{\partial x_{0k}\partial x_{0l}} P(\mathbf{x}, t|\mathbf{x}_0, t_0), \; P(\mathbf{x}, t|\mathbf{x}_0, t) = \delta(\mathbf{x} - \mathbf{x}_0). \qquad (5.137)\end{aligned}$$

In view of the special role that the Gaussian delta-correlated field $\mathbf{f}(\mathbf{x}, t)$ plays in physics, we give an alternative and more detailed discussion of this approximation commonly called the *approximation of the Gaussian delta-correlated field* in Part 3, page 184.

For random field $\mathbf{f}(\mathbf{x}, t)$ related to delta-correlated Poisson process (see Chapter 3, page 89) one can obtain the forward and backward equations of type of the Kolmogorov–Feller equation.

We illustrate the above general theory using several equations as examples.

5.5.1 One-dimensional nonlinear differential equation

Consider the one-dimensional stochastic equation

$$\frac{d}{dt}x(t) = f(x,t) + z(t)g(x,t), \quad x(0) = x_0, \tag{5.138}$$

where $f(x,t)$ and $g(x,t)$ are the deterministic functions and $z(t)$ is the random function of time. For indicator function $\varphi(x,t) = \delta(x(t) - x)$, we have the Liouville equation

$$\left(\frac{\partial}{\partial t} + \frac{\partial}{\partial x}f(x,t)\right)\varphi(x,t) = -z(t)\frac{\partial}{\partial x}\left\{g(x,t)\varphi(x,t)\right\},$$

so that the equation for the one-time probability density $P(x,t)$ has the form

$$\left(\frac{\partial}{\partial t} + \frac{\partial}{\partial x}f(x,t)\right)P(x,t) = \left\langle \dot{\Theta}_t\left[t, \frac{\delta}{i\delta z(\tau)}\right]\varphi(x,t)\right\rangle.$$

In the case of the delta-correlated random process $z(t)$, the equality

$$\dot{\Theta}_t[t, v(\tau)] = \dot{\Theta}_t[t, v(t)]$$

holds. Taking into account the equality

$$\frac{\delta}{\delta z(t-0)}\varphi(x,t) = -\frac{\partial}{\partial x}\left\{g(x,t)\varphi(x,t)\right\},$$

we obtain the closed operator equation

$$\left(\frac{\partial}{\partial t} + \frac{\partial}{\partial x}f(x,t)\right)P(x,t) = \dot{\Theta}_t\left[t, i\frac{\partial}{\partial x}g(x,t)\right]P(x,t). \tag{5.139}$$

For the Gaussian delta-correlated process, we have

$$\Theta[t, v(\tau)] = -\frac{1}{2}\int_0^t d\tau B(\tau)v^2(\tau), \tag{5.140}$$

and Eq. (5.139) assumes the form of the Fokker–Planck equation

$$\left(\frac{\partial}{\partial t} + \frac{\partial}{\partial x}f(x,t)\right)P(x,t) = \frac{1}{2}B(t)\frac{\partial}{\partial x}g(x,t)\frac{\partial}{\partial x}g(x,t)P(x,t). \tag{5.141}$$

For the Poisson delta-correlated process $z(t)$, we have

$$\Theta[t, v(\tau)] = \nu\int_0^t d\tau\left\{\int_{-\infty}^{\infty} d\xi p(\xi)e^{i\xi v(\tau)} - 1\right\}, \tag{5.142}$$

and Eq. (5.139) reduces to the form

$$\left(\frac{\partial}{\partial t} + \frac{\partial}{\partial x}f(x,t)\right)P(x,t) = \nu\left\{\int_{-\infty}^{\infty} d\xi p(\xi)e^{-\xi\frac{\partial}{\partial x}g(x,t)} - 1\right\}P(x,t). \tag{5.143}$$

If we set $g(x,t) = 1$, Eq. (5.138) assumes the form

$$\frac{d}{dt}x(t) = f(x,t) + z(t), \quad x(0) = x_0.$$

In this case, the operator in the right-hand side of Eq. (5.143) is the shift operator, and Eq. (5.143) assumes the form of the Kolmogorov–Feller equation

$$\left(\frac{\partial}{\partial t} + \frac{\partial}{\partial x}f(x,t)\right) P(x,t) = \nu \int_{-\infty}^{\infty} d\xi p(\xi) P(x-\xi,t) - \nu P(x,t).$$

Define now $g(x,t) = x$, so that Eq. (5.138) reduces to the form

$$\frac{d}{dt}x(t) = f(x,t) + z(t)x(t), \quad x(0) = x_0.$$

In this case, Eq. (5.143) assumes the form

$$\left(\frac{\partial}{\partial t} + \frac{\partial}{\partial x}f(x,t)\right) P(x,t) = \nu \left\{ \int_{-\infty}^{\infty} d\xi p(\xi) e^{-\xi \frac{\partial}{\partial x} x} - 1 \right\} P(x,t). \tag{5.144}$$

To determine the action of the operator in the right-hand side of Eq. (5.144), we expand it in series in ξ

$$\left\{ e^{-\xi \frac{\partial}{\partial x} x} - 1 \right\} P(x,t) = \sum_{n=1}^{\infty} \frac{(-\xi)^n}{n!} \left(\frac{\partial}{\partial x} x\right)^n P(x,t)$$

and consider the action of every term.

Representing then x in the form $x = e^{\varphi}$, we can transform this formula as follows (the fact that x is the alternating quantity is insignificant here)

$$\begin{aligned}
&\sum_{n=1}^{\infty} \frac{(-\xi)^n}{n!} e^{-\varphi} \frac{\partial^n}{\partial \varphi^n} e^{\varphi} P(e^{\varphi},t) \\
&= e^{-\varphi} \left\{ e^{-\xi \frac{\partial}{\partial \varphi}} - 1 \right\} e^{\varphi} P(e^{\varphi},t) = e^{-\xi} P(e^{\varphi-\xi},t) - P(e^{\varphi},t).
\end{aligned}$$

Reverting to variable x, we can represent Eq. (5.144) in the final form of the integro-differential equation similar to the Kolmogorov–Feller equation

$$\left(\frac{\partial}{\partial t} + \frac{\partial}{\partial x}f(x,t)\right) P(x,t) = \nu \int_{-\infty}^{\infty} d\xi p(\xi) e^{-\xi} P(xe^{-\xi},t) - \nu P(x,t).$$

In Chapter 3, we mentioned that formula

$$x(t) = \int_0^t d\tau g(t-\tau) z(\tau)$$

relates the Poisson process $x(t)$ with arbitrary impulse function $g(t)$ to the Poisson delta-correlated random process $z(t)$. Let $g(t) = e^{-\lambda t}$. In this case, process $x(t)$ satisfies the stochastic differential equation

$$\frac{d}{dt}x(t) = -\lambda x(t) + z(t)$$

and, consequently, both transition probability density and one-point probability density of this process satisfy, according to Eq. (5.143), the equations

$$\frac{\partial}{\partial t}p(\tilde{z},t|\tilde{z}_0,t_0)=\hat{L}_{\tilde{z}}p(\tilde{z},t|\tilde{z}_0,t_0),\quad \frac{\partial}{\partial t}P(\tilde{z},t)=\hat{L}_{\tilde{z}}P(\tilde{z},t),$$

where operator

$$\hat{L}_{\tilde{z}}=\lambda\frac{\partial}{\partial x}x+\nu\left\{\int\limits_{-\infty}^{\infty}d\xi p(\xi)e^{-\xi\frac{\partial}{\partial x}}-1\right\}. \tag{5.145}$$

5.5.2 Linear operator equation

Consider now the linear operator equation

$$\frac{d}{dt}\mathbf{x}(t)=\hat{A}(t)\mathbf{x}(t)+z(t)\hat{B}(t)\mathbf{x}(t),\quad \mathbf{x}(0)=\mathbf{x}_0, \tag{5.146}$$

where $\hat{A}(t)$ and $\hat{B}(t)$ are the deterministic operators (e.g., differential operators with respect to auxiliary variable or regular matrixes). We will assume that function $z(t)$ is the random delta-correlated function.

Averaging system (5.146), we obtain, according to general formulas,

$$\frac{d}{dt}\left\langle\mathbf{x}(t)\right\rangle=\hat{A}(t)\left\langle\mathbf{x}(t)\right\rangle+\left\langle\dot{\Theta}_t\left[t,\frac{\delta}{i\delta z(t)}\right]\mathbf{x}(t)\right\rangle. \tag{5.147}$$

Then, taking into account the equality

$$\frac{\delta}{\delta z(t-0)}\mathbf{x}(t)=\hat{B}(t)\mathbf{x}(t)$$

that follows immediately from Eq. (5.146), we can rewrite Eq. (5.147) in the form

$$\frac{d}{dt}\left\langle\mathbf{x}(t)\right\rangle=\hat{A}(t)\left\langle\mathbf{x}(t)\right\rangle+\dot{\Theta}_t\left[t,-i\hat{B}\right]\left\langle\mathbf{x}(t)\right\rangle. \tag{5.148}$$

Thus, in the case of linear system (5.146), equations for average values also are the linear equations.

We can expand the logarithm of the characteristic functional $\Theta[t;v(\tau)]$ of delta-correlated processes in the functional Fourier series

$$\Theta[t;v(\tau)]=\sum_{n=1}^{\infty}\frac{i^n}{n!}\int\limits_0^t d\tau K_n(\tau)v^n(\tau), \tag{5.149}$$

where $K_n(t)$ determine the cumulant functions of process $z(t)$. Substituting Eq. (5.149) in Eq. (5.148), we obtain the equation

$$\frac{d}{dt}\left\langle\mathbf{x}(t)\right\rangle=\hat{A}(t)\left\langle\mathbf{x}(t)\right\rangle+\sum_{n=1}^{\infty}\frac{1}{n!}K_n(t)\left[\hat{B}(t)\right]^n\left\langle\mathbf{x}(t)\right\rangle. \tag{5.150}$$

If there exists power l such that $\hat{B}^l(t)=0$, then Eq. (5.150) assumes the form

$$\frac{d}{dt}\left\langle\mathbf{x}(t)\right\rangle=\hat{A}(t)\left\langle\mathbf{x}(t)\right\rangle+\sum_{n=1}^{l-1}\frac{1}{n!}K_n(t)\left[\hat{B}(t)\right]^n\left\langle\mathbf{x}(t)\right\rangle. \tag{5.151}$$

In this case, the equation for average value depends only on a finite number of cumulants of process $z(t)$. This means that there is no necessity in knowledge of probability distribution of function $z(t)$ in the context of the equation for average value; sufficient information includes only certain cumulants of process and knowledge of the fact that process $z(t)$ can be considered as the delta-correlated random process. Statistical description of an oscillator with fluctuating frequency is a good example of such system in physics.

Stochastic parametric resonance

Consider statistical description of an oscillator with fluctuating frequency (1.15), page 10 as an example of simple linear dynamic system that allows a sufficiently complete analysis. The problem on such an oscillator is formulated in terms of the initial value problem for the second-order differential equation

$$\frac{d^2}{dt^2}x(t) + \omega_0^2[1 + z(t)]x(t) = 0, \quad x(0) = x_0, \quad \frac{d}{dt}x(0) = y_0, \tag{5.152}$$

which is equivalent to the system of equations

$$\begin{gathered}\frac{d}{dt}x(t) = y(t), \quad \frac{d}{dt}y(t) = -\omega_0^2[1 + z(t)]x(t), \\ x(0) = x_0, \quad y(0) = y_0.\end{gathered} \tag{5.153}$$

For system (5.153), indicator function

$$\Phi(t; x, y) = \delta\left(x(t) - x\right)\delta\left(y(t) - y\right)$$

satisfies the Liouville equation

$$\left(\frac{\partial}{\partial t} + y\frac{\partial}{\partial x} - \omega_0^2 x\frac{\partial}{\partial y}\right)\Phi(t; x, y) = \omega_0^2 z(t) x\frac{\partial}{\partial y}\Phi(t; x, y).$$

The joint one-time probability density of solutions to system (5.153) is defined by the equality $P(t; x, y) = \langle\Phi(t; x, y)\rangle$ and satisfies the operator equation

$$\left(\frac{\partial}{\partial t} + y\frac{\partial}{\partial x} - \omega_0^2 x\frac{\partial}{\partial y}\right)P(t; x, y) = \left\langle\dot{\Theta}_t\left[t; \frac{\delta}{i\delta z(\tau)}\right]\Phi(t; x, y)\right\rangle, \tag{5.154}$$

where $\dot{\Theta}_t\left[t; v(\tau)\right] = \frac{d}{dt}\Theta\left[t; v(\tau)\right]$, and $\Theta\left[t; v(\tau)\right]$ is the logarithm of the characteristic functional of process $z(t)$,

$$\Theta\left[t; v(\tau)\right] = \ln\left\langle\exp\left\{i\int\limits_0^t d\tau z(\tau)v(\tau)\right\}\right\rangle.$$

In the case of delta-correlated process $z(t)$, the joint one-time probability density of solutions to system (5.153) satisfies the simplified operator equation

$$\left(\frac{\partial}{\partial t} + y\frac{\partial}{\partial x} - \omega_0^2 x\frac{\partial}{\partial y}\right)P(t; x, y) = \left\langle\dot{\Theta}_t\left[t; \frac{\delta}{i\delta z(t)}\right]\Phi(t; x, y)\right\rangle, \tag{5.155}$$

which, in view of the equality

$$\frac{\delta}{\delta z(t-0)}\Phi(t; x, y) = \omega_0^2 x\frac{\partial}{\partial y}\Phi(t; x, y)$$

immediately following from the Liouville equation, can be represented as the closed operator equation

$$\left(\frac{\partial}{\partial t}+y\frac{\partial}{\partial x}-\omega_0^2 x\frac{\partial}{\partial y}\right)P(t;x,y)=\dot{\Theta}_t\left[t;-i\omega_0^2 x\frac{\partial}{\partial y}\right]P(t;x,y). \tag{5.156}$$

Equation (5.156) offers a possibility of deriving closed systems of equations for moments of arbitrary orders. This possibility follows from the fact that the operator in the right-hand side of Eq. (5.156) depends only on homogeneous combination $x\frac{\partial}{\partial y}$ whose action cannot increase the order of the moment under consideration, which is, of cause, a consequence of linearity of the initial system of equations (5.153). Hence, the equations for moments will depend only on the process $z(t)$ cumulants whose orders are smaller or equal to the order of the moment of interest.

Indeed, consider the vector quantity

$$A_k(t)=x^k(t)y^{N-k}(t) \quad (k=0,\,\ldots,N).$$

One can derive from system (5.153) that this quantity satisfies the stochastic equation

$$\frac{d}{dt}A_k(t)=kA_{k-1}(t)-\omega_0^2(N-k)[1+z(t)]A_{k+1}(t) \quad (k=0,\,\ldots,N),$$

which corresponds to the linear operator equation (5.146), page 134 with constant matrixes

$$\hat{A}_{ij}=i\delta_{i,j+1}-\omega_0^2(N-i)\delta_{i,j-1}, \quad \hat{B}_{ij}=-\omega_0^2(N-i)\delta_{i,j-1}.$$

It is obvious that the square of matrix $\hat{B}_{ij}$ is

$$\hat{B}^2_{ij}=-\omega_0^4(N-i)(N-j+i)\delta_{i,j-2}$$

and so on for higher powers; consequently, for power $N+1$ we have

$$\hat{B}^{N+1}\equiv 0.$$

According to (5.150), page 134, averages $\langle A_k(t)\rangle$ $(k=0,\,\ldots,N)$ satisfy the equation

$$\frac{d}{dt}\langle A_k(t)\rangle=k\langle A_{k-1}(t)\rangle-\omega_0^2(N-k)\langle A_{k+1}(t)\rangle+\sum_{n=1}^{N}\frac{1}{n!}K_n\left[\hat{B}^n\right]_{kl}\langle A_l(t)\rangle, \tag{5.157}$$

where K_n are the cumulants of random process $z(t)$ and the summation is truncated at $n=N$ because, as was mentioned, the equation for average value can depend only on the process $z(t)$ cumulants whose orders are smaller or equal to N. In particular, the first moments of the solution to the system of stochastic equations (5.153) for the delta-correlated process $z(t)$ are independent of fluctuations of system parameters in view of the equality $K_1=0$, and the second moments satisfy the system of equations that coincides with the system derived for the Gaussian fluctuations of system parameters.

In the case of the delta-correlated process $z(t)$, we can additionally obtain the correlation functions of solutions to system of equations (5.153). Indeed, multiplying system (5.153) by $x(t')$, where $t'<t$, and averaging the result over an ensemble of realizations of process $z(t)$, we obtain the closed system

$$\frac{d}{dt}\langle x(t)x(t')\rangle=\langle y(t)x(t')\rangle, \quad \frac{d}{dt}\langle y(t)x(t')\rangle=-\omega_0^2\langle x(t)x(t')\rangle, \tag{5.158}$$

because

$$\frac{\delta}{\delta z(t-0)} x(t)x(t') = 0.$$

The initial values of this system are as follows

$$\langle x(t)x(t')\rangle_{t=t'} = \left\langle x^2(t')\right\rangle, \quad \langle y(t)x(t')\rangle_{t=t'} = \langle x(t')y(t')\rangle. \tag{5.159}$$

The system of equations for the other pair of correlation functions for $t > t'$ is derived similarly

$$\frac{d}{dt}\langle x(t)y(t')\rangle = \langle y(t)y(t')\rangle, \quad \frac{d}{dt}\langle y(t)y(t')\rangle = -\omega_0^2 \langle x(t)y(t')\rangle. \tag{5.160}$$

The corresponding boundary conditions are

$$\langle x(t)y(t')\rangle_{t=t'} = \langle x(t')y(t')\rangle, \quad \langle y(t)y(t')\rangle_{t=t'} = \left\langle y^2(t')\right\rangle. \tag{5.161}$$

Solutions to systems of equations (5.158) and (5.160) with the respective initial values (5.159) and (5.161) have the form

$$\begin{aligned}
\langle x(t)x(t')\rangle &= \left\langle x^2(t')\right\rangle \cos\omega_0(t-t') + \frac{1}{\omega_0}\langle x(t')y(t')\rangle \sin\omega_0(t-t'),\\
\langle y(t)x(t')\rangle &= -\omega_0 \left\langle x^2(t')\right\rangle \sin\omega_0(t-t') + \langle x(t')y(t')\rangle \cos\omega_0(t-t'),\\
\langle x(t)y(t')\rangle &= \langle x(t')y(t')\rangle \cos\omega_0(t-t') + \frac{1}{\omega_0}\left\langle y^2(t')\right\rangle \sin\omega_0(t-t'),\\
\langle y(t)y(t')\rangle &= -\omega_0 \langle x(t')y(t')\rangle \sin\omega_0(t-t') + \left\langle y^2(t')\right\rangle \cos\omega_0(t-t').
\end{aligned} \tag{5.162}$$

Gaussian delta-correlated fluctuations of parameters For the Gaussian stationary delta-correlated process $z(t)$, functional $\Theta[t; v(\tau)]$ is given by the formula

$$\Theta[t; v(\tau)] = -\sigma^2\tau_0 \int\limits_0^t d\tau v^2(\tau) \quad \left(\langle z(t)\rangle = 0, \quad \langle z(t)z(t')\rangle = 2\sigma^2\tau_0\delta(t-t')\right),$$

where σ^2 is the variance and τ_0 is the temporal correlation radius of process $z(t)$, so that Eq. (5.156) assumes the form of the Fokker–Planck equation

$$\begin{aligned}
&\left(\frac{\partial}{\partial t} + y\frac{\partial}{\partial x} - \omega_0^2 x\frac{\partial}{\partial y}\right) P(t; x, y) = D\omega_0^2 x^2 \frac{\partial^2}{\partial y^2} P(t; x, y),\\
&P(0; x, y) = \delta(x - x_0)\,\delta(y - y_0),
\end{aligned} \tag{5.163}$$

where $D = \sigma^2\tau_0\omega_0^2$ is the diffusion coefficient in space $\{x, y/\omega_0\}$.

Let us derive the equations for the two first moments of solutions to system (5.153).

For average values of $x(t)$ and $y(t)$, we obtain the system of equations

$$\frac{d}{dt}\langle x(t)\rangle = \langle y(t)\rangle, \quad \frac{d}{dt}\langle y(t)\rangle = -\omega_0^2\langle x(t)\rangle, \quad x(0) = x_0, \quad y(0) = y_0 \tag{5.164}$$

that coincides with system (5.153) without fluctuations, which agrees with the above discussion. Consequently, we have

$$\begin{aligned}
\langle x(t)\rangle &= x_0 \cos\omega_0(t-t') + \frac{1}{\omega_0} y_0 \sin\omega_0(t-t'),\\
\langle y(t)\rangle &= -\omega_0 x_0 \sin\omega_0(t-t') + y_0 \cos\omega_0(t-t').
\end{aligned} \tag{5.165}$$

The second moments of quantities $x(t)$ and $y(t)$ satisfy the system of equations

$$\begin{aligned}\frac{d}{dt}\left\langle x^2(t)\right\rangle &= 2\left\langle x(t)y(t)\right\rangle, \quad \frac{d}{dt}\left\langle x(t)y(t)\right\rangle = \left\langle y^2(t)\right\rangle - \omega_0^2\left\langle x^2(t)\right\rangle, \\ \frac{d}{dt}\left\langle y^2(t)\right\rangle &= -2\omega_0^2\left\langle x(t)y(t)\right\rangle + D\omega_0^2\left\langle x^2(t)\right\rangle. \end{aligned} \tag{5.166}$$

From this system, we can derive the closed third-order equation for any particular moment. For example, for quantity $\langle U(t)\rangle = \langle x^2(t)\rangle$ that describes the average potential energy of the oscillator, we obtain the equation

$$\frac{d^3}{dt^3}\langle U(t)\rangle + 4\omega_0^2\frac{d}{dt}\langle U(t)\rangle - 4D\omega_0^2\langle U(t)\rangle = 0, \tag{5.167}$$

which corresponds to the following stochastic initial value problem for $U(t) = x^2(t)$

$$\begin{aligned}&\frac{d^3}{dt^3}U(t) + 4\omega_0^2\frac{d}{dt}U(t) + 2\omega_0^2\left(z(t)\frac{d}{dt}U(t) + \frac{d}{dt}z(t)U(t)\right) = 0, \\ &U(0) = x_0^2, \quad \left.\frac{d}{dt}U(t)\right|_{t=0} = 2x_0y_0, \\ &\left.\frac{d^2}{dt^2}U(t)\right|_{t=0} = 2y_0^2 - \omega_0^2[1+z(0)]x_0^2 \end{aligned} \tag{5.168}$$

that can also be obtained immediately from system (5.153).

To simplify the calculations, we will assume that the initial values of system (5.153) have the form

$$x(0) = 0, \quad y(0) = \omega_0. \tag{5.169}$$

Assuming that the problem has a small parameter related to the intensity of process $z(t)$ fluctuations, we can approximately (to terms of order of D/ω_0) represent the solution to system (5.166) in the form

$$\begin{aligned}\left\langle x^2(t)\right\rangle &= \frac{1}{2}\left\{e^{Dt} - e^{-\frac{Dt}{2}}\left[\cos(2\omega_0t) + \frac{3D}{4\omega_0}\sin(2\omega_0t)\right]\right\}, \\ \left\langle x(t)y(t)\right\rangle &= \frac{\omega_0}{4}\left\{2e^{-\frac{Dt}{2}}\sin(2\omega_0t) + \frac{D}{\omega_0}\left[e^{Dt} - e^{-\frac{Dt}{2}}\cos(2\omega_0t)\right]\right\}, \\ \left\langle y^2(t)\right\rangle &= \frac{\omega_0^2}{2}\left\{e^{Dt} + e^{-\frac{Dt}{2}}\left[\cos(2\omega_0t) - \frac{D}{4\omega_0}\sin(2\omega_0t)\right]\right\} \end{aligned} \tag{5.170}$$

Thus, solution (5.170) of system of equations (5.166) has terms increasing with time, which corresponds to statistical parametric build-up of fluctuations in dynamic system (5.153) at the expense of frequency fluctuations. In the case of weak fluctuations, the increment of fluctuations is

$$\mu = D \quad (D/\omega_0 \ll 1).$$

From Eqs. (5.170) follows that solutions to statistical problem (5.153) have two characteristic temporal scales $t_1 \sim 1/\omega_0$ and $t_2 \sim 1/D$. The first temporal scale corresponds to the period of oscillations in system (5.153) without fluctuations (fast processes), and the second scale characterizes slow variations of statistical characteristics, which appear due to fluctuations (slow processes). The ratio of these scales is small:

$$t_1/t_2 = D/\omega_0 \ll 1.$$

We can explicitly obtain slow variations of statistical characteristics of processes $x(t)$ and $y(t)$ by excluding fast motions by means of averaging the corresponding quantities over the period $T = 2\pi/\omega_0$. Denoting such averaging with the overbar, we have

$$\overline{\langle x^2(t)\rangle} = \frac{1}{2}e^{Dt}, \quad \overline{\langle x(t)y(t)\rangle} = 0, \quad \overline{\langle y^2(t)\rangle} = \frac{\omega_0^2}{2}e^{Dt}.$$

Stochastic problem with linear friction If we add the linear friction to the system of equations (5.153), i.e., if we consider the dynamic system

$$\frac{d}{dt}x(t) = y(t), \quad \frac{d}{dt}y(t) = -2\gamma y(t) - \omega_0^2[1 + z(t)]x(t), \tag{5.171}$$

then the corresponding Fokker–Planck equation will have the form

$$\left(\frac{\partial}{\partial t} - 2\gamma\frac{\partial}{\partial y}y + y\frac{\partial}{\partial x} - \omega_0^2 x\frac{\partial}{\partial y}\right) P(t;x,y) = D\omega_0^2 x^2 \frac{\partial^2}{\partial y^2}P(t;x,y),$$

and the system of equations for the second moments assumes, instead of (5.166), the form

$$\begin{aligned}
\frac{d}{dt}\left\langle x^2(t)\right\rangle &= 2\left\langle x(t)y(t)\right\rangle, \\
\frac{d}{dt}\left\langle x(t)y(t)\right\rangle &= \left\langle y^2(t)\right\rangle - 2\gamma\left\langle x(t)y(t)\right\rangle - \omega_0^2\left\langle x^2(t)\right\rangle, \\
\frac{d}{dt}\left\langle y^2(t)\right\rangle &= -4\gamma\left\langle y^2(t)\right\rangle - 2\omega_0^2\left\langle x(t)y(t)\right\rangle + D\omega_0^2\left\langle x^2(t)\right\rangle.
\end{aligned}$$

For this system, we will seek the solution proportional to $e^{\lambda t}$. The corresponding characteristic equation for λ assumes then the form

$$\lambda^3 + 6\gamma\lambda^2 + 4(\omega_0^2 + 2\gamma^2)\lambda + 4\omega_0^2(2\gamma - D) = 0.$$

As is known, the necessary and sufficient conditions of solution stability (which means the absence of roots λ_k with positive real parts) is formulated as the Routh–Hurwitz condition, which is equivalent in our case to the inequality $D < 2\gamma$. Thus, if this condition is not satisfied, i.e., if

$$2\gamma < D, \tag{5.172}$$

the second moments grow in time exponentially, which means the occurrence of the statistical parametric excitation of second moments. Note that conditions of the statistical parametric excitation differ for different moments. For example, the condition of exciting the fourth moments appears weaker than condition (5.172) and has the form [268]

$$D > \frac{2\gamma}{3}\frac{\omega_0^2 + 3\gamma^2}{\omega_0^2 + 6\gamma^2}.$$

For the stochastic parametric oscillator with friction, we can consider the problem on the steady-state regime that steadies under the action of random forces statistically independent of frequency fluctuations. This problem is formulated as the stochastic system of equations

$$\frac{d}{dt}x(t) = y(t), \quad \frac{d}{dt}y(t) = -2\gamma y(t) - \omega_0^2[1 + z(t)]x(t) + f(t), \tag{5.173}$$

where $f(t)$ is the Gaussian process statistically independent of process $z(t)$; it is assumed that $f(t)$ is the delta-correlated process with the following parameters

$$\langle f(t)\rangle = 0, \quad \langle f(t)f(t')\rangle = 2\sigma_f^2\tau_f\delta(t-t'),$$

where σ_f^2 is the variance and τ_f is the temporal correlation radius of process $f(t)$.

The one-time probability density of the solutions to stochastic system (5.173) satisfies the Fokker–Planck equation

$$\begin{aligned}\left(\frac{\partial}{\partial t} - 2\gamma\frac{\partial}{\partial y}y + y\frac{\partial}{\partial x} - \omega_0^2 x\frac{\partial}{\partial y}\right)P(t;x,y)\\ = D\omega_0^2 x^2\frac{\partial^2}{\partial y^2}P(t;x,y) + \sigma_f^2\tau_f\frac{\partial^2}{\partial y^2}P(t;x,y),\end{aligned} \tag{5.174}$$

and, consequently, we have

$$\langle x(t)\rangle = 0, \quad \langle y(t)\rangle = 0.$$

Equations for the second moments form in this case the system

$$\begin{aligned}&\frac{d}{dt}\left\langle x^2(t)\right\rangle = 2\left\langle x(t)y(t)\right\rangle,\\ &\frac{d}{dt}\left\langle x(t)y(t)\right\rangle = \left\langle y^2(t)\right\rangle - 2\gamma\left\langle x(t)y(t)\right\rangle - \omega_0^2\left\langle x^2(t)\right\rangle,\\ &\frac{d}{dt}\left\langle y^2(t)\right\rangle = -4\gamma\left\langle y^2(t)\right\rangle - 2\omega_0^2\left\langle x(t)y(t)\right\rangle + D\omega_0^2\left\langle x^2(t)\right\rangle + 2\sigma_f^2\tau_f,\end{aligned} \tag{5.175}$$

whose steady-state solution exists for $t \to \infty$ if the condition (5.172) is satisfied. This solution behaves as follows

$$\langle x(t)y(t)\rangle = 0, \quad \left\langle x^2(t)\right\rangle = \frac{\sigma_f^2\tau_f}{\omega_0^2(D-2\gamma)}, \quad \left\langle y^2(t)\right\rangle = \frac{\sigma_f^2\tau_f}{D-2\gamma}.$$

Poisson delta-correlated fluctuations of parameters

Functional $\Theta[t;v(\tau)]$ of the Poisson delta-correlated random process $z(t)$ is given by Eq. (3.43)

$$\Theta[t;v(\tau)] = \nu\int_0^t d\tau\int_{-\infty}^{\infty} d\xi p(\xi)\left[e^{i\xi v(\tau)} - 1\right],$$

so that Eq. (5.155) assumes the form of the Kolmogorov–Feller equation

$$\left(\frac{\partial}{\partial t} + y\frac{\partial}{\partial x} - \omega_0^2 x\frac{\partial}{\partial y}\right)P(t;x,y) = \nu\int_{-\infty}^{\infty} d\xi p(\xi)P(t;x,y+\xi\omega_0^2 x) - \nu P(t;x,y). \tag{5.176}$$

For sufficiently small parameter ξ, the logarithm of the characteristic functional grades into the expression

$$\Theta[t;v(\tau)] = -\nu\left\langle\xi^2\right\rangle\int_0^t d\tau v^2(\tau),$$

and Eq. (5.176) grades into the Fokker–Planck equation (5.163) with the diffusion coefficient

$$D = \frac{1}{2}\nu\left\langle\xi^2\right\rangle\omega_0^2 x.$$

5.5.3 Partial differential equations

Statistical interpretation of solutions to stochastic equations

In a number of cases, solutions to many deterministic problems can be treated as a result of averaging certain functionals over random trajectories. Such interpretation appears useful in the context of various applications.

Let us derive the conditions under which such interpretation is applicable to some simple equations.

Consider the problem formulated as the initial value problem for the partial differential equation

$$\frac{\partial}{\partial t}u(t,\mathbf{r}) = -q(t,\mathbf{r})u(t,\mathbf{r}) + Q(t,\boldsymbol{\nabla})u(t,\mathbf{r}), \quad u(0,\mathbf{r}) = u_0(\mathbf{r}). \tag{5.177}$$

Along with Eq. (5.177), we consider the first-order partial differential equation

$$\frac{\partial}{\partial t}\phi(t,\mathbf{r}) = -q(t,\mathbf{r})\phi(t,\mathbf{r}) + \mathbf{z}(t)\boldsymbol{\nabla}\phi(t,\mathbf{r}), \quad \phi(0,\mathbf{r}) = u_0(\mathbf{r}) \tag{5.178}$$

whose solution has the form

$$\phi[t,\mathbf{r};\mathbf{z}(\tau)] = u_0\left(\mathbf{r}+\int\limits_0^t d\tau\mathbf{z}(\tau)\right)\exp\left\{-\int\limits_0^t d\tau q\left(\tau,\mathbf{r}+\int\limits_\tau^t d\tau'\mathbf{z}(\tau')\right)\right\}. \tag{5.179}$$

We will assume that $z(t)$ is the random function delta-correlated in time t with characteristic functional $\Phi[t;v(\tau)]$. Averaging Eq. (5.178) over an ensemble of realizations $z(t)$, we obtain the equation

$$\frac{\partial}{\partial t}\langle\phi(t,\mathbf{r})\rangle = -q(t,\mathbf{r})\langle\phi(t,\mathbf{r})\rangle + \left\langle\dot{\Theta}_t\left[t,\frac{\delta}{i\delta\mathbf{z}(t)}\right]\phi(t,\mathbf{r})\right\rangle, \quad \langle\phi(0,\mathbf{r})\rangle = u_0(\mathbf{r}). \tag{5.180}$$

Taking into account the equality

$$\frac{\delta}{\delta\mathbf{z}(t-0)}\phi(t,\mathbf{r}) = \boldsymbol{\nabla}\phi(t,\mathbf{r}),$$

which is a consequence of the initial dynamic equation (5.178), we can rewrite Eq. (5.180) in the form

$$\frac{\partial}{\partial t}\langle\phi(t,\mathbf{r})\rangle = -q(t,\mathbf{r})\langle\phi(t,\mathbf{r})\rangle + \dot{\Theta}_t\left[t,-i\boldsymbol{\nabla}\right]\langle\phi(t,\mathbf{r})\rangle, \quad \langle\phi(0,\mathbf{r})\rangle = u_0(\mathbf{r}). \tag{5.181}$$

Comparing now Eq. (5.181) with Eq. (5.177), we can see that

$$u(t,\mathbf{r}) = \langle\phi[t,\mathbf{r};\mathbf{z}(\tau)]\rangle_{\mathbf{z}} \tag{5.182}$$

if

$$Q(t,\boldsymbol{\nabla}) = \dot{\Theta}_t[t,-i\boldsymbol{\nabla}]. \tag{5.183}$$

In this case, we can treat Eq. (5.182) as the solution to Eq. (5.177) written in the form of the continual integral.

In addition, we can give the operator form of Eq. (5.182) by introducing the functional shift operator

$$u(t,\mathbf{r}) = \langle\phi[t,\mathbf{r};\mathbf{z}(\tau)+\mathbf{v}(\tau)]\rangle_{\mathbf{z}}\big|_{\mathbf{v}=0} = \Phi\left[t,\frac{\delta}{i\delta\mathbf{v}(\tau)}\right]\phi[t,\mathbf{r};\mathbf{v}(\tau)]\bigg|_{\mathbf{v}=0}, \tag{5.184}$$

where $\Phi[t;\mathbf{v}(\tau)]$ is the characteristic functional of process $\mathbf{z}(t)$.

For the Gaussian process $z(t)$, we have

$$\Theta[t,\mathbf{v}(\tau)] = -\frac{1}{2}\int_0^t d\tau B(\tau)\mathbf{v}^2(\tau) \quad \left(Q(t,\boldsymbol{\nabla}) = \frac{1}{2}B(t)\Delta, \quad B(t)>0\right).$$

As a consequence, we obtain the well-known result that the solution to the diffusion equation

$$\frac{\partial}{\partial t}u(t,\mathbf{r}) = -q(t,\mathbf{r})u(t,\mathbf{r}) + \frac{1}{2}B(t)\Delta u(t,\mathbf{r}), \quad u(0,\mathbf{r}) = u_0(\mathbf{r}) \tag{5.185}$$

can be treated as the result of averaging the functional $\phi[t,\mathbf{r};\mathbf{z}(\tau)]$ over the Gaussian delta-correlated process $z(t)$, i.e.

$$u(t,\mathbf{r}) = \left\langle u_0\left(\mathbf{r}+\int_0^t d\tau\mathbf{z}(\tau)\right)\exp\left\{-\int_0^t d\tau q\left(\tau,\mathbf{r}+\int_\tau^t d\tau'\mathbf{z}(\tau')\right)\right\}\right\rangle_{\mathbf{z}}$$

$$= e^{-\frac{1}{2}\int_0^t d\tau B(\tau)\frac{\delta^2}{\delta\mathbf{v}^2(\tau)}}\left\{u_0\left(\mathbf{r}+\int_0^t d\tau\mathbf{v}(\tau)\right)\exp\left[-\int_0^t d\tau q\left(\tau,\mathbf{r}+\int_\tau^t d\tau'\mathbf{v}(\tau')\right)\right]\right\}_{\mathbf{v}=0}.$$

For the Poisson random process of type of the shot noise $\mathbf{z}(t) = \sum_{i=1}^{\infty}\boldsymbol{\xi}_i\delta(t-t_i)$, we have

$$\dot{\Theta}_t[t,\mathbf{v}(t)] = \nu\left\{\int d\boldsymbol{\xi}p(\boldsymbol{\xi})e^{i\boldsymbol{\xi}\mathbf{v}(t)} - 1\right\}, \quad Q(t,\boldsymbol{\nabla}) = \nu\left\{\int d\boldsymbol{\xi}p(\boldsymbol{\xi})e^{\boldsymbol{\xi}\boldsymbol{\nabla}} - 1\right\}.$$

Consequently, the solution to the integro-differential equation

$$\frac{\partial}{\partial t}u(t,\mathbf{r}) = -q(t,\mathbf{r})u(t,\mathbf{r}) + \nu\int d\boldsymbol{\xi}p(\boldsymbol{\xi})u(t,\mathbf{r}+\boldsymbol{\xi}) - \nu u(t,\mathbf{r}), \quad u(0,\mathbf{r}) = u_0(\mathbf{r}) \tag{5.186}$$

can be represented as the result of averaging the functional $\phi[t,\mathbf{r};\mathbf{z}(\tau)]$, i.e., in the form

$$u(t,\mathbf{r}) = \left\langle u_0\left(\mathbf{r}+\int_0^t d\tau\mathbf{z}(\tau)\right)\exp\left\{-\int_0^t d\tau q\left(\tau,\mathbf{r}+\int_\tau^t d\tau'\mathbf{z}(\tau')\right)\right\}\right\rangle_{\mathbf{z}},$$

if function $p(\boldsymbol{\xi})$ in Eq. (5.186) can be interpreted as the probability density of random quantity $\boldsymbol{\xi}$.

If $p(\boldsymbol{\xi}) = \delta(\boldsymbol{\xi}-\mathbf{r}_0)$, then Eq. (5.186) assumes simpler form

$$\frac{\partial}{\partial t}u(t,\mathbf{r}) = -q(t,\mathbf{r})u(t,\mathbf{r}) + \nu\left[u(t,\mathbf{r}+\mathbf{r}_0)-u(t,\mathbf{r})\right], \quad u(0,\mathbf{r}) = u_0(\mathbf{r}). \tag{5.187}$$

In this case, Eqs. (5.186) and (5.187) have the form of the transport equation.

If a dynamic system is formulated in terms of partial differential equations of orders higher than the first and fluctuations of system parameters are delta-correlated, we can obtain closed equations, but for characteristic functionals in functional spaces. Consider two examples of such equations.

Parabolic equation of quasi-optics

If we assume that field $\varepsilon(x,\mathbf{R})$ in the linear parabolic equation (5.13), page 100

$$\frac{\partial}{\partial x}u(x,\mathbf{R}) = \frac{i}{2k}\Delta_{\mathbf{R}}u(x,\mathbf{R}) + i\frac{k}{2}\varepsilon(x,\mathbf{R})u(x,\mathbf{R}), \quad u(0,\mathbf{R}) = u_0(\mathbf{R})$$

is the homogeneous delta-correlated random field, then Eq. (5.21) for the characteristic functional $\Phi[x;v,v^*]$ of the solution to this equation assumes the form

$$\frac{\partial}{\partial x}\Phi[x;v,v^*] = \left\langle \dot{\Theta}_x\left[x;\frac{\delta}{i\delta\varepsilon(x,\mathbf{R}')}\right]\varphi[x;v,v^*]\right\rangle$$
$$+\frac{i}{2k}\left\{\int d\mathbf{R}'\left[v(\mathbf{R}')\Delta_{\mathbf{R}'}\frac{\delta}{\delta v(\mathbf{R}')} - v^*(\mathbf{R}')\Delta_{\mathbf{R}'}\frac{\delta}{\delta v^*(\mathbf{R}')}\right]\right\}\Phi[x;v,v^*].$$

Taking into account Eq. (5.20), we can represent the last equation in the closed operator form [132, 134, 135]

$$\frac{\partial}{\partial x}\Phi[x;v,v^*] = \dot{\Theta}_x\left[x,\frac{k}{2}\hat{M}(\mathbf{R}')\right]\Phi[x;v,v^*)]$$
$$+\frac{i}{2k}\left\{\int d\mathbf{R}'\left[v(\mathbf{R}')\Delta_{\mathbf{R}'}\frac{\delta}{\delta v(\mathbf{R}')} - v^*(\mathbf{R}')\Delta_{\mathbf{R}'}\frac{\delta}{\delta v^*(\mathbf{R}')}\right]\right\}\Phi[x;v,v^*], \quad (5.188)$$

where$\hat{M}(\mathbf{R}')$ is given by the formula

$$\hat{M}(\mathbf{R}') = v(\mathbf{R}')\frac{\delta}{\delta v(\mathbf{R}')} - v^*(\mathbf{R}')\frac{\delta}{\delta v^*(\mathbf{R}')}$$

and functional

$$\dot{\Theta}_x\left[x;\psi(\xi,\mathbf{R}')\right] = \frac{d}{dx}\ln\left\langle \exp\left\{i\int\limits_0^x d\xi\int d\mathbf{R}'\varepsilon(\xi,\mathbf{R}')\psi(\xi,\mathbf{R}')\right\}\right\rangle$$

is the derivative of the logarithm of the characteristic functional of filed $\varepsilon(x,\mathbf{R})$.

Equation (5.188) yields the equations for the moment functions of filed $u(x,\mathbf{R})$,

$$M_{m,n}(x;\mathbf{R}_1,...,\mathbf{R}_m;\mathbf{R}'_1,...,\mathbf{R}'_n) = \langle u(x,\mathbf{R}_1)...u(x,\mathbf{R}_m)u^*(x,\mathbf{R}'_1)...u^*(x,\mathbf{R}'_n)\rangle$$

(for $m=n$, these functions are usually called the *coherence functions* of order $2n$), which are a consequence of the linearity of the initial dynamic equation (5.13). These equations have the form

$$\frac{\partial}{\partial x}M_{m,n} = \frac{i}{2k}\left(\sum_{p=1}^{m}\Delta_{\mathbf{R}_p} - \sum_{q=1}^{n}\Delta_{\mathbf{R}'_q}\right)M_{m,n}$$
$$+\dot{\Theta}_x\left[x,\frac{1}{k}\left(\sum_{p=1}^{m}\delta(\mathbf{R}'-\mathbf{R}_p) - \sum_{q=1}^{n}\delta(\mathbf{R}'-\mathbf{R}'_q)\right)\right]M_{m,n}, \quad (5.189)$$

If we assume now that $\varepsilon(x,\mathbf{R})$ is the homogeneous Gaussian delta-correlated field with the correlation function

$$B_\varepsilon(x,\mathbf{R}) = A(\mathbf{R})\delta(x), \quad A(\mathbf{R}) = \int\limits_{-\infty}^{\infty}dx B_\varepsilon(x,\mathbf{R}),$$

so that functional $\Theta\left[x;\psi(\xi,\mathbf{R}')\right]$ has the form

$$\Theta\left[x;\psi(\xi,\mathbf{R}')\right] = -\frac{1}{2}\int_0^x d\xi \int d\mathbf{R}' \int d\mathbf{R} A(\mathbf{R}'-\mathbf{R})\psi(\xi,\mathbf{R}')\psi(\xi,\mathbf{R}),$$

then Eq. (5.188) assumes the closed operator form [132, 134, 135, 268, 295]

$$\begin{aligned}\frac{\partial}{\partial x}\Phi[x;v,v^*] &= -\frac{k^2}{8}\int d\mathbf{R}' \int d\mathbf{R} A(\mathbf{R}'-\mathbf{R})\hat{M}(\mathbf{R}')\hat{M}(\mathbf{R})\Phi[x;v,v^*] \\ &+\frac{i}{2k}\left\{\int d\mathbf{R}'\left[v(\mathbf{R}')\Delta_{\mathbf{R}'}\frac{\delta}{\delta v(\mathbf{R}')} - v^*(\mathbf{R}')\Delta_{\mathbf{R}'}\frac{\delta}{\delta v^*(\mathbf{R}')}\right]\right\}\Phi[x;v,v^*], \qquad (5.190)\end{aligned}$$

and Eqs. (5.189) for the moment functions of wavefield $u(x,\mathbf{R})$ assume the form

$$\frac{\partial}{\partial x}M_{m,n} = \frac{i}{2k}\left(\sum_{p=1}^{m}\Delta_{\mathbf{R}_p} - \sum_{q=1}^{n}\Delta_{\mathbf{R}'_q}\right)M_{m,n} - \frac{k^2}{8}Q(\mathbf{R}_1,...,\mathbf{R}_m;\mathbf{R}'_1,...,\mathbf{R}'_m)M_{m,n}, \qquad (5.191)$$

where

$$\begin{aligned}&Q(\mathbf{R}_1,...,\mathbf{R}_m;\mathbf{R}'_1,...,\mathbf{R}'_m) \\ &= \sum_{i=1}^{m}\sum_{j=1}^{m}A(\mathbf{R}_i-\mathbf{R}_j) - 2\sum_{i=1}^{m}\sum_{j=1}^{n}A(\mathbf{R}_i-\mathbf{R}'_j) + \sum_{i=1}^{n}\sum_{j=1}^{n}A(\mathbf{R}'_i-\mathbf{R}'_j). \qquad (5.192)\end{aligned}$$

Remark 2 ***Another derivation of Eqs.*** (5.189) ***and*** (5.191).

In the case of the delta-correlated fluctuations of medium parameters, there is another, physically more clear way of deriving Eqs. (5.189) and (5.191) for the moment functions of wavefield $u(x,\mathbf{R})$ [130, 132].

As was mentioned earlier, field $u(x,\mathbf{R})$ depends functionally only on the preceding values of field $\varepsilon(\xi,\mathbf{R}')$, i.e., for $\xi \leqslant x$. However, in the general case, there is statistical relationship between field $u(x,\mathbf{R})$ and subsequent values of field $\varepsilon(\xi,\mathbf{R}')$ for $\xi \geqslant x$. In the approximation of the delta-correlated fluctuations of medium parameters, this statistical relationship disappears, and fields $u(\xi_i,\mathbf{R})$ for $\xi_i < x$ are independent of $\varepsilon(\eta_j,\mathbf{R}')$ for $\eta_j > x$ not only functionally, but also statistically; i.e., for $\xi_i < x; \eta_j > x$, the following equality holds:

$$\left\langle\prod_{i,j}u(\xi_i,\mathbf{R}_i)\varepsilon(\eta_j,\mathbf{R}_j)\right\rangle = \left\langle\prod_{i}u(\xi_i,\mathbf{R}_i)\right\rangle\left\langle\prod_{j}\varepsilon(\eta_j,\mathbf{R}_j)\right\rangle. \qquad (5.193)$$

Using Eq. (5.193), we can easily obtain the equations for statistical moments of field $u(x,\mathbf{R})$. We derive the equation for average field $\langle u(x,\mathbf{R})\rangle$ as an example. With this goal in view, we rewrite the initial stochastic equation (5.13) in the form of the integral equation

$$\begin{aligned}u(x,\mathbf{R}) &= u_0(\mathbf{R})\exp\left\{i\frac{k}{2}\int_0^x d\xi\varepsilon(\xi,\mathbf{R})\right\} \\ &+\frac{i}{2k}\int_0^x d\xi \exp\left\{i\frac{k}{2}\int_\xi^x d\eta\varepsilon(\eta,\mathbf{R})\right\}\Delta_{\mathbf{R}}u(\xi,\mathbf{R}). \qquad (5.194)\end{aligned}$$

Averaging Eq. (5.194) over an ensemble of realizations of random field $\varepsilon(\xi, \mathbf{R})$, we take into account Eq. (5.193) to obtain the closed integral equation

$$\langle u(x,\mathbf{R})\rangle = u_0(\mathbf{R})\left\langle \exp\left\{i\frac{k}{2}\int\limits_0^x d\xi\varepsilon(\xi,\mathbf{R})\right\}\right\rangle$$
$$+\frac{i}{2k}\int\limits_0^x d\xi\left\langle \exp\left\{i\frac{k}{2}\int\limits_\xi^x d\eta\varepsilon(\eta,\mathbf{R})\right\}\right\rangle \Delta_{\mathbf{R}}\langle u(\xi,\mathbf{R})\rangle. \quad (5.195)$$

To transform the integral equation into the differential equation, we use the fact that the equality

$$\left\langle \exp\left\{i\frac{k}{2}\int\limits_0^x d\xi\varepsilon(\xi,\mathbf{R})\right\}\right\rangle = \left\langle \exp\left\{i\frac{k}{2}\int\limits_0^\xi d\eta\varepsilon(\eta,\mathbf{R})\right\}\right\rangle\left\langle \exp\left\{i\frac{k}{2}\int\limits_\xi^x d\eta\varepsilon(\eta,\mathbf{R})\right\}\right\rangle,$$

holds in the case of the delta-correlated fluctuations of medium parameter for any point $0 \leqslant \xi \leqslant x$. Thus, introducing function

$$\Phi(x,\mathbf{R}) = \left\langle \exp\left\{i\frac{k}{2}\int\limits_0^x d\eta\varepsilon(\eta,\mathbf{R})\right\}\right\rangle,$$

we can rewrite Eq. (5.195) in the form

$$\langle u(x,\mathbf{R})\rangle = u_0(\mathbf{R})\Phi(x,\mathbf{R}) + \frac{i}{2k}\int\limits_0^x d\xi\frac{\Phi(x,\mathbf{R})}{\Phi(\xi,\mathbf{R})}\Delta_{\mathbf{R}}\langle u(\xi,\mathbf{R})\rangle, \quad (5.196)$$

from which easily follows the differential equation for $\langle u(x,\mathbf{R})\rangle$

$$\frac{\partial}{\partial x}\langle u(x,\mathbf{R})\rangle = \frac{i}{2k}\Delta_{\mathbf{R}}\langle u(x,\mathbf{R})\rangle + \langle u(x,\mathbf{R})\rangle\frac{\partial}{\partial x}\ln\Phi(x,\mathbf{R}), \quad u(0,\mathbf{R}) = u_0(\mathbf{R})$$

coinciding with Eq. (5.189) for $m = 1, n = 0$. Equations for the higher-order moments of field $u(x,\mathbf{R})$ can be derived similarly. ♦

Random forces in hydrodynamic turbulence

In the case of the hydrodynamic equation (5.22) under the assumption that random field $\mathbf{f}(\mathbf{x},t)$ is homogeneous in space and stationary and delta-correlated in time, Eq. (5.25), page 103 for the characteristic functional of the Fourier transform of the velocity field

$$\Phi[t;\mathbf{z}(\mathbf{k}')] = \Phi[t;\mathbf{z}] = \langle\varphi[t;\mathbf{z}(\mathbf{k}')]\rangle = \left\langle \exp\left\{i\int d\mathbf{k}'\mathbf{z}(\mathbf{k}')\hat{\mathbf{u}}(\mathbf{k}',t)\right\}\right\rangle$$

assumes the form

$$\frac{\partial}{\partial t}\Phi[t;\mathbf{z}] = \left\langle \dot{\Theta}_t\left[t;\frac{\delta}{i\delta\mathbf{f}(\boldsymbol{\kappa},t)}\right]\varphi[t;\mathbf{z}]\right\rangle$$
$$-\int d\mathbf{k}z_i(\mathbf{k})\left\{\frac{1}{2}\int d\mathbf{k}_1\int d\mathbf{k}_2\Lambda_{i,\alpha\beta}(\mathbf{k}_1,\mathbf{k}_2,\mathbf{k})\frac{\delta^2}{\delta z_\alpha(\mathbf{k}_1)\delta z_\beta(\mathbf{k}_2)} + \nu k^2\frac{\delta}{\delta z_i(\mathbf{k})}\right\}\Phi[t;\mathbf{z}], \quad (5.197)$$

where

$$\dot{\Theta}_t\left[t;\boldsymbol{\psi}(\boldsymbol{\kappa},\tau)\right] = \frac{d}{dt}\ln\left\langle \exp\left\{ i\int\limits_0^t d\tau\int d\boldsymbol{\kappa}\,\hat{\mathbf{f}}(\boldsymbol{\kappa},\tau)\boldsymbol{\psi}(\boldsymbol{\kappa},\tau)\right\}\right\rangle$$

is the derivative of the logarithm of the characteristic functional of external forces $\hat{\mathbf{f}}(\mathbf{k},t)$. By virtue of equality (5.24), page 103

$$\frac{\delta}{\delta\hat{\mathbf{f}}(\mathbf{k},t-0)}\varphi[t;\mathbf{z}] = iz(k)\varphi[t;\mathbf{z}],$$

we can rewrite Eq. (5.197) in the form of the closed equation

$$\begin{aligned}\frac{\partial}{\partial t}\Phi[t;\mathbf{z}] &= \dot{\Theta}_t\left[t;\mathbf{z}(\mathbf{k})\right]\Phi[t;\mathbf{z}] \\ &- \int d\mathbf{k} z_i(\mathbf{k})\left\{\frac{1}{2}\int d\mathbf{k}_1\int d\mathbf{k}_2\Lambda_{i,\alpha\beta}(\mathbf{k}_1,\mathbf{k}_2,\mathbf{k})\frac{\delta^2}{\delta z_\alpha(\mathbf{k}_1)\delta z_\beta(\mathbf{k}_2)} + \nu k^2\frac{\delta}{\delta z_i(\mathbf{k})}\right\}\Phi[t;\mathbf{z}].\end{aligned} \tag{5.198}$$

If we assume now that $\mathbf{f}(\mathbf{x},t)$ is the Gaussian random field homogeneous and isotropic in space and stationary in time with the correlation tensor

$$B_{ij}(\mathbf{x}_1-\mathbf{x}_2,t_1-t_2) = \left\langle f_i(\mathbf{x}_1,t_1)f_j(\mathbf{x}_2,t_2)\right\rangle,$$

then the field $\hat{\mathbf{f}}(\mathbf{k},t)$ will also be the Gaussian stationary random field with the correlation tensor

$$\left\langle \hat{f}_i(\mathbf{k},t+\tau)\hat{f}_j(\mathbf{k}',t)\right\rangle = \frac{1}{2}F_{ij}(\mathbf{k},\tau)\delta(\mathbf{k}+\mathbf{k}),$$

where $F_{ij}(\mathbf{k},\tau)$ is the external force spatial spectrum given by the formula

$$F_{ij}(\mathbf{k},\tau) = 2(2\pi)^3\int d\mathbf{x}B_{ij}(\mathbf{x},\tau)e^{-i\mathbf{k}\mathbf{x}}.$$

In view of the fact that forces are spatially isotropic, we have

$$F_{ij}(\mathbf{k},\tau) = F(k,\tau)\Delta_{ij}(\mathbf{k}).$$

As long as field $\hat{\mathbf{f}}(\mathbf{k},t)$ is delta-correlated in time, we have

$$F(k,\tau) = F(k)\delta(\tau),$$

so that functional $\Theta\left[t;\boldsymbol{\psi}(\boldsymbol{\kappa},\tau)\right]$ is given by the formula

$$\Theta\left[t;\boldsymbol{\psi}(\boldsymbol{\kappa},\tau)\right] = -\frac{1}{4}\int\limits_0^t d\tau\int d\boldsymbol{\kappa}\,F(\kappa)\Delta_{ij}(\boldsymbol{\kappa})\psi_i(\boldsymbol{\kappa},\tau)\psi_j(-\boldsymbol{\kappa},\tau),$$

and Eq. (5.198) assumes the closed form [255]

$$\begin{aligned}\frac{\partial}{\partial t}\Phi[t;\mathbf{z}] &= -\frac{1}{4}\int d\mathbf{k}F(k)\Delta_{ij}(\mathbf{k})z_i(\mathbf{k})z_j(-\mathbf{k})\Phi[t;\mathbf{z}] \\ &- \int d\mathbf{k} z_i(\mathbf{k})\left\{\frac{1}{2}\int d\mathbf{k}_1\int d\mathbf{k}_2\Lambda_{i,\alpha\beta}(\mathbf{k}_1,\mathbf{k}_2,\mathbf{k})\frac{\delta^2}{\delta z_\alpha(\mathbf{k}_1)\delta z_\beta(\mathbf{k}_2)} + \nu k^2\frac{\delta}{\delta z_i(\mathbf{k})}\right\}\Phi[t;\mathbf{z}].\end{aligned} \tag{5.199}$$

Equation (5.199) plays the role of the Fokker–Planck equation of the problem under consideration. The unknown in this equation is the characteristic functional, and this fact distinguishes this equation from the standard equation of this type, where the unknown is the probability density expressed as the Fourier transform of this functional.

Another distinction consists in the fact that Eq. (5.199) is the diffusion equation in the infinite-dimensional space, because of which it is the variational differential equation. The diffusion coefficient can be different for different wave components; it is given by the spectral tensor of external forces $F(k)\Delta_{ij}(\mathbf{k})$.

Remark 3 ***Equilibrium distributions for hydrodynamic flows.***

In the conditions of absent molecular viscosity and random external forces, the problem on evolution of the velocity field specified at the initial moment becomes meaningful. In the context of this problem, the characteristic functional of velocity satisfies the equation

$$\frac{\partial}{\partial t}\Phi[t;\mathbf{z}] = -\frac{1}{2}\int d\mathbf{k} z_i(\mathbf{k})\int d\mathbf{k}_1\int d\mathbf{k}_2 \Lambda_i^{\alpha\beta}(\mathbf{k}_1,\mathbf{k}_2,\mathbf{k})\frac{\delta^2\Phi[t;\mathbf{z}]}{\delta z_\alpha(\mathbf{k}_1)\,\delta z_\beta(\mathbf{k}_2)}. \tag{5.200}$$

Note that this equation was considered by E. Hopf in his the classic paper [116] and is called now the *Hopf equation* (see also [117, 118]). The integro-differential equation

$$\frac{\partial}{\partial t}\widehat{u}_i(\mathbf{k},t) + \frac{i}{2}\int d\mathbf{k}_1\int d\mathbf{k}_2 \Lambda_i^{\alpha\beta}(\mathbf{k}_1,\mathbf{k}_2,\mathbf{k})\,\widehat{u}_\alpha(\mathbf{k}_1,t)\,\widehat{u}_\beta(\mathbf{k}_2,t) = 0,$$

which is the input equation for this problem, describes the motion of the ideal liquid. It can have a number of integrals of motion, which may result in the existence of the solution to Eq. (5.200) steady-state for $t\to\infty$ and independent of initial values. Such a solution is called the equilibrium distribution. For the two-dimensional and three-dimensional velocity fields, these distributions appear significantly different.

Consider Eq. (5.200). In view of multiple nonlinear interactions between different harmonics of random velocity field, we can expect that the steady-state distribution of the velocity field exists for $t\to\infty$ and satisfies the steady-state Hopf equation

$$\int d\mathbf{k} z_i(\mathbf{k})\int d\mathbf{k}_1\int d\mathbf{k}_2 \Lambda_i^{\alpha\beta}(\mathbf{k}_1,\mathbf{k}_2,\mathbf{k})\frac{\delta^2\Phi[\mathbf{z}]}{\delta z_\alpha(\mathbf{k}_1)\,\delta z_\beta(\mathbf{k}_2)} = 0.$$

As was shown in [119], the unique solution to this equation in the class of the Gaussian functionals is the functional

$$\Phi[\mathbf{z}(\mathbf{k})] = \exp\left\{-\frac{\gamma}{2}\int d\mathbf{k}\Delta_{ij}(\mathbf{k})z_i(\mathbf{k})z_j(-\mathbf{k})\right\} \tag{5.201}$$

corresponding to the uniform energy distribution over wave numbers (the white noise).

Note that solution (5.201) can satisfy the initial equation (5.199) if random forces are specially fit to compensate the molecular viscosity. Indeed, substituting functional (5.201) in Eq. (5.199), we see that the term with the second variational derivative vanishes (which is a consequence of the fact that the integral of motion — energy — exists in the case of the ideal liquid) and other terms — they correspond to the linearized initial value problem — are mutually cancelled only if

$$F_{ij}(\mathbf{k}) = 4\nu\gamma k^2\Delta_{ij}(\mathbf{k}). \tag{5.202}$$

This relationship corresponds to the so-called *fluctuation–dissipation theorem* for hydrodynamic flows.

In the case of the two-dimensional perfect liquid, the second integral of motion quadratic in velocities appears available in addition to the energy integral; it is the square of the vorticity of the velocity field. In this case, there appears the equilibrium distribution different from the white noise (5.202) and characterized by a number of features, the main of which consists in the existence of coherent structures whose energy is described by spectral density proportional to the delta-function [129].

In the simplest case, the incompressible liquid flow in the two-dimensional plane $\mathbf{R} = (x, y)$ is described by the stream function $\psi(\mathbf{R}, t)$ satisfying Eq. (1.101), page 35 that has, in the absence of the Coriolis forces and topographic inhomogeneities of underlying surface, the following form

$$\frac{\partial}{\partial t}\Delta\psi(\mathbf{R}, t) = J\left\{\Delta\psi(\mathbf{R}, t); \psi(\mathbf{R}, t)\right\}, \quad \psi(\mathbf{R}, 0) = \psi_0(\mathbf{R}), \tag{5.203}$$

where

$$J\left\{\psi(\mathbf{R}, t); \varphi(\mathbf{R}, t)\right\} = \frac{\partial\psi(\mathbf{R}, t)}{\partial x}\frac{\partial\varphi(\mathbf{R}, t)}{\partial y} - \frac{\partial\varphi(\mathbf{R}, t)}{\partial x}\frac{\partial\psi(\mathbf{R}, t)}{\partial y}$$

is the Jacobian of two functions.

Nonlinear interactions must bring the hydrodynamic system (5.203) to statistical equilibrium. In view of the fact that establishing this equilibrium requires a great number of interactions between the disturbances of different scales, we can suppose that, in the simplest case of statistically homogeneous and isotropic initial random field $\psi_0(\mathbf{R})$, this distribution will be the Gaussian distribution, so that our task consists in the determination of this distribution parameters. During the evolution, random stream function $\psi(\mathbf{R}, t)$ remains a homogeneous and isotropic function. Because the stream function is defined to an additive constant, we can describe its statistical characteristics by the one-time structure function

$$D_{\psi}(\mathbf{R} - \mathbf{R}', t) = \left\langle \left[\psi(\mathbf{R}, t) - \psi(\mathbf{R}', t)\right]^2 \right\rangle = 2\left[B_{\psi}(0, t) - B_{\psi}(\mathbf{R} - \mathbf{R}', t)\right],$$

where

$$B_{\psi}(\mathbf{R} - \mathbf{R}', t) = \langle \psi(\mathbf{R}, t)\psi(\mathbf{R}', t) \rangle$$

is the spatial correlation function of field $\psi(\mathbf{R}, t)$.

Under the assumption that we seek the steady-state (equilibrium) distribution on the class of the Gaussian distributions of statistically homogeneous and isotropic field $\psi(\mathbf{R}, t)$ described by the structure function $D_{\psi}(R) = \lim_{t\to\infty} D_{\psi}(\mathbf{R}, t)$, we can obtain the equation for this structure function

$$\left(\Delta_q + \lambda\right)\Delta_q^2 D_{\psi}(q) = 0, \tag{5.204}$$

where λ is the separation constant with the dimension of the inverse square of length and Δ_q is the radial part of the Laplace operator.

There are two possible solutions to Eq. (5.204), depending on whether constant λ is positive ($\lambda = k_0^2 > 0$) or negative ($\lambda = -k_0^2 < 0$).

If $\lambda = k_0^2 > 0$, Eq. (5.204) can be reduced to the equation

$$\Delta_q D_{\psi}(q) = CJ_0(k_0 q),$$

where $J_0(z)$ is the Bessel function of the first kind. In this case, structure function $D_\psi(q)$ is determined as the solution to the Laplace equation, and we obtain the spectral density of energy in the form

$$E(k) = E\delta(k - k_0).$$

The delta-like behavior of spectral density is evidence of the fact that fields $\psi(\mathbf{R}, t)$ are highly correlated, which suggests that coherent structures can exist in the developed turbulent flow of the two-dimensional liquid (in the sense of the existence of the corresponding eigenfunctions slowly decaying with distance).

In the case $(\lambda = -k_0^2 < 0)$, Eq. (5.204) can be reduced to the similar equation

$$\Delta_q D_\psi(q) = CK_0(k_0 q).$$

However, the right-hand side of this equation is proportional to the McDonalds function $K_0(z)$ with the dimensional parameters k_0 and C. The corresponding spectral density of energy is now given by the formula [201, 202, 203, 247]

$$E(k) = \frac{E_0}{k^2 + k_0^2}.$$

The behavior of density $E(k)$ is characterized by the logarithmic divergence of the average kinetic energy, which is not surprising because our model neglects the viscous dissipation.

The steady-state solution to the initial dynamic equation (5.203) satisfies the equation

$$\Delta\psi(\mathbf{R}) = F(\psi(\mathbf{R})),$$

where $F(\psi(\mathbf{R}))$ is the arbitrary function determined from boundary conditions at infinity. In the simplest case of the *Fofonoff flow* [66] corresponding to the linear function $F(\psi(\mathbf{R})) = -\lambda\psi(\mathbf{R})$, this equation assumes the form

$$\Delta\psi(\mathbf{R}) = -\lambda\psi(\mathbf{R}). \tag{5.205}$$

Considering formally Eq. (5.205) as the stochastic equation, we can easily obtain that the structure function of field $\psi(\mathbf{R})$ satisfies the equation coinciding with Eq. (5.204). This means that the Gaussian equilibrium state is statistically equivalent to the stochastic Fofonoff flow of the liquid. Of course, the realizations of dynamic systems (5.203) and (5.205) are different. Thus, despite strong nonlinearity of the input equation (5.203), the equilibrium regime (for $t \to \infty$) appears statistically equivalent to the linear equation in which the nonlinear interactions are absent.

Equilibrium states for quasi-geostrophic flows described by Eqs. (1.101) and (1.102), page 35 that includes the random topography of underlying surface can be considered similarly [144, 155, 156].

A characteristic feature of all above solutions consists in the fact that they predict the possibility for coherent states to exist in the developed turbulent flow. Nothing can be said about the stability of these states. However, we note that the above Gaussian equilibrium ensemble forms the natural noise in a number of geophysical systems described in the quasi-geostrophic approximation and is similar to the thermal noise in the statistical physics. For this reason, this noise may play very important and sometimes determinative role in the statistical theory of quasi-geostrophic flows of liquid. ♦

Chapter 6

Stochastic equations with the Markovian fluctuations of parameters

In the preceding chapter, we dealt with the statistical description of dynamic systems in terms of the general methods that assumed the knowledge of the characteristic functional of fluctuating parameters. However, this functional is unknown in most cases, and we are forced to resort either to assumptions on the model of parameter fluctuations, or to asymptotic approximations.

The methods based on approximating the fluctuating parameters with the Markovian random processes and fields with a finite temporal correlation radius are widely used. Such approximations can be found, for example, as solutions to the dynamic equations with delta-correlated parameter fluctuations. Consider such methods in greater detail using the Markovian random processes as an example [133]–[135].

Consider stochastic equations of the form

$$\frac{d}{dt}\mathbf{x}(t) = f\left(t, \mathbf{x}, \mathbf{z}(t)\right), \quad \mathbf{x}(0) = \mathbf{x}_0, \tag{6.1}$$

where $f\left(t, \mathbf{x}, \mathbf{z}(t)\right)$ is the deterministic function of its arguments and $\mathbf{z}(t) = \{z_1(t), ..., z_n(t)\}$ is the Markovian vector process whose transition probability density satisfies the equation (see Chapter 3, page 73)

$$\frac{\partial}{\partial t}p(\mathbf{z}, t|\mathbf{z}_0, t_0) = \hat{L}(\mathbf{z})p(\mathbf{z}, t|\mathbf{z}_0, t_0).$$

In this equation, operator $\hat{L}(\mathbf{z})$ is called the *kinetic operator*.

Our task consists in the determination of statistical characteristics of the solution to Eq. (6.1) from known statistical characteristics of process $\mathbf{z}(t)$, for example, from the kinetic operator $\hat{L}(\mathbf{z})$.

In the general case of arbitrary Markovian process $\mathbf{z}(t)$, we cannot judge about process $\mathbf{x}(t)$. We can only assert that the joint process $\{\mathbf{x}(t), \mathbf{z}(t)\}$ is the Markovian process. Indeed, as we showed in Chapter 4, page 87 the following differentiation formula

$$\begin{aligned}
&\frac{d}{dt}\left\langle \delta(\mathbf{z}(t)-\mathbf{z})R[t;\mathbf{z}(\tau)]\right\rangle \\
&\quad = \left\langle \delta(\mathbf{z}(t)-\mathbf{z})\frac{d}{dt}R[t;\mathbf{z}(\tau)]\right\rangle + \hat{L}(\mathbf{z})\left\langle \delta(\mathbf{z}(t)-\mathbf{z})R[t;\mathbf{z}(\tau)]\right\rangle,
\end{aligned} \tag{6.2}$$

holds for arbitrary functional $R[t;\mathbf{z}(\tau)]$, $\tau \leq t$ if $\mathbf{z}(t)$ is the Markovian process. Multiplying Eq. (6.2) by arbitrary function $F(\mathbf{z})$ and integrating the result over $\mathbf{z}$, we obtain another representation of the differentiation formula

$$\frac{d}{dt}\left\langle F\left(\mathbf{z}(t)\right) R[t;\mathbf{z}(\tau)]\right\rangle$$
$$= \left\langle F\left(\mathbf{z}(t)\right)\frac{d}{dt}R[t;\mathbf{z}(\tau)]\right\rangle + \left\langle R[t;\mathbf{z}(\tau)]\left[\hat{L}^{+}(\mathbf{z})F\left(\mathbf{z}(t)\right)\right]\right\rangle, \tag{6.3}$$

where $\hat{L}^{+}(\mathbf{z})$ is the operator conjugated to operator $\hat{L}(\mathbf{z})$.

Now, we specify functional $R[t;\mathbf{z}(\tau)]$ in the form of the indicator function

$$R[t,\mathbf{x};\mathbf{z}(\tau)] = \delta(\mathbf{x}(t) - \mathbf{x}),$$

where $\mathbf{x}(t)$ is the solution to Eq. (6.1). In this case, function $R[t,\mathbf{x};\mathbf{z}(\tau)]$ satisfies the equation

$$\frac{\partial}{\partial t}R[t,\mathbf{x};\mathbf{z}(\tau)] = -\frac{\partial}{\partial x_i}f_i(t,\mathbf{x},\mathbf{z})R[t,\mathbf{x};\mathbf{z}(\tau)],$$

which is the stochastic Liouville equation for our problem. Note that the correlator

$$\langle\delta(\mathbf{z}(t) - \mathbf{z})R[t,\mathbf{x};\mathbf{z}(\tau)]\rangle = P(\mathbf{x},\mathbf{z},t)$$

appears in this case the one-point joint probability density of processes $\mathbf{x}(t)$ and $\mathbf{z}(t)$. Consequently, the differentiation formula (6.2) assumes the form of the closed equation for the one-point probability density

$$\frac{\partial}{\partial t}P(\mathbf{x},\mathbf{z},t) = -\frac{\partial}{\partial x_i}f_i(t,\mathbf{x},\mathbf{z})P(\mathbf{x},\mathbf{z},t) + \hat{L}(\mathbf{z})P(\mathbf{x},\mathbf{z},t). \tag{6.4}$$

It is obvious that the transition probability density of the joint process $\{\mathbf{x}(t),\, \mathbf{z}(t)\}$ also satisfies Eq. (6.4), which means that process $\{\mathbf{x}(t),\, \mathbf{z}(t)\}$ is the Markovian process. If we would able to solve Eq. (6.4), then we could integrate the solution over $\mathbf{z}$ to obtain the probability density of the solution to Eq. (6.1), i.e., function $P(\mathbf{x},t)$. In this case, process $\mathbf{x}(t)$ would not be the Markovian process.

There are several types of processes $\mathbf{z}(t)$ that allow obtaining equations for density $P(\mathbf{x},t)$ without solving Eq. (6.4) for $P(\mathbf{x},\mathbf{z},t)$. Among these processes, we mention first of all the telegrapher's and generalized telegrapher's processes, Markovian processes with finite number of states, and Gaussian Markovian processes. Below, we discuss these processes in more detail as examples of processes widely used in different applications.

6.1 Telegrapher's processes

Recall that telegrapher's random process $z(t)$ (the two-state, or binary process) is defined by the equality

$$z(t) = a(-1)^{n(0,t)},$$

where random quantity a assumes values $a = \pm a_0$ with probabilities $1/2$ and $n(t_1,t_2)$, $t_1 < t_2$ is the Poisson integer-valued process with average value $\overline{n(t_1,t_2)} = \nu|t_1 - t_2|$.

Telegrapher's process $z(t)$ is stationary in time and its correlation function

$$\langle z(t)z(t')\rangle = a_0^2 e^{-2\nu|t-t'|}$$

has the temporal correlation radius $\tau_0 = 1/(2\nu)$.

For splitting the correlation between telegrapher's process $z(t)$ and arbitrary functional $R[t; z(\tau)]$, where $\tau \le t$, we obtained the relationship (4.31), page 83

$$\langle z(t) R[t; z(\tau)] \rangle = a_0^2 \int_0^t dt_1 e^{-2\nu(t-t_1)} \left\langle \frac{\delta}{\delta z(t_1)} \tilde{R}[t; z(\tau)] \right\rangle, \qquad (6.5)$$

where functional $\tilde{R}[t; z(\tau)]$ is given by the formula

$$\tilde{R}[t, t_1; z(\tau)] = R[t; z(\tau)\theta(t_1 - \tau + 0)] \quad (t_1 < t). \qquad (6.6)$$

Formula (6.5) is appropriate for analyzing stochastic equations linear in process $z(t)$. Let functional $R[t; z(\tau)]$ is the solution to a system of differential equations of the first order in time with initial values at $t = 0$. Functional $\tilde{R}[t, t_1; z(\tau)]$ will also satisfy the same system of equations with product $z(t)\theta(t_1 - t)$ instead of $z(t)$. Consequently, we obtain that functional $\tilde{R}[t, t_1; z(\tau)] = R[t; 0]$ for all times $t > t_1$; moreover, it satisfies the same system of equations for absent fluctuations (i.e., at $z(t) = 0$) with the initial value $\tilde{R}[t_1, t_1; z(\tau)] = R[t_1; z(\tau)]$.

Another formula convenient in the context of stochastic equations linear in random telegrapher's process $z(t)$ concerns the differentiation of the correlation of this process with arbitrary functional $R[t; z(\tau)]$ $(\tau \le t)$ (4.37), page 85

$$\frac{d}{dt} \langle z(t) R[t; z(\tau)] \rangle = -2\nu \langle z(t) R[t; z(\tau)] \rangle + \left\langle z(t) \frac{d}{dt} R[t; z(\tau)] \right\rangle. \qquad (6.7)$$

In addition, we have the equality

$$\langle z(t') R[t; z(\tau)] \rangle = e^{-2\nu|t-t'|} \langle z(t) R[t; z(\tau)] \rangle, \quad t' \geqslant t, \quad \tau \leqslant t. \qquad (6.8)$$

Formula (6.7) determines the rule of factoring the differentiation operation out of averaging brackets

$$\left\langle z(t) \frac{d^n}{dt^n} R[t; z(\tau)] \right\rangle = \left(\frac{d}{dt} + 2\nu \right)^n \langle z(t) R[t; z(\tau)] \rangle. \qquad (6.9)$$

We consider some special examples to show the usability of these formulas. It is evident that both methods give the same result. However, the method based on the differentiation formula appears more practicable.

6.1.1 System of linear operator equations

The first example concerns the system of linear operator equations

$$\frac{d}{dt} \mathbf{x}(t) = \hat{A}(t) \mathbf{x}(t) + z(t) \hat{B}(t) \mathbf{x}(t), \quad \mathbf{x}(0) = \mathbf{x}_0, \qquad (6.10)$$

where $\hat{A}(t)$ and $\hat{B}(t)$ are certain differential operators (they may include differential operators with respect to auxiliary variables). If operators $\hat{A}(t)$ and $\hat{B}(t)$ are matrixes, then Eqs. (6.10) describe the linear dynamic system.

Average Eqs. (6.10) over an ensemble of random functions $z(t)$. The result will be the equation

$$\frac{d}{dt} \langle \mathbf{x}(t) \rangle = \hat{A}(t) \langle \mathbf{x}(t) \rangle + \hat{B}(t) \boldsymbol{\psi}(t), \quad \langle \mathbf{x}(0) \rangle = \mathbf{x}_0, \qquad (6.11)$$

where we introduced new functions

$$\boldsymbol{\psi}(t) = \langle z(t)\mathbf{x}(t)\rangle .$$

We can use formula (6.7) for these functions; as a result, we obtain the equality

$$\frac{d}{dt}\boldsymbol{\psi}(t) = -2\nu\boldsymbol{\psi}(t) + \left\langle z(t)\frac{d}{dt}\mathbf{x}(t)\right\rangle . \tag{6.12}$$

Substituting now derivative $d\mathbf{x}/dt$ (6.10) in Eq. (6.12), we obtain the equation for the vector function $\boldsymbol{\psi}(t)$

$$\left(\frac{d}{dt} + 2\nu\right)\boldsymbol{\psi}(t) = \hat{A}(t)\boldsymbol{\psi}(t) + \hat{B}(t)\left\langle z^2(t)\mathbf{x}(t)\right\rangle . \tag{6.13}$$

Because $z^2(t) \equiv a_0^2$ for telegrapher's process, we obtain finally the closed system of linear equations for vectors $\langle \mathbf{x}(t)\rangle$ and $\boldsymbol{\psi}(t)$

$$\begin{aligned}
&\frac{d}{dt}\langle \mathbf{x}(t)\rangle = \hat{A}(t)\langle \mathbf{x}(t)\rangle + \hat{B}(t)\boldsymbol{\psi}(t), \quad \langle \mathbf{x}(0)\rangle = \mathbf{x}_0,\\
&\left(\frac{d}{dt} + 2\nu\right)\boldsymbol{\psi}(t) = \hat{A}(t)\boldsymbol{\psi}(t) + a_0^2\hat{B}(t)\langle \mathbf{x}(t)\rangle , \quad \boldsymbol{\psi}(0) = (0).
\end{aligned} \tag{6.14}$$

If operators $\hat{A}(t)$ and $\hat{B}(t)$ are the time-independent matrixes A and B, we can solve system (6.14) using the Laplace transform. After the Laplace transform, system (6.14) becomes the algebraic system of equations

$$\begin{aligned}
&(pE - A)\langle \mathbf{x}\rangle_p - B\boldsymbol{\psi}_p = \mathbf{x}_0,\\
&[(p + 2\nu)E - A]\boldsymbol{\psi}_p - a_0^2 B\langle \mathbf{x}\rangle_p = 0,
\end{aligned} \tag{6.15}$$

where E is the unit matrix. From this system, we obtain solution $\langle \mathbf{x}\rangle_p$ in the form

$$\langle \mathbf{x}\rangle_p = \left[(pE - A) - a_0^2 B\frac{1}{(p + 2\nu)E - A}B\right]^{-1}\mathbf{x}_0. \tag{6.16}$$

Stochastic parametric resonance Consider the problem on the statistical description of an oscillator with fluctuating frequency as a simple example of the linear dynamic system (6.10). This problem is formulated as the second-order equation (1.15), page 10 with initial values

$$\begin{aligned}
&\frac{d^2}{dt^2}x(t) + \omega_0^2[1 + z(t)]x(t) = 0,\\
&x(0) = x_0, \quad \left.\frac{d}{dt}x(t)\right|_{t=0} = y_0,
\end{aligned} \tag{6.17}$$

which is equivalent to the system of equations

$$\begin{aligned}
&\frac{d}{dt}x(t) = y(t), \quad \frac{d}{dt}y(t) = -\omega_0^2[1 + z(t)]x(t),\\
&x(0) = x_0, \quad y(0) = y_0.
\end{aligned} \tag{6.18}$$

If our interest concerns only the average value of the solution to statistical problem (6.17), we can deal without rewriting it in the form of the system of equations (6.18). Averaging Eq. (6.17) over an ensemble of realizations $z(t)$, we obtain the unclosed equation

$$\left(\frac{d^2}{dt^2}+\omega_0^2\right)\langle x(t)\rangle+\omega_0^2\langle z(t)x(t)\rangle=0. \tag{6.19}$$

To split the correlator in the right-hand side of Eq. (6.19), we multiply Eq. (6.17) by function $z(t)$ and average the result to obtain the equation

$$\left\langle z(t)\left(\frac{d^2}{dt^2}+\omega_0^2\right)x(t)\right\rangle+\omega_0^2a_0^2\langle x(t)\rangle=0. \tag{6.20}$$

Deriving Eq. (6.20), we took into account that quantity $z^2(t)\equiv a_0^2$ is not random in the case of telegrapher's process.

Then, we use the rule of factoring the derivative out of averaging brackets (6.9), page 152 to rewrite Eq. (6.20) in the form

$$\left[\left(\frac{d}{dt}+2\nu\right)^2+\omega_0^2\right]\langle z(t)x(t)\rangle+\omega_0^2a_0^2\langle x(t)\rangle=0. \tag{6.21}$$

Now, Eqs. (6.19) and (6.21) form the closed system of equations.

From Eq. (6.21), we obtain

$$\langle z(t)x(t)\rangle=\omega_0a_0^2\int_0^t dt'e^{-2\nu(t-t')}\sin\omega_0(t-t')\langle x(t')\rangle.$$

Consequently, Eq. (6.19) can be represented in the form of the integro-differential equation

$$\left(\frac{d^2}{dt^2}+\omega_0^2\right)\langle x(t)\rangle+\omega_0^3a_0^2\int_0^t dt'e^{-2\nu(t-t')}\sin\omega_0(t-t')\langle x(t')\rangle=0. \tag{6.22}$$

We can again use the Laplace transform to solve either the system of equations (6.19) and (6.21) or Eq. (6.22); in both cases, the solution has the form

$$\langle x\rangle_p=F(p)\frac{L(p+2\nu)}{L(p)L(p+2\nu)-\omega_0^4a_0^2}=F(p)\frac{1}{p^2+\omega_0^2-\frac{\omega_0^4a_0^2}{(p+2\nu)^2+\omega_0^2}}, \tag{6.23}$$

where

$$F(p)=px_0+y_0,\quad L(p)=p^2+\omega_0^2.$$

Under the conditions

$$\omega_0\ll 2\nu,\quad \frac{\omega_0^2a_0^2}{4\nu^2}\ll 1,$$

solution (6.23) grades into the Laplace transform of Eq. (5.165), i.e., corresponds to the Gaussian random process $z(t)$ delta-correlated in time.

Consider now the problem on the second moments of the solution to Eq. (6.17). Here, the use of system of equations (6.18) appears necessary. In a way similar to the above derivation of the system of equations (6.19) and (6.21), we obtain the system of six equations for second moments

$$\begin{aligned}
&\frac{d}{dt}\left\langle x^2(t)\right\rangle = 2\left\langle x(t)y(t)\right\rangle, \\
&\frac{d}{dt}\left\langle x(t)y(t)\right\rangle = \left\langle y^2(t)\right\rangle - \omega_0^2\left\langle x^2(t)\right\rangle - \omega_0^2\left\langle z(t)x^2(t)\right\rangle, \\
&\frac{d}{dt}\left\langle y^2(t)\right\rangle = -2\omega_0^2\left\langle x(t)y(t)\right\rangle - 2\omega_0^2\left\langle z(t)x(t)y(t)\right\rangle; \\
&\left(\frac{d}{dt}+2\nu\right)\left\langle z(t)x^2(t)\right\rangle = 2\left\langle z(t)x(t)y(t)\right\rangle, \\
&\left(\frac{d}{dt}+2\nu\right)\left\langle z(t)x(t)y(t)\right\rangle = \left\langle z(t)y^2(t)\right\rangle - \omega_0^2\left\langle z(t)x^2(t)\right\rangle - \omega_0^2 a_0^2\left\langle x^2(t)\right\rangle, \\
&\left(\frac{d}{dt}+2\nu\right)\left\langle z(t)y^2(t)\right\rangle = -2\omega_0^2\left\langle z(t)x(t)y(t)\right\rangle - 2\omega_0^2 a_0^2\left\langle x(t)y(t)\right\rangle.
\end{aligned} \tag{6.24}$$

System of equations (6.24) allows one to obtain closed systems for every unknown function $\langle x^2(t)\rangle$, $\langle x(t)y(t)\rangle$, and $\langle y^2(t)\rangle$. For example, the average value of the potential energy $\langle U(t)\rangle$, where $U(t) = x^2(t)$, satisfies the closed system of two equations (every of which is the third-order equation)

$$\begin{aligned}
&\frac{d^3}{dt^3}\left\langle U(t)\right\rangle + 4\omega_0^2\frac{d}{dt}\left\langle U(t)\right\rangle + 4\omega_0^2\left(\frac{d}{dt}+\nu\right)\left\langle z(t)U(t)\right\rangle = 0, \\
&\left(\frac{d}{dt}+2\nu\right)\left\{\left(\frac{d}{dt}+2\nu\right)^2 + 4\omega_0^2\right\}\left\langle z(t)U(t)\right\rangle + 4\omega_0^2 a_0^2\left(\frac{d}{dt}+\nu\right)\left\langle U(t)\right\rangle = 0.
\end{aligned} \tag{6.25}$$

It is clear that we could obtain system (6.25) without deriving the complete system of equations (6.24). Indeed, random quantity $U(t)$ satisfies the stochastic third-order equation (5.168)

$$\frac{d^3}{dt^3}U(t) + 4\omega_0^2\frac{d}{dt}U(t) + 2\omega_0^2\left(z(t)\frac{d}{dt}U(t) + \frac{d}{dt}z(t)U(t)\right) = 0 \tag{6.26}$$

with the initial value that can generally depend on process $z(t)$ and its derivatives. Averaging Eq. (6.26) over an ensemble of random process realizations and using rule (6.9), page 152 to factor the derivative out of averaging brackets, we obtain the first equation of system (6.25). Then, multiplying Eq. (6.26) by $z(t)$ and using again the rule (6.9), we obtain the second equation of system (6.25).

Systems of equations (6.24) and (6.25) can be solved using the Laplace transform. For example, in the case of the conditions $x(0) = 0, y(0) = y_0$, we have

$$U(0) = x_0^2, \quad \left.\frac{d}{dt}U(t)\right|_{t=0} = 0, \quad \left.\frac{d^2}{dt^2}U(t)\right|_{t=0} = 2y_0^2, \tag{6.27}$$

and we obtain the solution of Eqs. (6.25) in the form

$$\begin{aligned}
\langle U\rangle_p &= 2y_0^2\frac{L(p+2\nu)}{L(p)L(p+2\nu) - a_0^2 M^2(p)}, \\
L(p) &= p\left(p^2+4\omega_0^2\right), \quad M(p) = 4\omega_0^2\left(p^2+\nu\right).
\end{aligned} \tag{6.28}$$

In the limiting case of great parameters ν and a_0^2, but finite ratio $a_0^2/2\nu = \sigma^2\tau_0$, we obtain from the second equation of system (6.25)

$$\langle z(t)U(t)\rangle = -\frac{\omega_0^2\sigma^2\tau_0}{\nu}\langle U(t)\rangle .$$

Consequently, average potential energy $\langle U(t)\rangle$ satisfies in this limiting case the closed third-order equation

$$\frac{d^3}{dt^3}\langle U(t)\rangle + 4\omega_0^2\frac{d}{dt}\langle U(t)\rangle - 4\omega_0^4\sigma^2\tau_0\langle U(t)\rangle = 0,$$

which coincides with Eq. (5.167) and corresponds to the Gaussian delta-correlated process $z(t)$.

The system of equations for correlation functions $\langle x(t)x(t')\rangle$ and $\langle y(t)x(t')\rangle$ for $t > t'$ can be obtained in a way similar to the derivation of Eqs. (6.24); it has, obviously, the form

$$\begin{aligned}
&\frac{d}{dt}\langle x(t)x(t')\rangle = \langle y(t)x(t')\rangle ,\\
&\frac{d}{dt}\langle y(t)x(t')\rangle = -\omega_0^2\langle x(t)x(t')\rangle - \omega_0^2\langle z(t)x(t)x(t')\rangle ,\\
&\left(\frac{d}{dt}+2\nu\right)\langle z(t)x(t)x(t')\rangle = \langle z(t)y(t)x(t')\rangle ,\\
&\left(\frac{d}{dt}+2\nu\right)\langle z(t)y(t)x(t')\rangle = -\omega_0^2\langle z(t)x(t)x(t')\rangle - \omega_0^2a_0^2\langle x(t)x(t')\rangle .
\end{aligned}$$

The initial values for this system are obtained as the solution to system (6.24) at $t = t'$. In a similar way, one can derive the second pair of equations for correlation functions $\langle x(t)y(t')\rangle$, $\langle y(t)y(t')\rangle$ for $t > t'$. In the limit $\nu \to \infty$, $a_0^2 \to \infty$, but finite ratio $a_0^2/2\nu = \sigma^2\tau_0$, we revert to systems of equations (5.158) and (5.160), which correspond to the Gaussian delta-correlated process $z(t)$.

6.1.2 One-dimension nonlinear differential equation

Consider now the nonlinear one-dimensional equation

$$\frac{d}{dt}x(t) = f(x,t) + z(t)g(x,t), \quad x(0) = x_0. \tag{6.29}$$

In this case, the indicator function $\varphi(x,t) = \delta(x(t) - x)$ satisfies the stochastic Liouville equation

$$\frac{\partial}{\partial t}\varphi(x,t) = -\frac{\partial}{\partial x}f(x,t)\varphi(x,t) - z(t)\frac{\partial}{\partial x}g(x,t)\varphi(x,t). \tag{6.30}$$

Averaging Eq. (6.30) over an ensemble of realizations of functions $z(t)$ yields the equation for the probability density of solutions to Eq. (6.29) $P(x,t) = \delta\langle\varphi(x,t)\rangle$ in the form

$$\frac{\partial}{\partial t}P(x,t) = -\frac{\partial}{\partial x}f(x,t)P(x,t) - \frac{\partial}{\partial x}g(x,t)\Psi(x,t), \tag{6.31}$$

where we introduced new function

$$\Psi(x,t) = \langle z(t)\varphi(x,t)\rangle .$$

Since solution to Eq. (6.30) is a functional of process $z(t)$, we can apply formula (6.7), page 152 to $\Psi(x,t)$ to obtain the equality

$$\left(\frac{d}{dt}+2\nu\right)\Psi(x,t)=\left\langle z(t)\frac{\partial}{\partial t}\varphi(x,t)\right\rangle. \tag{6.32}$$

Substitution of the right-hand side of Eq. (6.30) in Eq (6.32) yields the equation

$$\left(\frac{d}{dt}+2\nu\right)\Psi(x,t)=-\frac{\partial}{\partial x}f(x,t)\Psi(x,t)-\frac{\partial}{\partial x}g(x,t)\left\langle z^2(t)\varphi(x,t)\right\rangle, \tag{6.33}$$

and we obtain the closed system of equations

$$\begin{aligned}
&\frac{\partial}{\partial t}P(x,t)=-\frac{\partial}{\partial x}f(x,t)P(x,t)-\frac{\partial}{\partial x}g(x,t)\Psi(x,t),\\
&\left(\frac{d}{dt}+2\nu\right)\Psi(x,t)=-\frac{\partial}{\partial x}f(x,t)\Psi(x,t)-a_0^2\frac{\partial}{\partial x}g(x,t)P(x,t).
\end{aligned} \tag{6.34}$$

If functions $f(x,t)$ and $g(x,t)$ are independent of time, the steady-state probability distribution satisfies (if it exists) the equations

$$\begin{aligned}
&f(x)P(x)=-\frac{d}{dx}g(x)\Psi(x),\\
&\left(2\nu+\frac{d}{dx}f(x)\right)\Psi(x)=-a_0^2\frac{d}{dx}g(x)P(x).
\end{aligned} \tag{6.35}$$

Eliminating function $\Psi(x)$, we obtain the first-order differential equation [133]–[135]

$$\left(2\nu+\frac{d}{dx}f(x)\right)\frac{f(x)}{g(x)}P(x)=a_0^2\frac{d}{dx}g(x)P(x),$$

whose solution can be represented in the form of the quadrature ($|f(x)|<a_0|g(x)|$)

$$P(x)=\frac{C|g(x)|}{a_0^2g^2(x)-f^2(x)}\exp\left\{\frac{2\nu}{a_0^2}\int dx\frac{f(x)}{a_0^2g^2(x)-f^2(x)}\right\}, \tag{6.36}$$

where the positive constant C is determined from the normalization condition.

Note that, in the limit $\nu\to\infty$ and $a_0^2\to\infty$ under the condition that $a_0^2\tau_0=\text{const}$ ($\tau_0=1/(2\nu)$), probability distribution (6.36) grades into the expression

$$P(x)=\frac{C}{|g(x)|}\exp\left\{\frac{2\nu}{a_0^2}\int dx\frac{f(x)}{g^2(x)}\right\}$$

corresponding to the Gaussian delta-correlated process $z(t)$, i.e., the Gaussian process with the correlation function $\langle z(t)z(t')\rangle=2a_0^2\tau_0\delta(t-t')$.

To obtain an idea of system dynamics under the condition of finite correlation radius of process $z(t)$, we consider the simple example with $g(x)=1$, $f(x)=-x$ and $a_0=1$. In this case, we obtain from Eq. (6.36) the probability distribution

$$P(x)=\frac{1}{B(\nu,1/2)}(1-x^2)^{\nu-1}\quad(|x|<1), \tag{6.37}$$

where $B(\nu,1/2)$ is the beta-function. This distribution has essentially different behaviors for $\nu>1$, $\nu=1$ and $\nu<1$, which is schematically shown in Fig. 6.1.

One can easily see that this system resides mainly near state $x=0$ if $\nu>1$, and near states $x=\pm1$ if $\nu<1$. In the case $\nu=1$, we obtain the uniform probability distribution on segment $[-1,1]$.

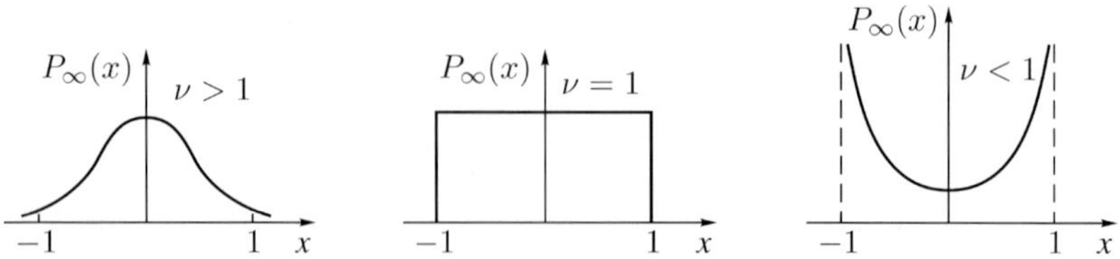

Figure 6.1: Steady-state probability distribution (6.37) of the solution to the equation $\frac{d}{dt}x(t) = -x + z(t)$ versus parameter ν.

6.1.3 Particle in the one-dimension potential field

Another example of nonlinear system concerns the one-dimensional motion of a particle in the filed $U(x)$ under the condition that random forces have a finite temporal correlation radius. We will describe the motion of the particle by the stochastic system of equations

$$\frac{d}{dt}x(t) = y(t), \quad \frac{d}{dt}y(t) = -\frac{dU(x)}{dx} - \lambda y(t) + \mu z(t), \tag{6.38}$$

where function $z(t)$ is assumed to be telegrapher's process ($z^2(t) = 1$). Similarly to the derivation of Eq. (6.34), we obtain the operator equation for the steady-state joint probability density of particle coordinate x and velocity y

$$\hat{L}(x,y)P(x,y) = \mu^2 \frac{\partial}{\partial y}\frac{1}{2\nu + \hat{L}(x,y)}\frac{\partial}{\partial y}P(x,y), \tag{6.39}$$

where $\hat{L}(x,y)$ is the Liouville operator,

$$\hat{L}(x,y) = y\frac{\partial}{\partial x} - \frac{dU(x)}{dx}\frac{\partial}{\partial y} - \lambda\frac{\partial}{\partial y}y.$$

For $\nu \to \infty$, Eq. (6.39) grades into the steady-state Fokker–Planck equation

$$\hat{L}(x,y)P(x,y) = \frac{\mu^2}{2\nu}\frac{\partial^2}{\partial y^2}P(x,y),$$

whose solution is the Gibbs distribution

$$P(x,y) = C\exp\left\{-\beta\left(\frac{y^2}{2} + U(x)\right)\right\} \quad \left(\beta = \frac{2\nu\lambda}{\mu^2}\right). \tag{6.40}$$

But in the general case, Eq. (6.39) describes the deformation of distribution (6.40) because of finite correlation time $\tau_0 = 1/(2\nu)$ of process $z(t)$. Equation (6.39) can be rewritten in the form of the partial differential equation

$$\begin{aligned}&\left\{\left(2\nu + \hat{L}\right)^2\hat{L} - \lambda\left(2\nu + \hat{L}\right)\hat{L}\right\}P(x,y)\\ &+\left\{-\mu^2\left(2\nu + \hat{L}\right)\frac{\partial^2}{\partial y^2} + \mu^2\frac{\partial^2}{\partial x\partial y} + \frac{\partial^2 U(x)}{\partial x^2}\hat{L}\right\}P(x,y) = 0.\end{aligned} \tag{6.41}$$

Deriving Eq. (6.41), we used the differentiation formula for the inverse operator $\hat{L}^{-1}(\alpha)$

$$\frac{\partial}{\partial\alpha}\hat{L}^{-1}(\alpha) = -\hat{L}^{-1}(\alpha)\frac{\partial\hat{L}(\alpha)}{\partial\alpha}\hat{L}^{-1}(\alpha).$$

Equation (6.41) is rather complicated, and it hardly can be solved for arbitrary field $U(x)$. However, one can easily see that solution to Eq. (6.41) will not be a function of sole particle energy as it is the case in Eq. (6.40); in addition, particle coordinate and velocity will be statistically dependent quantities.

6.1.4 Ordinary differential equation of the n-th order

Let now operators $\hat{A}(t)$ and $\hat{B}(t)$ in Eq. (6.10) be the matrixes, i.e.,

$$\frac{d}{dt}\mathbf{x}(t) = A(t)\mathbf{x}(t) + z(t)B(t)\mathbf{x}(t), \quad \mathbf{x}(0) = \mathbf{x}_0.$$

If only one component is of our interest, we can obtain for it the operator equation

$$\hat{L}\left(\frac{d}{dt}\right)x(t) + \sum_{i+j=0}^{n-1} b_{ij}(t)\frac{d^i}{dt^i}z(t)\frac{d^j}{dt^j}x(t) = f(t), \tag{6.42}$$

where

$$\hat{L}\left(\frac{d}{dt}\right) = \sum_{i=0}^{n} a_i(t)\frac{d^i}{dt^i}$$

and n is the order of matrixes A and B in Eq. (6.10).

The initial values for x are included in function $f(t)$ through the corresponding derivatives of the delta function. Note that function $f(t)$ can depend additionally on derivatives of random process $z(t)$ at $t = 0$, i.e., $f(t)$ is also the random function statistically related to process $z(t)$.

Averaging Eq. (6.42) over an ensemble of realizations of process $z(t)$ with the use of formula (6.9), we obtain the equation

$$\hat{L}\left(\frac{d}{dt}\right)\langle x(t)\rangle + M\left[\frac{d}{dt}, \frac{d}{dt} + 2\nu\right]\langle z(t)x(t)\rangle = \langle f(t)\rangle, \tag{6.43}$$

where

$$M[p,q] = \sum_{i+j=0}^{n-1} b_{ij}(t)p^i q^j.$$

However, Eq. (6.43) is unclosed because of the presence of correlator $\langle z(t)x(t)\rangle$. Multiplying Eq. (6.42) by $z(t)$ and averaging the result, we obtain the equation for correlator $\langle z(t)x(t)\rangle$

$$\hat{L}\left(\frac{d}{dt} + 2\nu\right)\langle z(t)x(t)\rangle + a_0^2 M\left[\frac{d}{dt} + 2\nu, \frac{d}{dt}\right]\langle x(t)\rangle = \langle z(t)f(t)\rangle. \tag{6.44}$$

System of equations (6.43) and (6.44) is the closed system. If functions $a_i(t)$ and $b_{ij}(t)$ are independent of time, this system can be solved using the Laplace transform. This solution is as follows:

$$\langle x\rangle_p = \frac{\hat{L}(p+2\nu)\langle f\rangle_p - M[p,p+2\nu]\langle zf\rangle_p}{\hat{L}(p)\hat{L}(p+2\nu) - a_0^2 M[p+2\nu,p]M[p,p+2\nu]}. \tag{6.45}$$

Note that Eq. (6.26) considered earlier is a special case of Eq. (6.42) and, consequently, Eq. (6.28) is a special case of Eq. (6.45), page 159.

6.1.5 Statistical interpretation of telegrapher's equation

In the preceding chapter, page 141, we showed that solutions to certain class of partial differential equations can be treated as the result of averaging certain functional over the random process delta-correlated in time. A similar situation occurs for telegrapher's random process.

Consider the initial value problem for the wave equation with linear friction

$$\left(\frac{\partial^2}{\partial t^2} + 2\nu\frac{\partial}{\partial t} - v^2\frac{\partial^2}{\partial x^2}\right) F(x,t) = 0,$$

$$F(x,0) = \varphi(x), \quad \frac{\partial}{\partial t}F(x,t)\Big|_{t=0} = \psi(x). \tag{6.46}$$

We can rewrite Eq. (6.46) as the integro-differential equation

$$\frac{\partial}{\partial t}F(x,t) = \psi(x)e^{-2\nu t} + v^2\int\limits_0^t dt_1 e^{-2\nu(t-t_1)}\frac{\partial^2}{\partial x^2}F(x,t). \tag{6.47}$$

Introduce now the auxiliary stochastic equation

$$\frac{\partial}{\partial t}f(x,t) = \psi(x)e^{-2\nu t} + vz(t)\frac{\partial}{\partial x}f(x,t), \quad f(x,0) = \varphi(x), \tag{6.48}$$

where $z(t)$ is telegrapher's process ($z^2 = 1$). From the above material obviously follows that

$$F(x,t) = \langle f(x,t)\rangle_z .$$

The solution to Eq. (6.48) has the form

$$f(x,t) = \varphi\left(x + v\int\limits_0^t d\tau z(\tau)\right) + \int\limits_0^t dt_1 e^{-2\nu t_1}\psi\left(x + v\int\limits_{t_1}^t d\tau z(\tau)\right).$$

Consequently, the solution to Eq. (6.46) can be represented as the statistical average over random process $z(t)$:

$$F(x,t) = \left\langle \varphi\left(x + v\int\limits_0^t d\tau z(\tau)\right)\right\rangle_z + \int\limits_0^t dt_1 e^{-2\nu t_1}\left\langle \psi\left(x + v\int\limits_{t_1}^t d\tau z(\tau)\right)\right\rangle_z .$$

6.2 Generalized telegrapher's process

Generalized telegrapher's process is defined by the formula

$$z(t) = a_{n(0,t)}, \tag{6.49}$$

where $n(t_1, t_2)$, $t_1 < t_2$ is the integer-valued Poisson random process statistically independent of random quantities a_i, which are also statistically independent and have probability density $p(a)$.

Generalized telegrapher's process $z(t)$ is stationary in time and its correlation function

$$\langle z(t)z(t')\rangle = \left\langle a^2\right\rangle e^{-\nu|t-t'|}$$

is characterized by the temporal correlation radius $\tau_0 = 1/\nu$.

As in the case of telegrapher's process, two alternative methods are appropriate for analyzing stochastic equations whose parameter fluctuations can be described by generalized telegrapher's process.

The first method immediately deals with the formula (4.40), page 86 for splitting the correlation of process $z(t)$ with arbitrary functional $R[t; z(\tau)]$ of this process:

$$\langle z(t)R[t;z(\tau)]\rangle = \langle \tilde{a}R[t;\tilde{a}]\rangle\, e^{-\upsilon t} + a_0^2 \int\limits_0^t dt_1 e^{-\upsilon(t-t_1)} \left\langle \tilde{a}\tilde{R}[t,t_1;\tilde{a},z(\tau)]\right\rangle, \qquad (6.50)$$

where functional $\tilde{R}[t,t_1;\tilde{a},z(\tau)]$ is given by the formula

$$\tilde{R}[t,t_1;\tilde{a},z(\tau)] = R[t;\tilde{a}\theta(\tau - t_1 + 0) + z(\tau)\theta(t_1 - \tau + 0)], \qquad (6.51)$$

and random quantity $\tilde{a}$ is statistically independent of process $z(t)$.

In contrast to telegrapher's process, the second method appears here more formal and deals with the differentiation formula (4.47), (4.49), page 88 that has the form

$$\left(\frac{d}{dt} + \nu\right) \langle F\left(z(t)\right) R[t;z(\tau)]\rangle = \left\langle F\left(z(t)\right) \frac{d}{dt} R[t;z(\tau)]\right\rangle, \quad (\tau \le t) \qquad (6.52)$$

under the condition that $\langle F(z(t))\rangle = \langle F(a)\rangle_a = 0$. In particular, we have the formula

$$\left(\frac{d}{dt} + \nu\right)^n \langle F\left(z(t)\right) R[t;z(\tau)]\rangle = \left\langle F\left(z(t)\right) \frac{d^n}{dt^n} R[t;z(\tau)]\right\rangle \qquad (6.53)$$

defining the rule of factoring the differential operator out of averaging brackets.

The further analysis becomes simpler if we define function $F(z(t))$ as follows:

$$F(t) = F(z(t)) = \frac{1}{1+\lambda z(t)} - C_0(\lambda), \qquad (6.54)$$

where

$$C_k(\lambda) = \left\langle \frac{a^k}{1+\lambda a}\right\rangle_a \qquad (6.55)$$

and λ is arbitrary parameter. This function $F(t)$ satisfies the identity

$$z(t)F(t) = -\frac{1}{\lambda}F(t) + C_1(t) - z(t)C_0(\lambda). \qquad (6.56)$$

Now, we consider several examples of working according to the above formalisms.

6.2.1 Stochastic linear equation

First of all, we consider, as in the previous section, the stochastic linear equation (6.10), page 152 assuming that linear operators $\hat{A}(t)$ and $\hat{B}(t)$ in this equation are constant matrixes A and B. In this case, the equation for the mean value $\langle \mathbf{x}(t)\rangle$ is

$$\frac{d}{dt}\langle \mathbf{x}(t)\rangle = A\langle \mathbf{x}(t)\rangle + B\langle z(t)\mathbf{x}(t)\rangle. \qquad (6.57)$$

Using Eq. (6.50), we can rewrite this equation in the form

$$\frac{d}{dt}\langle \mathbf{x}(t)\rangle = A\langle \mathbf{x}(t)\rangle + B\langle \tilde{a}\mathbf{x}[t;\tilde{a}]\rangle e^{-\nu t} + \nu B\int_0^t dt_1 e^{-\nu(t-t_1)}\langle \tilde{a}\mathbf{x}[t,t_1;\tilde{a},z(\tau)]\rangle . \quad (6.58)$$

According to Eq. (6.51), functional $\mathbf{x}[t,t_1;\tilde{a},z(\tau)]$ satisfies the equation

$$\frac{d}{dt}\mathbf{x}(t) = A\mathbf{x}(t) + \tilde{a}B\mathbf{x}(t) \quad (t > t_1) \quad (6.59)$$

with the initial value

$$\mathbf{x}[t_1,t_1;\tilde{a},z(\tau)] = \mathbf{x}[t_1;z(\tau)]. \quad (6.60)$$

Hence, functional $\mathbf{x}[t,t_1;\tilde{a},z(\tau)]$ has the form

$$\mathbf{x}[t,t_1;\tilde{a},z(\tau)] = e^{(A+\tilde{a}B)(t-t_1)}\mathbf{x}(t_1),$$

and Eq. (6.58) turns into the integro-differential equation

$$\begin{aligned}\frac{d}{dt}\langle \mathbf{x}(t)\rangle &= A\langle \mathbf{x}(t)\rangle + e^{-\nu t}B\left\langle \tilde{a}e^{(A+\tilde{a}B)t}\right\rangle \mathbf{x}_0 \\ &+\nu B\int_0^t dt_1 e^{-\nu(t-t_1)}\left\langle \tilde{a}e^{(A+\tilde{a}B)(t-t_1)}\right\rangle \langle \mathbf{x}(t_1)\rangle \end{aligned} \quad (6.61)$$

with the initial value $\langle \mathbf{x}(0)\rangle = \mathbf{x}_0$.

We can easily solve Eq. (6.61) using the Laplace transform. The solution has the form

$$\langle \mathbf{x}\rangle_p = (E-\nu C)^{-1}C\mathbf{x}_0, \quad (6.62)$$

where

$$C = \left\langle \{(p+\nu)E - A - \tilde{a}B\}^{-1}\right\rangle_a$$

and E is the unit matrix.

Use now the alternative method for splitting the correlator in Eq. (6.57).

According to the differentiation formula (6.52), we have

$$\left(\frac{d}{dt}+\nu\right)\langle F(t)\mathbf{x}(t)\rangle = \left\langle F(t)\frac{d}{dt}\mathbf{x}(t)\right\rangle = A\langle F(t)\mathbf{x}(t)\rangle + B\langle z(t)F(t)\mathbf{x}(t)\rangle . \quad (6.63)$$

Using then Eq. (6.56), we can rewrite Eq. (6.63) as the identity

$$\left[\left(\frac{d}{dt}+\nu\right)E - A + \frac{1}{\lambda}B\right]\langle F(t)\mathbf{x}(t)\rangle = BC_1(\lambda)\langle \mathbf{x}(t)\rangle - BC_0(\lambda)\langle z(t)\mathbf{x}(t)\rangle . \quad (6.64)$$

Performing the Laplace transform of Eqs. (6.57) and (6.64), we obtain the unclosed system of equations

$$\begin{aligned}&(pE-A)\langle \mathbf{x}\rangle_p - B\langle z\mathbf{x}\rangle_p = \mathbf{x}_0, \\ &\left[(p+\nu)E - A + \frac{1}{\lambda}B\right]\langle F(t)\mathbf{x}(t)\rangle_p = BC_1(\lambda)\langle \mathbf{x}\rangle_p - BC_0(\lambda)\langle z\mathbf{x}\rangle_p ,\end{aligned} \quad (6.65)$$

which is valid for arbitrary λ. For $\frac{1}{\lambda} = [A-(p+\nu)E]B^{-1}$, the left-hand side of the second equation vanishes, and we obtain the algebraic relationship between $\langle \mathbf{x}\rangle_p$ and $\langle z\mathbf{x}\rangle_p$; together with the first equation of system (6.65), this relationship yields the solution that coincides with Eq. (6.62).

Stochastic parametric resonance

We consider the statistical description of solution to problem (6.17), page 153 as a specific example. Averaging Eq. (6.17) over an ensemble of realizations of generalized telegrapher's process $z(t)$, we obtain the unclosed equation

$$\left(\frac{d^2}{dt^2}+\omega_0^2\right)\langle x(t)\rangle+\omega_0^2\langle z(t)x(t)\rangle=0, \tag{6.66}$$

with the initial values

$$\langle x(0)\rangle=x_0,\quad \left\langle\frac{dx(t)}{dt}\right\rangle_{t=0}=y_0.$$

To split the correlator appeared in (6.66), we multiply Eq. (6.17), page 153 by function $F(z(t))$ and average the result. Using then formula (6.52), page 161 that defines the rule of factoring the derivative out of averaging brackets, we obtain the equation

$$\left[\left(\frac{d}{dt}+\nu\right)^2+\omega_0^2\right]\langle F(z(t))\,x(t)\rangle+\omega_0^2\langle z(t)F(z(t))\,x(t)\rangle=0 \tag{6.67}$$

with zero-valued initial values

$$\langle F(z(t))\,x(0)\rangle=0,\quad \left\langle F(z(t))\frac{dx(t)}{dt}\right\rangle_{t=0}=0.$$

The further analysis becomes simpler if we use function $F(z(t))$ in form (6.54), page 161 and rewrite Eq. (6.67) as follows:

$$\begin{aligned}&\left[\left(\frac{d}{dt}+\nu\right)^2+\omega_0^2\left(1-\frac{1}{\lambda}\right)\right]\langle F(z(t))\,x(t)\rangle\\&+\omega_0^2C_1(\lambda)\langle x(t)\rangle-\omega_0^2C_0(\lambda)\langle z(t)x(t)\rangle=0.\end{aligned} \tag{6.68}$$

Performing the Laplace transform of Eqs. (6.66) and (6.68), we obtain

$$\begin{aligned}&\left(p^2+\omega_0^2\right)\langle x\rangle_p+\omega_0^2\langle zx\rangle_p=y_0+px_0,\\&\left[(p+\nu)^2+\omega_0^2\left(1-\frac{1}{\lambda}\right)\right]\langle Fx\rangle_p+\omega_0^2C_1(\lambda)\langle x\rangle_p-\omega_0^2C_0(\lambda)\langle zx\rangle_p=0.\end{aligned} \tag{6.69}$$

In Eqs. (6.68) and (6.69), parameter λ is arbitrary parameter. Now, we specify it as follows:

$$\lambda=\lambda_p=\frac{\omega_0^2}{L(p+\nu)},\quad L(p)=p^2+\omega_0^2. \tag{6.70}$$

In this case, the first term in the second equation of system (6.69) vanishes, and we obtain the relationship between correlator $\langle zx\rangle_p$ and average solution to Eq. (5.152) $\langle x\rangle_p$ in the form

$$\langle zx\rangle_p=\frac{C_1(p)}{C_0(p)}\langle x\rangle_p, \tag{6.71}$$

where

$$C_k(p)=\left\langle\frac{a^k}{L(p+\nu)+a\omega_0^2}\right\rangle_a.$$

Substituting Eq. (6.71) in the first equation of system (6.69), we obtain the solution in the form

$$\langle x\rangle_p = \frac{y_0 + p x_0}{L(p) + \omega_0^2 \frac{C_1(p)}{C_0(p)}}. \tag{6.72}$$

As was noted earlier, the mean value of the solution to problem (6.17), page 153 can be obtained with the use of the other — alternative and more intuitive — method. Using Eq. (6.50), page 161, we can rewrite Eq. (6.66) in the form

$$\begin{aligned}\left(\frac{d^2}{dt^2} + \omega_0^2\right)\langle x(t)\rangle &= -\omega_0^2 \langle a\widetilde{x}[t;a]\rangle_a e^{-\nu t} \\ &-\nu\omega_0^2 \int\limits_0^t dt_1 e^{-\nu(t-t_1)} \langle a\widetilde{x}[t,t_1;a,z(\tau)]\rangle_a,\end{aligned} \tag{6.73}$$

where functional $\widetilde{x}[t,t_1;a,z(\tau)]$ satisfies the equation

$$\left(\frac{d^2}{dt^2} + \omega_0^2\right)\widetilde{x}(t) + \omega_0^2 a\widetilde{x}(t) = 0$$

with the initial values

$$\widetilde{x}[t,t_1;a,z(\tau)]_{t=t_1} = x(t_1), \quad \frac{d}{dt}\widetilde{x}[t,t_1;a,z(\tau)]\Big|_{t=t_1} = \frac{d}{dt_1}x(t_1)$$

and $\widetilde{x}[t;a] = \widetilde{x}[t,0;a,z(\tau)]$.

The solution to this equation is as follows

$$\begin{aligned}&\widetilde{x}[t,t_1;a,z(\tau)] \\ &\quad = x(t_1)\cos\left[\omega_0\sqrt{1+a}(t-t_1)\right] + \frac{dx(t_1)}{dt_1}\frac{\sin\left[\omega_0\sqrt{1+a}(t-t_1)\right]}{\omega_0\sqrt{1+a}},\end{aligned}$$

and, consequently, Eq. (6.28) can be rewritten in the closed form

$$\begin{aligned}\left(\frac{d^2}{dt^2} + \omega_0^2\right)\langle x(t)\rangle &= -x_0\omega_0^2 e^{-\nu t}\left\langle a\cos\left(\omega_0\sqrt{1+a t}\right)\right\rangle_a \\ &-y_0\omega_0^2 e^{-\nu t}\left\langle a\frac{\sin\left(\omega_0\sqrt{1+a t}\right)}{\omega_0\sqrt{1+a}}\right\rangle_a \\ &-\nu\omega_0^2\int\limits_0^t dt_1 e^{-\nu(t-t_1)}\langle x(t_1)\rangle\left\langle a\cos\left[\omega_0\sqrt{1+a}(t-t_1)\right]\right\rangle_a \\ &-\nu\omega_0^2\int\limits_0^t dt_1 e^{-\nu(t-t_1)}\frac{d\langle x(t_1)\rangle}{dt_1}\left\langle a\frac{\sin\left[\omega_0\sqrt{1+a}(t-t_1)\right]}{\omega_0\sqrt{1+a}}\right\rangle_a\end{aligned} \tag{6.74}$$

Equation (6.74) can be easily solved using the Laplace transform; the result coincides with Eq. (6.72).

6.2.2 One-dimensional nonlinear differential equation

Consider now the one-dimensional equation (6.29), page 156

$$\frac{d}{dt}x(t) = f(x) + z(t)g(x), \quad x(0) = x_0$$

assuming that $z(t)$ is generalized telegrapher's process and functions $f(x)$ and $g(x)$ are independent of time. In this case, the indicator function satisfies the Liouville equation (6.30), page 156 that assumes here the form

$$\frac{\partial}{\partial t}\varphi(x,t) = -\frac{\partial}{\partial x}f(x)\varphi(x,t) - z(t)\frac{\partial}{\partial x}g(x)\varphi(x,t). \tag{6.75}$$

Averaging Eq. (6.75), we obtain the equation for the one-time probability density

$$\begin{aligned}\frac{\partial}{\partial t}P(x,t) + \frac{\partial}{\partial x}f(x)P(x,t) &= -\frac{\partial}{\partial x}g(x)\left\langle z(t)\varphi(x,t)\right\rangle \\ = -e^{-\nu t}\frac{\partial}{\partial x}g(x)\left\langle \tilde{a}\varphi[x,t;\tilde{a}]\right\rangle &- \nu\frac{\partial}{\partial x}g(x)\int\limits_0^t dt_1 e^{-\nu(t-t_1)}\left\langle \tilde{a}\tilde{\varphi}[x,t,t_1;\tilde{a},z(\tau)]\right\rangle.\end{aligned} \tag{6.76}$$

Functional $\tilde{\varphi}[x,t,t_1;\tilde{a},z(\tau)]$ will satisfy now the equation

$$\frac{\partial}{\partial t}\tilde{\varphi}(x,t) = -\frac{\partial}{\partial x}f(x)\tilde{\varphi}(x,t) - \tilde{a}\frac{\partial}{\partial x}g(x)\tilde{\varphi}(x,t)$$

with the initial value $\tilde{\varphi}(x,t_1) = \varphi(x,t_1)$. In the operator form, the solution to this equation will be

$$\tilde{\varphi}(x,t) = e^{-(t-t_1)\frac{\partial}{\partial x}[f(x)+\tilde{a}g(x)]}\varphi(x,t_1).$$

Hence, we can rewrite the equation for the probability density (6.76) in the closed operator form

$$\begin{aligned}\left(\frac{\partial}{\partial t} + \frac{\partial}{\partial x}f(x)\right)P(x,t) &= -e^{-\nu t}\frac{\partial}{\partial x}g(x)\left\langle \tilde{a}e^{-t\frac{\partial}{\partial x}[f(x)+\tilde{a}g(x)]}\right\rangle\varphi(x,0) \\ -\nu\frac{\partial}{\partial x}g(x)\int\limits_0^t dt_1 e^{-\nu(t-t_1)}&\left\langle \tilde{a}e^{-(t-t_1)\frac{\partial}{\partial x}[f(x)+\tilde{a}g(x)]}\right\rangle P(x,t_1).\end{aligned} \tag{6.77}$$

The steady-state probability distribution (if it exists) satisfies the operator equation

$$f(x)P(x) = -\nu g(x)\int\limits_0^\infty d\tau e^{-\nu\tau}\left\langle \tilde{a}e^{-\tau\frac{d}{dx}[f(x)+\tilde{a}g(x)]}\right\rangle P(x)$$

that can be rewritten as follows:

$$f(x)P(x) = -\nu g(x)\left\langle \frac{\tilde{a}}{\nu + \frac{d}{dx}[f(x)+\tilde{a}g(x)]}\right\rangle P(x). \tag{6.78}$$

To convert Eq. (6.78) to the differential equation, we must specify the probability distribution of random quantity a.

Assume for example that quantity a is characterized by sufficiently small intensity of fluctuations and $\langle a\rangle = 0$. Then, expanding the operator in the right-hand side of Eq. (6.78) in powers of a and neglecting all terms higher than $\langle a^2\rangle$, we obtain the operator equation

$$f(x)P(x) = -\nu\left\langle a^2\right\rangle g(x)\frac{1}{\nu+\frac{d}{dx}f(x)}\frac{d}{dx}\frac{1}{\nu+\frac{d}{dx}f(x)}P(x). \tag{6.79}$$

If we represent now function $P(x)$ in the form

$$P(x) = \left[\nu + \frac{d}{dx}f(x)\right]\psi(x),$$

then we obtain the second-order differential equation for function $\psi(x)$

$$\left[\nu+\frac{d}{dx}f(x)\right]\frac{F(x)}{g(x)}\left[\nu+\frac{d}{dx}f(x)\right]\psi(x) = -\nu\left\langle a^2\right\rangle\frac{d}{dx}g(x)\psi(x). \tag{6.80}$$

For $\nu \to \infty$, we can expand the mean value in the right-hand side of Eq. (6.78) in powers of $1/\nu$ and obtain the steady-state Fokker–Planck equation

$$f(x)P(x) = g(x)\frac{\nu}{\langle a^2\rangle}\frac{d}{dx}g(x)P(x)$$

corresponding to the Gaussian delta-correlated process $z(t)$.

6.2.3 Ordinal differential equation of the n-th order

Consider now Eq. (6.42), page 159

$$\hat{L}\left(\frac{d}{dt}\right)x(t) + \sum_{i+j=0}^{n-1} b_{ij}(t)\frac{d^i}{dt^i}z(t)\frac{d^j}{dt^j}x(t) = f(t)$$

with generalized telegrapher's process $z(t)$. For simplicity, we will assume that initial values for Eq. (6.42) are independent of $z(t)$ and coefficients a_i and b_{ij} are constants.

Averaging Eq. (6.42) with the use of formula (6.53), page 161, we obtain the equation

$$\hat{L}\left(\frac{d}{dt}\right)\langle x(t)\rangle + M\left[\frac{d}{dt}, \frac{d}{dt}+\nu\right]\langle z(t)x(t)\rangle = f(t), \tag{6.81}$$

where $M[p,q] = \sum\limits_{i+j=0}^{n-1} b_{ij}(t)p^iq^j$, as before.

Consider now correlator $\langle F(t)x(t)\rangle$, where $F(t)$ is given by Eq. (6.54). In accordance with the differentiation formula (6.53), this function satisfies the equation

$$\begin{aligned}\hat{L}\left(\frac{d}{dt}+\nu\right)\langle F(t)x(t)\rangle &= \left\langle F(t)\hat{L}\left(\frac{d}{dt}\right)x(t)\right\rangle \\ = -\sum_{i+j=0}^{n-1} b_{ij}(t)&\left\langle F(t)\frac{d^i}{dt^i}z(t)\frac{d^j}{dt^j}x(t)\right\rangle - \sum_{i+j=0}^{n-1} b_{ij}(t)\left(\frac{d}{dt}+\nu\right)^i\left\langle z(t)F(t)\frac{d^j}{dt^j}x(t)\right\rangle.\end{aligned} \tag{6.82}$$

Using now Eq. (6.56), we can rewrite the right-hand side of Eq. (6.82) in the form

$$
\begin{aligned}
&-\sum_{i+j=0}^{n-1} b_{ij}(t)\left(\frac{d}{dt}+\nu\right)^{i} - \frac{1}{\lambda}\left\langle F(t)\frac{d^j}{dt^j}x(t)\right\rangle \\
&+C_1(\lambda)\frac{d^j}{dt^j}\left\langle x(t)\right\rangle - C_0(\lambda)\left\langle z(t)\frac{d^j}{dt^j}x(t)\right\rangle \\
&\quad= \frac{1}{\lambda}M\left[\frac{d}{dt}+\nu, \frac{d}{dt}+\nu\right]\left\langle F(t)x(t)\right\rangle \\
&-C_1(\lambda)M\left[\frac{d}{dt}+\nu, \nu\right]\left\langle x(t)\right\rangle + C_0(\lambda)M\left[\frac{d}{dt}+\nu, \frac{d}{dt}+\nu\right]\left\langle z(t)x(t)\right\rangle .
\end{aligned}
\tag{6.83}
$$

Consequently, Eq. (6.82) assumes the form

$$
\begin{aligned}
&\left\{\hat{L}\left(\frac{d}{dt}+\nu\right) - \frac{1}{\lambda}M\left[\frac{d}{dt}+\nu, \frac{d}{dt}+\nu\right]\right\}\left\langle F(t)x(t)\right\rangle \\
=\;& -C_1(\lambda)M\left[\frac{d}{dt}+\nu, \frac{d}{dt}\right]\left\langle x(t)\right\rangle + C_0(\lambda)M\left[\frac{d}{dt}+\nu, \frac{d}{dt}+\nu\right]\left\langle z(t)x(t)\right\rangle
\end{aligned}
\tag{6.84}
$$

with the initial value $\langle F(t)x(t)\rangle\,|_{t=0} = 0$.

Perform now the Laplace transform of Eqs. (6.81) and (6.84). As a result, we obtain the algebraic system of equations

$$
\begin{aligned}
&\hat{L}\,(p)\left\langle x\right\rangle_p + M\,[p, p+\nu]\left\langle zx\right\rangle_p = f(p), \\
&\left\{\hat{L}\,(p+\nu) - \frac{1}{\lambda}M\,[p+\nu, p+\nu]\right\}\left\langle Fx\right\rangle_p \\
=\;& -C_1(\lambda)M\,[p+\nu, p]\left\langle x\right\rangle_p + C_0(\lambda)M\,[p+\nu, p+\nu]\left\langle zx\right\rangle_p .
\end{aligned}
\tag{6.85}
$$

Equations (6.85) are valid for arbitrary λ. If we set

$$
\lambda = \lambda_p = M[p+\nu, p+v]/L(p+\nu),
\tag{6.86}
$$

the second equation of system (6.85) becomes the algebraic relationship between $\langle zx\rangle_p$ and $\langle x\rangle_p$:

$$
\left\langle z(t)x(t)\right\rangle_p = \frac{\tilde{C}_1(p)}{\tilde{C}_0(p)}\frac{M\,[p+\nu, p]}{M\,[p+\nu, p+\nu]}\left\langle x\right\rangle_p ,
\tag{6.87}
$$

where

$$
\tilde{C}_k(p) = \left\langle \frac{a^k}{\hat{L}(p+\nu) + aM\,[p+\nu, p+\nu]}\right\rangle_a .
$$

Substituting (6.87) in the first equation of system (6.85), we obtain the algebraic equation for $\langle x\rangle_p$, whose solution has the form [216]

$$
\left\langle x\right\rangle_p = f(p)\left[\hat{L}(p) + \frac{M\,[p+\nu, p]\,M\,[p, p+\nu]}{M\,[p+\nu, p+\nu]}\frac{\tilde{C}_1(p)}{\tilde{C}_0(p)}\right]^{-1} .
\tag{6.88}
$$

6.3 Gaussian Markovian processes

Here, we consider several examples associated with the Gaussian Markovian processes. Define random process $z(t)$ by the formula

$$z(t) = z_1(t) + ... + z_N(t), \tag{6.89}$$

where $z_i(t)$ are statistically independent telegrapher's processes with correlation functions

$$\langle z_i(t) z_j(t')\rangle = \delta_{ij} \left\langle z^2 \right\rangle e^{-\alpha|t-t'|} \quad (\alpha = 2\nu).$$

If we set $\langle z^2 \rangle = \sigma^2/N$, then this process passes for $N \to \infty$ into the Gaussian Markovian process with correlation function

$$\langle z_i(t) z_j(t')\rangle = \left\langle z^2 \right\rangle e^{-\alpha|t-t'|}.$$

Thus, process $z(t)$ (6.89) approximates the Gaussian Markovian process in terms of the Markovian process with a finite number of states.

It is evident that the differentiation formula and the rule of factoring the derivative out of averaging brackets assume for process $z(t)$ the forms

$$\left(\frac{d}{dt} + \alpha k\right) \langle z_1(t)...z_k(t) R[t; z(\tau)]\rangle = \left\langle z_1(t)...z_k(t) \frac{d}{dt} R[t; z(\tau)]\right\rangle,$$

$$\left\langle z_1(t)...z_k(t) \frac{d^n}{dt^n} R[t; z(\tau)]\right\rangle = \left(\frac{d}{dt} + \alpha k\right)^n \langle z_1(t)...z_k(t) R[t; z(\tau)]\rangle, \tag{6.90}$$

where $R[t; z(\tau)]$ is arbitrary functional of process $z(t)$ $(\tau \leq t)$.

6.3.1 Stochastic linear equation

Consider again Eq. (6.10), which we represent here in the form

$$\frac{d}{dt}\mathbf{x}(t) = \hat{A}(t)\mathbf{x}(t) + [z_1(t) + ... + z_N(t)]\hat{B}(t)\mathbf{x}(t), \quad \mathbf{x}(0) = \mathbf{x}_0, \tag{6.91}$$

and introduce vector-function

$$\mathbf{X}_k(t) = \langle z_1(t)...z_k(t)\mathbf{x}(t)\rangle, \quad k = 1, ..., N; \quad \mathbf{X}_0(t) = \langle \mathbf{x}(t)\rangle. \tag{6.92}$$

Using formula (6.90) for differentiating correlations (6.92) and Eq. (6.91), we obtain the recurrence equation for $\mathbf{x}_k(t), k = 0, 1, \ldots, N$

$$\begin{aligned}\frac{d}{dt}\mathbf{X}_k(t) &= \hat{A}(t)\mathbf{X}_k(t) + \left\langle z_1(t)\ldots z_k(t)[z_1(t) + \ldots + z_N(t)]\hat{B}(t)\mathbf{x}(t)\right\rangle \\ &= \hat{A}(t)\mathbf{X}_k(t) + k\left\langle z^2\right\rangle \hat{B}(t)\mathbf{X}_{k-1}(t) + (N-k)\hat{B}(t)\mathbf{X}_{k+1}(t),\end{aligned} \tag{6.93}$$

with the initial value

$$\mathbf{X}_k(0) = \mathbf{x}_0 \delta_{k,0}.$$

Thus, the mean value of the solution to system (6.91) satisfies the closed system of $(N+1)$ vector equations. If operators $\hat{A}(t)$, $\hat{B}(t)$ are time-independent matrixes, system (6.93) can be easily solved using the Laplace transform. It is clear that such a solution will have the form of a finite segment of continued fraction. If we set $\langle z^2 \rangle = \sigma^2/N$ and proceed to the limit $N \to \infty$, then random process $z(t) = z_1(t) + ... + z_N(t)$ will grade, as was mentioned earlier, into the Gaussian Markovian process and solution to system (6.93) will assume the form of the infinite continued fraction.

6.3.2 Ordinal differential equation of the n-th order

Consider stochastic equation (6.42), page 159

$$\hat{L}\left(\frac{d}{dt}\right)x(t)+\sum_{i+j=0}^{n-1}b_{ij}(t)\frac{d^i}{dt^i}z(t)\frac{d^j}{dt^j}x(t)=f(t)$$

with random process $z(t)$ given by Eq. (6.89) and introduce, as in the previous example, functions

$$X_k(t)=\langle z_1(t)...z_k(t)x(t)\rangle\,,\quad k=1,...,N;\quad X_0(t)=\langle x(t)\rangle\,.$$

To obtain equations for these functions, we multiply Eq. (6.42) by $z_1(t)...z_k(t)$ and average the result over an ensemble of realizations of $z_i(t)$. Using Eqs. (6.90), we obtain that function $X_k(t)$ satisfies the closed system of recurrence equations

$$\begin{aligned}&\hat{L}\left(\frac{d}{dt}+\alpha k\right)X_k(t)=F_k(t)-k\left\langle z^2\right\rangle M\left[\frac{d}{dt}+\alpha k,\frac{d}{dt}+\alpha(k-1)\right]X_{k-1}(t)\\&-(N-k)\left\langle z^2\right\rangle M\left[\frac{d}{dt}+\alpha k,\frac{d}{dt}+\alpha(k+1)\right]X_{k+1}(t),\end{aligned}\tag{6.94}$$

where

$$F_k(t)=\langle z_1(t)...z_k(t)f(t)\rangle\,.$$

If operator $\hat{L}$ and functions b_{ij} are independent of time t, the Laplace transform reduces system (6.94) to the algebraic system

$$\begin{aligned}&\hat{L}\left(p+\alpha k\right)X_k(p)=F_k(p)-k\left\langle z^2\right\rangle M\left[p+\alpha k,p+\alpha(k-1)\right]X_{k-1}(p)\\&-(N-k)\left\langle z^2\right\rangle M\left[p+\alpha k,p+\alpha(k+1)\right]X_{k+1}(p).\end{aligned}\tag{6.95}$$

In the special case of function $f(t)$ independent of $z_k(t)$, when $F_k(p)=f(p)\delta_{k,0}$, Eq. (6.95) can be easily solved, and the solution has the form of the finite segment of continued fraction

$$X_0(p)=f(p)K_0(p),\quad K_l(p)=\frac{1}{A_l(p)-B_l(p)K_{l+1}(p)},\tag{6.96}$$

where

$$\begin{aligned}A_l(p)&=\hat{L}\left(p+\alpha l\right),\\B_l(p)&=\left\langle z^2\right\rangle(N-l)(l+1)M\left[p+\alpha l,p+\alpha(l+1)\right]M\left[p+\alpha(l+1),p+\alpha l\right].\end{aligned}\tag{6.97}$$

If $N=1$, i.e., if we deal with only one telegrapher's process, the solution (6.96), (6.97) assumes the form of Eq. (6.45), page 159, which corresponds to the two-level continued fraction.

If we set $\langle z^2\rangle=\sigma^2/N$ and proceed to the limit $N\to\infty$, we obtain solution $\langle x(p)\rangle$ in the form if the infinite continued fraction (6.96) with parameters [216]

$$\begin{aligned}A_l(p)&=\hat{L}\left(p+\alpha l\right),\\B_l(p)&=\sigma^2(l+1)M\left[p+\alpha l,p+\alpha(l+1)\right]M\left[p+\alpha(l+1),p+\alpha l\right],\end{aligned}\tag{6.98}$$

which corresponds to the Gaussian Markovian process $z(t)$.

Stochastic parametric resonance

We illustrate the above material using statistical description of solution to problem (6.17), page 153 for the Gaussian Markovian process $z(t)$ as an example.

We introduce function

$$X_l(t) = \langle z_1(t) \ldots z_l(t) x(t) \rangle \,, \tag{6.99}$$

where $x(t)$ is the solution to problem (6.17). Multiplying then Eq. (6.17) by product $z_1(t) \ldots z_l(t)$, averaging the result over an ensemble of realizations of all processes $z_i(t)$, and using Eq. (6.90), we obtain the recurrent equality

$$\hat{L}\left(\frac{d}{dt} + \alpha l\right) X_l(t) + \omega_0^2 \left\langle z^2 \right\rangle l X_{l-1} + \omega_0^2 (N - l) X_{l+1} = 0, \quad (l = 0, \ldots, .N), \tag{6.100}$$

where

$$\hat{L}\left(\frac{d}{dt} + \alpha l\right) = \left(\frac{d}{dt} + \alpha l\right)^2 + \omega_0^2.$$

Equality (6.100) can be considered as the closed system of N equations with the initial values

$$X_0(0) = 0, \quad \left.\frac{d}{dt} X_0(t)\right|_{t=0} = y_0.$$

Performing the Laplace transform, we obtain recurrent algebraic system of equations

$$L(p + \alpha l) X_l(p) + \omega_0^2 \left\langle z^2 \right\rangle X_{l-1}(p) + \omega_0^2 (N - l) X_{l+1}(p) = F(p) \delta_{l,0}, \tag{6.101}$$

where $F(p) = y_0 + p x_0$. Now, we set

$$X_l(p) = -\omega_0^2 \left\langle z^2 \right\rangle l K_l(p) X_{l-1} \tag{6.102}$$

for $l \neq 0$. Substituting Eq. (6.102) in Eq. (6.101), we obtain the finite segment of continued fraction

$$K_l(p) = \frac{1}{L(p + \alpha l) - \omega_0^4 \langle z^2 \rangle (N - l)(l + 1) K_{l+1}(p)}, \tag{6.103}$$

and the solution to problem (5.152), page 135 is

$$\langle x \rangle_p = X_0(p) = F(p) K_0(p). \tag{6.104}$$

At $N = 1$, equality (6.104) grades into Eq. (6.23) for single telegrapher's process and corresponds to the two-level continued fraction (6.103).

Setting $\langle z^2 \rangle = \sigma^2 / N$ and proceeding to the limit $N \to \infty$, we obtain the solution for the Gaussian Markovian process in the form of the infinite continued fraction (6.104), where

$$K_l(p) = \frac{1}{L(p + \alpha l) - \omega_0^4 \sigma^2 (l + 1) K_{l+1}(p)}. \tag{6.105}$$

The second moments of the solution to problem (5.152), page 135 can be considered similarly. For example, considering potential energy $U(t) = x^2(t)$ that satisfies dynamic equation (6.26) with initial values (6.27), page 155, we obtain the mean value $\langle U(t) \rangle$ in the form of the finite segment of continued fraction (in the case of N telegrapher's processes)

$$\langle U \rangle_p = 2 y_0^2 K_0(p), \quad K_l(p) = \frac{1}{A_l(p) - B_l(p) K_{l+1}(p)}, \tag{6.106}$$

where

$$\begin{aligned} A_l(p) &= (p+\alpha l)\left[(p+\alpha l)^2+4\omega_0^2\right], \\ B_l(p) &= 4\left\langle z^2\right\rangle \omega_0^4(l+1)(N-l)\left[2p+\alpha(2l+1)^2\right]. \end{aligned}$$

At $N=1$, we obtain the solution (6.28), page 155 corresponding to single telegrapher's process. Setting $\langle z^2\rangle=\sigma^2/N$ and proceeding to the limit $N\to\infty$, we obtain the solution for the Gaussian Markovian process in the form of the infinite continued fraction (6.106), where

$$\begin{aligned} A_l(p) &= (p+\alpha l)\left[(p+\alpha l)^2+4\omega_0^2\right], \\ B_l(p) &= 4\sigma^2\omega_0^4(l+1)\left[2p+\alpha(2l+1)^2\right]. \end{aligned}$$

6.3.3 The square of the Gaussian Markovian process

The finite-dimensional approximation of the Gaussian Markovian process (6.89), page 168 is practicable for describing fluctuations of dynamic systems of the form $F(z(t))$, where $z(t)$ is the Gaussian Markovian process, too.

For example, for system $F(z(t))=z^2(t)-\langle z^2(t)\rangle$, the finite-dimensional approximation has the form

$$F(z(t))=\sum_{i\neq k}^{N} z_i(t)z_k(t).$$

In this case, the mean value of the solution to system of equations (6.10), page 152 (we assume that operators $\hat{A}(t)$ and $\hat{B}(t)$ are matrixes)

$$\frac{d}{dt}\mathbf{x}(t)=A\mathbf{x}(t)+\sum_{i\neq k}^{N} z_i(t)z_k(t)B\mathbf{x}(t) \tag{6.107}$$

will satisfy the closed system of $([N/2]-1)$ vector equations for functions

$$\mathbf{X}_n(t)=\langle z_1(t)...z_{2n}(t)\mathbf{x}(t)\rangle, \quad n=1,...,[N/2]; \quad \mathbf{X}_0(t)=\langle\mathbf{x}(t)\rangle.$$

Here, $[N/2]$ is the integer part of $N/2$.

It is obvious that functions $\mathbf{X}_n(t)$ satisfy the equations

$$\left(\frac{d}{dt}+2\alpha n\right)\mathbf{X}_n(t)-A\mathbf{X}_n(t)=B\left\langle z_1(t)...z_{2n}(t)\sum_{i\neq k}^{N} z_i(t)z_k(t)\mathbf{x}(t)\right\rangle. \tag{6.108}$$

The further analysis is similar to the derivation of system of equations (6.94). Divide the sum over i and k in the right-hand side of Eq. (6.108) into four regions (Fig. 6.2).

In region (1), both functions $z_i(t)$ and $z_k(t)$ will be extinguished by the corresponding functions of product $z_1(t)...z_{2n}(t)$. The number of such terms is $2n(2n-1)$; consequently, in region (1), the right-hand side of Eq. (6.108) assumes the form

$$2n(2n-1)\left\langle z^2\right\rangle^2 B\mathbf{X}_{n-1}(t)$$

In region (2), none of functions $z_i(t)$ and $z_k(t)$ is extinguished, and we obtain that the corresponding term in the right-hand side of Eq. (6.108) has the form

$$(N-2n)(N-2n-1)B\mathbf{X}_{n+1}(t)$$

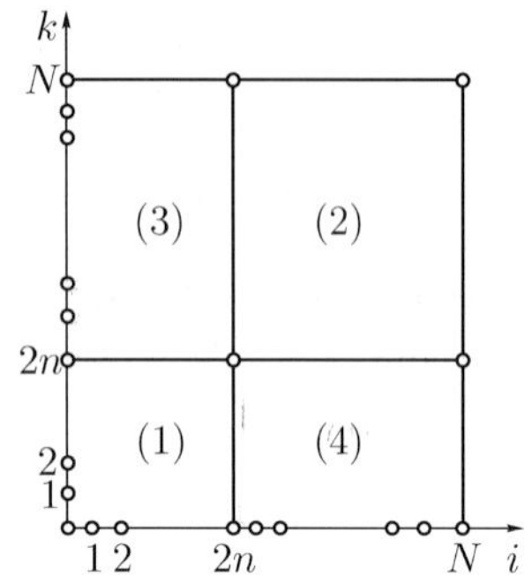

Figure 6.2: Schematic of index division in sum (6.108).

In regions (3) and (4), only one of functions $z_i(t)$ and $z_k(t)$ is extinguished. The number of such terms is $4n(N-2n)$, so that the corresponding term in the right-hand side of Eq. (6.108) has the form

$$4n(N-2n)\left\langle z^2\right\rangle B\mathbf{X}_n(t)$$

As a result, Eq. (6.108) assumes the form of the closed system of recurrence equations

$$\left[\left(\frac{d}{dt}+2\alpha n\right)E-A-4n(N-2n)\left\langle z^2\right\rangle B\right]\mathbf{X}_n(t)$$
$$=2n(2n-1)\left\langle z^2\right\rangle^2 B\mathbf{X}_{n-1}(t)+(N-2n)(N-2n-1)B\mathbf{X}_{n+1}(t),$$

where $n=0,1,...,[N/2]$. It is obvious that, for constant matrixes A and B, the solution to this system again has the form of a finite segment of continued fraction. The simplest approximations with $N=2$ and $N=3$ give the closed systems of only two vector equations.

6.4 Markovian processes with finite-dimensional phase space

All considered processes — telegrapher's and generalized telegrapher's processes and process $z(t)=z_1(t)+...+z_N(t)$, where $z_i(t)$ are statistically independent telegrapher's processes, — are special cases of the Markovian processes $z(t)$ with finite number of states (or with finite-dimensional phase space). We assume that possible values of process $z(t)$ are in the general case $z_1,...,z_n$. As a result, all realizations of process $z(t)$ satisfy the identity

$$(z(t)-z_1)(z(t)-z_2)...(z(t)-z_n)\equiv 0,$$

and, consequently,

$$z^n(t)=(z_1+...+z_n)z^{n-1}(t)+(-1)^{n-1}z_1...z_n. \tag{6.109}$$

In this case, the mean value of the solution to the system of equations

$$\frac{d}{dt}\mathbf{x}(t)=\hat{A}(t)\mathbf{x}(t)+z(t)\hat{B}(t)\mathbf{x}(t),\quad \mathbf{x}(0)=\mathbf{x}_0 \tag{6.110}$$

will again satisfy a closed system of equations. Indeed, averaging Eq. (6.110) and repeatedly using the differentiation formula (6.3), page 151 for correlators

$$\left\langle z^k(t)\mathbf{x}(t)\right\rangle \quad (k=1,...,n-1),$$

we reach function $\langle z^n(t)\mathbf{x}(t)\rangle$ at the last step. Because this function is expressed in terms of the functions of preceding steps (see Eq. (6.109)), we obtain the closed system of vector equations of the nth order.

6.4.1 Two-state process

Consider the process with two states z_1, z_2 and respective transition probabilities ν and μ as an example. In this case, Eq. (6.109) assumes the form

$$z^2(t) = (z_1+z_2)z(t) - z_1 z_2. \tag{6.111}$$

Averaging Eq. (6.110), we obtain

$$\left(E\frac{d}{dt} - \hat{A}(t)\right)\langle\mathbf{x}(t)\rangle = \hat{B}(t)\langle z(t)\mathbf{x}(t)\rangle, \quad \langle\mathbf{x}(0)\rangle = \mathbf{x}_0. \tag{6.112}$$

According to Eq. (6.3), page 151, correlation $\langle z(t)\mathbf{x}(t)\rangle$ is given by the formula

$$\frac{d}{dt}\langle z(t)\mathbf{x}(t)\rangle = \left\langle z(t)\frac{d}{dt}\mathbf{x}(t)\right\rangle + \left\langle \mathbf{x}(t)\left[\hat{L}^{+}(z)z(t)\right]\right\rangle,$$

where the kinetic and conjugated operators are the following matrixes (see page 202)

$$\hat{L}(z) = \begin{pmatrix} -\nu & \mu \\ \nu & -\mu \end{pmatrix}, \quad \hat{L}^{+}(z) = \begin{pmatrix} -\nu & \nu \\ \mu & -\mu \end{pmatrix}. \tag{6.113}$$

Because the action of operator $\hat{L}^{+}(z)$ on $z(t)$ is representable in the form

$$\begin{aligned}\hat{L}^{+}(z)z(t) &= \begin{pmatrix} -\nu & \nu \\ \mu & -\mu \end{pmatrix}\begin{pmatrix} z_1 \\ z_2 \end{pmatrix} = \begin{pmatrix} -\nu(z_1-z_2) \\ \mu(z_1-z_2) \end{pmatrix} \\ &= (\nu z_2 + \mu z_1) - (\nu+\mu)z(t),\end{aligned}$$

we can rewrite the equation for correlation $\langle z(t)\mathbf{x}(t)\rangle$ as follows

$$\begin{aligned}&\left\{E\left(\frac{d}{dt}+\mu+\nu\right) - \hat{A}(t) - (z_1+z_2)\hat{B}(t)\right\}\langle z(t)\mathbf{x}(t)\rangle \\ &\quad = \left\{(\nu z_2+\mu z_1)E - z_1 z_2\hat{B}(t)\right\}\langle\mathbf{x}(t)\rangle. \end{aligned} \tag{6.114}$$

Equations (6.112) and (6.114) form the closed system of two vector equations.

Note that, in the special case of the scalar equation with parameters $A = 0$ and $B(t) = iv(t)$, the solution to Eq. (6.110) is

$$x(t) = \exp\left\{i\int_0^t d\tau z(\tau)v(\tau)\right\},$$

so that the mean value of this solution coincides with the characteristic functional of random process $z(t)$

$$\Phi[t; v(\tau)] = \langle x(t) \rangle = \left\langle \exp \left\{ i \int_0^t d\tau z(\tau) v(\tau) \right\} \right\rangle .$$

In this case, we can obtain the differential equation for functional $\Phi[t; v(\tau)]$ by eliminating function $\langle z(t) x(t) \rangle$ from Eqs. (6.112) and (6.114):

$$\begin{aligned} &\frac{d^2}{dt^2} \Phi[t; v(\tau)] + \left[\mu + \nu - \frac{1}{v(t)} \frac{dv(t)}{dt} - iv(t)(z_1 + z_2) \right] \frac{d}{dt} \Phi[t; v(\tau)] \\ &\quad - \left[iv(t)(\nu z_2 + \mu z_1) + z_1 z_2 v^2(t) \right] \Phi[t; v(\tau)] = 0. \end{aligned} \tag{6.115}$$

6.5 Causal stochastic integral equations

Consider the simplest one-dimensional integral equation

$$S(t, t') = g(t - t') \theta(t - t') + \Lambda \int_0^t d\tau g(t - \tau) z(\tau) S(\tau, t'), \tag{6.116}$$

where $z(t)$ is the random function of time, $g(t - t')$ is the deterministic function, Λ is a constant parameter, and $\theta(t)$ is the Heaviside step function. Iterating this equation, we can see that its solution $S(t, t')$ depends on random function $z(\tau)$ only for $t' \leq \tau \leq t$, which means that the causality condition

$$\frac{\delta}{\delta z(\tau)} S(t, t') = 0 \quad \text{for} \quad \tau < t', \quad \tau > t$$

holds. In addition, $S(t, t') \sim \theta(t - t')$.

Average Eq. (6.116) over an ensemble of realizations of function $z(t)$. In the case of stationary process $z(t)$, function

$$\langle S(t, t') \rangle = \langle S(t - t') \rangle ,$$

and the result of averaging assumes the form

$$\langle S(t - t') \rangle = g(t - t') \theta(t - t') + \Lambda \int_0^t d\tau g(t - \tau) \int_0^\tau d\tau' Q(\tau - \tau') \langle S(\tau' - t') \rangle , \tag{6.117}$$

where $Q(t) \sim \theta(t)$ is the mass function defined by the equality

$$\Lambda \langle z(t) S(t, t') \rangle = \int_0^t d\tau Q(t - \tau) \langle S(\tau - t') \rangle .$$

Performing the Laplace transform in Eq. (6.117) with respect to $t - t'$, we obtain

$$\langle S \rangle_p = g(p) + g(p) Q(p) \langle S \rangle_p , \tag{6.118}$$

where

$$\Lambda \langle z(t)S(t,t')\rangle_p = Q(p)\langle S\rangle_p . \tag{6.119}$$

Note that if the integral equation (6.116) can be reduced to the differential equation

$$L\left(\frac{d}{dt}\right)S(t,t') = \Lambda z(t)S(t,t') + \delta(t-t'),$$

then $g(p) = L^{-1}(p)$.

According to the material of Sect. 5.3, the structure of function $Q(p)$ can be determined from the auxiliary equation

$$\tilde{S}(t,t') = g(t-t')\theta(t-t') + \Lambda \int_0^t d\tau g(t-\tau)\left[z(\tau)+\eta(\tau)\right]\tilde{S}(\tau,t'), \tag{6.120}$$

where $\eta(\tau)$ is arbitrary deterministic function. If we average Eq. (6.120) and denote the solution to averaged equation $G[t,t';\eta(\tau)]$, then vertex function $\Gamma(t,t_1,t') = \Gamma(t-t_1,t_1-t')$ will be given by the equality

$$\Gamma(t,t_1,t') = -\left.\frac{\delta}{\delta\eta(t_1)}G^{-1}[t,t';\eta]\right|_{\eta=0}.$$

The variational derivative $\delta G/\delta\eta$ at $\eta(\tau)=0$ can be expressed in terms of vertex function $\Gamma(t,t_1,t')$ and average Green's function by the relationship (5.37), page 107

$$\left.\frac{\delta}{\delta\eta(t_1)}G[t,t';\eta]\right|_{\eta=0} = \int d\tau_1 \int d\tau_2 \langle S(t-\tau_1)\rangle \Gamma(\tau_1-t_1,t_1-\tau_2)\langle S(\tau_2-t')\rangle, \tag{6.121}$$

where the domain of integration is defined by the condition of positiveness of all arguments. Performing the Laplace transform in (6.121) with respect to $(t-t_1)$ and (t_1-t'), we obtain the equality

$$\frac{\delta G}{\delta\eta}(p,q) = \langle S\rangle_p \Gamma(p,q)\langle S\rangle_q \tag{6.122}$$

that makes it possible to determine the Laplace transform of the vertex function. In this case, the mass function is expressed through the characteristic functional of process $z(t)$.

Variational derivative $\delta G/\delta\eta$ in the right-hand side of Eq. (6.122) can be obtained by varying Eq. (6.120) with respect to $\eta(t_1)$ followed by setting $\eta(t)=0$ and averaging the obtained equation. If Eq. (6.120) can be averaged analytically, variational derivative $\delta G/\delta\eta$ can be obtained by varying the averaged equation with respect to $\eta(t)$.

Consider the realization of the above scheme for different processes $z(t)$.

6.5.1 Telegrapher's random process

Let $z(t)$ is telegrapher's process with correlation function

$$\langle z(t)z(t')\rangle = \left\langle z^2\right\rangle e^{-\alpha|t-t'|}.$$

Averaging Eq. (6.120), we obtain

$$G(t,t') = g(t-t')\theta(t-t')$$
$$+\Lambda \int_0^t d\tau g(t-\tau)\eta(\tau)G(\tau,t') + \Lambda \int_0^t d\tau g(t-\tau)\left\langle z(\tau)\tilde{S}(\tau,t')\right\rangle. \tag{6.123}$$

Equation (6.123) is unclosed because it contains new unknown function $\left\langle z(\tau)\tilde{S}(\tau,t')\right\rangle$. To obtain the equation for this function, we multiply Eq. (6.120) by $z(t)$ and average the result

$$\left\langle z(t)\tilde{S}(t,t')\right\rangle = \Lambda\int\limits_0^t d\tau g(t-\tau)\eta(\tau)\left\langle z(t)\tilde{S}(\tau,t')\right\rangle$$

$$+\Lambda\int\limits_0^t d\tau g(t-\tau)\left\langle z(t)z(\tau)\tilde{S}(\tau,t')\right\rangle. \tag{6.124}$$

Taking into account formula (4.32), page 84

$$\left\langle z(t)R[t';z(\tau)]\right\rangle = e^{-\alpha(t-t')}\left\langle z(t')R[t';z(\tau)]\right\rangle, \tag{6.125}$$

which is valid for arbitrary functional $R[t';z(\tau)]$ such that $t\geq t'$, we obtain the equation ($z^2(t)\equiv\langle z^2\rangle$)

$$\left\langle z(t)\tilde{S}(t,t')\right\rangle = \Lambda\int\limits_0^t d\tau g(t-\tau)\eta(\tau)e^{-\alpha(t-\tau)}\left\langle z(\tau)\tilde{S}(\tau,t')\right\rangle$$

$$+\Lambda\left\langle z^2\right\rangle\int\limits_0^t d\tau g(t-\tau)e^{-\alpha(t-\tau)}G(\tau,t'). \tag{6.126}$$

System of equations (6.123) and (6.126) is the closed system. Setting $\eta=0$ in this system and performing then the Laplace transform with respect to $(t-t')$, we obtain the algebraic system

$$\langle S\rangle_p = g(p)+\Lambda g(p)\langle zS\rangle_p, \quad \langle zS\rangle_p = \Lambda\left\langle z^2\right\rangle g(p)\langle S\rangle_p, \tag{6.127}$$

whose solution is as follows:

$$\langle S\rangle_p = \frac{g(p)}{1-\Lambda^2\langle z^2\rangle g(p)g(p+\alpha)}, \quad \langle zS\rangle_p = \frac{\Lambda\langle z^2\rangle g(p)g(p+\alpha)}{1-\Lambda^2\langle z^2\rangle g(p)g(p+\alpha)}. \tag{6.128}$$

According to Eq. (6.119), the mass function $Q(p)$ is

$$Q(p) = \Lambda^2\left\langle z^2\right\rangle g(p+\alpha). \tag{6.129}$$

In order to determine the vertex function, we vary Eqs. (6.123) and (6.126) with respect to $\eta(t_1)$, set $\eta(t)=0$, and perform the Laplace transform with respect to $(t-t_1)$ and (t_1-t'). As a result, we obtain the algebraic system

$$\begin{aligned}\frac{\delta G}{\delta\eta}(p,q) &= \Lambda g(p)\langle S\rangle_p + \Lambda g(p)\left\langle z\frac{\delta\tilde{S}}{\delta\eta}\right\rangle_{p,q},\\ \left\langle z\frac{\delta\tilde{S}}{\delta\eta}\right\rangle_{p,q} &= \Lambda g(p+\alpha)\langle zS\rangle_q + \Lambda\left\langle z^2\right\rangle g(p+\alpha)\frac{\delta G}{\delta\eta}(p,q),\end{aligned} \tag{6.130}$$

whose solution is

$$\frac{\delta G}{\delta\eta}(p,q) = \Lambda\langle S\rangle_p\left\{1+\Lambda^2\left\langle z^2\right\rangle g(p+\alpha)g(q+\alpha)\right\}\langle S\rangle_q, \tag{6.131}$$

where we used Eq. (6.127). Comparing Eq. (6.131) with Eq. (6.122), we obtain the vertex function in the form

$$\Gamma(p,q) = \Lambda\left\{1+\Lambda^2\left\langle z^2\right\rangle g(p+\alpha)g(q+\alpha)\right\}. \tag{6.132}$$

6.5.2 Generalized telegrapher's random process

Let $z(t)$ is the generalized telegrapher's process. Averaging Eq. (6.120), we obtain Eq. (6.123). Then we should derive the equation for function $\left\langle F_\lambda(t)\tilde{S}(t,t')\right\rangle$, where

$$F_\lambda(t) = \frac{1}{1+\lambda z(t)} - C_0(\lambda), \quad C_k(\lambda) = \left\langle \frac{a^k}{1+\lambda a}\right\rangle_a \quad \left(\langle F_\lambda(t)\rangle_a = 0\right),$$

and λ is the arbitrary parameter. Multiplying Eq. (6.120) by $F_\lambda(t)$ and averaging the result, we obtain

$$\begin{aligned}\left\langle F_\lambda(t)\tilde{S}(t,t')\right\rangle &= \Lambda \int_0^t d\tau g(t-\tau)e^{-\alpha(t-\tau)}\eta(\tau)\left\langle F_\lambda(t)\tilde{S}(\tau,t')\right\rangle \\ &-\Lambda \int_0^t d\tau g(t-\tau)e^{-\alpha(t-\tau)}\left\{\frac{1}{\lambda}\left\langle F_\lambda(\tau)\tilde{S}(\tau,t')\right\rangle\right. \\ &\left. - C_1(\lambda)G(\tau,t') + C_0(\lambda)\left\langle z(\tau)\tilde{S}(\tau,t')\right\rangle\right\}. \end{aligned} \tag{6.133}$$

Deriving Eq. (6.133), we used the equality

$$\langle F_\lambda(t)R[t';z(\tau)]\rangle = e^{-\alpha(t-t')}\langle F_\lambda(t')R[t';z(\tau)]\rangle,$$

which is valid for arbitrary functional $R[t;z(\tau)]$ of random process $z(t)$ for $t' \geqslant t$, and the identity (6.128)

$$z(t)F(t) = -\frac{1}{\lambda}F(t) + C_1(t) - z(t)C_0(\lambda).$$

To determine the mass function, we set $\eta(t) = 0$ in Eqs. (6.123) and (6.133) and perform the Laplace transform. As a result, we obtain the system of equations

$$\begin{aligned} &\langle S\rangle_p = g(p) + \Lambda g(p)\langle zS\rangle_p, \\ &\langle F_\lambda(t)S(t,t')\rangle_p\left\{1+\frac{\Lambda}{\lambda}g(p+\alpha)\right\} \\ &\quad = \Lambda g(p+\alpha)\left\{C_1(\lambda)\langle S\rangle_p - C_0(\lambda)\langle F_\lambda(t)S(t,t')\rangle_p\right\} \end{aligned}$$

valid for arbitrary λ. Setting

$$\lambda = \lambda_p = -\Lambda g(p+\alpha),$$

we obtain the algebraic relationship between $\langle zS\rangle_p$ and $\langle S\rangle_p$

$$\langle z(t)S(t,t')\rangle_p = \langle S\rangle_p \frac{C_1(\lambda_p)}{C_0(\lambda_p)} \tag{6.134}$$

and, consequently,

$$\langle S\rangle_p = \frac{g(p)}{1-\Lambda g(p)\frac{C_1(\lambda_p)}{C_0(\lambda_p)}}. \tag{6.135}$$

Using (6.134), function $\langle F_\lambda(t)S(t,t')\rangle_p$ for arbitrary λ can be represented in the form

$$\langle F_\lambda(t)S(t,t')\rangle_p = -\frac{\lambda\lambda_p\langle S\rangle_p\left[C_1(\lambda)C_0(\lambda_p) - C_0(\lambda)C_1(\lambda_p)\right]}{(\lambda-\lambda_p)\,C_0(\lambda_p)}. \tag{6.136}$$

In this case, the mass function, as follows from Eq. (6.134), is

$$Q(p) = \Lambda \frac{C_1(\lambda_p)}{C_0(\lambda_p)}. \tag{6.137}$$

To determine the vertex function, we vary Eqs. (6.123) and (6.133) with respect to $\eta(t_1)$, set $\eta(t) = 0$, and perform the Laplace transform. As a result, we obtain the system

$$\begin{aligned}
&\frac{\delta G}{\delta \eta}(p,q) = \Lambda g(p) \langle S \rangle_q + \Lambda g(p) \left\langle z \frac{\delta \tilde{S}}{\delta \eta} \right\rangle_{p,q}, \\
&\left\langle F_\lambda \frac{\delta \tilde{S}}{\delta \eta} \right\rangle_p = \left\{ 1 + \frac{\Lambda}{\lambda} g(p+\alpha) \right\} \Lambda g(p+\alpha) F_\lambda S_q \\
&\quad + \Lambda g(p+\alpha) \left\{ C_1(\lambda) \frac{\delta G}{\delta \eta}(p,q) - C_0(\lambda) \left\langle z \frac{\delta \tilde{S}}{\delta \eta} \right\rangle_{p,q} \right\}.
\end{aligned} \tag{6.138}$$

Then, setting $\lambda = \lambda_p$ in Eq. (6.138), we obtain the algebraic system for $\frac{\delta G}{\delta \eta}(p,q)$ and $\left\langle z \frac{\delta \tilde{S}}{\delta \eta} \right\rangle_{p,q}$, whose solution can be represented as follows

$$\frac{\delta G}{\delta \eta}(p,q) = \Lambda \langle S \rangle_p \left\{ 1 + \Lambda \frac{g(p+\alpha) g(q+\alpha)}{g(p+\alpha) - g(q+\alpha)} \left[\frac{C_1(\lambda_p)}{C_0(\lambda_p)} - \frac{C_1(\lambda_q)}{C_0(\lambda_q)} \right] \right\} \langle S \rangle_q, \tag{6.139}$$

where we used Eqs. (6.135), (6.136). Consequently, the vertex function is

$$\Gamma(p,q) = \Lambda \left\{ 1 + \Lambda \frac{g(p+\alpha) g(q+\alpha)}{g(p+\alpha) - g(q+\alpha)} \left[\frac{C_1(\lambda_p)}{C_0(\lambda_p)} - \frac{C_1(\lambda_q)}{C_0(\lambda_q)} \right] \right\}. \tag{6.140}$$

If the probability distribution of quantity a has the form

$$p(a) = \frac{1}{2} \left[\delta(a - a_0) + \delta(a + a_0) \right],$$

then $C_1(\lambda)/C_0(\lambda) = -\lambda a_0^2$, and we turn back to telegrapher's process with parameter $\langle z^2 \rangle = a_0^2$.

If a is the continuous random quantity with zero-valued mean and sufficiently small variance, then

$$C_0(\lambda) \approx 1, \quad C_1(\lambda) \approx -\lambda \left\langle a^2 \right\rangle,$$

and the vertex function assumes the form

$$\Gamma(p,q) = \Lambda \left\{ 1 + \Lambda^2 \left\langle a^2 \right\rangle g(p+\alpha) g(q+\alpha) \right\}. \tag{6.141}$$

However, Eq. (6.141) is valid only if obvious inequalities

$$|\lambda^2| \left\langle a^2 \right\rangle \ll 1 \quad (\lambda = \lambda_p, \lambda_q) \tag{6.142}$$

are satisfied.

6.5.3 Gaussian Markovian process

Let $z(t)$ is the Gaussian Markovian process with correlation function

$$\langle z(t)z(t)\rangle = \left\langle z^2\right\rangle e^{-\alpha|t-t'|}.$$

This process can be obtained from the process with a finite number of states

$$\xi_N = z_1(t) + ... + z_N(t),$$

where $z_i(t)$ are the statistically independent telegrapher's processes with $\langle z^2\rangle = \sigma^2/N$, using limit process $N \to \infty$.

So, we consider Eq. (6.120) with $z(t) = \xi_N(t)$ and introduce functions

$$G_l(t,t') = \left\langle z_1(t)...z_l(t)\tilde{S}(t,t')\right\rangle \quad (G_0(t,t') = G(t,t')) . \tag{6.143}$$

Multiplying Eq. (6.120) by product $z_1(t)...z_N(t)$, averaging the result, and using Eq. (6.125), we can obtain the recurrence equation ($l = 0, 1, ..., N$)

$$G_l(t,t') = g(t-t')\theta(t-t')\delta_{l,0} + \Lambda \int\limits_0^t d\tau g(t-\tau)\eta(\tau)e^{-\alpha l(t-\tau)}G_l(\tau,t')$$
$$+\Lambda \int\limits_0^t d\tau g(t-\tau)e^{-\alpha l(t-\tau)} \left\{l\left\langle z^2\right\rangle G_{l-1}(\tau,t') + (N-l)G_{l+1}(\tau,t')\right\}. \tag{6.144}$$

Setting $\eta(t) = 0$ and performing Laplace transform with respect to $(t-t')$, we obtain the algebraic recurrence equation

$$G_l(p) = g_0(p)\delta_{l,0} + \Lambda g_l(p)\left\{l\left\langle z^2\right\rangle G_{l-1}(p) + (N-l)G_{l+1}(p)\right\}, \tag{6.145}$$

where $g_l(p) = g(l + \alpha l)$.

The solution to Eq. (6.145) has the form of a finite segment of the continued fraction

$$G_l(p) = \Lambda g_l(p) l\left\langle z^2\right\rangle K_l(p)G_{l-1}(p), \quad l = 1, ..., N, \tag{6.146}$$

where

$$K_l(p) = \frac{1}{1-\gamma_l(p)K_{l+1}(p)}, \quad \gamma_l(p) = \Lambda^2\left\langle z^2\right\rangle (l+1)(N-l)g_l(p)g_{l+1}(p). \tag{6.147}$$

Consequently,

$$G_l(p) = \Lambda^l \left\langle z^2\right\rangle^l l!\left\{g_l(p)K_l(p)\right\}! g_0(p)K_0(p), \tag{6.148}$$

where $f_l!$ stands for the product $f_1...f_l$. Taking into account the fact that

$$\langle \xi_N(t)S(t,t')\rangle_p = NG_1(p),$$

we obtain the expression for the mass function:

$$Q_N(p) = \Lambda^2 N\left\langle z^2\right\rangle g_1(p)K_1(p). \tag{6.149}$$

Setting now $\langle z^2 \rangle = \sigma^2/N$ and proceeding to the limit $N \to \infty$, we obtain the mass function for the Gaussian Markovian process:

$$Q(p) = \Lambda^2 \sigma^2 g_1(p) K_1(p), \tag{6.150}$$

where $K_1(p)$ is the infinite continued fraction (6.147) with parameter

$$\tilde{\gamma}_l(p) = \Lambda^2 \sigma^2 (l+1) g_l(p) g_{l+1}(p). \tag{6.151}$$

Calculating the vertex function in the cases of telegrapher's and generalized telegrapher's processes, we straightforwardly followed the procedure valid for arbitrary integral equations. The goal of that consideration was to illustrate the general procedure. In the case of Eq. (6.116), we can immediately obtain the expression for the vertex function if only the solution to the Dyson equation is known. Indeed, according to Eqs. (6.121) and (5.30), page 105, we have the following relationship for Eq. (6.116)

$$\Lambda \langle S(t,t_0) S(t_0,t') \rangle = \int\int d\tau_1 d\tau_2 \langle S(t-\tau_1) \rangle \Gamma(\tau_1 - t_1, t_1 - \tau_2) \langle S(\tau_2 - t') \rangle . \tag{6.152}$$

Let now random process $z(t)$ is a function of process $\xi_N(t)$. Then, we can split the correlator in the left-hand side of Eq. (6.152) using formula (4.27), which assumes in this case the form

$$\begin{aligned} &\langle S(t,t_0) S(t_0,t') \rangle \\ &= \sum_{k=0}^{N} C_N^k \frac{1}{\langle z^2 \rangle^k} \langle z_1(t_0) \ldots z_k(t_0) S(t,t_0) \rangle \langle z_1(t_0) \ldots z_k(t_0) S(t_0,t') \rangle . \end{aligned} \tag{6.153}$$

Performing the Laplace transform with respect to $(t - t_0)$ and $(t_0 - t')$, we obtain the equality

$$\langle SS \rangle_{p,q} = \sum_{k=0}^{N} C_N^k \frac{1}{\langle z^2 \rangle^k} G_k(p) G_k(q), \tag{6.154}$$

where function $G_k(p)$ is given by Eq. (6.143). Consequently, we obtain the following expression for the vertex function $\Gamma(p,q)$

$$\Gamma(p,q) = \frac{1}{\Lambda} \sum_{k=0}^{N} C_N^k \frac{1}{\langle z^2 \rangle^k} \frac{G_k(p) G_k(q)}{G_0(p) G_0(q)}. \tag{6.155}$$

For $z(t) = \xi_N(t)$, functions $G_k(p)$ are given by Eq. (6.148), and we obtain

$$\Gamma_N(p,q) = \Lambda \left[1 + \sum_{k=0}^{N} \Lambda^{2k} \left\langle z^2 \right\rangle^k \frac{N!k!}{(N-k)!} \{ g_k(p) g_k(q) K_k(p) K_k(q) \}! \right] . \tag{6.156}$$

At $N = 1$, we turn back to the case of a single telegrapher's process, and Eq. (6.156) grades into (6.132).

Setting $\langle z^2 \rangle = \sigma^2/N$ and proceeding to the limit $N \to \infty$, we obtain the vertex function for the Gaussian Markovian process in the form of the infinite series

$$\Gamma(p,q) = \Lambda \left[1 + \sum_{k=0}^{\infty} k! \Lambda^{2k} \sigma^{2k} \{ g_k(p) g_k(q) K_k(p) K_k(q) \}! \right] , \tag{6.157}$$

whose terms include infinite continued fractions (6.147) with parameter (6.151). Two first terms of series (6.157) are as follows

$$\Gamma(p,q) = \Lambda\left[1 + \Lambda^2\sigma^2 g_1(p)g_1(q)K_1(p)K_1(q) + ...\right].$$

Now, we dwell on approximations commonly used in analyzing stochastic integral equations.

First of all, we consider the Gaussian Markovian process. In this case, the mass function is related to the vertex function by the formula

$$Q(t-t') = \Lambda\sigma^2 \int\int d\tau_1 d\tau_2 e^{-\alpha(t-\tau)} \langle S(t-\tau_1)\rangle \Gamma(\tau_1-\tau, \tau-t'), \tag{6.158}$$

where the domain of integration is defined by the condition of positiveness of all arguments. Performing the Laplace transform in Eq. (6.158) with respect to $(t-t')$, we obtain the equality

$$Q(p) = \Lambda\sigma^2 \langle S\rangle_{p+\alpha} \Gamma(p+\alpha, p). \tag{6.159}$$

The Kraichnan approximation corresponds to the replacement of vertex function $\Gamma(p+\alpha, p)$ in Eq. (6.158) by Λ, and the Bourret approximation assumes additionally substitution of $\langle S\rangle_{p+\alpha}$ with $g_1(p)$.

The solution to the Dyson equation (6.118) depends primarily on the poles and other significant singularities of function $g_1(p)$. Denote p_0 the singular point of this function. Then, if the condition

$$\Lambda^2\sigma^2|g_1(p_0)|^2|K_1(p_0)|^2 \ll 1 \tag{6.160}$$

holds, we can neglect all terms of series (6.157) excluding the first one. Functions $K_1(p_0)$ themselves depend on parameter $\beta^2 = \Lambda^2|g_1(p_0)|^2$, and $|K_1(p_0)| \sim 1$ for $\beta^2 \ll 1$.

Thus, we can replace vertex function $\Gamma(p,q)$ by Λ under the condition that

$$\beta^2 = \Lambda^2|g_1(p_0)|^2 \ll 1. \tag{6.161}$$

Earlier, we showed that function $\langle S\rangle_{p_0+\alpha}$ also has a small parameter. In the first approximation with respect to this small parameter, the mass function is

$$Q(p) = \Lambda^2 g_1(p), \tag{6.162}$$

which corresponds to the Bourret approximation. Thus, the Kraichnan approximation fails in the context of this problem, whereas the Bourret approximation represents the first term of the asymptotic expansion of the solution in the above small parameter.

Note that the mass function in the Bourret approximation (6.162) coincides with the mass function for telegrapher's process (6.129).

The limit process $\alpha \to \infty$ in the solutions obtained for all above processes results in the Gaussian delta-correlated process with correlation function

$$\langle z(t)z(t')\rangle = 2\sigma^2\tau_0\delta(t-t'), \quad \tau_0 = 1/\alpha.$$

It is clear that this solution can be obtained immediately from Eq. (6.116).

Part III

Asymptotic and approximate methods for analyzing stochastic equations

Chapter 7

Gaussian random field delta-correlated in time (ordinary differential equations)

In the foregoing chapters, we considered in detail the general methods of analyzing stochastic equations. Here, we give an alternative and more detailed consideration of the approximation of the Gaussian random delta-correlated (in time) field in the context of stochastic equations and discuss the physical meaning of this widely used approximation.

7.1 The Fokker–Planck equation

Let vector function $\mathbf{x}(t) = \{x_1(t), x_2(t), ..., x_n(t)\}$ satisfies the dynamic equation

$$\frac{d}{dt}\mathbf{x}(t) = \mathbf{v}(\mathbf{x}, t) + \mathbf{f}(\mathbf{x}, t), \qquad \mathbf{x}(t_0) = \mathbf{x}_0, \tag{7.1}$$

where $v_i(\mathbf{x}, t)$ $(i = 1, ..., n)$ are the deterministic functions and $f_i(\mathbf{x}, t)$ are the random functions of $(n + 1)$ variable that have the following properties:

(a) $f_i(\mathbf{x}, t)$ is the Gaussian random field in the $(n + 1)$-dimensional space $(\mathbf{x}, t)$;

(b) $\langle f_i(\mathbf{x}, t)\rangle = 0$.

For definiteness, we assume that t is the temporal variable and $\mathbf{x}$ is the spatial variable.

Statistical characteristics of field $f_i(\mathbf{x}, t)$ are completely described by the correlation tensor

$$B_{ij}(\mathbf{x}, t; \mathbf{x}', t') = \langle f_i(\mathbf{x}, t) f_j(\mathbf{x}', t')\rangle .$$

Because Eq. (7.1) is the first-order equation with the initial value, its solution satisfies the dynamic causality condition

$$\frac{\delta}{\delta f_j(\mathbf{x}', t')} x_i(t) = 0 \quad \text{for} \quad t' < t_0 \quad \text{and} \quad t' > t, \tag{7.2}$$

which means that solution $\mathbf{x}(t)$ depends only on values of function $f_j(\mathbf{x}, t')$ for times t' preceding time t, i.e., $t_0 \le t' \le t$. In addition, we have the following equality for the variational derivative

$$\frac{\delta}{\delta f_j(\mathbf{x}', t - 0)} x_i(t) = \delta_{ij}\delta(\mathbf{x}(t) - \mathbf{x}'). \tag{7.3}$$

Nevertheless, the statistical relationship between $\mathbf{x}(t)$ and function values $f_j(\mathbf{x}, t'')$ for consequent times $t'' > t$ can exist, because such function values $f_j(\mathbf{x}, t'')$ are correlated with values $f_j(\mathbf{x}, t')$ for $t' \le t$. It is obvious that the correlation between function $\mathbf{x}(t)$ and consequent values $f_j(\mathbf{x}, t'')$ is appreciable only for $t'' - t \le \tau_0$, where τ_0 is the correlation radius of field $\mathbf{f}(\mathbf{x}, t)$ with respect to variable t.

For many actual physical processes, characteristic temporal scale T of function $\mathbf{x}(t)$ significantly exceeds correlation radius τ_0 ($T \gg \tau_0$); in this case, the problem has the small parameter τ_0/T that can be used for constructing an approximate solution.

In the first approximation in this small parameter, one can consider the asymptotic solution for $\tau_0 \to 0$. In this case values of function $\mathbf{x}(t')$ for $t' < t$ will be independent of values $\mathbf{f}(\mathbf{x}, t'')$ for $t'' > t$ not only functionally, but also statistically. This approximation is equivalent to the replacement of correlation tensor B_{ij} with the effective tensor

$$B^{\mathrm{eff}}_{ij}(\mathbf{x}, t; \mathbf{x}', t') = 2\delta(t - t')F_{ij}(\mathbf{x}, \mathbf{x}'; t). \tag{7.4}$$

Here, quantity $F_{ij}(\mathbf{x}, \mathbf{x}'; t)$ is determined from the condition that integrals of $B_{ij}(\mathbf{x}, t; \mathbf{x}', t')$ and $B^{\mathrm{eff}}_{ij}(\mathbf{x}, t; \mathbf{x}', t')$ over t' coincide

$$F_{ij}(\mathbf{x}, \mathbf{x}'; t) = \frac{1}{2}\int\limits_{-\infty}^{\infty} dt' B_{ij}(\mathbf{x}, t; \mathbf{x}', t'),$$

which just corresponds to the passage to the Gaussian random field delta-correlated in time t.

Introduce the indicator function

$$\varphi(\mathbf{x}, t) = \delta(\mathbf{x}(t) - \mathbf{x}), \tag{7.5}$$

where $\mathbf{x}(t)$ is the solution to Eq. (7.1), which satisfies the Liouville equation

$$\left(\frac{\partial}{\partial t} + \frac{\partial}{\partial \mathbf{x}}\mathbf{v}(\mathbf{x}, t)\right)\varphi(\mathbf{x}, t) = -\frac{\partial}{\partial \mathbf{x}}\mathbf{f}(\mathbf{x}, t)\varphi(\mathbf{x}, t) \tag{7.6}$$

and the equality

$$\frac{\delta}{\delta f_j(\mathbf{x}', t - 0)}\varphi(\mathbf{x}, t) = -\frac{\partial}{\partial x_j}\left\{\delta(\mathbf{x} - \mathbf{x}')\varphi(\mathbf{x}, t)\right\}. \tag{7.7}$$

The equation for the probability density of the solution to Eq. (7.1)

$$P(\mathbf{x}, t) = \langle\varphi(\mathbf{x}, t)\rangle = \langle\delta(\mathbf{x}(t) - \mathbf{x})\rangle$$

can be obtained by averaging Eq. (7.6) over an ensemble of realizations of field $\mathbf{f}(\mathbf{x}, t)$,

$$\left(\frac{\partial}{\partial t} + \frac{\partial}{\partial \mathbf{x}}\mathbf{v}(\mathbf{x}, t)\right)P(\mathbf{x}, t) = -\frac{\partial}{\partial \mathbf{x}}\langle\mathbf{f}(\mathbf{x}, t)\varphi(\mathbf{x}, t)\rangle. \tag{7.8}$$

We rewrite Eq. (7.8) in the form

$$\left(\frac{\partial}{\partial t} + \frac{\partial}{\partial \mathbf{x}}\mathbf{v}(\mathbf{x}, t)\right)P(\mathbf{x}, t) = -\frac{\partial}{\partial x_i}\int d\mathbf{x}'\int\limits_{t_0}^{t} dt' B_{ij}(\mathbf{x}, t; \mathbf{x}', t')\left\langle\frac{\delta\varphi(\mathbf{x}, t)}{\delta f_j(\mathbf{x}', t')}\right\rangle, \tag{7.9}$$

where we used the Furutsu–Novikov formula

$$\langle f_k(\mathbf{x},t)R[t;\mathbf{f}(\mathbf{y},\tau)]\rangle = \int d\mathbf{x}' \int dt' B_{kl}(\mathbf{x},t;\mathbf{x}',t') \left\langle \frac{\delta R[t;\mathbf{f}(\mathbf{y},\tau)]}{\delta f_l(\mathbf{x}',t')} \right\rangle \tag{7.10}$$

for the correlator of the Gaussian random field $\mathbf{f}(\mathbf{x},t)$ with arbitrary functional $R[t;\mathbf{f}(\mathbf{y},\tau)]$ of it and the dynamic causality condition (7.2).

Equation (7.9) shows that the one-time probability density of solution $\mathbf{x}(t)$ at instant t is governed by functional dependence of solution $\mathbf{x}(t)$ on field $\mathbf{f}(\mathbf{x}',t)$ for all times in the interval (t_0,t).

In the general case, there is no closed equation for the probability density $P(\mathbf{x},t)$. However, if we use approximation (7.4) for the correlation function of field $\mathbf{f}(\mathbf{x},t)$, there appear terms related to variational derivatives $\delta\varphi[\mathbf{x},t;\mathbf{f}(\mathbf{y},\tau)]/\delta f_j(\mathbf{x}',t')$ at coincident temporal arguments $t'=t-0$,

$$\left(\frac{\partial}{\partial t}+\frac{\partial}{\partial \mathbf{x}}\mathbf{v}(\mathbf{x},t)\right)P(\mathbf{x},t) = -\frac{\partial}{\partial x_i}\int d\mathbf{x}' F_{ij}(\mathbf{x},\mathbf{x}';t)\left\langle \frac{\delta\varphi(\mathbf{x},t)}{\delta f_j(\mathbf{x}',t-0)}\right\rangle.$$

According to Eq. (7.7), these variational derivatives can be expressed immediately in terms of quantity $\varphi[\mathbf{x},t;\mathbf{f}(\mathbf{y},\tau)]$. Thus, we obtain the closed Fokker-Planck equation

$$\left(\frac{\partial}{\partial t}+\frac{\partial}{\partial x_k}\left[v_k(\mathbf{x},t)+A_k(\mathbf{x},t)\right]\right)P(\mathbf{x},t) = \frac{\partial^2}{\partial x_k \partial x_l}\left[F_{kl}(\mathbf{x},\mathbf{x};t)P(\mathbf{x},t)\right], \tag{7.11}$$

where

$$A_k(\mathbf{x},t) = \left.\frac{\partial}{\partial x'_l}F_{kl}(\mathbf{x},\mathbf{x}';t)\right|_{\mathbf{x}'=\mathbf{x}}.$$

Equation (7.11) should be solved with the initial condition $P(\mathbf{x},t_0)=\delta(\mathbf{x}-\mathbf{x}_0)$, or with a more general initial condition $P(\mathbf{x},t_0)=W_0(\mathbf{x})$ if the initial conditions are also random, but statistically independent of field $\mathbf{f}(\mathbf{x},t)$.

The Fokker-Planck equation (7.11) is a partial differential equation and its further analysis essentially depends on boundary conditions with respect to $\mathbf{x}$ whose form can vary depending on the problem at hand.

Consider the quantities appeared in Eq. (7.11). In this equation, the terms containing $A_k(\mathbf{x},t)$ and $F_{kl}(\mathbf{x},\mathbf{x}';t)$ are stipulated by fluctuations of field $\mathbf{f}(\mathbf{x},t)$. If field $\mathbf{f}(\mathbf{x},t)$ is stationary in time, quantities $A_k(\mathbf{x})$ and $F_{kl}(\mathbf{x},\mathbf{x}')$ are independent of time. If field $\mathbf{f}(\mathbf{x},t)$ is additionally homogeneous and isotropic in all spatial coordinates, then $F_{kl}(\mathbf{x},\mathbf{x},t)=\text{const}$, which corresponds to the constant tensor of diffusion coefficients, and $A_k(\mathbf{x},t)=0$ (note however that quantities $F_{kl}(\mathbf{x},\mathbf{x}';t)$ and $A_k(\mathbf{x},t)$ can depend on $\mathbf{x}$ because of the use of a curvilinear coordinate systems).

7.2 Transitional probability distributions

Turn back to dynamic system (7.1) and consider the m-time probability density

$$P_m(\mathbf{x}_1,t_1;...;\mathbf{x}_m,t_m) = \langle\delta(\mathbf{x}(t_1)-\mathbf{x}_1)...\delta(\mathbf{x}(t_m)-\mathbf{x}_m)\rangle \tag{7.12}$$

for m different instants $t_1<t_2<...<t_m$. Differentiating Eq. (7.12) with respect to time t_m and using then dynamic equation (7.1), dynamic causality condition (7.2), definition of

function $F_{kl}(\mathbf{x}, \mathbf{x}'; t)$, and the Furutsu–Novikov formula (7.10), one can obtain the equation similar to the Fokker–Planck equation (7.11),

$$\begin{aligned}
&\frac{\partial}{\partial t_m} P_m(\mathbf{x}_1, t_1; ...; \mathbf{x}_m, t_m) \\
&+ \sum_{k=1}^{n} \frac{\partial}{\partial x_{mk}} \left[v_k(\mathbf{x}_m, t_m) + A_k(\mathbf{x}_m, t_m) \right] P_m(\mathbf{x}_1, t_1; ...; \mathbf{x}_m, t_m) \\
&= \sum_{k=1}^{n} \sum_{l=1}^{n} \frac{\partial^2}{\partial x_{mk} \partial x_{ml}} \left[F_{kl}(\mathbf{x}_m, \mathbf{x}_m; t_m) P_m(\mathbf{x}_1, t_1; ...; \mathbf{x}_m, t_m)) \right]. \qquad (7.13)
\end{aligned}$$

No summation over index m is performed here. The initial value to Eq. (7.13) can be determined from Eq. (7.12). Setting $t_m = t_{m-1}$ in (7.12), we obtain

$$P_m(\mathbf{x}_1, t_1; ...; \mathbf{x}_m, t_{m-1}) = \delta(\mathbf{x}_m - \mathbf{x}_{m-1}) P_{m-1}(\mathbf{x}_1, t_1; ...; \mathbf{x}_{m-1}, t_{m-1}). \qquad (7.14)$$

Equation (7.13) assumes the solution in the form

$$P_m(\mathbf{x}_1, t_1; ...; \mathbf{x}_m, t_m) = p(\mathbf{x}_m, t_m | \mathbf{x}_{m-1}, t_{m-1}) P_{m-1}(\mathbf{x}_1, t_1; ...; \mathbf{x}_{m-1}, t_{m-1}). \qquad (7.15)$$

Because all differential operations in Eq. (7.13) concern only t_m and $\mathbf{x}_m$, we can find the equation for the *transitional probability density* by substituting Eq. (7.15) in Eqs. (7.13) and (7.14):

$$\begin{aligned}
\left(\frac{\partial}{\partial t} + \frac{\partial}{\partial x_k} \left[v_k(\mathbf{x}, t) + A_k(\mathbf{x}, t) \right] \right) p(\mathbf{x}, t | \mathbf{x}_0, t_0) &= \frac{\partial^2}{\partial x_k \partial x_l} \left[F_{kl}(\mathbf{x}, \mathbf{x}; t) p(\mathbf{x}, t | \mathbf{x}_0, t_0) \right], \\
p(\mathbf{x}, t | \mathbf{x}_0, t_0)|_{t \to t_0} &= \delta(\mathbf{x} - \mathbf{x}_0), \qquad (7.16)
\end{aligned}$$

where

$$p(\mathbf{x}, t | \mathbf{x}_0, t_0) = \langle \delta(\mathbf{x}(t) - \mathbf{x}) | \mathbf{x}(t_0) = \mathbf{x}_0 \rangle .$$

In Eq. (7.16) we denoted variables $\mathbf{x}_m$ and t_m as $\mathbf{x}$ and t, and variables $\mathbf{x}_{m-1}$ and t_{m-1} as $\mathbf{x}_0$ and t_0.

Using formula (7.15) $(m-1)$ times, we obtain the relationship

$$P_m(\mathbf{x}_1, t_1; ...; \mathbf{x}_m, t_m) = p(\mathbf{x}_m, t_m | \mathbf{x}_{m-1}, t_{m-1}) \ \ldots p(\mathbf{x}_2, t_2 | \mathbf{x}_1, t_1) P(\mathbf{x}_1, t_1), \qquad (7.17)$$

where $P(\mathbf{x}_1, t_1)$ is the one-time probability density governed by Eq. (7.11). Equality (7.17) expresses the multi-time probability density as the product of transitional probability densities, which means that random process $\mathbf{x}(t)$ is the Markovian process.

Equation (7.11) is usually called the *forward Fokker–Planck equation*. The *backward Fokker–Planck equation* (it describes the transitional probability density as a function of the initial parameters t_0 and $\mathbf{x}_0$) can also be easily derived.

Indeed, we obtained the backward Liouville equation (2.4), page 39 for indicator function

$$\left(\frac{\partial}{\partial t_0} + \mathbf{v}(\mathbf{x}_0, t_0) \frac{\partial}{\partial \mathbf{x}_0} \right) \varphi(\mathbf{x}, t | \mathbf{x}_0, t_0) = -\mathbf{f}(\mathbf{x}_0, t_0) \frac{\partial}{\partial \mathbf{x}_0} \varphi(\mathbf{x}, t | \mathbf{x}_0, t_0), \qquad (7.18)$$

with the initial value

$$\varphi(\mathbf{x}, t | \mathbf{x}_0, t) = \delta(\mathbf{x} - \mathbf{x}_0).$$

This equation describes the dynamic system evolution in terms of initial parameters t_0 and $\mathbf{x}_0$. From Eq. (7.18) follows the equality similar to Eq. (7.7),

$$\frac{\delta}{\delta f_j(\mathbf{x}', t_0 + 0)}\varphi(\mathbf{x}, t|\mathbf{x}_0, t_0) = \delta(\mathbf{x} - \mathbf{x}')\frac{\partial}{\partial x_{0j}}\varphi(\mathbf{x}, t|\mathbf{x}_0, t_0). \tag{7.19}$$

Averaging now the backward Liouville equation (7.18) over an ensemble of realizations of random field $\mathbf{f}(\mathbf{x}, t)$ with the effective correlation tensor (7.4), using the Furutsu–Novikov formula (7.10), and relationship (7.19) for the variational derivative, we obtain the backward Fokker–Planck equation (see also [72])

$$\begin{aligned}&\left(\frac{\partial}{\partial t_0} + [v_k(\mathbf{x}_0, t_0) + A_k(\mathbf{x}_0, t_0)]\frac{\partial}{\partial x_{0k}}\right)p(\mathbf{x}, t|\mathbf{x}_0, t_0)\\ &= -F_{kl}(\mathbf{x}_0, \mathbf{x}_0; t_0)\frac{\partial^2}{\partial x_{0k}\partial x_{0l}}p(\mathbf{x}, t|\mathbf{x}_0, t_0), \quad p(\mathbf{x}, t|\mathbf{x}_0, t) = \delta(\mathbf{x} - \mathbf{x}_0).\end{aligned} \tag{7.20}$$

The forward and backward Fokker–Planck equations are equivalent. The forward equation is more convenient for analyzing the temporal behavior of statistical characteristics of the solution to Eq. (7.1). The backward equation appears more convenient for studying statistical characteristics related to initial values, such as the time during which process $\mathbf{x}(t)$ resides in certain spatial region and the time at which the process arrives at region's boundary. In this case the probability of the fact that random process $\mathbf{x}(t)$ resides in spatial region V is given by the integral

$$G(t; \mathbf{x}_0, t_0) = \int\limits_V d\mathbf{x}p(\mathbf{x}, t|\mathbf{x}_0, t_0),$$

which, according to Eq. (7.20), satisfies the closed equation

$$\begin{aligned}&\left(\frac{\partial}{\partial t_0} + [v_k(\mathbf{x}_0, t_0) + A_k(\mathbf{x}_0, t_0)]\frac{\partial}{\partial x_{0k}}\right)G(t; \mathbf{x}_0, t_0)\\ &= -F_{kl}(\mathbf{x}_0, \mathbf{x}_0; t_0)\frac{\partial^2}{\partial x_{0k}\partial x_{0l}}G(t; \mathbf{x}_0, t_0), \quad G(t; \mathbf{x}_0, t_0) = \begin{cases} 1 & (\mathbf{x}_0 \in V), \\ 0 & (\mathbf{x}_0 \notin V). \end{cases}\end{aligned} \tag{7.21}$$

For Eq. (7.21), we must formulate additional boundary conditions, which depend on characteristics of both region V and its boundaries.

7.3 Applicability range of the Fokker–Planck equation

To estimate the applicability range of the Fokker–Planck equation, we must include into consideration the finite-valued correlation radius τ_0 of field $\mathbf{f}(\mathbf{x}, t)$ with respect to time. In this case, the equation for the probability density (7.11) is replaced with the equation

$$\widehat{E}P(\mathbf{x}, t) = -\frac{\partial}{\partial x_k}S'(\mathbf{x}, t),$$

where $\widehat{E}$ — is the operator appeared in the left-hand side of Eq. (7.11) in which quantity $F_{kl}(\mathbf{x}, \mathbf{x}', t)$ is replaced with

$$\widetilde{F}_{kl}(\mathbf{x}, \mathbf{x}', t) = \int\limits_0^t dt' B_{kl}(\mathbf{x}, \mathbf{x}', t)$$

and term $S'(\mathbf{x}, t)$ includes corrections to the factor of the probability flux density because of finiteness of τ_0. For $\tau_0 \to 0$, we turn back to Eq. (7.11). Thus, smallness of parameter τ_0/T is the necessary, but generally not sufficient condition in order that one can describe statistical characteristics of the solution to Eq. (7.1) using the approximation of the delta-correlated random field of which a consequence is the Fokker–Planck equation. Every particular problem requires more detailed investigations. Below, we give a more physical method called the *diffusion approximation*. This method also leads to the Markovian property of the solution to Eq. (7.1); however, it considers to some extent the finite value of the temporal correlation radius.

Here, we emphasize that the approximation of the delta-correlated random field does not reduce to the formal replacement of random field $\mathbf{f}(\mathbf{x}, t)$ in Eq. (7.1) with the random field with correlation function (7.4). This approximation corresponds to the construction of an asymptotic expansion for temporal correlation radius τ_0 of filed $\mathbf{f}(\mathbf{x}, t)$ approaching to zero. It is in such limit process that exact average quantities like

$$\langle \mathbf{f}(\mathbf{x}, t) R[t; \mathbf{f}(\mathbf{x}', \tau)] \rangle$$

grade into the expressions obtained by the formal replacement of the correlation tensor of field $\mathbf{f}(\mathbf{x}, t)$ with the effective tensor (7.4).

7.3.1 Langevin equation

We illustrate the above material by the example of the *Langevin equation* that allows an exhaustive statistical analysis [140]. This equation has the form

$$\frac{d}{dt} x(t) = -\lambda x(t) + f(t), \quad x(t_0) = 0 \tag{7.22}$$

and assumes that the sufficiently fine smooth function $f(t)$ is the stationary Gaussian process with zero-valued average and correlation function

$$\langle f(t) f(t') \rangle = B_f(t - t').$$

For any particular realization of random force $f(t)$, the solution to Eq. (7.22) has the form

$$x(t) = \int_{t_0}^{t} d\tau f(\tau) e^{-\lambda(t-\tau)}.$$

Consequently, this solution $x(t)$ is also the Gaussian process with the parameters

$$\langle x(t) \rangle = 0, \quad \langle x(t) x(t') \rangle = \int_{t_0}^{t} d\tau_1 \int_{t_0}^{t'} d\tau_2 B_f(\tau_1 - \tau_2) e^{-\lambda(t + t' - \tau_1 - \tau_2)}.$$

In addition, we have, for example,

$$\langle f(t) x(t) \rangle = \int_{0}^{t - t_0} d\tau B_f(\tau) e^{-\lambda\tau}.$$

Note that the one-point probability density $P(x,t) = \langle \delta(x(t)-x) \rangle$ for Eq. (7.22) satisfies the equation

$$\left(\frac{\partial}{\partial t} - \lambda \frac{\partial}{\partial x} x \right) P(x,t) = \int_0^{t-t_0} d\tau B_f(\tau) e^{-\lambda\tau} \frac{\partial^2}{\partial x^2} P(x,t), \quad P(x,t_0) = \delta(x),$$

which rigorously follows from Eq. (5.93), page 119. As a consequence, we obtain

$$\frac{d}{dt} \left\langle x^2(t) \right\rangle = -2\lambda \left\langle x^2(t) \right\rangle + 2 \int_0^{t-t_0} d\tau B_f(\tau) e^{-\lambda\tau}.$$

For $t_0 \to -\infty$, process $x(t)$ grades into the stationary Gaussian process with the following one-time statistical parameters

$$\langle x(t) \rangle = 0, \quad \sigma_x^2 = \left\langle x^2(t) \right\rangle = \frac{1}{\lambda} \int_0^{\infty} d\tau B_f(\tau) e^{-\lambda\tau}, \quad \langle f(t)x(t) \rangle = \int_0^{\infty} d\tau B_f(\tau) e^{-\lambda\tau}.$$

In particular, for exponential correlation function $B_f(t)$,

$$B_f(t) = \sigma_f^2 e^{-|\tau|/\tau_0},$$

we obtain the expressions

$$\langle x(t) \rangle = 0, \quad \left\langle x^2(t) \right\rangle = \frac{\sigma_f^2 \tau_0}{\lambda(1+\lambda\tau_0)}, \quad \langle f(t)x(t) \rangle = \frac{\sigma_f^2 \tau_0}{1+\lambda\tau_0}, \tag{7.23}$$

which grade into the asymptotic expressions

$$\left\langle x^2(t) \right\rangle = \frac{\sigma_f^2 \tau_0}{\lambda}, \quad \langle f(t)x(t) \rangle = \sigma_f^2 \tau_0 \tag{7.24}$$

for $\tau_0 \to 0$.

Multiply now Eq. (7.22) by $x(t)$. Assuming that function $x(t)$ is sufficiently fine function, we obtain the equality

$$x(t) \frac{d}{dt} x(t) = \frac{1}{2} \frac{d}{dt} x^2(t) = -\lambda x^2(t) + f(t)x(t).$$

Averaging this equation over an ensemble of realizations of function $f(t)$, we obtain the equation

$$\frac{1}{2} \frac{d}{dt} \left\langle x^2(t) \right\rangle = -\lambda \left\langle x^2(t) \right\rangle + \langle f(t)x(t) \rangle, \tag{7.25}$$

whose steady-state solution (it corresponds to the limit process $t_0 \to -\infty$ and $\tau_0 \to 0$)

$$\left\langle x^2(t) \right\rangle = \frac{1}{\lambda} \langle f(t)x(t) \rangle$$

coincides with Eqs. (7.23) and (7.24).

Taking into account the fact that $\delta x(t)/\delta f(t-0) = 1$, we obtain the same result for correlation $\langle f(t)x(t) \rangle$ by using the formula

$$\langle f(t)x(t) \rangle = \int_{-\infty}^{t} d\tau B_f(t-\tau) \left\langle \frac{\delta}{\delta f(\tau)} x(t) \right\rangle \tag{7.26}$$

with the effective correlation function

$$B_f^{\rm eff}(t) = 2\sigma_f^2\tau_0\delta(t).$$

Earlier, we mentioned that statistical characteristics of solutions to dynamic problems in the approximation of the delta-correlated random process (field) coincide with the statistical characteristics of the Markovian processes. However, one should clearly understand that this is the case only for statistical averages and equations for these averages. In particular, realizations of process $x(t)$ satisfying the Langevin equation (7.22) drastically differ from realizations of the corresponding Markovian process. The latter satisfies Eq. (7.22) in which function $f(t)$ in the right-hand side is the ideal white noise with the correlation function $B_f(t) = 2\sigma_f^2\tau_0\delta(t)$; moreover, this equation must be treated in the sense of generalized functions, because the Markovian processes are not differentiable in the ordinary sense. At the same time, process $x(t)$ — whose statistical characteristics coincide with characteristics of the Markovian process — behaves as sufficiently fine function and is differentiable in the ordinary sense. For example,

$$x(t)\frac{d}{dt}x(t) = \frac{1}{2}\frac{d}{dt}x^2(t),$$

and we have for $t_0 \to -\infty$ in particular

$$\left\langle x(t)\frac{d}{dt}x(t)\right\rangle = 0. \tag{7.27}$$

On the other hand, in the case of the ideal Markovian process $x(t)$ satisfying (in the sense of generalized functions) the Langevin equation (7.22) with the white noise in the right-hand side, Eq. (7.27) makes no sense at all, and the meaning of the relationship

$$\left\langle x(t)\frac{d}{dt}x(t)\right\rangle = -\lambda\left\langle x^2(t)\right\rangle + \langle f(t)x(t)\rangle \tag{7.28}$$

depends on the definition of averages. Indeed, if we will mean Eq. (7.28) as the limit of the equality

$$\left\langle x(t+\Delta)\frac{d}{dt}x(t)\right\rangle = -\lambda\langle x(t)x(t+\Delta)\rangle + \langle f(t)x(t+\Delta)\rangle \tag{7.29}$$

for $\Delta \to 0$, the result will be essentially different depending on whether we use limit processes $\Delta \to +0$, or $\Delta \to -0$. For limit process $\Delta \to +0$, we have

$$\lim_{\Delta\to+0}\langle f(t)x(t+\Delta)\rangle = 2\sigma_f^2\tau_0,$$

and, taking into account Eq. (7.26), we can rewrite Eq. (7.29) in the form

$$\left\langle x(t+0)\frac{d}{dt}x(t)\right\rangle = \sigma_f^2\tau_0. \tag{7.30}$$

On the contrary, for limit process $\Delta \to -0$, we have $\langle f(t)x(t-0)\rangle = 0$ because of the dynamic causality condition, and Eq. (7.29) assumes the form

$$\left\langle x(t-0)\frac{d}{dt}x(t)\right\rangle = -\sigma_f^2\tau_0. \tag{7.31}$$

Comparing Eq. (7.27) with Eqs. (7.30) and (7.31), we see that, for the ideal Markovian process described by the solution to the Langevin equation with the white noise in the right-hand side and commonly called the *Ohrnstein–Ulenbeck process*, we have

$$\left\langle x(t+0)\frac{d}{dt}x(t)\right\rangle \neq \left\langle x(t-0)\frac{d}{dt}x(t)\right\rangle \neq \frac{1}{2}\frac{d}{dt}\left\langle x^2(t)\right\rangle.$$

Note that equalities (7.30) and (7.31) can also be obtained from the correlation function

$$\langle x(t)x(t+\tau)\rangle = \frac{\sigma_f^2\tau_0}{\lambda}e^{-\lambda|\tau|}$$

of process $x(t)$.

To conclude with the discussion of the approximation of the delta-correlated random process (field), we emphasize that, in all further examples, we will treat the sentence like 'dynamic system (equation) with the delta-correlated parameter fluctuations' as the asymptotic limit in which these parameters have temporal correlation radii small in comparison with all characteristic temporal scales of the problem under consideration.

Chapter 8

Methods for solving and analyzing the Fokker-Planck equation

The Fokker–Planck equations for the one-point probability density (7.11), page 186 and for the transitional probability density (7.16), page 187 are the partial differential equations of parabolic type, so that we can use methods of the theory of mathematical physics equations to solve them. In this context, the basic methods are such as the method of separation of variables, the Fourier transformation with respect to spatial coordinates, and other integral transformations.

However, there are only few Fokker–Planck equations that allow an exact solution. First of all, among them are the Fokker–Planck equations corresponding to the stochastic equations that are themselves solvable in the analytic form. Such problems often allow determination of not only the one-point and transitional probability densities, but also the characteristic functional and other statistical characteristics important for practice.

8.1 System of linear equations

Consider the system of linear equations for the components of vector function $\mathbf{x}(t)$

$$\frac{d}{dt}\mathbf{x}(t) = A\mathbf{x}(t) + \mathbf{f}(t), \quad \mathbf{x}(t_0) = \mathbf{x}_0 \tag{8.1}$$

with constant matrix A. In Sect. 7.3, we considered in detail a special one-dimensional case of this equation — the Langevin equation. We will assume functions $f_i(t)$ the Gaussian functions delta-correlated in time, i.e., we set

$$\langle f_i(t) f_j(t') \rangle = 2B_{ij}\delta(t - t').$$

The solution to system of equations (8.1) has the form

$$\mathbf{x}(t) = e^{(t-t_0)A}\mathbf{x}_0 + \int\limits_{t_0}^{t} d\tau e^{(t-t_0)A}\mathbf{f}(\tau),$$

so that quantity $\mathbf{x}(t)$ is the Gaussian vector function with the parameters

$$\begin{aligned}\langle \mathbf{x}(t)\rangle &= e^{(t-t_0)A}\mathbf{x}_0,\\ \sigma_{ij}^2(t,t') &= \langle [x_i(t)-\langle x_i(t)\rangle][x_j(t')-\langle x_j(t')\rangle]\rangle\\ &= \int\limits_{t_0}^{t} d\tau \left\{e^{(t-t_0)A}Be^{(t-t_0)A^T}\right\}_{ij},\end{aligned} \tag{8.2}$$

where A^T is the transposed matrix of A.

We can easily see in this case that the Gaussian probability distribution with parameters (8.2) satisfies the Fokker–Planck equation for the transitional probability density $p(\mathbf{x},t|\mathbf{x}_0,t_0)$,

$$\left(\frac{\partial}{\partial t}+\frac{\partial}{\partial x_i}A_{ik}x_k\right)p(\mathbf{x},t|\mathbf{x}_0,t_0)=\frac{\partial}{\partial x_i}B_{ik}\frac{\partial}{\partial x_k}p(\mathbf{x},t|\mathbf{x}_0,t_0), \tag{8.3}$$

corresponding to stochastic system (8.1).

We note that Eq. (8.3) by itself also can be easily solved by the Fourier transform with respect to spatial coordinates.

The simplest special case of Eq. (8.1) is the equation that defines the *Wiener random process*. In view of the significant role that such processes plays in physics (for example, they describe the *Brownian motion of particles*), we consider the Wiener process in detail.

8.1.1 Wiener random process

The Wiener random process is defined as the solution to the stochastic equation

$$\frac{d}{dt}w(t)=z(t),\quad w(0)=0,$$

where $z(t)$ is the Gaussian process delta-correlated in time and described by the parameters

$$\langle z(t)\rangle=0,\quad \langle z(t)z(t')\rangle=2\sigma^2\tau_0\delta(t-t').$$

The solution to this equation

$$w(t)=\int\limits_0^t d\tau z(\tau)$$

is the continuous Gaussian nonstationary random process with the parameters

$$\langle w(t)\rangle=0,\quad \langle w(t)w(t')\rangle=2\sigma^2\tau_0\min(t,t').$$

As a consequence, its characteristic functional has the form

$$\Phi[t;v(\tau)]=\left\langle \exp\left\{i\int\limits_0^t d\tau w(\tau)v(\tau)\right\}\right\rangle = e^{-\sigma^2\tau_0\int\limits_0^t d\tau_1\int\limits_0^t d\tau_2 v(\tau_1)v(\tau_2)\min(\tau_1,\tau_2)} \tag{8.4}$$

Note that the increment of process $w(t)$ on the temporal interval (t_1,t_2)

$$w(t_1;t_2)=w(t_2)-w(t_1)=\int\limits_{t_1}^{t_2} d\tau z(\tau)$$

has, like process $w(t)$ itself, the Gaussian statistics with the parameters

$$\langle w(t_1;t_2)\rangle = 0, \quad \left\langle [w(t_1;t_2)]^2 \right\rangle = 2\sigma^2\tau_0|t_2 - t_1|.$$

The Wiener random process $w(t)$ is the Gaussian continuous process with independent increments. This means that increments of process $w(t)$ on the nonoverlapping intervals $(t_1;t_2)$ and $(t_3;t_4)$ are statistically independent.

The characteristic functional of process

$$w(t_0;t_0+t) = \int\limits_{t_0}^{t_0+t} d\tau z(\tau)$$

coincides with the characteristic functional of process $w(t)$. This means that realizations of processes $w(t)$ and $w(t_0;t_0+t)$ are *statistically equivalent* for any given parameter t_0. Thus, dealing solely with process realizations, we cannot decide to which process these realizations belong. In addition, processes $w(t)$ and $w(-t)$ are also statistically equivalent, which means that the Wiener random process is the time-reversible process in the sense specified above.

An additional — *fractal* — property inheres in realizations of the Wiener process. According to this property, realizations of the Wiener process $w(at)$ (compressed in time for $a > 1$) are statistically equivalent to realizations of process $a^{1/2}w(t)$ (elongated in amplitude). The fractal property of the Wiener process can be treated also as statistical equivalence of realizations of process $w(t)$ and realizations of process $w(at)/a^{1/2}$, which is compressed both in time t and amplitude, because their characteristic functionals coincide.

Consider a more general process that includes additionally the drift dependent on parameter α

$$w(t;\alpha) = -\alpha t + w(t), \quad \alpha > 0.$$

Process $w(t;\alpha)$ is the Markovian process, and its probability density

$$P(w,t;\alpha) = \langle \delta(w(t;\alpha) - w)\rangle$$

satisfies the Fokker–Planck equation

$$\left(\frac{\partial}{\partial t} - \alpha\frac{\partial}{\partial w}\right) P(w,t;\alpha) = D\frac{\partial^2}{\partial w^2}P(w,t;\alpha), \quad P(w,0;\alpha) = \delta(w), \tag{8.5}$$

where $D = \sigma^2\tau_0$ is the diffusion coefficient. The solution to this equation has the form of the Gaussian distribution

$$P(w,t;\alpha) = \frac{1}{2\sqrt{\pi D t}} \exp\left\{-\frac{(w+\alpha t)^2}{4Dt}\right\}. \tag{8.6}$$

The corresponding integral distribution function defined as the probability of the event that $w(t;\alpha) < w$ is given by the formula

$$F(w,t;\alpha) = \int\limits_{-\infty}^{w} dw P(w,t;\alpha) = \Phi\left(\frac{w}{\sqrt{2Dt}} + \alpha\sqrt{\frac{t}{2D}}\right), \tag{8.7}$$

where

$$\Phi(z) = \frac{1}{\sqrt{2\pi}} \int_{-\infty}^{z} dy \exp\left\{-\frac{y^2}{2}\right\} \tag{8.8}$$

is the error function.

In addition to the initial value, supplement Eq. (8.5) with the boundary condition

$$P(w,t;\alpha)|_{w=h} = 0, \quad (t > 0). \tag{8.9}$$

This condition breaks down realizations of process $w(t;\alpha)$ at the instant they reach boundary h. For $w < h$, the solution to the boundary-value problem (8.5), (8.9) (we denote it as $P(w,t;\alpha,h)$) describes the probability distribution of those realizations of process $w(t;\alpha)$ that survived instant t, i.e., never reached boundary h during the whole temporal interval. Correspondingly, the norm of the probability density appears not unity, but the probability of the event that $t < t^*$, where t^* is the instant at which process $w(t;\alpha)$ reaches boundary h for the first time

$$\int_{-\infty}^{h} dwP(w,t;\alpha,h) = P(t < t^*). \tag{8.10}$$

Introduce the integral distribution function and probability density of random instant at which the process reaches boundary h

$$\begin{aligned} F(t;\alpha,h) &= P(t^* < t) = 1 - P(t < t^*) = 1 - \int_{-\infty}^{h} dwP(w,t;\alpha,h), \\ P(t;\alpha,h) &= \frac{\partial}{\partial t}F(t;\alpha,h) = -D\frac{\partial}{\partial w}P(w,t;\alpha,h)\bigg|_{w=h}. \end{aligned} \tag{8.11}$$

If $\alpha > 0$, process $w(t;\alpha)$ moves on average out of boundary h; as a result, probability $P(t < t^*)$ (8.10) tends for $t \to \infty$ to the probability of the event that process $w(t;\alpha)$ never reaches boundary h. In other words, limit

$$\lim_{t\to\infty} \int_{-\infty}^{h} dwP(w,t;\alpha,h) = P\left(w_{\max}(\alpha) < h\right) \tag{8.12}$$

is equal to the probability of the event that the process absolute maximum

$$w_{\max}(\alpha) = \max_{t\in(0,\infty)} w(t;\alpha)$$

is less than h. Thus, from Eq. (8.12) follows that the integral distribution function of the absolute maximum $w_{\max}(\alpha)$ is given by the formula

$$F(h;\alpha) = P\left(w_{\max}(\alpha) < h\right) = \lim_{t\to\infty} \int_{-\infty}^{h} dwP(w,t;\alpha,h). \tag{8.13}$$

After we solve boundary-value problem (8.5), (8.9) by using, for example, the reflection method, we obtain

$$P(w,t;\alpha,h) = \frac{1}{2\sqrt{\pi Dt}}\left\{\exp\left[-\frac{(w+\alpha t)^2}{4Dt}\right] - \exp\left[-\frac{h\alpha}{D} - \frac{(w-2h+\alpha t)^2}{4Dt}\right]\right\}. \tag{8.14}$$

Substituting this expression in Eq. (8.11), we obtain the probability density of instant t^* at which process $w(t;\alpha)$ reaches boundary h for the first time

$$P(t;\alpha,h) = \frac{1}{2Dt\sqrt{\pi Dt}} \exp\left\{-\frac{(h+\alpha t)^2}{4Dt}\right\}.$$

Finally, integrating Eq. (8.14) over w and setting $t \to \infty$, we obtain, in accordance with Eq. (8.13), the integral distribution function of absolute maximum $w_{\max}(\alpha)$ of process $w(t;\alpha)$ in the form [142, 166]

$$F(h;\alpha) = P\left(w_{\max}(\alpha) < h\right) = 1 - \exp\left\{-\frac{h\alpha}{D}\right\}. \tag{8.15}$$

Consequently, the absolute maximum of the Wiener process has the exponential probability density

$$P(h;\alpha) = \langle\delta\left(w_{\max}(\alpha) - h\right)\rangle = \frac{\alpha}{D}\exp\left\{-\frac{h\alpha}{D}\right\}.$$

The Wiener random process offers a possibility of constructing different other processes convenient for modeling different physical phenomena. In the case of positive quantities, the simplest approximation of such kind is the logarithmic-normal (lognormal) process. Consider this process in greater detail.

8.1.2 Logarithmic-normal random process

We define the lognormal process (logarithmic-normal process) by the formula

$$y(t;\alpha) = e^{w(t;\alpha)} = \exp\left\{-\alpha t + \int\limits_0^t d\tau z(\tau)\right\}, \tag{8.16}$$

where $z(t)$ is the Gaussian white noise process with the parameters

$$\langle z(t)\rangle = 0, \quad \langle z(t)z(t')\rangle = 2\sigma^2\tau_0\delta(t-t').$$

The lognormal process satisfies the stochastic equation0

$$\frac{d}{dt}y(t;\alpha) = \{-\alpha + z(t)\}\, y(t;\alpha), \qquad y(0;\alpha) = 1.$$

The one-time probability density of the lognormal process is given by the formula

$$P(y,t;\alpha) = \left\langle\delta\left(e^{w(t;\alpha)} - y\right)\right\rangle = \frac{1}{y}\, P(w,t;\alpha)\big|_{w=\ln y},$$

where $P(w,t;\alpha)$ is the one-time probability density of the Wiener process with a drift, which is given by Eq. (8.6), so that

$$P(y,t;\alpha) = \frac{1}{2y\sqrt{\pi Dt}}\exp\left\{-\frac{(\ln y + \alpha t)^2}{4Dt}\right\} = \frac{1}{2y\sqrt{\pi Dt}}\exp\left\{-\frac{\ln^2\left(ye^{\alpha t}\right)}{4Dt}\right\}, \tag{8.17}$$

where $D = \sigma^2\tau_0$.

Figure 8.1 shows the curves of the lognormal probability density (8.17) for $\alpha/D = 1$ and dimensionless times $\tau = Dt = 0.1$ and 1. One can see the long flat *tail* that appears

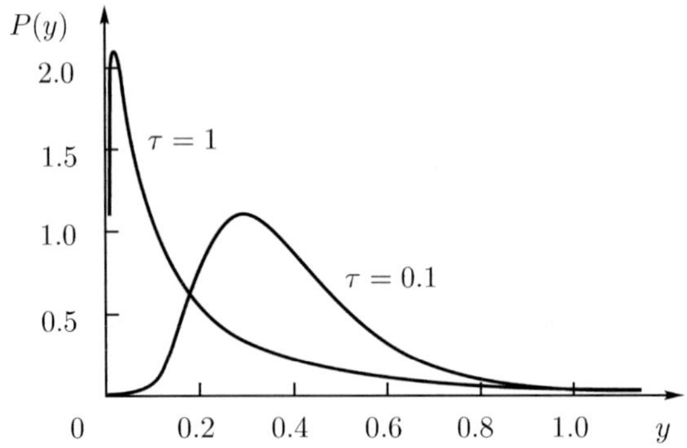

Figure 8.1: Logarithmic-normal probability density (8.17) for $\alpha/D = 1$ and dimensionless times $\tau = 0.1$ and 1.

for the curve at $\tau = 1$; this tail increases the role of high peaks of process $y(t;\alpha)$ in the formation of the one-time statistics. Correspondingly, the integral distribution function is given, in accordance with Eqs. (8.7), (8.8), by the expression

$$F(y,t;\alpha) = P\left(y(t;\alpha) < y\right) = \Phi\left(\frac{1}{\sqrt{2Dt}}\ln\left(ye^{\alpha t}\right)\right). \tag{8.18}$$

Having only the one-point statistical characteristics of process $y(t;\alpha)$, one can obtain a deeper insight into the behavior of realizations of process $y(t;\alpha)$ on the whole interval of times $(0,\infty)$ [142, 166]. In particular,

(**1**) From the integral distribution function, one can calculate the typical realization curve of lognormal process $y(t;\alpha)$ (see Chapter 3, page 56); this distribution function appears the exponentially decaying curve

$$y^*(t;\alpha) = e^{-\alpha t}. \tag{8.19}$$

(**2**) The lognormal process $y(t;\alpha)$ is the Markovian process and its one-time probability density (8.17) satisfies the Fokker–Planck equation

$$\left(\frac{\partial}{\partial t} - \alpha\frac{\partial}{\partial y}y\right)P(y,t;\alpha) = D\frac{\partial}{\partial y}y\frac{\partial}{\partial y}yP(y,t;\alpha), \quad P(y,0;\alpha) = \delta(y-1). \tag{8.20}$$

From Eq. (8.20), one can easily derive the equations for the moment functions of process $y(t;\alpha)$; solutions to these equations are given by the formulas

$$\langle y^n(t;\alpha)\rangle = e^{n(n-\alpha/D)Dt}, \quad \left\langle\frac{1}{y^n(t;\alpha)}\right\rangle = e^{n(n+\alpha/D)Dt}, \quad n = 1,2,... \tag{8.21}$$

from which follows that moments exponentially grow with time. Consequently, the exponential increase of moments is caused by deviations of process $y(t;\alpha)$ from the typical realization curve $y^*(t;\alpha)$ towards both large and small values of y.

At $\alpha/D = 1$, the average value of process $y(t;D)$ is independent of time and is equal to unity. Despite this fact, according to Eq. (8.18), the probability of the event that $y < 1$

for $Dt \gg 1$ rapidly approaches the unity by the law

$$P\left(y(t;D) < 1\right) = \Phi\left(\sqrt{\frac{Dt}{2}}\right) = 1 - \frac{1}{\sqrt{\pi Dt}} e^{-Dt/4},$$

i.e., the curves of process realizations run mainly below the level of the process average $\langle y(t;D) \rangle = 1$, though namely large peaks of the process govern the behavior of statistical moments of process $y(t;D)$.

Here, we have a clear contradiction between the behavior of statistical characteristics of process $y(t;\alpha)$ and the behavior of process realizations.

(**3**) The behavior of realizations of process $y(t;\alpha)$ on the whole temporal interval can also be evaluated with the use of the p-majorant curves $M_p(t,\alpha)$ whose definition is as follows [142, 166]. We call the majorant curve the curve $M_p(t,\alpha)$ for which inequality $y(t;\alpha) < M_p(t,\alpha)$ is satisfied for all times t with probability p, i.e.,

$$P\left\{y(t;\alpha) < M_p(t,\alpha) \text{ for all } t \in (0,\infty)\right\} = p.$$

The above statistics (8.15) of the absolute maximum of the Wiener process with a drift $w(t;\alpha)$ makes it possible to outline a wide enough class of the majorant curves. Indeed, let p be the probability of the event that the absolute maximum $w_{\max}(\beta)$ of the auxiliary process $w(t;\beta)$ with arbitrary parameter β in the interval $0 < \beta < \alpha$ satisfies inequality $w(t;\beta) < h = \ln A$. It is clear that the whole realization of process $y(t;\alpha)$ will run in this case below the majorant curve

$$M_p(t,\alpha,\beta) = Ae^{(\beta-\alpha)t}. \tag{8.22}$$

with the same probability p. As may be seen from Eq. (8.15), the probability of the event that process $y(t;\alpha)$ never exceeds majorant curve (8.22) depends on this curve parameters according to the formula

$$p = 1 - A^{-\beta/D}.$$

This means that we derived the one-parameter class of exponentially decaying majorant curves

$$M_p(t,\alpha,\beta) = \frac{1}{(1-p)^{D/\beta}} e^{(\beta-\alpha)t}. \tag{8.23}$$

Notice the remarkable fact that, despite statistical average $\langle y(t;D) \rangle$ remains constant ($\langle y(t;D) \rangle = 1$) and higher-order moments of process $y(t;D)$ are exponentially increasing functions, one can always select an exponentially decreasing majorant curve (8.23) such that realizations of process $y(t;D)$ will run below it with arbitrary predetermined probability $p < 1$. In particular, inequality ($\tau = Dt$)

$$y(t;D) < M_{1/2}(t,D,D/2) = M(\tau) = 4e^{-\tau/2} \tag{8.24}$$

is satisfied with probability $p = 1/2$ for any instant t from interval $(0,\infty)$.

Figure 8.2 schematically shows the behaviors of a realization of process $y(t;D)$ and the majorant curve (8.24). This schematic is an additional fact in favor of our conclusion that the exponential growth of moments of process $y(t;D)$ with time is the purely statistical effect caused by averaging over the whole ensemble of realizations.

Note that the area below the exponentially decaying majorant curves has a finite value. Consequently, high peaks of process $y(t;\alpha)$, which are the reason of the exponential growth

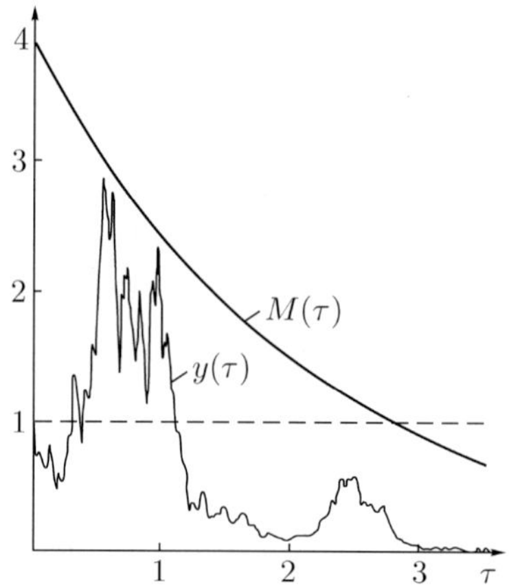

Figure 8.2: Schematic behaviors of a realization of process $y(t; D)$ and majorant curve $M(\tau)$ (8.24).

of higher moments, only insignificantly contribute to the area below realizations; this area appears finite for almost all realizations, which means that the peaks of the lognormal process $y(t;\alpha)$ are sufficiently narrow.

(**4**) In this connection, it is of interest to investigate immediately the statistics of random area below realizations of process $y(t;\alpha)$

$$S_n(t;\alpha) = \int_0^t d\tau y^n(\tau;\alpha). \tag{8.25}$$

This function satisfies the system of stochastic equations

$$\begin{aligned} \frac{d}{dt} S_n(t;\alpha) &= y^n(t;\alpha), \quad S_n(0;\alpha) = 0, \\ \frac{d}{dt} y(t;\alpha) &= \{-\alpha + z(t)\}\, y(t;\alpha), \quad y(0;\alpha) = 1, \end{aligned} \tag{8.26}$$

so that the two-component process $\{y(t;\alpha), S_n(t;\alpha)\}$ is the Markovian process whose one-point probability density

$$P(S_n, y, t; \alpha) = \langle \delta\left(S_n(t;\alpha) - S_n\right) \delta\left(y(t;\alpha) - y\right) \rangle$$

and transition probability density satisfy the Fokker–Planck equation

$$\begin{aligned} \left(\frac{\partial}{\partial t} + y^n \frac{\partial}{\partial S_n} - \alpha \frac{\partial}{\partial y} y\right) P(S_n, y, t; \alpha) &= D \frac{\partial}{\partial y} y \frac{\partial}{\partial y} y P(S_n, y, t; \alpha), \\ P(S_n, y, 0; \alpha) &= \delta(S_n)\delta(y-1). \end{aligned} \tag{8.27}$$

Unfortunately, Eq. (8.27) cannot be solved analytically, which prevents from studying the statistics of process $S_n(t;\alpha)$ exhaustively. However, for the one-time statistical averages of process $S_n(t;\alpha)$, i.e., averages at a fixed instant, the corresponding statistics can be studied in sufficient detail.

With this goal in view, we rewrite Eq. (8.25) in the form

$$S_n(t;\alpha) = \int\limits_0^t d\tau e^{-n\alpha\tau + n\int\limits_0^\tau d\tau_1 z(\tau_1)} = \int\limits_0^t d\tau e^{-n\alpha(t-\tau) + n\int\limits_0^{t-\tau} d\tau_1 z(t-\tau-\tau_1)},$$

from which follows that quantity $S_n(t;\alpha)$ in the context of the one-time statistics is statistically equivalent to the quantity

$$S_n(t;\alpha) = \int\limits_0^t d\tau e^{-n\alpha(t-\tau) + n\int\limits_0^{t-\tau} d\tau_1 z(\tau+\tau_1)}. \tag{8.28}$$

Differentiating now Eq. (8.28) with respect to time, we obtain the statistically equivalent stochastic equation

$$\frac{d}{dt} S_n(t;\alpha) = 1 - n\{\alpha - z(t)\} S_n(t;\alpha), \quad S_n(0;\alpha) = 0,$$

whose one-time statistical characteristics are described by the one-time probability density $P(S_n, t;\alpha) = \langle \delta\left(S_n(t;\alpha) - S_n\right)\rangle$ that satisfies the Fokker–Planck equation

$$\left(\frac{\partial}{\partial t} + \frac{\partial}{\partial S_n} - n\alpha \frac{\partial}{\partial S_n} S_n\right) P(S_n, t;\alpha) = n^2 D \frac{\partial}{\partial S_n} S_n \frac{\partial}{\partial S_n} S_n P(S_n, y, t;\alpha). \tag{8.29}$$

As may be seen from Eq. (8.29), random integrals $S_n(\alpha) = \int_0^\infty d\tau y^n(\tau;\alpha)$ are distributed according to the steady-state probability density

$$P(S_n;\alpha) = \frac{1}{(n^2 D)^{\alpha/nD} \Gamma\left(\frac{\alpha}{D}\right) S_n^{1+\alpha/D}} \exp\left\{-\frac{1}{n^2 D S_n}\right\},$$

where $\Gamma(z)$ is the *gamma function*. In the special case $n = 1$, quantity

$$S(\alpha) = S_1(\alpha) = \int\limits_0^\infty d\tau y(\tau;\alpha)$$

has the following probability density

$$P(S;\alpha) = \frac{1}{D^{\alpha/D} \Gamma\left(\frac{\alpha}{D}\right) S^{1+\alpha/D}} \exp\left\{-\frac{1}{DS}\right\}. \tag{8.30}$$

If we set now $\alpha = D$, then the steady-state probability density and the corresponding integral distribution function will have the form

$$P(S;D) = \frac{1}{DS^2} \exp\left\{-\frac{1}{DS}\right\}, \quad F(S;D) = \exp\left\{-\frac{1}{DS}\right\}. \tag{8.31}$$

The time-dependent behavior of the probability density of random process

$$\tilde{S}(t,\alpha) = \int\limits_t^\infty d\tau y(\tau;\alpha) \tag{8.32}$$

gives an additional information about the behavior of realizations of process $y(t;\alpha)$ with time t. The integral in the right-hand side of Eq. (8.32) can be represented in the form

$$\tilde{S}(t,\alpha) = y(t;\alpha)\int_0^\infty d\tau \exp\left\{-\alpha\tau + \int_0^\tau d\tau_1 z(\tau_1 + t)\right\}. \tag{8.33}$$

In Eq. (8.33), random process $y(t;\alpha)$ is statistically independent of the integral factor, because they depend functionally on process $z(\tau)$ for nonoverlapping intervals of times τ; in addition, the integral factor by itself appears statistically equivalent to random quantity $S(\alpha)$. Consequently, the one-time probability density $P(\tilde{S},t;\alpha) = \left\langle\delta\left(\tilde{S}(t;\alpha) - \tilde{S}\right)\right\rangle$ of random process $\tilde{S}(t,\alpha)$ is described by the expression

$$P(\tilde{S},t;\alpha) = \int_0^\infty\int_0^\infty dy dS \delta(yS - \tilde{S}) P(y,t;\alpha) = \int_0^\infty \frac{dy}{y} P(y,t;\alpha) P(\tilde{S}/y;\alpha), \tag{8.34}$$

where $P(y,t;\alpha)$ is the one-time probability density of lognormal process $y(t;\alpha)$ (8.17) and $P(\tilde{S}/y;\alpha)$ is the probability density (8.30) of random area.

The corresponding integral distribution function

$$F(\tilde{S},t;\alpha) = P\left(P(\tilde{S},t;\alpha) < \tilde{S}\right) = \int_0^{\tilde{S}} d\tilde{S} P(\tilde{S},t;\alpha)$$

is given by the integral

$$F(\tilde{S},t;\alpha) = \int_0^\infty dy P(y,t;\alpha) F(\tilde{S}/y;\alpha),$$

where $F(S;\alpha)$ is the integral distribution function of random area $S(t;\alpha)$. In the special case $\alpha = D$, we obtain, according to Eqs. (8.17) and (8.31), the expression

$$F(\tilde{S},t;\alpha) = \frac{1}{2\sqrt{\pi D t}} \int_0^\infty \frac{dy}{y} \exp\left\{-\frac{\ln^2\left(ye^{Dt}\right)}{4Dt} - \frac{y}{D\tilde{S}}\right\}$$

from which follows that the probability of the event that inequality $\tilde{S}(t;\alpha) < \tilde{S}$ is satisfied monotonously tends to unity with increasing Dt for any predetermined value of $D\tilde{S}$. This is an additional evidence in favor of the fact that every separate realization of the lognormal process tends to zero with increasing Dt, though moment functions of process $y(t;\alpha)$ show the exponential growth caused by large spikes.

8.2 Integral transformations

Integral transformations are very practicable for solving the Fokker–Planck equation. Indeed, earlier we mentioned the convenience of the Fourier transformation in (7.11) if the diffusion coefficient tensor $F_{kl}(\mathbf{x},\mathbf{x};t)$ is independent of $\mathbf{x}$. Different integral transformations related to eigenfunctions of the diffusion operator

$$\hat{L} = \frac{\partial^2}{\partial x_k \partial x_l} F_{kl}(\mathbf{x},\mathbf{x};t)$$

can be used in other situations.

For example, in the case of the Legendre operator

$$\hat{L} = \frac{\partial}{\partial x}(x^2 - 1)\frac{\partial}{\partial x},$$

it is quite natural to use the integral transformation related to the *Legendre functions.* This transformation is called the *Meller–Fock* transform (see, e.g., [56]) and is defined by the formula

$$F(\mu) = \int_1^\infty dx f(x) P_{-1/2+i\mu}(x) \quad (\mu > 0), \tag{8.35}$$

where $P_{-1/2+i\mu}(x)$ is the *complex index Legendre function of the first kind,* which satisfies the equation

$$\frac{d}{dx}(x^2 - 1)\frac{d}{dx}P_{-1/2+i\mu}(x) = -\left(\mu^2 + \frac{1}{4}\right)P_{-1/2+i\mu}(x). \tag{8.36}$$

The inversion of the transform (8.35) has the form

$$f(x) = \int_0^\infty d\mu\, \mu \tanh(\pi\mu) F(\mu) P_{-1/2+i\mu}(x) \quad (1 \le x \le \infty), \tag{8.37}$$

where $F(\mu)$ is given by formula (8.35).

Another integral transformation called the Kantorovich–Lebedev transform (see, e.g., [56]), is related to diffusion operator $\hat{L} = \frac{\partial}{\partial x}x^2\frac{\partial}{\partial x}$ and has the form

$$F(\tau) = \int_0^\infty dx f(x) K_{i\tau}(x) \quad (\tau > 0), \tag{8.38}$$

where $K_{i\tau}(x)$ is the *imaginary index McDonalds function of the first kind,* which satisfies the equations

$$\left(x^2\frac{d^2}{dx^2} + x\frac{d}{dx} - x^2 + \tau^2\right)K_{i\tau}(x) = 0,$$

$$\left(\frac{d}{dx}x^2\frac{d}{dx} - x\frac{d}{dx}\right)K_{i\tau}(x) = \left(x^2 - \tau^2\right)K_{i\tau}(x). \tag{8.39}$$

The corresponding inversion has the form

$$f(x) = \frac{2}{\pi^2 x}\int_0^\infty d\tau \sinh(\pi\tau) F(\tau) K_{i\tau}(x). \tag{8.40}$$

As a concrete example, consider the Fokker–Planck equation ($x \ge 1$)

$$\frac{\partial}{\partial t}p(x,t|x_0,t_0) = D\frac{\partial}{\partial x}(x^2 - 1)\frac{\partial}{\partial x}p(x,t|x_0,t_0), \quad p(x,t_0|x_0,t_0) = \delta(x - x_0). \tag{8.41}$$

Multiplying Eq. (8.41) by $P_{-1/2+i\mu}(x)$, integrating the result over x from 1 to ∞, and introducing function

$$p(t,\mu) = \int_1^\infty dx p(x,t|x_0,t_0) P_{-1/2+i\mu}(x),$$

we obtain the equation

$$\frac{\partial}{\partial t}p(t,\mu) = D\int\limits_{1}^{\infty} dx P_{-1/2+i\mu}(x)\frac{\partial}{\partial x}(x^2-1)\frac{\partial}{\partial x}p(x,t|x_0,t_0) \tag{8.42}$$

with the initial value

$$p(t_0,\mu) = P_{-1/2+i\mu}(x_0). \tag{8.43}$$

Integrating two times by parts in the right-hand side of Eq. (8.42) and using the differential Legendre equation for function $P_{-1/2+i\mu}(x)$ (8.36), we obtain the ordinary differential equation in $p(t,\mu)$

$$\frac{d}{dt}p(t,\mu) = -D\left(\mu^2+\frac{1}{4}\right)p(t,\mu),$$

whose solution satisfying the initial value (8.43) has the form

$$p(t,\mu) = P_{-1/2+i\mu}(x_0)e^{-D\left(\mu^2+\frac{1}{4}\right)(t-t_0)}.$$

Using now inversion (8.37), we obtain the solution to Eq. (8.41) in terms of the Meller–Fock integral

$$p(x,t|x_0,t_0) = \int\limits_{0}^{\infty} d\mu\,\mu\tanh(\pi\mu)e^{-D\left(\mu^2+\frac{1}{4}\right)(t-t_0)}P_{-1/2+i\mu}(x)P_{-1/2+i\mu}(x_0). \tag{8.44}$$

If $x_0 = 1$ at the initial instant $t_0 = 0$, we obtain the equation

$$P(x,t) = \int\limits_{0}^{\infty} d\mu\,\mu\tanh(\pi\mu)e^{-D\left(\mu^2+\frac{1}{4}\right)(t-t_0)}P_{-1/2+i\mu}(x) \tag{8.45}$$

corresponding to the solution of the Fokker–Planck equation for the one-point probability density (8.41) with the initial value

$$P(x,0) = \delta(x-1).$$

8.3 Steady-state solutions of the Fokker–Planck equation

In previous sections, we discussed the general methods of solving the Fokker–Planck equation for both transition and one-point probability densities. However, the problem on the one-point probability density can have peculiarities related to possible existence of the steady-state solution; in a number of cases, such a solution can be obtained immediately. The steady-state solution, if it exists, is independent of the initial values and is the solution of the Fokker–Planck equation in the limit $t\to\infty$.

There are two classes of problems for which the steady-state solution of the Fokker–Planck equation can be easily found. These classes deal with one-dimensional differential equations and with the Hamiltonian systems of equations. Consider these cases in greater detail.

8.3.1 One-dimensional nonlinear differential equation

The one-dimensional nonlinear systems are described by the stochastic equation

$$\frac{d}{dt}x(t) = f(x) + z(t)g(x), \quad x(0) = x_0, \tag{8.46}$$

where $z(t)$ is, as earlier, the Gaussian delta-correlated process with the parameters

$$\langle z(t)\rangle = 0, \quad \langle z(t)z(t')\rangle = 2D\delta(t-t') \quad (D = \sigma_z^2\tau_0).$$

The corresponding Fokker–Planck equation has the form

$$\left(\frac{\partial}{\partial t} + \frac{\partial}{\partial x}f(x)\right)P(x,t) = D\frac{\partial}{\partial x}g(x)\frac{\partial}{\partial x}g(x)P(x,t). \tag{8.47}$$

The steady-state probability distribution $P(x)$, if it exists, satisfies the equation

$$f(x)P(x) = Dg(x)\frac{d}{dx}g(x)P(x) \tag{8.48}$$

(we assume that $P(x)$ is distributed over the whole space, i.e., for $-\infty < x < \infty$) whose solution is as follows

$$P(x) = \frac{C}{|g(x)|}\exp\left\{\frac{1}{D}\int dx\frac{f(x)}{g^2(x)}\right\}, \tag{8.49}$$

where constant C is determined from the normalization condition

$$\int\limits_{-\infty}^{\infty} dxP(x) = 1.$$

In the special case of the Langevin equation (8.1), page 193 ($f(x) = -\lambda x$, $g(x) = 1$), Eq. (8.49) grades into the Gaussian probability distribution

$$P(x) = \sqrt{\frac{\lambda}{2\pi D}}\exp\left\{-\frac{\lambda}{2D}x^2\right\}. \tag{8.50}$$

8.3.2 Hamiltonian systems

Another type of dynamic systems that allow obtaining the steady-state probability distribution is described by the Hamiltonian system with linear friction

$$\begin{aligned}\frac{d}{dt}\mathbf{r}_i(t) &= \frac{\partial}{\partial \mathbf{p}_i}H(\{\mathbf{r}_i\},\{\mathbf{p}_i\}),\\ \frac{d}{dt}\mathbf{p}_i(t) &= -\frac{\partial}{\partial \mathbf{r}_i}H(\{\mathbf{r}_i\},\{\mathbf{p}_i\}) - \lambda\mathbf{p}_i + \mathbf{f}_i(t),\end{aligned} \tag{8.51}$$

where $i = 1, 2, ..., N$,

$$H(\{\mathbf{r}_i\},\{\mathbf{p}_i\}) = \frac{\mathbf{p}_i^2}{2} + U(\mathbf{r}_1, \ldots, \mathbf{r}_N)$$

is the Hamiltonian, λ is a constant coefficient (friction), and random forces $\mathbf{f}_i(t)$ are the Gaussian delta-correlated random vector functions with the correlation tensor

$$\left\langle f_i^\alpha(t)f_j^\beta(t')\right\rangle = 2D\delta_{ij}\delta_{\alpha\beta}\delta(t-t'), \quad D = \sigma_f^2\tau_0. \tag{8.52}$$

Here, α and β are the vector indexes.

System of equations (8.51) describes the Brownian motion of a system of N interacting particles. The Fokker–Planck equation for the joint probability density of the solution to system (8.51) has the form

$$\begin{aligned}\frac{\partial}{\partial t}P(\{\mathbf{r}_i\},\{\mathbf{p}_i\},t)+\sum_{k=1}^{N}\{H,P\}_{(k)}-\lambda\sum_{k=1}^{N}\frac{\partial}{\partial\mathbf{p}_k}\{\mathbf{p}_kP\}\\ =D\sum_{k=1}^{N}\frac{\partial^2}{\partial\mathbf{p}_k^2}P(\{\mathbf{r}_i\},\{\mathbf{p}_i\},t),\end{aligned} \tag{8.53}$$

where

$$\{\varphi,\psi\}_{(k)}=\frac{\partial\varphi}{\partial\mathbf{p}_k}\frac{\partial\psi}{\partial\mathbf{r}_k}-\frac{\partial\psi}{\partial\mathbf{p}_k}\frac{\partial\varphi}{\partial\mathbf{r}_k}$$

is the Poisson bracket for the k-th particle.

One can easily check that the steady-state solution to Eq. (8.53) is the *canonical Gibbs distribution*

$$P(\{\mathbf{r}_i\},\{\mathbf{p}_i\})=C\exp\left\{-\frac{\lambda}{D}H(\{\mathbf{r}_i\},\{\mathbf{p}_i\})\right\}. \tag{8.54}$$

The specificity of this distribution consists in the Gaussian behavior with respect to momenta and statistical independence of particle coordinates and momenta.

Integrating Eq. (8.54) over all $\mathbf{r}$, we can obtain the *Maxwell distribution* that describes velocity fluctuations of the Brownian particles. The case $U(\mathbf{r}_1,\ldots,\mathbf{r}_N)=0$ corresponds to the Brownian motion of a system of free particles (8.50).

If we integrate probability distribution (8.54) over momenta (velocities), we obtain the *Boltzmann distribution* of particle coordinates

$$P(\{\mathbf{r}_i\})=C\exp\left\{-\frac{\lambda}{D}U(\{\mathbf{r}_i\})\right\}. \tag{8.55}$$

In the case of sufficiently strong friction, the equilibrium distribution (8.54) is formed in two stages. First, the Gaussian momentum distribution (the Maxwell distribution) is formed relatively quickly and then, the spatial distribution (the Boltzmann distribution) is formed at far slower rate. The latter stage is described by the Fokker–Planck equation

$$\frac{\partial}{\partial t}P(\{\mathbf{r}_i\},t)=\frac{1}{\lambda}\sum_{k=1}^{N}\frac{\partial}{\partial\mathbf{r}_k}\left(\frac{\partial U(\{\mathbf{r}_i\})}{\partial\mathbf{r}_k}P(\{\mathbf{r}_i\},t)\right)+\frac{D}{\lambda^2}\sum_{k=1}^{N}\frac{\partial^2}{\partial\mathbf{r}_k^2}P(\{\mathbf{r}_i\},t), \tag{8.56}$$

which is usually called the *Einstein–Smolukhovsky equation*. Derivation of Eq. (8.56) from the Fokker–Planck equation (8.53) is called the *Kramers problem* (see, e.g., [303] and the corresponding discussion in Sect. 5.4.1, where dynamics of particles under a random force is considered as an example). Note that Eq. (8.56) statistically corresponds to the stochastic equation

$$\frac{d}{dt}\mathbf{r}_i(t)=-\frac{1}{\lambda}\frac{\partial}{\partial\mathbf{r}_i}U(\{\mathbf{r}_i\})+\frac{1}{\lambda}\mathbf{f}_i(t),$$

which, nevertheless, cannot be considered as the limit of Eq. (8.51) for $\lambda\to\infty$.

In the one-dimensional case, Eqs. (8.51) are simplified and assume the form of the system of two equations

$$\frac{d}{dt}x(t)=y(t),\quad \frac{d}{dt}y(t)=-\frac{\partial}{\partial x}U(x)-\lambda y(t)+f(t). \tag{8.57}$$

The corresponding steady-state probability distribution has the form

$$P(x,y) = C \exp\left\{-\frac{\lambda}{D} H(x,y)\right\}, \quad H(x,y) = \frac{y^2}{2} + U(x). \tag{8.58}$$

8.3.3 Systems of hydrodynamic type

Note that system of equations (8.57) can appear in problems having no concern with the Brownian motion.

As an example, we consider the simplest hydrodynamic-type system formulated in terms of the following stochastic equation

$$\begin{aligned} \frac{d}{dt} v_0(t) &= -v_1^2(t) - v_0(t) + R + f(t), \\ \frac{d}{dt} v_1(t) &= v_0(t) v_1(t) - v_1(t). \end{aligned} \tag{8.59}$$

This system describes the motion of a triplet (gyroscope) with linear friction under exciting force combined of regular (R) and random ($f(t)$) components that act on the instable mode.

If $R < 1$ and random component of force is absent ($f(t) = 0$), the system has the stable steady-state solution

$$v_1 = 0, \quad v_0 = R, \tag{8.60}$$

and fluctuations of component $v_0(t)$ caused by the random force satisfy the stochastic equation

$$\frac{d}{dt} \tilde{v}_0(t) = -\tilde{v}_0(t) + f(t) \qquad (\tilde{v}_0(t) = v_0(t) - R). \tag{8.61}$$

Thus, for $R < 1$, the steady-state probability distribution of component $\tilde{v}_0(t)$ will be, according to Eq. (8.50), the Gaussian distribution.

For $R > 1$, we have drastically another situation. In this case, two stable equilibrium states occur for $f(t) = 0$

$$v_0 = 1, \quad v_1 = \pm\sqrt{R-1}. \tag{8.62}$$

Represent component $v_0(t)$ as $v_0(t) = 1 + \tilde{v}_0(t)$. Then, system of equations (8.59) assumes the form

$$\begin{aligned} \frac{d}{dt} \tilde{v}_0(t) &= -v_1^2(t) - \tilde{v}_0(t) + (R-1) + f(t), \\ \frac{d}{dt} v_1(t) &= \tilde{v}_0(t) v_1(t), \end{aligned} \tag{8.63}$$

and temporal evolution of component $v_1(t)$ depends on its initial value. If $v_1(0) > 0$, then $v_1(t) > 0$ too. In this case, we can represent $v_1(t)$ in the form

$$v_1(t) = e^{\varphi(t)}$$

and rewrite system of equations (8.63) in the Hamiltonian form (8.57)

$$\frac{d}{dt} \tilde{v}_0(t) = -\frac{\partial U(\varphi)}{\partial \varphi} - \tilde{v}_0(t) + f(t), \quad \frac{d}{dt} \varphi(t) = \tilde{v}_0(t), \tag{8.64}$$

where

$$U(\varphi) = \frac{1}{2} e^{2\varphi} - (R-1)\varphi.$$

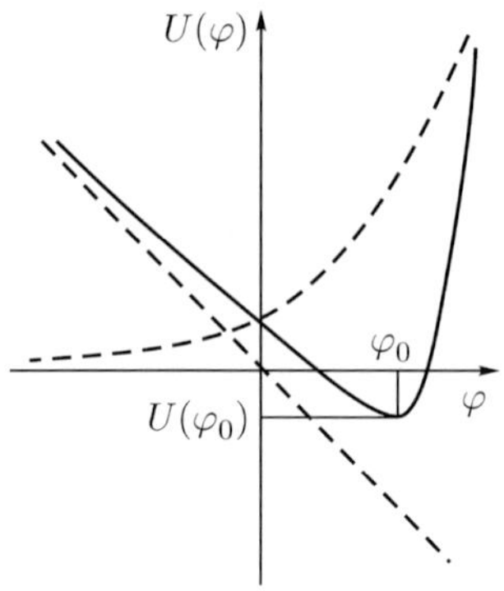

Figure 8.3: Potential function $U(\varphi)$. The dashed lines show curve $U(\varphi) = \frac{1}{2}\exp\{2\varphi\}$ and straight line $U(\varphi) = -(R-1)\varphi$.

Here, variable $\varphi(t)$ plays the role of particle's coordinate and variable $v_0(t)$ plays the role of particle's velocity.

The solid line in Fig. 8.3 shows the behavior of function $U(\varphi)$. At point $\varphi_0 = \ln\sqrt{R-1}$, this function has a minimum $U(\varphi_0) = \frac{1}{2}(R-1)\,[1 - \ln(R-1)]$ corresponding to the stable equilibrium state $v_1 = \sqrt{R-1}$. Thus, the steady-state probability distribution of $\varphi(t)$ and $\tilde{v}_0(t)$ is similar to the Gibbs distribution (8.58)

$$P(\tilde{v}_0,\varphi) = C\exp\left\{-\frac{1}{D}H(\tilde{v}_0,\varphi)\right\}, \quad H(\tilde{v}_0,\varphi) = \frac{\tilde{v}_0^2}{2} + U(\varphi). \tag{8.65}$$

From Eq. (8.65) follows that, for $R > 1$, the steady-state probability distribution is composed of two independent steady-state distributions, of which the distribution of component $v_0(t)$ of system (8.59) is the Gaussian distribution

$$P(v_0) = \frac{1}{\sqrt{2\pi D}}\exp\left\{-\frac{(v_0-1)^2}{2D}\right\},$$

and the distribution of quantity $\varphi(t)$ is the non-Gaussian distribution. If we turn back to variable $v_1(t)$, we obtain the corresponding steady-state probability distribution in the form

$$P(v_1) = \text{const}\, v_1^{\frac{R-1}{D}-1}\exp\left\{-\frac{v_1^2}{2D}\right\}. \tag{8.66}$$

As may be seen from Eq. (8.66), no steady-state probability distribution of component $v_1(t)$ exists in the critical regime ($R = 1$). Note that, if we include an additional random force acting on component $v_1(t)$, the steady-state probability density will exist even in the critical regime. In this case, the intensity of fluctuations of component $v_1(t)$ increases, and, for example, $\langle v_1^2(t)\rangle \sim \sqrt{D}$ (see, e.g., [132]).

8.4 Boundary-value problems for the Fokker-Planck equation (transfer phenomena)

The Fokker–Planck equations are the partial differential equations and they generally require boundary conditions whose particular form depends on the problem under consideration. We can proceed from both forward and backward Fokker–Planck equations, which are equivalent. Consider several examples.

8.4.1 Transfer phenomena in regular systems

Consider the nonlinear oscillator with friction described by the equation

$$\frac{d^2}{dt^2}x(t) + \lambda\frac{d}{dt}x(t) + \omega_0^2 x(t) + \beta x^3(t) = f(t) \quad (\beta,\, \lambda > 0) \tag{8.67}$$

and assume that random force $f(t)$ is the delta-correlated random function with the parameters

$$\langle f(t)\rangle = 0, \quad \langle f(t)f(t')\rangle = 2D\delta(t-t') \quad (D = \sigma_f^2\tau_0).$$

At $\lambda = 0$ and $f(t) = 0$, this equation is called the *Duffing equation*.

We can rewrite Eq. (8.67) in the standard form of the Hamiltonian system in functions $x(t)$ and $v(t) = \frac{d}{dt}x(t)$,

$$\frac{d}{dt}x(t) = \frac{\partial}{\partial v}H(x,v), \quad \frac{d}{dt}v(t) = -\frac{\partial}{\partial x}H(x,v) - \lambda v + f(t),$$

where

$$H(x,v) = \frac{v^2}{2} + U(x), \quad U(x) = \frac{\omega_0^2 x^2}{2} + \beta\frac{x^4}{4}$$

is the Hamiltonian.

According to Eq. (8.58), the steady-state solution to the corresponding Fokker–Planck equation has the form

$$P(x,v) = C\exp\left\{-\frac{\lambda}{D}H(x,v)\right\}. \tag{8.68}$$

It is clear that this distribution is the product of two independent distributions, of which one — the steady-state probability distribution of quantity $v(t)$ — is the Gaussian distribution and the other — the steady-state probability distribution of quantity $x(t)$ — is the non-Gaussian distribution. Integrating Eq. (8.68) over v, we obtain the steady-state probability distribution of $x(t)$

$$P(x,v) = C\exp\left\{-\frac{\lambda}{D}\left(\frac{\omega_0^2 x^2}{2} + \beta\frac{x^4}{4}\right)\right\}.$$

This distribution is maximum at the stable equilibrium point $x = 0$.

Consider now the equation

$$\frac{d^2}{dt^2}x(t) + \lambda\frac{d}{dt}x(t) - \omega_0^2 x(t) + \beta x^3(t) = f(t) \quad (\beta,\, \lambda > 0). \tag{8.69}$$

In this case again, the steady-state probability distribution has the form (8.68), where now

$$H(x,v) = \frac{v^2}{2} + U(x), \quad U(x) = -\frac{\omega_0^2 x^2}{2} + \beta\frac{x^4}{4}.$$

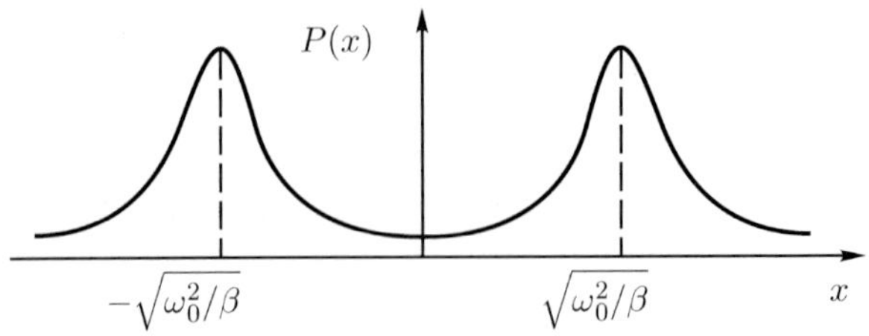

Figure 8.4: Probability distribution (8.70).

The steady-state probability distribution of $x(t)$ assumes now the form

$$P(x,v) = C\exp\left\{-\frac{\lambda}{D}\left(-\frac{\omega_0^2 x^2}{2} + \beta\frac{x^4}{4}\right)\right\} \tag{8.70}$$

and has maxima at points $x = \pm\sqrt{\omega_0^2/\beta}$ and a minimum at point $x = 0$; the maxima correspond to the stable equilibrium points of problem (8.69) for $f(t) = 0$ and the minimum, to the instable equilibrium point. Figure 8.4 shows the behavior of the probability distribution (8.70).

As we mentioned earlier, the formation of distribution (8.70) is described by the Einstein–Smolukhovsky equation (8.56), which has in this case the form

$$\frac{\partial}{\partial t}P(x,t) = \frac{1}{\lambda}\frac{\partial}{\partial x}\left(\frac{\partial U(x)}{\partial x}P(x,t)\right) + \frac{1}{\lambda^2}\frac{\partial^2}{\partial x^2}P(x,t). \tag{8.71}$$

This equation is statistically equivalent to the dynamic equation

$$\frac{d}{dt}x(t) = -\frac{1}{\lambda}\frac{\partial U(x)}{\partial x} + \frac{1}{\lambda}f(t). \tag{8.72}$$

Probability distribution (8.70) corresponds to averaging over an ensemble of realizations of random process $f(t)$. If we deal with a single realization, the system arrives at one of states corresponding to the distribution maxima with a probability of 1/2. In this case, averaging over time will form the probability distribution around the maximum position. However, after a lapse of certain time T (the longer, the smaller D), the system will be transferred in the vicinity of the other maximum due to the fact that function $f(t)$ can assume sufficiently large values. For this reason, temporal averaging will form probability distribution (8.71) only if averaging time $t \gg T$.

Introducing dimensionless coordinate $x \to \sqrt{\frac{\omega_0^2}{\beta}}x$ and time $t \to \frac{\lambda}{\omega_0^2}t$, we can rewrite Eq. (8.71) in the form

$$\frac{\partial}{\partial t}P(x,t) = \frac{\partial}{\partial x}\left(\frac{\partial U(x)}{\partial x}P(x,t)\right) + \mu\frac{\partial^2}{\partial x^2}P(x,t), \tag{8.73}$$

where

$$\mu = \frac{\beta D}{\lambda\omega_0^4}, \quad U(x) = -\frac{x^2}{2} + \frac{x^4}{4}.$$

In this case, the equivalent stochastic equation (8.72) assumes the form of Eq. (1.10), page 6

$$\frac{d}{dt}x(t) = -\frac{\partial U(x)}{\partial x} + f(t). \tag{8.74}$$

Estimate the time required for the system to switch from a most probable state $x = -1$ to the other $x = 1$.

Let the system described by stochastic equation (8.74) was at a point from the interval (a, b) at instant t_0. The corresponding probability for the system to leave this interval

$$G(t; x_0, t_0) = 1 - \int_a^b dx p(x, t|x_0, t_0)$$

satisfies Eq. (7.21) following from the backward Fokker–Planck equation (7.20), page 188, i.e., the equation

$$\frac{\partial}{\partial t_0}G(t; x_0, t_0) = \frac{\partial U(x_0)}{\partial x_0}\frac{\partial}{\partial x_0}G(t; x_0, t_0) - \mu\frac{\partial^2}{\partial x_0^2}G(t; x_0, t_0)$$

with the boundary conditions

$$G(t; x_0, t) = 0, \quad G(t; a, t_0) = G(t; b, t_0) = 1.$$

Taking into account the fact that $G(t; x_0, t_0) = G(t - t_0; x_0)$ in our problem, we can denote $(t - t_0) = \tau$ and rewrite the boundary-value problem in the form

$$\begin{aligned} \frac{\partial}{\partial \tau}G(\tau; x_0) &= \frac{\partial U(x_0)}{\partial x_0}\frac{\partial}{\partial x_0}G(\tau; x_0) - \mu\frac{\partial^2}{\partial x_0^2}G(\tau; x_0), \\ G(0; x_0) &= 0,\ G(\tau; a) = G(\tau; b) = 1 \quad \left(\lim_{\tau\to\infty} G(\tau; x_0) = 0\right). \end{aligned} \tag{8.75}$$

From Eq. (8.75), one can easily see that average time required for the system to leave interval (a, b)

$$T(x_0) = \int_0^\infty d\tau\, \tau \frac{\partial G(\tau; x_0)}{\partial \tau}$$

satisfies the boundary-value problem

$$\mu\frac{d^2T(x_0)}{dx_0^2} - \frac{dU(x_0)}{dx_0}\frac{dT(x_0)}{dx_0} = -1, \quad T(a) = T(b) = 0. \tag{8.76}$$

Equation (8.76) can be easily solved, and we obtain that the average time required for the system under random force to switch its state from $x_0 = -1$ to $x_0 = 1$ (this time is usually called the *Kramers time*) is given by the expression

$$T = \frac{1}{\mu}\int_{-1}^{1} d\xi \int_{-\infty}^{\xi} d\eta \exp\left\{\frac{1}{\mu}\left[U(\xi) - U(\eta)\right]\right\} = \frac{C(\mu)}{\mu}\int_0^1 d\xi \exp\left\{\frac{1}{\mu}U(\xi)\right\}, \tag{8.77}$$

where $C(\mu) = \int_{-\infty}^{\infty} d\xi e^{\frac{1}{\mu}U(\xi)}$. For $\mu \ll 1$, we obtain

$$T \approx \sqrt{2}\pi e^{\frac{1}{4\mu}},$$

i.e., the average switching time increases exponentially with decreasing the intensity of fluctuations of the force.

Remark 4 ***Stochastic resonance.***

In addition to the Duffing stochastic equation (8.69), a great attention is given recently to the equation

$$\frac{d^2}{dt^2}x(t) + \lambda\frac{d}{dt}x(t) - \omega_0^2 x(t) + \beta x^3(t) = f(t) + A\cos\omega_0 t \quad (\beta,\, \lambda > 0)$$

and, in particular, to the effect of an additional (except the noise) periodic impact on the statistical characteristics of the solution to Eq. (8.69) (see, e.g., reviews [8] and [127]). In this case, there sometimes occurs the phenomenon commonly called the stochastic resonance. In the context of this problem, the physical meaning of the term 'resonance' differs from the generally accepted one. Here, it reflects the fact that the response of the nonlinear stochastic oscillator on an external action appears a non-monotonous (i.e., 'resonance') function of the intensity of stochastic noise $f(t)$. In the case of the above problem, such a stochastic resonance occurs if the periodic signal frequency ω_0 coincides with the frequency of system switching between two stable states $\omega \sim 1/T$, which is called the *Kramers frequency*. ♦

8.4.2 Transfer phenomena in singular systems

Consider now the singular stochastic problem described by Eq. (1.14), page 9. We rewrite this equation in the form ($\lambda = 1$)

$$\frac{d}{dt}x(t) = -x^2(t) + f(t), \quad x(0) = x_0, \tag{8.78}$$

where we assume as earlier that random process $f(t)$ is the Gaussian delta-correlated process with the parameters

$$\langle f(t)\rangle = 0, \quad \langle f(t)f(t')\rangle = 2D\delta(t-t') \quad (D = \sigma_f^2\tau_0).$$

In the absence of fluctuations, the solution to Eq. (8.78) has the form

$$x(t) = \frac{1}{t-t_0}, \quad t_0 = -\frac{1}{x_0}.$$

If $x_0 > 0$, the solution monotonously tends to zero. But if $x_0 < 0$, the solution arrives at the infinite value within a finite time t_0.

The solution of the statistical problem (8.78) is described by the forward and backward Fokker–Planck equations ($t - t_0 = \tau$)

$$\begin{aligned}\frac{\partial}{\partial\tau}p(x,\tau|x_0) &= \frac{\partial}{\partial x}x^2 p(x,\tau|x_0) + D\frac{\partial^2}{\partial x^2}p(x,\tau|x_0),\\ \frac{\partial}{\partial\tau}p(x,\tau|x_0) &= -x_0^2\frac{\partial}{\partial x_0}p(x,\tau|x_0) + D\frac{\partial^2}{\partial x_0^2}p(x,\tau|x_0).\end{aligned} \tag{8.79}$$

Note that respective dimensions of quantities $x, p(x,\tau|x_0)$ and D are

$$[x] = \tau^{-1}, \quad [D] = \tau^{-3}, \quad [p] = \tau.$$

Consequently, we can reduce Eqs. (8.79) to the following dimensionless form

$$\begin{aligned}\frac{\partial}{\partial \tau}p(x,\tau|x_0) &= \frac{\partial}{\partial x}x^2 p(x,\tau|x_0)+\frac{\partial^2}{\partial x^2}p(x,\tau|x_0),\\ \frac{\partial}{\partial \tau}p(x,\tau|x_0) &= -x_0^2\frac{\partial}{\partial x_0}p(x,\tau|x_0)+\frac{\partial^2}{\partial x_0^2}p(x,\tau|x_0).\end{aligned} \tag{8.80}$$

Now, we must formulate boundary conditions to Eqs. (8.80). Two types of problem boundary conditions are of the first-hand interest.

Boundary conditions of the first type correspond to the assumption that curve $x(t)$ stops at point t_0 where it becomes equal to $-\infty$. This means that probability flux density

$$J(\tau,x)=x^2p(x,\tau|x_0)+\frac{\partial}{\partial x}p(x,\tau|x_0) \tag{8.81}$$

must vanish for $x\to\infty$, i.e.,

$$\begin{aligned}J(\tau,x) &\to 0, \quad \text{for} \quad x\to\infty;\\ p(x,\tau|x_0) &\to 0, \quad \text{for} \quad x\to-\infty.\end{aligned}$$

In this case, quantity $G(\tau|x_0)=\int_{-\infty}^{\infty}dxp(x,\tau|x_0)\neq 1$ is the probability of the event that function $x(t)$ remain finite along the whole axis $(-\infty,\infty)$; in other words, this quantity is the probability of the absence of singular point at instant t: $G(\tau|x_0)=P(t<t_0)$. Consequently, the probability of the appearance of singular point at instant t is given by the equality

$$P(t>t_0)=1-\int_{-\infty}^{\infty}dxp(x,\tau|x_0),$$

and the corresponding probability density

$$p(\tau|x_0)=\frac{\partial}{\partial t}P(t>t_0)=-\frac{\partial}{\partial \tau}\int_{-\infty}^{\infty}dxp(x,\tau|x_0) \tag{8.82}$$

satisfies the equation

$$\frac{\partial}{\partial \tau}p(\tau|x_0)=-x_0^2\frac{\partial}{\partial x_0}p(\tau|x_0)+\frac{\partial^2}{\partial x_0^2}p(\tau|x_0), \qquad \lim_{\tau\to 0,\,\tau\to\infty}p(\tau|x_0)\to 0 \tag{8.83}$$

following from the backward Fokker–Planck equation (8.80).

Estimate the average time

$$\langle T(x_0)\rangle=\int_0^{\infty}\tau d\tau p(\tau|x_0)$$

during which the system switches from state x_0 to state $(-\infty)$. This time is described by the equation following from Eq. (8.83)

$$-1=-x_0^2\frac{d}{dx_0}\langle T(x_0)\rangle+\frac{d^2}{dx_0^2}\langle T(x_0)\rangle \tag{8.84}$$

the boundary conditions to which are formulated as $\langle T(x_0)\rangle \to 0$ for $x_0 \to -\infty$ and finiteness of $\langle T(x_0)\rangle$ for $x_0 \to \infty$. This equation can be easily integrated, and the result has the form

$$\langle T(x_0)\rangle = \int\limits_{-\infty}^{x_0} d\xi \int\limits_{\xi}^{\infty} d\eta \exp\left\{\frac{1}{3}\left(\xi^3 - \eta^3\right)\right\}. \qquad (8.85)$$

From Eq. (8.85), we obtain for the average time of switching between two singular points ($x_0 \to \infty$)

$$\langle T(\infty)\rangle = \sqrt{\pi}\frac{12^{1/6}}{3}\Gamma\left(\frac{1}{6}\right) \approx 4.976.$$

Additionally, we note that quantity $\langle T(0)\rangle = \frac{2}{3}\langle T(\infty)\rangle$ is the average time of switching from state $x_0 = 0$ to state $x_0 = -\infty$.

Drastically different boundary conditions appear under the assumption that function $x(t)$ is discontinuous and defined for all times t. If we assume additionally that function value $-\infty$ at instant $t \to t_0 - 0$ is immediately followed by value ∞ at instant $t \to t_0 + 0$, the boundary condition to Eq. (8.80) will be the condition of continuity of probability density flux (8.81), i.e., the condition

$$J(\tau, x)|_{x=-\infty} = J(\tau, x)|_{x=+\infty}.$$

In this case, the steady-state probability density exists and is independent of x_0,

$$P(x) = J \int\limits_{-\infty}^{x} d\xi \exp\left\{\frac{1}{3}\left(\xi^3 - x^3\right)\right\}, \qquad (8.86)$$

where

$$J = \frac{1}{\langle T(\infty)\rangle}$$

is the steady-state probability flux density.

From (8.86) follows the asymptotic formula

$$P(x) \approx \frac{1}{\langle T(\infty)\rangle\, x^2} \qquad (8.87)$$

for great x. This asymptotic is formed by discontinuities of function $x(t)$. Indeed, function $x(t)$ behaves near the discontinuity as

$$x(t) = \frac{1}{t - t_k},$$

and the effect of randomness appears insignificant. In this case, we have for sufficiently great t ($t \gg \langle T(\infty)\rangle$) and x

$$\begin{aligned} p(x,t|x_0) &= \sum_{k=0}^{\infty}\left\langle \delta\left(x - \frac{1}{t-t_k}\right)\right\rangle = \frac{1}{x^2}\sum_{k=0}^{\infty}\left\langle \delta\left(t - t_k\right)\right\rangle \\ &= \frac{1}{2\pi x^2}\int\limits_{-\infty}^{\infty} d\omega e^{-i\omega t}\sum_{k=0}^{\infty}\left\langle e^{i\omega t_k}\right\rangle = \frac{1}{2\pi x^2}\int\limits_{-\infty}^{\infty} d\omega e^{-i\omega t}\frac{\Phi_0(\omega)}{1-\Phi(\omega)}, \end{aligned}$$

where $\Phi_0(\omega) = \langle e^{i\omega t_0} \rangle$ is the characteristic function of the first singular point, and $\Phi(\omega) = \left\langle e^{i\omega T} \right\rangle$ is the characteristic function of the temporal interval between the singularities. As a result, for $t \to \infty$, we obtain the asymptotic

$$P(x) = -\frac{1}{2\pi i x^2 \langle T(\infty) \rangle} \int\limits_{-\infty}^{\infty} d\omega e^{-i\omega t} \frac{1}{\omega + i0} = \frac{J}{x^2},$$

coincident with Eq. (8.87).

8.5 Asymptotic and approximate methods of solving the Fokker-Plank equation

If parameter fluctuations of the dynamic system are sufficiently small, the Fokker–Planck equation can be analyzed using different asymptotic and approximate methods. Consider in more detail three methods most used in statistical analysis.

8.5.1 Asymptotic expansion

First of all, one can formulate some convergence method with respect to small parameters related to fluctuating quantities. This is the standard procedure for partial differential equations in which the small parameter appears as a factor of the highest derivative. The schematic of such a method is as follows (see, e.g., [72]).

Rewrite the Fokker–Planck equation in the form

$$\begin{aligned} &\frac{\partial}{\partial t} P(\mathbf{x}, t) + A(\mathbf{x}, t; \varepsilon) P(\mathbf{x}, t) + B_i(\mathbf{x}, t; \varepsilon) \frac{\partial}{\partial x_i} P(\mathbf{x}, t) \\ &= \varepsilon^2 D_{ij}(\mathbf{x}, t; \varepsilon) \frac{\partial^2}{\partial x_i \partial x_j} P(\mathbf{x}, t), \quad P(\mathbf{x}, 0) = p_0(\mathbf{x}), \end{aligned} \tag{8.88}$$

where we introduced parameter ε^2 that characterizes the intensity of fluctuations of dynamic system parameters. Representing the solution to Eq. (8.88) in the form

$$P(\mathbf{x}, t) = C(\varepsilon) \exp\left\{ -\frac{1}{\varepsilon^2} \phi(\mathbf{x}, t; \varepsilon) \right\}, \tag{8.89}$$

we obtain the nonlinear equation for function $\phi(\mathbf{x}, t; \varepsilon)$

$$\begin{aligned} &\frac{\partial}{\partial t} \phi(\mathbf{x}, t; \varepsilon) - \varepsilon^2 A(\mathbf{x}, t; \varepsilon) + B_i(\mathbf{x}, t; \varepsilon) \frac{\partial}{\partial x_i} \phi(\mathbf{x}, t; \varepsilon) \\ &- \varepsilon^2 D_{ij}(\mathbf{x}, t; \varepsilon) \frac{\partial^2}{\partial x_i \partial x_j} \phi(\mathbf{x}, t; \varepsilon) + D_{ij}(\mathbf{x}, t; \varepsilon) \left(\frac{\partial}{\partial x_i} \phi(\mathbf{x}, t; \varepsilon) \right) \left(\frac{\partial}{\partial x_j} \phi(\mathbf{x}, t; \varepsilon) \right) = 0, \end{aligned} \tag{8.90}$$

whose solution can be sought in the form of the series in powers of ε^2

$$\phi(\mathbf{x}, t; \varepsilon) = \phi_0(\mathbf{x}, t) + \varepsilon^2 \phi_1(\mathbf{x}, t) + \ldots.$$

To derive the convergence method for Eq. (8.90), we substitute this expansion in Eq. (8.90), expand the equation coefficients in series in ε^2, and group the terms with the corresponding powers of ε^2. In particular, for function $\phi_0(\mathbf{x},t)$, we obtain the equation

$$\left(\frac{\partial}{\partial t}+B_i(\mathbf{x},t;0)\frac{\partial}{\partial x_i}\right)\phi_0(\mathbf{x},t)+D_{ij}(\mathbf{x},t;0)\left(\frac{\partial}{\partial x_i}\phi_0(\mathbf{x},t)\right)\left(\frac{\partial}{\partial x_j}\phi_0(\mathbf{x},t)\right)=0,$$

which is the first-order partial differential equation and can be solved by the method of characteristics, for example. Function $\phi_0(\mathbf{x},t)$ is the first term of the convergence method series; it describes the main singularity of the Fokker–Planck equation. The next term $\phi_1(\mathbf{x},t)$ describes the preexponential factor and constant $C(\varepsilon^2)$ in (8.89) can be obtained from the behavior of the solution to Eq. (8.88) for $t\to 0$ and the corresponding initial value.

This convergence method holds only for a finite-duration initial stage of evolution and fails in the limit $t\to\infty$. To analyze Eq. (8.88) in this limit, this equation is usually rearranged to the form containing the self-adjoint operator with respect to spatial variables, which has the discrete spectrum.

Consider now two approximate methods of solving the Fokker–Planck equation.

8.5.2 Method of cumulant expansions

The first method is called the method of cumulant expansions [235]. If we perform the Fourier transform of the Fokker–Planck equation (7.11), page 186 with respect to spatial variables $\mathbf{x}$, i.e., turn from the probability density of the solution to stochastic equations (7.1), page 184 to the characteristic function

$$\Phi(\boldsymbol{\lambda},t)=\left\langle e^{i\boldsymbol{\lambda}\mathbf{x}(t)}\right\rangle=e^{\Theta(\boldsymbol{\lambda},t)} \tag{8.91}$$

and expand this function in the Taylor series in powers of $\boldsymbol{\lambda}$, we obtain that the expansion coefficients (i.e., one-point cumulants of random process $\mathbf{x}(t)$) satisfy the infinite system of nonlinear equations. The method of cumulant expansions considers this system neglecting all higher-order cumulants beginning from some certain order (if this order is three, we arrive at the Gaussian approximation, if four, the excess approximation, and so on). The retained cumulants satisfy the closed nonlinear system of ordinary differential equations whose solution determines the time-dependent behavior of cumulants. Note that monograph [235] suggests the general approach for deriving these equations directly from stochastic equations (7.1), without considering the Fokker–Planck equation (7.11), page 186 or the equation for the characteristic function. A disadvantage of this method consists in the fact that the neglect of the infinite number of cumulants, as is known, impairs the probability distribution. In particular, such impaired distribution appears negative in certain regions of spatial variables. Nevertheless, examples show that the method of cumulant expansions adequately describes time-dependent behavior of certain cumulants for a wide class of problems. It seems that this class of problems is limited to the problems for which statistical characteristics of the solution are analytic functions with respect to the intensity of random actions. Most likely, this method will fail for problems characterized by the nonanalytic behavior with respect to this parameter (such as problems on the escape of the trajectory of a system out of certain spatial region and the problem on the arrival at a given boundary).

8.5.3 Method of fast oscillation averaging

Another approximate method widely used in the context of stochastic oscillating systems is called the method of averaging over fast parameters. For example, let a stochastic system is described by the dynamic equations

$$\begin{aligned} \frac{d}{dt}x(t) &= A(x,\widetilde{\phi}) + z(t)B(x,\widetilde{\phi}), \\ \frac{d}{dt}\phi(t) &= C(x,\widetilde{\phi}) + z(t)D(x,\widetilde{\phi}), \end{aligned} \tag{8.92}$$

where

$$\widetilde{\phi}(t) = \omega_0 t + \phi(t),$$

functions $A(x,\widetilde{\phi})$, $B(x,\widetilde{\phi})$, $C(x,\widetilde{\phi})$, and $D(x,\widetilde{\phi})$ are the periodic functions of variable $\widetilde{\phi}$, and $z(t)$ is the Gaussian delta-correlated process with the parameters

$$\langle z(t)\rangle = 0, \quad \langle z(t)z(t')\rangle = 2D\delta(t-t'), \quad D = \sigma^2\tau_0.$$

Variables $x(t)$ and $\phi(t)$ can mean the vector module and phase, respectively. The Fokker–Planck equation corresponding to system of equations (8.92) has the form

$$\begin{aligned} \frac{\partial}{\partial t}P(x,\phi,t) &= -\frac{\partial}{\partial x}A(x,\widetilde{\phi})P(x,\phi,t) - \frac{\partial}{\partial\phi}C(x,\widetilde{\phi})P(x,\phi,t) \\ &+D\left[\frac{\partial}{\partial x}B(x,\widetilde{\phi}) + \frac{\partial}{\partial\phi}D(x,\widetilde{\phi})\right]^2 P(x,\phi,t). \end{aligned} \tag{8.93}$$

Commonly, Eq. (8.93) is very complicated to immediately analyze the joint probability density. We rewrite this equation in the form

$$\begin{aligned} &\frac{\partial}{\partial t}P(x,\phi,t) = -\frac{\partial}{\partial x}A(x,\widetilde{\phi})P(x,\phi,t) - \frac{\partial}{\partial\phi}C(x,\widetilde{\phi})P(x,\phi,t) \\ &-D\frac{\partial}{\partial x}\left(\frac{\partial B^2(x,\widetilde{\phi})}{2\partial x} + \frac{\partial B(x,\widetilde{\phi})}{\partial\widetilde{\phi}}D(x,\widetilde{\phi})\right)P(x,\phi,t) \\ &-D\frac{\partial}{\partial\phi}\left(\frac{\partial D(x,\widetilde{\phi})}{\partial x}B(x,\widetilde{\phi}) + \frac{\partial D^2(x,\widetilde{\phi})}{2\partial\widetilde{\phi}})\right)P(x,\phi,t) \\ &+D\left\{\frac{\partial^2}{\partial x^2}B^2(x,\widetilde{\phi}) + 2\frac{\partial^2}{\partial x\partial\phi}B(x,\widetilde{\phi})D(x,\widetilde{\phi}) + \frac{\partial^2}{\partial\phi^2}D^2(x,\widetilde{\phi})\right\}P(x,\phi,t). \end{aligned} \tag{8.94}$$

Now, we assume that functions $A(x,\widetilde{\phi})$ and $C(x,\widetilde{\phi})$ are sufficiently small and fluctuation intensity of process $z(t)$ is also small. In this case, statistical characteristics of system of equations (8.92) only slightly vary during times $\sim 1/\omega_0$. To study these small variations (accumulated effects), we can average Eq. (8.94) over the period of all oscillating functions. Assuming that function $P(x,\phi,t)$ remains intact under averaging, we obtain the equation

$$\begin{aligned} &\frac{\partial}{\partial t}\overline{P(x,\phi,t)} = -\frac{\partial}{\partial x}\overline{A(x,\widetilde{\phi})}\,\overline{P(x,\phi,t)} - \frac{\partial}{\partial\phi}\overline{C(x,\widetilde{\phi})}\,\overline{P(x,\phi,t)} \\ &-D\left\{\frac{\partial}{\partial x}\left(\overline{\frac{\partial B^2(x,\widetilde{\phi})}{2\partial x}} + \overline{\frac{\partial B(x,\widetilde{\phi})}{\partial\widetilde{\phi}}D(x,\widetilde{\phi})}\right) + \frac{\partial}{\partial\phi}\overline{\frac{\partial D(x,\widetilde{\phi})}{\partial x}B(x,\widetilde{\phi})}\right\}\overline{P(x,\phi,t)} \\ &+D\left\{\frac{\partial^2}{\partial x^2}\overline{B^2(x,\widetilde{\phi})} + 2\frac{\partial^2}{\partial x\partial\phi}\overline{B(x,\widetilde{\phi})D(x,\widetilde{\phi})} + \frac{\partial^2}{\partial\phi^2}\overline{D^2(x,\widetilde{\phi})}\right\}\overline{P(x,\phi,t)}, \end{aligned} \tag{8.95}$$

where the overbar denotes quantities averaged over the oscillation period.

Integrating Eq. (8.95) over ϕ, we obtain the Fokker–Planck equation for function $\overline{P(x,t)}$

$$\frac{\partial}{\partial t}\overline{P(x,\phi,t)} = -\frac{\partial}{\partial x}\overline{A(x,\widetilde{\phi})}\,\overline{P(x,\phi,t)}$$
$$+D\frac{\partial}{\partial x}\left(\overline{\frac{\partial B^2(x,\widetilde{\phi})}{2\partial x}} + \overline{\frac{\partial B(x,\widetilde{\phi})}{\partial\widetilde{\phi}}D(x,\widetilde{\phi})}\right)\overline{P(x,\phi,t)} + D\frac{\partial}{\partial x}\overline{B^2(x,\widetilde{\phi})}\frac{\partial}{\partial x}\overline{P(x,\phi,t)}. \tag{8.96}$$

Note that quantity $x(t)$ appears the one-dimensional Markovian random process in this approximation.

If we assume that

$$\overline{B(x,\widetilde{\phi})D(x,\widetilde{\phi})} = \overline{\frac{\partial D(x,\widetilde{\phi})}{\partial x}B(x,\widetilde{\phi})} = 0, \text{ and } \overline{C(x,\widetilde{\phi})} = \text{const}\,,\ \overline{D^2(x,\widetilde{\phi})} = \text{const}$$

in Eq. (8.95), then processes $x(t)$ and $\phi(t)$ become statistically independent, and process $\phi(t)$ becomes the Markovian Gaussian process whose variance is the linear increasing function of time t. This means that probability distribution of quantity $\phi(t)$ on segment $[0, 2\pi]$ becomes uniform for large t (at $\overline{C(x,\widetilde{\phi})} = 0$).

As an illustration of using the above technique, we consider the problem on the stochastic parametric resonance.

Stochastic parametric resonance

Consider the stochastic second-order equation equivalent to system of the first-order equations (5.171), page 139

$$\frac{d}{dt}x(t) = y(t), \quad \frac{d}{dt}y(t) = -2\gamma y(t) - \omega_0^2[1 + z(t)]x(t). \tag{8.97}$$

In Chapter 5, we discussed the general approach to this problem in the case of the delta-correlated fluctuations of frequency. Here, we will assume that $z(t)$ is the Gaussian random process with the parameters

$$\langle z(t)\rangle = 0, \quad \langle z(t)z(t')\rangle = 2\sigma^2\tau_0\delta(t - t').$$

Replace functions $x(t)$ and $y(t)$ with the variables — oscillation amplitude $A(t)$ and phase $\phi(t)$ — defined by the formulas

$$x(t) = A(t)\sin\left(\omega_0 t + \phi(t)\right), \quad y(t) = \omega_0 A(t)\cos\left(\omega_0 t + \phi(t)\right). \tag{8.98}$$

Substituting Eqs. (8.98) in system of equations (8.97), we obtain the system of equations for functions $A(t)$ and $\phi(t)$

$$\begin{aligned}\frac{d}{dt}A(t) &= -2\gamma A(t)\cos^2\psi(t) - \frac{\omega_0}{2}z(t)A(t)\sin\left(2\psi(t)\right), \\ \frac{d}{dt}\phi(t) &= 2\gamma\sin\left(2\psi(t)\right) + \omega_0 z(t)\sin^2\psi(t),\end{aligned} \tag{8.99}$$

where $\psi(t) = \omega_0 t + \phi(t)$. Representing amplitude $A(t)$ as $A(t) = e^{u(t)}$, we can rewrite system (8.99) in the form

$$\begin{aligned}\frac{d}{dt}u(t) &= -2\gamma\cos^2\psi(t) - \frac{\omega_0}{2}z(t)\sin\left(2\psi(t)\right),\\ \frac{d}{dt}\phi(t) &= \gamma\sin\left(2\psi(t)\right) + \omega_0 z(t)\sin^2\psi(t).\end{aligned} \tag{8.100}$$

Consider now the joint probability density of the solution to system of equations (8.99) $P(t;u,\phi) = \langle\varphi(t;u,\phi)\rangle$, where the indicator function

$$\varphi(t;u,\phi) = \delta\left(u(t)-u\right)\delta\left(\phi(t)-\phi\right)$$

satisfies the Liouville equation

$$\begin{aligned}\frac{\partial}{\partial t}\varphi(t;u,\phi) &= \gamma\left\{2\frac{\partial}{\partial u}\cos^2\psi(t) - \frac{\partial}{\partial\phi}\sin\left(2\psi(t)\right)\right\}\varphi(t;u,\phi)\\ &+z(t)\omega_0\left\{\frac{1}{2}\frac{\partial}{\partial u}\sin\left(2\psi(t)\right) - \frac{\partial}{\partial\phi}\sin^2\psi(t)\right\}\varphi(t;u,\phi).\end{aligned} \tag{8.101}$$

Averaging now Eq. (8.101) over an ensemble of realizations of random delta-correlated process $z(t)$, using the Furutsu–Novikov formula and the equality

$$\frac{\delta}{\delta z(t-0)}\varphi(t;u,\phi) = \omega_0\left\{\frac{1}{2}\frac{\partial}{\partial u}\sin\left(2\psi(t)\right) - \frac{\partial}{\partial\phi}\sin^2\psi(t)\right\}\varphi(t;u,\phi)$$

following from Eq. (8.101), we obtain the Fokker–Planck equation for the probability density

$$\begin{aligned}\frac{\partial}{\partial t}P(t;u,\phi) &= \gamma\left\{2\frac{\partial}{\partial u}\cos^2\psi(t) - \frac{\partial}{\partial\phi}\sin\left(2\psi(t)\right)\right\}P(t;u,\phi)\\ &+D\left\{\frac{1}{2}\frac{\partial}{\partial u}\sin\left(2\psi(t)\right) - \frac{\partial}{\partial\phi}\sin^2\psi(t)\right\}^2 P(t;u,\phi),\end{aligned}$$

where $D = \sigma^2\tau_0\omega_0^2$. This equation can be rewritten in the form

$$\begin{aligned}\frac{\partial}{\partial t}P(t;u,\phi) &= \gamma\left\{2\frac{\partial}{\partial u}\cos^2\psi(t) - \frac{\partial}{\partial\phi}\sin\left(2\psi(t)\right)\right\}P(t;u,\phi)\\ &+D\left\{\frac{\partial}{\partial u}\cos\left(2\psi(t)\right)\sin^2\psi(t) - 2\frac{\partial}{\partial\phi}\sin^3\psi(t)\cos\psi(t)\right\}P(t;u,\phi)\\ &+D\left\{\frac{1}{4}\frac{\partial^2}{\partial u^2}\sin^2\left(2\psi(t)\right) - \frac{\partial^2}{\partial u\partial\phi}\sin\left(2\psi(t)\right)\sin^2\psi(t) + \frac{\partial^2}{\partial\phi^2}\sin^4\psi(t)\right\}P(t;u,\phi).\end{aligned} \tag{8.102}$$

Assuming that absorption parameter γ is small in comparison with oscillation frequency ω_0 ($\gamma \ll \omega_0$), we can average Eq. (8.102) over oscillation period $T = 2\pi/\omega_0$ (the assumption that statistical characteristics only slightly vary during times $\sim T$ allows us to average solely trigonometric functions appeared in the right-hand side of Eq. (8.102)) to obtain the equation for the averaged (i.e., describing slow variations of statistical characteristics) probability density

$$\begin{aligned}\frac{\partial}{\partial t}\overline{P(t;u,\phi)} &= \gamma\frac{\partial}{\partial u}\overline{P(t;u,\phi)} - \frac{D}{4}\frac{\partial}{\partial u}\overline{P(t;u,\phi)}\\ &+\frac{D}{8}\frac{\partial^2}{\partial u^2}\overline{P(t;u,\phi)} + \frac{3D}{8}\frac{\partial^2}{\partial\phi^2}\overline{P(t;u,\phi)}\end{aligned} \tag{8.103}$$

with the initial value

$$\overline{P(0;u,\phi)} = \delta(u-u_0)\delta(\phi-\phi_0).$$

For example, in the case of initial values $u_0 = 0$, $\phi_0 = 0$ corresponding to $x(0) = 0$, $y(0) = \omega_0$, from Eq. (8.103) follows that statistical characteristics of amplitude and phase (averaged over the oscillation period) are statistically independent and the corresponding probability densities are the Gaussian densities,

$$\overline{P(t;u)} = \frac{1}{\sqrt{2\pi\sigma_u^2(t)}} e^{-\frac{(u-\langle u(t)\rangle)^2}{2\sigma_u^2(t)}}, \quad \overline{P(t;\phi)} = \frac{1}{\sqrt{2\pi\sigma_\phi^2(t)}} e^{-\frac{(\phi-\phi_0)^2}{2\sigma_\phi^2(t)}}, \tag{8.104}$$

where

$$\langle u(t)\rangle = u_0 - \gamma t + \frac{D}{4}t, \quad \sigma_u^2(t) = \frac{D}{4}t, \quad \langle \phi(t)\rangle = \phi_0, \quad \sigma_\phi^2(t) = \frac{3D}{4}t.$$

As an example of using the above expressions, consider expressions for $\langle x(t)\rangle$ and $\langle x^2(t)\rangle$ corresponding to initial values $u_0 = 0$ and $\phi_0 = 0$.

For the average value, we have the expression

$$\begin{aligned} \langle x(t)\rangle &= \langle A(t)\rangle \langle \sin(\omega_0 t + \phi(t))\rangle \\ &= \frac{1}{2i}\left\langle e^{u(t)}\right\rangle \left\langle e^{i\omega_0 t + i\phi(t)} - e^{-i\omega_0 t - i\phi(t)}\right\rangle \\ &= \exp\left\{\langle u(t)\rangle + \frac{1}{2}\sigma_u^2(t) - \frac{1}{2}\sigma_\phi^2(t)\right\}\sin(\omega_0 t) = e^{-\gamma t}\sin(\omega_0 t) \end{aligned} \tag{8.105}$$

coinciding with the problem solution in the case of absent fluctuations.

For quantity $\langle x^2(t)\rangle$, we obtain the expression

$$\begin{aligned} \left\langle x^2(t)\right\rangle &= \left\langle e^{2u(t)}\right\rangle \left\langle \sin^2(\omega_0 t + \phi(t))\right\rangle = \frac{1}{2}\left\langle e^{2u(t)}\right\rangle \{1 - \langle \cos 2(\omega_0 t + \phi(t))\rangle\} \\ &= \frac{1}{2}e^{2\langle u(t)\rangle + 2\sigma_u^2(t)}\left\{1 - e^{-2\sigma_\phi^2(t)}\cos(\omega_0 t)\right\} = \frac{1}{2}e^{(D-2\gamma)t}\left\{1 - e^{-\frac{3D}{2}t}\cos(2\omega_0 t)\right\} \end{aligned} \tag{8.106}$$

that coincides (in the absence of absorption) with Eq. (5.170), page 138 to terms of order $D/\omega_0 \ll 1$, and statistical parametric excitation of the system occurs if the condition

$$D > 2\gamma$$

is satisfied.

As was mentioned earlier, the random amplitude has the lognormal probability distribution; consequently, its moment functions are given by the expression

$$\langle A^n(t)\rangle = \left\langle e^{nu(t)}\right\rangle = A_0^n \exp\left\{-n\gamma t + \frac{1}{8}n(n+2)Dt\right\}. \tag{8.107}$$

Under the condition

$$8\gamma < (n+2)D,$$

stochastic dynamic system (8.97) is statistically excited beginning from the moment function of order n. Nevertheless, the typical realization curve of the random amplitude has the form

$$A^*(t) = A_0 e^{-(\gamma - \frac{D}{4})t},$$

and, under sufficiently weak absorption, namely if

$$1 < 4\frac{\gamma}{D} < 1 + \frac{1}{2}n,$$

which is the case if n is sufficiently great, the typical realization curve decreases exponentially with time, whereas all moment functions of random amplitude $A(t)$ of order n and higher are exponentially increasing functions of time. This means that statistics of random amplitude $A(t)$ is formed by high peaks above the exponentially decreasing typical realization curve, which is a consequence of the fact that random amplitude $A(t)$ is the lognormal quantity.

Chapter 9

Gaussian delta-correlated random field (causal integral equations)

In problems discussed in the previous chapter, we succeeded in deriving the closed statistical description in the approximation of the delta-correlated random field due to the fact that every of these problems corresponded to a system of the first-order (in temporal coordinate) differential equations with given initial values at $t = 0$. The solutions to such systems satisfy the dynamic causality condition, which means that the solution at instant t depends only on system parameter fluctuations for preceding times and is independent of fluctuations for consequent times.

However, problems described in terms of integral equations that generally cannot be reduced to a system of differential equations also can satisfy the causality condition.

In this case, the parent stochastic equation is the linear integral equation for Green's function

$$S(\mathbf{r},\mathbf{r}') = S_0(\mathbf{r},\mathbf{r}') + \int d\mathbf{r}_1 d\mathbf{r}_2 d\mathbf{r}_3 S_0(\mathbf{r},\mathbf{r}_1)\Lambda(\mathbf{r}_1,\mathbf{r}_2,\mathbf{r}_3) f(\mathbf{r}_2) S(\mathbf{r}_3,\mathbf{r}'), \tag{9.1}$$

where $\mathbf{r}$ denotes all arguments of functions $S(\mathbf{r},\mathbf{r}')$ and $f(\mathbf{r})$ including the index arguments that assume summation instead of integration. It is assumed here that function $f(\mathbf{r})$ is the random field and function $S_0(\mathbf{r},\mathbf{r}')$ is Green's function for the problem with absent parameter fluctuations, i.e., for $f(\mathbf{r}) = 0$. We will assume additionally that quantity $\Lambda(\{\mathbf{r}_i\})$ is a function.

The solution to Eq. (9.1) is a functional of field $f(\mathbf{r})$, i.e. $S(\mathbf{r},\mathbf{r}') = S[\mathbf{r},\mathbf{r}'; f(\widetilde{\mathbf{r}})]$ and Eq. (9.1) appears equivalent to the functional equation that contains the variational derivative in functional space $\{f(\widetilde{\mathbf{r}})\}$ (5.30), page 105

$$\frac{\delta}{\delta f(\mathbf{r}_0)} S[\mathbf{r},\mathbf{r}'; f(\widetilde{\mathbf{r}})] = \int d\mathbf{r}_1 d\mathbf{r}_2 S[\mathbf{r},\mathbf{r}_1; f(\widetilde{\mathbf{r}})]\Lambda(\mathbf{r}_1,\mathbf{r}_0,\mathbf{r}_2) S[\mathbf{r}_2,\mathbf{r}'; f(\widetilde{\mathbf{r}})], \tag{9.2}$$

and satisfies the initial value

$$S[\mathbf{r},\mathbf{r}'; f(\widetilde{\mathbf{r}})]_{f=0} = S_0(\mathbf{r},\mathbf{r}').$$

Now, we select the temporal coordinate t in Eq. (9.1), i.e., rewrite it in the form

$$\begin{aligned} &S(\mathbf{r},t;\mathbf{r}',t') = S_0(\mathbf{r},t;\mathbf{r}',t') \\ &+ \int d\mathbf{r}_1 d\mathbf{r}_2 d\mathbf{r}_3 \int d\tau S_0(\mathbf{r},t;\mathbf{r}_1,\tau)\Lambda(\mathbf{r}_1,\mathbf{r}_2,\mathbf{r}_3) f(\mathbf{r}_2,\tau) S(\mathbf{r}_3,\tau;\mathbf{r}',t') \end{aligned} \tag{9.3}$$

9.1 Causal integral equation

In what follows, we will assume that

$$S_0(\mathbf{r},t;\mathbf{r}',t') = g(\mathbf{r},t;\mathbf{r}',t')\theta(t-t'),$$

where $\theta(t)$ is the Heaviside step function. In this case, the solution to Eq. (9.3) also has the form

$$S(\mathbf{r},t;\mathbf{r}',t') = G(\mathbf{r},t;\mathbf{r}',t')\theta(t-t'),$$

where function $G(\mathbf{r},t;\mathbf{r}',t')$ is described by the causal (in time) integral equation

$$G(\mathbf{r},t;\mathbf{r}',t') = g(\mathbf{r},t;\mathbf{r}',t') + \int d\mathbf{r}_1 d\mathbf{r}_2 d\mathbf{r}_3 \int\limits_{t'}^{t} d\tau g(\mathbf{r},t;\mathbf{r}_1,\tau)\Lambda(\mathbf{r}_1,\mathbf{r}_2,\mathbf{r}_3)f(\mathbf{r}_2,\tau)G(\mathbf{r}_3,\tau;\mathbf{r}',t'). \tag{9.4}$$

Function $G(\mathbf{r},t;\mathbf{r}',t')$ is a functional of field $f(\tilde{\mathbf{r}},\tau)$, i.e.,

$$G(\mathbf{r},t;\mathbf{r}',t') = G[\mathbf{r},t;\mathbf{r}',t';f(\tilde{\mathbf{r}},\tau)].$$

Consequently, the dynamic causality condition has the form

$$\frac{\delta}{\delta f(\mathbf{r}_0,\tau)}G(\mathbf{r},t;\mathbf{r}',t') = 0 \quad \text{for} \quad \tau < t' \quad \text{and} \quad \tau > t.$$

In this case, the variational derivative is given, in view of Eq. (9.2), by the expression

$$\frac{\delta}{\delta f(\mathbf{r}_0,t_0)}G(\mathbf{r},t;\mathbf{r}',t') = \int d\mathbf{r}_1 d\mathbf{r}_2 G(\mathbf{r},t;\mathbf{r}_1,t_0)\Lambda(\mathbf{r}_1,\mathbf{r}_0,\mathbf{r}_2)G(\mathbf{r}_2,t_0;\mathbf{r}',t'), \tag{9.5}$$

from which follows that

$$\frac{\delta}{\delta f(\mathbf{r}_0,t-0)}G(\mathbf{r},t;\mathbf{r}',t') = \int d\mathbf{r}_1 d\mathbf{r}_2 g(\mathbf{r},t;\mathbf{r}_1,t)\Lambda(\mathbf{r}_1,\mathbf{r}_0,\mathbf{r}_2)G(\mathbf{r}_2,t;\mathbf{r}',t'). \tag{9.6}$$

9.2 Statistical averaging

Let now random field $f(\mathbf{r},t)$ is the Gaussian random field whose average value is zero. In this case, assignment of the correlation function

$$B(\mathbf{r},t;\mathbf{r}',t') = \langle f(\mathbf{r},t)f(\mathbf{r}',t')\rangle .$$

describes all statistical characteristics of the field. For statistically homogeneous and stationary random field $f(\mathbf{r},t)$, we have

$$B(\mathbf{r},t;\mathbf{r}',t') = B(\mathbf{r}-\mathbf{r}',t-t').$$

Averaging Eq. (9.4) over an ensemble of realizations of field $f(\mathbf{r},t)$, we obtain the equation

$$\langle G(\mathbf{r},t;\mathbf{r}',t')\rangle = g(\mathbf{r},t;\mathbf{r}',t') + \int d\mathbf{r}_1 d\mathbf{r}_2 d\mathbf{r}_3 \int\limits_{t'}^{t} d\tau g(\mathbf{r},t;\mathbf{r}_1,\tau)\Lambda(\mathbf{r}_1,\mathbf{r}_2,\mathbf{r}_3)\langle f(\mathbf{r}_2,\tau)G(\mathbf{r}_3,\tau;\mathbf{r}',t')\rangle . \tag{9.7}$$

Using the Furutsu–Novikov formula (7.10), page 186 to split the correlation in the right-hand side of Eq. (9.7), we obtain

$$\langle G(\mathbf{r},t;\mathbf{r}',t')\rangle = g(\mathbf{r},t;\mathbf{r}',t') + \int d\mathbf{r}_1 d\mathbf{r}_2 d\mathbf{r}_3 \int\limits_{t'}^{t} d\tau g(\mathbf{r},t;\mathbf{r}_1,\tau)\Lambda(\mathbf{r}_1,\mathbf{r}_2,\mathbf{r}_3)$$
$$\times \int d\mathbf{r}_0 \int\limits_{t'}^{\tau} dt_0 B(\mathbf{r}_2-\mathbf{r}_0;\tau-t_0)\left\langle \frac{\delta}{\delta f(\mathbf{r}_0,t_0)} G(\mathbf{r}_3,\tau;\mathbf{r}',t')\right\rangle. \tag{9.8}$$

If we use Eq. (9.5) for the variational derivative, we obtain that Eq. (9.8) assumes the form of the equality

$$\langle G(\mathbf{r},t;\mathbf{r}',t')\rangle = g(\mathbf{r},t;\mathbf{r}',t') + \int d\mathbf{r}_1 d\mathbf{r}_2 d\mathbf{r}_3 \int\limits_{t'}^{t} d\tau g(\mathbf{r},t;\mathbf{r}_1,\tau)\Lambda(\mathbf{r}_1,\mathbf{r}_2,\mathbf{r}_3)$$
$$\times \int d\mathbf{r}_0 \int\limits_{t'}^{\tau} dt_0 B(\mathbf{r}_2-\mathbf{r}_0;\tau-t_0) \int d\mathbf{r}' d\mathbf{r}'' \left\langle G(\mathbf{r}_3,\tau;\mathbf{r}',t_0)\Lambda(\mathbf{r}',\mathbf{r}_0,\mathbf{r}'')G(\mathbf{r}'',t_0;\mathbf{r}',t')\right\rangle. \tag{9.9}$$

Now, the correlation function of field $G(\mathbf{r},t;\mathbf{r}',t')$ appears in the right-hand side of Eq. (9.9).

If we tend the temporal correlation radius of random field $f(\mathbf{r},t)$ to zero, $\tau_0 \to 0$, then Eq. (9.9) is simplified and assumes, for $t \gg \tau_0$, the form of the closed integral equation

$$\langle G(\mathbf{r},t;\mathbf{r}',t')\rangle = g(\mathbf{r},t;\mathbf{r}',t') \int d\mathbf{r}_1 d\mathbf{r}_2 d\mathbf{r}_3 \int\limits_{t'}^{t} d\tau g(\mathbf{r},t;\mathbf{r}_1,\tau)\Lambda(\mathbf{r}_1,\mathbf{r}_2,\mathbf{r}_3)$$
$$\times \int d\mathbf{r}_0 F(\mathbf{r}_2-\mathbf{r}_0) \int d\mathbf{r}' d\mathbf{r}' g(\mathbf{r}_3,\tau;\mathbf{r}',\tau)\Lambda(\mathbf{r}',\mathbf{r}_0,\mathbf{r}') \langle G(\mathbf{r}',\tau;\mathbf{r}',t')\rangle,$$

where

$$F(\mathbf{r}) = \int\limits_0^\infty dt B(\mathbf{r};t).$$

This result is equivalent to the introduction of the effective correlation function of random field $f(\mathbf{r},t)$ in Eq. (9.8)

$$B(\mathbf{r};t) = 2F(\mathbf{r})\delta(t), \quad F(\mathbf{r}) = \int\limits_0^\infty dt B(\mathbf{r};t)$$

and the use of Eq. (9.6) instead of (9.5), which just corresponds to the delta-correlated approximation for random field $f(\mathbf{r},t)$ in time.

The equation for the correlation function of solution to Eq. (9.4) can be derived in a similar way. For short, we illustrate this derivation by the simplest example of the one-dimensional causal equation ($t > t'$)

$$G(t;t') = g(t;t') + \Lambda \int\limits_{t'}^{t} d\tau g(t;\tau)z(\tau)G(\tau;t'), \tag{9.10}$$

where we assume that $z(t)$ is the Gaussian delta-correlated random function with the parameters

$$\langle z(t)\rangle = 0, \quad \langle z(t)z(t')\rangle = 2D\delta(t-t') \quad (D = \sigma_z^2\tau_0).$$

Averaging then Eq. (9.10) over an ensemble of realizations of random function $z(t)$, we obtain the equation

$$\langle G(t;t')\rangle = g(t;t') + \Lambda \int_{t'}^{t} d\tau g(t;\tau)\langle z(\tau)G(\tau;t')\rangle . \tag{9.11}$$

Taking into account Eq. (9.6) that assumes here the form

$$\frac{\delta}{\delta z(t)} G(t;t') = g(t;t)\Lambda G(t;t'), \tag{9.12}$$

we can rewrite the correlation in the right-hand side of Eq. (9.11) in the from

$$\langle z(\tau)G(\tau;t')\rangle = D\left\langle \frac{\delta}{\delta z(\tau)} G(\tau;t')\right\rangle = \Lambda D g(\tau;\tau)\langle G(\tau;t')\rangle .$$

As a consequence, Eq. (9.11) grades into the closed integral equation for average Green's function

$$\langle G(t;t')\rangle = g(t;t') + \Lambda^2 D \int_{t'}^{t} d\tau g(t;\tau) g(\tau;\tau)\langle G(\tau;t')\rangle , \tag{9.13}$$

which, according to the general-form derivation technique, has the form of the *Dyson equation*

$$\begin{aligned}
\langle G(t;t')\rangle &= g(t;t') + \Lambda \int_{t'}^{t} d\tau g(t;\tau) \int_{t'}^{\tau} d\tau' Q(\tau;\tau')\langle G(\tau';t')\rangle ,\\
\langle G(t;t')\rangle &= g(t;t') + \Lambda \int_{t'}^{t} d\tau \langle G(t;\tau)\rangle \int_{t'}^{\tau} d\tau' Q(\tau;\tau') g(\tau';t'),
\end{aligned} \tag{9.14}$$

with the *mass function*

$$Q(\tau;\tau') = \Lambda^2 D g(\tau;\tau)\delta(\tau-\tau'). \tag{9.15}$$

Derive now the equation for the correlation function

$$\Gamma(t,t';t_1,t_1') = \langle G(t;t')G^*(t_1;t_1')\rangle \quad (t > t', \quad t_1 > t_1'),$$

where $G^*(t;t')$ is complex conjugated Green's function. With this goal in view, we multiply Eq. (9.10) by $G^*(t_1;t_1')$ and average the result over an ensemble of realizations of random function $z(t)$. The result is the equation that can be symbolically represented as

$$\Gamma = g\langle G^*\rangle + \Lambda g\langle zGG^*\rangle . \tag{9.16}$$

Taking into account the Dyson equation (9.14)

$$\langle G\rangle = \{1 + \langle G\rangle Q\} g,$$

we apply operator $\{1 + \langle G\rangle Q\}$ to Eq. (9.16). As the result, we obtain the symbolic-form equation

$$\Gamma = \langle G\rangle \langle G^*\rangle + \langle G\rangle \Lambda \{\langle zGG^*\rangle - Q\Gamma\},$$

which can be represented in common variables in the form

$$\begin{aligned}\Gamma(t,t';t_1,t_1') &= \langle G(t;t')\rangle \langle G^*(t_1;t_1')\rangle \\ &+\Lambda D\int_0^t d\tau \langle G(t;\tau)\rangle \left[\left\langle \frac{\delta G(\tau;t')}{\delta z(\tau)} G^*(t_1;t_1') + 2G(\tau;t')\frac{\delta G^*(t_1;t_1')}{\delta z(\tau)}\right\rangle\right] \\ &-\Lambda^2 D\int_0^t d\tau \langle G(t;\tau)\rangle g(\tau;\tau)\Gamma(\tau,t';t_1,t_1'). \end{aligned} \tag{9.17}$$

Deriving Eq. (9.17), we used additionally Eq. (4.58), page 89 for splitting correlators between the Gaussian delta-correlated process $z(t)$ and functionals of this process

$$\langle z(t')R[t;z(\tau)]\rangle >= \begin{cases} D\left\langle \frac{\delta}{\delta z(t)} R[t;z(\tau)]\right\rangle & (t'=t, \quad \tau<t), \\ 2D\left\langle \frac{\delta}{\delta z(t')} R[t;z(\tau)]\right\rangle & (t'<t, \quad \tau<t). \end{cases}$$

Taking into account formulas (9.12) and (9.2) of which the latter assumes in our case the form

$$\frac{\delta}{\delta z(\tau)} G^*(t_1;t_1') = \Lambda G^*(t_1;\tau)G^*(\tau;t_1'),$$

we can rewrite Eq. (9.17) as

$$\begin{aligned}\Gamma(t,t';t_1,t_1') &= \langle G(t;t')\rangle \langle G^*(t_1;t_1')\rangle \\ &+2|\Lambda|^2 D\int_0^t d\tau \langle G(t;\tau)\rangle \langle G^*(t_1;\tau)G(\tau;t')G^*(\tau;t_1')\rangle . \end{aligned} \tag{9.18}$$

Now, we take into account the fact that function $G^*(t_1;\tau)$ functionally depends on random process $z(\widetilde{\tau})$ for $\widetilde{\tau} \geq \tau$ while functions $G(\tau;t')$ and $G^*(\tau;t_1')$ depend on it for $\widetilde{\tau} \leq \tau$. Consequently, these functions are statistically independent in the case of the delta-correlated process $z(\widetilde{\tau})$, and we can rewrite Eq. (9.18) in the form of the closed equation $(t_1 \geq t)$

$$\Gamma(t,t';t_1,t_1') = \langle G(t;t')\rangle \langle G^*(t_1;t_1')\rangle + 2|\Lambda|^2 D\int_0^t d\tau \langle G(t;\tau)\rangle \langle G^*(t_1;\tau)\rangle \Gamma(\tau;t';\tau;t_1'), \tag{9.19}$$

which corresponds to the *Bete–Salpeter equation* (5.55), page 110 with the *intensity operator kernel*

$$K(\tau_1,\tau';\tau_2,\tau'') = 2|\Lambda|^2 D\delta(\tau_1-\tau')\delta(\tau_2-\tau'')\delta(\tau_1-\tau_2). \tag{9.20}$$

Thus, for the one-dimensional causal equation (9.10), the *ladder approximation* appears the exact equality in the case of the delta-correlated process $z(t)$.

Chapter 10

Diffusion approximation

10.1 General remarks

Applicability of the approximation of the delta-correlated random field $\mathbf{f}(\mathbf{x},t)$ (i.e., applicability of the Fokker–Planck equation) is restricted by the smallness of the temporal correlation radius τ_0 of random field $\mathbf{f}(\mathbf{x},t)$ with respect to all temporal scales of the problem under consideration. The effect of the finite-valued temporal correlation radius of random field $\mathbf{f}(\mathbf{x},t)$ can be considered within the framework of the diffusion approximation (see, e.g., [140, 142]). The diffusion approximation appears more obvious and physical than the formal mathematical derivation of the approximation of the delta-correlated random field. This approximation also holds for sufficiently weak parameter fluctuations of the stochastic dynamic system and allows describing new physical effects caused by the finite-valued temporal correlation radius of random parameters, rather than only obtaining the applicability range of the delta-correlated approximation. The diffusion approximation assumes that the effect of random actions is insignificant during temporal scales about τ_0, i.e., the system behaves during these times as the free system.

Again, let vector function $\mathbf{x}(t)$ satisfies the dynamic equation (7.1), page 184

$$\frac{d}{dt}\mathbf{x}(t) = \mathbf{v}(\mathbf{x},t) + \mathbf{f}(\mathbf{x},t), \;\; \mathbf{x}(t_0) = \mathbf{x}_0, \tag{10.1}$$

where $\mathbf{v}(\mathbf{x},t)$ is the deterministic vector function and $\mathbf{f}(\mathbf{x},t)$ is the random statistically homogeneous and stationary Gaussian vector field with the statistical characteristics

$$\langle f(\mathbf{x},t)\rangle = 0, \quad B_{ij}(\mathbf{x},t;\mathbf{x}',t') = B_{ij}(\mathbf{x}-\mathbf{x}';t-t') = \langle f_i(\mathbf{x},t)f_j(\mathbf{x}',t')\rangle .$$

Introduce the indicator function

$$\varphi(\mathbf{x},t) = \delta(\mathbf{x}(t)-\mathbf{x}), \tag{10.2}$$

($\mathbf{x}(t)$ is the solution to Eq. (10.1)) satisfying the Liouville equation (7.6)

$$\left(\frac{\partial}{\partial t} + \frac{\partial}{\partial \mathbf{x}}\mathbf{v}(\mathbf{x},t)\right)\varphi(\mathbf{x},t) = -\frac{\partial}{\partial \mathbf{x}}\mathbf{f}(\mathbf{x},t)\varphi(\mathbf{x},t). \tag{10.3}$$

As earlier, we obtain the equation for the probability density of the solution to Eq. (10.1)

$$P(\mathbf{x}(t) = \langle\varphi(\mathbf{x},t)\rangle = \langle\delta(\mathbf{x}(t)-\mathbf{x})\rangle$$

by averaging Eq. (10.3) over an ensemble of realizations of field $\mathbf{f}(\mathbf{x},t)$

$$\left(\frac{\partial}{\partial t}+\frac{\partial}{\partial \mathbf{x}}\mathbf{v}(\mathbf{x},t)\right)P(\mathbf{x},t)=-\frac{\partial}{\partial \mathbf{x}}\left\langle \mathbf{f}(\mathbf{x},t)\varphi(\mathbf{x},t)\right\rangle, \quad P(\mathbf{x},t_0)=\delta(\mathbf{x}-\mathbf{x}_0). \tag{10.4}$$

Using the Furutsu–Novikov formula (7.10), page 186

$$\langle f_k(\mathbf{x},t)R[t;\mathbf{f}(\mathbf{y},\tau)]\rangle=\int d\mathbf{x}'\int dt' B_{kl}(\mathbf{x},t;\mathbf{x}',t')\left\langle \frac{\delta}{\delta f_l(\mathbf{x}',t')}R[t;\mathbf{f}(\mathbf{y},\tau)]\right\rangle$$

valid for the correlation between the Gaussian random field $\mathbf{f}(\mathbf{x},t)$ and arbitrary functional $R[t;\mathbf{f}(\mathbf{y},\tau)]$ of this field, we can rewrite Eq. (10.4) in the form

$$\left(\frac{\partial}{\partial t}+\frac{\partial}{\partial \mathbf{x}}\mathbf{v}(\mathbf{x},t)\right)P(\mathbf{x},t)=-\frac{\partial}{\partial x_i}\int d\mathbf{x}'\int_{t_0}^{t} dt' B_{ij}(\mathbf{x},t;\mathbf{x}',t')\left\langle \frac{\delta}{\delta f_j(\mathbf{x}',t')}\varphi(\mathbf{x},t)\right\rangle. \tag{10.5}$$

In the diffusion approximation, Eq. (10.5) is the exact equation, and the variational derivative and indicator function satisfy, within temporal scales of about temporal correlation radius τ_0 of random field $\mathbf{f}(\mathbf{x},t)$, the system of dynamic equations

$$\begin{aligned}\frac{\partial}{\partial t}\frac{\delta\varphi(\mathbf{x},t)}{\delta f_i(\mathbf{x}',t')}&=-\frac{\partial}{\partial \mathbf{x}}\left\{\mathbf{v}(\mathbf{x},t)\frac{\delta\varphi(\mathbf{x},t)}{\delta f_i(\mathbf{x}',t')}\right\},\\ \left.\frac{\delta\varphi(\mathbf{x},t)}{\delta f_i(\mathbf{x}',t')}\right|_{t=t'}&=-\frac{\partial}{\partial x_i}\left\{\delta(\mathbf{x}-\mathbf{x}')\varphi(\mathbf{x},t')\right\},\\ \frac{\partial}{\partial t}\varphi(\mathbf{x},t)=-\frac{\partial}{\partial \mathbf{x}}&\left\{\mathbf{v}(\mathbf{x},t)\varphi(\mathbf{x},t)\right\}, \quad \varphi(\mathbf{x},t)|_{t=t'}=\varphi(\mathbf{x},t').\end{aligned} \tag{10.6}$$

The solution to problem (10.5), (10.6) holds for all times t. In this case, the solution $\mathbf{x}(t)$ to problem (10.1) cannot be considered as the Markovian vector random process because its multi-time probability density cannot be factorized in terms of the transition probability density. However, in asymptotic limit $t \gg \tau_0$, the diffusion-approximation solution to the initial dynamic system (10.1) will be the Markovian random process, and the corresponding conditions of applicability are formulated as smallness of all statistical effects within temporal scales of about temporal correlation radius τ_0.

10.2 Dynamics of a particle

The use of the diffusion approximation in concrete physical problems will be discussed in the Part 4. Here, we illustrate this approximation by the example of the dynamics of a particle with linear friction under random forces, which is described by the stochastic system (1.12), page 8

$$\begin{aligned}\frac{d}{dt}\mathbf{r}(t)&=\mathbf{v}(t), \quad \frac{d}{dt}\mathbf{v}(t)=-\lambda\mathbf{v}(t)+\mathbf{f}(\mathbf{r},t),\\ \mathbf{r}(0)&=\mathbf{r}_0, \quad \mathbf{v}(0)=\mathbf{v}_0.\end{aligned} \tag{10.7}$$

Introduce the indicator function for particle position and velocity

$$\varphi(\mathbf{r},\mathbf{v},t)=\delta(\mathbf{r}(t)-\mathbf{r})\delta(\mathbf{v}(t)-\mathbf{v}).$$

This function satisfies the stochastic Liouville equation

$$\left(\frac{\partial}{\partial t}+\mathbf{v}\frac{\partial}{\partial \mathbf{r}}-\lambda\frac{\partial}{\partial \mathbf{v}}\mathbf{v}\right)\varphi(\mathbf{r},\mathbf{v},t)=-\mathbf{f}(\mathbf{r},t)\frac{\partial}{\partial \mathbf{v}}\varphi(\mathbf{r},\mathbf{v},t). \tag{10.8}$$

Averaging Eq. (10.8) over an ensemble of realizations of field $\mathbf{f}(\mathbf{r},t)$, we obtain that the one-time probability density

$$P(\mathbf{r},\mathbf{v},t)=\langle\delta(\mathbf{r}(t)-\mathbf{r})\delta(\mathbf{v}(t)-\mathbf{v})\rangle$$

satisfies the equation

$$\left(\frac{\partial}{\partial t}+\mathbf{v}\frac{\partial}{\partial \mathbf{r}}-\lambda\frac{\partial}{\partial \mathbf{v}}\mathbf{v}\right)P(\mathbf{r},\mathbf{v},t)=-\frac{\partial}{\partial \mathbf{v}}\left\langle\mathbf{f}(\mathbf{r},t)\varphi(\mathbf{r},\mathbf{v},t)\right\rangle. \tag{10.9}$$

Taking into account the Furutsu–Novikov formula (7.10), page 186 we can rewrite this equation in the form

$$\left(\frac{\partial}{\partial t}+\mathbf{v}\frac{\partial}{\partial \mathbf{r}}-\lambda\frac{\partial}{\partial \mathbf{v}}\mathbf{v}\right)P(\mathbf{r},\mathbf{v},t)=-\frac{\partial}{\partial v_i}\int d\mathbf{r}'\int\limits_0^t dt' B_{ij}(\mathbf{r}-\mathbf{r}',t-t')\left\langle\frac{\delta\varphi(\mathbf{r},\mathbf{v},t)}{\delta f_j(\mathbf{r}',t')}\right\rangle. \tag{10.10}$$

If we express the variational derivative in the right-hand side of Eq. (10.10) in terms of the variational derivatives of functions $\mathbf{r}(t)$ and $\mathbf{v}(t)$, the equation will assume the form

$$\left(\frac{\partial}{\partial t}+\mathbf{v}\frac{\partial}{\partial \mathbf{r}}-\lambda\frac{\partial}{\partial \mathbf{v}}\mathbf{v}\right)P(\mathbf{r},\mathbf{v},t)=\frac{\partial}{\partial v_i}\int d\mathbf{r}'\int\limits_0^t dt' B_{ij}(\mathbf{r}-\mathbf{r}',t-t')$$
$$\times\left\langle\left[\frac{\partial}{\partial r_k}\frac{\delta r_k(t)}{\delta f_j(\mathbf{r}',t')}+\frac{\partial}{\partial v_k}\frac{\delta v_k(t)}{\delta f_j(\mathbf{r}',t')}\right]\varphi(\mathbf{r},\mathbf{v},t)\right\rangle. \tag{10.11}$$

The variational derivatives of functions $\mathbf{r}(t)$ and $\mathbf{v}(t)$ appeared in Eq. (10.11) satisfy the system of equations following from Eq. (10.7) for $t'<t$,

$$\begin{aligned}\frac{d}{dt}\frac{\delta r_k(t)}{\delta f_j(\mathbf{r}',t')}&=\frac{\delta v_k(t)}{\delta f_j(\mathbf{r}',t')},\\ \frac{d}{dt}\frac{\delta v_k(t)}{\delta f_j(\mathbf{r}',t')}&=-\lambda\frac{\delta v_k(t)}{\delta f_j(\mathbf{r}',t')}+\frac{\partial f_k(\mathbf{r},t)}{\partial r_l}\frac{\delta r_l(t)}{\delta f_j(\mathbf{r}',t')},\end{aligned} \tag{10.12}$$

with the initial values

$$\left.\frac{\delta r_k(t)}{\delta f_j(\mathbf{r}',t')}\right|_{t-t'}=0,\qquad \left.\frac{\delta v_k(t)}{\delta f_j(\mathbf{r}',t')}\right|_{t=t'}=\delta_{kj}\delta(\mathbf{r}(t')-\mathbf{r}'). \tag{10.13}$$

The integral with respect to time in the right-hand side of Eq. (10.11), depends mainly on variational derivative behaviors within the temporal interval $t-t'\sim\tau_0$. Assuming that the effect of random forces is insignificant within this temporal scale, we can omit the last term in Eq. (10.12) to obtain the deterministic system of equations

$$\frac{d}{dt}\frac{\delta r_k(t)}{\delta f_j(\mathbf{r}',t')}=\frac{\delta v_k(t)}{\delta f_j(\mathbf{r}',t')},\quad \frac{d}{dt}\frac{\delta v_k(t)}{\delta f_j(\mathbf{r}',t')}=-\lambda\frac{\delta v_k(t)}{\delta f_j(\mathbf{r}',t')}. \tag{10.14}$$

Nevertheless, the initial values (10.13) to this system remain random because $\mathbf{r}(t')$ is the stochastic function.

The solution to system (10.14) with initial values (10.13) is given by the formula

$$\frac{\delta v_k(t)}{\delta f_j(\mathbf{r}',t')} = \delta_{kj} e^{-\lambda(t-t')}\delta(\mathbf{r}(t')-\mathbf{r}'), \quad \frac{\delta r_k(t)}{\delta f_j(\mathbf{r}',t')} = \frac{1}{\lambda}\delta_{kj}\left[1-e^{-\lambda(t-t')}\right]\delta(\mathbf{r}(t')-\mathbf{r}'). \tag{10.15}$$

Assuming further that the dynamics of the particle is also only slightly affected by random forces, we can express function $\mathbf{r}(t')$ in Eq. (10.15) through function $\mathbf{r}(t)$ that satisfies the simplified system of equations (10.7)

$$\frac{d}{dt}\mathbf{r}(t) = \mathbf{v}(t), \quad \frac{d}{dt}\mathbf{v}(t) = -\lambda \mathbf{v}(t) \tag{10.16}$$

with the initial values

$$\mathbf{r}(t)|_{t=t'} = \mathbf{r}(t'), \quad \mathbf{v}(t)|_{t=t'} = \mathbf{v}(t'), \tag{10.17}$$

from which follows that

$$\mathbf{r}(t') = \mathbf{r}(t) - \frac{1}{\lambda}\mathbf{v}\left[e^{\lambda(t-t')}-1\right], \quad \mathbf{v}(t') = e^{\lambda(t-t')}\mathbf{v}(t). \tag{10.18}$$

The above simplifying procedures of passing from Eqs. (10.12), (10.13) to Eq. (10.15) and from Eq. (10.7) to Eq. (10.18) form the basis of the diffusion approximation in the context of the problem under consideration.

Using now Eqs. (10.15) and (10.18), we can rewrite Eq. (10.11) in the closed form

$$\begin{aligned}\left(\frac{\partial}{\partial t} + \mathbf{v}\frac{\partial}{\partial \mathbf{r}} - \lambda\frac{\partial}{\partial \mathbf{v}}\mathbf{v}\right) P(\mathbf{r},\mathbf{v},t) &= -\frac{\partial}{\partial v_i}\int d\mathbf{r}' \int_0^t d\tau B_{ij}(\mathbf{r}-\mathbf{r}',\tau) \\ \times \left\{\frac{\partial}{\partial r_j}\frac{1}{\lambda}\left[1-e^{-\lambda\tau}\right] + \frac{\partial}{\partial v_j}e^{-\lambda\tau}\right\} &\delta\left(\mathbf{r}-\mathbf{r}'-\frac{1}{\lambda}\left[e^{\lambda\tau}-1\right]\mathbf{v}\right) P(\mathbf{r},\mathbf{v},t).\end{aligned} \tag{10.19}$$

The operator in braces commutes with the delta-function, and integration over $\mathbf{r}'$ results in the equation

$$\begin{aligned}&\left(\frac{\partial}{\partial t} + \mathbf{v}\frac{\partial}{\partial \mathbf{r}} - \lambda\frac{\partial}{\partial \mathbf{v}}\mathbf{v}\right) P(\mathbf{r},\mathbf{v},t) \\ &\quad = \frac{\partial}{\partial v_i}\left\{D_{ij}^{(1)}(\mathbf{v},t)\frac{\partial}{\partial v_j} + D_{ij}^{(2)}(\mathbf{v},t)\frac{\partial}{\partial r_j}\right\} P(\mathbf{r},\mathbf{v},t).\end{aligned} \tag{10.20}$$

where we introduced the diffusion coefficients

$$\begin{aligned} D_{ij}^{(1)}(\mathbf{v},t) &= \int_0^t d\tau e^{-\lambda\tau} B_{ij}\left(\frac{1}{\lambda}\left[e^{\lambda\tau}-1\right]\mathbf{v},\tau\right), \\ D_{ij}^{(2)}(\mathbf{v},t) &= \frac{1}{\lambda}\int_0^t d\tau \left[1-e^{-\lambda\tau}\right] B_{ij}\left(\frac{1}{\lambda}\left[e^{\lambda\tau}-1\right]\mathbf{v},\tau\right). \end{aligned} \tag{10.21}$$

Equation (10.21) for the one-point probability density is correct even for times $t < \tau_0$. In this case, the solution $\{\mathbf{r}(t),\mathbf{v}(t)\}$ to problem (10.7) will not be the Markovian vector process because its multi-time probability density cannot be factorized in terms of the transition probability density. Nevertheless, it will be the Markovian random process in asymptotic limit $t \gg \tau_0$. In this limit, we can replace the upper limits of integrals in Eq. (10.21) with infinity. This replacement results in the Fokker–Planck equation for the one-time probability density

$$\left(\frac{\partial}{\partial t}+\mathbf{v}\frac{\partial}{\partial \mathbf{r}}-\lambda\frac{\partial}{\partial \mathbf{v}}\mathbf{v}\right)P(\mathbf{r},\mathbf{v},t)=\frac{\partial}{\partial v_i}\left\{D_{ij}^{(1)}(\mathbf{v})\frac{\partial}{\partial v_j}+D_{ij}^{(2)}(\mathbf{v})\frac{\partial}{\partial r_j}\right\}P(\mathbf{r},\mathbf{v},t), \quad (10.22)$$

with the diffusion coefficients

$$\begin{aligned} D_{ij}^{(1)}(\mathbf{v}) &= \int_0^\infty d\tau e^{-\lambda\tau}B_{ij}\left(\frac{1}{\lambda}\left[e^{\lambda\tau}-1\right]\mathbf{v},\tau\right), \\ D_{ij}^{(2)}(\mathbf{v}) &= \frac{1}{\lambda}\int_0^\infty d\tau\left[1-e^{-\lambda\tau}\right]B_{ij}\left(\frac{1}{\lambda}\left[e^{\lambda\tau}-1\right]\mathbf{v},\tau\right). \end{aligned} \quad (10.23)$$

Note that the approximation of the delta-correlated random field corresponds to Eq. (10.20) with the diffusion coefficients

$$D_{ij}^{(1)}(\mathbf{v})=\int_0^\infty d\tau B_{ij}(0,\tau), \quad D_{ij}^{(2)}(\mathbf{v})=0.$$

Integrating Eq. (10.22) over $\mathbf{r}$, we arrive at the Fokker–Planck equation for the one-time probability density of particle velocity

$$\left(\frac{\partial}{\partial t}-\lambda\frac{\partial}{\partial \mathbf{v}}\mathbf{v}\right)P(\mathbf{v},t)=\frac{\partial}{\partial v_i}D_{ij}^{(1)}(\mathbf{v})\frac{\partial}{\partial v_j}P(\mathbf{v},t).$$

The steady-state probability density corresponding to the limit process $t \to \infty$ satisfies the equation

$$-\lambda\frac{\partial}{\partial \mathbf{v}}\mathbf{v}P(\mathbf{v})=\frac{\partial}{\partial v_i}D_{ij}^{(1)}(\mathbf{v})\frac{\partial}{\partial v_j}P(\mathbf{v}), \quad (10.24)$$

whose solution essentially depends on the behavior of the diffusion coefficient, i.e., on the correlation function of random vector field $\mathbf{f}(\mathbf{r},t)$. For example, if we consider the one-dimensional case and specify the correlation function $B_f(x,t)$ in the form

$$B_f(x,t)=\sigma_f^2\exp\left\{-\frac{|x|}{l_0}-\frac{|t|}{\tau_0}\right\},$$

where l_0 and τ_0 are the spatial and temporal correlation radii, respectively, then we obtain that, for a sufficiently small friction ($\lambda\tau_0 \ll 1$), the solution to Eq. (10.24) has the form [140]

$$P(v)=C\exp\left\{-\frac{\lambda v^2}{2\sigma_f^2\tau_0}\left[1+\frac{2}{3}\frac{|v|\tau_0}{l_0}\right]\right\}. \quad (10.25)$$

For small particle velocity $|v|\tau_0 \ll l_0$, probability distribution (10.25) grades into the Gaussian distribution corresponding to the approximation of the delta-correlated (in time) random field $f(x,t)$. However, in the opposite limiting case $|v|\tau_0 \gg l_0$, probability distribution (10.25) decreases significantly faster than in the case of the approximation of the delta-correlated (in time) random field $f(x,t)$, namely,

$$P(v) = C \exp\left\{-\frac{\lambda v^2|v|}{3\sigma_f^2 l_0}\right\}, \tag{10.26}$$

which corresponds to the diffusion coefficient decreasing according to the law $D^{(1)} \sim 1/|v|$ for great particle velocities. Physically, it means that the effect of random force $f(x,t)$ on faster particles is significantly smaller than on slower ones.

Thus, the diffusion approximation lifts the basic restriction on smallness of the temporal correlation radius τ_0 remaining within the framework of the Markovian process.

Remark 5 ***Diffusion in fast random wavefields of velocity.***

In some cases, the diffusion coefficients can vanish in both approximation of the delta-correlated random field and diffusion approximation. Such a situation occurs, for example, when a particle moves in fast random wavefields of velocity [171] (see also [306]–[308]).

In this case, particle diffusion is described by the equation

$$\frac{d}{dt}\mathbf{r}(t) = \mathbf{u}(\mathbf{r},t), \qquad \mathbf{r}(0) = \mathbf{r}_0, \tag{10.27}$$

where $\mathbf{u}(\mathbf{r},t)$ is the statistically homogeneous and stationary random wave vector field such that $\langle \mathbf{u}(\mathbf{r},t)\rangle = 0$ and the correlation tensor has the form

$$B_{ij}(\mathbf{r},t) = \int d\mathbf{k} F_{ij}(\mathbf{k}) \cos\{\mathbf{kr} - \omega(\mathbf{k})t\}. \tag{10.28}$$

The spectral function $F_{ij}(\mathbf{k})$ is such that $\int d\mathbf{k} F_{ii}(\mathbf{k}) = \sigma_{\mathbf{u}}^2$ and $\omega = \omega(\mathbf{k}) > 0$ is the equation of the dispersion curve for wave motions. For conventional wave motions, the spectral functions of the velocity satisfies the condition $\Phi_{ij}(0) = 0$, where $\Phi_{ij}(\omega) = \int d\mathbf{k} F_{ij}(\mathbf{k})\delta[\omega - \omega(\mathbf{k})]$, so that the tensor diffusion coefficient in the Fokker–Planck equation vanishes, i.e.,

$$D_{ij} = \int_0^\infty B_{ij}(0,t)dt = 0.$$

The same diffusion coefficient appears in the diffusion approximation for $t \gg \tau_0$, where τ_0 is the temporal correlation radius of the velocity field. Consequently, both approximation of the delta-correlated field of velocity and diffusion approximation give no finite result, and one needs to take into account higher-order terms [171].

Let the maximum of spectral function $F_{ij}(\mathbf{k})$ corresponds to a certain wave number k_m, and the maximum of spectral function $\Phi_{ij}(\omega)$, to frequency ω_m. The corresponding spatial and temporal scales are $l = 1/k_m$ and $\tau_0 = 1/\omega_m$. Quantity $\varepsilon = \sigma_{\mathbf{u}}\tau_0/l$ appears usually small for actual wavefields and can be used as the basic small parameter of the problem, i.e., $\varepsilon \ll 1$. ♦

Part IV

Coherent phenomena in stochastic dynamic systems

Chapter 11

Passive tracer diffusion and clustering in random hydrodynamic flows

One of concerns of statistical hydrodynamics is the problem on spreading a passive tracer in random velocity field, which is of significant importance in ecological problems of tracer diffusion in Earth's atmosphere and oceans [51, 222, 241, 251, 258], in the diffusion in porous media [52], and in the problem on the large-scale mass distribution at the last stage of the formation of universe [275]. This problem is extensively investigated beginning from pioneer works [24, 25, 300, 301]. Further, many researchers obtained different equations for describing passive tracer statistical characteristics in both Eulerian and Lagrangian descriptions. Derivation of such type equations (for both moment functions of the tracer concentration field and tracer concentration probability density) for different models of fluctuating parameters in different approximations and their analysis was actively continued even in the last decade (see, e.g., surveys [146, 149]).

11.1 General remarks

The evolution of the density (concentration) of a passive tracer moving in velocity field $\mathbf{U}(\mathbf{r}, t)$ is described by the equation

$$\left(\frac{\partial}{\partial t} + \frac{\partial}{\partial \mathbf{r}}\mathbf{U}(\mathbf{r}, t)\right) \rho(\mathbf{r}, t) = \mu\Delta\rho(\mathbf{r}, t), \quad \rho(\mathbf{r}, 0) = \rho_0(\mathbf{r}). \tag{11.1}$$

where $\mathbf{U}(\mathbf{r}, t) = \mathbf{u}_0(\mathbf{r}, t) + \mathbf{u}(\mathbf{r}, t)$, $\mathbf{u}_0(\mathbf{r}, t)$ is the deterministic component of the velocity field (mean flow), and $\mathbf{u}(\mathbf{r}, t)$ is the random component. In the general case, random field $\mathbf{u}(\mathbf{r}, t)$ can be composed of both solenoidal (for which $\operatorname{div} \mathbf{u}(\mathbf{r}, t) = 0$) and potential (for which $\operatorname{div} \mathbf{u}(\mathbf{r}, t) \neq 0$) components. The right-hand side of Eq. (11.1) takes into account the molecular diffusion with the diffusion coefficient μ; it is assumed that the total tracer mass is conserved during the evolution process, i.e.,

$$M = M(t) = \int d\mathbf{r}\rho(\mathbf{r}, t) = \int d\mathbf{r}\rho_0(\mathbf{r}) = \text{const}.$$

The effect of the molecular diffusion can be neglected during the initial stages of diffusion development. In this case, Eq. (11.1) becomes simpler and assumes the form

$$\left(\frac{\partial}{\partial t}+\mathbf{U}(\mathbf{r},t)\frac{\partial}{\partial \mathbf{r}}\right)\rho(\mathbf{r},t)+\frac{\partial \mathbf{U}(\mathbf{r},t)}{\partial \mathbf{r}}\rho(\mathbf{r},t)=0. \tag{11.2}$$

A more complete analysis requires that the field of gradient of tracer concentration $\mathbf{p}(\mathbf{r},t)=\boldsymbol{\nabla}\rho(\mathbf{r},t)$ was included into consideration. This field satisfies the equation

$$\begin{aligned}\left(\frac{\partial}{\partial t}+\frac{\partial}{\partial \mathbf{r}}\mathbf{U}(\mathbf{r},t)\right)p_i(\mathbf{r},t) &= -p_k(\mathbf{r},t)\frac{\partial U_k(\mathbf{r},t)}{\partial r_i}-\rho(\mathbf{r},t)\frac{\partial^2 \mathbf{U}(\mathbf{r},t)}{\partial r_i \partial \mathbf{r}},\\ \mathbf{p}(\mathbf{r},0) &= \mathbf{p}_0(\mathbf{r})=\boldsymbol{\nabla}\rho_0(\mathbf{r}).\end{aligned} \tag{11.3}$$

The above equations correspond to the *Eulerian description* of the concentration evolution.

Equation (11.2) is the first-order partial differential equation and can be solved by the method of characteristics. Introducing characteristic curves $\mathbf{r}(t)$ satisfying the equations of particle motion

$$\frac{d}{dt}\mathbf{r}(t)=\mathbf{U}(\mathbf{r},t), \qquad \mathbf{r}(0)=\mathbf{r}_0, \tag{11.4}$$

we can change over from Eq. (11.2) to the ordinary differential equation

$$\frac{d}{dt}\rho(t)=-\frac{\partial \mathbf{U}(\mathbf{r},t)}{\partial \mathbf{r}}\rho(t),\ \rho(0)=\rho_0(\mathbf{r}_0). \tag{11.5}$$

Solutions to Eqs. (11.4) and (11.5) have an obvious geometric interpretation. They describe the concentration behavior around a fixed tracer particle moving along trajectory $\mathbf{r}=\mathbf{r}(t)$. As may be seen from Eq. (11.5), the concentration in divergent flows varies: it increases in regions where the medium is compressed and decreases in regions where the medium is rarefied.

Solutions to system (11.4), (11.5) depend on characteristic parameter $\mathbf{r}_0$ (the initial coordinate of the particle)

$$\mathbf{r}(t)=\mathbf{r}(t|\mathbf{r}_0), \quad \rho(t)=\rho(t|\mathbf{r}_0), \tag{11.6}$$

which we will separate by the bar. Components of vector $\mathbf{r}_0$ are called the *Lagrangian coordinates* of the particle; they unambiguously specify the position of arbitrary particle. Equations (11.4), (11.5) correspond in this case to the *Lagrangian description* of the concentration evolution. The first of the equalities (11.6) specify the relationship between the Eulerian and Lagrangian descriptions. Solving it in $\mathbf{r}_0$, we obtain the relationship that expresses the Lagrangian coordinates in terms of the Eulerian ones

$$\mathbf{r}_0=\mathbf{r}_0(\mathbf{r},t). \tag{11.7}$$

Then, using Eq. (11.7), to eliminate $\mathbf{r}_0$ in the last equality in (11.6), we turn back to the concentration in the *Eulerian description*

$$\rho(\mathbf{r},t)=\rho(t|\mathbf{r}_0(\mathbf{r},t))=\int d\mathbf{r}_0\rho(t|\mathbf{r}_0)j(t|\mathbf{r}_0)\delta\left(\mathbf{r}(t|\mathbf{r}_0)-\mathbf{r}\right), \tag{11.8}$$

where we introduced new function called *divergence*

$$j(t|\mathbf{r}_0)=\det\|j_{ik}(t|\mathbf{r}_0)\|=\det\left\|\frac{\partial r_i(t|\mathbf{r}_0)}{\partial r_{0k}}\right\|.$$

Differentiating Eq. (11.4) with respect to components of vector $\mathbf{r}_0$, we arrive at the equations for elements of the Jacobian matrix $j_{ik}(t|\mathbf{r}_0)$

$$\frac{d}{dt} j_{ik}(t|\mathbf{r}_0) = \frac{\partial U_i(\mathbf{r},t)}{\partial r_l} j_{lk}(t|\mathbf{r}_0), \quad j_{ik}(0|\mathbf{r}_0) = \delta_{ik},$$

from which follows that the determinant of this matrix satisfies the equation

$$\frac{d}{dt} j(t|\mathbf{r}_0) = \frac{\partial \mathbf{U}(\mathbf{r},t)}{\partial \mathbf{r}} j(t|\mathbf{r}_0), \quad j(0|\mathbf{r}_0) = 1. \tag{11.9}$$

Function $j(t|\mathbf{r}_0)$ is the quantitative measure of the degree of compression (extension) of physically infinitely small liquid particles. Comparing Eq. (11.5) with Eq. (11.9), we see that

$$\rho(t|\mathbf{r}_0) = \frac{\rho_0(\mathbf{r}_0)}{j(t|\mathbf{r}_0)}. \tag{11.10}$$

Thus, we can rewrite Eq. (11.8) as the equality

$$\rho(\mathbf{r},t) = \int d\mathbf{r}_0 \rho_0(\mathbf{r}_0) \delta\left(\mathbf{r}(t|\mathbf{r}_0) - \mathbf{r}\right) \tag{11.11}$$

specifying the relationship between the Lagrangian and Eulerian characteristics. Delta function in the right-hand side of Eq. (11.11) is the *indicator function* for the position of the Lagrangian particle; as a consequence, after averaging Eq. (11.11) over an ensemble of realizations of random velocity field, we arrive at the well-known relationship between the average concentration in the Eulerian description and the one-time probability density

$$P(t,\mathbf{r}|\mathbf{r}_0) = \left\langle \delta\left(\mathbf{r}(t|\mathbf{r}_0) - \mathbf{r}\right) \right\rangle$$

of the Lagrangian particle (see, e.g., [251])

$$\langle \rho(\mathbf{r},t) \rangle = \int d\mathbf{r}_0 \rho_0(\mathbf{r}_0) P\left(t,\mathbf{r}|\mathbf{r}_0\right).$$

The relationship between the spatial correlation function of the density field in the Eulerian description

$$\Gamma(\mathbf{r}_1,\mathbf{r}_2,t) = \langle \rho(\mathbf{r}_1,t)\rho(\mathbf{r}_2,t) \rangle$$

and the joint probability density of positions of two particles

$$P(t,\mathbf{r}_1,\mathbf{r}_2|\mathbf{r}_{01},\mathbf{r}_{02}) = \left\langle \delta\left(\mathbf{r}_1(t|\mathbf{r}_{01}) - \mathbf{r}_1\right) \delta\left(\mathbf{r}_2(t|\mathbf{r}_{02}) - \mathbf{r}_2\right) \right\rangle$$

can be obtained similarly

$$\Gamma(\mathbf{r}_1,\mathbf{r}_2,t) = \int d\mathbf{r}_{01} \int d\mathbf{r}_{02} \rho_0(\mathbf{r}_{01}) \rho_0(\mathbf{r}_{02}) P(t,\mathbf{r}_1,\mathbf{r}_2|\mathbf{r}_{01},\mathbf{r}_{02}).$$

For the divergence-free velocity field ($\operatorname{div} \mathbf{U}(\mathbf{r},t) = 0$), both particle divergence and particle concentration are invariant, i.e.,

$$j(t|\mathbf{r}_0) = 1, \quad \rho(t|\mathbf{r}_0) = \rho_0(\mathbf{r}_0)$$

so that the solution to Eq. (11.2) has in this case the following structure

$$\rho(\mathbf{r},t) = \rho_0(\mathbf{r}_0(\mathbf{r},t)).$$

Remark 6 *Inclusion of molecular diffusion.*

Note that, as we mentioned in Part 2, page 141, the statistical interpretation of the solution to the stochastic equation with first-order derivatives can appear useful even in the general case of Eq. (11.1). Namely, if we consider the auxiliary equation

$$\left(\frac{\partial}{\partial t}+\frac{\partial}{\partial \mathbf{r}}\mathbf{U}(\mathbf{r},t)+\boldsymbol{\eta}(t)\frac{\partial}{\partial \mathbf{r}}\right)\tilde{\rho}(\mathbf{r},t)=0, \quad \tilde{\rho}(\mathbf{r},0)=\rho_0(\mathbf{r}), \tag{11.12}$$

where $\mathbf{V}(t)$ is the vector Gaussian white-noise process with the characteristics

$$\langle \eta_i(t)\rangle = 0, \quad \left\langle \eta_i(t)\eta_j(t')\right\rangle = 2\mu\delta_{ij}\delta(t-t'), \tag{11.13}$$

then

$$\rho(\mathbf{r},t)=\langle\tilde{\rho}(\mathbf{r},t)\rangle_{\boldsymbol{\eta}}.$$

According to Eq. (11.11), we can represent the solution to Eq. (11.12) in the form

$$\tilde{\rho}(\mathbf{r},t)=\int d\mathbf{r}_0\rho_0(\mathbf{r}_0)\delta\left(\mathbf{r}(t|\mathbf{r}_0)-\mathbf{r}\right),$$

so that

$$\rho(\mathbf{r},t)=\int d\mathbf{r}_0\rho_0(\mathbf{r}_0)\left\langle\delta\left(\mathbf{r}(t|\mathbf{r}_0)-\mathbf{r}\right)\right\rangle_{\boldsymbol{\eta}}, \tag{11.14}$$

where the characteristic curve (particle trajectory) satisfies the dynamic equation

$$\frac{d}{dt}\mathbf{r}(t)=\mathbf{U}(\mathbf{r},t)+\boldsymbol{\eta}(t), \qquad \mathbf{r}(0)=\mathbf{r}_0. \tag{11.15}$$

Averaging now Eq. (11.14) over an ensemble of realizations of random field $\mathbf{U}(\mathbf{r},t)$, we obtain the final equality

$$\langle\rho(\mathbf{r},t)\rangle=\int d\mathbf{r}_0\rho_0(\mathbf{r}_0)P(t,\mathbf{r}|\mathbf{r}_0), \tag{11.16}$$

where the one-time probability density of the position of the Lagrangian particle is given now by the formula

$$P(t,\mathbf{r}|\mathbf{r}_0)=\left\langle\left\langle\delta\left(\mathbf{r}(t|\mathbf{r}_0)-\mathbf{r}\right)\right\rangle_{\boldsymbol{\eta}}\right\rangle_{\mathbf{U}}. \tag{11.17}$$

Thus, we can deal with the Lagrangian description based on the dynamic equation (11.12) even in the case of the equation with the second-order partial derivatives (11.1). In a similar way, the spatial correlation function of the concentration field in the Eulerian description with allowance for the molecular diffusion effect

$$\Gamma(\mathbf{r}_1,\mathbf{r}_2,t)=\langle\rho(\mathbf{r}_1,t)\rho(\mathbf{r}_2,t)\rangle$$

can be related to the joint probability density of positions of two particles

$$P(t,\mathbf{r}_1,\mathbf{r}_2|\mathbf{r}_{01},\mathbf{r}_{02})=\left\langle\delta\left(\mathbf{r}_1(t|\mathbf{r}_{01})-\mathbf{r}_1\right)\delta\left(\mathbf{r}_2(t|\mathbf{r}_{02})-\mathbf{r}_2\right)\right\rangle$$

through the relationship

$$\Gamma(\mathbf{r}_1,\mathbf{r}_2,t)=\int d\mathbf{r}_{01}\int d\mathbf{r}_{02}\rho_0(\mathbf{r}_{01})\rho_0(\mathbf{r}_{02})P(t,\mathbf{r}_1,\mathbf{r}_2|\mathbf{r}_{01},\mathbf{r}_{02}),$$

where the joint probability density of positions of two particles

$$P(t, \mathbf{r}_1, \mathbf{r}_2|\mathbf{r}_{01}, \mathbf{r}_{02}) = \langle \delta\left(\mathbf{r}_1(t|\mathbf{r}_{01}) - \mathbf{r}_1\right) \delta\left(\mathbf{r}_2(t|\mathbf{r}_{02}) - \mathbf{r}_2\right)\rangle_{\{\mathbf{V}\},\mathbf{U}}$$

is determined from the statistical analysis of dynamics of two particles whose trajectories satisfy now the equations

$$\begin{aligned} \frac{d}{dt}\mathbf{r}_1(t) &= \mathbf{U}(\mathbf{r}_1, t) + \mathbf{V}_1(t), \qquad \mathbf{r}_1(0) = \mathbf{r}_{01}, \\ \frac{d}{dt}\mathbf{r}_2(t) &= \mathbf{U}(\mathbf{r}_2, t) + \mathbf{V}_2(t), \qquad \mathbf{r}_2(0) = \mathbf{r}_{02}, \end{aligned} \tag{11.18}$$

where $\mathbf{V}_1(t)$ and $\mathbf{V}_2(t)$ are the statistically independent vector processes with the parameters (11.13). ♦

Thus, in the Lagrangian representation, the behavior of passive tracer is described in terms of ordinary differential equations (11.4), (11.5), and (11.9). We can easily pass on from these equations to the linear Liouville equation in the corresponding phase space. With this goal in view, introduce the indicator function

$$\varphi_{\text{Lag}}(t; \mathbf{r}, \rho, j|\mathbf{r}_0) = \delta(\mathbf{r}(t|\mathbf{r}_0) - \mathbf{r})\delta(\rho(t|\mathbf{r}_0) - \rho)\delta(j(t|\mathbf{r}_0) - j), \tag{11.19}$$

where we explicitly emphasized the fact that the solution to the initial dynamic equations depends on the Lagrangian coordinates $\mathbf{r}_0$. Differentiating Eq. (11.19) with respect to time and using Eqs. (11.4), (11.5), and (11.9), we arrive at the Liouville equation equivalent to the initial value problem

$$\begin{aligned} &\left(\frac{\partial}{\partial t} + \frac{\partial}{\partial \mathbf{r}}\mathbf{U}(\mathbf{r}, t)\right) \varphi_{\text{Lag}}(t; \mathbf{r}, \rho, j|\mathbf{r}_0) = \frac{\partial \mathbf{U}(\mathbf{r}, t)}{\partial \mathbf{r}} \left(\frac{\partial}{\partial \rho}\rho - \frac{\partial}{\partial j}j\right) \varphi_{\text{Lag}}(t; \mathbf{r}, \rho, j|\mathbf{r}_0), \\ &\varphi_{\text{Lag}}(0; \mathbf{r}, \rho, j|\mathbf{r}_0) = \delta(\mathbf{r}_0 - \mathbf{r})\delta(\rho_0(\mathbf{r}_0) - \rho)\delta(j - 1). \end{aligned} \tag{11.20}$$

The one-time probability density of the solutions to dynamic problems (11.4), (11.5), and (11.9) coincides with the indicator function averaged over an ensemble of realizations

$$P(t; \mathbf{r}, \rho, j|\mathbf{r}_0) = \left\langle \varphi_{\text{Lag}}(t; \mathbf{r}, \rho, j|\mathbf{r}_0) \right\rangle.$$

In order to describe the concentration field in the Eulerian representation, we introduce the indicator function similar to function (11.19)

$$\varphi(t, \mathbf{r}; \rho) = \delta(\rho(\mathbf{r}, t) - \rho), \tag{11.21}$$

which is defined on surface $\rho(\mathbf{r}, t) = \rho = \text{const}$ in the three-dimensional case or on a contour in the two-dimensional case. It satisfies the equation (2.9), page 40

$$\begin{aligned} \left(\frac{\partial}{\partial t} + \mathbf{U}(\mathbf{r}, t)\frac{\partial}{\partial \mathbf{r}}\right) \varphi(t, \mathbf{r}; \rho) &= \frac{\partial \mathbf{U}(\mathbf{r}, t)}{\partial \mathbf{r}} \frac{\partial}{\partial \rho} \left[\rho \varphi(t, \mathbf{r}; \rho)\right], \\ \varphi(0, \mathbf{r}; \rho) &= \delta(\rho_0(\mathbf{r}) - \rho). \end{aligned} \tag{11.22}$$

For divergence-free velocity fields, Eqs. (11.22) and (11.2) coincide. Essential differences appear only for divergent velocity fields.

In this case, the one-point probability density of the solution to dynamic equation (11.2) coincides with the indicator function averaged over an ensemble of realizations

$$P(t, \mathbf{r}; \rho) = \langle \delta(\rho(\mathbf{r}, t) - \rho) \rangle.$$

As a result, the one-point probability density of the concentration field in the Eulerian representation is related to the one-point probability density in the Lagrangian representation through the equality

$$P(t,\mathbf{r};\rho)=\int d\mathbf{r}_0\int\limits_0^\infty jdjP(t;\mathbf{r},\rho,j|\mathbf{r}_0). \tag{11.23}$$

In addition, the indicator function provides reach quantitative and qualitative data on the geometry of random fields (see page 55).

Like ordinary topography of mountain ranges, statistical topography deals mainly with the system of contours (in the two-dimensional case) or surfaces (in the three-dimensional case) corresponding to constant values, which are defined by the equality

$$\rho(\mathbf{r},t)=\rho=\text{const}\,.$$

In analyzing the system of contours (for simplicity we will deal here with the two-dimensional case), it appears convenient to introduce the singular indicator function (11.21) lumped on these contours, which is a functional of medium parameters.

In terms of function (11.21), one can express various quantities, such as the total area of regions located inside level lines (where $\rho(\mathbf{r},t)>\rho$)

$$S(t,\rho)=\int\limits_\rho^\infty d\tilde{\rho}\int d\mathbf{r}\,\varphi(t,\mathbf{r};\tilde{\rho}) \tag{11.24}$$

and the total field mass present in these regions

$$M(t,\rho)=\int\limits_\rho^\infty \tilde{\rho}d\tilde{\rho}\int d\mathbf{r}\,\varphi(t,\mathbf{r};\tilde{\rho}). \tag{11.25}$$

Indeed, in the context of the passive tracer dynamics described by the Liouville equation (11.22), we can obtain the following expressions

$$\begin{aligned}\frac{\partial}{\partial t}S(t,\rho) &= \int d\mathbf{r}\int\limits_\rho^\infty d\tilde{\rho}\frac{\partial\mathbf{U}(\mathbf{r},t)}{\partial\mathbf{r}}\left(\frac{\partial}{\partial\tilde{\rho}}\tilde{\rho}+1\right)\varphi(t,\mathbf{r};\tilde{\rho}),\\ \frac{\partial}{\partial t}M(t,\rho) &= \int d\mathbf{r}\int\limits_\rho^\infty d\tilde{\rho}\frac{\partial\mathbf{U}(\mathbf{r},t)}{\partial\mathbf{r}}\tilde{\rho}\left(\frac{\partial}{\partial\tilde{\rho}}\tilde{\rho}+1\right)\varphi(t,\mathbf{r};\tilde{\rho})\end{aligned}$$

by differentiating Eqs. (11.24) and (11.25) with respect to time. Consequently, the total area of the region lying within the contour $\rho(\mathbf{r},t)=\rho=\text{const}$ and the total mass present in this region remain invariant for divergence-free velocity field. In this case, an additional invariant quantity—the number of closed contours—appears; these contours cannot appear and disappear in the medium, they can only vary in time depending on their initial spatial distribution specified by the equality $\rho_0(\mathbf{r})=\rho=\text{const}$.

For the velocity field with the non-zero potential component, the above quantities are no more invariant in time.

Ensemble averages of Eqs. (11.24) and (11.25) can be immediately calculated using the one-point probability density.

Additional structural details of field $\rho(\mathbf{r},t)$ can be obtained by considering the spatial gradient $\mathbf{p}(\mathbf{r},t) = \boldsymbol{\nabla}\rho(\mathbf{r},t)$. For example, quantity

$$l(t,\rho) = \int d\mathbf{r}\,|\mathbf{p}(\mathbf{r},t)|\,\delta(\rho(\mathbf{r},t)-\rho) = \oint dl \tag{11.26}$$

describes the total length of contours $\rho(\mathbf{r},t) = \rho = \text{const}$.

Description of Eq. (11.26) requires the extended indicator function

$$\varphi(t,\mathbf{r};\rho,\mathbf{p}) = \delta\left(\rho(\mathbf{r},t)-\rho\right)\delta\left(\mathbf{p}(\mathbf{r},t)-\mathbf{p}\right), \tag{11.27}$$

which satisfies (in the case of tracer in random velocity field) the Liouville equation following from Eqs. (11.2) and (11.3)

$$\begin{aligned}
&\left(\frac{\partial}{\partial t} + \mathbf{U}(\mathbf{r},t)\frac{\partial}{\partial \mathbf{r}}\right)\varphi(t,\mathbf{r};\rho,\mathbf{p})\\
&= \left[\frac{\partial U_k(\mathbf{r},t)}{\partial r_i}\frac{\partial}{\partial p_i}p_k + \frac{\partial \mathbf{U}(\mathbf{r},t)}{\partial \mathbf{r}}\left(\frac{\partial}{\partial \rho}\rho + \frac{\partial}{\partial \mathbf{p}}\mathbf{p}\right) + \frac{\partial^2 \mathbf{U}(\mathbf{r},t)}{\partial r_i \partial \mathbf{r}}\frac{\partial}{\partial p_i}\rho\right]\varphi(t,\mathbf{r};\rho,\mathbf{p}),\\
&\varphi(0,\mathbf{r};\rho,\mathbf{p}) = \delta\left(\rho_0(\mathbf{r})-\rho\right)\delta\left(\mathbf{p}_0(\mathbf{r})-\mathbf{p}\right).
\end{aligned} \tag{11.28}$$

A consequence of Eq. (11.28) is, for example, the evolution equation for contour length (11.26)

$$\begin{aligned}
&\frac{\partial}{\partial t}l(t,\rho) = \int d\mathbf{r}\int d\mathbf{p}\,p\frac{\partial}{\partial t}\varphi(t,\mathbf{r};\rho,\mathbf{p})\\
&= \int d\mathbf{r}\int d\mathbf{p}\left[-\frac{\partial U_k(\mathbf{r},t)}{\partial r_i}\frac{p_i p_k}{p} + \frac{\partial U_k(\mathbf{r},t)}{\partial r_k}p\frac{\partial}{\partial \rho}\rho - \frac{\partial^2 U_k(\mathbf{r},t)}{\partial r_i \partial r_k}\frac{p_i}{p}\rho\right]\varphi(t,\mathbf{r};\rho,\mathbf{p}),
\end{aligned} \tag{11.29}$$

from which follows that the contour length is not invariant in time even in the case of divergence-free velocity field.

Note that averages of formulas (11.26) are (11.29) are related to the joint one-point probability density $P(t,\mathbf{r};\rho,\mathbf{p})$ of field $\rho(\mathbf{r},t)$ and its gradient $\mathbf{p}(\mathbf{r},t)$; this probability density is determined by averaging the indicator function (11.27) over an ensemble of realizations

$$P(t,\mathbf{r};\rho,\mathbf{p}) = \left\langle \delta\left(\rho(\mathbf{r},t)-\rho\right)\delta\left(\mathbf{p}(\mathbf{r},t)-\mathbf{p}\right)\right\rangle.$$

11.2 Statistical description

Consider now the problem of statistical description of passive tracer diffusion in the random velocity field.

We assume that the random component of the velocity field is in the general case the divergent ($\operatorname{div}\mathbf{u}(\mathbf{r},t) \neq 0$), statistically homogeneous and isotropic, stationary Gaussian random field whose average is zero-valued ($\langle \mathbf{u}(\mathbf{r},t)\rangle = 0$) and correlation and spectral tensors are given by the formulas

$$\begin{aligned}
&\langle u_i(\mathbf{r},t)u_j(\mathbf{r}',t')\rangle = B_{ij}(\mathbf{r}-\mathbf{r}',t-t') = \int d\mathbf{k}E_{ij}(\mathbf{k},t-t')e^{i\mathbf{k}(\mathbf{r}-\mathbf{r}')},\\
&B_{ij}(\mathbf{r},t) = B^{\mathrm{s}}_{ij}(\mathbf{r},t) + B^{\mathrm{p}}_{ij}(\mathbf{r},t), \quad E_{ij}(\mathbf{k},t) = E^{\mathrm{s}}_{ij}(\mathbf{k},t) + E^{\mathrm{p}}_{ij}(\mathbf{k},t)\\
&E^{\mathrm{s}}_{ij}(\mathbf{k},t) = \frac{1}{(2\pi)^d}\int d\mathbf{r}B^{\mathrm{s}}_{ij}(\mathbf{r},t)e^{-i\mathbf{k}\mathbf{r}}, \quad E^{\mathrm{p}}_{ij}(\mathbf{k},t) = \frac{1}{(2\pi)^d}\int d\mathbf{r}B^{\mathrm{p}}_{ij}(\mathbf{r},t)e^{-i\mathbf{k}\mathbf{r}},
\end{aligned} \tag{11.30}$$

where d is the dimension of space and the spectral tensor of the velocity field assumes the following structure

$$E_{ij}^{\mathrm{s}}(\mathbf{k},t) = E^{\mathrm{s}}(k,t)\left(\delta_{ij} - \frac{k_i k_j}{k^2}\right), \quad E_{ij}^{\mathrm{p}}(\mathbf{k},t) = E^{\mathrm{p}}(k,t)\frac{k_i k_j}{k^2}. \qquad (11.31)$$

Here, $E^{\mathrm{s}}(k,t)$ and $E^{\mathrm{p}}(k,t)$ are the solenoidal and potential components of the spectral density of the velocity field, respectively.

The following cases are of immediate practical importance:

- A divergence-free hydrodynamic flow ($\operatorname{div}\mathbf{u}(\mathbf{r},t) = 0$, $E^{\mathrm{p}}(k,t) = 0$);
- A potential velocity field ($E^{\mathrm{s}}(k,t) = 0$);
- A mixed situation. This case corresponds to the diffusion of buoyant tracer and the diffusion of low-inertia particles.

Calculating statistical properties of the concentration field and its gradient, we will approximate the velocity field $\mathbf{u}(\mathbf{r},t)$ by the random process delta correlated in time. In the framework of this approximation, correlation tensor (11.30) is approximated by the expression

$$B_{ij}(\mathbf{r},t) = 2B_{ij}^{\mathrm{eff}}(\mathbf{r})\delta(t), \qquad (11.32)$$

where

$$B_{ij}^{\mathrm{eff}}(\mathbf{r}) = \frac{1}{2}\int_{-\infty}^{\infty} dt B_{ij}(\mathbf{r},t) = \int_{0}^{\infty} dt B_{ij}(\mathbf{r},t).$$

In view of homogeneity and isotropy of the velocity field $\mathbf{u}(\mathbf{r},t)$, we have the equalities

$$\begin{aligned}
&B_{kl}^{\mathrm{eff}}(0) = D_0\delta_{kl}, \quad \frac{\partial}{\partial r_i}B_{kl}^{\mathrm{eff}}(0) = 0,\\
&-\frac{\partial^2}{\partial r_i \partial r_j}B_{kl}^{\mathrm{eff}}(0) = \frac{D^{\mathrm{s}}}{d(d+2)}\left[(d+1)\delta_{kl}\delta_{ij} - \delta_{ki}\delta_{lj} - \delta_{kj}\delta_{li}\right] + \\
&\qquad + \frac{D^{\mathrm{p}}}{d(d+2)}\left[\delta_{kl}\delta_{ij} + \delta_{ki}\delta_{lj} + \delta_{kj}\delta_{li}\right],\\
&\frac{\partial^4}{\partial r_i \partial r_k \partial r_j \partial r_l}B_{kl}^{\mathrm{eff}}(0) = D_4^{\mathrm{p}}\delta_{ij}.
\end{aligned} \qquad (11.33)$$

As usual, the repeated indexes assume summation. In addition, we introduced here the following notations:

$$\begin{aligned}
D_0 &= \frac{1}{d}\int_0^{\infty} dt \int d\mathbf{k}\left[(d-1)E^{\mathrm{s}}(k,t) + E^{\mathrm{p}}(k,t)\right],\\
D^{\mathrm{s}} &= \int_0^{\infty} dt \int d\mathbf{k}\, k^2 E^{\mathrm{s}}(k,t), \quad D^{\mathrm{p}} = \int_0^{\infty} dt \int d\mathbf{k}\, k^2 E^{\mathrm{p}}(k,t),\\
D_4^{\mathrm{p}} &= \frac{1}{d}\int_0^{\infty} dt \int d\mathbf{k}\, k^4 E^{\mathrm{p}}(k,t).
\end{aligned} \qquad (11.34)$$

Tracer diffusion in random velocity field is described by the Liouville equation (11.20) in the Lagrangian representation, and by Eqs. (11.22) and (11.28) in the Eulerian representation. If we average these equations over an ensemble of realizations of the velocity field $\mathbf{u}$, we obtain the equations for the one-time Lagrangian probability density $P(t;\mathbf{r},\rho,j|\mathbf{r}_0)$ and the one-point Eulerian probability distributions $P(t,\mathbf{r};\rho)$ and $P(t,\mathbf{r};\rho,\mathbf{p})$.

11.2.1 Lagrangian description (particle diffusion)

One-point statistical characteristics

Averaging Eq. (11.20) over an ensemble of realizations of random field $\mathbf{u}(\mathbf{r},t)$, using the Furutsu–Novikov formula, and taking into account the equality

$$\frac{\delta}{\delta u_\beta(\mathbf{r}',t-0)}\varphi_{\mathrm{Lag}}(t;\mathbf{r},\rho,j|\mathbf{r}_0)$$
$$=\left\{-\frac{\partial}{\partial r_\beta}\delta(\mathbf{r}-\mathbf{r}')+\frac{\partial\delta(\mathbf{r}-\mathbf{r}')}{\partial r_\beta}\left(\frac{\partial}{\partial\rho}\rho-\frac{\partial}{\partial j}j\right)\right\}\varphi_{\mathrm{Lag}}(t;\mathbf{r},\rho,j|\mathbf{r}_0),$$

and relationships (11.33), we arrive at the Fokker–Planck equation for the one-time Lagrangian probability density

$$\left(\frac{\partial}{\partial t}-D_0\Delta\right)P(t;\mathbf{r},\rho,j|\mathbf{r}_0)=D^{\mathrm{p}}\left(\frac{\partial}{\partial\rho}\rho^2\frac{\partial}{\partial\rho}-2\frac{\partial^2}{\partial\rho\partial j}\rho j+\frac{\partial^2}{\partial j^2}j^2\right)P(t;\mathbf{r},\rho,j|\mathbf{r}_0),$$
$$P(0;\mathbf{r},\rho,j|\mathbf{r}_0)=\delta(\mathbf{r}-\mathbf{r}_0)\delta(\rho_0(\mathbf{r}_0)-\rho)\delta(j-1). \tag{11.35}$$

The solution to Eq. (11.35) is as follows

$$P(t;\mathbf{r},\rho,j|\mathbf{r}_0)=P(t;\mathbf{r}|\mathbf{r}_0)P(t;j|\mathbf{r}_0)\delta\left(\rho-\frac{\rho_0(\mathbf{r}_0)}{j}\right), \tag{11.36}$$

where

$$P(t;\mathbf{r}|\mathbf{r}')=e^{D_0t\Delta}\delta(\mathbf{r}-\mathbf{r}')=\frac{1}{(4\pi D_0dt)^{d/2}}\exp\left\{\frac{(\mathbf{r}-\mathbf{r}')^2}{4D_0t}\right\} \tag{11.37}$$

is the probability distribution of coordinates of passive tracer particle and

$$P(t;j|\mathbf{r}_0)=e^{D^{\mathrm{p}}t\frac{\partial^2}{\partial j^2}j^2}\delta(j-1)=\frac{1}{2j\sqrt{\pi\tau}}\exp\left\{-\frac{\ln^2(je^{\tau})}{4\tau}\right\} \tag{11.38}$$

is the probability distribution of the divergence field in the vicinity of this particle. In Eq. (11.38) and below, we use the dimensionless time $\tau=D^{\mathrm{p}}t$. We emphasize that solution (11.36) expresses the fact that coordinates $\mathbf{r}(t|\mathbf{r}_0)$ and divergence $j(t|\mathbf{r}_0)$ are statistically independent in the vicinity of the particle with the Lagrangian coordinates $\mathbf{r}_0$. From the lognormal distribution (11.38) follows that quantity $\chi(t|\mathbf{r}_0)=\ln j(t|\mathbf{r}_0)$ is distributed according to the Gaussian law with the parameters

$$\langle\chi(t|\mathbf{r}_0)\rangle=-\tau,\quad \sigma_\chi^2(t)=2\tau. \tag{11.39}$$

In particular, Eq. (11.38) (and immediately Eq. (11.35), too) results in the following expressions for moments of the random divergence field

$$\langle j^n(t|\mathbf{r}_0)\rangle=e^{n(n-1)\tau},\quad n=\pm1,\pm2,\ldots. \tag{11.40}$$

We emphasize that average divergence remains constant $\langle j(t|\mathbf{r}_0)\rangle = 1$, while its higher moments exponentially grow in time.

In addition, we note that, according to Eqs. (11.10) and (11.40), the Lagrangian moments of concentration can be represented in the form

$$\langle \rho^n(t|\mathbf{r}_0)\rangle = \rho_0^n(\mathbf{r}_0)e^{n(n+1)\tau},$$

from which follows that both average concentration and its higher moments are exponentially increasing functions in the Lagrangian representation. In this case, the probability density of particle concentration has the form

$$P(t;\rho|\mathbf{r}_0) = \frac{1}{2\rho\sqrt{\pi\tau}}\exp\left\{-\frac{\ln^2\left(\rho e^{-\tau}/\rho_0(\mathbf{r}_0)\right)}{4\tau}\right\}. \tag{11.41}$$

This probability density can be obtained also as the solution to the Fokker–Planck equation following from Eq. (11.35)

$$\frac{\partial}{\partial t}P(t;\rho|\mathbf{r}_0) = D^{\mathrm{p}}\frac{\partial}{\partial\rho}\rho^2\frac{\partial}{\partial\rho}P(t;\rho|\mathbf{r}_0), \quad P(0;\rho|\mathbf{r}_0) = \delta(\rho_0(\mathbf{r}_0)-\rho).$$

The above paradoxical behavior of statistical characteristics of the divergence and concentration (simultaneous growth of all moment functions in time) is a consequence of the lognormal probability distribution. Indeed, the typical realization curve of random divergence is the exponentially decaying curve

$$j^*(t) = e^{-\tau}.$$

Moreover, realizations of the lognormal process satisfy certain majorant estimates. For example, with probability $p = 1/2$, we have

$$j(t|\mathbf{r}_0) < 4e^{-\tau/2}$$

throughout the whole temporal interval $t \in (t_1, t_2)$.

Similarly, the typical realization curve of concentration and its minorant estimate have the following form

$$\rho^*(t) = \rho_0 e^{\tau}, \quad \rho(t|\mathbf{r}_0) > \frac{\rho_0}{4}e^{\tau/2}.$$

We emphasize that the above Lagrangian statistical properties of a particle in flows containing the potential random component are qualitatively different from the statistical properties of a particle in divergence-free flows where $j(t|\mathbf{r}_0) \equiv 1$ and particle concentration remains invariant in the vicinity of a fixed particle $\rho(t|\mathbf{r}_0) = \rho_0(\mathbf{r}_0) = \text{const}$. The above statistical estimates mean that the statistics of random processes $j(t|\mathbf{r}_0)$ and $\rho(t|\mathbf{r}_0)$ is formed by the realization spikes relative typical realization curves.

At the same time, probability distributions of particle coordinates in essence coincide for both divergent and divergence-free velocity fields.

Plane-parallel mean shear Above, we considered the statistical description of particle dynamics in the conditions of absent mean flow of liquid. The case of the two-dimensional plane-parallel mean flow in which case

$$\mathbf{u}_0(\mathbf{r},t) = v(y)\mathbf{l},$$

where $\mathbf{r} = (x, y)$ and $\mathbf{l} = (1, 0)$, is also of certain interest. In these conditions, vector equation (11.4) reduces to two scalar equations

$$\frac{d}{dt}x(t) = v(y) + u_1(\mathbf{r}, t), \quad \frac{d}{dt}y(t) = u_2(\mathbf{r}, t). \tag{11.42}$$

The following types of flows are of practical importance:

- The linear shear flow described by function $v(y) = \alpha y$;
- The tangential gap described by function $v(y) = v_0\theta(y - y_0) - v_0\theta(y_0 - y)$, where $\theta(y)$ is the Heaviside step function equal to unity for $y > 0$ and zero otherwise;
- The Kolmogorov flow described by function $v(y) = v_0 \sin \beta y$;
- The jet flow described by function $v(y) = \tilde{v}(y)\theta(|y_0| - y)$.

Results concerning stability of such flows can be found, for example, in monographs [58, 75, 251].

In the context of problem (11.42), the stochastic Liouville equation for the indicator function

$$\varphi(t; x, y) = \delta(x(t) - x)\delta(y(t) - y)$$

is simplified and assumes in this case the form

$$\left(\frac{\partial}{\partial t} + v(y)\frac{\partial}{\partial x}\right) \varphi(t; \mathbf{r}) = -\left[\frac{\partial}{\partial x}u_1(\mathbf{r}, t) + \frac{\partial}{\partial y}u_2(\mathbf{r}, t)\right] \varphi(t; \mathbf{r}). \tag{11.43}$$

Averaging now Eq. (11.43) over an ensemble of realizations of random field $\mathbf{u}(\mathbf{r}, t)$, we obtain the Fokker–Planck equation

$$\left(\frac{\partial}{\partial t} + v(y)\frac{\partial}{\partial x}\right) P(t; \mathbf{r}) = D_0 \Delta P(t; \mathbf{r}), \quad P(0; \mathbf{r}) = \delta(x - x_0)\, \delta(y - y_0). \tag{11.44}$$

In this case, Eq. (11.44) can be associated with the stochastically equivalent particle whose behavior is governed by the equations

$$\frac{d}{dt}x(t) = v(y) + u_1(t), \quad \frac{d}{dt}y(t) = u_2(t),$$

where $u_i(t)$, $i = 1, 2$ are the statistically independent Gaussian white-noise processes with the statistical characteristics

$$\langle \mathbf{u}(t) \rangle = 0, \quad \langle u_i(t) u_j(t') \rangle = 2D_0 \delta(t - t').$$

These equations can be easily integrated:

$$y(t) = y_0 + w_2(t), \quad x(t) = x_0 + w_1(t) + \int\limits_0^t d\tau v\left(y + w_2(\tau)\right), \tag{11.45}$$

where

$$w_i(t) = \int\limits_0^t d\tau u_i(\tau)$$

are the independent Wiener processes with the characteristics

$$\langle \mathbf{w}(t)\rangle = 0, \quad \langle w_i(t)w_j(t')\rangle = 2D_0\delta_{ij}\min\{t,t'\}.$$

From Eqs. (11.45) follows in particular that coordinate $y(t)$ has the Gaussian probability density with the parameters

$$\langle y(t)\rangle = y_0, \quad \left\langle y^2(t)\right\rangle = y_0^2 + 2D_0 t,$$

which corresponds to the ordinary Brownian motion with the diffusion coefficient D_0.

In addition, Eqs. (11.45) make it possible to easily calculate arbitrary moment functions $\langle x^n(t)\rangle$ and correlations $\langle x^n(t)y^m(t)\rangle$ for particle trajectories.

For example, in the simplest example of the linear shear

$$v_x = \alpha y, \quad v_y = 0,$$

Eqs. (11.45) correspond to the joint Gaussian probability density with the parameters [51, 89, 317]

$$\begin{aligned}
\langle y(t)\rangle &= y_0, \quad \langle x(t)\rangle = x_0 + \alpha y_0 t, \\
\sigma_{xx}^2(t) &= 2D_0 t\left(1 + \alpha t + \frac{1}{3}\alpha^2 t^2\right), \quad \sigma_{yy}^2(t) = 2D_0 t, \quad \sigma_{xy}^2(t) = 2D_0 t\,(1+\alpha t)\,.
\end{aligned}$$

In the case of the Kolmogorov flow, we have [143]

$$\langle y(t)\rangle = y_0, \quad \langle x(t)\rangle = x_0 + \frac{v_0}{\beta^2 D_0}\left[1 - e^{-\beta^2 D_0 t}\right]\sin(\beta y_0),$$

and, under the condition $t \gg 1/(D_0\beta^2)$,

$$\langle x(t)\rangle = x_0 + \frac{v_0}{\beta^2 D_0}\sin(\beta y_0),$$

which means that the particle is on average located in the finite part of space. In this case, the correlation of $x(t)$ and $y(t)$ appears also independent of time:

$$\langle (x(t)-x_0)(y(t)-y_0)\rangle_{t\to\infty} = x_0 + \frac{4v_0}{\beta^3 D_0}\cos(\beta y_0).$$

However, in this limit, quantity $x(t)$ behaves like the Brownian particle with the diffusion coefficient D_0, i.e., $\sigma_{xx}^2 \sim 2D_0 t$.

Note that the Kolmogorov flow becomes the quasi-periodic flow in plane (x, y) after it looses stability. For tracer diffusion in flows of such type with $\mathbf{u}_0(\mathbf{r},t) = \{B\cos y,\ A\sin x\}$, see papers [50, 62].

Remark 7 ***Diffusion of tracer cloud.***

Above, we considered particle diffusion in the presence of mean plane-parallel flow of liquid. In this case, average concentration in the Eulerian description also satisfies the equation

$$\left(\frac{\partial}{\partial t} + v(y)\frac{\partial}{\partial x}\right)\langle\rho(\mathbf{r},t)\rangle = D_0\Delta\langle\rho(\mathbf{r},t)\rangle\,, \quad \langle\rho(\mathbf{r},0)\rangle = \rho_0(\mathbf{r}),$$

and the problem differs from problem (11.44) only in the initial condition. In terms of the Eulerian description of average tracer concentration, the moment functions $\langle x^n(t) y^m(t)\rangle$ obtained above characterize spreading of a tracer cloud. For example, quantity

$$\langle \mathbf{r}(t)\rangle = \frac{1}{M}\int d\mathbf{r}\,\mathbf{r}\,\langle \rho(\mathbf{r},t)\rangle\,,$$

where $M = \int dr\,\langle \rho(\mathbf{r},t)\rangle = \int dr \rho_0(r)$ is the total mass of tracer, defines the time-dependent position of the tracer cloud center of gravity, while higher moments like

$$\langle r_i(t) r_j(t)\rangle = \frac{1}{M}\int d\mathbf{r}\, r_i r_j\,\langle \rho(\mathbf{r},t)\rangle$$

characterize cloud's deformation. ♦

Two-point statistical characteristics

Consider now the joint dynamics of two particles in the absence of mean flow. In this case, the indicator function

$$\varphi(t;\mathbf{r}_1,\mathbf{r}_2) = \delta\left(\mathbf{r}_1(t)-\mathbf{r}_1\right)\delta\left(\mathbf{r}_2(t)-\mathbf{r}_2\right)$$

satisfies the Liouville equation

$$\frac{\partial}{\partial t}\varphi(t;\mathbf{r}_1,\mathbf{r}_2) = -\left[\frac{\partial}{\partial \mathbf{r}_1}\mathbf{u}_1(\mathbf{r},t) + \frac{\partial}{\partial \mathbf{r}_2}\mathbf{u}_2(\mathbf{r},t)\right]\varphi(t;\mathbf{r}_1,\mathbf{r}_2).$$

If we average the indicator function over an ensemble of realizations of field $\mathbf{u}(\mathbf{r},t)$, use the Furutsu–Novikov formula (7.10), page 186, and the equality

$$\frac{\delta}{\delta u_j(\mathbf{r}',t-0)}\varphi(t;\mathbf{r}_1,\mathbf{r}_2) = -\left[\frac{\partial}{\partial r_{1j}}\delta(\mathbf{r}_1-\mathbf{r}') + \frac{\partial}{\partial r_{2j}}\delta(\mathbf{r}_2-\mathbf{r}')\right]\varphi(t;\mathbf{r}_1,\mathbf{r}_2),$$

then we obtain that the joint probability density of positions of two particles

$$P(t;\mathbf{r}_1,\mathbf{r}_2) = \langle \varphi(t;\mathbf{r}_1,\mathbf{r}_2)\rangle$$

satisfies the Fokker–Planck equation

$$\begin{aligned}\frac{\partial}{\partial t}P(t;\mathbf{r}_1,\mathbf{r}_2) &= \left[\frac{\partial^2}{\partial_{1i}\partial r_{1j}} + \frac{\partial^2}{\partial r_{2i}\partial r_{2j}}\right] B^{\text{eff}}_{ij}(0) P(t;\mathbf{r}_1,\mathbf{r}_2) \\ &+2\frac{\partial^2}{\partial_{1i}\partial r_{2j}} B^{\text{eff}}_{ij}(\mathbf{r}_1-\mathbf{r}_2) P(t;\mathbf{r}_1,\mathbf{r}_2).\end{aligned} \tag{11.46}$$

Multiplying now Eq. (11.46) by function $\delta(\mathbf{r}_1-\mathbf{r}_2-\mathbf{l})$ and integrating over $\mathbf{r}_1$ and $\mathbf{r}_2$, we obtain that the probability density of relative diffusion of two particles

$$P(t;\mathbf{l}) = \langle \delta(\mathbf{r}_1(t)-\mathbf{r}_2(t)-\mathbf{l})\rangle$$

satisfies the Fokker–Planck equation

$$\frac{\partial}{\partial t}P(t;\mathbf{l}) = \frac{\partial^2}{\partial l_\alpha \partial l_\beta} D_{\alpha\beta}(\mathbf{l}) P(t;\mathbf{l}), \quad P(0;\mathbf{l}) = \delta(\mathbf{l}-\mathbf{l}_0), \tag{11.47}$$

$$\begin{aligned}
&\langle u_i(\mathbf{r},t)u_j(\mathbf{r}',t')\rangle = B_{ij}(\mathbf{r}-\mathbf{r}',t-t') = \int d\mathbf{k}E_{ij}(\mathbf{k},t-t')e^{i\mathbf{k}(\mathbf{r}-\mathbf{r}')}, \\
&B_{ij}(\mathbf{r},t) = B_{ij}^{\mathrm{s}}(\mathbf{r},t) + B_{ij}^{\mathrm{p}}(\mathbf{r},t), \quad E_{ij}(\mathbf{k},t) = E_{ij}^{\mathrm{s}}(\mathbf{k},t) + E_{ij}^{\mathrm{p}}(\mathbf{k},t) \\
&E_{ij}^{\mathrm{s}}(\mathbf{k},t) = \frac{1}{(2\pi)^d}\int d\mathbf{r}B_{ij}^{\mathrm{s}}(\mathbf{r},t)e^{-i\mathbf{k}\mathbf{r}}, \quad E_{ij}^{\mathrm{p}}(\mathbf{k},t) = \frac{1}{(2\pi)^d}\int d\mathbf{r}B_{ij}^{\mathrm{p}}(\mathbf{r},t)e^{-i\mathbf{k}\mathbf{r}},
\end{aligned} \tag{11.30}$$

$$B_{ij}(\mathbf{r},t) = 2B^{\mathrm{eff}}_{ij}(\mathbf{r})\delta(t), \tag{11.32}$$

$$B^{\mathrm{eff}}_{ij}(\mathbf{r}) = \frac{1}{2}\int\limits_{-\infty}^{\infty} dt B_{ij}(\mathbf{r},t) = \int\limits_{0}^{\infty} dt B_{ij}(\mathbf{r},t).$$

where

$$D_{\alpha\beta}(\mathbf{l}) = 2\left[B^{\mathrm{eff}}_{\alpha\beta}(0) - B^{\mathrm{eff}}_{\alpha\beta}(\mathbf{l})\right]$$

is the structure matrix of vector field $\mathbf{u}(\mathbf{r},t)$ and $\mathbf{l}_0$ is the initial distance between the particles.

In the general case, Eq. (11.47) cannot be solved analytically. However, if the initial distance between particles l_0 is sufficiently small, namely, if $l_0 \ll l_{\mathrm{cor}}$, where l_{cor} is the spatial correlation radius of the velocity field $\mathbf{u}(\mathbf{r},t)$, we can expand functions $D_{\alpha\beta}(\mathbf{l})$ in the Taylor series to obtain in the first approximation

$$D_{\alpha\beta}(\mathbf{l}) = -\left.\frac{\partial^2 B^{\mathrm{eff}}_{\alpha\beta}(\mathbf{l})}{\partial l_i \partial l_j}\right|_{\mathbf{l}=0} l_i l_j.$$

The use of representation (11.31) – (11.34) simplifies the diffusion tensor reducing it to the form

$$D_{\alpha\beta}(\mathbf{l}) = \frac{1}{d(d+2)}\left[\left(D^{\mathrm{s}}(d+1) + D^{\mathrm{p}}\right)\delta_{\alpha\beta} l^2 - 2\left(D^{\mathrm{s}} - D^{\mathrm{p}}\right) l_\alpha l_\beta\right], \tag{11.48}$$

where d is the dimension of space.

Substituting now Eq. (11.48) in Eq. (11.47), multiplying both sides of the resulting equation by l^n, and integrating over $\mathbf{l}$, we obtain the closed equation

$$\begin{aligned}
&\frac{d}{dt}\ln\langle l^n(t)\rangle = \\
&\quad = \frac{1}{d(d+2)}\left[\left(D^{\mathrm{s}}(d+1) + D^{\mathrm{p}}\right) n\,(d+n-2) - 2\left(D^{\mathrm{s}} - D^{\mathrm{p}}\right) n(n-1)\right],
\end{aligned}$$

whose solution shows the exponential growth of all moment functions ($n = 1, 2, ...$) in time. In this case, the probability distribution of random process $l(t)/l_0$ will be logarithmic-normal. As a consequence, the typical realization curve of the distance between two particles will be the exponential function of time

$$l^*(t) = l_0 \exp\left\{\frac{1}{d(d+2)}\left\{D^{\mathrm{s}} d\,(d-1) - D^{\mathrm{p}}\,(4-d)\right\} t\right\}. \tag{11.49}$$

It appears that this expression in the two-dimensional case ($d = 2$)

$$l^*(t) = l_0 \exp\left\{\frac{1}{4}\left(D^{\mathrm{s}} - D^{\mathrm{p}}\right) t\right\}$$

significantly depends on the sign of the difference $(D^{\mathrm{s}} - D^{\mathrm{p}})$. In particular, for the divergence-free velocity field ($D^{\mathrm{p}} = 0$), we have the exponentially increasing typical realization curve, which means that particle scatter is exponentially fast for small distances between them. This result is valid for times

$$\frac{1}{4} D^{\mathrm{s}} t \ll \ln\left(\frac{l_{\mathrm{cor}}}{l_0}\right),$$

for which expansion (11.48) holds. In another limiting case of the potential velocity field ($D^{\mathrm{s}} = 0$), the typical realization curve is the exponentially decreasing curve, which means that particles tend to join. In view of the fact that liquid particles themselves are compressed during this process, we arrive at the conclusion that particles must form *clusters*, i.e., compact particle concentration zones located merely in rarefied regions, which agrees with the evolution of the realization (see Fig. 1.1b, page 4) obtained by simulating the behavior of the initially homogeneous particle distribution in random potential velocity field (though, for drastically other statistical model of the velocity field). This means that the phenomenon of clustering by itself is independent of the model of the velocity field, although statistical parameters characterizing this phenomenon surely depend on this model.

In the three-dimensional case ($d = 3$) Eq. (11.49) grades into

$$l^*(t) = l_0 \exp\left\{\frac{1}{15}\left(6D^{\mathrm{s}} - D^{\mathrm{p}}\right)t\right\},$$

and typical realization curve will exponentially decay in time under the condition

$$D^{\mathrm{p}} > 6D^{\mathrm{s}},$$

which is stronger than in the two-dimensional case.

In the one-dimensional case, we have

$$l^*(t) = l_0 e^{-D^{\mathrm{p}}t},$$

and typical realization curve always decays in time because the velocity is always divergent in this case.

Remark 8 ***Probability density of vector modulus*** $l(t) = |\mathbf{l}(t)|$.

Note that, multiplying Eq. (11.47) by $\delta(l(t) - l)$ and integrating the result over $\mathbf{l}$, we can easily obtain that the probability density of the modulus of vector $\mathbf{l}(t)$

$$P(t;l) = \langle \delta(|\mathbf{l}(t)| - l)\rangle = \int d\mathbf{l}\delta(|\mathbf{l}(t)| - l)P(t,\mathbf{l})$$

satisfies the equation

$$\frac{\partial}{\partial t}P(t,l) = -\frac{\partial}{\partial l}\frac{D_{ii}(\mathbf{l})}{l}P(t,l) + \frac{\partial}{\partial l}N(l)P(t,l) + \frac{\partial^2}{\partial l^2}N(l)P(t,l),$$

where $N(l) = l_j l_i D_{ij}(\mathbf{l})/l^2$. ♦

11.2.2 Eulerian description

First of all we note that, in the case of the delta-correlated random velocity field, linear equation (11.1) in the absence of mean flow allows a relatively simple passage to the closed equations for both buoyant tracer average concentration and its higher multipoint correlation functions. For example, averaging Eq. (11.1), using the Furutsu–Novikov formula (7.10), page 186, and the expression for variational derivative

$$\frac{\delta\rho(\mathbf{r},t)}{\delta u_\alpha(\mathbf{r}',t-0)} = -\frac{\partial}{\partial r_\alpha}\delta(\mathbf{r}-\mathbf{r}')\rho(\mathbf{r},t)$$

following from Eq. (11.1), page 234, we obtain the equation for the average tracer concentration

$$\frac{\partial}{\partial t}\langle\rho(\mathbf{r},t)\rangle = (D_0 + \mu)\,\Delta\,\langle\rho(\mathbf{r},t)\rangle\,. \tag{11.50}$$

Under the condition $D_0 \gg \mu$ ($\mu \ll \sigma_u^2 l_{\text{cor}}^2$), where σ_u^2 is the variance of the random velocity field and l_{cor} is the correlation radius of this field, Eq. (11.50) coincides with the equation for the probability distribution of particle coordinate (11.37); consequently, diffusion coefficient D_0 characterizes only the scales of the region in which tracer is concentrated as a whole and provides no information on the local structure of concentration realization, which is similar to the diffusion in the divergence-free random velocity field.

In a similar way, we obtain that the spatial correlation function of the concentration field

$$\Gamma(\mathbf{r}_1, \mathbf{r}_2, t) = \langle \rho(\mathbf{r}_1, t)\rho(\mathbf{r}_2, t)\rangle$$

satisfies the equation

$$\frac{\partial}{\partial t}\Gamma(\mathbf{r}_1, \mathbf{r}_2, t) = \left[\frac{\partial^2}{\partial r_{1i}\partial r_{1j}} + \frac{\partial^2}{\partial r_{2i}\partial r_{2j}}\right]\left[B_{ij}^{\text{eff}}(0) + \mu\delta_{ij}\right]\Gamma(\mathbf{r}_1, \mathbf{r}_2, t)$$
$$+ \frac{\partial^2}{\partial r_{1i}\partial r_{2j}} B_{ij}^{\text{eff}}(\mathbf{r}_1 - \mathbf{r}_2)\Gamma(\mathbf{r}_1, \mathbf{r}_2, t)$$

coinciding with the equation for the two-particle probability density in the absence of the molecular diffusion ($\mu = 0$).

In the special case of constant initial distribution of the concentration field ($\rho_0(\mathbf{r}) = \rho_0 = \text{const}$), random field $\rho(\mathbf{r}, t)$ will be the homogeneous isotropic random field. In this case, $\langle \rho(\mathbf{r}, t)\rangle = \rho_0$, and the equation for correlation function becomes simpler and assumes the form

$$\frac{\partial}{\partial t}\Gamma(\mathbf{r}, t) = 2\mu\Delta\Gamma(\mathbf{r}, t) + \frac{\partial^2}{\partial r_i \partial r_j} D_{ij}(\mathbf{r})\Gamma(\mathbf{r}, t), \quad \Gamma(\mathbf{r}, 0) = \rho_0^2,$$

where $\mathbf{r} = \mathbf{r}_1 - \mathbf{r}_2$ and

$$D_{ij}(\mathbf{r}) = 2\left[B_{ij}^{\text{eff}}(0) - B_{ij}^{\text{eff}}(\mathbf{r})\right]$$

is the structure matrix of vector field $\mathbf{u}(\mathbf{r}, t)$. In the absence of molecular diffusion, this equation coincides with the equation for the probability density of relative diffusion of two particles.

Correlation function $\Gamma(\mathbf{r}, t)$ will now depend on the modulus of vector $\mathbf{r}$ ($\Gamma(\mathbf{r}, t) = \Gamma(r, t)$) and, as a function of variables $\{r, t\}$, will satisfy the equation

$$\frac{\partial}{\partial t}\Gamma(r, t) = \frac{1}{r^{d-1}}\frac{\partial}{\partial r} r^{d-1}\left[\frac{\partial D_{ii}^{\text{p}}(r, t)}{\partial r} + \left(2\mu + \frac{r_i r_j}{r^2}D_{ij}(\mathbf{r})\right)\frac{\partial}{\partial r}\right]\Gamma(r, t),$$

where d is the dimension of space, as earlier. This equation has the steady-state solution $\Gamma(r) = \Gamma(r, \infty)$ [20, 21],

$$\Gamma(r) = \rho_0^2 \exp\left\{\int\limits_r^\infty dr' \frac{\partial D_{ii}^{\text{p}}(r')/\partial r'}{2\mu + r_i' r_j' D_{i,j}(\mathbf{r}')/r'^2}\right\},$$

which corresponds to the boundary-value condition

$$\Gamma(\infty) = \rho_0^2.$$

For incompressible turbulent liquid flow, this equation was analyzed in paper [232].

To describe the local behavior of tracer realizations in random velocity field, we need the probability distribution of tracer concentration, which is possible only in the absence of molecular diffusion. The equation for the Eulerian probability density can be easily derived in view of formula (11.23) by multiplying Eq. (11.35) by j and integrating the result over all possible values of j and $\mathbf{r}_0$. As a result, we arrive at the equation for the probability density of the concentration field in the form

$$\left(\frac{\partial}{\partial t} - D_0\Delta\right) P(t,\mathbf{r};\rho) = D^{\mathrm{p}}\frac{\partial^2}{\partial\rho^2}\rho^2 P(t,\mathbf{r};\rho), \quad P(0,\mathbf{r};\rho) = \delta(\rho_0(\mathbf{r}) - \rho). \tag{11.51}$$

From this equation follows in particular that moment functions of the concentration field satisfy the equation

$$\left(\frac{\partial}{\partial t} - D_0\Delta\right) \langle\rho^n(\mathbf{r},t)\rangle = D^{\mathrm{p}} n(n-1) \langle\rho^n(\mathbf{r},t)\rangle, \quad \langle\rho^n(\mathbf{r},0)\rangle = \rho_0^n(\mathbf{r}).$$

Its solution can be represented in the form

$$\langle\rho^n(\mathbf{r},t)\rangle = e^{n(n-1)\tau}\int d\mathbf{r}' P(t;\mathbf{r}|\mathbf{r}')\rho_0^n(\mathbf{r}'),$$

where function $P(t;\mathbf{r}|\mathbf{r}')$ is defined by Eq. (11.37).

For example, in the case of uniform initial tracer concentration ($\rho_0(\mathbf{r}) = \rho_0 = \text{const}$), the tracer concentration is the lognormal quantity whose probability distribution is independent of $\mathbf{r}$; the corresponding probability density and integral distribution function are as follows

$$P(t;\rho) = \frac{1}{2\rho\sqrt{\pi\tau}}\exp\left\{-\frac{\ln^2(\rho e^{\tau}/\rho_0)}{4\tau}\right\}, \quad F(t;\rho) = \Phi\left(\frac{\ln(\rho e^{\tau}/\rho_0)}{2\sqrt{\tau}}\right), \tag{11.52}$$

where $\Phi(z)$ is the error function,

$$\Phi(z) = \frac{1}{\sqrt{2\pi}}\int\limits_{-\infty}^{z} dy \exp\left\{-\frac{y^2}{2}\right\}.$$

In this case, all moment functions beginning from the second one appear the exponentially increasing functions of time

$$\langle\rho(\mathbf{r},t)\rangle = \rho_0, \quad \langle\rho^n(\mathbf{r},t)\rangle = \rho_0^n e^{n(n-1)\tau},$$

and the typical realization curve of the concentration field exponentially decays with time at any fixed spatial point

$$\rho^*(t) = \rho_0 e^{-\tau},$$

which is evidence of cluster behavior of medium density fluctuations in arbitrary divergent flows. The Eulerian concentration statistics at any fixed point is formed due to concentration fluctuations about this curve.

Even the above discussion of the one-point probability density of tracer concentration in the Eulerian representation revealed several regularities concerning the temporal behavior of concentration field realizations at fixed spatial points. Now we show that this distribution additionally allows us to reveal certain features in the space-time structure of concentration field realizations.

For simplicity, we content ourselves with the two-dimensional case. As was mentioned earlier, the analysis of level lines defined by the equality

$$\rho(\mathbf{r},t) = \rho = \text{const}$$

can give important data on the spatial behavior of realizations, in particular, different functionals of the concentration field, such as the total area $S(t,\rho)$ of regions where $\rho(\mathbf{r},t) > \rho$ and the total tracer mass within these regions $M(t,\rho)$. Average values of these functionals can be expressed in terms of the one-point probability density:

$$\langle S(t,\rho)\rangle = \int\limits_{\rho}^{\infty} d\tilde{\rho} \int d\mathbf{r}\, P(t,\mathbf{r};\tilde{\rho}), \quad \langle M(t,\rho)\rangle = \int\limits_{\rho}^{\infty} \tilde{\rho} d\tilde{\rho} \int d\mathbf{r}\, P(t,\mathbf{r};\tilde{\rho}). \tag{11.53}$$

Substituting the solution to Eq. (11.51) in these expressions and performing some rearrangement, we easily obtain explicit expressions for these quantities

$$\begin{aligned}
\langle S(t,\rho)\rangle &= \int d\mathbf{r}\, \Phi\left(\frac{1}{2\sqrt{\tau}} \ln\left(\frac{\rho_0(\mathbf{r})e^{-\tau}}{\rho}\right)\right), \\
\langle M(t,\rho)\rangle &= \int d\mathbf{r}\, \rho_0(\mathbf{r})\, \Phi\left(\frac{1}{2\sqrt{\tau}} \ln\left(\frac{\rho_0(\mathbf{r})e^{\tau}}{\rho}\right)\right).
\end{aligned} \tag{11.54}$$

These expressions show in particular that, for $\tau \gg 1$, the average area of regions where concentration exceeds level ρ decreases in time according to the law

$$\langle S(t,\rho)\rangle \approx \frac{1}{\sqrt{\pi\tau\rho}} e^{-\tau/4} \int d\mathbf{r} \sqrt{\rho_0(\mathbf{r})}, \tag{11.55}$$

while the average tracer mass within these regions

$$\langle M(t,\rho)\rangle \approx M - \sqrt{\frac{\rho}{\pi\tau}} e^{-\tau/4} \int d\mathbf{r} \sqrt{\rho_0(\mathbf{r})}. \tag{11.56}$$

monotonously tends to the total mass $M = \int d\mathbf{r}\rho_0(\mathbf{r})$.

This is an additional evidence in favor of the above conclusion that tracer particles tend to join in clusters, i.e., in compact regions of enhanced concentration surrounded with rarefied regions.

We illustrate the dynamics of cluster formation by the example of the initially uniform distribution of buoyant tracer over the plane, in which case $\rho_0(\mathbf{r}) = \rho_0 = \text{const}$. In this case, the average specific area (per unit surface area) of regions within which $\rho(\mathbf{r},t) > \rho$ is

$$s(t,\rho) = \int\limits_{\rho}^{\infty} d\tilde{\rho} P(t;\tilde{\rho}) = \Phi\left(\frac{\ln\left(\rho_0\, e^{-\tau}/\rho\right)}{2\sqrt{\tau}}\right), \tag{11.57}$$

where $P(t;\rho)$ is the solution to Eq. (11.51) independent of $\mathbf{r}$ (i.e., function (11.52)), and average specific tracer mass (per unit surface area) within these regions is given by the expression

$$m(t,\rho) = \frac{1}{\rho_0} \int\limits_{\rho}^{\infty} \tilde{\rho} d\tilde{\rho} P(t;\tilde{\rho}) = \Phi\left(\frac{\ln\left(\rho_0\, e^{\tau}/\rho\right)}{2\sqrt{\tau}}\right). \tag{11.58}$$

From Eqs. (11.57) and (11.58) follows that, for sufficiently large times, average specific area of such regions decreases according to the law

$$s(t,\rho) = \Phi(-\sqrt{\tau}/2) \approx \frac{1}{\sqrt{\pi\tau}} e^{-\tau/4} \tag{11.59}$$

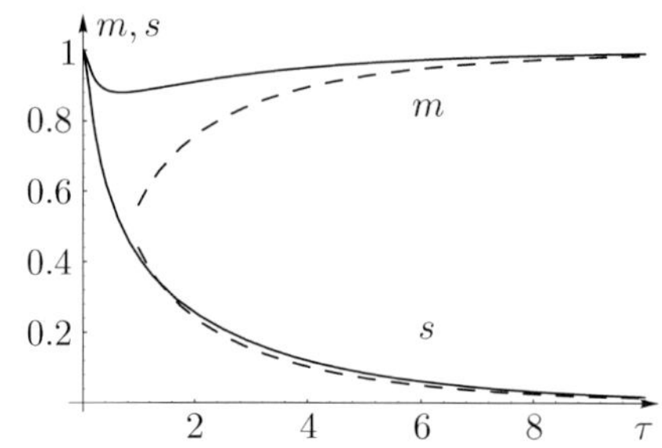

Figure 11.1: Cluster formation dynamics for $\rho/\rho_0 = 0.5$.

irrespective of ratio ρ/ρ_0; at the same time, these regions concentrate almost all tracer mass

$$m(t, \rho) = \Phi(\sqrt{\tau}/2) \approx 1 - \frac{1}{\sqrt{\pi\tau}} e^{-\tau/4}. \tag{11.60}$$

Nevertheless, the time-dependent behavior of the formation of cluster structure essentially depends on ratio ρ/ρ_0. If $\rho/\rho_0 < 1$, then $s(0, \rho) = 1$ and $m(0, \rho) = 1$ at the initial instant. Then, in view of the fact that particles of buoyant tracer initially tend to scatter, there appear small areas within which $\rho(\mathbf{r}, t) < \rho$ and which concentrate only insignificant portion of the total mass. These regions rapidly grow with time and their mass flows into cluster region relatively quickly approaching asymptotic expressions (11.59) and (11.60) (Fig. 11.1). Note that $s(t^*, \rho) = 1/2$ at instant $\tau^* = \ln(\rho/\rho_0)$.

In the opposite, more interesting case $\rho/\rho_0 > 1$, we have $s(0, \rho) = 0$ and $m(0, \rho) = 0$ at the initial instant. In view of initial scatter of particles, there appear small cluster regions within which $\rho(\mathbf{r}, t) > \rho$; these regions remain at first almost invariable in time and intensively absorb a significant portion of total mass. With time, the area of these regions begins to decrease and the mass within them begins to increase according to asymptotic expressions (11.59) and (11.60) (Fig. 11.2*a* and Fig. 11.2*b*).

As we mentioned earlier, a more detailed description of tracer concentration field in random velocity field requires that spatial gradient $\mathbf{p}(\mathbf{r}, t) = \nabla\rho(\mathbf{r}, t)$ (and, generally, higher-order derivatives) was additionally included in the consideration.

The concentration gradient satisfies dynamic equation (11.3); consequently, the extended indicator function

$$\varphi(t, \mathbf{r}; \rho, \mathbf{p}) = \delta(\rho(\mathbf{r}, t) - \rho)\, \delta(\mathbf{p}(\mathbf{r}, t) - \mathbf{p})$$

satisfies Eq. (11.28). Averaging now Eq. (11.28) over an ensemble of velocity field realizations in the approximation of the delta-correlated velocity field, we obtain the equation for the one-point joint probability density of the concentration and gradient $P(t, \mathbf{r}; \rho, \mathbf{p}) = \langle\varphi(t, \mathbf{r}; \rho, \mathbf{p})\rangle$ (this probability density is a function of the space-time point

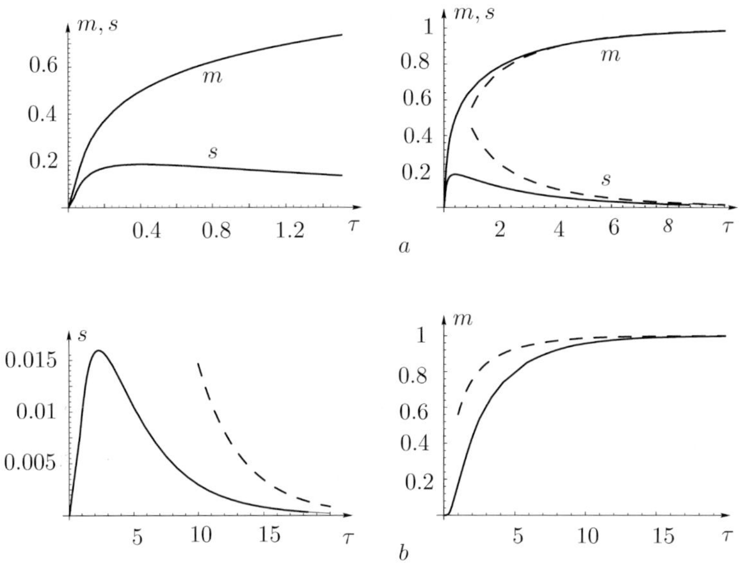

Figure 11.2: Cluster formation dynamics for (a) $\rho/\rho_0 = 1.5$ and (b) $\rho/\rho_0 = 10$.

$(\mathbf{r}, t))$

$$\frac{\partial}{\partial t}P(t,\mathbf{r};\rho,\mathbf{p}) = \left[D_0\Delta + \frac{2}{d}D^{\mathrm{p}}\frac{\partial^2}{\partial\mathbf{r}\partial\mathbf{p}}\rho + D^{\mathrm{p}}\frac{\partial^2}{\partial\rho^2}\rho^2 \right.$$
$$+\frac{1}{d(d+2)}D^{\mathrm{s}}\hat{L}^{\mathrm{s}}(\mathbf{p}) + \frac{1}{d(d+2)}D^{\mathrm{p}}\hat{L}^{p}(\mathbf{p})$$
$$\left.+\frac{2(d+1)}{d}D^{\mathrm{p}}\frac{\partial}{\partial\mathbf{p}}\mathbf{p}\frac{\partial}{\partial\rho}\rho + D_4^{\mathrm{p}}\frac{\partial^2}{\partial\mathbf{p}^2}\rho^2\right]P(t,\mathbf{r};\rho,\mathbf{p}), \tag{11.61}$$

where we introduced the operators

$$\begin{aligned}\hat{L}^{\mathrm{s}}(\mathbf{p}) &= (d+1)\frac{\partial^2}{\partial\mathbf{p}^2}\mathbf{p}^2 - 2\frac{\partial}{\partial\mathbf{p}}\mathbf{p} - 2\left(\frac{\partial}{\partial\mathbf{p}}\mathbf{p}\right)^2 \\ &= (d+1)\frac{\partial^2}{\partial\mathbf{p}^2}\mathbf{p}^2 - 2\frac{\partial^2}{\partial p_k\partial p_l}p_kp_l, \\ \hat{L}^{\mathrm{p}}(\mathbf{p}) &= \frac{\partial^2}{\partial\mathbf{p}^2}\mathbf{p}^2 + (d^2+4d+6)\left(\frac{\partial}{\partial\mathbf{p}}\mathbf{p}\right)^2 + (d^2+2d+2)\frac{\partial}{\partial\mathbf{p}}\mathbf{p}.\end{aligned}$$

Investigation of Eq. (11.61) is hardly possible in the general case. Such investigation appears possible only for the divergence-free velocity field, in which case Eq. (11.61) assumes the form

$$\frac{\partial}{\partial t}P(t,\mathbf{r};\rho,\mathbf{p}) = D_0\Delta P(t,\mathbf{r};\rho,\mathbf{p})$$
$$+\frac{1}{d(d+2)}D^{\mathrm{s}}\left((d+1)\frac{\partial^2}{\partial\mathbf{p}^2}\mathbf{p}^2 - 2\frac{\partial^2}{\partial p_k\partial p_l}p_kp_l\right)P(t,\mathbf{r};\rho,\mathbf{p}). \tag{11.62}$$

In view of the fact that the random velocity field is divergence-free, we can represent the solution to Eq. (11.62) in the form

$$P(t, \mathbf{r}; \rho, \mathbf{p}) = \int d\mathbf{r}_0 P(t, \mathbf{r}|\mathbf{r}_0) P(t, \mathbf{p}|\mathbf{r}_0), \tag{11.63}$$

where $P(t, \mathbf{r}|\mathbf{r}_0)$ and $P(t, \mathbf{p}|\mathbf{r}_0)$ are the corresponding Lagrangian probability densities of particle position and gradient. The first density is given by Eq. (11.37), and the second satisfies the equation

$$\frac{\partial}{\partial t} P(t, \mathbf{p}|\mathbf{r}_0) = \frac{1}{d(d+2)} D^{\mathrm{s}} \left((d+1) \frac{\partial^2}{\partial \mathbf{p}^2} \mathbf{p}^2 - 2 \frac{\partial^2}{\partial p_k \partial p_l} p_k p_l \right) P(t, \mathbf{p}|\mathbf{r}_0). \tag{11.64}$$

From Eq. (11.64) follows that $\langle \mathbf{p}(\mathbf{r}, t) \rangle = \mathbf{p}_0(\mathbf{r}_0)$, i.e., the average tracer concentration gradient is invariant. As regards the moment functions of the concentration gradient modulus, they satisfy the equations

$$\frac{\partial}{\partial t} \langle p^n(t|\mathbf{r}_0) \rangle = \frac{n(d+n)(d-1)}{d(d+2)} D^{\mathrm{s}} \langle p^n(t|\mathbf{r}_0) \rangle, \quad \langle p^n(0|\mathbf{r}_0) \rangle = p_0^n(\mathbf{r}_0). \tag{11.65}$$

that follow from (11.64).

Consequently, concentration gradient modulus in the Lagrangian representation is the logarithmic-normal quantity whose typical realization curve and all moment functions increase exponentially in time. In particular, the first and second moments in the two-dimensional case are given by the equalities

$$\langle |\mathbf{p}(t|\mathbf{r}_0)| \rangle = |\mathbf{p}_0(\mathbf{r}_0)| e^{3D^{\mathrm{s}}t/8}, \quad \left\langle \mathbf{p}^2(t|\mathbf{r}_0) \right\rangle = \mathbf{p}_0^2(\mathbf{r}_0) e^{D^{\mathrm{s}}t}. \tag{11.66}$$

Note that lognormal distribution for the concentration gradient modulus was for the first time suggested in paper [102] and agrees with atmospheric experimental data [53, 125].

In addition, from Eq. (11.62) with an allowance for Eq. (11.26) follows that the average total length of contour $\rho(\mathbf{r}, t) = \rho = \text{const}$ (remember that we deal with the two-dimensional case) also exponentially increases in time according to the law

$$\langle l(t, \rho) \rangle = l_0 e^{D^{\mathrm{s}}t},$$

where l_0 is the initial length of the contour [180, 181, 271]. Remind that, in the case of the divergence-free velocity field, the number of contours remains unchanged; the contours cannot appear and disappear in the medium, they only evolve in time depending on their spatial distribution at the initial instant.

Thus, the initially smooth tracer distribution becomes with time increasingly inhomogeneous in space; its spatial gradients sharpen and level lines acquire the fractal behavior. We observed such pattern in Fig. 1.1*a*, page 4 that shows simulated results (for quite other model of velocity field fluctuations). This means that the above general behavioral characteristics only slightly depend on the fluctuation model.

Remark 9 ***Diffusion of nonconservative tracer.***

Above, we studied statistical characteristics of the solution to Eq. (11.2), page 235 in the Lagrangian and Eulerian representations and showed that both particle dynamics and Eulerian concentration field show clustering if the velocity field has a nonzero potential component. Along with dynamic equation (11.2), there is certain interest to the equation describing transfer of nonconservative tracer (see, e.g., [217])

$$\left(\frac{\partial}{\partial t}+\mathbf{U}(\mathbf{r},t)\frac{\partial}{\partial\mathbf{r}}\right)\rho(\mathbf{r},t)=0,\quad \rho(\mathbf{r},0)=\rho_0(\mathbf{r}).$$

In this case, particle dynamics in the Lagrangian representation is described by the equation coinciding with Eq. (11.4), page 235; consequently, particles are clustered. However, in the Eulerian representation, no clustering occurs. In this case, as in the case of the divergence-free velocity field, average number of contours, average area of regions within which $\rho(\mathbf{r},t)>\rho$, and average tracer mass $\int dS\rho(\mathbf{r},t)$ within these regions remain invariant. ♦

Thus, we obtained that occurrence of tracer field clustering requires that the velocity field of hydrodynamic flow be with necessity the divergent field. In the context of problems concerning Earth's atmosphere and ocean, the medium is usually assumed *incompressible*, which means that it is generally described by the divergence-free velocity field. There are two cases in which clustering can occur in this case:

(1) Diffusion of buoyant tracer [157, 167] and

(2) Diffusion of low-inertia tracer [151, 153, 154] (see, also [48, 49] for experiments and numerical modelling).

In the first case, the motion of the passive tracer with the horizontal and vertical velocities $\mathbf{u}=(\mathbf{U},w)$ along surface $z=0$ in noncompressible medium ($\operatorname{div}\mathbf{u}(\mathbf{r},t)=0$) with absent mean flow creates on this surface the effective two-dimensional compressible flow with the two-dimensional divergence $\nabla_{\mathbf{R}}\mathbf{U}(\mathbf{R},t)=-\partial w(\mathbf{r},t)/\partial z|_{z=0}$. We assume that spatial spectral tensor of the velocity field $\mathbf{u}(\mathbf{r},t)$ has the form

$$E_{ij}(\mathbf{k},t)=E(k,t)\left(\delta_{ij}-\frac{k_ik_j}{k^2}\right).$$

Represent now the field of buoyant tracer in the form

$$\rho(\mathbf{r},t)=\rho(\mathbf{R},t)\delta(z),\quad \mathbf{r}=(\mathbf{R},z),\quad \mathbf{R}=(x,y).$$

Substituting this expression in Eq. (11.2), page 235 and integrating the result over z, we obtain the equation

$$\left(\frac{\partial}{\partial t}+\frac{\partial}{\partial\mathbf{R}}\mathbf{U}(\mathbf{R},t)\right)\rho(\mathbf{R},t)=0,\quad \rho(\mathbf{R},0)=\rho_0(\mathbf{R}).$$

Field $\mathbf{U}(\mathbf{R},t)$ is the homogeneous and isotropic Gaussian field with the spectral tensor

$$E_{\alpha\beta}(\mathbf{k}_\perp,t)=\int\limits_{-\infty}^{\infty}dk_zE\left(\mathbf{k}_\perp^2+k_z^2,t\right)\left(\delta_{\alpha\beta}-\frac{k_{\perp\alpha}k_{\perp\beta}}{\mathbf{k}_\perp^2+k_z^2}\right),\quad \alpha,\beta=1,2. \tag{11.67}$$

Correlating now Eq. (11.67) with Eqs. (11.30) and (11.31), we obtain the expressions for the solenoidal and potential components of velocity $\mathbf{U}(\mathbf{R},t)$ in plane $z=0$ [167]

$$E^s(\mathbf{k}_\perp,t)=\int\limits_{-\infty}^{\infty}dk_zE\left(\mathbf{k}_\perp^2+k_z^2,t\right),\quad E^p(\mathbf{k}_\perp,t)=\int\limits_{-\infty}^{\infty}dk_zE\left(\mathbf{k}_\perp^2+k_z^2\right)\frac{k_z^2}{\mathbf{k}_\perp^2+k_z^2}. \tag{11.68}$$

As a consequence, the probability density of concentration field $\rho(\mathbf{R},t)$ will satisfy the two-dimensional equation

$$\left(\frac{\partial}{\partial t} - D_0\Delta\right) P(t,\mathbf{R};\rho) = D^{\mathrm{p}}\frac{\partial^2}{\partial\rho^2}\rho^2 P(t,\mathbf{R};\rho), \quad P(0,\mathbf{R};\rho) = \delta(\rho_0(\mathbf{R}) - \rho) \tag{11.69}$$

with the diffusion coefficients given, in accordance with Eqs. (11.33), (11.34), and (11.68), by the equalities

$$\begin{aligned} D_0 &= 2\pi\int_0^\infty d\tau\int_0^\infty k^2 dk E(k,\tau), \\ D^{\mathrm{s}} &= \frac{4\pi}{3}\int_0^\infty d\tau\int_0^\infty k^4 dk E(k,\tau), \quad D^{\mathrm{p}} = \frac{4\pi}{5}\int_0^\infty d\tau\int_0^\infty k^4 dk E(k,\tau). \end{aligned} \tag{11.70}$$

Thus, we see that clustering of the concentration field in the Eulerian representation must occur for the diffusion of inertialess buoyant tracer concentration. At the same time, no clustering will occur for the diffusion of inertialess buoyant tracer particles, because in this case we have, according to (11.70), the inequality $D^{\mathrm{s}} > D^{\mathrm{p}}$ (see page **??**).

The second case of the diffusion of low-inertia tracer is of importance for applications. Beginning from classical work by Stokes, 1851 [292] (see also classical book [218]), research of the dynamics of inertial particles in hydrodynamic flows attracts attention of many researchers in view of its importance for different problems of the ecology of Earth's atmosphere and ocean and due to multiple technical applications (see, e.g., [273], [304]). Note that Maxey [240] was seemingly the first who noticed that the velocity field of inertial particles in the divergence-free velocity field of hydrodynamic flow appears divergent, as distinct from the inertialess passive particles.

Diffusion of the number density of particles (particles per unit volume) $n(\mathbf{r},t)$ moving in random hydrodynamic flows is described by the equation of continuity (1.60), page 25

$$\left(\frac{\partial}{\partial t} + \frac{\partial}{\partial\mathbf{r}}\mathbf{V}(\mathbf{r},t)\right) n(\mathbf{r},t) = 0, \quad n(\mathbf{r},0) = n_0(\mathbf{r}). \tag{11.71}$$

We will assume that velocity field $\mathbf{V}(\mathbf{r},t)$ satisfies the equation (1.59) (see, e.g., [240])

$$\left(\frac{\partial}{\partial t} + \mathbf{V}(\mathbf{r},t)\frac{\partial}{\partial\mathbf{r}}\right)\mathbf{V}(\mathbf{r},t) = -\lambda\left[\mathbf{V}(\mathbf{r},t) - \mathbf{u}(\mathbf{r},t)\right] \tag{11.72}$$

that we will consider the phenomenological equation. Parameter $\tau = 1/\lambda$ is the well-known Stokes time dependent on particle size and molecular viscosity.

The total number of particles remains invariant during evolution; this means that

$$N_0 = \int n(\mathbf{r},t)d\mathbf{r} = \int n_0(\mathbf{r})d\mathbf{r} = \mathrm{const}.$$

If velocity field $\mathbf{V}(\mathbf{r},t)$ is the random Gaussian field statistically homogeneous and isotropic in space and stationary in time with the zero-valued mean and correlation tensor

$$\langle V_i(\mathbf{r},t)V_j(\mathbf{r}',t')\rangle = B_{ij}^{(\mathbf{V})}(\mathbf{r}-\mathbf{r}',t-t'),$$

then the one-point probability density of the solution $P(t,\mathbf{r};n)$ to dynamic equation (11.71) satisfies, in the approximation of the delta-correlated (in time) field $\mathbf{V}(\mathbf{r},t)$, Eq. (11.51)

$$\begin{aligned}\left(\frac{\partial}{\partial t}-D_0\frac{\partial^2}{\partial\mathbf{r}^2}\right)P(t,\mathbf{r};n) &= D^{(\mathbf{V})}\frac{\partial^2}{\partial n^2}n^2P(t,\mathbf{r};n),\\ P(0,\mathbf{r};n) &= \delta\left(n_0(\mathbf{r})-n\right),\end{aligned} \tag{11.73}$$

where diffusion coefficients

$$\begin{aligned}D_0 &= \frac{1}{d}\int\limits_0^\infty d\tau\left\langle\mathbf{V}(\mathbf{r},t+\tau)\mathbf{V}(\mathbf{r},t)\right\rangle=\frac{1}{d}\tau_{\mathbf{V}}\left\langle\mathbf{V}^2(\mathbf{r},t)\right\rangle,\\ D^{(\mathbf{V})} &= \int\limits_0^\infty d\tau\left\langle\frac{\partial\mathbf{V}(\mathbf{r},t+\tau)}{\partial\mathbf{r}}\frac{\partial\mathbf{V}(\mathbf{r},t)}{\partial\mathbf{r}}\right\rangle=\tau_{\mathrm{div}\mathbf{V}}\left\langle\left(\frac{\partial\mathbf{V}(\mathbf{r},t)}{\partial\mathbf{r}}\right)^2\right\rangle\end{aligned} \tag{11.74}$$

characterize spatial scatter of the number density of particles $n(\mathbf{r},t)$ and characteristic time of the formation of cluster structures, $\tau_{\mathbf{V}}$ and $\tau_{\mathrm{div}\mathbf{V}}$ are the temporal correlation radii of random fields $\mathbf{V}(\mathbf{r},t)$ and $\partial\mathbf{V}(\mathbf{r},t)/\partial\mathbf{r}$, and d is the dimension of space.

Thus, our task consists in the evaluation of diffusion coefficients (11.74) from stochastic equation (11.72), i.e., in the calculation of temporal correlation radii $\tau_{\mathbf{V}}$ and $\tau_{\mathrm{div}\mathbf{V}}$ of random fields $\mathbf{V}(\mathbf{r},t)$ and $\partial\mathbf{V}(\mathbf{r},t)/\partial\mathbf{r}$, spatial correlation scales and variances of these fields [151, 153].

We will assume that velocity field $\mathbf{u}(\mathbf{r},t)$ is the divergence-free field

$$\mathrm{div}\,\mathbf{u}(\mathbf{r},t)=\frac{\partial}{\partial\mathbf{r}}\mathbf{u}(\mathbf{r},t)=0.$$

Additionally, we will assume it the Gaussian random field homogeneous and isotropic in space and stationary in time with the zero-valued mean and the correlation tensor

$$B_{ij}(\mathbf{r}-\mathbf{r}',t-t')=\left\langle u_i(\mathbf{r},t)u_j(\mathbf{r}',t')\right\rangle.$$

Moreover, we will assume that variance of random velocity field $\sigma_{\mathbf{u}}^2=\left\langle\mathbf{u}^2(\mathbf{r},t)\right\rangle$ is sufficiently small and can be used as the fundamental small parameter of the problem. In this case, we can linearize Eq. (11.72) in quantity $\mathbf{V}(\mathbf{r},t)\approx\mathbf{u}(\mathbf{r},t)$.

As earlier, we define the temporal correlation radius of field $\mathbf{u}(\mathbf{r},t)$ by the equality

$$\tau_0B_{ii}(0,0)=\int\limits_0^\infty d\tau B_{ii}(0,\tau)=\int\limits_0^\infty d\tau\left\langle\mathbf{u}(\mathbf{r},t+\tau)\mathbf{u}(\mathbf{r},t)\right\rangle.$$

Within the framework of such a model, the spatial spectral and space-time spectral functions of field $\mathbf{u}(\mathbf{r},t)$ have the forms

$$B_{ij}(\mathbf{r},t)=\int d\mathbf{k}E_{ij}(\mathbf{k},t)e^{i\mathbf{kr}},\quad B_{ij}(\mathbf{r},t)=\int d\mathbf{k}\int\limits_{-\infty}^{\infty}d\omega\Phi_{ij}(\mathbf{k},\omega)e^{i\mathbf{kr}+i\omega t},$$

where

$$E_{ij}(\mathbf{k},t)=E(k,t)\left(\delta_{ij}-\frac{k_ik_j}{\mathbf{k}^2}\right),\quad \Phi_{ij}(\mathbf{k},\omega)=\Phi(k,\omega)\left(\delta_{ij}-\frac{k_ik_j}{k^2}\right). \tag{11.75}$$

In this case,

$$B_{ij}(0,t) = \frac{d-1}{d} \int d\mathbf{k} E(k,t)\delta_{ij}, \tag{11.76}$$

where d is the dimension of space and the fourth-order tensor $\frac{\partial^2 B_{ij}(0,\tau)}{\partial r_k \partial r_l}$ has the following representation

$$-\frac{\partial^2 B_{ij}(0,\tau)}{\partial r_k \partial r_l} = \frac{D(\tau)}{d(d+2)} \left[(d+1)\delta_{kl}\delta_{ij} - \delta_{ki}\delta_{lj} - \delta_{kj}\delta_{li}\right]. \tag{11.77}$$

Coefficient $D(\tau)$ in (11.77) has the form

$$D(\tau) = \int d\mathbf{k}\, \mathbf{k}^2 E(k,\tau) = -\frac{1}{d-1} \left\langle \mathbf{u}(\mathbf{r},t+\tau)\Delta\mathbf{u}(\mathbf{r},t)\right\rangle,$$

so that quantity

$$D(0) = -\frac{1}{d-1} \left\langle \mathbf{u}(\mathbf{r},t)\Delta\mathbf{u}(\mathbf{r},t)\right\rangle$$

is related to the vortex structure of random divergence-free field $\mathbf{u}(\mathbf{r},t)$.

Coefficients (11.74) in the equation for the probability density (11.73) were calculated in paper [153]; they are expressed as follows

$$\begin{aligned} D_0 &= \frac{1}{d}\tau_{\mathbf{V}} \left\langle \mathbf{V}^2(\mathbf{r},t)\right\rangle = \frac{1}{d}\tau_0 B_{ii}(0,0) = \frac{d-1}{d}\tau_0 \int d\mathbf{k} E(k,0), \\ D^{(\mathbf{V})} &= \tau_{\mathrm{div}\mathbf{V}} \left\langle \left(\frac{\partial \mathbf{V}(\mathbf{r},t)}{\partial \mathbf{r}}\right)^2 \right\rangle = \frac{4}{\lambda} \frac{d^2-1}{d(d+2)} D_1 D_2(\lambda), \end{aligned} \tag{11.78}$$

where coefficients

$$\begin{aligned} D_1 &= \int\limits_0^\infty d\tau D(\tau) = \int\limits_0^\infty d\tau \int d\mathbf{k}\mathbf{k}^2 E(k,\tau) \\ D_2(\lambda) &= \int\limits_0^\infty d\tau e^{-\lambda\tau} D(\tau) = \int\limits_0^\infty d\tau e^{-\lambda\tau} \int d\mathbf{k}\mathbf{k}^2 E(k,\tau). \end{aligned}$$

In particular, we have

$$\begin{aligned} D_0 &= \frac{1}{3}\tau_{\mathbf{V}} \left\langle \mathbf{V}^2(\mathbf{r},t)\right\rangle = \frac{1}{3}\tau_0 B_{ii}(0,0) = \frac{2}{3}\tau_0 \int d\mathbf{k} E(k,0), \\ D^{(\mathbf{V})} &= \tau_{\mathrm{div}\mathbf{V}} \left\langle \left(\frac{\partial \mathbf{V}(\mathbf{r},t)}{\partial \mathbf{r}}\right)^2 \right\rangle = \frac{8}{15} \frac{\tau_0}{\lambda^2} \left\langle \mathbf{u}(\mathbf{r},t)\Delta\mathbf{u}(\mathbf{r},t)\right\rangle^2 \end{aligned} \tag{11.79}$$

for low-inertia particles under the condition $\lambda\tau_0 \gg 1$ in the three-dimensional case. In the two-dimensional case, we have for $\lambda\tau_0 \gg 1$

$$\begin{aligned} D_0 &= \frac{1}{2}\tau_{\mathbf{V}} \left\langle \mathbf{V}^2(\mathbf{r},t)\right\rangle = \frac{1}{2}\tau_0 B_{ii}(0,0) = \tau_0 \int d\mathbf{k} E(k,0), \\ D^{(\mathbf{V})} &= \tau_{\mathrm{div}\mathbf{V}} \left\langle \left(\frac{\partial \mathbf{V}(\mathbf{r},t)}{\partial \mathbf{r}}\right)^2 \right\rangle = \frac{3}{2} \frac{\tau_0}{\lambda^2} \left\langle \mathbf{u}(\mathbf{r},t)\Delta\mathbf{u}(\mathbf{r},t)\right\rangle^2. \end{aligned} \tag{11.80}$$

Thus, one can see that coefficient $D^{(\mathbf{V})} \sim \sigma_{\mathbf{u}}^4$. Consequently, the vortex component of field $\mathbf{u}(\mathbf{r},t)$ first generates the vortex component of field $\mathbf{V}(\mathbf{r},t)$ by the direct linear mechanism without advection, and then the vortex component of field $\mathbf{V}(\mathbf{r},t)$ generates the divergent component of field $\mathbf{V}(\mathbf{r},t)$ by way of the advection mechanism.

The obtained expressions are applicable under the condition

$$\frac{\sigma_{\mathbf{u}}^2\tau_0^2}{l_0^2} \ll 1.$$

By the same way it is possible to solve the problem of diffusion of the settling tracer in random hydrodynamic flows [151].

Discuss now the two-dimensional hydrodynamic flow with allowance for rotation. Such a flow is described by the equation

$$\left(\frac{\partial}{\partial t} + \mathbf{V}(\mathbf{r},t)\frac{\partial}{\partial \mathbf{r}}\right) V_i(\mathbf{r},t) = -\lambda\left[V_i(\mathbf{r},t) - u_i(\mathbf{r},t)\right] + 2\Omega\Gamma_{i\mu}V_\mu(\mathbf{r},t),$$

where

$$\Gamma = \left\|\begin{array}{cc} 0 & 1 \\ -1 & 0 \end{array}\right\|, \quad \Gamma^2 = -E,$$

and E is the unit matrix. This equation can be written in the form

$$\left(\frac{\partial}{\partial t} + \mathbf{V}(\mathbf{r},t)\frac{\partial}{\partial \mathbf{r}}\right) \mathbf{V}(\mathbf{r},t) = -\Lambda\left[\mathbf{V}(\mathbf{r},t) - \mathbf{U}(\mathbf{r},t)\right], \tag{11.81}$$

where $\Lambda = (\lambda E - 2\Omega\Gamma)$ and random velocity field assumes the form

$$\mathbf{U}(\mathbf{r},t) = \lambda\Lambda^{-1}\mathbf{u}(\mathbf{r},t), \quad \Lambda^{-1} = \frac{\lambda E + 2\Omega\Gamma}{\lambda^2 + 4\Omega^2}. \tag{11.82}$$

In the case of large λ or ($\lambda \to \infty$, or $\Omega \to \infty$), we obtain approximately

$$\mathbf{V}(\mathbf{r},t) \approx \mathbf{U}(\mathbf{r},t). \tag{11.83}$$

Note that one can introduce new vector $\mathbf{W}(\mathbf{r},t) = \Gamma\mathbf{V}(\mathbf{r},t)$, in terms of which quantity

$$\xi(\mathbf{r},t) = \frac{\partial W_i(\mathbf{r},t)}{\partial r_i} = \frac{\partial \mathbf{W}(\mathbf{r},t)}{\partial \mathbf{r}} = \frac{\partial V_2(\mathbf{r},t)}{\partial r_1} - \frac{\partial V_1(\mathbf{r},t)}{\partial r_2}$$

describes the vortex component of velocity field $\mathbf{V}(\mathbf{r},t)$.

The difference between Eqs. (11.81) and (11.72) consists in the fact that parameter Λ is now the tensor. Furthermore, field $\mathbf{U}(\mathbf{r},t)$ in Eq. (11.81) is the divergent field, and, for divergence-free field $\mathbf{u}(\mathbf{r},t)$, quantity

$$\operatorname{div}\mathbf{U}(\mathbf{r},t) = \frac{\partial \mathbf{U}(\mathbf{r},t)}{\partial \mathbf{r}} = \lambda\frac{\partial}{\partial r_k}\Lambda_{k\mu}^{-1}u_\mu(\mathbf{r},t) = \frac{2\lambda\Omega}{\lambda^2 + 4\Omega^2}\Gamma_{k\mu}\frac{\partial u_\mu(\mathbf{r},t)}{\partial r_k}$$

is related to the vortex component of field $\mathbf{u}(\mathbf{r},t)$.

As earlier, we will assume that variance $\sigma_{\mathbf{u}}^2 = \langle \mathbf{u}^2(\mathbf{r},t)\rangle$ is small, and we can linearize Eq. (11.81) in flow (11.83) for large λ or Ω. In this case, the spatial diffusion coefficient D_0 in Eq. (11.74) is independent of parameter λ and is given by the expression

$$D_0 = \frac{1}{2}\tau_{\mathbf{V}}\left\langle \mathbf{V}^2(\mathbf{r},t)\right\rangle = \frac{1}{2}\int\limits_0^\infty d\tau B_{ii}(0,\tau)\cos 2\Omega\tau = \frac{\pi}{2}\int d\mathbf{k}\Phi(k,2\Omega), \tag{11.84}$$

where $\Phi(k,\omega)$ is the space-time spectral function (11.75) of field $\mathbf{u}(\mathbf{r},t)$. As regards the diffusion coefficient $D^{(\mathbf{V})}$, it can be expressed in the form [153]:

$$D^{(\mathbf{V})} = \frac{4\lambda^2\Omega^2}{\left(\lambda^2+4\Omega^2\right)^2}\int\limits_0^\infty d\tau e^{-\lambda\tau}\cos 2\Omega\tau D(\tau). \tag{11.85}$$

If $\{\lambda,\Omega\}\tau_0 \gg 1$, then

$$D^{(\mathbf{V})} = \frac{4\lambda^3\Omega^2 D(0)}{\left(\lambda^2+4\Omega^2\right)^3} = \begin{cases} 4\Omega^2 D(0)/\lambda^3, & \text{if} \quad \lambda \gg \Omega, \\ \lambda^3 D(0)/16\Omega^4, & \text{if} \quad \lambda \ll \Omega, \end{cases} \tag{11.86}$$

where, as earlier,

$$D(0) = \int d\mathbf{k}k^2 E(k,0) = -\left\langle \mathbf{u}(\mathbf{r},t)\Delta\mathbf{u}(\mathbf{r},t)\right\rangle.$$

Thus, in the framework of the problem under consideration, the generation of the divergent component of field $\mathbf{V}(\mathbf{r},t)$ is described under the conditions $\{\lambda,\Omega\}\tau_0 \gg 1$ in terms of the linear equation without allowance for advective terms. If $\lambda \gg \Omega$ in addition to the above conditions, then one should take into account consequent corrections whose order of magnitude is $\sigma_{\mathbf{u}}^4$ (11.80) and which can appear sometimes comparable with (11.86); in this case, we obtain the expression

$$\begin{aligned} D^{(\mathbf{V})} &= \frac{3}{2}\frac{\tau_0}{\lambda^2}\left\langle \mathbf{u}(\mathbf{r},t)\Delta\mathbf{u}(\mathbf{r},t)\right\rangle^2 - \frac{4\Omega^2}{\lambda^3}\left\langle \mathbf{u}(\mathbf{r},t)\Delta\mathbf{u}(\mathbf{r},t)\right\rangle \\ &= -\frac{4\Omega^2}{\lambda^3}\left\langle \mathbf{u}(\mathbf{r},t)\Delta\mathbf{u}(\mathbf{r},t)\right\rangle\left\{1-\frac{3\lambda\tau_0}{2\Omega^2}\left\langle \mathbf{u}(\mathbf{r},t)\Delta\mathbf{u}(\mathbf{r},t)\right\rangle\right\}. \end{aligned} \tag{11.87}$$

11.3 Additional factors

Above, we considered the simplest statistical problems on tracer diffusion in random velocity field in the absence of regular flows and molecular diffusion. Moreover, our statistical description used the approximation of random field delta-correlated in time. All unaccounted factors act beginning from certain time, so that the above results hold only during the initial stage of diffusion. Furthermore, these factors can give rise to new physical effects. In this section, we outline these additional problems for the divergence-free (noncompressible) velocity field.

11.3.1 Plane-parallel mean shear

Particle dynamics in the presence of the plane-parallel hydrodynamic flow was considered in Sect. 11.2.1.

If we include in consideration the field of tracer concentration gradient, we obtain a more complete pattern of tracer diffusion. In this case, Eq. (11.62) is replaced for the two-dimensional problem with allowance for linear shear $\mathbf{u}_0(\mathbf{r},t) = \alpha y\mathbf{l}$, $\mathbf{l} = (1,0)$ with the equation [180, 181]

$$\begin{aligned} \frac{\partial}{\partial t}P(t,\mathbf{r};\rho,\mathbf{p}) &= \left[-\alpha y\frac{\partial}{\partial x} + D_0\Delta\right]P(t,\mathbf{r};\rho,\mathbf{p}) \\ &+ \left\{\alpha p_x\frac{\partial}{\partial p_y} + \frac{1}{8}D^{\mathrm{s}}\left(3\frac{\partial^2}{\partial\mathbf{p}^2}\mathbf{p}^2 - 2\frac{\partial^2}{\partial p_k\partial p_l}p_k p_l\right)\right\}P(t,\mathbf{r};\rho,\mathbf{p}). \end{aligned} \tag{11.88}$$

Solution to Eq. (11.88) can be again represented in the form of integral (11.63), where the Lagrangian probability densities $P(t,\mathbf{r}|\mathbf{r}_0)$ and $P(t,\mathbf{p}|\mathbf{r}_0)$ satisfy the equations

$$\frac{\partial}{\partial t}P(t,\mathbf{r}|\mathbf{r}_0) = \left[-\alpha y\frac{\partial}{\partial x}+D_0\Delta\right]P(t,\mathbf{r}|\mathbf{r}_0), \quad P(0,\mathbf{r}|\mathbf{r}_0)=\delta(\mathbf{r}-\mathbf{r}_0); \tag{11.89}$$

$$\begin{aligned}\frac{\partial}{\partial t}P(t,\mathbf{p}|\mathbf{r}_0) &= \left\{\alpha p_x\frac{\partial}{\partial p_y}+\frac{1}{8}D^{\mathrm{s}}\left(3\frac{\partial^2}{\partial\mathbf{p}^2}\mathbf{p}^2-2\frac{\partial^2}{\partial p_k\partial p_l}p_kp_l\right)\right\}P(t,\mathbf{p}|\mathbf{r}_0),\\ P(0,\mathbf{p}|\mathbf{r}_0) &= \delta\left(\mathbf{p}-\mathbf{p}_0(\mathbf{r}_0)\right).\end{aligned} \tag{11.90}$$

Particle diffusion described by Eq. (11.89) was considered earlier. From Eq. (11.90) follows that the average gradient of tracer concentration field is no more invariant; instead, it varies in accordance with the solution to the problem in the absence of velocity field fluctuations

$$\langle p_x(t)\rangle = p_x(0), \quad \langle p_y(t)\rangle = p_y(0)-\alpha p_x(0)t.$$

As regards the second moments of the gradient, they satisfy the system of equations

$$\begin{aligned}\frac{d}{dt}\left\langle p^2(t)\right\rangle &= D^{\mathrm{s}}\left\langle p^2(t)\right\rangle-2\alpha\left\langle p_x(t)p_y(t)\right\rangle,\\ \frac{d}{dt}\left\langle p_x(t)p_y(t)\right\rangle &= -\frac{1}{2}D^{\mathrm{s}}\left\langle p_x(t)p_y(t)\right\rangle-\alpha\left\langle p_x^2(t)\right\rangle,\\ \frac{d}{dt}\left\langle p_x^2(t)\right\rangle &= \frac{3}{4}D^{\mathrm{s}}\left\langle p^2(t)\right\rangle-\frac{1}{2}\left\langle p_x^2(t)\right\rangle,\end{aligned} \tag{11.91}$$

following from Eq. (11.90). Assuming that solution to system (11.91) has the exponential form $e^{\lambda t}$, we obtain the characteristic equation in λ

$$\left(\lambda+\frac{1}{2}D^{\mathrm{s}}\right)^2\left(\lambda-D^{\mathrm{s}}\right)=\frac{3}{2}\alpha^2D^{\mathrm{s}}, \tag{11.92}$$

whose roots essentially depend of parameter α/D^{s}.

If this parameter is small $\alpha/D^{\mathrm{s}}\ll 1$, the roots of Eq. (11.92) can be approximately represented by the formulas

$$\lambda_1 = D_2^s+\frac{2\alpha^2}{3D_2^s}, \quad \lambda_2=-\frac{1}{2}D_2^s+i|\alpha|, \quad \lambda_3=-\frac{1}{2}D_2^s-i|\alpha|.$$

Consequently, for times t satisfying the condition $D^{\mathrm{s}}t\gg 2$, the random factor completely governs the solution to the problem. This means that the effect of velocity field fluctuations completely predominates the effect of the weak gradient of linear shear.

In the other limiting case $\alpha/D^{\mathrm{s}}\gg 1$, the roots of Eq. (11.92) are

$$\lambda_1=\left(\frac{3}{2}\alpha^2D^{\mathrm{s}}\right)^{1/3}, \quad \lambda_2=\left(\frac{3}{2}\alpha^2D^{\mathrm{s}}\right)^{1/3}e^{i(2/3)\pi}, \quad \lambda_2=\left(\frac{3}{2}\alpha^2D^{\mathrm{s}}\right)^{1/3}e^{-i(2/3)\pi}.$$

Real parts of roots λ_2 and λ_3 are negative; for this reason, the asymptotic solution to system (11.91) for $\left(\frac{3}{2}\alpha^2D^{\mathrm{s}}\right)^{1/3}t\gg 1$ has the form

$$\left\langle p^2(t)\right\rangle \sim \exp\left\{\left(\frac{3}{2}\alpha^2D_2^s\right)^{1/3}t\right\}.$$

Consequently, even small fluctuations of the velocity field appear the defining factor in the presence of strong gradient of shearing flow.

11.3.2 Effect of molecular diffusion

As we mentioned earlier, in the presence of velocity field fluctuations, the initially smooth distribution of tracer becomes increasingly inhomogeneous with time, there appear changes on increasingly shorter scales, and concentration spatial gradients sharpen. In actuality, molecular diffusion smooths these processes, so that the mentioned behavior of tracer concentration holds only for limited temporal intervals.

In the presence of molecular diffusion, the tracer diffusion is described in terms of the second-order partial differential equation (11.1) that do not allow deriving the equation for the one-point probability density. In this case, one must resort to different approximate methods (see, e.g., [41, 71, 126, 290]) or to numerical simulations. The first attempt of simulating the effect of molecular diffusion on the tracer field cluster structure is described in paper [198].

Applicability condition of the neglect of molecular diffusion

We estimate the time during which the effect of molecular diffusion remains insignificant by the simplest example of the two-dimensional divergence-free flow [180, 181].

A consequence of Eq. (11.1), page 234 is the circumstance that quantity $\rho^n(\mathbf{r},t)$, $n = 1, 2, ...$ will now satisfy the unclosed equation

$$\left(\frac{\partial}{\partial t} + \frac{\partial}{\partial \mathbf{r}}\mathbf{u}(\mathbf{r},t)\right)\rho^n(\mathbf{r},t) = \mu\Delta\rho^n(\mathbf{r},t) - \mu n(n-1)\rho^{n-2}(\mathbf{r},t)\mathbf{p}^2(\mathbf{r},t).$$

Averaging this equation over an ensemble of velocity field realizations, we obtain the unclosed equation

$$\frac{\partial}{\partial t}\left\langle\rho^n(\mathbf{r},t)\right\rangle = (D_0+\mu)\,\Delta\left\langle\rho^n(\mathbf{r},t)\right\rangle - \mu n(n-1)\left\langle\rho^{n-2}(\mathbf{r},t)\mathbf{p}^2(\mathbf{r},t)\right\rangle. \tag{11.93}$$

Under the condition $\mu \ll D_0$, we can rewrite Eq. (11.93) in the integral form

$$\left\langle\rho^n(\mathbf{r},t)\right\rangle = e^{D_0 t\Delta}\rho_0^n(\mathbf{r}) - \mu n(n-1)\int\limits_0^t d\tau e^{D_0(t-\tau)\Delta}\left\langle\rho^{n-2}(\mathbf{r},\tau)\mathbf{p}^2(\mathbf{r},\tau)\right\rangle. \tag{11.94}$$

To estimate the last term in Eq. (11.94), we use Eq. (11.64) that was derived for the case of absent molecular diffusion. In this way, we can derive the closed equation for quantity $\left\langle\rho^{n-2}(\mathbf{r},\tau)\mathbf{p}^2(\mathbf{r},\tau)\right\rangle$; the solution to this equation has the form

$$\left\langle\rho^{n-2}(\mathbf{r},t)\mathbf{p}^2(\mathbf{r},t)\right\rangle = e^{D^{\mathrm{s}}t + D_0 t\Delta}\rho_0^{n-2}(\mathbf{r})\mathbf{p}_0^2(\mathbf{r}). \tag{11.95}$$

Substituting Eq. (11.95) in Eq. (11.94), we can obtain the conditions under which the last term in the right-hand side of Eq. (11.94) plays only insignificant role. These conditions impose restrictions on the characteristic spatial scale of the initial concentration distribution $\mathbf{r}_0^2$ and the temporal interval. These restrictions are as follows

$$D^{\mathrm{s}}\mathbf{r}_0^2 \gg \mu n(n-1), \qquad D^{\mathrm{s}}t \ll \ln\frac{D^{\mathrm{s}}\mathbf{r}_0^2}{\mu n^2}.$$

Nonzero mean concentration gradient

Problems in which mean concentration gradient assumes nonzero values allow a more complete analysis [169, 180, 181]. This case corresponds to solving Eq. (11.1) with the following initial conditions (here, we again content ourselves with the two-dimensional case)

$$\rho_0(\mathbf{r}) = \mathbf{G}\mathbf{r}, \quad \mathbf{p}_0(\mathbf{r}) = \mathbf{G}.$$

Substituting the concentration field in the form

$$\rho(\mathbf{r}, t) = \mathbf{G}\mathbf{r} + \widetilde{\rho}(\mathbf{r}, t),$$

we obtain the equation for fluctuating portion $\widetilde{\rho}(\mathbf{r}, t)$ of the concentration field

$$\left(\frac{\partial}{\partial t} + \frac{\partial}{\partial \mathbf{r}}\mathbf{u}(\mathbf{r}, t)\right)\widetilde{\rho}(\mathbf{r}, t) = -\mathbf{G}\mathbf{u}(\mathbf{r}, t) + \mu\Delta\widetilde{\rho}(\mathbf{r}, t), \quad \widetilde{\rho}(\mathbf{r}, 0) = 0. \tag{11.96}$$

Unlike the problems discussed earlier, the solution to this problem is the random field statistically homogeneous in space, so that all one-point statistical averages are independent of $\mathbf{r}$ and steady-state probability densities exist for $t \to \infty$ for both concentration field and its gradient. Recently, this problem attracted considerable attention of both theorists and experimenters (see, e.g., [84, 114, 115, 125, 264, 265, 289]). Using simulations and phenomenological models, they discovered that the steady-state distribution has slowly decaying exponential tails. Note that authors of paper [289] discovered that the steady-state probability density of the concentration field also has slowly decaying tails.

In this case, from Eq. (11.96) follows not Eq. (11.93), but the equation

$$\frac{d}{dt}\left\langle\tilde{\rho}^n(\mathbf{r}, t)\right\rangle = n(n-1)D_0\mathbf{G}^2\left\langle\tilde{\rho}^{n-2}(\mathbf{r}, t)\right\rangle - \mu n(n-1)\left\langle\tilde{\rho}^{n-2}(\mathbf{r}, t)\widetilde{\mathbf{p}}^2(\mathbf{r}, t)\right\rangle, \tag{11.97}$$

where

$$\widetilde{\mathbf{p}}(\mathbf{r}, t) = \frac{\partial}{\partial \mathbf{r}}\widetilde{\rho}(\mathbf{r}, t) = \mathbf{p}(\mathbf{r}, t) - \mathbf{G}.$$

In the steady-state regime (for $t \to \infty$), we obtain from Eq. (11.97) that

$$\left\langle\tilde{\rho}^{n-2}(\mathbf{r}, t)\widetilde{\mathbf{p}}^2(\mathbf{r}, t)\right\rangle = \frac{D_0\mathbf{G}^2}{\mu}\left\langle\tilde{\rho}^{n-2}(\mathbf{r}, t)\right\rangle. \tag{11.98}$$

For $n = 2$ in particular, we obtain the expression for the variance of fluctuations of the concentration gradient [169]

$$\lim_{t\to\infty}\left\langle\widetilde{\mathbf{p}}^2(\mathbf{r}, t)\right\rangle = \frac{D_0\mathbf{G}^2}{\mu}. \tag{11.99}$$

Consequently, Eq. (11.98) can be rewritten in the form

$$\left\langle\tilde{\rho}^{n-2}(\mathbf{r}, t)\widetilde{\mathbf{p}}^2(\mathbf{r}, t)\right\rangle = \left\langle\widetilde{\mathbf{p}}^2(\mathbf{r}, t)\right\rangle\left\langle\tilde{\rho}^{n-2}(\mathbf{r}, t)\right\rangle, \tag{11.100}$$

i.e., quantities $\tilde{\rho}(\mathbf{r}, t)$ and $\widetilde{\mathbf{p}}^2(\mathbf{r}, t)$ appear statistically independent in the steady-state regime.

Rewrite now Eq. (11.97) in the form

$$\frac{d}{dt}\left\langle\tilde{\rho}^n(\mathbf{r}, t)\right\rangle = n(n-1)D_0\mathbf{G}^2\left\langle f(\mathbf{r}, t)\tilde{\rho}^{n-2}(\mathbf{r}, t)\right\rangle, \tag{11.101}$$

where

$$f(\mathbf{r},t) = 1 - \frac{\mu}{D_0 \mathbf{G}^2} \tilde{\mathbf{p}}^2(\mathbf{r},t).$$

Consequently, the variance of concentration is given by the expression ($\langle \tilde{\rho}(\mathbf{r},t) \rangle = 0$)

$$\left\langle \tilde{\rho}^2(\mathbf{r},t) \right\rangle = 2D_0 \mathbf{G}^2 \int\limits_0^t d\tau \left\langle f(\mathbf{r},\tau) \right\rangle . \tag{11.102}$$

In the absence of molecular diffusion, we have $f(\mathbf{r},t) \equiv 1$, so that

$$\left\langle \tilde{\rho}^2(\mathbf{r},t) \right\rangle = 2D_0 \mathbf{G}^2 t. \tag{11.103}$$

In this case, the one-point distribution of field $\tilde{\rho}(\mathbf{r},t)$ is the Gaussian distribution and field $\tilde{\rho}(\mathbf{r},t)$ and its spatial gradient are uncorrelated quantities. In the general case, Eq. (11.103) holds for sufficiently short times.

Note that the correlation function of field $\tilde{\rho}(\mathbf{r},t)$,

$$\Gamma(\mathbf{r},t) = \left\langle \tilde{\rho}(\mathbf{r}_1,t)\tilde{\rho}(\mathbf{r}_2,t) \right\rangle , \quad \mathbf{r} = \mathbf{r}_1 - \mathbf{r}_2,$$

satisfies the equation

$$\frac{\partial}{\partial t}\Gamma(\mathbf{r},t) = 2G_i G_j B_{ij}^{\text{eff}}(\mathbf{r}) + 2\left[B_{ij}^{\text{eff}}(0) - B_{ij}^{\text{eff}}(\mathbf{r}) + \mu\delta_{ij}\right] \frac{\partial^2}{\partial r_i \partial r_j}\Gamma(\mathbf{r},t)$$

that follows from Eq. (11.96); consequently, its steady-state limit

$$\Gamma(\mathbf{r}) = \lim_{t\to\infty} \Gamma(\mathbf{r},t)$$

satisfies the equation

$$G_i G_j B_{ij}^{\text{eff}}(\mathbf{r}) = -\left[B_{ij}^{\text{eff}}(0) - B_{ij}^{\text{eff}}(\mathbf{r}) + \mu\delta_{ij}\right] \frac{\partial^2}{\partial r_i \partial r_j}\Gamma(\mathbf{r}). \tag{11.104}$$

Setting $\mathbf{r} = 0$ in this equation and taking into account Eqs. (11.33) and (11.34), we arrive at equality (11.99). If we twice differentiate Eq. (11.104) with respect to $\mathbf{r}$ and then set $\mathbf{r} = 0$, we obtain the equality

$$\mu^2 \left\langle (\Delta\tilde{\rho}(\mathbf{r},t))^2 \right\rangle = \frac{1}{2} D_2^s (D_0 + \mu)\mathbf{G}^2. \tag{11.105}$$

Exact equalities (11.99) and (11.105) can appear practicable for testing different numerical schemes and checking simulated results. However, the calculation of the steady-state limit $\langle \tilde{\rho}^2(\mathbf{r},t) \rangle$ for $t \to \infty$ requires the knowledge of the time-dependent behavior of the second moment $\langle \tilde{\mathbf{p}}^2(\mathbf{r},t) \rangle$, which can be obtained only if molecular diffusion is absent. In this case, the probability density of the concentration gradient satisfies Eq. (11.62); in the problem under consideration, this equation assumes the form

$$\begin{aligned} \frac{\partial}{\partial t} P(t,\mathbf{r};\mathbf{p}) &= \frac{1}{8} D^{\text{s}} \left(3\frac{\partial^2}{\partial \mathbf{p}^2}\mathbf{p}^2 - 2\frac{\partial^2}{\partial p_k \partial p_l} p_k p_l \right) P(t,\mathbf{r};\mathbf{p}), \\ P(0,\mathbf{r};\mathbf{p}) &= \delta\left(\mathbf{p} - \mathbf{G}\right). \end{aligned} \tag{11.106}$$

Consequently, according to Eq. (11.66), we have

$$\left\langle |\widetilde{\mathbf{p}}(\mathbf{r},t)|^2 \right\rangle = \mathbf{G}^2 \left\{ e^{D^{\mathrm{s}}t} - 1 \right\}. \tag{11.107}$$

The exact formula (11.99) and Eq. (11.107) give a possibility of estimating the time required for quantity $\langle \widetilde{\mathbf{p}}^2(\mathbf{r},t) \rangle$ to approach at the steady-state regime for $t \to \infty$; namely,

$$D^{\mathrm{s}} T_0 \sim \ln\left(\frac{D_0 + \mu}{\mu}\right).$$

As a consequence, we obtain from Eq. (11.102) the following estimate of the steady-state variance of concentration field fluctuations

$$\lim_{t\to\infty} \left\langle \widetilde{\rho}^2(\mathbf{r},t) \right\rangle \sim 2\frac{D_0}{D^{\mathrm{s}}} \mathbf{G}^2 \ln\left(\frac{D_0 + \mu}{\mu}\right).$$

Taking into account the fact that $D_0 \sim \sigma_u^2 \tau_0$ and $D_0/D^{\mathrm{s}} \sim l_0^2$ (σ_u^2 is the variance of velocity field fluctuations and τ_0 and l_0 are this field temporal and spatial correlation radii, respectively), we see that time T_0, in view of its logarithmic dependence on molecular diffusion coefficient μ, can be not very long

$$T_0 \approx \frac{l_0^2}{\sigma_u^2 \tau_0} \ln\left(\frac{\sigma_u^2 \tau_0}{\mu}\right),$$

and quantity

$$\left\langle \widetilde{\rho}^2 \right\rangle \sim \mathbf{G}^2 l_0^2 \ln\left(\frac{\sigma_u^2 \tau_0}{\mu}\right) \quad \text{for} \quad \mu \ll \sigma_u^2 \tau_0.$$

11.3.3 Consideration of finite temporal correlation radius

In previous consideration, we used the approximation of the delta-correlated random field $\mathbf{u}(\mathbf{r},t)$, which is applicable under the condition that correlation radius τ_0 of random field $\mathbf{u}(\mathbf{r},t)$ is small in comparison with all temporal scales of the problem, i.e., under the condition that $\tau_0 \ll \tau_1 = L/v$, where parameter L represents the typical spatial scale. This scale can depend, for example, on characteristics of the mean flow ($L = u_0/|\boldsymbol{\nabla} u_0|$ is the typical size of eddies) or on characteristics of tracer concentration ($L = \rho/|\boldsymbol{\nabla}\rho|$). In any case, this scale decreases with time due to the appearance of small-scale structures. As a result, this scale becomes comparable with correlation radius τ_0, and the approximation of the delta-correlated random field fails. In this situation, we must take into consideration the finiteness of temporal correlation radius τ_0. The general formal mathematical classification of parameter regions in which different approximate schemes can be used is given in paper [261].

As was mentioned earlier, consideration of the finite temporal correlation radius of random field $\mathbf{u}(\mathbf{r},t)$ can be performed within the framework of the diffusion approximation. This approximation assumes that the effect of random action is insignificant on temporal scales about τ_0, i.e., the system behaves on these scales as the free system, which, of cause, imposes its own restrictions on parameters of the statistical problem.

The limiting case of steady-state random velocity field $\mathbf{u}(\mathbf{r})$ (it corresponds to the limit process $\tau_0 \to \infty$) cannot be described in the diffusion approximation. Being convenient for simulating, this case is very difficult for analytical treatment, though certain results were obtained even in this case (see, e.g., [12, 233, 234]).

As an illustration, we use the diffusion approximation to derive the equations for the settling tracer diffusion [151, 165], tracer diffusion in the plane-parallel hydrodynamic flow [143], and equations for average tracer concentration and for tracer with a constant concentration gradient.

Diffusion of settling tracer

The spread of foreign particles and inclusions whose velocities relative to the surrounding medium (rest on average) are appreciable due to buoyancy and gravity forces is important for ecology and climatology. Among such inclusions are fine dust of industrial objects, dust generated in ecological catastrophe centers, and artificial condensation/scattering centers. Velocity of inclusion sedimentation/flotation $\mathbf{v}$ is usually directed along the vertical. Its magnitude can be determined from the balance of the buoyancy and viscous friction forces; the result can be represented, for example, in the form

$$\frac{4}{3}\pi r^3 g(\rho_i - \rho_m) = 6\pi\eta r v,$$

where r is the inclusion radius; g is the acceleration of gravity; ρ_i and ρ_m are the densities of inclusion and medium, respectively; and η is the medium dynamic viscosity. If particles are additionally involved in the chaotic motion of the medium, the constant sedimentation velocity can significantly change the particle diffusion coefficient. Consider this problem in detail assuming the medium incompressible and taking into account the effect of molecular diffusion.

We start with the stochastic equation

$$\frac{d}{dt}\mathbf{r}(t) = \mathbf{v} + \mathbf{u}(\mathbf{r}, t) + \boldsymbol{\eta}(t), \quad \mathbf{r}(0) = \mathbf{r}_0, \tag{11.108}$$

where $\boldsymbol{\eta}(t)$ is the random vector delta-correlated process with the statistical characteristics (see Remark 6, page 237)

$$\langle \eta_i(t) \rangle = 0, \quad \left\langle \eta_i(t)\eta_j(t')\right\rangle = 2\mu\delta_{ij}\delta(t - t'),$$

and $\mathbf{v} = \text{const}$. We assume that field $\mathbf{u}(\mathbf{r}, t)$ is the random divergence-free Gaussian field homogeneous in space and stationary in time with the correlation and spectral tensors

$$B_{ij}(\mathbf{r}, t) = \int d\mathbf{k} E_{ij}(\mathbf{k}, t) e^{i\mathbf{kr}},$$

where

$$E_{ij}(\mathbf{k}, t) = E^s(k, t)\Delta_{ij}(\mathbf{k}), \quad \Delta_{ij}(\mathbf{k}) = \left(\delta_{ij} - \frac{k_i k_j}{k^2}\right).$$

We denote l_0 and τ_0 the spatial and temporal correlation radii of field $\mathbf{u}(\mathbf{r}, t)$, respectively.

The indicator function of particle coordinate

$$\varphi(t; \mathbf{r}) = \delta\left(\mathbf{r}(t) - \mathbf{r}\right),$$

satisfies the Liouville equation

$$\left(\frac{\partial}{\partial t} + \left[\mathbf{v} + \mathbf{u}(\mathbf{r}, t) + \boldsymbol{\eta}(t)\right]\frac{\partial}{\partial \mathbf{r}}\right)\varphi(t; \mathbf{r}) = 0, \quad \varphi(0; \mathbf{r}) = \delta\left(\mathbf{r} - \mathbf{r}_0\right). \tag{11.109}$$

Averaging Eq. (11.109) over an ensemble of realizations of random process $\boldsymbol{\eta}(t)$, we obtain that function

$$\widetilde{\varphi}(t;\mathbf{r}) = \langle \delta\left(\mathbf{r}(t) - \mathbf{r}\right)\rangle_{\boldsymbol{\eta}(t)}$$

satisfies the equation

$$\left(\frac{\partial}{\partial t} + \left[\mathbf{v} + \mathbf{u}(\mathbf{r},t)\right]\frac{\partial}{\partial \mathbf{r}}\right)\widetilde{\varphi}(t;\mathbf{r}) = \mu\Delta\widetilde{\varphi}(t;\mathbf{r}), \quad \widetilde{\varphi}(0;\mathbf{r}) = \delta\left(\mathbf{r} - \mathbf{r}_0\right). \tag{11.110}$$

Equation (11.110) is still the stochastic equation and describes the passive tracer diffusion in random velocity field with allowance for molecular diffusion.

Average now Eq. (11.110) over an ensemble of realizations of random field $\mathbf{u}(\mathbf{r},t)$. Using the Furutsu–Novikov formula, we obtain that the one-particle probability density

$$P(t;\mathbf{r}) = \left\langle \langle \delta\left(\mathbf{r}(t) - \mathbf{r}\right)\rangle_{\boldsymbol{\eta}(t)}\right\rangle_{\mathbf{u}}$$

satisfies the equation containing the variational derivative

$$\left(\frac{\partial}{\partial t} + \mathbf{v}\frac{\partial}{\partial \mathbf{r}}\right)P(t;\mathbf{r}) = \mu\Delta P(t;\mathbf{r}) - \int d\mathbf{r}'\int\limits_0^t dt' B_{ij}(\mathbf{r}-\mathbf{r}',t-t')\frac{\partial}{\partial r_i}\left\langle\frac{\delta\widetilde{\varphi}(t;\mathbf{r})}{\delta u_j(\mathbf{r}',t')}\right\rangle. \tag{11.111}$$

In the diffusion approximation, the variational derivative satisfies for $t' < t$ the equation

$$\left(\frac{\partial}{\partial t} + \mathbf{v}\frac{\partial}{\partial \mathbf{r}}\right)\left\langle\frac{\delta\widetilde{\varphi}(t;\mathbf{r})}{\delta u_j(\mathbf{r}',t')}\right\rangle = \mu\Delta\left\langle\frac{\delta\widetilde{\varphi}(t;\mathbf{r})}{\delta u_j(\mathbf{r}',t')}\right\rangle$$

with the initial condition

$$\left\langle\frac{\delta\widetilde{\varphi}(t;\mathbf{r})}{\delta u_j(\mathbf{r}',t')}\right\rangle_{t=t'} = -\delta(\mathbf{r}-\mathbf{r}')\frac{\partial}{\partial r_j}\widetilde{\varphi}(t';\mathbf{r}),$$

whose solution is

$$\left\langle\frac{\delta\widetilde{\varphi}(t;\mathbf{r})}{\delta u_j(\mathbf{r}',t')}\right\rangle = -e^{(t-t')\left(\mu\Delta - \mathbf{v}\frac{\partial}{\partial \mathbf{r}}\right)}\left[\delta(\mathbf{r}-\mathbf{r}')\frac{\partial}{\partial r_j}\widetilde{\varphi}(t';\mathbf{r})\right]. \tag{11.112}$$

On temporal scales about temporal correlation radius τ_0, function $\widetilde{\varphi}(t;\mathbf{r})$ itself is described by the initial-value problem

$$\left(\frac{\partial}{\partial t} + \mathbf{v}\frac{\partial}{\partial \mathbf{r}}\right)\widetilde{\varphi}(t;\mathbf{r}) = \mu\Delta\widetilde{\varphi}(t;\mathbf{r}), \quad \widetilde{\varphi}(t;\mathbf{r})|_{t=t'} = \widetilde{\varphi}\left(t';\mathbf{r}\right). \tag{11.113}$$

Consequently,

$$\widetilde{\varphi}(t';\mathbf{r}) = e^{-(t-t')\left(\mu\Delta - \mathbf{v}\frac{\partial}{\partial \mathbf{r}}\right)}\widetilde{\varphi}(t;\mathbf{r}). \tag{11.114}$$

Substituting Eqs. (11.112) and (11.114) in Eq. (11.111), we obtain the closed equation for the one-particle probability density

$$\begin{aligned}&\left(\frac{\partial}{\partial t} + \mathbf{v}\frac{\partial}{\partial \mathbf{r}}\right)P(t;\mathbf{r}) = \mu\Delta P(t;\mathbf{r})\\ &+ \int d\mathbf{r}'\int\limits_0^t d\tau B_{ij}(\mathbf{r}-\mathbf{r}',\tau)\frac{\partial}{\partial r_i}e^{\tau\left(\mu\Delta - \mathbf{v}\frac{\partial}{\partial \mathbf{r}}\right)}\left[\delta(\mathbf{r}-\mathbf{r}')\frac{\partial}{\partial r_j}e^{-\tau\left(\mu\Delta - \mathbf{v}\frac{\partial}{\partial \mathbf{r}}\right)}P(t;\mathbf{r})\right]\end{aligned} \tag{11.115}$$

with the initial condition $P(0;\mathbf{r}) = \delta(\mathbf{r}-\mathbf{r}_0)$. Performing the shift operations in Eq. (11.115), we obtain the operator equation in the final form

$$\left(\frac{\partial}{\partial t}+\mathbf{v}\frac{\partial}{\partial \mathbf{r}}\right)P(t;\mathbf{r}) = \mu\Delta P(t;\mathbf{r})$$
$$+\frac{\partial}{\partial r_i}\int d\mathbf{r}'\int\limits_0^t d\tau B_{ij}(\mathbf{r}-\mathbf{r}',\tau)e^{\tau\mu\Delta}\left[\delta(\mathbf{r}-\mathbf{r}'-\mathbf{v}\tau)\frac{\partial}{\partial r_j}e^{-\tau\mu\Delta}P(t;\mathbf{r})\right]. \tag{11.116}$$

Equation (11.116) allows an explicit solution. With this goal in view, we introduce the Fourier transform with respect to variable $\mathbf{r}$

$$P(t;\mathbf{r}) = \int d\mathbf{q}\widehat{P}(t;\mathbf{q})e^{i\mathbf{q}\mathbf{r}}, \quad \widehat{P}(t;\mathbf{q}) = \frac{1}{(2\pi)^d}\int d\mathbf{r}P(t;\mathbf{r})e^{-i\mathbf{q}\mathbf{r}},$$

so that function $\widehat{P}(t;\mathbf{q})$ is the characteristic function of random process $\mathbf{r}(t)$. Then, from Eq. (11.116), we obtain the equation for the Fourier transform $\widehat{P}(t;\mathbf{q})$

$$\left(\frac{\partial}{\partial t}+i\mathbf{q}\mathbf{v}\right)\widehat{P}(t;\mathbf{q}) = -\left\{\mu q^2 + q_i q_j\int\limits_0^t d\tau D_{ij}(\tau,\mathbf{q};\mathbf{v})\right\}\widehat{P}(t;\mathbf{q}), \tag{11.117}$$

where

$$D_{ij}(\tau,\mathbf{q};\mathbf{v}) = \int d\mathbf{k}E_{ij}(\mathbf{k},\tau)\exp\left\{-\mu\tau\left(k^2-2\mathbf{k}\mathbf{q}\right)+i\tau\mathbf{k}\mathbf{v}\right\}. \tag{11.118}$$

Here, we introduced the spatial spectral function of the velocity field. Integrating Eq. (11.118), we obtain the expression

$$\widehat{P}(t;\mathbf{q}) = \widehat{P}_0(\mathbf{q})\exp\left\{-\mu q^2 t - i\mathbf{q}\mathbf{v}t + q_i q_j\int\limits_0^t d\tau(t-\tau)D_{ij}(\tau,\mathbf{q};\mathbf{v})\right\}, \tag{11.119}$$

so that

$$P(t;\mathbf{r}) = \frac{1}{(2\pi)^d}\int d\mathbf{r}'P_0(\mathbf{r}')$$
$$\times\int d\mathbf{q}\exp\left\{i\mathbf{q}\left(\mathbf{r}-\mathbf{r}'-\mathbf{v}t\right)-\mu q^2 t + q_i q_j\int\limits_0^t d\tau(t-\tau)D_{ij}(\tau,\mathbf{q};\mathbf{v})\right\}. \tag{11.120}$$

For sufficiently large t, from this equation follows the asymptotic formula

$$P(t;\mathbf{r}) = \frac{1}{(2\pi)^d}\int d\mathbf{r}'P_0(\mathbf{r}')\int d\mathbf{q}\exp\left\{i\mathbf{q}\left(\mathbf{r}-\mathbf{r}'-\mathbf{v}t\right)-\mu q^2 t + q_i q_j t D_{ij}(\mathbf{v})\right\}, \tag{11.121}$$

where

$$D_{ij}(\mathbf{v}) = \int\limits_0^\infty d\tau D_{ij}(\tau,0;\mathbf{v}) = \int\limits_0^\infty d\tau\int d\mathbf{k}E_{ij}(\mathbf{k},\tau)e^{-\mu\tau k^2+i\tau\mathbf{k}\mathbf{v}}. \tag{11.122}$$

Expression (11.122) shows that there appears natural anisotropy of the diffusion tensor, which is related to the direction of tracer sedimentation vector $\mathbf{v}$.

Note that definition of the Fourier transform of characteristic function $\widehat{P}(t;\mathbf{q})$ immediately yields the expressions for statistical moments of particle coordinates

$$\langle r_k(t)\rangle = i(2\pi)^d \frac{\partial}{\partial q_k}\widehat{P}(t;\mathbf{q})\bigg|_{\mathbf{q}=0}, \quad \langle r_k(t) r_l(t)\rangle = -(2\pi)^d \frac{\partial^2}{\partial q_k \partial q_l}\widehat{P}(t;\mathbf{q})\bigg|_{\mathbf{q}=0}, \ldots,$$

where $(2\pi)^d \widehat{P}(t;0) = 1$. If we now repeatedly differentiate (11.117) and set in the result $\mathbf{q} = 0$, we obtain the equations for the moment functions of particle coordinates. In particular, the average particle trajectory $\langle \mathbf{r}(t)\rangle$ and its variance

$$\sigma_{ij}^2(t) = \langle [r_i(t) - \langle r_i(t)\rangle]\,[r_j(t) - \langle r_j(t)\rangle]\rangle$$

satisfy the equations

$$\begin{aligned}
\langle \mathbf{r}(t)\rangle &= \mathbf{r}_0 + \mathbf{v}t, \\
\frac{d}{dt}\sigma_{ij}^2(t) &= 2\left[\mu\delta_{ij} + \int_0^t d\tau D_{ij}(\tau, 0; \mathbf{v})\right],
\end{aligned}$$

from which follows that the combined turbulent diffusion coefficient is governed, for large times t, by quantity $D_{ij}(\mathbf{v})$

$$D_{ij}^{\text{tur}}(\mathbf{v}) = \lim_{t\to\infty} \frac{d}{dt}\sigma_{ij}^2(t) = 2\left[\mu\delta_{ij} + D_{ij}(\mathbf{v})\right],$$

where

$$D_{ij}(\mathbf{v}) = A(v)\delta_{ij} + B(v)\Delta_{ij}(\mathbf{v}).$$

Here,

$$A(v) = \frac{v_i v_j}{\mathbf{v}^2} D_{ij}(\mathbf{v}), \quad B(v) = \frac{1}{d-1}\left[\delta_{ij} - d\frac{v_i v_j}{\mathbf{v}^2}\right] D_{ij}(\mathbf{v}).$$

From this representation follows that, if we direct one axis of the coordinate system (the x-axis) along vector $\mathbf{v}$, then particle diffusion along different axes will be statistically independent with the diffusion coefficient

$$D_{\|}(v) = A(v)$$

along vector $\mathbf{v}$ and

$$D_{\perp} = A(v) + B(v)$$

in the transverse plane $(\mathbf{r})$. This property is related to the finite temporal correlation radius of random velocity field $\mathbf{u}(\mathbf{r}, t)$; in the approximation of the delta-correlated random field, no such property occurs. In this coordinate system, Eq. (11.121) assumes the form

$$\begin{aligned}
P(t;\mathbf{r}) = &\frac{1}{(2\pi)^d}\int d\mathbf{r}' P_0(\mathbf{r}')\int d\mathbf{q}\int d\mathbf{q}_\perp \\
&\times \exp\left\{-q(x - x' - vt) - i\mathbf{q}_\perp(\mathbf{r} - \mathbf{r}') - q^2 t\left[\mu + D_{\|}(\mathbf{v})\right] - \mathbf{q}_\perp^2 t\left[\mu + D_\perp(\mathbf{v})\right]\right\},
\end{aligned}$$

or

$$\begin{aligned}
P(t;\mathbf{r}) &= \frac{1}{\left[4\pi t\left(\mu + A(v) + B(v)\right)\right]^{d/2}}\sqrt{\frac{\mu + A(v) + B(v)}{\mu + A(v)}} \\
&\times \int d\mathbf{r}' P_0(\mathbf{r}')\exp\left\{-\frac{(x - x' - vt)^2}{4t\left[\mu + D_{\|}(\mathbf{v})\right]} - \frac{(\mathbf{r} - \mathbf{r}')^2}{4t\left[\mu + D_\perp(\mathbf{v})\right]}\right\}, \quad (11.123)
\end{aligned}$$

For particles initially distributed according to the probability density

$$P_0(\mathbf{r}) = \delta(\mathbf{r} - \mathbf{r}_0),$$

Eq. (11.123) becomes simpler and assumes the form of the Gaussian distribution

$$\begin{aligned} P(t;\mathbf{r}) &= \frac{1}{[4\pi t\,(\mu + A(v) + B(v))]^{d/2}}\sqrt{\frac{\mu + A(v) + B(v)}{\mu + A(v)}} \\ &\times \exp\left\{-\frac{(x - x_0 - vt)^2}{4t\left[\mu + D_{\parallel}(\mathbf{v})\right]} - \frac{(\mathbf{r} - \mathbf{r}_0)^2}{4t\left[\mu + D_{\perp}(\mathbf{v})\right]}\right\}, \end{aligned} \tag{11.124}$$

We estimate diffusion coefficients using the velocity field fluctuation model in which the spectral function has the form

$$E(k,t) = E(k)e^{-|t|/\tau_0},$$

where τ_0 is the temporal correlation radius of the velocity field. In this case,

$$D_{ij}(\mathbf{v}) = \frac{1}{v}\int d\mathbf{k}E(k)\Delta_{ij}(\mathbf{k})\frac{p(k,v)}{k}\frac{1}{1 + p^2(k,v)\cos^2\theta},$$

where $\cos\theta = \mathbf{k}\mathbf{v}/kv$ and we introduced the function

$$p(k,v) = \frac{kv\tau_0}{1 + \mu k^2\tau_0}.$$

Consequently, in the three-dimensional case ($d = 3$), this tensor projections on the tracer sedimentation direction and the transverse plane can be represented in the form

$$D_{\parallel}(v) = \frac{4\pi}{v}\int\limits_0^{\infty} kdkE(k)f_{\parallel}(k,v), \quad D_{\perp}(v) = \frac{4\pi}{v}\int\limits_0^{\infty} kdkE(k)f_{\perp}(k,v),$$

where

$$\begin{aligned} f_{\parallel}(k,v) &= \arctan p(k,v) + \frac{1}{p(k,v)}\left(\frac{1}{p(k,v)}\arctan p(k,v) - 1\right), \\ f_{\perp}(k,v) &= \arctan p(k,v) - \frac{1}{p(k,v)}\left(\frac{1}{p(k,v)}\arctan p(k,v) - 1\right). \end{aligned} \tag{11.125}$$

If parameter p is small (i.e., $v\tau_0 \ll l_0$, where l_0 is the spatial correlation radius of velocity field), functions $f_{\parallel}(k,v)$ and $f_{\perp}(k,v)$ appear about $2p/3$, which corresponds to isotropic diffusion independent of the sedimentation velocity; conversely, for large parameter p ($v\tau_0 \gg l_0$), we have $f_{\parallel}(k,v) = 2f_{\perp}(k,v) \cong \pi/2$. This diffusion anisotropy can be explained by the fact that tracer diffusion relative to medium turbulent motions decreases the time during which a tracer particle dwells in the region of correlated velocities. Moreover, in the isotropic divergence-free random velocity field, the transverse correlation radius is two times shorter that the longitudinal correlation radius (see, e.g., [251]), which just explains the above anisotropy of the diffusion coefficient (for $\mu\tau_0 \ll l_0^2$ diffusion tensor $D_{ij}(\mathbf{v})$ is independent of parameter μ). Emphasize once more that this anisotropy is related to the finite temporal correlation radius of the velocity field.

As was mentioned, the diffusion of tracer cloud in the Eulerian description can be considered in a similar way, including the problem with tracer source. It is obvious, the tracer diffusion in this case will be characterized by the same diffusion coefficients (11.134).

Effect of plane-parallel mean shear

Consider now the diffusion approximation of the two-dimensional problem with the plane-parallel mean flow, which is described by the dynamic system (11.42), page 244 supplemented with random terms responsible for molecular diffusion

$$\frac{d}{dt}\mathbf{r}(t) = v(y)\mathbf{l} + \mathbf{u}(\mathbf{r},t) + \boldsymbol{\eta}(t), \quad \mathbf{l} = (1,0),$$

where $\boldsymbol{\eta}(t)$ is the random vector delta-correlated process with the statistical characteristics (see Remark 6 on page 237)

$$\langle \eta_i(t) \rangle = 0, \quad \left\langle \eta_i(t)\eta_j(t') \right\rangle = 2\mu\delta_{ij}\delta(t-t'),$$

or

$$\frac{d}{dt}x(t) = v(y) + u_1(\mathbf{r},t) + \eta_1(t), \quad \frac{d}{dt}y(t) = u_2(\mathbf{r},t) + \eta_2(t) \tag{11.126}$$

in the scalar form.

In problem (11.126), the indicator function

$$\varphi(t;\mathbf{r}) = \varphi(t;x,y) = \delta(x(t)-x)\delta(y(t)-y)$$

satisfies the vector equation

$$\left(\frac{\partial}{\partial t} + v(y)\mathbf{l}\frac{\partial}{\partial \mathbf{r}}\right)\varphi(t;\mathbf{r}) = -\left[\mathbf{u}(\mathbf{r},t) + \boldsymbol{\eta}(t)\right]\frac{\partial}{\partial \mathbf{r}}\varphi(t;\mathbf{r}) \tag{11.127}$$

rather than the stochastic Liouville equation (11.43). In scalar form, this equation is as follows

$$\begin{aligned} &\left(\frac{\partial}{\partial t} + v(y)\frac{\partial}{\partial x}\right)\varphi(t;x,y) \\ &= -\left\{\left[u_1(\mathbf{r},t) + \eta_1(t)\right]\frac{\partial}{\partial x} + \left[u_2(\mathbf{r},t) + \eta_2(t)\right]\frac{\partial}{\partial y}\right\}\varphi(t;x,y). \end{aligned} \tag{11.128}$$

Averaging now Eq. (11.127) over an ensemble of realizations of random function $\boldsymbol{\eta}(t)$, we obtain that function

$$\widetilde{\varphi}(t;\mathbf{r}) = \widetilde{\varphi}(t;x,y) = \langle \delta(x(t)-x)\delta(y(t)-y) \rangle_{\boldsymbol{\eta}(t)}$$

satisfies the equation

$$\left(\frac{\partial}{\partial t} + \left[v(y)\mathbf{l} + \mathbf{u}(\mathbf{r},t)\right]\frac{\partial}{\partial \mathbf{r}}\right)\widetilde{\varphi}(t;\mathbf{r}) = \mu\Delta\widetilde{\varphi}(t;\mathbf{r}). \tag{11.129}$$

Equation (11.129) is still the stochastic equation. Averaging it over an ensemble of realizations of random field $\mathbf{u}(\mathbf{r},t)$, we obtain the equation

$$\begin{aligned} &\left(\frac{\partial}{\partial t} + v(y)\mathbf{l}\frac{\partial}{\partial \mathbf{r}}\right)P(t;\mathbf{r}) = \mu\Delta P(t;\mathbf{r}) \\ &- \int d\mathbf{r}' \int\limits_0^t dt' B_{ij}(\mathbf{r}-\mathbf{r}',t-t')\frac{\partial}{\partial r_i}\left\langle \frac{\delta\widetilde{\varphi}(t;\mathbf{r})}{\delta u_j(\mathbf{r}',t')} \right\rangle_{\mathbf{u}}. \end{aligned} \tag{11.130}$$

for the one-point probability density

$$P(t;\mathbf{r}) = \langle\widetilde{\varphi}(t;x,y)\rangle_{\mathbf{u}} = \left\langle\langle\delta(x(t)-x)\delta(y(t)-y)\rangle_{\boldsymbol{\eta}}\right\rangle_{\mathbf{u}}$$

In the diffusion approximation, the variational derivative in Eq. (11.130) satisfies for $t' < t$ the equation

$$\left(\frac{\partial}{\partial t} + v(y)\mathbf{l}\frac{\partial}{\partial \mathbf{r}}\right)\left\langle\frac{\delta\widetilde{\varphi}(t;\mathbf{r})}{\delta u_j(\mathbf{r}',t')}\right\rangle_{\mathbf{u}} = \mu\Delta\left\langle\frac{\delta\widetilde{\varphi}(t;\mathbf{r})}{\delta u_j(\mathbf{r}',t')}\right\rangle_{\mathbf{u}} \tag{11.131}$$

with the initial value

$$\left\langle\frac{\delta\widetilde{\varphi}(t;\mathbf{r})}{\delta u_j(\mathbf{r}',t')}\right\rangle_{t=t'} = -\delta(\mathbf{r}-\mathbf{r}')\frac{\partial}{\partial r_j}\widetilde{\varphi}(t';\mathbf{r}). \tag{11.132}$$

In geophysical problems, the effect of molecular diffusion coefficient μ is usually small within temporal scales about temporal correlation radius τ_0, so that we can omit the terms proportional to μ in Eqs. (11.130) and (11.131) (in any case, only the limit $\mu \to 0$ is of interest here). Consequently, we can consider that the variational derivative satisfies the simpler equation

$$\left(\frac{\partial}{\partial t} + v(y)\mathbf{l}\frac{\partial}{\partial \mathbf{r}}\right)\left\langle\frac{\delta\widetilde{\varphi}(t;\mathbf{r})}{\delta u_j(\mathbf{r}',t')}\right\rangle_{\mathbf{u}} = 0,$$

$$\left\langle\frac{\delta\widetilde{\varphi}(t;\mathbf{r})}{\delta u_j(\mathbf{r}',t')}\right\rangle_{t=t'} = -\delta(\mathbf{r}-\mathbf{r}')\frac{\partial}{\partial r_j}\widetilde{\varphi}(t';\mathbf{r}). \tag{11.133}$$

Nevertheless, we retain the term proportional to μ in Eq. (11.130), because it can be sometimes the regularizing factor. In this case, the solution to Eq. (11.133) has the form

$$\left\langle\frac{\delta\widetilde{\varphi}(t;\mathbf{r})}{\delta u_j(\mathbf{r}',t')}\right\rangle = -e^{-(t-t')v(y)\mathbf{l}\frac{\partial}{\partial \mathbf{r}}}\delta(\mathbf{r}-\mathbf{r}')\frac{\partial}{\partial r_j}\widetilde{\varphi}(t';\mathbf{r}). \tag{11.134}$$

In the diffusion approximation, quantity $\widetilde{\varphi}(t';\mathbf{r})$ in the right-hand side of Eq. (11.134) can be determined from the initial dynamic system (11.129) with absent fluctuating term and the term proportional to μ

$$\left(\frac{\partial}{\partial t} + v(y)\mathbf{l}\frac{\partial}{\partial \mathbf{r}}\right)\widetilde{\varphi}(t;\mathbf{r}) = 0, \quad \widetilde{\varphi}(0;\mathbf{r}) = \delta\left(\mathbf{r}-\mathbf{r}_0\right). \tag{11.135}$$

We have, consequently,

$$\widetilde{\varphi}(t';\mathbf{r}) = e^{(t-t')v(y)\mathbf{l}\frac{\partial}{\partial \mathbf{r}}}\widetilde{\varphi}(t;\mathbf{r}). \tag{11.136}$$

Substituting now Eqs. (11.134) and (11.136) in Eq. (11.130), we obtain the desired equation

$$\left(\frac{\partial}{\partial t} + v(y)\mathbf{l}\frac{\partial}{\partial \mathbf{r}}\right)P(t;\mathbf{r}) = \mu\Delta P(t;\mathbf{r})$$
$$+\int d\mathbf{r}'\int\limits_0^t d\tau B_{ij}(\mathbf{r}-\mathbf{r}',\tau)\frac{\partial}{\partial r_i}\left\{e^{-\tau v(y)\mathbf{l}\frac{\partial}{\partial \mathbf{r}}}\delta(\mathbf{r}-\mathbf{r}')\frac{\partial}{\partial r_j}\left[e^{\tau v(y)\mathbf{l}\frac{\partial}{\partial \mathbf{r}}}P(t;\mathbf{r})\right]\right\}. \tag{11.137}$$

We can perform the integration over $\mathbf{r}'$ in Eq. (11.137) to obtain the equation

$$\left(\frac{\partial}{\partial t} + v(y)\mathbf{l}\frac{\partial}{\partial \mathbf{r}}\right) P(t;\mathbf{r}) = \mu \Delta P(t;\mathbf{r})$$
$$+\frac{\partial}{\partial r_i}\int\limits_0^t d\tau B_{ij}(\tau v(y)\mathbf{l},\tau) e^{-\tau v(y)\mathbf{l}\frac{\partial}{\partial \mathbf{r}}}\frac{\partial}{\partial r_j}\left[e^{\tau v(y)\mathbf{l}\frac{\partial}{\partial \mathbf{r}}} P(t;\mathbf{r})\right]. \tag{11.138}$$

Note that the operator in the right-hand side of Eq. (11.138) can be represented in the form

$$e^{-\tau v(y)\mathbf{l}\frac{\partial}{\partial \mathbf{r}}}\frac{\partial}{\partial r_j} e^{\tau v(y)\mathbf{l}\frac{\partial}{\partial \mathbf{r}}} = \frac{\partial}{\partial r_j} + \tau \frac{dv(y)}{dy}\delta_{j,2}\mathbf{l}\frac{\partial}{\partial \mathbf{r}},$$

so that Eq. (11.138) assumes the form

$$\left(\frac{\partial}{\partial t} + v(y)\mathbf{l}\frac{\partial}{\partial \mathbf{r}}\right) P(t;\mathbf{r}) = \mu \Delta P(t;\mathbf{r})$$
$$+\frac{\partial}{\partial r_i}\left(D_{ij}^{(1)}(\mathbf{r},t)\frac{\partial}{\partial r_j} + D_{i2}^{(2)}(\mathbf{r},t)\frac{\partial}{\partial x}\right) P(t;\mathbf{r}), \tag{11.139}$$

where we introduced diffusion coefficients

$$D_{ij}^{(1)}(\mathbf{r},t) = \int\limits_0^t d\tau B_{ij}(\tau v(y)\mathbf{l},\tau), \quad D_{i2}^{(2)}(\mathbf{r},t) = \int\limits_0^t d\tau\, \tau B_{i2}(\tau v(y)\mathbf{l},\tau)\frac{dv(y)}{dy}. \tag{11.140}$$

Equation (11.139) adequately describes the behavior of the one-point probability density $P(t;\mathbf{r})$ even for times $t \le \tau_0$, where τ_0 is the temporal correlation radius of random field $\mathbf{u}(\mathbf{r},t)$. However, in this case, the statistical solution to Eq. (11.126) will not have the Markovian property. If we reduce the problem to the consideration of system behavior only for times $t \gg \tau_0$, we can replace the upper limits of integrals in Eq. (11.140) by infinity and rewrite Eq. (11.139) in the form

$$\left(\frac{\partial}{\partial t} + v(y)\mathbf{l}\frac{\partial}{\partial \mathbf{r}}\right) P(t;\mathbf{r}) = \mu \Delta P(t;\mathbf{r})$$
$$+\frac{\partial}{\partial r_i}\left(D_{ij}^{(1)}(\mathbf{r})\frac{\partial}{\partial r_j} + D_{i2}^{(2)}(\mathbf{r})\frac{\partial}{\partial x}\right) P(t;\mathbf{r}), \tag{11.141}$$

where the diffusion coefficients are now given by the formulas

$$D_{ij}^{(1)}(\mathbf{r}) = \int\limits_0^\infty d\tau B_{ij}(\tau v(y)\mathbf{l},\tau), \quad D_{i2}^{(2)}(\mathbf{r}) = \int\limits_0^\infty d\tau\, \tau B_{i2}(\tau v(y)\mathbf{l},\tau)\frac{dv(y)}{dy}. \tag{11.142}$$

In this case, the solution to Eq. (11.126) will be the Markovian process whose probability density will also satisfy Eq. (11.141).

Mean concentration field in the diffusion approximation

Averaging of Eq. (11.1), page 234 over an ensemble of realizations of the Gaussian random velocity field with the use of the Furutsu–Novikov formula (7.10), page 186 results in the equation

$$\left(\frac{\partial}{\partial t}+\frac{\partial}{\partial \mathbf{r}}\mathbf{u}_0(\mathbf{r},t)\right)\langle\rho(\mathbf{r},t)\rangle$$
$$=\mu\Delta\langle\rho(\mathbf{r},t)\rangle-\frac{\partial}{\partial r_i}\int d\mathbf{r}'\int_0^t dt' B_{ij}(\mathbf{r}-\mathbf{r},t-t')\left\langle\frac{\delta\rho(\mathbf{r},t)}{\delta u_j(\mathbf{r}',t')}\right\rangle. \quad (11.143)$$

In the diffusion approximation, Eq. (11.143) is the exact equation and the variational derivative satisfies, for times smaller or about the temporal correlation radius τ_0, the dynamic equation with the initial value

$$\left(\frac{\partial}{\partial t}+\frac{\partial}{\partial \mathbf{r}}\mathbf{u}_0(\mathbf{r},t)\right)\frac{\delta\rho(\mathbf{r},t)}{\delta u_j(\mathbf{r}',t')}=\mu\Delta\frac{\delta\rho(\mathbf{r},t)}{\delta u_j(\mathbf{r}',t')},$$
$$\left.\frac{\delta\rho(\mathbf{r},t)}{\delta u_j(\mathbf{r}',t')}\right|_{t=t'}=-\frac{\partial}{\partial r_j}\left\{\delta\left(\mathbf{r}-\mathbf{r}'\right)\rho(\mathbf{r},t')\right\}. \quad (11.144)$$

Within these temporal scales, concentration field $\rho(\mathbf{r},t)$ itself also satisfies the dynamic equation with the initial value

$$\left(\frac{\partial}{\partial t}+\frac{\partial}{\partial \mathbf{r}}\mathbf{u}_0(\mathbf{r},t)\right)\rho(\mathbf{r},t)=\mu\Delta\rho(\mathbf{r},t),\quad \rho(\mathbf{r},t)|_{t=t'}=\rho(\mathbf{r},t'). \quad (11.145)$$

Eliminating function $\rho(\mathbf{r},t')$ from Eqs. (11.144) and (11.145), we obtain the relationship between the variational derivative and function $\rho(\mathbf{r},t)$; as a result, Eq. (11.143) reduces to the closed equation for function $\langle\rho(\mathbf{r},t)\rangle$.

In particular,

$$\rho(\mathbf{r},t')=e^{-\mu(t-t')\Delta}\rho(\mathbf{r},t)$$

in the case of absent mean flow, and the variational derivative assumes the form

$$\frac{\delta\rho(\mathbf{r},t)}{\delta u_j(\mathbf{r}',t')}=-e^{\mu(t-t')\Delta}\frac{\partial}{\partial r_j}\left\{\delta\left(\mathbf{r}-\mathbf{r}'\right)e^{-\mu(t-t')\Delta}\rho(\mathbf{r},t)\right\}.$$

Consequently, in this case, Eq. (11.143) reduces to the following closed equation for the mean concentration field

$$\frac{\partial}{\partial t}\langle\rho(\mathbf{r},t)\rangle=\mu\Delta\langle\rho(\mathbf{r},t)\rangle$$
$$+\frac{\partial}{\partial r_i}\int d\mathbf{r}'\int_0^t d\tau B_{ij}(\mathbf{r}-\mathbf{r}',\tau)e^{\mu\tau\Delta}\frac{\partial}{\partial r_j}\left\{\delta\left(\mathbf{r}-\mathbf{r}'\right)e^{-\mu\tau\Delta}\langle\rho(\mathbf{r},t)\rangle\right\}.$$

For $t\gg\tau_0$, we can replace the upper limit of the integral with infinity; thus, we arrive at the final form of the equation

$$\frac{\partial}{\partial t}\langle\rho(\mathbf{r},t)\rangle=\mu\Delta\langle\rho(\mathbf{r},t)\rangle$$
$$+\frac{\partial}{\partial r_i}\int d\mathbf{r}'\int_0^\infty d\tau B_{ij}(\mathbf{r}-\mathbf{r}',\tau)e^{\mu\tau\Delta}\frac{\partial}{\partial r_j}\left\{\delta\left(\mathbf{r}-\mathbf{r}'\right)e^{-\mu\tau\Delta}\langle\rho(\mathbf{r},t)\rangle\right\}, \quad (11.146)$$

which can be easily solved with the use of the Fourier transformation with respect to the spatial coordinate.

Tracer with constant concentration gradient in the diffusion approximation

Using again the Furutsu–Novikov formula, we obtain in this case not Eq. (11.97), but the following equation

$$\frac{d}{dt}\langle\tilde{\rho}^n(\mathbf{r},t)\rangle = -\mu n(n-1)\left\langle\tilde{\rho}^{n-2}(\mathbf{r},t)\tilde{\mathbf{p}}^2(\mathbf{r},t)\right\rangle$$

$$+n(n-1)G_i\int d\mathbf{r}'\int_0^t dt' B_{ij}(\mathbf{r}-\mathbf{r}',t-t')\left\langle\tilde{\rho}^{n-2}(\mathbf{r},t)\frac{\delta\tilde{\rho}(\mathbf{r},t)}{\delta u_j(\mathbf{r}',t')}\right\rangle \qquad (11.147)$$

that contains the variational derivative. In the diffusion approximation, the variational derivative is given by the expression

$$\frac{\delta\tilde{\rho}(\mathbf{r},t)}{\delta u_j(\mathbf{r}',t')} = --e^{\mu(t-t')\Delta}\delta\left(\mathbf{r}-\mathbf{r}'\right)\left[\frac{\partial}{\partial r_j}\tilde{\rho}(\mathbf{r},t')-G_j\right],$$

where

$$\tilde{\rho}(\mathbf{r},t') = e^{-\mu(t-t')\Delta}\tilde{\rho}(\mathbf{r},t).$$

Consequently, Eq. (11.147) can be rewritten in the closed form

$$\frac{d}{dt}\langle\tilde{\rho}^n(\mathbf{r},t)\rangle = n(n-1)D_0(t;\mu)\mathbf{G}^2\left\langle\tilde{\rho}^{n-2}(\mathbf{r},t)\right\rangle - \mu n(n-1)\left\langle\tilde{\rho}^{n-2}(\mathbf{r},t)\tilde{\mathbf{p}}^2(\mathbf{r},t)\right\rangle, \quad (11.148)$$

where

$$D_0(t;\mu) = \frac{1}{2}\int_0^t d\tau\int d\mathbf{k}E^s(k,t)e^{-\mu\tau k^2}. \qquad (11.149)$$

The condition of applicability range of the diffusion approximation is

$$D_0(t;\mu)\mathbf{G}^2\tau_0 \ll 1. \qquad (11.150)$$

If the conditions

$$\mu\tau_0/l_0^2 \ll 1 \quad \text{and} \quad t \gg \tau_0 \qquad (11.151)$$

are satisfied in addition, then Eq. (11.148) grades into Eq. (11.97) corresponding to the approximation of the delta-correlated random field. In this case, condition (11.150) can be rewritten in the form

$$\sigma_u^2\mathbf{G}^2\tau_0 \ll 1. \qquad (11.152)$$

When we derived the estimator of steady-state value $\langle\tilde{\rho}^2(\mathbf{r},t)\rangle$, we had need for data on the random field of the concentration gradient; in particular, we used formula (11.107). The condition of its applicability is obviously the condition

$$D^s\tau_0 \ll 1,$$

which is equivalent to the condition

$$\sigma_u^2\tau_0^2 \ll l_0^2. \qquad (11.153)$$

Thus, the applicability range of the approximation of the delta-correlated (in time) random velocity field is restricted by the conditions (11.151)–(11.153). These conditions restrict both velocity field variance and molecular diffusion coefficient. In the context of geophysical flows, these restrictions appear not very strong.

Remark 10 *Tracer diffusion in random wavefields.*

As was mentioned in Remark 5, page 232, the feature of tracer diffusion in random wavefields consists in the fact that the diffusion coefficient vanishes in both approximation of the delta-correlated velocity field and diffusion approximation. In this case, one is forced to resort higher-order approximations [171] (see, also, [306]–[308]). ♦

To conclude with this chapter, we list general deductions following from the foregoing consideration.

- Statistical characteristics of the solution to the problem on passive tracer diffusion in random divergent fields may have little in common with the behavior of separate realizations. The tradition approach based on moment description appears inadequate for such problems. They require the statistical description in terms of probability density (at least the one-time or one-point probability density).

- The feature of problems on passive tracer diffusion in random divergent fields consists in the appearance of coherent statistical physical phenomena occurring with a probability of unity, such as clustering of particles and tracer concentration in the divergent velocity field. This means that the coherent phenomena occur in almost all realizations of the random concentration field.

- By themselves, the coherent phenomena are almost independent of the fluctuation model of dynamic system parameters; as a result, their temporal behavior can be described in terms of the one-time and one-point probability densities with the use of methods of statistical topography. Of cause, particular parameters characteristic of this phenomena (such as characteristic times of cluster structure formation and characteristic spatial scales) can significantly depend on the fluctuation model.

For example, in Chapter 1, page 3 we considered examples of the formation of tracer field cluster structure in random velocity field $\mathbf{u}(\mathbf{r},t) = \mathbf{v}(t)f(\mathbf{r})$, where $f(\mathbf{r})$ is the deterministic function and $\mathbf{v}(t)$ is the vector Gaussian random field. Assume that $\mathbf{v}(t)$ is the stationary Gaussian random process with the parameters

$$\langle \mathbf{v}(t)\rangle = 0, \quad B_{ij}(t-t') = \langle v_i(t)v_j(t')\rangle \quad \left(B_{ii}(0) = \left\langle \mathbf{v}^2(t)\right\rangle\right).$$

In the approximation of the delta-correlated random process $\mathbf{v}(t)$, we have

$$B_{ij}(t-t') = 2\sigma^2\delta_{ij}\tau_0\delta(t-t') \quad \left(\sigma^2\delta_{ij}\tau_0 = \int\limits_0^\infty d\tau B_{ij}(\tau)\right), \tag{11.154}$$

where σ^2 is the variance of velocity field fluctuation and τ_0 is the temporal correlation radius.

For example, in the Eulerian representation, the indicator function $\varphi(t,\mathbf{r};\rho) = \delta(\rho(\mathbf{r},t)-\rho)$ satisfies the Liouville equation

$$\left(\frac{\partial}{\partial t}+\mathbf{v}(t)f(\mathbf{r})\frac{\partial}{\partial \mathbf{r}}\right)\varphi(t,\mathbf{r};\rho) = \mathbf{v}(t)\frac{\partial f(\mathbf{r})}{\partial \mathbf{r}}\frac{\partial}{\partial \rho}\rho\varphi(t,\mathbf{r};\rho), \quad \varphi(0,\mathbf{r};\rho) = \delta(\rho_0(\mathbf{r})-\rho)$$

that we rearrange to the form

$$\frac{\partial}{\partial t}\varphi(t,\mathbf{r};\rho) = -\mathbf{v}(t)\left\{\frac{\partial}{\partial \mathbf{r}}f(\mathbf{r}) - \frac{\partial f(\mathbf{r})}{\partial \mathbf{r}}\left(1+\frac{\partial}{\partial \rho}\rho\right)\right\}\varphi(t,\mathbf{r};\rho). \tag{11.155}$$

Averaging Eq. (11.155) over an ensemble of realizations of random process $\mathbf{v}(t)$, we obtain the equation for the probability density

$$\frac{\partial}{\partial t}P(t,\mathbf{r};\rho) = \sigma^2\tau_0\left\{\frac{\partial^2}{\partial \mathbf{r}^2}f^2(\mathbf{r}) - \left(3+2\frac{\partial}{\partial\rho}\rho\right)\frac{\partial}{\partial \mathbf{r}}f(\mathbf{r})\frac{\partial f(\mathbf{r})}{\partial \mathbf{r}}\right.$$
$$\left. + f(\mathbf{r})\frac{\partial^2 f(\mathbf{r})}{\partial \mathbf{r}^2}\left(1+\frac{\partial}{\partial\rho}\rho\right) + \frac{\partial f(\mathbf{r})}{\partial \mathbf{r}}\frac{\partial f(\mathbf{r})}{\partial \mathbf{r}}\left(1+\frac{\partial}{\partial\rho}\rho\right)^2\right\}P(t,\mathbf{r},\rho). \tag{11.156}$$

If characteristic scale of function $f(\mathbf{r})$ is $\tilde{k}^{-1}$ and the function is a kind of periodic function (fast function), then we can additionally average Eq. (11.156) over this scale to obtain the equation describing slow spatial variations

$$\frac{\partial}{\partial t}P(t,\mathbf{r};\rho) = \sigma^2\tau_0\left\{\overline{f^2(\mathbf{r})}\Delta + \overline{\frac{\partial f(\mathbf{r})}{\partial \mathbf{r}}\frac{\partial f(\mathbf{r})}{\partial \mathbf{r}}}\frac{\partial^2}{\partial\rho^2}\rho^2\right\}P(t,\mathbf{r},\rho). \tag{11.157}$$

Equation (11.157) coincides with Eq. (11.51); this means that, in the context of the one-point statistical characteristics of concentration field, the above model of the velocity field is equivalent to the model of the Gaussian delta-correlated field $\mathbf{u}(\mathbf{r},t)$. Consequently, this model of velocity field fluctuations also must give rise to tracer clustering if $\mathbf{v}(t)\partial f(\mathbf{r})/\partial\mathbf{r} \neq 0$, which was observed in simulations for the simplest case of function $f(\mathbf{r}) = \sin 2(\mathbf{kr})$.

In the special case of uniform (independent of $\mathbf{r}$) initial concentration distribution $\rho_0(\mathbf{r}) = \rho_0$, this model results in the concentration field

$$\rho(\mathbf{r},t)/\rho_0 = \frac{1}{e^{T(t)}\cos^2(kx) + e^{-T(t)}\sin^2(kx)}.$$

As a consequence, the concentration averaged over fast spatial variables appears independent of random factor $T(t)$

$$\overline{\rho(\mathbf{r},t)/\rho_0} = 1.$$

In a similar way, we obtain that

$$\overline{(\rho(\mathbf{r},t)/\rho_0)^2} = \frac{1}{2}\left(e^{T(t)} + e^{-T(t)}\right),$$

so that we have

$$\left\langle\overline{(\rho(\mathbf{r},t)/\rho_0)^2}\right\rangle = \left\langle e^{T(t)}\right\rangle = \exp\left\{\frac{1}{2}\left\langle T^2(t)\right\rangle\right\}$$

for the Gaussian random process $v_x(t)$, which agrees with the lognormal probability distribution of field $\rho(\mathbf{r},t)$.

Chapter 12

Wave localization in randomly layered media

The problem on plane wave propagation in layered media is formulated in terms of the one-dimensional boundary-value problem. It attracts attention of many researchers because it is much simpler in comparison with the corresponding two- and three-dimensional problems and provides a deep insight into wave propagation in random media. In view of the fact that the one-dimensional problem allows an exact asymptotic solution, we can use it for tracing the effect of different models, medium parameters, and boundary conditions on statistical characteristics of the wavefield.

The problem in the one-dimensional statement was given in Part 1.

12.1 General remarks

Let the layer of inhomogeneous medium occupies the portion of space $L_0 < x < L$. The unit-amplitude plane wave is incident on this layer from region $x > L$. The wavefield in the inhomogeneous layer satisfies the Helmholtz equation

$$\frac{d^2}{dx^2}u(x) + k^2(x)u(x) = 0, \tag{12.1}$$

where

$$k^2(x) = k^2[1 + \varepsilon(x)]$$

and function $\varepsilon(x)$ describes inhomogeneities of the medium. In the simplest case of unmatched boundary, we assume that $k(x) = k$, i.e., $\varepsilon(x) = 0$ outside the layer and $\varepsilon(x) = \varepsilon_1(x) + i\gamma$ inside the layer, where $\varepsilon_1(x)$ is the real part responsible for wave scattering in the medium and $\gamma \ll 1$ is the imaginary part responsible for wave absorption in the medium.

The boundary conditions for Eq. (12.1) are formulated as the continuity of function $u(x)$ and derivative $\frac{d}{dx}u(x)$ at layer boundaries; these conditions can we written in the form

$$\left(1 + \frac{i}{k}\frac{d}{dx}\right)u(x)\bigg|_{x=L} = 2, \qquad \left(1 - \frac{i}{k}\frac{d}{dx}\right)u(x)\bigg|_{x=L_0} = 0. \tag{12.2}$$

For $x < L$, from Eq. (12.1) follows the equality

$$k\gamma I(x) = \frac{d}{dx}S(x), \tag{12.3}$$

where $S(x)$ is the energy flux density,

$$S(x)=\frac{i}{2k}\left[u(x)\frac{d}{dx}u^*(x)-u^*(x)\frac{d}{dx}u(x)\right],$$

and $I(x)$ is the wavefield intensity, $I(x)=|u(x)|^2$. In addition,

$$S(L)=1-|R_L|^2,\quad S(L_0)=|T_L|^2,$$

where R_L is the complex reflection coefficient from the medium layer and T_L is the complex transmission coefficient of the wave. Integrating Eq. (12.3) over the inhomogeneous layer, we obtain the equality

$$|R_L|^2+|T_L|^2+k\gamma\int\limits_{L_0}^{L}dxI(x)=1. \tag{12.4}$$

If the medium causes no wave attenuation ($\gamma=0$), then conservation of the energy flux density is expressed by the equality

$$|R_L|^2+|T_L|^2=1.$$

The imbedding method provides a possibility of reformulating boundary-value problem (12.1), (12.2) in terms of the dynamic initial value problem with respect to parameter L (geometric position of the right-hand boundary of the layer) by considering the solution of the problem as a function of this parameter. For example, reflection coefficient R_L satisfies the Riccati equation (see Appendix C, page 445)

$$\frac{d}{dL}R_L=2ikR_L+\frac{ik}{2}\varepsilon(L)\left(1+R_L\right)^2,\quad R_{L_0}=0 \tag{12.5}$$

and the wavefield in medium layer $u(x)\equiv u(x;L)$ satisfies the linear equation

$$\frac{\partial}{\partial L}u(x;L)=iku(x;L)+\frac{ik}{2}\varepsilon(L)\left(1+R_L\right)u(x;L),\quad u(x;x)=1+R_x. \tag{12.6}$$

From Eqs. (12.5) and (12.6) follow the equations for the squared modulus of the reflection coefficient $W_L=|R_L|^2$ and the wavefield intensity $I(x;L)=|u(x;L)|^2$

$$\begin{aligned}\frac{d}{dL}W_L&=-\frac{k\gamma}{2}\left[4W_L+\left(R_L+R_L^*\right)\left(1+W_L\right)\right]-\frac{ik}{2}\varepsilon_1(L)\left(R_L-R_L^*\right)\left(1-W_L\right),\\ W_{L_0}&=0,\\ \frac{\partial}{\partial L}I(x;L)&=-\frac{k\gamma}{2}\left(2+R_L+R_L^*\right)I(x;L)-\frac{ik}{2}\varepsilon_1(L)\left(R_L-R_L^*\right)I(x;L),\\ I(x;x)&=|1+R_x|^2,\end{aligned} \tag{12.7}$$

or, after rearrangement,

$$\begin{aligned}\frac{d}{dL}\ln\left(1-W_L\right)&=-\frac{k\gamma}{2}\frac{4W_L+\left(R_L+R_L^*\right)\left(1+W_L\right)}{1-W_L}-\frac{ik}{2}\varepsilon_1(L)\left(R_L-R_L^*\right),\\ \frac{\partial}{\partial L}\ln I(x;L)&=-\frac{k\gamma}{2}\left(2+R_L+R_L^*\right)-\frac{ik}{2}\varepsilon_1(L)\left(R_L-R_L^*\right).\end{aligned} \tag{12.8}$$

Excluding from Eqs. (12.8) terms containing $\varepsilon_1(L)$, we obtain the equality

$$\frac{\partial}{\partial L}\ln\frac{I(x;L)}{1-W_L}=-k\gamma\frac{|1+R_L|^2}{1-W_L}.$$

Consequently, the wavefield intensity is related to the reflection coefficient by the expression

$$I(x;L) = \frac{|1+R_x|^2(1-W_L)}{1-W_x} \exp\left\{-k\gamma \int\limits_x^L d\xi \frac{|1+R_\xi|^2}{1-W_\xi}\right\}. \tag{12.9}$$

Setting $x = L_0$ in Eq. (12.9), we express the modulus of the transmission coefficient in terms of the reflection coefficient

$$|T_L|^2 = (1-W_L) \exp\left\{-k\gamma \int\limits_{L_0}^L d\xi \frac{|1+R_\xi|^2}{1-W_\xi}\right\}. \tag{12.10}$$

In the case of non-absorptive medium, from Eq. (12.9) follows the expression

$$I(x;L) = \frac{|1+R_x|^2(1-W_L)}{1-W_x}. \tag{12.11}$$

Thus, in the case of non-absorptive medium, Eq. (12.7) can be integrated in analytic form; the resulting wavefield intensity inside the inhomogeneous layer is explicitly expressed in terms of the layer reflection coefficient.

Similarly, the field of the point source located in the layer of random medium is described in terms of the boundary-value problem for Green's function of the Helmholtz equation

$$\begin{gathered}\frac{d^2}{dx^2}G(x;x_0) + k^2[1+\varepsilon(x)]G(x;x_0) = 2ik\delta(x-x_0),\\ \left(\frac{d}{dx}+ik\right)G(x;x_0)\Big|_{x=L_0} = 0, \quad \left(\frac{d}{dx}-ik\right)G(x;x_0)\Big|_{x=L} = 0.\end{gathered} \tag{12.12}$$

Note that the problem on the source at layer boundary $x_0 = L$ coincides with boundary-value problem (12.1), (12.2) on wave incidence on the layer, i.e.,

$$G(x;L) = u(x;L).$$

The solution of boundary-value problem (12.12) for $x < x_0$ can be represented in the form (1.35), page 17

$$G(x;x_0) = \frac{[1+R_1(x_0)][1+R_2(x_0)]}{1-R_1(x_0)R_2(x_0)} \exp\left[ik\int\limits_x^{x_0} d\xi \frac{1-R_1(\xi)}{1+R_1(\xi)}\right], \tag{12.13}$$

where $R_1(L) = R_L$ is the reflection coefficient of the plane wave incident on the layer from region $x > L$ and $R_2(x_0)$ is the reflection coefficient of the wave incident on layer (x_0, L) from the homogeneous half-space $x < x_0$ (where $\varepsilon = 0$).

Problems with perfectly reflecting boundaries (at which either $G(x;x_0)$ or $\frac{d}{dx}G(x;x_0)$ vanishes) are of great interest for applications. Indeed, in the latter case, we have $R_2(x_0) = 1$ for the source located at this boundary; consequently,

$$G_{\text{ref}}(x;x_0) = \frac{2}{1-R_1(x_0)} \exp\left[ik\int\limits_x^{x_0} d\xi \frac{1-R_1(\xi)}{1+R_1(\xi)}\right], \quad x \le x_0. \tag{12.14}$$

In addition, the expression for wavefield intensity $I(x;x_0) = |G(x;x_0)|^2$ follows from Eq. (12.12)

$$k\gamma I(x;x_0) = \frac{d}{dx} S(x;x_0), \tag{12.15}$$

for $x < x_0$, where energy flux density $S(x;x_0)$ is given by the expression

$$S(x;x_0) = \frac{i}{2k}\left[G(x;x_0)\frac{d}{dx}G^*(x;x_0) - G^*(x;x_0)\frac{d}{dx}G(x;x_0)\right].$$

Using Eq. (12.13), we can represent $S(x;x_0)$ in the form ($x \le x_0$)

$$S(x;x_0) = S(x_0;x_0)\exp\left[-k\gamma\int\limits_x^{x_0} d\xi \frac{|1+R_1(\xi)|^2}{1-|R_1(\xi)|^2}\right],$$

where the energy flux density at the point of source location

$$S(x_0;x_0) = \frac{\left[1-|R_1(x_0)|^2\right]|1+R_2(x_0)|^2}{|1-R_1(x_0)R_2(x_0)|^2}. \tag{12.16}$$

Below, our concern will be with statistical problems on waves incident on random half-space ($L_0 \to -\infty$) and source-generated waves in infinite space ($L_0 \to -\infty$, $L \to \infty$) for sufficiently small absorption ($\gamma \to 0$). One can see from Eq. (12.15) that these limit processes are not commutable in the general case. Indeed, if $\gamma = 0$, then energy flux density $S(x;x_0)$ is conserved in the whole half-space $x < x_0$. However, integrating Eq. (12.15) over half-space $x < x_0$ in the case of small but finite absorption, we obtain the restriction on the energy confined in this half-space

$$k\gamma\int\limits_{-\infty}^{x_0} dx I(x;x_0) = S(x_0;x_0) = \frac{\left[1-|R_1(x_0)|^2\right]|1+R_2(x_0)|^2}{|1-R_1(x_0)R_2(x_0)|^2}. \tag{12.17}$$

Three simple statistical problems are of interest:

- Wave incidence on medium layer (of finite and infinite thickness);
- Wave source in the medium layer or infinite medium;
- Effect of boundaries on statistical characteristics of the wavefield.

All these problems can be exhaustively solved in analytic form. One can easily simulate these problems numerically and compare the simulated and analytic results.

We will assume that $\varepsilon_1(x)$ is the Gaussian delta-correlated random process with the parameters

$$\langle\varepsilon_1(L)\rangle = 0, \quad \langle\varepsilon_1(L)\varepsilon_1(L')\rangle = B_\varepsilon(L-L') = 2\sigma_\varepsilon^2 l_0\delta(L-L'), \tag{12.18}$$

where $\sigma_\varepsilon^2 \ll 1$ is the variance and l_0 is the correlation radius of random function $\varepsilon_1(L)$. This approximation means that asymptotic limit process to asymptotic case $l_0 \to 0$ in the exact problem solution with a finite correlation radius l_0 must give the result coinciding with the solution to the statistical problem with parameters (12.18).

In view of smallness of parameter σ_ε^2, all statistical effects can be divided into two types, local and accumulated due to multiple wave reflections in the medium. Our concern will be with the latter.

The statement of boundary wave problems in terms of the imbedding method clearly shows that two types of wavefield characteristics are of immediate interest. The first type of characteristics deals with quantities, such as values of the wavefield at layer boundaries (reflection and transmission coefficients R_L and T_L), field at the point of source location $G(x_0; x_0)$, and energy flux density at the point of source location $S(x_0; x_0)$. The second type of characteristics deals with statistical characteristics of wavefield intensity in the medium layer, which is the subject matter of the statistical theory of radiative transfer.

12.2 Statistics of scattered field at layer boundaries

12.2.1 Reflection and transmission coefficients

Complex coefficient of wave reflection from a medium layer satisfies the closed Riccati equation (12.5).

Represent reflection coefficient in the form $R_L = \rho_L e^{i\phi_L}$, where ρ_L is the modulus and ϕ_L is the phase. Then, starting from Eq. (12.5), we obtain the system of equations for squared modulus of the reflection coefficient $W_L = \rho_L^2 = |R_L|^2$

$$\begin{aligned} \frac{d}{dL} W_L &= -2k\gamma W_L + k\varepsilon_1(L)\sqrt{W_L}\,(1 - W_L)\sin\phi_L, \quad W_{L_0} = 0, \\ \frac{d}{dL}\phi_L &= 2k + k\varepsilon_1(L)\left\{1 + \frac{1 + W_L}{2\sqrt{W_L}}\cos\phi_L\right\}, \quad \phi_{L_0} = 0. \end{aligned} \tag{12.19}$$

Fast functions producing only little contribution to accumulated effects are omitted in the dissipative terms of system (12.19) (cf. with Eq (12.7)).

Introduce the indicator function $\varphi(L; W) = \delta(W_L - W)$ that satisfies the Liouville equation

$$\frac{\partial}{\partial L}\varphi(L; W) = 2k\gamma\frac{\partial}{\partial W}\{W\varphi(L; W)\}$$
$$-k\varepsilon_1(L)\frac{\partial}{\partial W}\left\{\sqrt{W}\,(1 - W)\sin\phi_L\varphi(L; W)\right\}, \quad \varphi(L_0; W) = \delta(W - 1). \tag{12.20}$$

Averaging this equation over an ensemble of realizations of function $\varepsilon_1(L)$ and using the Furutsu–Novikov formula (7.10), page 186, we obtain the equation for the probability density of reflection coefficient squared modulus $P(L; W) = \langle\varphi(L; W)\rangle$

$$\begin{aligned} \frac{\partial}{\partial L}P(L; W) &= 2k\gamma\frac{\partial}{\partial W}\{WP(L; W)\} \\ &-k\frac{\partial}{\partial W}\int\limits_{L_0}^{L} dL' B_\varepsilon(L - L')\sqrt{W}\,(1 - W) \\ &\times\left\langle\cos\phi_L\frac{\delta\phi_L}{\delta\varepsilon_1(L')}\varphi(L; W) + \sin\phi_L\frac{\delta\varphi(L; W)}{\delta\varepsilon_1(L')}\right\rangle, \end{aligned} \tag{12.21}$$

where $B_\varepsilon(L - L')$ is the correlation function of random process $\varepsilon_1(L)$. Substituting the correlation function (12.18) in this equation and taking into account the equalities

$$\begin{aligned} \frac{\delta\varphi(L;W)}{\delta\varepsilon_1(L-0)} &= -k\frac{\partial}{\partial W}\left\{\sqrt{W}\,(1-W)\sin\phi_L\varphi(L;W)\right\}, \\ \frac{\delta\phi_L}{\delta\varepsilon_1(L-0)} &= k\left\{1+\frac{1+W_L}{2\sqrt{W_L}}\cos\phi_L\right\} \end{aligned}$$

following immediately from Eqs. (12.20) and (12.19), we obtain the unclosed equation for probability density $P(L;W)$

$$\begin{aligned} &\frac{\partial}{\partial L}P(L;W) = 2k\gamma\frac{\partial}{\partial W}\left\{WP(L;W)\right\} \\ &-k^2\sigma_\varepsilon^2 l_0\frac{\partial}{\partial W}(1-W)\left\langle\left[\sqrt{W}\cos\phi_L+\frac{1}{2}(1+W)\cos^2\phi_L\right]\varphi(L;W)\right\rangle \\ &+k^2\sigma_\varepsilon^2 l_0\frac{\partial}{\partial W}\left\{\sqrt{W}\,(1-W)\frac{\partial}{\partial W}\left[\sqrt{W}\,(1-W)\left\langle\sin^2\phi_L\varphi(L;W)\right\rangle\right]\right\}. \end{aligned}$$

In view of the fact that the phase of the reflection coefficient

$$\phi_L = k(L-L_0)+\widetilde{\phi}_L,$$

rapidly varies on distances about the wavelength, we can additionally average this equation over fast oscillations, which will be valid under the natural restriction $k/D \gg 1$. Thus we arrive at the Fokker–Planck equation

$$\begin{aligned} &\frac{\partial}{\partial L}P(L;W) = 2k\gamma\frac{\partial}{\partial W}WP(L;W) - 2D\frac{\partial}{\partial W}W(1-W)P(L;W) \\ &+D\frac{\partial}{\partial W}W(1-W)^2\frac{\partial}{\partial W}P(L;W), \quad P(L_0,W)=\delta(W-1) \end{aligned} \tag{12.22}$$

with the diffusion coefficient

$$D = \frac{k^2\sigma_\varepsilon^2 l_0}{2}.$$

Representation of quantity W_L in the form

$$W_L = \frac{u_L-1}{u_L+1}, \quad u_L = \frac{1+W_L}{1-W_L}, \quad u_L \ge 1 \tag{12.23}$$

appears more convenient in some cases. Quantity u_L satisfies the stochastic system of equations

$$\begin{aligned} \frac{d}{dL}u_L &= -k\gamma\left(u_L^2-1\right)+k\varepsilon_1(L)\sqrt{u_L^2-1}\sin\phi_L, \quad u_{L_0}=1, \\ \frac{d}{dL}\phi_L &= 2k+k\varepsilon_1(L)\left\{1+\frac{u_L}{\sqrt{u_L^2-1}}\cos\phi_L\right\}, \quad \phi_{L_0}=0, \end{aligned}$$

and we obtain that probability density $P(L;u) = \langle\delta(u_L-u)\rangle$ of random quantity u_L satisfies the Fokker–Planck equation

$$\frac{\partial}{\partial L}P(L;u) = k\gamma\frac{\partial}{\partial u}\left(u^2-1\right)P(L;u)+D\frac{\partial}{\partial u}\left(u^2-1\right)\frac{\partial}{\partial u}P(L;u). \tag{12.24}$$

Note that quantity inverse to the diffusion coefficient defines the natural spatial scale related to medium inhomogeneities and is usually called the *localization length*

$$l_{\text{loc}} = 1/D.$$

In further analysis of wavefield statistics, we will see that this quantity determines the scale of the *dynamic wave localization* in separate realizations of the wavefield, although the *statistical localization* related to statistical characteristics of the wavefield may not occur in some cases.

Nondissipative medium (normal wave incidence)

If the medium is non-absorptive (i.e., if $\gamma = 0$), then Eq. (12.24) for the dimensionless layer thickness $\eta = D(L - L_0)$ assumes the form

$$\frac{\partial}{\partial \eta} P(\eta; u) = \frac{\partial}{\partial u} \left(u^2 - 1\right) \frac{\partial}{\partial u} P(\eta; u). \tag{12.25}$$

The solution to this equation can be easily obtained using the integral *Meller–Fock transform* (see Sect. 8.2, page 203). This solution has the form (8.45), page 204

$$P(\eta, u) = \int_0^\infty d\mu \, \mu \tanh(\pi\mu) \exp\left\{-\left(\mu^2 + \frac{1}{4}\right)\eta\right\} P_{-\frac{1}{2}+i\mu}(u), \tag{12.26}$$

where $P_{-1/2+i\mu}(x)$ is the first-order *complex index Legendre function* (*conal function*).

In view of the formula

$$\int_1^\infty \frac{dx}{(1+x)^n} P_{-\frac{1}{2}+i\mu}(x) = \frac{\pi}{\cosh(\mu\pi)} K_n(\mu),$$

where

$$K_{n+1}(\mu) = \frac{1}{2n}\left[\mu^2 + \left(n - \frac{1}{2}\right)^2\right] K_n(\mu), \quad K_1(\mu) = 1,$$

representation (12.26) offers a possibility of calculating statistical characteristics of reflection and transmission coefficients $W_L = |R_L|^2$ and $|T_L|^2 = 1 - |R_L|^2 = 2/(1 + u_L)$; in particular, we obtain the following expression for the moments of the transmission coefficient squared modulus [134]–[136]

$$\left\langle |T_L|^{2n} \right\rangle = 2^n \pi \int_0^\infty d\mu \frac{\mu \sinh(\mu\pi)}{\cosh^2(\mu\pi)} K_n(\mu) e^{-(\mu^2+1/4)\eta}. \tag{12.27}$$

Figure 12.1 shows coefficients $\langle W_L \rangle = \langle |R_L|^2 \rangle$ and $\langle |T_L|^2 \rangle = 1 - \langle |R_L|^2 \rangle$ as functions of layer thickness.

For sufficiently thick layers, namely, for $\eta = D(L - L_0) \gg 1$, from Eq. (12.27) follows the asymptotic formula for the moments of the reflection coefficient squared modulus

$$\left\langle |T_L|^{2n} \right\rangle \approx \frac{[(2n-3)!!]^2 \, \pi^2 \sqrt{\pi}}{2^{2n-1}(n-1)!} \frac{1}{\eta\sqrt{\eta}} e^{-\eta/4}.$$

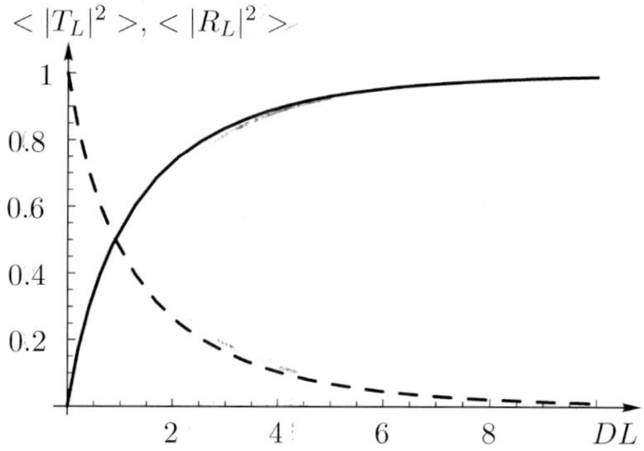

Figure 12.1: Quantities $\langle |R_L|^2 \rangle$ and $\langle |T_L|^2 \rangle$ versus layer thickness.

As may be seen, all moments of the reflection coefficient modulus $|T_L|$ vary with layer thickness according to the universal law (only the numerical factor is changed).

The fact that all moments of quantity $|T_L|$ tend to zero with increasing layer thickness means that $|R_L| \to 1$ with a probability of unity, i.e., *the half-space of randomly layered nondissipative medium completely reflects the incident wave.* It is clear that this phenomenon is independent of the statistical model of medium and the condition of applicability of the description based on the additional averaging over fast oscillations associated with the reflection coefficient phase.

In the approximation of the delta-correlated random process $\varepsilon_1(L)$, random processes W_L and u_L are obviously the Markovian processes with respect to parameter L. It is obvious that the transition probability density

$$p(u, L|u', L') = \langle \delta(u_L - u|u_{L'} = u') \rangle$$

also satisfies in this case Eq. (12.25), i.e.,

$$\frac{\partial}{\partial L} p(u, L|u', L') = D \frac{\partial}{\partial u} \left(u^2 - 1\right) \frac{\partial}{\partial u} p(u, L|u', L')$$

with the initial value

$$p(u, L''|u', L') = \delta(u - u').$$

The corresponding solution has the form (8.44), page 204, i.e.,

$$p(u, L|u', L') = \int_0^\infty d\mu\, \mu \tanh(\pi\mu) e^{-D(\mu^2+1/4)(L-L')} P_{-\frac{1}{2}+i\mu}(u) P_{-\frac{1}{2}+i\mu}(u'). \qquad (12.28)$$

At $L' = L_0$ and $u' = 1$, expression (12.28) grades into the one-point probability density (12.26).

Nondissipative medium (oblique wave incidence)

The situation remains the same even if the wave is incident on the half-space of random medium obliquely, at angle θ relative the x-axis. In this case, the reflection coefficient and wavefield in the medium satisfy the imbedding equations (C.43) derived in Appendix C, page 451

$$\begin{aligned} &\frac{d}{dL}R_L = 2ik(\cos\theta)R_L + \frac{ik}{2(\cos\theta)}\varepsilon(L)(1+R_L)^2, \quad R_{L_0}=0, \\ &\frac{\partial}{\partial L}u(x;L) = ik(\cos\theta)u(x;L) + \frac{ik}{2(\cos\theta)}\varepsilon(L)(1+R_L)u(x;L), \\ &u(x;x) = 1+R_x. \end{aligned} \tag{12.29}$$

From the first equation (12.29) follows that quantity $W_L = |R_L|^2$ in nondissipative medium satisfies the equation

$$\frac{d}{dL}W_L = \frac{k}{\cos\theta}\varepsilon_1(L)\sqrt{W_L}(1-W_L)\sin\phi_L, \quad W_{L_0}=0, \tag{12.30}$$

where ϕ_L is the phase of the reflection coefficient. It is quite obvious that, in the limit of the half-space filled with random medium ($L\to\infty$), quantity $W_L\to 1$ with a probability of unity for any random process $\varepsilon_1(L)$ and arbitrary angle of incidence θ.

In this case, reflection coefficient has the form $R_L = e^{i\phi_L}$, where phase ϕ_L satisfies the imbedding equation following from Eq. (12.29)

$$\frac{d}{dL}\phi_L = 2k(\cos\theta) + \frac{k}{\cos\theta}\varepsilon_1(L)(1+\cos\phi_L), \quad \phi_{L_0}=0. \tag{12.31}$$

Our interest here is the probability density of random quantity ϕ_L. Solution to Eq. (12.31) defines this distribution along the whole ϕ_L-axis, i.e., in interval $(-\infty,\infty)$. However, from the application viewpoint, the probability distribution in interval $(-\pi,\pi)$ appears more practicable. Such a distribution must naturally be independent of L in the limit of the half-space. To derive this distribution, it appears convenient to introduce singular function $z_L = \tan(\phi_L/2)$. This function satisfies the equation

$$\frac{d}{dL}z_L = k\cos\theta\left(1+z_L^2\right) + \frac{k}{\cos\theta}\varepsilon_1(L), \quad z_{L_0}=0. \tag{12.32}$$

Assuming that $\varepsilon_1(L)$ is the Gaussian delta-correlated random function with the parameters (12.18), we obtain that probability density

$$P(L,z) = \langle\delta(z_L - z)\rangle$$

defined on the whole axis $(-\infty,\infty)$ satisfies the Fokker–Planck equation

$$\frac{\partial}{\partial L}P(L,z) = -k\cos\theta\frac{\partial}{\partial z}\left(1+z_L^2\right)P(L,z) + \frac{2D}{\cos^2\theta}\frac{\partial^2}{\partial z^2}P(L,z). \tag{12.33}$$

In the limit of the half-space of random medium ($L_0\to-\infty$), the corresponding steady-state (independent of L) solution to the Fokker–Planck equation

$$P(z) = \lim_{L_0\to-\infty} P(L,z)$$

is described by the equation

$$-\kappa\frac{d}{dz}\left(1+z^2\right)P(z) + \frac{d^2}{dz^2}P(z) = 0, \tag{12.34}$$

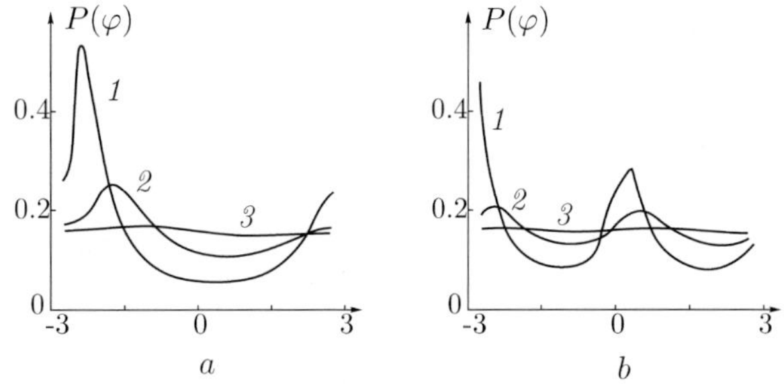

Figure 12.2: Steady-state probability density $P(\varphi)$ for (a) unmatched and (b) matched boundaries. Curves *1* to *3* correspond to $\kappa = 0.1$, 1, and 10, respectively.

where

$$\kappa = \frac{\alpha}{2}\cos^3\theta, \quad \alpha = \frac{k}{D}, \quad D = \frac{k^2\sigma_\varepsilon^2 l_0}{2}.$$

Note that, in the case of normal wave incidence ($\theta = 0$), parameter $\kappa = \alpha/2$ describes the effect of the wave number on problem statistical characteristics [262, 263, 311].

Under the condition of constant probability flux density, the solution to Eq. (12.34) has the form [112, 142]

$$P(z) = J(\kappa)\int\limits_z^\infty d\xi \exp\left\{-\kappa\xi\left[1+\frac{\xi^3}{3}+z(z+\xi)\right]\right\}, \tag{12.35}$$

where

$$J^{-1}(\kappa) = \sqrt{\frac{\pi}{\kappa}}\int\limits_0^\infty \xi^{-1/2}d\xi \exp\left\{-\kappa\left(\xi+\frac{\xi^3}{12}\right)\right\}$$

is the steady-state probability flux density. Figure 12.2a shows the corresponding probability density of the wave phase in interval $(-\pi, \pi)$

$$P(\phi) = \left.\frac{1+z^2}{2}P(z)\right|_{z=\mathrm{tang}(\phi/2)}$$

for different $\kappa = 0.1$, 1.0, and 10.

For $\kappa \gg 1$, we have asymptotically

$$P(z) = \frac{1}{\pi(1+z^2)},$$

which corresponds to the uniform distribution of the reflection coefficient phase

$$P(\phi) = \frac{1}{2\pi}, \quad -\pi < \phi < \pi.$$

In the opposite limiting case $\kappa \ll 1$, which corresponds to grazing wave incidence on the half-space ($\theta \to \pi/2$), we obtain

$$P(z) = \kappa^{1/3}\left(\frac{3}{4}\right)^{1/6}\frac{1}{\sqrt{\pi}\Gamma(1/6)}\Gamma\left(\frac{1}{3}, \frac{\kappa z^3}{3}\right),$$

where $\Gamma(\mu, z)$ is the *incomplete gamma function*. From this expression follows that

$$P(z) = \kappa^{1/3} \left(\frac{3}{4}\right)^{1/6} \frac{1}{\sqrt{\pi}\Gamma(1/6)} \left(\frac{3}{\kappa z^3}\right)^{2/3}$$

for $\kappa|z|^3 \gg 3$ and $|z| \to \infty$.

Probability distribution (12.35) offers a possibility of calculating different statistical characteristics related to the reflection coefficient. In particular, the average intensity of the wavefield at layer boundary $x = L$ is described by the asymptotic expressions

$$\langle I(L; L\rangle = 2\left\langle 1 + \cos\phi_L\right\rangle = \begin{cases} 2, & \kappa \gg 1, \\ 2(3)^{1/6}\Gamma(2/3)\kappa^{1/3}, & \kappa \ll 1. \end{cases}$$

Thus, in the case of grazing incidence of the wave, i.e., for $\theta \to \pi/2$, quantity $R_L \to -1$, so that at layer boundary $x = L$ wavefield $u(L, L) = 1 + R_L$ tends to zero. This result shows that, in the case of grazing incidence, random medium behaves as if it were a mirror. This effect is essentially a consequence of discontinuity of function $\varepsilon_1(x)$ at layer boundary $x = L$. This small step only slightly contributes to the statistics for small angles of incidence (normal incidence); however, in the case of grazing incidence, this step acts as an infinite barrier, and statistics is drastically changed. Consequently, probability distribution of the reflection coefficient phase(12.35) is informative of both wave scattering on random inhomogeneities of the medium and wave scattering on the discontinuity of function $\varepsilon_1(x)$ at layer boundary without distinguishing these effects. These effects can be distinguished by considering the problem with matched boundary within the framework of the diffusion approximation, which will be done below.

Dissipative medium

In the case of absorptive medium, Eqs. (12.22) and (12.24) cannot be solved analytically for the layer of finite thickness. Nevertheless, in the limit of half-space ($L_0 \to -\infty$), quantities W_L and u_L have the steady-state probability density [1, 184] independent of L and satisfying the equations

$$\begin{aligned} &2\left(\beta - 1 + W\right)P(W) + (1 - W)^2 \frac{d}{dW} P(W) = 0, \quad 0 < W < 1, \\ &\beta P(u) + \frac{d}{du} P(u) = 0, \quad u > 1, \end{aligned} \tag{12.36}$$

where $\beta = k\gamma/D$ is the dimensionless absorption coefficient.

Solutions to Eqs. (12.36) have the form

$$P(W) = \frac{2\beta}{(1 - W)^2} \exp\left\{-\frac{2\beta W}{1 - W}\right\}, \quad P(u) = \beta e^{-\beta(u-1)}, \tag{12.37}$$

and Fig. 12.3 shows function $P(W)$ for different values of parameter β.

The physical meaning of probability density (12.37) is obvious. It describes the statistics of the reflection coefficient from the random layer sufficiently thick for the incident wave could not reach its end because of dynamic absorption in the medium.

Using distributions (12.37), we can calculate all moments of quantity $W_L = |R_L|^2$. For example, we have for the average square of reflection coefficient modulus

$$\langle W\rangle = \int\limits_0^1 dW\, W P(W) = \int\limits_1^\infty du \frac{u - 1}{u + 1} P(u) = 1 + 2\beta e^{2\beta}\, \mathrm{Ei}(-2\beta),$$

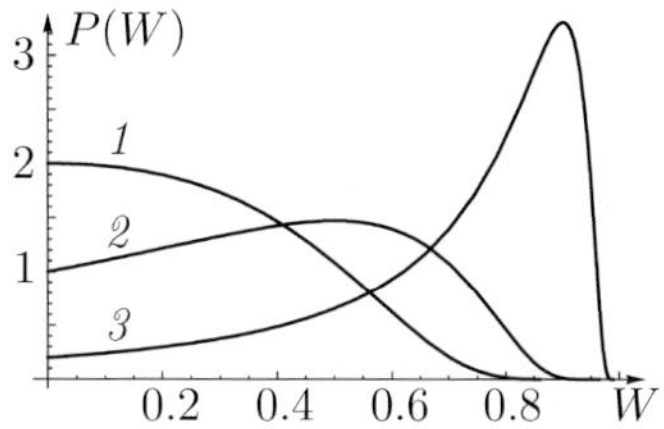

Figure 12.3: Probability density of squared reflection coefficient modulus $P(W)$. Curves *1* to *3* correspond to $\beta = 1$, 0.5, and 0.1, respectively.

where $\mathrm{Ei}(-x) = -\int\limits_x^\infty \frac{dt}{t} e^{-t}$ $(x > 0)$ is the *integral exponent*. Using asymptotic expansions of function $\mathrm{Ei}(-x)$ (see, e.g., [2])

$$\mathrm{Ei}(-x) = \begin{cases} \ln x & (x \ll 1), \\ -e^{-x}\frac{1}{x}\left(1 - \frac{1}{x}\right) & (x \gg 1), \end{cases}$$

we obtain the asymptotic expansions of quantity $\langle W \rangle = \langle |R_L|^2 \rangle$

$$\langle W \rangle \approx \begin{cases} 1 - 2\beta \ln(1/\beta), & \beta \ll 1, \\ 1/2\beta, & \beta \gg 1. \end{cases} \tag{12.38}$$

To determine higher moments of quantity $W_L = |R_L|^2$, we multiply the first equation in (12.36) by W^n and integrate the result over W from 0 to 1. As a result, we obtain the recurrence equation

$$n\left\langle W^{n+1}\right\rangle - 2(\beta + n)\left\langle W^n\right\rangle + n\left\langle W^{n-1}\right\rangle = 0 \quad (n = 1, 2, ...). \tag{12.39}$$

Using this equation, we can recursively calculate all higher moments. For example, we have for $n = 1$

$$\left\langle W^2\right\rangle = 2(\beta + 1)\left\langle W\right\rangle - 1.$$

The steady-state probability distribution can be obtained not only by limiting process $L_0 \to -\infty$, but also $L \to \infty$. Equation (12.22) was solved numerically at $\beta = 1.0$ and $\beta = 0.08$ for different initial values [135, 136]. Figure 12.4 shows moments $\langle W_L \rangle$ and $\langle W_L^2 \rangle$ calculated from the obtained solutions versus dimensionless layer thickness $\eta = D(L - L_0)$. The curves show that the probability distribution approaches the steady-state behavior relatively rapidly ($\eta \sim 1.5$) for $\beta \geq 1$ and much slower ($\eta \geq 5$) for strongly stochastic problem at $\beta = 0.08$.

Note that, for the problem under consideration, energy flux density and wavefield intensity at layer boundary $x = L$ can be expressed in terms of the reflection coefficient. Consequently, we have for $\beta \ll 1$

$$\langle S(L, L)\rangle = 1 - \langle W_L \rangle = 2\beta \ln(1/\beta), \quad \langle I(L, L)\rangle = 1 + \langle W_L \rangle = 2. \tag{12.40}$$

Taking into account that $|T_L| = 0$ in the case of the random half-space and using Eq. (12.4), we obtain that the wavefield energy contained in this half-space

$$E = D \int\limits_{-\infty}^{L} dx I(x; L),$$

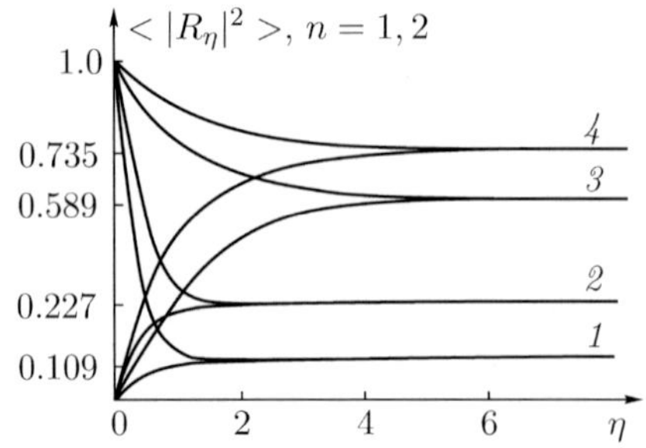

Figure 12.4: Fig. 4.6. Statistical characteristics of quantity $W_L = |R_L|^2$. Curves *1* and *2* show the second and first moments at $\beta = 1$, and curves *3* and *4* show the second and first moments at $\beta = 0.08$.

has the probability distribution

$$P(E) = \beta P(W)_{W=(1-\beta E)} = \frac{2}{E^2} \exp\left\{-\frac{2}{E}(1-\beta E)\right\} \theta(1-\beta E), \tag{12.41}$$

so that we have, in particular,

$$\langle E \rangle = 2 \ln(1/\beta) \tag{12.42}$$

for $\beta \ll 1$.

Note that probability distribution (12.41) allows limit process $\beta \to 0$; as a result, we obtain the limiting probability density

$$P(E) = \frac{2}{E^2} \exp\left\{-\frac{2}{E}\right\} \tag{12.43}$$

that decays according to the power law for large energies E. The corresponding integral distribution function has the form

$$F(E) = \exp\left\{-\frac{2}{E}\right\}.$$

A consequence of Eq. (12.43) is the fact that all moments of the total wave energy appear infinite. Nevertheless, the total energy in separate wavefield realizations can be limited to arbitrary value with a finite probability.

One can also show [134]–[136] that the expression

$$D \int\limits_{-\infty}^{L} dx \left\langle I^2(x;L) \right\rangle = \frac{1}{\beta}.$$

holds in the case of the half-space ($L_0 \to -\infty$) for $\beta \ll 1$.

Remark 11 ***Correlation function of reflection coefficient.***

Above, we considered in detail statistics of the squared modulus of refraction coefficient. Correlations of complex function R_L can be considered similarly. Consider function $\langle R_L R_{L'}^* \rangle$ with $L' < L$ as an example.

Multiplying Eq. (12.5) by $R^*_{L'}$ and averaging the result over an ensemble of realizations of random process $\varepsilon_1(L)$, we obtain the equation

$$\frac{d}{dL}\langle R_L R^*_{L'}\rangle = 2ik\langle R_L R^*_{L'}\rangle + \frac{ik}{2}\left\langle \varepsilon_1(L)(1+R_L)^2 R^*_{L'}\right\rangle$$
$$-\frac{k\gamma}{2}\left\langle (1+R_L)^2 R^*_{L'}\right\rangle, \quad \langle R_L R^*_{L'}\rangle_{L=L'} = \left\langle |R_{L'}|^2\right\rangle.$$

Using then the Furutsu–Novikov formula and the expressions for variational derivatives

$$\frac{\delta}{\delta\varepsilon_1(L)}R_L = \frac{ik}{2}(1+R_L)^2, \quad \frac{\delta}{\delta\varepsilon_1(L)}R^*_{L'} = 0,$$

we obtain, after an additional averaging over fast oscillations, the closed equation

$$\frac{d}{dL}\langle R_L R^*_{L'}\rangle = [2ik - D(3+\beta)]\langle R_L R^*_{L'}\rangle, \quad \langle R_L R^*_{L'}\rangle_{L=L'} = \left\langle |R_{L'}|^2\right\rangle,$$

whose solution is

$$\langle R_L R^*_{L'}\rangle = \left\langle |R_{L'}|^2\right\rangle \exp\left\{[2ik - D(3+\beta)](L-L')\right\}. \tag{12.44}$$

Note that quantity $\langle R_L R^*_{L'}\rangle$ by itself has no physical meaning. It describes the correlation of solutions to two different boundary-value problems corresponding to layers of thickness $(L - L_0)$ and $(L' - L_0)$. Nevertheless this quantity is convenient for comparing with simulations and, in particular, for checking ergodicity of the reflection coefficient with respect to parameter L. ♦

12.2.2 Source inside the medium layer

If the source of plane waves is located inside the medium layer, the wavefield and energy flux density at the point of source location are given by Eqs. (12.13) and (12.16). Quantities $R_1(x_0)$ and $R_2(x_0)$ are statistically independent within the framework of the model of the delta-correlated fluctuations of $\varepsilon_1(x)$, because they satisfy dynamic equations (1.34), page 17 for nonoverlapping space portions. In the case of the infinite space ($L_0 \to -\infty, L \to \infty$), probability densities of quantities $R_1(x_0)$ and $R_2(x_0)$ are given by Eq. (12.37); as a result, average intensity of the wavefield and average energy flux density at the point of source location are given by the expressions [134]–[136]

$$\langle I(x_0;x_0)\rangle = 1 + \frac{1}{\beta}, \quad \langle S(x_0;x_0)\rangle = 1. \tag{12.45}$$

The infinite increase of the average intensity at the point of source location for $\beta \to 0$ is evidence of the accumulation of wave energy in a randomly layered medium; at the same time, average energy flux density at the point of source location is independent of medium parameter fluctuations and coincides with energy flux density in free space.

For the source located at perfectly reflecting boundary $x_0 = L$, we obtain from Eqs. (12.14) and (12.16)

$$\langle I_{\text{ref}}(L;L)\rangle = 4\left(1+\frac{2}{\beta}\right), \quad \langle S_{\text{ref}}(L;L)\rangle = 4, \tag{12.46}$$

i.e., average energy flux density of the source located at the reflecting boundary is also independent of medium parameter fluctuations and coincides with energy flux density in free space.

Note the singularity of the above formulas (12.45) and (12.46) for $\beta \to 0$ from which follows that absorption (even arbitrarily small) serves regularizing factor in the problem on the point source.

Using Eq. (12.17), we can obtain the probability distribution of wavefield energy in the half-space

$$E = D \int_{-\infty}^{x_0} dx I(x; x_0).$$

In particular, for the source located at reflecting boundary, we obtain the expression

$$P_{\text{ref}}(E) = \sqrt{\frac{2}{\pi}} \frac{1}{E\sqrt{E}} \exp\left\{-\frac{2}{E}\left(1 - \frac{\beta E}{4}\right)^2\right\},$$

that allows limiting process $\beta \to 0$, which is similar to the case of wave incidence on the half-space of random medium.

12.2.3 Statistical localization of energy

In view of Eq. (12.17), the obtained results related to wavefield at fixed spatial points (at layer boundaries and at the point of source location) offer a possibility of making certain general conclusions about the behavior of the wavefield average intensity inside the random medium.

For example, from Eq. (12.17) follows the expression for average energy contained in the half-space $(-\infty, x)$

$$\langle E \rangle = D \int_{-\infty}^{x_0} dx \, \langle I(x; x_0) \rangle = \frac{1}{\beta} \langle S(x_0; x_0) \rangle \, . \qquad (12.47)$$

In the case of the plane wave ($x_0 = L$) incident on the half-space $x \le L$, Eqs. (12.40) and (12.47) result, for $\beta \ll 1$, in the expressions

$$\langle E \rangle = 2 \ln(1/\beta), \quad \langle I(L; L\rangle = 2. \qquad (12.48)$$

Consequently, the space portion

$$D l_\beta \cong \ln(1/\beta),$$

concentrates the most portion of average energy, which means that there occurs the *wavefield statistical localization* caused by wave absorption. Note that, in the absence of medium parameter fluctuations, energy localization occurs on scales about absorption length $Dl_{\text{abs}} \cong 1/\beta$. However, we have $l_{\text{abs}} \gg l_\beta$ for $\beta \ll 1$. If $\beta \to 0$, then $l_\beta \to \infty$, and statistical localization of the wavefiled disappears in the limiting case of non-absorptive medium.

In the case of the source in unbounded space, we have

$$\langle E \rangle = \frac{1}{\beta}, \quad \langle I(x_0; x_0) \rangle = 1 + \frac{1}{\beta},$$

and average energy localization is characterized, as distinct from the foregoing case, by spatial scale $D|x - x_0| \cong 1$ for $\beta \to 0$.

In a similar way, we have for the source located at reflecting boundary

$$\langle E\rangle = \frac{4}{\beta}, \quad \langle I_{\text{ref}}(L;L)\rangle = 4\left(1+\frac{2}{\beta}\right),$$

from which follows that average energy localization is characterized by the half spatial scale $D(L-x)| \cong 1/2$ for $\beta \to 0$.

In the considered problems, wavefield average energy essentially depends on parameter β and tends to infinity for $\beta \to 0$. However, this is the case only for average quantities. In our further analysis of the wavefield in random medium, we will show that the field is localized in separate realization due to the *dynamic localization* even in non-absorptive media, which corresponds to the so-called *Anderson localization* [7].

12.2.4 Diffusion approximation

Unmatched boundary

Deriving Eqs. (12.22) and (12.24), we used the delta-correlated approximation for function $\varepsilon_1(x)$ and an additional averaging over fast oscillations, which restricts the spatial correlation radius l_0 of random process $\varepsilon_1(x)$ to small values. Note that the case of the medium characterized by two spatial sales was considered in paper [108]. The effect of finite correlation radius can be estimated in the diffusion approximation. This approximation assumes that the effect of fluctuations of process $\varepsilon_1(x)$ on the wavefield dynamics is small within spatial scales about correlation radius l_0; in other words, it assumes that the wave propagates within scales about l_0 as if it would propagate in free space.

We start from the exact equation (12.21). Within the framework of the diffusion approximation, variational derivatives $\delta\phi_L/\delta\varepsilon_1(L')$ and $\delta\varphi(L;W)/\delta\varepsilon_1(L')$ within scales about l_0 satisfy the equations with initial values (wave absorption in the medium is again assumed small here)

$$\frac{\partial}{\partial L}\frac{\delta\varphi(L;W)}{\delta\varepsilon_1(L')} = 0, \quad \left.\frac{\delta\varphi(L;W)}{\delta\varepsilon_1(L')}\right|_{L=L'} = -k\frac{\partial}{\partial W}\left\{\sqrt{W}(1-W)\sin\phi_{L'}\varphi(L';W)\right\},$$
$$\frac{d}{dL}\frac{\delta\phi_L}{\delta\varepsilon_1(L')} = 0, \quad \left.\frac{\delta\phi_L}{\delta\varepsilon_1(L')}\right|_{L=L'} = k\left\{1+\frac{1+W_{L'}}{2\sqrt{W_{L'}}}\cos\phi_{L'}\right\}.$$

In addition, functions $\varphi(L;W)$, W_L, and ϕ_L satisfy, within scales about l_0, the equations

$$\frac{\partial}{\partial L}\varphi(L;W) = 0, \quad \varphi(L;W)|_{L=L'} = \varphi(L';W),$$
$$\frac{d}{dL}W_L = 0, \quad W_L|_{L=L'} = W_{L'}, \quad \frac{d}{dL}\phi_L = 2k, \quad \phi_L|_{L=L'} = \phi_{L'}.$$

Consequently, within the framework of the diffusion approximation, we have

$$\varphi(L';W) = \varphi(L;W), \quad W_{L'} = W_L, \quad \phi_{L'} = \phi_L - 2k(L-L'),$$

and variational derivatives $\delta\phi_L/\delta\varepsilon_1(L')$ and $\delta\varphi(L;W)/\delta\varepsilon_1(L')$ assume the form

$$\frac{\delta\varphi(L;W)}{\delta\varepsilon_1(L')} = -k\frac{\partial}{\partial W}\left\{\sqrt{W}(1-W)\sin\left[\phi_L - 2k(L-L')\right]\varphi(L;W)\right\},$$
$$\frac{\delta\phi_L}{\delta\varepsilon_1(L')} = k\left\{1+\frac{1+W_{L'}}{2\sqrt{W_{L'}}}\cos\left[\phi_L - 2k(L-L')\right]\right\}. \tag{12.49}$$

Substituting Eqs. (12.49) in Eq. (12.21), additionally averaging over fast oscillations, and assuming that the thickness of random layer significantly exceeds scale l_0 and wavelength, we arrive at the Fokker–Planck equation (12.22) with the diffusion coefficient

$$D(k, l_0) = \frac{k^2}{4} \int_{-\infty}^{\infty} d\xi B_\varepsilon(\xi) \cos(2k\xi) = \frac{k^2}{4} \Phi_\varepsilon(2k), \tag{12.50}$$

where $\Phi_\varepsilon(q) = \int_{-\infty}^{\infty} d\xi B_\varepsilon(\xi) e^{iq\xi}$ is the spectral function of random process $\varepsilon_1(x)$. Argument $2k$ of the spectrum of function $\varepsilon_1(x)$ physically follows from the well-known *Bragg condition* for diffraction on spatial structures (see, e.g., [28]).

The diffusion approximation assumes the smallness of the effect of fluctuations of process $\varepsilon_1(x)$ on wavefield dynamics within scales about correlation radius l_0. Under this assumption, the wavefield as a function of parameter L is the Markovian random process, which is the case if the conditions

$$D(k, l_0) l_0 \ll 1, \quad \alpha = \frac{k}{D(k, l_0)} \gg 1$$

are satisfied.

Structurally, diffusion coefficient $D(k, l_0)$ depends on parameter kl_0. If $kl_0 \ll 1$, then the delta-correlated approximation of process $\varepsilon_1(x)$ holds, in which the diffusion coefficient is independent of the model of medium and is given by the formula

$$D(k, l_0) = \frac{k^2}{4} \int_{-\infty}^{\infty} d\xi B_\varepsilon(\xi) = \frac{k^2}{4} \Phi_\varepsilon(0).$$

In the opposite limiting case $kl_0 \gg 1$, the diffusion coefficient can significantly depend on the model of medium.

Thus, the diffusion approximation holds for sufficiently small parameters $\sigma_\varepsilon^2 \ll 1$.

Matched boundary

As we noted earlier, in the case of unmatched boundary $x = L$, wave reflection occurs not only due to inhomogeneities of medium, but also due to the discontinuity of function $\varepsilon_1(x)$ at this boundary. We can separate these effects by considering the matched boundary, in which case no discontinuity of function $\varepsilon_1(x)$ is present at layer boundary $x = L$, i.e., when the wave number in free half-space $x > L$ is equal to $k_L = k\sqrt{1 + \varepsilon_1(L)}$. In this case, the wavefield is described by the boundary-value problem

$$\begin{gathered} \frac{d^2}{dx^2} u(x) + k^2(x) u(x) = 0, \\ \left(1 + \frac{i}{k(x)} \frac{d}{dx}\right) u(x)\Big|_{x=L} = 2, \quad \left(1 - \frac{i}{k(x)} \frac{d}{dx}\right) u(x)\Big|_{x=L_0} = 0, \end{gathered} \tag{12.51}$$

where

$$k^2(x) = k^2[1 + \varepsilon(x)].$$

Again, the imbedding method makes it possible to reformulate boundary-value problem (12.51) into the initial value problem with respect to parameter L, whose meaning is the position of the layer right-hand side the wave is incident on (see Appendix C, page 455). In the case of small fluctuations of function $\varepsilon_1(L)$, reflection coefficient R_L and wavefield in the layer $u(x) \equiv u(x;L)$ satisfy the equations

$$\frac{d}{dL}R_L = 2ikR_L - k\gamma R_L + \frac{\xi(L)}{2}\left(1 - R_L^2\right), \quad R_{L_0} = 0, \tag{12.52}$$

$$\begin{aligned} &\frac{\partial}{\partial L}u(x;L) = iku(x;L) - \frac{k\gamma}{2}u(x;L) + \frac{\xi(L)}{2}(1 - R_L)\,u(x;L), \\ &\quad u(x;x) = 1 + R_x. \end{aligned} \tag{12.53}$$

where $\xi(L) = \varepsilon_1'(L)$. One can see that the nonlinear term in the equation for reflection coefficient has now another structure; moreover, random inhomogeneities are described here in terms of the spatial derivative of function $\varepsilon_1(L)$. For this reason, the approximation of the delta-correlated process is inapplicable, so that the diffusion approximation appears the simplest approximation for this problem.

In the case of the Gaussian process $\varepsilon_1(x)$ with correlation function $B_\varepsilon(x)$, random process $\xi(x)$ is also the Gaussian process with the correlation function

$$B_\xi(x - x') = \langle \xi(x)\xi(x') \rangle = -\frac{\partial^2}{\partial x^2}B_\varepsilon(x - x'). \tag{12.54}$$

As earlier, consider quantity $W_L = |R_L|^2$. For this quantity, we obtain the dynamic equation

$$\frac{d}{dL}W_L = -2k\gamma W_L + \frac{\xi(L)}{2}(1 - W_L)(R_L + R_L^*), \quad W_{L_0} = 0. \tag{12.55}$$

Introduce the indicator function $\varphi(L;W) = \delta(W_L - W)$ satisfying the Liouville equation

$$\left(\frac{\partial}{\partial L} - 2k\gamma\frac{\partial}{\partial W}W\right)\varphi(L;W) = -\frac{\xi(L)}{2}\frac{\partial}{\partial W}\left\{(1 - W)(R_L + R_L^*)\,\varphi(L;W)\right\}. \tag{12.56}$$

Averaging this equation over an ensemble of realizations of function $\xi(L)$ and using the Furutsu–Novikov formula (7.10), page 186, we obtain that probability density of reflection coefficient squared modulus $P(L;W) = \langle \varphi(L;W) \rangle$ satisfies the equation

$$\begin{aligned} \left(\frac{\partial}{\partial L} - 2k\gamma\frac{\partial}{\partial W}W\right)P(L;W) &= -\frac{1}{2}\frac{\partial}{\partial W}\int\limits_{L_0}^{L} dL' B_\xi(L - L')(1 - W) \\ &\times \left\langle \left(\frac{\delta R_L}{\delta \xi(L')} + \frac{\delta R_L^*}{\delta \xi(L')}\right)\varphi(L;W) + (R_L + R_L^*)\frac{\delta \varphi(L;W)}{\delta \xi(L')} \right\rangle. \end{aligned} \tag{12.57}$$

where $B_\xi(L - L')$ is the correlation function of random process $\xi(L)$.

In the diffusion approximation, variational derivatives $\delta R_L/\delta\xi(L')$ and $\delta\varphi(L;W)/\delta\xi(L')$ within scales about l_0 satisfy the equations with initial values (wave absorption is again assumed small)

$$\begin{aligned} &\frac{\partial}{\partial L}\frac{\delta \varphi(L;W)}{\delta \xi(L')} = 0, \\ &\quad \left.\frac{\delta \varphi(L;W)}{\delta \xi(L')}\right|_{L=L'} = -\frac{1}{2}\frac{\partial}{\partial W}\left\{(1 - W)(R_{L'} + R_{L'}^*)\,\varphi(L';W)\right\}, \\ &\frac{d}{dL}\frac{\delta R_L}{\delta \xi(L')} = 2ik\frac{\delta R_L}{\delta \xi(L')}, \quad \left.\frac{\delta R_L}{\delta \xi(L')}\right|_{L=L'} = \frac{1}{2}\left(1 - R_{L'}^2\right). \end{aligned}$$

Moreover, functions $\varphi(L;W)$ and W_L themselves satisfy, within scales about l_0 the equations

$$\frac{\partial}{\partial L}\varphi(L;W) = 0, \quad \varphi(L;W)|_{L=L'} = \varphi(L';W),$$
$$\frac{d}{dL}R_L = 2ikR_L, \quad R_L|_{L=L'} = R_{L'}.$$

Consequently, we have

$$\varphi(L';W) = \varphi(L;W), \quad R_{L'} = R_L e^{-2ik(L-L')}$$

in the framework of the diffusion approximation, and variational derivatives $\delta R_L/\delta\xi(L')$ and $\delta\varphi(L;W)/\delta\xi(L')$ assume the form

$$\frac{\delta\varphi(L;W)}{\delta\xi(L')} = -\frac{1}{2}\frac{\partial}{\partial W}\left\{\left(R_L e^{-2ik(L-L')} + R_L^* e^{2ik(L-L')}\right)(1-W)\varphi(L;W)\right\},$$
$$\frac{\delta R_L}{\delta\xi(L')} = \frac{1}{2}\frac{\partial}{\partial W}\left(e^{2ik(L-L')} - R_L^2 e^{-2ik(L-L')}\right)(1-W)\varphi(L;W). \tag{12.58}$$

Substituting Eqs. (12.58) in Eq. (12.57), additionally averaging the result over fast oscillations, and assuming that the thickness of random layer significantly exceeds scale l_0 and wavelength, we arrive at the Fokker–Planck equation (12.22) with the diffusion coefficient

$$D(k,l_0) = \frac{1}{16}\int_{-\infty}^{\infty} d\eta B_\xi(\eta)\cos(2k\eta) = \frac{1}{16}\Phi_\xi(2k) = \frac{k^2}{4}\Phi_\varepsilon(2k). \tag{12.59}$$

Thus, statistics of the reflection coefficient modulus in the problem with matched boundary coincides with the corresponding statistics for unmatched boundary. This is quite natural because the step of function $k(L)$ at this boundary is small for normal wave incidence. One might expect to observe the difference only in the case of oblique wave incidence, or in the situation when averaging over fast oscillation is impossible.

In the case of non-absorptive medium, from Eqs. (12.52) and (12.55) follows that $W_L = 1$ for random half-space $x < L$ ($L_0 \to -\infty$), i.e., random half-space totally reflects the wave. A similar situation occurs when the wave is incident on the layer obliquely, at angle θ relative x-axis. In this case, Eq. (12.52) is replaced with the following equations for reflection coefficient and wavefield in the medium

$$\frac{d}{dL}R_L = 2ik(\cos\theta)R_L + \frac{\xi(L)}{2\cos^2\theta}\left(1-R_L^2\right), \quad R_{L_0} = 0,$$
$$\frac{\partial}{\partial L}u(x;L) = ik(\cos\theta)u(x;L) + \frac{\xi(L)}{2\cos^2\theta}(1-R_L)u(x;L),$$
$$u(x;x) = 1 + R_x. \tag{12.60}$$

For half-space $x < L$, we have $R_L^2 \to 1$ for grazing wave incidence, so that $R_L \to \pm 1$. Consequently, these values produce the main contribution to the statistics of the reflection coefficient phase.

Representing reflection coefficient in the form $R_L = e^{i\phi_L}$, we obtain that the phase satisfies the equation

$$\frac{d}{dL}\phi_L = 2k(\cos\theta) - \frac{\xi(L)}{2\cos^2\theta}\sin\phi_L, \quad \phi_{L_0} = 0. \tag{12.61}$$

As in the case of unmatched boundary, introduce new function $z_L = \tan(\phi_L/2)$ having singular points. It satisfies the dynamic equation

$$\frac{d}{dL} z_L = 2k(\cos\theta)\left(1 + z_L^2\right) - \frac{\xi(L)}{2\cos^2\theta} z_L, \qquad z_{L_0} = 0. \tag{12.62}$$

If we proceed with stochastic equation (12.56) as earlier, then we obtain that probability density $P(L, z) = \langle \delta(z_L - z) \rangle$ defined along the whole axis $(-\infty, \infty)$ satisfies in the diffusion approximation the Fokker–Planck equation [109]

$$\frac{\partial}{\partial L} P(L, z) = -k\cos\theta \frac{\partial}{\partial z}\left(1 + z_L^2\right) P(L, z) + \frac{2D}{\cos^2\theta} \frac{\partial}{\partial z} z \frac{\partial}{\partial z} z P(L, z), \tag{12.63}$$

where $D = k^2 \sigma_\varepsilon^2 l_0 / 2$ is, as earlier, the diffusion coefficient for the wave normally incident on the medium layer.

In the case of half-space ($L_0 \to -\infty$), steady-state (independent of L) probability density $P(z)$ satisfies the equation

$$-\kappa \frac{d}{dz}\left(1 + z_L^2\right) P(z) + \frac{2D}{\cos^2\theta} \frac{d}{dz} z \frac{d}{dz} z P(z), \tag{12.64}$$

where

$$\kappa = \frac{\alpha}{2}\cos^3\theta, \qquad \alpha = \frac{k}{D},$$

as earlier. Under the condition of constant probability flux density, the solution to this equation has the form of the following quadrature

$$P(z) = -\frac{J(\kappa)}{z} \int\limits_{z_0}^{z} \frac{dz_1}{z_1} \exp\left\{\kappa\left(z - \frac{1}{z} - z_1 + \frac{1}{z_1}\right)\right\}.$$

Constant $J(\kappa)$ is determined from the normalization condition $\int\limits_{-\infty}^{\infty} dz P(z) = 1$ and arbitrary parameter z_0 must be determined from the condition of finiteness of the quadrature for all z from interval $(-\infty, \infty)$.

As a result, we obtain

$$\begin{aligned} P(z) &= \theta(z) P_+(z) + \theta(-z) P_-(z), \\ P_+(z) &= \frac{J(\kappa)}{z} \int\limits_0^\infty \frac{ds}{1+s} \exp\left\{-\kappa s\left(z + \frac{1}{z(1+s)}\right)\right\}, \quad z > 0, \\ P_-(z) &= -\frac{J(\kappa)}{z} \int\limits_0^1 \frac{ds}{1-s} \exp\left\{\kappa s\left(z + \frac{1}{z(1-s)}\right)\right\}, \quad z < 0. \end{aligned} \tag{12.65}$$

Probability density $P(z)$ is the continuous function and

$$P_+(z = +0) = P_-(z = -0) = \frac{J(\kappa)}{\kappa},$$

where

$$\frac{1}{J(\kappa)} = \pi^2\left[J_0^2(2\kappa) + N_0^2(2\kappa)\right] = \begin{cases} \pi, & \kappa \gg 1, \\ \left[\pi^2 + 4(\ln\kappa + C)^2\right], & \kappa \ll 1. \end{cases}$$

Here, $J_0(x)$ is the *Bessel function*, $N_0(x)$ is the *Neumann function*, and C is the *Euler constant*.

Under the condition $\kappa \gg 1$, we obtain the asymptotic solution in the form

$$P(z) = \frac{1}{\pi\left(1+z^2\right)}$$

that corresponds to the uniform distribution of the reflection coefficient phase

$$P(\phi) = \frac{1}{2\pi}$$

on interval $(-\pi, \pi)$. For $\kappa \ll 1$, there is no uniform asymptotic expression for $P(\phi)$. Figure 12.2*b*, page 287 shows numerical results for $\kappa = 0.1$, 1.0, and 10.

Consider now the wavefield at boundary $x = L$ and its statistical characteristics related to fluctuations of the reflection coefficient phase in the asymptotic case $\kappa \ll 1$. The average intensity of the wavefield at boundary $x = L$ is given by the expression

$$\langle I(L;L\rangle = 2 + \langle R_L + R_L^*\rangle = 2\left\langle(1+\cos\phi_L)\right\rangle = 4\int\limits_{-\infty}^{\infty}\frac{dz}{1+z^2}P(z).$$

Consequently, we have for $\kappa \ll 1$ the equality

$$\langle I(L;L)\rangle = 2,$$

which means that statistical weights of values $R_L = +1$ and $R_L = -1$ coincide, despite the probability density is essentially different from the uniform one.

12.3 Statistical description of a wavefield in random medium

Now, we dwell on the statistical description of a wavefield in random medium (statistical theory of radiative transfer). We consider two problems of which the first concerns the wave incident on the medium layer and the second concerns the waves generated by source located in the medium.

12.3.1 Normal wave incidence on the layer of random media

In the general case of absorptive medium, the wavefield is described by the boundary-value problem (12.1), (12.2), page 278. We introduce complex opposite waves

$$u(x) = u_1(x) + u_2(x), \quad \frac{d}{dx}u(x) = -ik[u_1(x) - u_2(x)],$$

related to the wavefield through the relationships (1.23), page 14

$$\begin{aligned} u_1(x) &= \frac{1}{2}\left[1 + \frac{i}{k}\frac{d}{dx}\right]u(x), \quad u_1(L) = 1, \\ u_2(x) &= \frac{1}{2}\left[1 - \frac{i}{k}\frac{d}{dx}\right]u(x), \quad u_2(L_0) = 0, \end{aligned}$$

and, consequently, the boundary-value problem (12.1), (12.2) can be rewritten as

$$\begin{aligned}\left(\frac{d}{dx}+ik\right)u_1(x) &= -\frac{ik}{2}\varepsilon(x)\left[u_1(x)+u_2(x)\right], \quad u_1(L)=1,\\ \left(\frac{d}{dx}-ik\right)u_2(x) &= -\frac{ik}{2}\varepsilon(x)\left[u_1(x)+u_2(x)\right], \quad u_2(L_0)=0.\end{aligned}$$

The wavefield as a function of parameter L satisfies imbedding equation (12.6), page 279. It is obvious that the opposite waves will also satisfy Eq. (12.6), but with different initial values:

$$\begin{aligned}\frac{\partial}{\partial L}u_1(x;L) &= ik\left(1+\frac{1}{2}\varepsilon(L)\left(1+R_L\right)\right)u_1(x;L), \quad u_1(x;x)=1,\\ \frac{\partial}{\partial L}u_2(x;L) &= ik\left(1+\frac{1}{2}\varepsilon(L)\left(1+R_L\right)\right)u_2(x;L), \quad u_2(x;x)=R_x,\end{aligned}$$

where reflection coefficient R_L satisfies Eq. (12.5), page 279.

Introduce now intensities of the opposite waves $W_1(x;L)=|u_1(x;L)|^2$ and $W_2(x;L)=|u_2(x;L)|^2$ satisfying the equations

$$\begin{aligned}\frac{\partial}{\partial L}W_1(x;L) &= -k\gamma W_1(x;L)+\frac{ik}{2}\varepsilon(L)\left(R_L-R_L^*\right)W_1(x;L),\\ \frac{\partial}{\partial L}W_2(x;L) &= -k\gamma W_2(x;L)+\frac{ik}{2}\varepsilon(L)\left(R_L-R_L^*\right)W_2(x;L),\\ W_1(x;x) &= 1, \qquad W_2(x;x)=|R_x|^2.\end{aligned} \tag{12.66}$$

Quantity $W_L=|R_L|^2$ appearing in the initial value of Eq. (12.66) satisfies Eq. (12.7), page 279, or the equation

$$\frac{d}{dL}W_L=-2k\gamma W_L-\frac{ik}{2}\varepsilon_1(L)\left(R_L-R_L^*\right)\left(1-W_L\right), \quad W_{L_0}=0. \tag{12.67}$$

In Eqs. (12.66) and (12.67), we omitted dissipative terms producing no contribution in accumulated effects.

As earlier, we will assume that $\varepsilon_1(x)$ is the Gaussian delta-correlated process with correlation function (12.18), page 281. In view of the fact that Eqs. (12.66), (12.67) are the first-order equations with initial values, we can use the standard procedure of deriving the Fokker–Planck equation for the joint probability density of quantities $W_1(x;L)$, $W_2(x;L)$, and W_L

$$P(x;L;W_1,W_2,W)=\left\langle\delta(W_1(x;L)-W_1)\delta(W_2(x;L)-W_2)\delta(W_L-W)\right\rangle.$$

As a result, we obtain the Fokker–Planck equation

$$\begin{aligned}&\frac{\partial}{\partial L}P(x;L;W_1,W_2,W)\\ &= k\gamma\left(\frac{\partial}{\partial W_1}W_1+\frac{\partial}{\partial W_2}W_2+2\frac{\partial}{\partial W}W\right)P(x;L;W_1,W_2,W)\\ &+D\left[\frac{\partial}{\partial W_1}W_1+\frac{\partial}{\partial W_2}W_2-\frac{\partial}{\partial W}(1-W)\right]P(x;L;W_1,W_2,W)\\ &+D\left[\frac{\partial}{\partial W_1}W_1+\frac{\partial}{\partial W_2}W_2-\frac{\partial}{\partial W}(1-W)\right]^2 WP(x;L;W_1,W_2)\end{aligned} \tag{12.68}$$

with the initial value

$$P(x; x; W_1, W_2, W) = \delta(W_1 - 1)\delta(W_2 - W)P(x; W),$$

where function $P(L; W)$ is the probability density of reflection coefficient squared modulus W_L, which satisfies Eq. (12.22), page 283. As earlier, the diffusion coefficient in Eq. (12.68) is $D = k^2\sigma_\varepsilon^2 l_0/2$. Deriving this equation, we used an additional averaging over fast oscillations ($u(x) \sim e^{\pm ikx}$) that appear in the solution of the problem for $\varepsilon = 0$.

In view of the fact that Eqs. (12.66) are linear in $W_n(x; L)$, we can introduce the generating function of moments of opposite wave intensities

$$Q(x; L; \mu, \lambda, W) = \int_0^1 dW_1 \int_0^1 dW_2\, W_1^{\mu-\lambda} W_2^{\lambda} P(x; L; W_1, W_2, W), \tag{12.69}$$

which satisfies the simpler equation

$$\begin{aligned}\frac{\partial}{\partial L} Q(x; L; \mu, \lambda, W) &= -k\gamma\left(\mu - 2\frac{\partial}{\partial W}W\right) Q(x; L; \mu, \lambda, W)\\ &-D\left[\mu + \frac{\partial}{\partial W}(1-W)\right] Q(x; L; \mu, \lambda, W)\\ &+\left[\mu - \frac{\partial}{\partial W}(1-W)\right]^2 W Q(x; L; \mu, \lambda, W),\end{aligned} \tag{12.70}$$

with the initial value

$$Q(x; x; \mu, \lambda, W) = W^{\lambda} P(x; W).$$

With function $Q(x; L; \mu, \lambda, W)$, we can determine the moment functions of opposite wave intensities by the formula

$$\left\langle W_1^{\mu-\lambda}(x; L) W_2^{\lambda}(x; L)\right\rangle = \int_0^1 dW\, Q(x; L; \mu, \lambda, W). \tag{12.71}$$

Equation (12.70) describes statistics of the wavefield in medium layer $L_0 \le x \le L$. In particular, if we set $x = L_0$, it describes the transmission coefficient of the wave.

In the limiting case of the half-space ($L_0 \to -\infty$), Eq. (12.70) grades into the equation

$$\begin{aligned}\frac{\partial}{\partial \xi} Q(\xi; \mu, \lambda, W) &= -\beta\left(\mu - 2\frac{\partial}{\partial W}W\right) Q(\xi; \mu, \lambda, W)\\ -\left[\mu + \frac{\partial}{\partial W}(1-W)\right] &Q(\xi; \mu, \lambda, W) + \left[\mu - \frac{\partial}{\partial W}(1-W)\right]^2 W Q(\xi; \mu, \lambda, W),\\ Q(0; \mu, \lambda, W) &= W^{\lambda} P(W),\end{aligned} \tag{12.72}$$

where $\xi = D(L - x) > 0$ is the dimensionless distance, and steady-state (independent of L) probability density of the reflection coefficient modulus $P(W)$ is given by Eq. (12.37). In this case, Eq. (12.71) assumes the form

$$\left\langle W_1^{\mu-\lambda}(\xi) W_2^{\lambda}(\xi)\right\rangle = \int_0^1 dW\, Q(\xi; \mu, \lambda, W). \tag{12.73}$$

Further discussion will be more convenient if we consider separately the cases of absorptive (dissipative) and non-absorptive (nondissipative) random medium.

Nondissipative medium (stochastic wave parametric resonance and dynamic wave localization)

For non-absorptive medium, imbedding equations (12.5) and (12.6), page 279 are simplified. In this case, Eq. (12.7) for wavefield intensity can be integrated analytically, and relationship (12.11) expresses the intensity immediately in terms of the reflection coefficient. Using reflection coefficient in representation (12.23), page 283, we can rewrite this relationship in the form

$$\frac{1}{2}I(x;L) = \frac{u_x + \sqrt{u_x^2 - 1}\cos\phi_x}{1+u_L}, \tag{12.74}$$

where reflection coefficient phase ϕ_x has the form $\phi_x = 2kx + \widetilde{\phi}_x$ and u_x and $\widetilde{\phi}_x$ are slow functions for distances about wavelength. For this reason, it is expedient to consider only slow variations of combinations of function $I(x;L)$ with respect to x, which corresponds to preliminary averaging over functions that rapidly vary within scales of about wavelength. We will use overbar to denote such averaging. For example, averaging of Eq. (12.74) gives

$$\frac{1}{2}\overline{I(x;L)} = \frac{u_x}{1+u_L}. \tag{12.75}$$

We have similarly

$$\frac{1}{4}\overline{I^2(x;L)} = \frac{3u_x^2 - 1}{2\left(1+u_L\right)^2}, \tag{12.76}$$

and so on.

As was mentioned earlier, function u_x appearing in equations like Eqs. (12.75) and (12.76) is the Markovian random process with transition probability density (12.28) and one-point probability density (12.26). Consequently, determination of statistical characteristics of wave intensity reduces simply to calculating a quadrature. For example, for quantity $\overline{I^n(x;L)}$, we obtain the expression

$$\frac{1}{2^n}\overline{I^n(x;L)} = \frac{g_n(u_x)}{\left(1+u_L\right)^n},$$

where $g_n(u_x)$ is the polynomial of power n in u_x, so that

$$\frac{1}{2^n}\left\langle \overline{I^n(x;L)} \right\rangle = \int\limits_1^\infty \frac{du_L}{\left(1+u_L\right)^n} \int\limits_1^\infty du_x g_n(u_x) p(u_L, L|u_x, x) P(x, u_x). \tag{12.77}$$

Substituting Eq. (12.28), page 285 for $p(u_L, L|u_x, x)$ in Eq. (12.77) and using formula

$$\int\limits_1^\infty \frac{dx}{\left(1+x\right)^n} P_{-\frac{1}{2}+i\mu}(x) = \frac{\pi}{\cosh(\mu\pi)} K_n(\mu), \tag{12.78}$$

where

$$K_{n+1}(\mu) = \frac{1}{2n}\left[\mu^2 + \left(n - \frac{1}{2}\right)^2\right] K_n(\mu), \quad K_1(\mu) = 1,$$

we can perform integration over u_L to obtain the two-fold (in appearance) integral

$$\frac{1}{2^n}\left\langle \overline{I^n(x;L)} \right\rangle = \pi \int\limits_0^\infty d\mu\, \mu \frac{\sinh(\mu\pi)}{\cosh^2(\mu\pi)} K_n(\mu) e^{-\left(\mu^2+\frac{1}{4}\right)(L-x)} \int\limits_1^\infty du g_n(u) P_{-\frac{1}{2}+i\mu}(u) P(x,u). \tag{12.79}$$

Here, we introduced dimensionless distances $DL \to L$ and $Dx \to x$. In addition, we will assume that $L_0 = 0$.

In view of the expression

$$I(0;L) = |T_L|^2 = \frac{2}{1+u_L},$$

the integral

$$\int\limits_1^\infty \frac{du_L}{(1+u_L)^n} \int\limits_1^\infty du_x g_k(u_x) p(u_L, L|u_x, x) P(x, u_x)$$

describes correlations of the wave transmission coefficient with the wave intensity in the layer.

Our further task consists in calculating the inner integral in Eq. (12.79), which reduces to the solution of a simple system of differential equations [134]–[136].

Indeed, consider the expressions

$$f_k(x) = \int\limits_1^\infty du u^k P_{-\frac{1}{2}+i\mu}(u) P(x,u) \quad (k = 0, 1, \ldots), \tag{12.80}$$

which are the Meller–Fock transforms of functions $u^k P(x;u)$ (see Sect. 8.2, page 202). Differentiating Eq. (12.80) with respect to x, using the Fokker–Planck equation (12.25) for function $P(x;u)$ and differential equation for the *Legendre function* $P_{-\frac{1}{2}+i\mu}(x)$, page 203

$$\frac{d}{dx}\left(x^2-1\right)\frac{d}{dx}P_{-\frac{1}{2}+i\mu}(x) = -\left(\mu^2+\frac{1}{4}\right)P_{-\frac{1}{2}+i\mu}(x),$$

and integrating the result by parts, we arrive at the equation

$$\frac{d}{dx}f_k(x) = -\left(\mu^2+\frac{1}{4}-k^2-k\right)f_k(x) + 2k\psi_k(x) - k(k-1)f_{k-2}(x), \tag{12.81}$$

where

$$\psi_k(x) = \int\limits_1^\infty du u^{k-1} P(x,u)\left(u^2-1\right)\frac{d}{du}P_{-\frac{1}{2}+i\mu}(u). \tag{12.82}$$

Differentiating now function $\psi_k(x)$ with respect to x, we similarly obtain that this function satisfies the equation

$$\begin{aligned}\frac{d}{dx}\psi_k(x) = &-\left(\mu^2+\frac{1}{4}-k^2+k\right)\psi_k(x) - 2k\left(\mu^2+\frac{1}{4}\right)f_k(x)\\ &-(k-1)(k-2)\psi_{k-2}(x) + 2(k-1)\left(\mu^2+\frac{1}{4}\right)f_{k-2}(x).\end{aligned} \tag{12.83}$$

The initial values for Eqs. (12.81) and (12.83) are, obviously, the conditions

$$f_k(0) = 1, \quad \psi_k(0) = 0.$$

Thus, functions $f_k(x)$ and $\psi_k(x)$ are mutually related and satisfy the closed recursive system of inhomogeneous second-order differential equations with constant coefficients, and this system can be easily solved.

Represent the solution to system (12.81), (12.83) in the form

$$f_k(x) = \widetilde{f}_k(x)e^{-(\mu^2+\frac{1}{4}-k^2)x}, \quad \psi_k(x) = \widetilde{\psi}_k(x)e^{-(\mu^2+\frac{1}{4}-k^2)x}. \tag{12.84}$$

The, for functions $\widetilde{f}_k(x)$ and $\widetilde{\psi}_k(x)$, we obtain the system of equations

$$\begin{aligned}
&\left(\frac{d}{dx} - k\right)\widetilde{f}_k(x) = 2k\widetilde{\psi}_k(x) - k(k-1)\widetilde{f}_{k-2}(x)e^{-4(k-1)x},\\
&\left(\frac{d}{dx} + k\right)\widetilde{\psi}_k(x) = -2k\left(\mu^2+\frac{1}{4}\right)\widetilde{f}_k(x)\\
&+(k-1)\left[2\left(\mu^2+\frac{1}{4}\right)\widetilde{f}_{k-2}(x) - (k-2)\widetilde{\psi}_{k-2}(x)\right]e^{-4(k-1)x}
\end{aligned} \tag{12.85}$$

with the initial values

$$\widetilde{f}_k(0) = 1, \quad \widetilde{\psi}_k(0) = 0.$$

We note that the corresponding solution to the homogeneous system has the form

$$\widetilde{f}_k(x) = A(\mu)\sin(2k\mu x) + B(\mu)\cos(2k\mu x).$$

Consider the simplest cases.

1. In the case $k = 0$, we have

$$\frac{d}{dx}\widetilde{f}_0(x) = 0, \quad \widetilde{f}_0(L_0) = 1,$$

so that

$$f_0(x) = \exp\left\{-\left(\mu^2+\frac{1}{4}\right)x\right\}.$$

Then, the integral

$$\begin{aligned}
\left\langle |T_L|^{2n}\right\rangle &= \int_1^\infty \frac{du_L}{(1+u_L)^n}\int_1^\infty du p(u_L, L|u,x)P(x,u)\\
&= 2^n\pi\int_0^\infty d\mu \frac{\mu\sinh(\mu\pi)}{\cosh^2(\mu\pi)}K_n(\mu)e^{-(\mu^2+\frac{1}{4})(L-L_0)}
\end{aligned}$$

describes the moments of the modulus of coefficient of wave transmission through the layer of random medium.

2. In the case $k = 1$, we have the system of equations

$$\begin{aligned}
\left(\frac{d}{dx} - 2\right)\widetilde{f}_1(x) &= 2\widetilde{\psi}_1(x),\\
\left(\frac{d}{dx} + 1\right)\widetilde{\psi}_1(x) &= -2\left(\mu^2+\frac{1}{4}\right)\widetilde{f}_1(x),
\end{aligned}$$

so that

$$f_1(x) = \exp\left\{-\left(\mu^2-\frac{3}{4}\right)x\right\}\left(\cos(2\mu x) + \frac{1}{2\mu}\sin(2\mu x)\right).$$

In this case, integral (12.77) at $n = 1$ describes the distribution of the wavefield average intensity in the layer of random medium [73]

$$\left\langle \overline{I(x;L)}\right\rangle = 2\pi\int_0^\infty d\mu\frac{\mu\sinh(\mu\pi)}{\cosh^2(\mu\pi)}e^{x-(\mu^2+\frac{1}{4})L}\left(\cos(2\mu x) + \frac{1}{2\mu}\sin(2\mu x)\right).$$

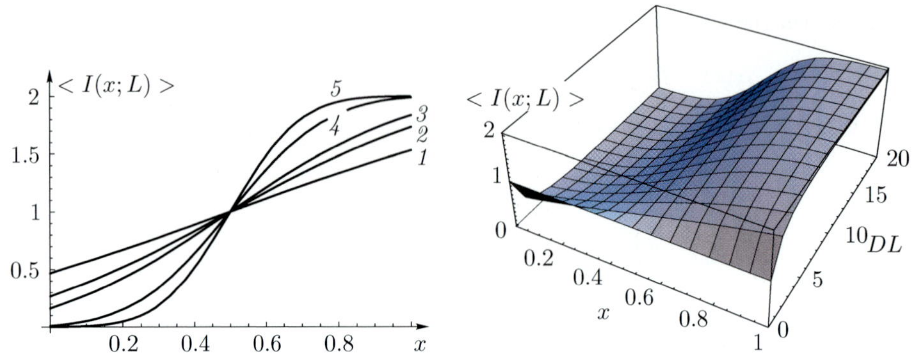

Figure 12.5: Wavefield average intensity in the problem on a wave incident on medium layer. Curves *1* to *5* correspond to parameter $DL = 1$, 2, 3, 10, and 20, respectively.

Figure 12.5 shows this intensity distribution for different layer thicknesses.

3. In the case $k = 2$, we have the system of equations

$$\begin{aligned} \left(\frac{d}{dx} - 2\right)\widetilde{f}_2(x) &= 4\widetilde{\psi}_2(x) - 2e^{-4x}, \\ \left(\frac{d}{dx} + 2\right)\widetilde{\psi}_2(x) &= -2\left(\mu^2 + \frac{1}{4}\right)\left[2\widetilde{f}_2(x) - e^{-4x}\right], \end{aligned}$$

so that

$$\widetilde{f}_2(x) = \frac{\mu^2 + 5/4}{2\left(1 + \mu^2\right)}\cos\left(4\mu\pi\right) + \frac{\mu^2 + 3/4}{2\mu\left(1 + \mu^2\right)}\sin\left(4\mu\pi\right) + \frac{\mu^2 + 3/4}{2\left(1 + \mu^2\right)}e^{-4x}.$$

In this case, integral (12.79) at $n = 2$ describes the distribution of the second moment of the intensity along the layer

$$\left\langle \overline{I^2(x; L)} \right\rangle = \pi \int\limits_0^\infty d\mu \frac{\mu \sinh(\mu\pi)}{\cosh^2(\mu\pi)} e^{-(\mu^2 + \frac{1}{4})L}\left(\mu^2 + \frac{1}{4}\right)\left(3e^{4x}\widetilde{f}_2(x) - 1\right).$$

Figure 12.6 shows this distribution for different layer thicknesses.

Thus, solving successively the recurrent system of equations (12.85), we can express the corresponding moment of intensity in terms of the sole quadrature.

Consider the structure of the obtained expressions. As we have seen earlier, moments of the wavefield intensity in the layer of medium are expressed in terms of the integrals

$$\begin{aligned} \left\langle \overline{I^n(x; L)} \right\rangle &\sim \int\limits_{-\infty}^{\infty} d\mu \frac{\sinh(\mu\pi)}{\cosh^2(\mu\pi)} \Phi(\mu) e^{n^2 x + 2in\mu x - \left(\mu^2 + \frac{1}{4}\right)L} \\ &= e^{-\frac{1}{4}L + n^2 L\xi(1-\xi)} \int\limits_{-\infty}^{\infty} d\mu \frac{\sinh(\mu\pi)}{\cosh^2(\mu\pi)} \Phi(\mu) e^{-(\mu - in\xi)^2 L}, \end{aligned} \qquad (12.86)$$

where $\xi = x/L$ and $\Phi(\mu)$ is the algebraic function of parameter μ. If we consider asymptotic limit $L \to \infty$ under the condition that ξ remains finite, then we obtain from Eq. (12.86)

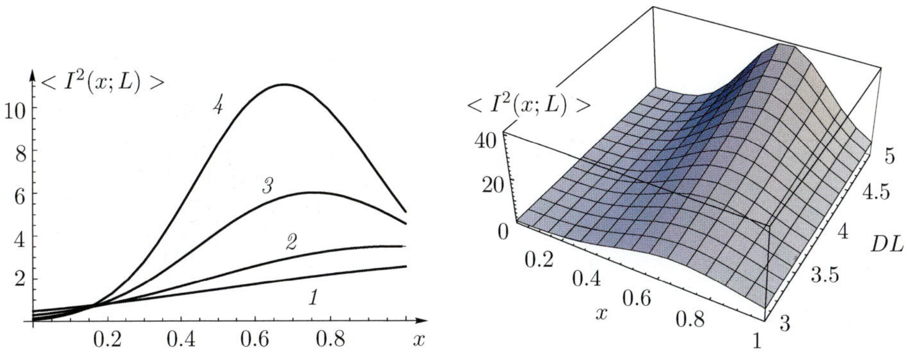

Figure 12.6: Second moment of wavefield intensity in the problem on a wave incident on medium layer. Curves *1* to *4* correspond to parameter $DL = 0.5$, 1, 2, and 3, respectively.

that two spatial scales

$$\xi_1 = \frac{1}{2}\left(1 - \sqrt{1 - \frac{1}{n^2}}\right) \quad \text{and} \quad \xi_2 = 1 - \frac{1}{2n}$$

exist such that quantity $\left\langle \overline{I^n(x;L)} \right\rangle$ is exponentially small for $0 \le \xi \le \xi_1$. For $\xi_1 \le \xi \le \xi_2$, quantity $\left\langle \overline{I^n(x;L)} \right\rangle$ is exponentially great and achieves its maximum in the vicinity of point $\xi \approx 1/2$, where $\left\langle \overline{I^n(x;L)} \right\rangle_{\max} \sim \exp\left\{\left(n^2 - 1\right) l/4\right\}$. For $1 \ge \xi \ge \xi_2$, quantity $\left\langle \overline{I^n(x;L)} \right\rangle$ exponentially tends to unity. The above behavior is pertinent to the case $n \ge 2$. The case $n = 1$ forms the exception; in this case, points ξ_1 and ξ_2 merge, and average intensity distribution appears monotonous.

The first scale follows from the relationship $n^2 \xi (1 - \xi) \sim 1/4$, and the second scale follows from the fact that, in view of limiting condition $\left\langle \overline{I^n(x;L)} \right\rangle \to 2^n$ for $L \to \infty$, integral (12.86) is contributed mainly by the pole $\mu_n = i(n - 1/2)$, so that integration contour must run above μ_n, i.e., $\mu_n < in\xi$. With increasing n, variable $\xi_1 \to 0$, and variable $\xi_2 \to 1$ (see Fig. 12.7).

The fact that moments of intensity behave in the layer as exponentially increasing functions is evidence of the *phenomenon of stochastic wave parametric resonance*, which is similar to the ordinary parametric resonance. The only difference consists in the fact that values of intensity moments at layer boundary are asymptotically predetermined; as a result, the wavefield intensity exponentially increases inside the layer and its maximum occurs approximately in the middle of the layer.

In the limit of the half-space ($L_0 \to -\infty$), the region of the exponential growth of all moments beginning from the second one occupies the whole of the half-space, and $\left\langle \overline{I(x;L)} \right\rangle = 2$.

Now, we turn back to the equation for moments of opposite wave intensities in non-absorptive medium, i.e., to Eq. (12.72) at $\beta = 0$ in the limit of the half-space ($L_0 \to -\infty$) filled with random medium. In this case, $W_L = 1$ with a probability of unity, and the solution to Eq. (12.72) has the form

$$Q(x, L; \mu, \lambda, W) = \delta(W - 1)e^{D\lambda(\lambda-1)(L-x)},$$

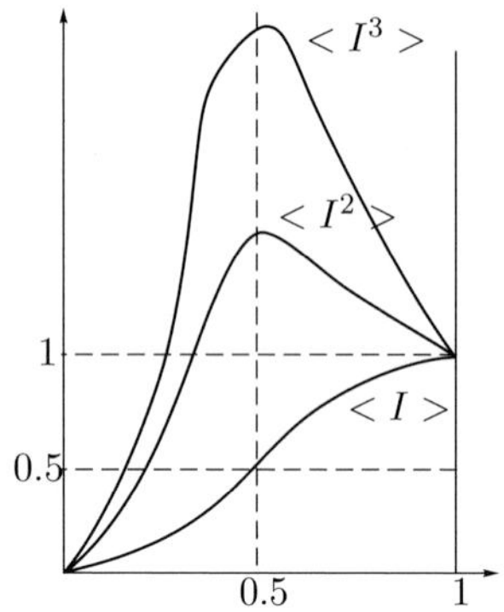

Figure 12.7: Schematic of the behavior of moments of wavefield intensity in the problem on a wave incident on medium layer (Stochastic parametric resonance).

so that

$$\left\langle W_1^{\lambda-\mu}(x;L) W_2^{\mu}(x;L) \right\rangle = e^{D\lambda(\lambda-1)(L-x)}. \tag{12.87}$$

In view of arbitrariness of parameters λ and μ, this means that

$$W_1(x;L) = W_2(x;L) = W(x;L)$$

with a probability of unity and quantity $W(x;L)$ has the lognormal probability density. In addition, the mean value of this quantity is equal to unity, and its higher moments beginning from the second one exponentially increase with the distance in medium

$$\langle W(x;L) \rangle = 1, \quad \langle W^n(x;L) \rangle = e^{Dn(n-1)(L-x)}, \quad n = 2, 3, \ldots. \tag{12.88}$$

Note that wavefield intensity $I(x;L)$ has in this case the form

$$I(x;L) = 2W(x;L)\left(1 + \cos\phi_x\right), \tag{12.89}$$

where ϕ_L is the phase of the reflection coefficient.

In view of the lognormal probability distribution, the typical realization curve of function $W(x;L)$ is the curve exponentially decaying with distance in the medium

$$W^*(x;L) = e^{-D(L-x)}. \tag{12.90}$$

For example, realizations of function $W(x;L)$ satisfy the inequality

$$W(x;L) < 4e^{-D(L-x)/2}$$

within the whole of the half-space with a probability of 1/2.

In physics of disordered systems, the exponential decay of typical realization curve (12.90) with increasing $\xi = D(L-x)$ is usually identified with the property of *dynamic localization* (see, e.g., [7, 68, 166, 223], [278] – [281]), and quantity

$$l_{\text{loc}} = \frac{1}{D}$$

is usually called the *localization length*. Here,

$$l_{\text{loc}}^{-1} = -\frac{\partial}{\partial L} \langle \varkappa(x;L) \rangle,$$

where

$$\varkappa(x;L) = \ln W(x;L).$$

Physically, the lognormal property of wavefield intensity $W(x;L)$ implies the existence of large spikes relative typical realization curve (12.90) towards both large and small intensities. This result agrees with the example of simulations given in Chapter 1 (see Fig. 1.7, page 12). However, these spikes of intensity contain only small energy, because random area below curve $W^n(x;L)$,

$$S_n(L) = D \int_{-\infty}^{L} dx W^n(x;L),$$

has, in accordance with the lognormal probability distribution attribute, the steady-state (independent of L) probability density

$$P_n(S) = \frac{1}{n^{2/n}\Gamma(1/n)} \frac{1}{S^{1+1/n}} \exp\left\{-\frac{1}{n^2 S}\right\}, \tag{12.91}$$

where $\Gamma(x)$ is the *Gamma function*. In particular, the area below curve $W(x;L)$

$$S_1(L) = D \int_{-\infty}^{L} dx W(x;L)$$

is distributed by the law

$$P_1(S) = \frac{1}{S^2} \exp\left\{-\frac{1}{S}\right\}$$

that coincides with the distribution of total energy of the wavefield in the half-space (12.43) if we set $E = 2S$. This means that the term dependent on fast phase oscillations of reflection coefficient in Eq. (12.89) only slightly contributes to total energy.

Thus, the knowledge of the one-point probability density provides an insight into the evolution of separate realizations of wavefield intensity in the whole space and allows estimating the parameters of this evolution in terms of statistical characteristics of fluctuating medium.

Dissipative medium

In the presence of a finite (even arbitrary small) absorption in the medium occupying the half-space, the exponential growth of moment functions must cease and give place to attenuation. If $\beta \gg 1$ (i.e., if the effect of absorption is great in comparison with the effect of diffusion), then $P(W) = 2\beta e^{-2\beta W}$, and, as can be easily seen from Eq. (12.72), opposite wave intensities $W_1(x;L)$ and $W_2(x;L)$ appear statistically independent, i.e., uncorrelated. In this case,

$$\langle W_1(\xi)\rangle = \exp\left\{-\beta\xi\left(1+\frac{1}{\beta}\right)\right\}, \quad \langle W_2(\xi)\rangle = \frac{1}{2\beta}\exp\left\{-\beta\xi\left(1+\frac{1}{\beta}\right)\right\}.$$

Figures 12.8–12.11 show the examples of moment functions of random processes obtained by numerical solution of Eq. (12.72) and calculation of quadrature (12.73) for different values of parameter β [15, 19, 135, 136, 173]. Different figures mark the curves corresponding to different values of parameter β. Figure 12.8 shows average intensities of

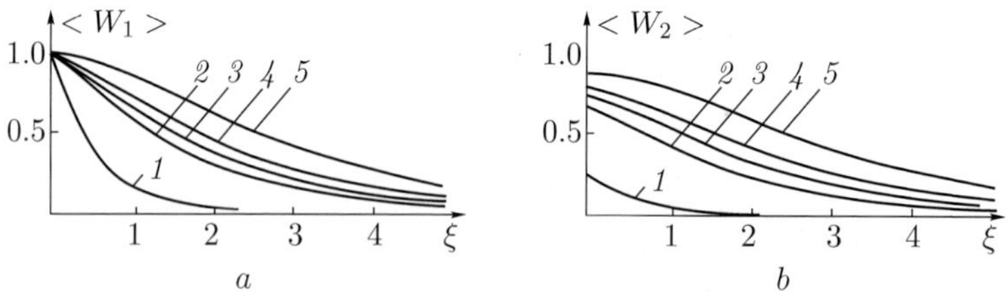

Figure 12.8: Distribution of wavefield average intensity along the medium; (a) the transmitted wave and (b) the reflected wave. Curves *1* to *5* correspond to parameter $\beta = 1$, 0.1, 0.06, 0.04 and 0.02, respectively.

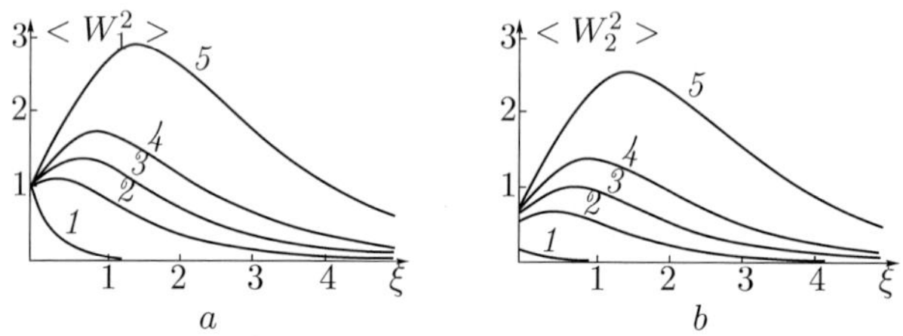

Figure 12.9: Distribution of the second moment of wavefield intensity along the medium; (a) the transmitted wave and (b) the reflected wave. Curves *1* to *5* correspond to parameter $\beta = 1$, 0.1, 0.06, 0.04 and 0.02, respectively.

the transmitted and reflected waves. The curves monotonically decrease with increasing ξ. Figure 12.9 shows the corresponding curves for second moments. We see that $\langle W_1^2(0)\rangle = 1$ and $\langle W_2^2(0)\rangle = \langle |R_L|^4\rangle$ at $\xi = 0$. For $\beta < 1$, the curves as functions of ξ become non-monotonic; the moments first increase, then pass the maximum, and finally monotonically decay. With decreasing parameter β, the position of the maximum moves to the right and the maximum value increases. Figure 12.10 shows the similar curves for the third moment $\langle W_1^3(\xi)\rangle$, and Fig. 12.11 shows curves for mutual correlation of intensities of the transmitted and reflected waves $\langle \Delta W_1(\xi)\Delta W_2(\xi)\rangle$ (here, $\Delta W_n(\xi) = W_n(\xi) - \langle W_n(\xi)\rangle$). For $\beta \geq 1$, this correlation disappears. For $\beta < 1$, the correlation is strong, and wave division into opposite waves appears physically senseless, but mathematically useful technique. For $\beta \geq 1$, such a division is justified in view of the lack of mutual correlation.

As was shown earlier, in the case of the half-space of random medium with $\beta = 0$, all wavefield moments beginning from the second one exponentially increase with the distance the wave travels in the medium. It is clear, that problem solution for small β ($\beta \ll 1$) must show the singular behavior in β in order to vanish the solution for sufficiently long distances. Consider this asymptotic case in more detail [106].

We introduce function

$$Q(x; L; \mu, \lambda, u) = \left\langle W_1^{\mu-\lambda}(x; L) W_2^{\lambda}(x; L)\delta(u_L - u)\right\rangle$$

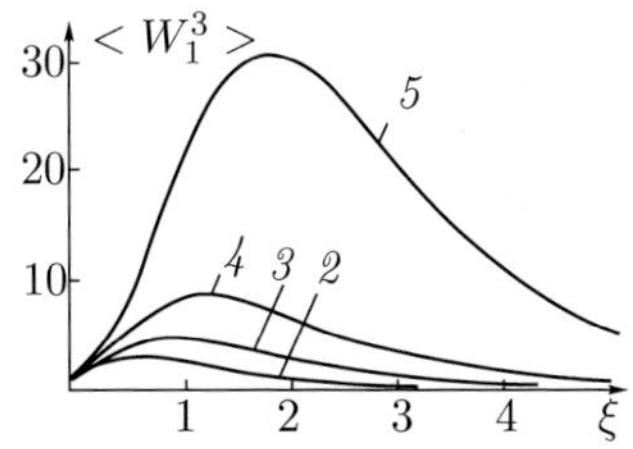

Figure 12.10: Distribution of the third moment of transmitted wave intensity. Curves *1* to *5* correspond to parameter $\beta = 1$, 0.1, 0.06, 0.04 and 0.02, respectively.

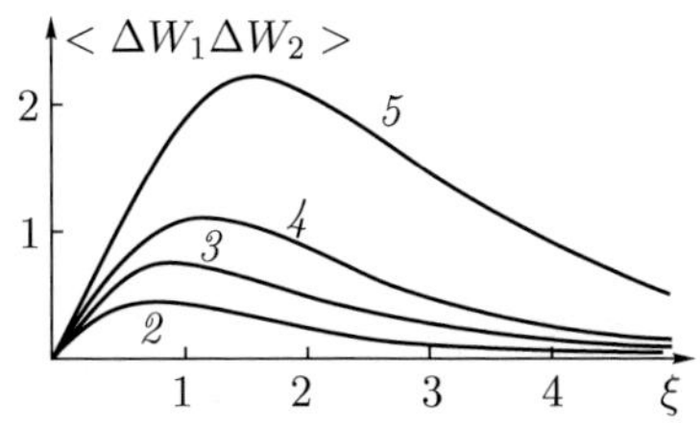

Figure 12.11: Correlation between the intensities of transmitted and reflected waves. Curves *1* to *5* correspond to parameter $\beta = 1$, 0.1, 0.06, 0.04 and 0.02, respectively.

satisfying in the case of the half-space the equation

$$\begin{aligned}
&\frac{\partial}{\partial \xi} Q(\xi;\mu,\lambda,u) \\
&= \left(-\beta\mu + \beta\frac{\partial}{\partial u}\left(u^2-1\right) + \mu(\mu+1) - \frac{2\mu^2}{u+1}\right) Q(\xi;\mu,\lambda,u) \\
&+ \left[2\mu(u-1)\frac{\partial}{\partial u} + \frac{\partial}{\partial u}\left(u^2-1\right)\frac{\partial}{\partial u}\right] Q(\xi;\mu,\lambda,u)
\end{aligned} \tag{12.92}$$

with the initial value

$$Q(0;\mu,\lambda,u) = \left(\frac{u-1}{u+1}\right)^{\lambda} P(u),$$

where $\xi = D(L-x) > 0$ and $P(u)$ is the steady-state probability density (12.37).

Our interest is in quantities

$$\left\langle W_1^{\mu-\lambda}(\xi) W_2^{\lambda}(\xi) \right\rangle = \int_1^{\infty} du Q(\xi;\mu,\lambda,u).$$

Replace variable $u \to \beta(u-1)$ in Eq. (12.92) and perform limit process $\beta \to 0$. As a result, we obtain a simpler equation

$$\begin{aligned}
&\frac{\partial}{\partial \xi} Q(\xi;\mu,\lambda,u) = \left(\mu(\mu+1) - \frac{2\mu^2\beta}{u} + 2\mu u\frac{\partial}{\partial u} + \frac{\partial}{\partial u}u^2\frac{\partial}{\partial u}\right) Q(\xi;\mu,\lambda,u), \\
&Q(0;\mu,\lambda,u) = e^{-u}.
\end{aligned} \tag{12.93}$$

The solution to this equation as a function of variable u (and, consequently, parameter β) has a singularity in the case of arbitrary small, but finite absorption in the medium. This solution can be obtained using the integral Kantorovich–Lebedev transform (see Sect. 8.2, page 203). As a result, in the case of integer parameters $\mu = n$, $\lambda = m$, we obtain the asymptotic representation in the form of the quadrature

$$\begin{aligned}\left\langle W_1^{n-m}(\xi)W_2^m(\xi)\right\rangle &= \frac{4}{\pi\,(\varepsilon n)^{2n-1}}\int\limits_0^\infty d\tau\,\tau\sinh\left(\frac{\pi\tau}{2}\right)e^{-\xi(1+\tau^2)/4}g_n(\tau)\psi_0(\tau),\\ \psi_0(\tau) &= \int\limits_0^\infty \frac{dy}{y^{2(n+1)}}\frac{e^{-1/y^2}}{(1+2\beta y^2)^m}K_{i\tau}(\varepsilon n y),\end{aligned} \tag{12.94}$$

where $\varepsilon = \sqrt{8\beta}$,

$$g_n(\tau) = \left[(2n-3)^2+\tau^2\right]g_{n-1}(\tau), \quad g_1(\tau) = 1,$$

and $K_{i\tau}(x)$ is the *imaginary index McDonalds function* of the first kind satisfying Eq. (8.39), page 203.

From Eq. (12.94), we see that, in asymptotic limit $\beta \ll 1$, intensities of opposite waves are equal with a probability of unity, and the solution for small distances from the boundary coincides with the solution corresponding to the stochastic parametric resonance.

For sufficiently great distances ξ, namely

$$\xi \gg 4\left(n-\frac{1}{2}\right)\ln\left(\frac{n}{\beta}\right),$$

quantities $\langle W^n(\xi)\rangle$ are characterized by the universal spatial localization behavior [106]

$$\langle W^n(\xi)\rangle \cong A_n\frac{1}{\beta^{n-1/2}}\ln\left(\frac{1}{\beta}\right)\frac{1}{\xi\sqrt{\xi}}e^{-\xi/4},$$

which coincides, to a numerical factor, with the asymptotic behavior of moments of the transmission coefficient of a wave passed through the layer of thickness ξ in the case $\beta = 0$.

Thus, the behavior of moments of opposite wave intensities appears essentially different in three regions. In the first region (it corresponds to the stochastic parametric resonance), the moments exponentially increase with the distance in medium and wave absorption plays only insignificant role. In the second region, absorption plays the most important role, because namely absorption ceases the exponential growth of moments. In the third region, the decrease of moment functions of opposite wave intensities is independent of absorption. The boundaries of these regions depend on parameter β and tend to infinity for $\beta \to 0$.

Note that, in the general case of arbitrary parameter β, mean logarithm of forward wave and its variance are given, in accordance with Eqs. (12.66), by the relationships [134, 136]

$$\langle \varkappa_1(x;L)\rangle = -(1+\beta)\,\xi, \quad \sigma^2_{\varkappa_1}(x;L) = 2\left\langle |R_L|^2\right\rangle\xi, \tag{12.95}$$

where $\langle |R_L|^2\rangle$ is given by Eq. (12.38), page 289.

12.3.2 Plane wave source located in random medium

In the previous section, we considered in detail the problem on the wave incidence on a layer (half-space) of random medium. We can consider similarly the problem on the wave generated by the plane wave source located in random medium. As earlier, let the layer of medium occupy a portion of space. Then, the wavefield in the layer is described by the solution to boundary-value problem (12.12), page 280. Considering this solution as a function of parameter L, we can obtain the imbedding equations (see Appendix C, page 447)

$$\begin{aligned}
&\frac{\partial}{\partial L}G(x;x_0;L) = i\frac{k}{2}\varepsilon(L)\,u(x_0;L)\,u(x;L),\\
&\quad G(x;x_0;L)_{L=\max(x,x_0)} = \begin{cases} u(x;x_0), & x \geq x_0 \\ u(x_0;x), & x \leq x_0 \end{cases},\\
&\frac{\partial}{\partial L}u(x;L) = ik\left\{1+\varepsilon(L)\,u(L;L)\right\}u(x;L), \quad u(x;x) = 1+R_x,\\
&\frac{d}{dL}u(L;L) = 2ik\left[u(L;L)-1\right] + i\frac{k}{2}\varepsilon(L)\,u^2(L;L), \quad R_{L_0} = 0.
\end{aligned} \tag{12.96}$$

Two last equations in Eqs. (12.96) describe the wavefield $u(x;L)$ in the problem on wave incidence on medium layer (L_0, L) and the field $u(L;L) = 1 + R_L$ (R_L is the reflection coefficient) at layer boundary $x = L$.

We introduce the intensity of the wavefield $I(x;x_0;L) = |G(x;x_0;L)|^2$ and consider its average value. Using Eq. (12.96), the corresponding complex conjugated equation and averaging over an ensemble of realizations of random function $\varepsilon_1(x)$ and fast oscillations, we obtain that average intensity satisfies the imbedding equation

$$\frac{\partial}{\partial L}\langle I(x;x_0;L)\rangle = D\,\langle I(x_0;L)\,I(x;L)\rangle, \tag{12.97}$$

where $I(x;L) = |u(x;L)|^2$ is the wavefield intensity in the problem on wave incidence on medium layer. As a result, we have (for definiteness, we assume that $x_0 > x$),

$$\langle I(x;x_0;L)\rangle = \langle I(x;x_0)\rangle + D\int\limits_{x_0}^{L} d\xi\,\langle I(x_0;\xi)\,I(x;\xi)\rangle, \tag{12.98}$$

so that this quantity is expressed in terms of the correlation function of wavefield intensity in the problem on wave incidence on medium layer.

Introduce functions

$$\begin{aligned}
\psi(x;x_0;L,W) &= \left\langle I(x_0;L)\,I(x;L)\,\delta\left(|R_L|^2 - W\right)\right\rangle,\\
\chi(x;L,W) &= \left\langle I(x;L)\,\delta\left(|R_L|^2 - W\right)\right\rangle.
\end{aligned} \tag{12.99}$$

It is obvious that these functions satisfy Eq. (12.70): for $\mu = 2$ and $x \neq x_0$ in the case of function ψ and for $\mu = 1$ in the case of function χ; in other words, they satisfy the following equations with the initial values

$$\begin{aligned}
&\frac{\partial}{\partial L}\psi(x;x_0;L,W) = -2k\gamma\left(1 - \frac{\partial}{\partial W}W\right)\psi(x;x_0;L,W)\\
&-D\left[2 + \frac{\partial}{\partial W}(1-W)\right]\psi(x;x_0;L,W) + D\left[2 - \frac{\partial}{\partial W}(1-W)\right]^2 W\psi(x;x_0;L,W),\\
&\quad \psi(x;x_0;x_0,W) = (1+W)\chi(x;L,W),
\end{aligned} \tag{12.100}$$

$$\begin{aligned}&\frac{\partial}{\partial L}\chi(x;L,W)=-k\gamma\left(1-2\frac{\partial}{\partial W}W\right)\chi(x;L,W)\\&-D\left[1+\frac{\partial}{\partial W}(1-W)\right]\chi(x;L,W)+D\left[1-\frac{\partial}{\partial W}(1-W)\right]^2W\chi(x;L,W),\\&\chi(x;x,W)=(1+W)P(x;W).\end{aligned}\tag{12.101}$$

At $x=x_0$, function $\psi(x;x;L,W)$ also satisfies Eq. (12.100), but with different initial value, namely

$$\begin{aligned}&\frac{\partial}{\partial L}\psi(x;x;L,W)=-2k\gamma\left(1-\frac{\partial}{\partial W}W\right)\psi(x;x;L,W)\\&-D\left[2+\frac{\partial}{\partial W}(1-W)\right]\psi(x;x;L,W)+D\left[2-\frac{\partial}{\partial W}(1-W)\right]^2W\psi(x;x;L,W),\\&\psi(x;x;x,W)=(1+4W+W^2)P(x;W).\end{aligned}\tag{12.102}$$

In Eqs. (12.101) and (12.102), function $P(L;W)=\langle\delta\left(|R_L|^2-W\right)\rangle$ is the probability density of the reflection coefficient squared modulus; it satisfies Eq. (12.22), page 283.

Infinite space of random medium

Perform limit process $L_0\to-\infty$ to determine average intensity of the wavefield generated by a source in the infinite space. We denote $D(L-x_0)=\eta$ and assume that quantity $D(x_0-x)=\xi$ is the fixed parameter. Then, Eq. (12.98) is replaced with the equality

$$\langle I(x;x_0;L)\rangle=\langle I(\xi)\rangle+S(\xi),$$

where

$$\langle I(\xi)\rangle=\int_0^1 dW\chi(\xi;W),\quad S(\xi)=\int_0^1 dW\int_0^\infty d\eta\psi(\xi;\eta;W),$$

and functions $\psi(\xi;\eta;W)$, $\chi(\xi;W)$ satisfy the equations

$$\begin{aligned}&\frac{\partial}{\partial\eta}\psi(\xi;\eta;W)=-2\beta\left(1-\frac{\partial}{\partial W}W\right)\psi(\xi;\eta;W)\\&-\left[2+\frac{\partial}{\partial W}(1-W)\right]\psi(\xi;\eta;W)+\left[2-\frac{\partial}{\partial W}(1-W)\right]^2W\psi(\xi;\eta;W),\\&\psi(\xi;0;W)=\begin{cases}(1+W)\chi(\xi;W),&(\xi\neq0),\\(1+4W+W^2)P(W),&(\xi=0),\end{cases}\end{aligned}\tag{12.103}$$

$$\begin{aligned}&\frac{\partial}{\partial\eta}\chi(\xi;W)=-\beta\left(1-2\frac{\partial}{\partial W}W\right)\chi(\xi;W)\\&-\left[1+\frac{\partial}{\partial W}(1-W)\right]\chi(\xi;W)+\left[1-\frac{\partial}{\partial W}(1-W)\right]^2W\chi(\xi;W),\\&\chi(0;W)=(1+W)P(W).\end{aligned}\tag{12.104}$$

Equation (12.103) can be rewritten in the form

$$\frac{\partial}{\partial\eta}\psi(\xi;\eta;W)=\left\{2\beta W+2W(1-W)+\frac{\partial}{\partial W}W(1-W)^2\right\}\frac{\partial}{\partial W}\psi(\xi;\eta;W).\tag{12.105}$$

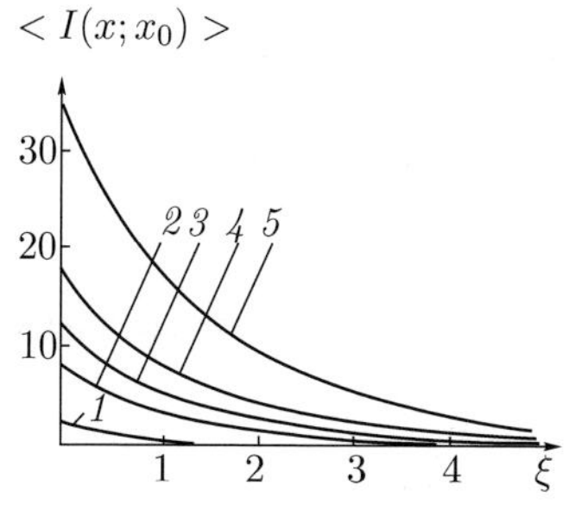

Figure 12.12: Distribution of average intensity of the field generated by a source in infinite space. Curves *1* to *5* correspond to parameter $\beta = 1$, 0.1, 0.06, 0.04 and 0.02, respectively.

Integrating Eq. (12.105) over η in limits $(0, \infty)$, we obtain that function

$$\tilde{\psi}(\xi; W) = \int_0^\infty d\eta \psi(\xi; \eta; W)$$

satisfies the following simple equation

$$-\psi(\xi; 0; W) = \left\{2\beta W + 2W(1-W) + \frac{\partial}{\partial W} W(1-W)^2\right\} \frac{\partial}{\partial W} \tilde{\psi}(\xi; W),$$

whose solution has the form

$$\tilde{\psi}(\xi; W) = \int_W^1 \frac{dW_1}{W_1} \int_0^{W_1} \frac{dW_2}{(1-W_2)^2} \psi(\xi; 0; W) \exp\left\{2\beta\left[\frac{1}{1-W_1} - \frac{1}{1-W_2}\right]\right\}. \quad (12.106)$$

Integrating then Eq. (12.106) over W, we obtain the final expression for function $S(\xi)$,

$$S(\xi) = \int_0^1 \frac{dW}{(1-W)^2} \psi(\xi; 0; W) \left[1 - W + 2\beta \exp\left(\frac{2\beta}{1-W}\right) \mathrm{Ei}\left(-\frac{2\beta}{1-W}\right)\right], \quad (12.107)$$

where $\mathrm{Ei}(-x) = -\int_x^\infty \frac{dt}{t} e^{-t}$ is the *integral exponent.*

Thus, average intensity of the wavefield generated by the source in infinite space satisfies the sole equation (12.104) and has the form $(x < x_0)$

$$\langle I(\xi)\rangle = \int_0^1 dW \left\{1 + \frac{1+W}{(1-W)^2}\left[1 - W + 2\beta e^{\frac{2\beta}{1-W}} \mathrm{Ei}\left(-\frac{2\beta}{1-W}\right)\right]\right\} \chi(\xi; W). \quad (12.108)$$

Figure 12.12 shows results of numerical integration of Eq. (12.108) for different values of parameter β.

For $\beta \gg 1$, from Eqs. (12.108) and (12.104) follows the expression

$$\langle I(\xi)\rangle = \left(1 + \frac{1}{\beta}\right) e^{-2\gamma(x_0 - x)\left(1+\frac{1}{\beta}\right)} \quad (12.109)$$

that corresponds to the linear phenomenological theory of radiative transfer.

The asymptotic case $\beta \ll 1$ will be considered in detail a little later.

Half-space of random medium

If the source of plane waves is located in region $L_0 < x_0 < \infty$, then average intensity $\langle I(x;x_0)\rangle$ as before will be given by Eq. (12.98) for $L \to \infty\,(x_0 \le x)$. In the case $x_0 \ge x$, one must interchange points x_0 and x in Eq. (12.98).

Introduce dimensionless variables $\tilde{x} = Dx$, $\tilde{x}_0 = Dx_0$, and $h = DL$. Replicating calculations of the foregoing subsection, we obtain that average intensity $\langle I(x;x_0)\rangle$ will be given by the expression (we omit here the tilde sign)

$$\langle I(x;x_0)\rangle = \int\limits_0^1 dW \left\{1 + \frac{1+W}{(1-W)^2}\right.$$
$$\left.\times\left[1 - W + 2\beta e^{\frac{2\beta}{1-W}}\,\mathrm{Ei}\left(-\frac{2\beta}{1-W}\right)\right]\right\}\chi(x;x_0;W), \tag{12.110}$$

where function $\chi(x;h;W)$ satisfies, as a function of variables h and W, the equation

$$\frac{\partial}{\partial h}\chi(x;h;W) = -\beta\left(1 - 2\frac{\partial}{\partial W}W\right)\chi(x;h;W) -$$
$$-(1-W)\frac{\partial}{\partial W}\left\{1 - (1-W)\frac{\partial}{\partial W}W\right\}\chi(x;h;W) \tag{12.111}$$

with the initial value

$$\chi(x;x;W) = (1+W)\,P(x;W). \tag{12.112}$$

Function $P(h;W)$ is the probability density of quantity $|R_h|^2$ and satisfies the equation

$$\frac{\partial}{\partial h}P(h;W) = -2\beta\frac{\partial}{\partial W}WP(h;W)$$
$$-\frac{\partial}{\partial W}(1-W)\left\{1 - \frac{\partial}{\partial W}(1-W)\,W\right\}P(h;W), \quad P(h_0;W) = \delta(W - |R_0|^2). \tag{12.113}$$

Introduce new variables $\xi = x_0 - x$ and $\eta = x - h_0$. In this case, we have $\chi(x;h;W) = \chi(\xi;\eta;W)$, and function $\chi(\xi;\eta;W)$ satisfies the equation

$$\frac{\partial}{\partial \xi}\chi(\xi;\eta;W) = -\beta\left(1 - 2\frac{\partial}{\partial W}W\right)\chi(\xi;\eta;W)$$
$$-(1-W)\frac{\partial}{\partial W}\left\{1 - (1-W)\frac{\partial}{\partial W}W\right\}\chi(\xi;\eta;W), \quad \chi(0;\eta;W) = (1+W)P(\eta;W), \tag{12.114}$$

where function $P(\eta;W)$ satisfies the equation

$$\frac{\partial}{\partial \eta}P(\eta;W) = -2\beta\frac{\partial}{\partial W}WP(\eta;W)$$
$$-\frac{\partial}{\partial W}(1-W)\left\{1 - \frac{\partial}{\partial W}(1-W)\,W\right\}P(\eta;W), \quad P(0;W) = \delta(W - |R_0|^2). \tag{12.115}$$

Thus, in the case $x_0 > x$, determination of the wavefield average intensity assumes solving Eqs. (12.114) and (12.115) and calculating quadrature (12.110). In the case $x_0 > x$,

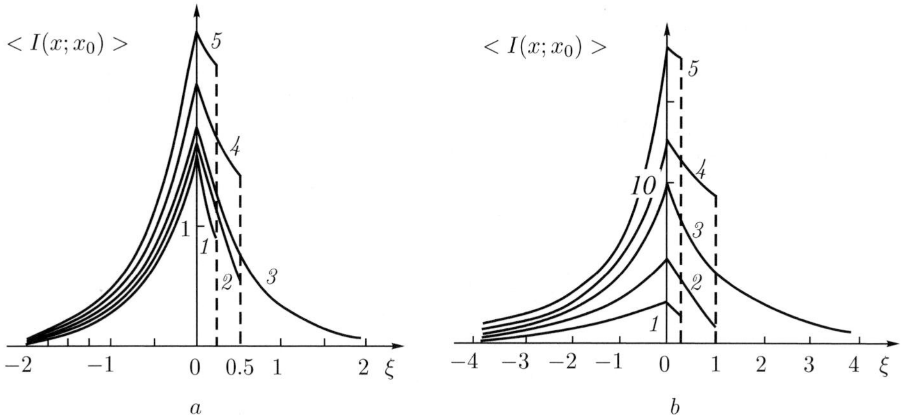

Figure 12.13: Distribution of average energy of the field generated by a source in the bounded medium for (a) $\beta = 1$ and (b) $\beta = 0.08$. Curves *1* and *2* correspond to the transmitting boundary, curve *3* corresponds to infinite space, and curves *4* and *5* correspond to the reflecting boundary.

Eqs. (12.114) and (12.115) remain valid, but we must replace variables ξ and η with the expressions $\xi = x - x_0$ and $\eta = x_0 - h$.

The case of the source in infinite space corresponds to the limit process $\eta \to \infty$ in Eq. (12.115). In this case, Eq. (12.115) has the steady-state solution, and the problem reduces to solving Eq. (12.114) with the initial value $\chi(0; \eta; W) = (1 + W)P(W)$. We analyzed this case in the foregoing subsection.

The magnitude or reflection coefficient $|R_0|^2$ appearing in Eq. (12.115) depends on medium parameters in region $x < L_0$. The case $R_0 = 0$ corresponds to the free wave penetration through the layer boundary. The limiting case of reflecting boundaries corresponds to $|R_0|^2 = 1$, and the above theory does not distinguish between the cases $R_0 = \pm 1$. The reason of this fact lies in averaging over fast oscillations. A similar situation is characteristic of the linear phenomenological theory of radiative transfer, which corresponds to the asymptotic case $\beta \gg 1$.

Numerical integration of Eqs. (12.114) and (12.115) was performed in paper [173] (see also [135, 136]). The calculations were carried out for $\beta = 1$ and $\beta = 0.08$. In the first case ($\beta = 1$), the expected result must nearly coincide with the result of the linear phenomenological theory of radiative transfer. The case $\beta = 0.08$ corresponds to a more stochastic problem. Figure 12.13a shows the curves of average wave intensity in the half-space calculated in the case of $\beta = 1$ for different positions of the boundary (the dashed lines) and different boundary conditions. In the case of penetrating boundary ($R_0 = 0$), the curves run below the corresponding curves for the case of the source in infinite space. In the case of reflecting boundary ($|R_0|^2 = 1$), the curves run above. Figure 12.13b shows the similar curves calculated for $\beta = 0.08$. The behavioral tendency of the curves remains unchanged; however, variations appear more prominent here.

Asymptotic case of small dissipation

Consider now the asymptotic solution of the problem on the plane wave source in infinite space ($L_0 \to -\infty$, $L \to \infty$) under the condition $\beta \to 0$. In this case, it appears convenient to calculate the average wavefield intensity in region $x < x_0$ using relationships (12.15) and (12.16), page 281

$$\beta \langle I(x;x_0)\rangle = \frac{1}{D}\frac{\partial}{\partial x}\langle S(x;x_0)\rangle = \frac{1}{D}\frac{\partial}{\partial x}\langle \psi(x;x_0)\rangle ,$$

where

$$\psi(x;x_0) = \exp\left\{-\beta D\int\limits_x^{x_0} d\xi \frac{|1+R_\xi|^2}{1-|R_\xi|^2}\right\},$$

so that this function satisfies, as a function of parameter x_0, the equation

$$\frac{\partial}{\partial x_0}\psi(x;x_0) = -\beta D\frac{|1+R_{x_0}|^2}{1-|R_{x_0}|^2}\psi(x;x_0), \quad \psi(x;x) = 1.$$

Introduce function

$$\Phi(x;x_0;u) = \psi(x;x_0)\delta(u_{x_0}-u), \tag{12.116}$$

where function $u_L = (1+W_L)/(1-W_L)$ satisfies the stochastic system of equations (12.23). Differentiating Eq. (12.116) with respect to x_0, we obtain the stochastic equation

$$\frac{\partial}{\partial x_0}\Phi(x;x_0;u) = -\beta D\left\{u+\sqrt{u^2-1}\cos\phi_{x_0}\right\}\Phi(x;x_0;u)$$
$$+\beta D\frac{\partial}{\partial u}\left\{\left(u^2-1\right)\Phi(x;x_0;u)\right\} - k\varepsilon_1(x_0)\frac{\partial}{\partial u}\left\{\sqrt{u^2-1}\sin\phi_{x_0}\Phi(x;x_0;u)\right\}. \tag{12.117}$$

Average now Eq. (12.117) over an ensemble of realizations of random process $\varepsilon_1(x_0)$ assuming it, as earlier, the Gaussian process delta-correlated in x_0. Using the Furutsu–Novikov formula (7.10), page 186, the following expression for the variational derivatives

$$\frac{\delta\Phi(x;x_0;u)}{\delta\varepsilon_1(x_0)} = -k\frac{\partial}{\partial u}\left\{\sqrt{u^2-1}\sin\phi_{x_0}\Phi(x;x_0;u)\right\},$$
$$\frac{\delta\phi_{x_0}}{\delta\varepsilon_1(x_0)} = k\left[1+\frac{u_{x_0}}{\sqrt{u_{x_0}^2-1}}\right]\cos\phi_{x_0},$$

and additionally averaging over fast oscillations (over the phase of the reflection coefficient), we obtain that function

$$\Phi(\xi;u) = \langle\Phi(x;x_0;u)\rangle = \langle\psi(x;x_0)\delta(u_{x_0}-u)\rangle ,$$

where $\xi = D|x-x_0|$, satisfies the equation

$$\frac{\partial}{\partial\xi}\Phi(\xi;u) = -\beta u\Phi(\xi;u)$$
$$+\beta\frac{\partial}{\partial u}\left(u^2-1\right)\Phi(\xi;u) + \frac{\partial}{\partial u}\left(u^2-1\right)\frac{\partial}{\partial u}\Phi(\xi;u),$$
$$\Phi(0;u) = P(u) = \beta e^{-\beta(u-1)}. \tag{12.118}$$

The average intensity can now be represented in the form

$$\beta\left\langle I(x;x_0)\right\rangle = -\frac{\partial}{\partial\xi}\int\limits_1^\infty du\Phi(\xi;u) = \beta\int\limits_1^\infty duu\Phi(\xi;u).$$

Equation (12.118) allows limiting process $\beta\to 0$. As a result, we obtain a simpler equation

$$\begin{aligned}\frac{\partial}{\partial\xi}\tilde{\Phi}(\xi;u) &= -u\tilde{\Phi}(\xi;u)+\frac{\partial}{\partial u}u^2\tilde{\Phi}(\xi;u)+\frac{\partial}{\partial u}u^2\frac{\partial}{\partial u}\tilde{\Phi}(\xi;u),\\ \tilde{\Phi}(0;u) &= e^{-u}.\end{aligned} \tag{12.119}$$

Consequently, localization of average intensity in space is described by the quadrature

$$\Phi_{\mathrm{loc}}(\xi) = \int\limits_1^\infty duu\tilde{\Phi}(\xi;u),$$

where

$$\Phi_{\mathrm{loc}}(\xi) = \lim_{\beta\to 0}\beta\left\langle I(x;x_0)\right\rangle = \lim_{\beta\to 0}\frac{\left\langle I(x;x_0)\right\rangle}{\left\langle I(x_0;x_0)\right\rangle}.$$

Thus, the average intensity of the wavefield generated by the point source has for $\beta\ll 1$ the following asymptotic behavior

$$\left\langle I(x;x_0)\right\rangle = \frac{1}{\beta}\Phi_{\mathrm{loc}}(\xi). \tag{12.120}$$

Equation (12.119) can be easily solved with the use of the Kantorovich–Lebedev transform (see Sect. 8.2, page 203); as a result, we obtain the expression for the *localization curve* [137]–[139]

$$\Phi_{\mathrm{loc}}(\xi) = 2\pi\int\limits_0^\infty d\tau\,\tau\left(\tau^2+\frac{1}{4}\right)\frac{\sinh(\pi\tau)}{\cosh^2(\pi\tau)}e^{-\left(\tau^2+\frac{1}{4}\right)\xi}. \tag{12.121}$$

Note that, structurally, Eq. (12.121) can be represented in the form

$$\Phi_{\mathrm{loc}}(\xi) = -\frac{\partial}{\partial\xi}|T_\xi|^2,$$

where $|T_\xi|^2$ is the squared modulus of the transmission coefficient of a wave incident on medium layer of thickness ξ (see Eq. (12.27), page 284).

For small distances ξ, the localization curve decays according to relatively fast law

$$\Phi_{\mathrm{loc}}(\xi)\approx e^{-2\xi}. \tag{12.122}$$

For great distances ξ (namely, for $\xi\gg\pi^2$), it decays significantly slower, according to the universal law

$$\Phi_{\mathrm{loc}}(\xi)\approx\frac{\pi^2\sqrt{\pi}}{8}\frac{1}{\xi\sqrt{\xi}}e^{-\xi/4}, \tag{12.123}$$

but for all that

$$\int\limits_0^\infty d\xi\Phi_{\mathrm{loc}}(\xi) = 1.$$

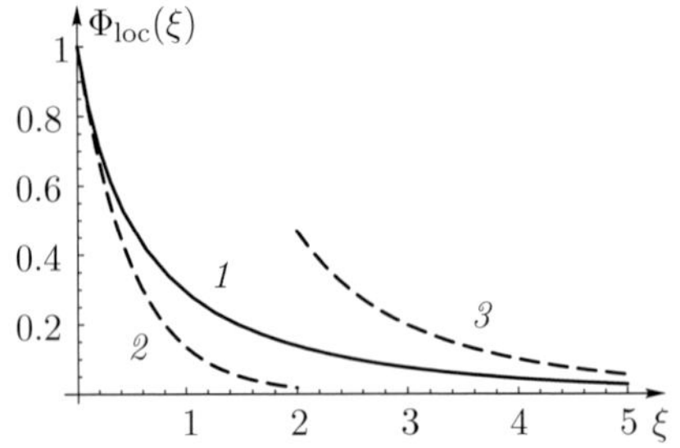

Figure 12.14: Localization curve for a source in infinite space (12.121) (curve *1*). Curves *2* and *3* correspond to asymptotic expressions for small and large distances from the source.

Function (12.121) is given in Fig. 12.14, where asymptotic curves (12.122) and (12.123) are also shown for comparison purposes.

Localization curve (12.121) corresponds to the double limit process

$$\Phi_{\rm loc}(\xi) = \lim_{\beta\to 0} \lim_{\substack{L_0\to -\infty \\ L\to\infty}} \frac{\langle I(x;x_0)\rangle}{\langle I(x_0;x_0)\rangle},$$

and one can easily see that these limit process are not permutable.

A similar situation occurs in the case of the plane wave source located at the reflecting boundary. In this case, we obtain the expression

$$\lim_{\beta\to 0} \frac{\langle I_{\rm ref}(x;L)\rangle}{\langle I_{\rm ref}(L;L)\rangle} = \frac{1}{2}\Phi_{\rm loc}(\xi), \quad \xi = D(L-x). \tag{12.124}$$

This result is valid in region $\xi > 1/3$, because it is obtained neglecting correlation $|\langle R_x R_L^* \rangle| = e^{-3\xi}$, unlike the case of the source in infinite space (see, Remark 11, page 290).

12.3.3 Numerical simulation

The above theory rests on two simplifications—on using the delta-correlated approximation of function $\varepsilon_1(x)$ (or the diffusion approximation) and extracting slow (within the scale of a wavelength) variations of statistical characteristics by averaging over fast oscillations. Averaging over fast oscillations is validated for statistical characteristics of the reflection coefficient only in the case of random medium occupying a half-space. For statistical characteristics of the wavefield intensity in medium, the corresponding validation appears very difficult if at all possible (this method is merely physical than mathematical). Numerical simulation of the exact problem offers a possibility of both verifying these simplifications and obtaining results concerning more difficult situations for which no analytic results exists.

In principle, such numerical simulation could be performed by way of multiply solving the problem for different realizations of medium parameters followed by averaging the obtained solutions over an ensemble of realizations (see, e.g., paper [183], where this procedure was carried out for the problem on the field of a point source). However, such an approach is not very practicable because it requires a vast body of realizations of medium

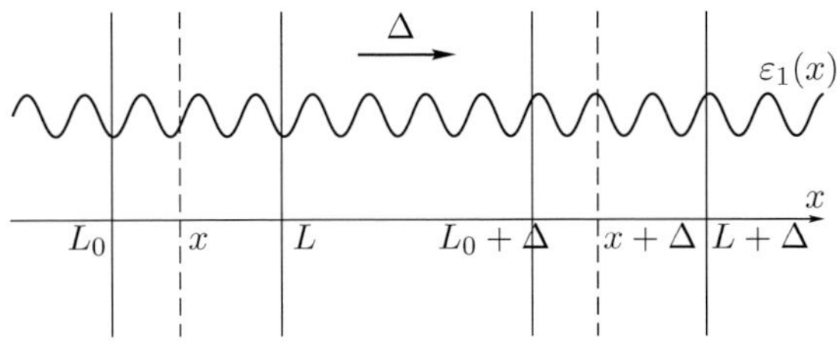

Figure 12.15: Averaging over parameter Δ by the procedure based on ergodicity of imbedding equations for a half-space of random medium.

parameters. Moreover, it is unsuitable for real physical problems, such as wave propagation in Earth's atmosphere and ocean, where only a single realization is usually available. A more practicable approach is based on the ergodic property of boundary-value problem solutions with respect to the displacement of the problem along the single realization of function $\varepsilon_1(x)$ defined along the half-axis (L_0, ∞) (see Fig. 12.15). This approach assumes that statistical characteristics are calculated by the formula

$$\langle F(L_0; x, x_0; L)\rangle = \lim_{\delta\to\infty} F_\delta(L_0; x, x_0; L),$$

where

$$F_\delta(L_0; x, x_0; L) = \frac{1}{\delta}\int\limits_0^\delta d\Delta F(L_0 + \Delta; x + \Delta, x_0 + \Delta; L + \Delta).$$

In the limit of a half-space ($L_0 \to -\infty$), statistical characteristics are independent of L_0, and, consequently, the problem possesses ergodic property with respect to the position of the right-hand layer boundary L (simultaneously, parameter L is the variable of the imbedding method), because this position is identified in this case with the displacement parameter. As a result, having solved the imbedding equation for the sole realization of medium parameters, we simultaneously obtain all desired statistical characteristics of this solution by using the obvious formula

$$\langle F(x, x_0; L)\rangle = \frac{1}{\delta}\int\limits_0^\delta d\xi F\left(\xi, \xi + x_0 - x; \xi + (L - x_0) + (x_0 - x)\right)$$

for sufficiently large interval $(0, \delta)$. This approach offers a possibility of calculating even the wave statistical characteristics that cannot be obtained within the framework of current statistical theory, and this calculation requires no additional simplifications.

In the case of the layer of finite thickness, the problem is not ergodic with respect to parameter L. However, the corresponding solution can be expressed in terms of two independent solutions of the problem on the half-space [160] and, consequently, it can be reduced to the problem ergodic with respect to L.

Systematically, the program of numerical simulation was implemented in paper [174] (see also [135, 136, 316]). In simulations, the following values of parameters $\alpha = k/D$ (dimensionless wave number) and $\beta = k\gamma/D$ (characteristics of the degree of stochasticity of the problem) were used

$$\alpha = 25, \quad \beta = 1;\ 0.08.$$

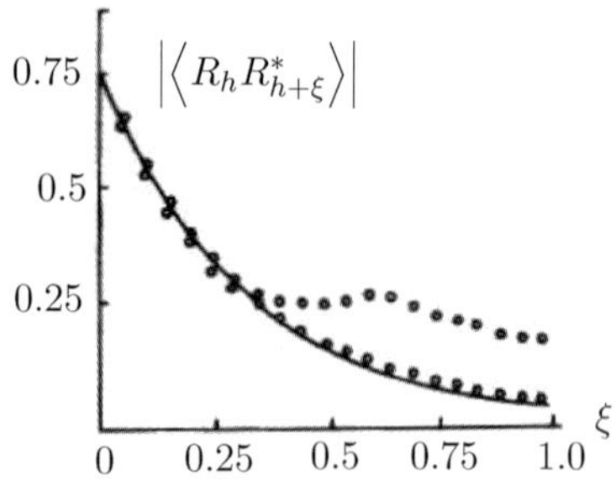

Figure 12.16: Modulus of reflection coefficient correlation function $\left|\left\langle R_h R^*_{h+\xi}\right\rangle\right|$ at $\beta = 0.08$ as a function of parameter ξ. The solid line corresponds to ensemble averaging, the circles ($\circ$) correspond to averaging over the realization of length $L = 10$, and dots ($\bullet$) correspond to averaging over the realization of length $L = 300$.

The values of parameter β were selected from the following considerations: for $\beta = 1$, the linear phenomenological theory of radiative transfer is approximately adequate, and $\beta = 0.08$ corresponds to a more stochastic problem in which case the linear theory fails. Moreover, some analytic curves are available for these values of parameter β (the curves obtained by analytic averaging over an ensemble of realizations), which offers a possibility of comparison between simulated and analytic results.

Consider several particular results obtained with numerical simulation.

Wave incident on the medium layer

The first stage of simulations consisted in studying the moments of the reflection coefficient. Figure 12.16 shows the modulus of reflection coefficient correlation function. Numerical simulation shows a good agreement with the results of Remark 11, page 290, and particularly with Eq. (12.44) in the case of sufficiently thick medium layer.

The second stage of simulations consisted in studying the first and second moments of the wavefield intensity $I(x; L)$ in the problem on the wave incident on random half-space. Simultaneously, we investigated the dependence of the result on the length of sampling used for averaging. Simulated results were compared with the above theoretical results.

Figure 12.17 shows moments $\langle I(x; L)\rangle$ and $\langle I^2(x; L)\rangle$ simulated with $\beta = 1$. The calculation showed that samplings of dimensionless length $L \sim 10 - 20$ are sufficient for obtaining satisfactory results. For $\beta = 0.08$, such a sampling appears insufficient, and obtaining the adequate result requires samplings of length $L \sim 300$ (Fig. 12.18).

Plane wave source in the medium layer

Figure 12.19 shows moment $\langle I(x, x_0)\rangle$ simulated in the case of the source in infinite space for sampling length $L = 10$ and $\beta = 0.08$. The solid line corresponds to the theoret-

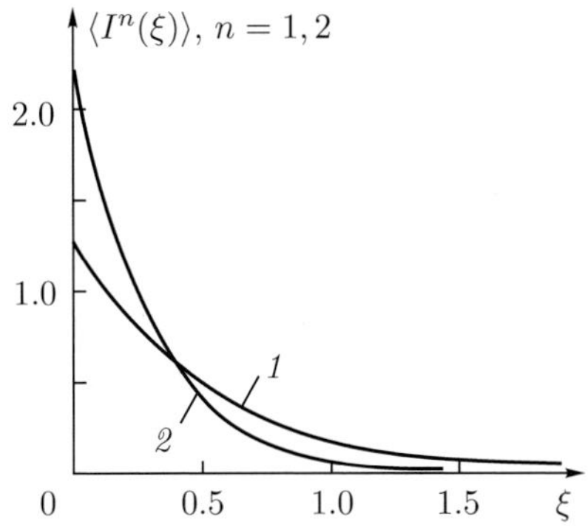

Figure 12.17: Moments of wavefield intensity in the problem on a wave incident on medium layer ($\beta = 1$). Curves *1* and *2* show average intensity $\langle I(x;L)\rangle$ and average square of the intensity $\left\langle I^2(x;L)\right\rangle$ calculated with the use of ensemble averaging.

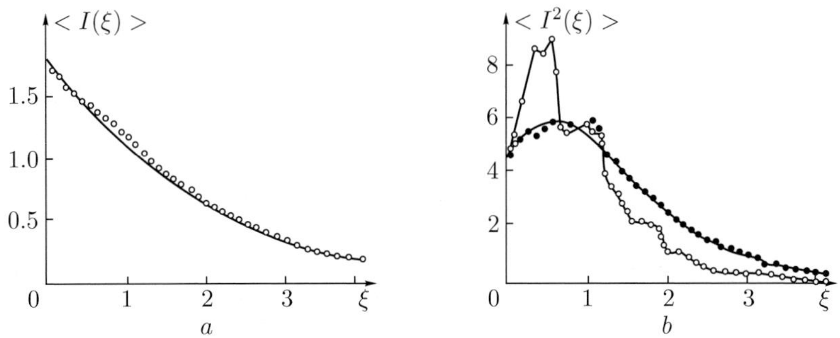

Figure 12.18: Moments of wavefield intensity in the problem on a wave incident on medium layer ($\beta = 0.08$). (a) Average intensity $\langle I(x;L)\rangle$ and (b) average square of the intensity $\left\langle I^2(x;L)\right\rangle$. The solid lines correspond to ensemble averaging, circles ($\circ$) correspond to averaging over a realization of length $L = 10$, and dots ($\bullet$) correspond to averaging over a realization of length $L = 300$.

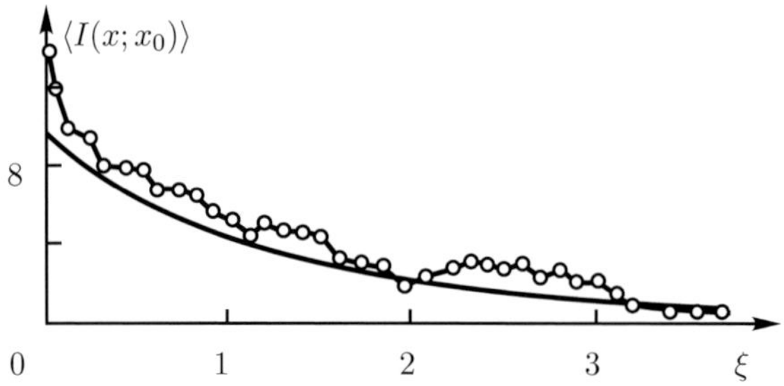

Figure 12.19: Average intensity $\langle I(x,x_0)\rangle$ of the field generated by a source in infinite space ($\beta = 0.08$). The solid line corresponds to ensemble averaging and circles ($\circ$) correspond to averaging over a realization of length $L = 10$.

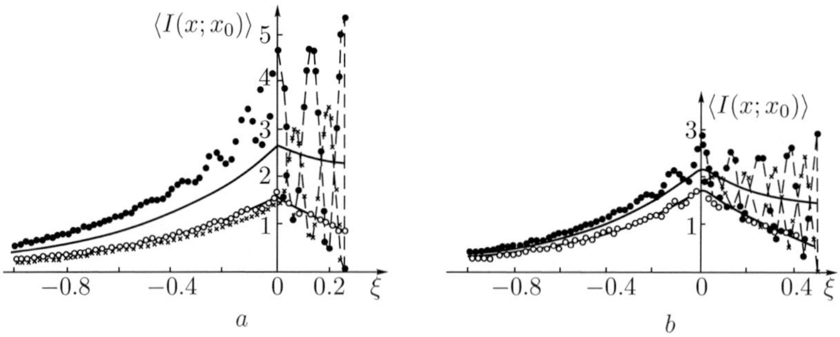

Figure 12.20: Average intensity of the source-generated field for $\beta = 1$ and boundary positions (a) $DH = 0.25$ and (b) $DH = 0.5$. The solid lines correspond to ensemble averaging, circles ($\circ$) correspond to simulations for free passage through the boundary, dots ($\bullet$) correspond to simulations for reflecting boundary with the condition $dG(H;x_0)/dx = 0$, and crosses ($\times$) correspond to simulations for reflecting boundary with the condition $G(H;x_0) = 0$.

ical result. We can see from this graph that even such short sampling adequately catches the behavioral tendency of average intensity of the field generated by a source in infinite space. All other curves were obtained with sampling length $L = 300 - 400$.

Figure 12.20 shows average intensity of the field generated by a source simulated with $\beta = 1$ for different boundary conditions. Again, the solid lines correspond to theoretical results. Figure 12.20 shows that simulated results are in adequate agreement with theoretical curves in the case of the penetrating boundary; at the same time, it shows that, in the case of reflecting boundary, average intensity is strongly oscillating, which indicates that the interference pattern of average intensity appears complicated even at $\beta = 1$. The amplitude of oscillations decreases with moving the source away from the boundary.

Figure 12.21 shows similar curves simulated with $\beta = 0.08$. This case is characterized by more intense variations of function $\langle I(x,x_0)\rangle$. Again, the amplitude of oscillations decreases with moving the source away from the boundary.

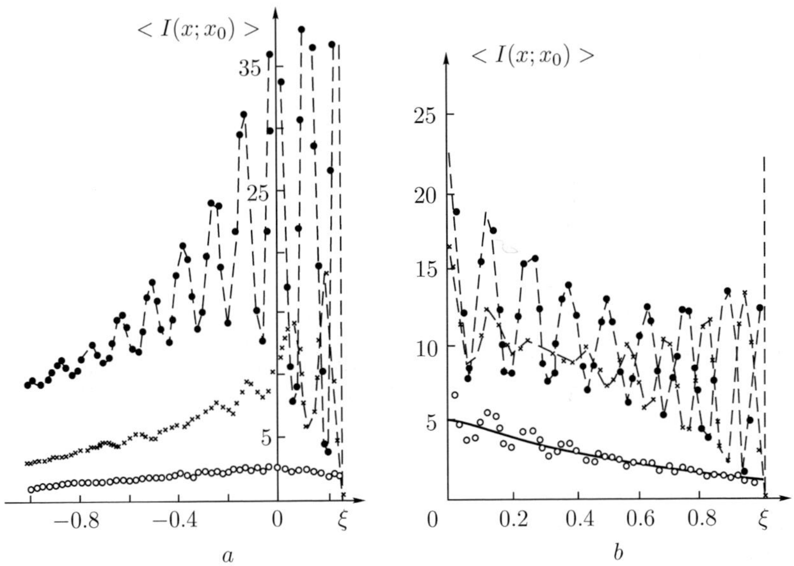

Figure 12.21: Average intensity of the source-generated field for $\beta = 0.08$ and boundary positions (*a*) $DH = 0.25$ and (*b*) $DH = 1$. The solid line corresponds to ensemble averaging, circles ($\circ$) correspond to simulations for free passage through the boundary, dots ($\bullet$) correspond to simulations for reflecting boundary with the condition $dG(H; x_0)/dx = 0$, and crosses ($\times$) correspond to simulations for reflecting boundary with the condition $G(H; x_0) = 0$.

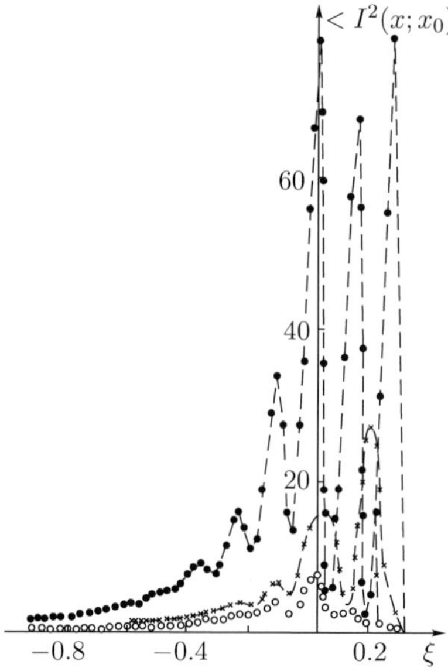

Figure 12.22: Second moment of intensity of the source-generated field for $\beta = 1$ and boundary position $DH = 0.25$. Circles ($\circ$) correspond to simulations for free passage through the boundary, dots ($\bullet$) correspond to simulations for reflecting boundary with the condition $dG(H; x_0)/dx = 0$, and crosses ($\times$) correspond to simulations for reflecting boundary with the condition $G(H; x_0) = 0$.

The method of numerical simulations enables us to find the statistical characteristics that cannot be determined theoretically yet. Figures 12.22 and 12.23 show the simulated second moments of intensity of the field generated by a source $\langle I^2(x, x_0)\rangle$ for $\beta = 1$ and $\beta = 0.08$ and different boundary conditions. The second moments oscillate with the same period, but oscillating amplitude significantly increases.

As can be seen from the above figures, the oscillations of period ~ 0.13 are characteristic of the moments of wavefield intensity in the presence of boundary. These oscillations are related to our choice of wave parameter $\alpha = 25$, because the corresponding period $T = \pi/\alpha = 0.126$.

The case of the point source located at reflecting boundary $x_0 = L$ with boundary condition $dG(x, x_0; L)/dx|_{x=L} = 0$ was considered in paper [138]. Figure 12.24 shows the quantity $\langle I_{\text{ref}}(x, x_0)\rangle$ simulated with $\beta = 0.08$ and $k/D = 25$. In region $\xi = D(L-x) < 0.3$, one can see oscillations of period $T = 0.13$. For larger ξ, simulated results agree well with localization curve (12.124).

Nonlinear problem on wave self-action in random media

Consider now the results of simulating the nonlinear problem on wave self-action in the medium whose parameter ε_1 is described by the model $\varepsilon_1(x, J(x, w)) = J(x, w) + \varepsilon_1(x)$, where $\varepsilon_1(x)$ is the Gaussian delta-correlated random process, $J(x, w) = w|u(x, w)|^2$, $u(x, w)$ is the wavefield in the nonlinear medium, and w is the intensity of incident wave [82], [312]–[315] (see Appendix C, page 470). A distinction of this problem in the case of

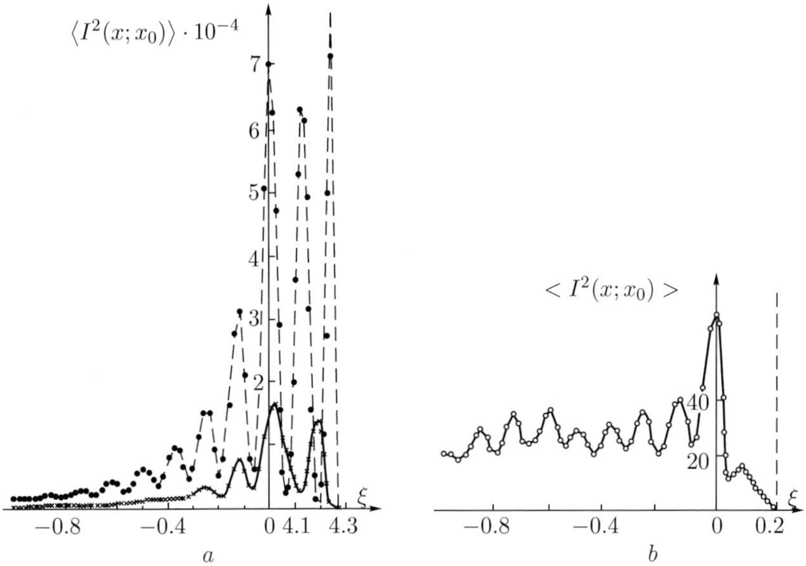

Figure 12.23: Second moment of intensity of the source-generated field for $\beta = 0.08$ and (a) reflecting boundary at $DH = 4.3$ and (d) freely penetrating boundary at $DH = 0.25$. Circles (∘) correspond to simulations for free passage through the boundary, dots (•) correspond to simulations for reflecting boundary with the condition $dG(H;x_0)/dx = 0$, and crosses (×) correspond to simulations for reflecting boundary with the condition $G(H;x_0) = 0$.

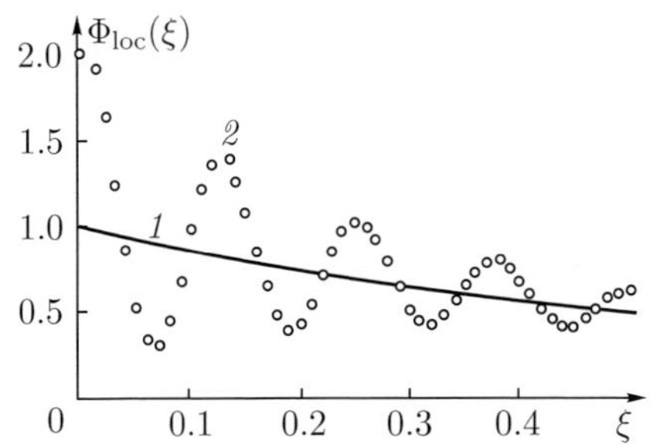

Figure 12.24: Average intensity $2\langle I_{\text{ref}}(x;L)\rangle / \langle I_{\text{ref}}(L;L)\rangle$ of the field of a source located at reflecting boundary ($\beta = 0.08$). Curve *1* shows the localization curve (12.121) and circles *2* show the simulated result.

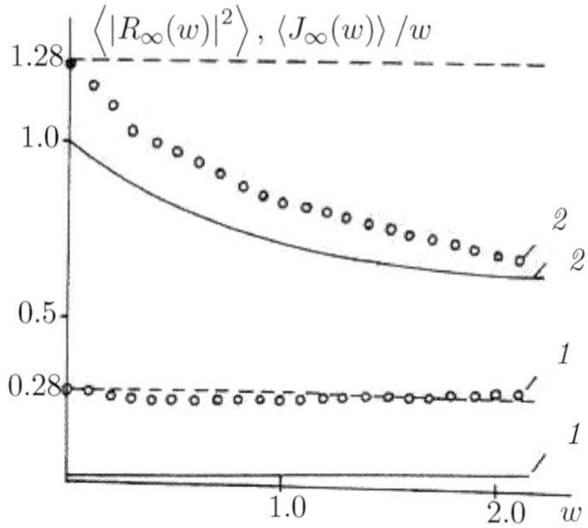

Figure 12.25: Simulated quantities $\langle |R|^2 \rangle$ and $\langle J(w) \rangle /w$ as functions of parameter w. Circles show the simulated quantities. Label *1* refers to quantity $\langle |R|^2 \rangle$ and label *2*, to $\langle J(w) \rangle /w$. The solid curves correspond to the solution for absent fluctuations ε and the dashed lines show the solution of the linear stochastic problem.

absent medium parameter fluctuations consists in the uniqueness and smoothness of the solution for arbitrary attenuation. However, in the presence of fluctuations, the solution may become nonunique, which depends on parameter β. Figure 12.25 shows the reflection coefficient squared modulus $\langle |R_\infty|^2 \rangle$ and normalized wavefield intensity at boundary $\langle J_\infty(w) \rangle /w$ as functions of incident wave intensity w, which were simulated for the case of random half-space ($L_0 \to -\infty$). This simulation was carried out with $\beta = 1$, which corresponds, from the one hand, to the moderate effect of statistics in the linear problem and, from the other hand, to the absence of non-uniquenesses in the nonlinear problem for the values of parameter w used in this simulation. As may be seen from Fig. 12.25, the medium only weakly reflects the incident wave for $w < 2$ (quantity $\langle |R_\infty|^2 \rangle$ is relatively small). In this case, reflecting properties of the medium are mainly governed by fluctuations of inhomogeneities, and quantity $\langle |R_\infty|^2 \rangle$ nearly coincide with that of the linear problem. Nevertheless, for wavefield intensity at the medium boundary, nonlinearity becomes significant even for small w, and quantity $\langle J_\infty(w) \rangle /w$ tends to the solution of the deterministic nonlinear problem with increasing w.

Figure 12.26 presents a more complete pattern of the effect of statistics and nonlinearity. It shows the simulated results for $\langle J(\xi, w) \rangle /w$ (circles, $\xi = D(L-x)$) and the corresponding solutions of both deterministic nonlinear and linear stochastic problems for incident waves of small ($w = 0.2$) and great ($w = 2$) intensities. In the case $w = 0.2$, the wavefield in the medium is first governed by fluctuations of medium inhomogeneities, and function $\langle J(\xi, w) \rangle /w$ is relatively close to the solution of the linear problem. In the case of incident wave with $w = 2$, near the boundary, function $\langle J(\xi, w) \rangle /w$ shows oscillations relative to the solution of the deterministic nonlinear problem (caused by the interference of the forward and reflected waves) and then decreases running between the two limiting solutions. This means that there is certain region in space (in which wavefield intensity is sufficiently great $\langle J(\xi, w) \rangle \gg 1$) where nonlinear effects dominate and statistical effects result only in

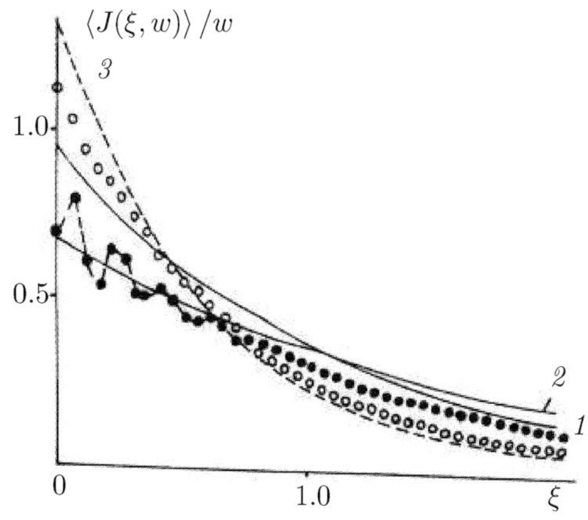

Figure 12.26: Simulated quantity $\langle J(x,w)\rangle/w$ (circles and dots). The solid lines show the solution of the deterministic problem (curve *1* corresponds to $w = 0.2$ and curve *2*, to $w = 2$), the dashed line shows the solution of the linear stochastic problem. Circles ($\circ$) correspond to $w = 0.2$ and dots ($\bullet$) correspond to $w = 2$.

interference phenomena. When the wave penetrates deep in medium and function $\langle J(\xi,w)\rangle$ becomes sufficiently small, the wavefield behavior in the medium is governed by medium parameter fluctuations.

12.4 Eigenvalue and eigenfunction statistics

In the foregoing section, we considered in detail statistical characteristics of the wavefield in random medium. We discussed the problems on the wave incident on a medium layer (half-space) and on the waves generates by a source in the medium. In parallel with the above problems, physics of disordered systems (see, e.g., [223]) places high emphasis on studying the statistics of *eigenvalues* of the Helmholtz equation (energy levels of the Schrödinger equation) for bounded randomly inhomogeneous systems. Wave propagation in different waveguides is an additional example of such problems (see, e.g., [269]). In the general case of many-dimensional systems, the analysis of eigenvalue and eigenfunction statistics faces great difficulties. However, in the one-dimensional case (plane layered media) the consideration appears significantly simpler.

In Appendix C we derived the system of dynamic equations that describes the behavior of eigenvalues (as functions of layer thickness) and appears quite appropriate for studying eigenvalue statistical characteristics.

12.4.1 General remarks

The eigenvalue analysis suggested in Appendix C, page 464, rests on analyzing zeros of the solution to the Riccati equation whose general form is as follows

$$\frac{d}{dL}f_L(\lambda) = a(L,\lambda) + b(L,\lambda)f_L(\lambda) + c(L,\lambda)f_L^2(\lambda). \tag{12.125}$$

To describe the indicator function of the solution to Eq (12.125) whose average over an ensemble of realizations of fluctuating parameters coincides with the probability density of the solution to Eq. (12.125), we introduce two functions

$$\varphi(L;\lambda;f)=\delta\left(f_L(\lambda)-f\right),\quad \Phi(L;\lambda;f)=A(L,\lambda)\delta\left(f_L(\lambda)-f\right), \tag{12.126}$$

where

$$A(L,\lambda)=\frac{\partial}{\partial\lambda}f_L(\lambda).$$

It is clear that these function are mutually related through the relationship

$$\frac{\partial}{\partial\lambda}\varphi(L;\lambda;f)=-\frac{\partial}{\partial f}\Phi(L;\lambda;f). \tag{12.127}$$

In view of the fact that function $f_L(\lambda)$ satisfies the initial value problem, the indicator function $\varphi(L;\lambda;f)=\delta\left(f_L(\lambda)-f\right)$ satisfies the Liouville equation (which is the stochastic equation if medium parameters fluctuate)

$$\frac{\partial}{\partial L}\varphi(L;\lambda;f)=-\frac{\partial}{\partial f}J(L;\lambda;f), \tag{12.128}$$

where

$$J(L;\lambda;f)=-\frac{df_L(\lambda)}{dL}\varphi(L;\lambda;f)$$

is the function whose average value is the probability flux density.

Consider derivative $\frac{\partial}{\partial L}\Phi(L;\lambda;f)$. Using relationship (12.127) and Eq. (12.128), we can express this derivative in the form

$$\begin{aligned}\frac{\partial}{\partial L}\Phi(L;\lambda;f) &= \frac{\partial A(L;\lambda)}{\partial L}\Phi(L;\lambda;f)-A(L;\lambda)\frac{\partial}{\partial f}\frac{df_L(\lambda)}{dL}\varphi(L;\lambda;f)\\ &= \frac{\partial A(L;\lambda)}{\partial L}\varphi(L;\lambda;f)-\frac{df_L(\lambda)}{dL}\frac{\partial}{\partial f}\Phi(L;\lambda;f)=\frac{\partial}{\partial\lambda}J(L;\lambda;f).\end{aligned}$$

This means that function $\Phi(L;\lambda;f)$ is simply (through a quadrature) related to function $J(L;\lambda;f)$; namely,

$$\Phi(L;\lambda;f)=\frac{\partial}{\partial\lambda}\int\limits_0^L d\xi J(\xi;\lambda;f). \tag{12.129}$$

The eigenvalues are defined as the roots of the equation

$$f_L(\lambda_L)=0. \tag{12.130}$$

Then, we have

$$\Phi(L;\lambda;0)=A(L;\lambda)\delta[f_L(\lambda)]=\sum_{n=1}^{\infty}\delta\left(\lambda-\lambda_L^{(n)}\right). \tag{12.131}$$

Consequently, *average eigenvalue density* (average number of eigenvalues per unit length) [223]

$$\rho(L;\lambda)=\frac{1}{L}\sum_{n=1}^{\infty}\delta\left(\lambda-\lambda_L^{(n)}\right)$$

is given by the expression

$$\rho(L;\lambda) = \frac{\partial}{\partial\lambda}\frac{1}{L}\int_0^L d\xi J(\xi;\lambda;0). \tag{12.132}$$

The average eigenvalue density is quite naturally a function of all eigenvalues of the initial value problem and gives no information about certain individual eigenvalue.

Differentiating Eq. (12.130) with respect to parameter L and taking into account Eq. (12.125), we obtain that eigenvalues as functions of parameter L satisfy the equation

$$a(L,\lambda_L) + A(L,\lambda_L)\frac{d}{dL}\lambda_L = 0 \tag{12.133}$$

with the additional condition that their behavior for $L \to L_0$ is predefined by system dynamics in the absence of parameter fluctuations. Consequently, the indicator function

$$\psi(L;\lambda) = \delta\left(\lambda_L^{(n)} - \lambda\right)$$

satisfies the Liouville equation

$$\frac{\partial}{\partial L}\psi(L;\lambda) = \frac{\partial}{\partial\lambda}\frac{a(L,\lambda)}{A(L,\lambda)}\psi(L;\lambda), \tag{12.134}$$

the initial value for which for $L \to L_0$ follows from the dynamics of the particular eigenvalue.

Another way of concrete definition of eigenvalue in the context of the one-dimensional problem consists in the use of the so-called *phase formalism.* The solution to the Riccati equation (12.125) varies from $-\infty$ to ∞, and we can use this fact to change variable according to one of the following formulas

$$f_L(\lambda) = \tan\phi_L(\lambda), \quad f_L(\lambda) = 1/\tan\phi_L(\lambda),$$

depending on the initial value to Eq. (12.125). In this case, eigenvalue $\lambda_L^{(n)}$ will be represented by either $\phi_L^{(n)} = n\pi$, or $\phi_L^{(n)} = \pi\,(n+1/2)$. Introducing the indicator function of quantity $\phi_L(\lambda)$

$$\psi(L;\lambda;\phi) = \delta\left(\phi_L(\lambda) - \phi\right),$$

and taking into account the fact that $\psi(L;\lambda;\phi)$ is related to function $\psi(L;\lambda) = \delta\left(\lambda_L^{(n)} - \lambda\right)$ through the relationship

$$\psi(L;\lambda;\phi_n)\frac{d}{dL}\phi_n(\lambda) = \psi(L;\lambda),$$

we can rewrite Eq. (12.134) in the form of the equality

$$\frac{\partial}{\partial L}\psi(L;\lambda) = \frac{\partial}{\partial\lambda}\frac{a(L,\lambda)}{\partial f/\partial\phi|_{\phi=\phi_n}}\psi(L;\lambda;\phi_n). \tag{12.135}$$

Thus, we expressed the indicator function of eigenvalues in terms of the indicator function of the solution to the Riccati equation (12.125) and this expression has the form of a quadrature.

12.4.2 Statistical averaging

If the medium has fluctuating parameters, all above expressions should be averaged over an ensemble of realizations of fluctuating parameters.

Consider the dynamic eigenvalue problem

$$\frac{d^2}{dx^2}u(x) + \lambda u(x) = \varepsilon(x)u(x), \quad u(0) = 0, \quad \left.\frac{du(x)}{dx}\right|_{x=L} = 0 \tag{12.136}$$

as an example.

Using the technique of the imbedding method described in the foregoing sections of this chapter, we consider instead of Eq. (12.136) the inhomogeneous problem

$$\frac{d}{dx}u(x) = v(x), \quad \frac{d}{dx}v(x) = [\varepsilon(x) - \lambda]\, u(x), \quad u(0) = 0, \quad v(L) = 1. \tag{12.137}$$

Considering now the solution to problem (12.137) as a function of parameter L, we obtain that quantity $u_L = u(L;L)$ satisfies the Riccati equation

$$\frac{d}{dL}u_L = 1 + [\lambda - \varepsilon(x)]\, u_L^2, \quad u_0 = 0. \tag{12.138}$$

The poles of the solution to this equation define the eigenvalues. Consequently, the eigenvalues correspond to zeros of function $f_L = 1/u_L$ satisfying the equation

$$\frac{d}{dL}f_L = -f_L^2 - \lambda + \varepsilon(x), \quad f_0 = \infty. \tag{12.139}$$

The indicator function $\varphi(L;\lambda;f) = \delta\left(f_L(\lambda) - f\right)$ satisfies now the stochastic Liouville equation

$$\frac{\partial}{\partial L}\varphi(L;\lambda;f) = \frac{\partial}{\partial f}\left(\lambda + f^2\right)\varphi(L;\lambda;f) - \varepsilon(L)\frac{\partial}{\partial f}\varphi(L;\lambda;f). \tag{12.140}$$

Assume now that $\varepsilon(x)$ is the delta-correlated Gaussian random process with the parameters

$$\langle\varepsilon(x)\rangle = 0, \quad \langle\varepsilon(x)\varepsilon(x')\rangle = 2\sigma_\varepsilon^2 l_0\delta(x - x').$$

Then, averaging Eq. (12.140) over an ensemble of realizations of process $\varepsilon(x)$, we obtain the Fokker–Planck equation for probability density $P(L;\lambda;f) = \langle\delta\left(f_L(\lambda) - f\right)\rangle$

$$\frac{\partial}{\partial L}P(L;\lambda;f) = \frac{\partial}{\partial f}\left(\lambda + f^2\right)P(L;\lambda;f) + D\frac{\partial^2}{\partial f^2}P(L;\lambda;f), \tag{12.141}$$

where $D = \sigma_\varepsilon^2 l_0$ is the diffusion coefficient.

For $L \to \infty$, the solution of Eq. (12.141) tends to the steady-state (independent of L) probability distribution that satisfies the equation

$$J_\infty(\lambda) = \left(\lambda + f^2\right)P_\infty(\lambda;f) + D\frac{d}{df}P_\infty(\lambda;f), \tag{12.142}$$

where $J_\infty(\lambda)$ is the constant of integration whose meaning is the steady-state probability flux density. The solution to Eq. (12.142) with the condition $P_\infty(\lambda;f) \to 0$ for $f \to -\infty$ has the form

$$P_\infty(\lambda;f) = \frac{J_\infty(\lambda)}{D}\exp\left\{-\frac{f}{D}\left(\frac{f^3}{3} + \lambda\right)\right\}\int\limits_{-\infty}^{f} d\xi \exp\left\{\frac{\xi}{D}\left(\frac{\xi^3}{3} + \lambda\right)\right\}. \tag{12.143}$$

From the normalization condition

$$\int_{-\infty}^{f} df P_{\infty}(\lambda; f) = 1,$$

we obtain the expression for the steady-state probability flux density

$$\frac{1}{J_{\infty}(\lambda)} = \frac{\sqrt{\pi}}{D^{1/3}} \int_{0}^{\infty} \frac{dx}{\sqrt{x}} \exp\left\{-\frac{x^3}{12} - \frac{\lambda x}{D^{2/3}}\right\}. \tag{12.144}$$

From Eq. (12.144), we obtain in particular the asymptotic formula for large λ, namely, for $\lambda \gg D^{2/3}$

$$J_{\infty}(\lambda) = \frac{\sqrt{\lambda}}{\pi} \quad \left(\lambda \gg D^{2/3}\right). \tag{12.145}$$

As a result, we obtain that average eigenvalue density (12.132) reduces for $L \to \infty$ and $\lambda \gg D^{2/3}$ to [223]

$$\rho(L; \lambda) = \frac{\partial}{\partial \lambda} \lim_{L \to \infty} \frac{1}{L} \int_{0}^{L} d\xi J(\xi; \lambda; 0) = \frac{\partial}{\partial \lambda} J_{\infty}(\lambda) = \frac{1}{2\pi\sqrt{\lambda}}. \tag{12.146}$$

It is obvious that this law of eigenvalue distribution is independent of the boundary condition at $x = L$ of problem (12.136). In particular, this law will hold for the boundary-value problem

$$\frac{d}{dx}u(x) = v(x), \quad \frac{d}{dx}v(x) = [\varepsilon(x) - \lambda]\, u(x), \quad u(0) = 0, \quad u(L) = 0. \tag{12.147}$$

In this case, eigenvalues coincide with zeros of function $f_L(\lambda)$ that satisfies the Riccati equation

$$\frac{d}{dL} f_L = 1 + [\lambda - \varepsilon(x)]\, f_L^2, \quad f_0 = 0. \tag{12.148}$$

Taking into account the fact that the solution to Eq. (12.148) with $\varepsilon(x) = 0$ has the form

$$f_L(\lambda) = \frac{1}{\sqrt{\lambda}} \tan\left(\sqrt{\lambda} L\right),$$

we change the variable according to the formula

$$f_L(\lambda) = \frac{1}{\sqrt{\lambda}} \tan \phi_L(\lambda). \tag{12.149}$$

Function $\phi_L(\lambda)$ satisfies then the equation

$$\frac{d}{dL}\phi_L(\lambda) = \sqrt{\lambda} - \frac{1}{\sqrt{\lambda}}\varepsilon(L) \sin^2 \phi_L(\lambda), \quad \phi_0(\lambda) = 0, \tag{12.150}$$

and eigenvalues correspond to the following values of function ϕ_L

$$\phi_L^{(n)} = n\pi \quad (n = 1, 2, \ldots). \tag{12.151}$$

In the case of the Gaussian delta-correlated random process $\varepsilon(L)$, the probability density of the solution to Eq. (12.150), i.e., function $P(L;\lambda;\phi) = \langle\psi(L;\lambda;\phi)\rangle$ satisfies the Fokker–Planck equation

$$\frac{\partial}{\partial L}P(L;\lambda;\phi) = -\sqrt{\lambda}\frac{\partial}{\partial\phi}P(L;\lambda;\phi) + \frac{D}{\lambda}\frac{\partial}{\partial\phi}\sin^2\phi\frac{\partial}{\partial\phi}\sin^2\phi P(L;\lambda;\phi). \tag{12.152}$$

Consequently, in the context of the problem under consideration, we can express probability density $P_n(L;\lambda)$ of eigenvalue $\lambda^{(n)}$ in terms of the solution to Eq. (12.152) (see Eq. (12.135)

$$\frac{\partial}{\partial L}P_n(L;\lambda) = \frac{\partial}{\partial\lambda}\sqrt{\lambda}P(L;\lambda;\phi_n). \tag{12.153}$$

Integrating this expression, we obtain

$$P_n(L;\lambda) = \frac{\partial}{\partial\lambda}\sqrt{\lambda}\int\limits_0^L d\xi P(\xi;\lambda;\phi_n). \tag{12.154}$$

Another expression for $P_n(L;\lambda)$ can be obtained by integrating Eq. (12.152) over ϕ in limits $(-\infty,\phi_n)$ with allowance for the condition $\sin\phi_n = 0$

$$P_n(L;\lambda) = -\frac{\partial}{\partial\lambda}\int\limits_0^{\phi_n} d\theta P(\xi;\lambda;\theta). \tag{12.155}$$

Of cause, this expression is equivalent to Eq. (12.154).

Thus, determination of function $P_n(L;\lambda)$ requires the knowledge of the solution to Eq. (12.152). It is hardly possible to solve Eq. (12.152) in the general case. If parameter λ assumes sufficiently large values, namely $\lambda \gg D^{2/3}$, we can use the approximate method of averaging over fast oscillations that appear in the problem solution for absent fluctuations in which case function $\phi_n = \sqrt{\lambda}L$.

In this case, we consider only slow variations of function $\phi_L(\lambda)$ caused by fluctuations and obtain the simpler equation for the probability density

$$\frac{\partial}{\partial L}P(L;\lambda;\phi) = -\sqrt{\lambda}\frac{\partial}{\partial\phi}P(L;\lambda;\phi) + \frac{3D}{8\lambda}\frac{\partial^2}{\partial\phi^2}P(L;\lambda;\phi). \tag{12.156}$$

The solution to this equation with the initial value $P(0;\lambda;\phi) = \delta(\phi)$ has the form of the Gaussian probability density

$$P(L;\lambda;\phi) = \sqrt{\frac{2\lambda}{3\pi DL}}\exp\left\{-\frac{2\lambda}{3DL}\left(\phi-\sqrt{\lambda}L\right)^2\right\}, \tag{12.157}$$

which means that $\phi_L(\lambda)$ as a function of parameter L is the Gaussian random function with the characteristics

$$\langle\phi_L(\lambda)\rangle = 0, \quad \sigma_\phi^2 = \frac{3}{4\lambda}DL.$$

Using Eq. (12.155), we obtain the expression for the probability density of the n-th eigenvalue

$$P_n(L;\phi) = \sqrt{\frac{L}{6\pi D}}\left[2-\sqrt{\frac{\lambda_{0n}}{\lambda}}\right]\exp\left\{-\frac{2\lambda L}{3D}\left(\sqrt{\lambda}-\sqrt{\lambda_{0n}}\right)^2\right\}, \tag{12.158}$$

where $\lambda_{0n} = n^2\pi^2/L^2$ is the eigenvalue of problem (12.147) in the absence of fluctuations [79], [80].

Note that, formally, Eq. (12.158) cannot be identified with the probability density, because it assumes negative values for $\lambda < \lambda_{0n}/4$, which is a consequence of averaging over fast oscillations.

In the case of sufficiently small variance σ_ε^2, function $P_n(L;\phi)$ is localized around $\lambda \approx \lambda_{0n}$; in this region, it can be represented in the form of the Gaussian distribution [269]

$$P_n(L;\phi) = \sqrt{\frac{L}{6\pi D}} \exp\left\{-\frac{L}{3D}(\lambda - \lambda_{0n})^2\right\}, \tag{12.159}$$

from which follows that

$$\langle\lambda_n\rangle = 0, \quad \sigma^2_{\lambda_n} = \frac{3D}{2L}.$$

This means that average value of quantity λ_n coincides with the value in the absence of medium parameter fluctuations, and the variance is independent of the eigenvalue number.

Thus, eigenvalue statistics is characterized by the dimensionless diffusion coefficient of the n-th eigenvalue

$$D_n = \frac{3DL}{8\lambda_{0n}},$$

and applicability range of all above expressions is limited by the condition

$$D_n \ll 1.$$

Note that probability distribution (12.159) coincides with the first approximation of the standard perturbation theory.

Above, we considered the specific boundary-value problem (12.147) in the context of eigenvalue statistics. However, one can easily see that all above results (except the expression for λ_{0n}) will hold for other boundary conditions. As an example, for the boundary-value problem

$$\frac{d^2}{dx^2}u(x) + \lambda u(x) = \varepsilon(x)u(x), \quad u(L) = 0, \quad \left.\frac{du(x)}{dx}\right|_{x=0} = 0 \tag{12.160}$$

all results remain in force, except the expression for λ_{0n}, which assumes here the form

$$\lambda_{0n} = \frac{\left(n + \frac{1}{2}\right)^2 \pi^2}{L^2}.$$

The above analytic results hold for sufficiently small dimensionless diffusion coefficient D_n. Such a situation occurs if either variance σ_ε^2 is sufficiently small, or number n of eigenvalues is sufficiently great. Numerical simulations offer a possibility of testing the validity of the obtained results even if $D_n > 1$. Such simulations were carried out in papers [78] and [79].

These papers dealt with boundary-value problem (12.160). The simulations showed that probability distribution (12.159) adequately describes eigenvalue statistics even if diffusion coefficient $D_0 \approx 5$. The only exception is the average value of the zeroth mode, in which case

$$\langle\lambda_0\rangle - \lambda_{00} \approx -D_0.$$

However, this result corresponds to the second-order perturbation theory, or to an additional expansion of Eq. (12.158) in $(\lambda - \lambda_{0n})$. Mutual correlation coefficients of different eigenvalues λ_n are close to a value of 2/3 that follows from the perturbation theory even for $D_0 \approx 5$.

Thus, the results of simulations show that the applicability range of the obtained asymptotic results significantly exceeds the range following from the restriction $D_n \ll 1$.

12.5 Multidimensional wave problems in layered random media

Consider now extensions of the stationary problem on plane waves in randomly layered media to the simplest multidimensional problems. Among these are the nonstationary problems on propagation of time-domain impulses in randomly layered media and the three-dimensional steady-state problem on the field of a point source in layered media.

12.5.1 Nonstationary problems

Formulation of boundary-value wave problems

Consider the nonstationary problem on plane wave $f[t+(x-L)/c_0]$ (c_0 is the velocity of the wave in free space) incident from region $x > L$ on medium layer occupying the portion of space $L_0 < x < L$. The wavefield in the layer satisfies the wave equation

$$\left[\frac{\partial^2}{\partial x^2} - \frac{1}{c^2(x)}\frac{\partial}{\partial t}\left(\frac{\partial}{\partial t} + \tilde{\gamma}\right)\right] u(x,t) = 0 \tag{12.161}$$

with the boundary conditions

$$\left(\frac{\partial}{\partial x} + \frac{\partial}{c_0 \partial t}\right) u(x,t)\bigg|_{x=L} = \frac{2}{c_0}\frac{\partial}{\partial t} f(t), \quad \left(\frac{\partial}{\partial x} - \frac{\partial}{c_0 \partial t}\right) u(x,t)\bigg|_{x=L_0} = 0. \tag{12.162}$$

Similarly, for the plane wave source located at point x_0 in the medium, we have the boundary-value problem

$$\begin{gathered}\left[\frac{\partial^2}{\partial x^2} - \frac{1}{c^2(x)}\frac{\partial}{\partial t}\left(\frac{\partial}{\partial t} + \tilde{\gamma}\right)\right] u(x;x_0;t) = -\frac{2}{c_0}\delta(x-x_0)\frac{\partial}{\partial t} f(t), \\ \left(\frac{\partial}{\partial x} + \frac{\partial}{c_0 \partial t}\right) u(x;x_0;t)\bigg|_{x=L} = 0, \quad \left(\frac{\partial}{\partial x} - 1\frac{\partial}{c_0 \partial t}\right) u(x;x_0;t)\bigg|_{x=L_0} = 0.\end{gathered} \tag{12.163}$$

Note that boundary-value problem (12.161), (12.162) coincides with boundary-value problem (12.163) for the source located at layer boundary, i.e., at $x_0 = L$. In this case, we have $u(x;L;t) = u(x;t)$.

The solution to problem (12.163) can be represented in the form of the Fourier integral (parameter $\tilde{\gamma}$ is assumed small)

$$u(x;x_0;t) = \frac{1}{2\pi}\int\limits_{-\infty}^{\infty} d\omega G_\omega(x;x_0) f(\omega) e^{-i\omega t}, \quad G_\omega(x;x_0) = \int\limits_{-\infty}^{\infty} dt G(x;x_0;t) e^{i\omega t}, \tag{12.164}$$

where

$$f(\omega) = \int_{-\infty}^{\infty} dt f(t) e^{i\omega t}.$$

Function $G_\omega(x; x_0)$ is the solution of the stationary problem on the field of the point source in randomly layered medium (12.12)

$$\begin{gathered} \frac{d^2}{dx^2} G_\omega(x; x_0) + k^2[1 + \varepsilon(x)] G_\omega(x; x_0) = 2ik\delta(x - x_0), \\ \left(\frac{d}{dx} + ik\right) G_\omega(x; x_0)\bigg|_{x=L_0} = 0, \quad \left(\frac{d}{dx} - ik\right) G_\omega(x; x_0)\bigg|_{x=L} = 0. \end{gathered} \tag{12.165}$$

where

$$\frac{1}{c^2(x)} = \frac{1}{c_0^2}[1 + \varepsilon(x)], \quad \varepsilon(x) = \varepsilon_1(x) + i\frac{\tilde{\gamma}}{\omega}, \quad k = \frac{\omega}{c_0}.$$

We considered this problem earlier. Parameter $\tilde{\gamma}$ characterizes wave absorption in the medium and is related to parameter γ introduced earlier through the relationship $\gamma = \tilde{\gamma}/2c_0$.

Introduce Green's nonstationary function $G(x; L; t)$. At the boundary $x = L$, wave $f[t + (x - L)/c_0]$ incident on the layer creates the distribution of sources $\tilde{f}(t_0)$ such that

$$f(t) = \frac{1}{2c_0} \int_{-\infty}^{\infty} dt_0 \theta(t - t_0) \tilde{f}(t_0), \quad \tilde{f}(t_0) = 2c_0 \frac{\partial}{\partial t_0} f(t_0).$$

Then, we can represent the wavefield in the layer in the form

$$u(x, t) = \int_{-\infty}^{\infty} dt_1 G(x; L; t - t_1) \frac{\partial}{\partial t_1} f(t_1),$$

where function $G(x; L; t - t_0)$ satisfies wave equation (12.161) with the boundary condition at $x = L$

$$\left(\frac{\partial}{\partial x} + \frac{\partial}{c_0 \partial t}\right) G(x; L; t - t_0)\bigg|_{x=L} = \frac{2}{c_0}\delta(t - t_0).$$

Using the imbedding method, we can reformulate the boundary-value problem of determining function $G(x; L; t)$ (for simplicity, we neglect wave absorption in the medium) into the with initia probleml with respect to parameter L (we assume that $t_0 = 0$) [17, 136]:

$$\begin{aligned} \left(\frac{\partial}{\partial L} + \frac{\partial}{c_0 \partial t}\right) G(x; L; t) &= -\frac{1}{2c_0}\varepsilon(L) \int_{-\infty}^{\infty} dt_1 \frac{\partial}{\partial t} G(x; L; t - t_1) \frac{\partial}{\partial t_1} H(L; t_1), \\ G(x; L; t)|_{L=x} &= H(x; t). \end{aligned} \tag{12.166}$$

Function $H(L; t) = G(L; L; t)$ is the wavefield at medium boundary; it satisfies the closed integro-differential equation with the initial value

$$\begin{gathered} \left(\frac{\partial}{\partial L} + \frac{2}{c_0}\frac{\partial}{\partial t}\right) H(L; t) = \frac{2}{c_0}\delta(t) \\ -\frac{1}{2c_0}\varepsilon(L) \int_{-\infty}^{\infty} dt_1 \frac{\partial}{\partial t} H(L; t - t_1) \frac{\partial}{\partial t_1} H(L; t_1), \quad H(L; t)|_{L=L_0} = \theta(t). \end{gathered} \tag{12.167}$$

Function $G(x;L;t)$ describes the wavefield in the medium under the condition that incident wave has the form $\theta(t+(x-L)/c_0)$. Function $H(L;t)$ also can be represented in the form

$$H(L;t) = \theta(t)H_L(t). \tag{12.168}$$

Substituting Eq. (12.168) in Eq. (12.167) and separating the singular ($\sim \delta(t)$) and regular ($\sim \theta(t)$) portions, we obtain the equation [32]

$$\begin{aligned}\left(\frac{\partial}{\partial L} + \frac{2}{c(L)}\frac{\partial}{\partial t}\right) H_L(t) &= -\frac{1}{2c_0}\varepsilon(L)\int\limits_0^t dt_1 \frac{\partial H_L(t-t_1)}{\partial t}\frac{\partial H_L(t_1)}{\partial t_1},\\ H_{L_0}(t) &= 1, \quad H_L(+0) = \frac{2c(L)}{c(L)+c_0}.\end{aligned} \tag{12.169}$$

Statistical description

Consider now statistical characteristics of the solution to the nonstationary problem on propagation of a time-domain impulse generated by a source located inside the layer of random medium. This problem is described by Eq. (12.161), and we can represent the solution in the form of the Fourier integral (12.164). Our interest is in limiting values of the wavefield average intensity

$$I(x;x_0;t) = u^2(x;x_0;t)$$

for $t\to\infty$ and $\tilde{\gamma}\to 0$. The average intensity can be represented in the form

$$\langle I(x;x_0;t\rangle = \frac{1}{(2\pi)^2}\int\limits_{-\infty}^{\infty} d\omega \int\limits_{-\infty}^{\infty} d\psi \left\langle I_{\omega,\psi}(x;x_0)\right\rangle f\left(\omega+\frac{\psi}{2}\right) f^*\left(\omega-\frac{\psi}{2}\right) e^{-i\psi t}.$$

For $t\to\infty$, the value of the integral is governed by the integrand behavior for small ψ, so that

$$\langle I(x;x_0;t\rangle = \frac{1}{(2\pi)^2}\int\limits_{-\infty}^{\infty} d\omega |f(\omega)|^2 \int\limits_{-\infty}^{\infty} d\psi \left\langle I_{\omega,\psi}(x;x_0)\right\rangle e^{-i\psi t}. \tag{12.170}$$

In Eq. (12.170), we introduced the two-frequency analog of the plane wave intensity

$$I_{\omega,\psi}(x;x_0) = G_{\omega+\psi/2}(x;x_0)G^*_{\omega-\psi/2}(x;x_0).$$

Note that, in the limit of small ψ and $\tilde{\gamma}$, one can obtain from Eq. (12.165) the following equality ($x\le x_0$)

$$\frac{d}{dx}S_{\omega,\psi}(x;x_0) = \frac{1}{c_0}(\tilde{\gamma}-i\psi)I_{\omega,\psi}(x;x_0), \tag{12.171}$$

where $S_{\omega,\psi}(x;x_0)$ is the two-frequency analog of the energy flux density

$$S_{\omega,\psi}(x;x_0) = \frac{c_0}{2i\psi}\left[G_{\omega+\psi/2}(x;x_0)\frac{d}{dx}G^*_{\omega-\psi/2}(x;x_0) - G^*_{\omega-\psi/2}(x;x_0)\frac{d}{dx}G_{\omega+\psi/2}(x;x_0)\right].$$

Integrating Eq. (12.171) over the whole half-space $-\infty<x<x_0$, we obtain

$$S_{\omega,\psi}(x;x_0) = \frac{1}{c_0}(\tilde{\gamma}-i\psi)\int\limits_{-\infty}^{x_0} dx I_{\omega,\psi}(x;x_0).$$

Consequently, after integrating Eq. (12.170) over the half-space, we obtain the expression for the average energy contained in this half-space

$$E(t)=\int\limits_{-\infty}^{x_0} dx\,\langle I(x;x_0;t)\rangle=\frac{c_0}{(2\pi)^2}\int\limits_{-\infty}^{\infty} d\omega|f(\omega)|^2\int\limits_{-\infty}^{\infty}\frac{d\psi}{\tilde{\gamma}-i\psi}\langle S_{\omega,\psi}(x;x_0)\rangle e^{-i\psi t}. \quad (12.172)$$

Now, we dwell on statistical description of quantities $S_{\omega,\psi}(x;x_0)$ and $I_{\omega,\psi}(x;x_0)$. In accordance with the corresponding expressions for the stationary problem, they are described in terms of the quantity

$$W_{\omega,\psi}(x_0)=R_{\omega+\psi/2}(x_0)R^*_{\omega-\psi/2}(x_0),$$

which is the two-frequency analog of the reflection coefficient squared modulus $W=|R|^2$.

At $\psi=0$, expressions for $S_{\omega,\psi}(x;x_0)$ and $I_{\omega,\psi}(x;x_0)$ grade into the corresponding expressions for the one-frequency characteristics of the stationary problem. Thus, the calculation of average values of $S_{\omega,\psi}(x;x_0)$ and $I_{\omega,\psi}(x;x_0)$ requires the knowledge of statistics of quantity $W_{\omega,\psi}(x_0)$.

Reflection coefficient $R_\omega(x)$ as a function of x satisfies the stochastic Riccati equation representable in the form

$$\frac{d}{dx}R_\omega(x)=\frac{2i}{c_0}\left(\omega+i\frac{\tilde{\gamma}}{2}\right)R_\omega(x)+\frac{\omega}{2c_0}\varepsilon_1(x)\left(1+R_\omega(x)\right)^2, \quad R_\omega(x)|_{x\to-\infty}=0.$$

Consequently, function $W_{\omega,\psi}(x)$ satisfies the equation

$$\begin{aligned}\frac{d}{dx}W_{\omega,\psi}(x)&=-\frac{2}{c_0}\left(\tilde{\gamma}-i\psi\right)W_{\omega,\psi}(x)\\&-i\frac{\omega}{2c_0}\varepsilon_1(x)\left(R_{\omega+\psi/2}(x)-R^*{}_{\omega-\psi/2}(x)\right)\left(1-W_{\omega,\psi}(x)\right),\end{aligned}$$

and, assuming as usually that $\varepsilon_1(x)$ is the Gaussian delta-correlated process with the parameters

$$\langle\varepsilon_1(x)\rangle=0, \quad \langle\varepsilon_1(x)\varepsilon_1(x')\rangle=2\sigma_\varepsilon^2 l_0\delta(x-x'),$$

we can use the standard procedure to derive for quantity $W^{(n)}_{\omega,\psi}(x)=\langle[W_{\omega,\psi}(x)]^n\rangle$ the recurrence equation

$$\begin{aligned}\frac{d}{dx}W^{(n)}_{\omega,\psi}(x)&=-\frac{2n}{c_0}\left(\tilde{\gamma}-i\psi\right)W^{(n)}_{\omega,\psi}(x)\\&+D(\omega)n^2\left\{W^{(n+1)}_{\omega,\psi}(x)-2W^{(n)}_{\omega,\psi}(x)+W^{(n-1)}_{\omega,\psi}(x)\right\},\end{aligned}$$

where

$$D(\omega)=\frac{\omega^2\sigma_\varepsilon^2 l_0}{2c_0^2},$$

as earlier.

As a result, we obtain that the solution independent of x and corresponding to the half-space of random medium satisfies the recurrence equation

$$\frac{2}{c_0}\left(\tilde{\gamma}-i\psi\right)W^{(n)}_{\omega,\psi}=D(\omega)n\left\{W^{(n+1)}_{\omega,\psi}-2W^{(n)}_{\omega,\psi}+W^{(n-1)}_{\omega,\psi}\right\}. \quad (12.173)$$

For $\psi = 0$, Eq. (12.173) grades into Eq. (12.39), page 289, to which probability density (12.37) corresponds. Equation (12.173) can be considered as analytic continuation of Eq. (12.39) to the complex region of parameter $\tilde{\gamma}$. This means that, being analytically continued to the complex region of attenuation parameter $\tilde{\gamma}$, all statistical characteristics obtained in the context of the stationary problem will grade into the corresponding two-frequency statistical characteristics [283].

Thus, in order to obtain the expressions of the two-frequency statistical characteristics for the problem with absent absorption in the medium ($\tilde{\gamma} = 0$), we must replace parameter $\tilde{\gamma}$ with $0 - i\psi$ in the corresponding statistical characteristics of the problem on plane waves, i.e., we must set

$$\beta(\omega) = \frac{1}{c_0 D(\omega)} (0 - i\psi) .$$

As a result, we obtain the expressions

$$\langle S_{\omega,\psi}(x_0; x_0) \rangle = 1, \quad \langle I_{\omega,\psi}(x_0; x_0) \rangle = i \frac{D(\omega) c_0}{\psi + i0}$$

valid for sufficiently small ψ in the limit $t \to \infty$ at $\tilde{\gamma} = 0$.

Consequently, after integration over ψ, formulas (12.170) and (12.172) grade into the expressions corresponding to asymptotic limit $t \to \infty$

$$\langle I(x_0; x_0; \infty) \rangle = \frac{c_0}{2\pi} \int\limits_{-\infty}^{\infty} d\omega D(\omega) |f(\omega)|^2, \quad E(\infty) = \frac{c_0}{2\pi} \int\limits_{-\infty}^{\infty} d\omega |f(\omega)|^2. \tag{12.174}$$

Thus, the average wavefield energy at the point of source location and the total energy in the whole half-space assume finite values (if the corresponding integrals exist). This fact confirms the existence of spatial statistical localization of average intensity; it is obvious that the corresponding localization length will be given by the formula

$$l_{\mathrm{loc}} = \frac{\int\limits_{-\infty}^{\infty} d\omega |f(\omega)|^2}{\int\limits_{-\infty}^{\infty} d\omega D(\omega) |f(\omega)|^2}.$$

The property of statistical localization follows from the finite-valuedness of the total energy concentrated in the half-space, which, in turn, follows from the independence of average energy flux of fluctuating medium parameters in the stationary problem on plane waves. The shape of localization curve can be obtained from Eq. (12.120), page 317

$$\langle I(x; x_0; \infty) \rangle = \frac{c_0}{2\pi} \int\limits_{-\infty}^{\infty} d\omega D(\omega) |f(\omega)|^2 \Phi_{\mathrm{loc}}(\xi) \quad (\xi = D(\omega)|x - x_0|), \tag{12.175}$$

where $\Phi_{\mathrm{loc}}(\xi)$ is the localization curve (12.121) of the steady-state problem. It depends on parameter ω only through diffusion coefficient $D(\omega)$.

If impulse $f(t)$ is characterized by one parameter (impulse width), then Eq. (12.175) gives for large $|x - x_0|$ the asymptotic dependence

$$\langle I(x; x_0; \infty) \rangle \sim |x - x_0|^{-3/2}.$$

If the impulse has high-frequency carrier, then the asymptotic dependence assumes the form

$$\langle I(x;x_0;\infty)\rangle \sim \Phi_{\rm loc}(\xi) \quad (\xi = D(\omega)|x-x_0|).$$

The corresponding expressions for the source located at reflecting boundary can be obtained similarly:

$$\begin{aligned}\langle I_{\rm ref}(x;x_0;\infty)\rangle &= \frac{2c_0}{\pi}\int\limits_{-\infty}^{\infty} d\omega D(\omega)|f(\omega)|^2\Phi_{\rm loc}(\xi) \quad (\xi = D(\omega)(L-x)),\\ \langle I_{\rm ref}(x_0;x_0;\infty)\rangle &= \frac{2c_0}{\pi}\int\limits_{-\infty}^{\infty} d\omega D(\omega)|f(\omega)|^2, \quad E(\infty) = \frac{2c_0}{\pi}\int\limits_{-\infty}^{\infty} d\omega |f(\omega)|^2.\end{aligned} \tag{12.176}$$

In this case, statistical localization is realized on the scale equal to a half of the localization length obtained in the previous case.

In the case of an impulse incident on the half-space of random medium $x < L$, we have for quantity $I(L;t) = u^2(L;t)$:

$$\begin{aligned}\langle I(L;t)\rangle &= \frac{1}{(2\pi)^2}\int\limits_{-\infty}^{\infty} d\omega |f(\omega)|^2 \int\limits_{-\infty}^{\infty} d\psi \left\{1 + W^{(1)}_{\omega,\psi}\right\} e^{-i\psi t},\\ E(t) = \int\limits_{-\infty}^{L} dx\, \langle I(x;L;t)\rangle &= \frac{c_0}{(2\pi)^2}\int\limits_{-\infty}^{\infty} d\omega |f(\omega)|^2 \int\limits_{-\infty}^{\infty} \frac{d\psi}{0 - i\psi}\left\{1 - W^{(1)}_{\omega,\psi}\right\} e^{-i\psi t},\end{aligned} \tag{12.177}$$

where

$$W^{(1)}_{\omega,\psi} = -\beta(\omega)\int\limits_0^{\infty} du \frac{u}{u+2} e^{-\beta(\omega)u}$$

is the analytic continuation of the corresponding expression for $\langle |R_L|^2\rangle$ with respect to parameter β. Performing integrations over ω and u in Eqs. (12.177), we obtain the following asymptotic expression [38]

$$\langle I(L;t)\rangle = \frac{c_0}{\pi}\int\limits_{-\infty}^{\infty} d\omega |f(\omega)|^2 \frac{D(\omega)}{[2 + D(\omega)c_0 t]^2}, \quad E(t) = \frac{2c_0}{\pi}\int\limits_{-\infty}^{\infty} d\omega |f(\omega)|^2 \frac{1}{2 + D(\omega)c_0 t} \tag{12.178}$$

valid for sufficiently large t.

Expressions (12.178) give the time-dependent asymptotic behavior of average intensity of impulse reflected from the half-space and average intensity contained in random medium. Asymptotically, we have in this case the relationships

$$\langle I(L;t)\rangle \sim t^{-2} \quad \text{and} \quad \langle I(L;t)\rangle \sim t^{-3/2}$$

for impulses with and without high-frequency carrier, respectively.

From Eqs. (12.178) follows additionally that the incident wave completely escapes from the random medium for $t \to \infty$.

In this section, we considered the statistical description of a wave impulse in random medium. The problem on a spatial wave pocket propagating in random medium can be considered similarly [9]–[11], [37], [185]–[188], [259]. It is clear that property of statistical localization will be inherent in this problem, too. In this case, the property of statistical localization can be treated as some kind of statistical waveguide in the direction perpendicular to the x-axis [67, 68, 87, 88].

12.5.2 Point source in randomly layered medium

Factorization of the wave equation in layered medium

Consider now the problem on the wavefield generated by the multidimensional point source located in randomly layered medium [107]. Green's function of this problem satisfies the equation

$$\left[\frac{\partial^2}{\partial z^2} + \Delta_{\mathbf{R}} + k^2\left[1 + \varepsilon(z, \mathbf{R})\right]\right] G(z, \mathbf{R}; z_0) = \delta(\mathbf{R})\delta(z - z_0), \tag{12.179}$$

where $\mathbf{R} = \{x, y\}$, $\Delta_{\mathbf{R}} = \frac{\partial^2}{\partial x^2} + \frac{\partial^2}{\partial y^2}$.

In this case, integral representations of Green's function have the following forms (see Appendix B, page 438)

$$\begin{aligned} G^{(1)}(z; z_0) &= \frac{1}{2ik}\int\limits_0^\infty dt e^{i\frac{kt}{2}}\psi(t, z; z_0), \\ G^{(2)}(x, z; z_0) &= \frac{1}{2ik}\sqrt{\frac{k}{2\pi i}}\int\limits_0^\infty \frac{dt}{\sqrt{t}} e^{i\frac{k}{2t}(x^2+t^2)}\psi(t, z; z_0), \\ G^{(3)}(z, \mathbf{R}; z_0) &= -\frac{1}{4\pi}\int\limits_0^\infty \frac{dt}{t} e^{i\frac{k}{2t}(\mathbf{R}^2+t^2)}\psi(t, z; z_0), \end{aligned} \tag{12.180}$$

for one-, two-, and three-dimensional spaces, respectively. Here, $\psi(t, z; z_0)$ is the solution (dependent on auxiliary parameter t) to the equation

$$\frac{\partial}{\partial t}\psi(t, z; z_0) = \frac{i}{2k}\left[\frac{\partial^2}{\partial z^2} + k^2\varepsilon(z)\right]\psi(t, z; z_0), \quad \psi(0, z; z_0) = \delta(z - z_0). \tag{12.181}$$

Formulas (12.180) and (12.181) express the factorization property of the Helmholtz equation in a layered medium.

Evolution problem (12.181) must be supplemented with a boundary condition in z. We will consider the following boundary-value problems:

(a) The source in infinite space and radiation condition for $z \to \pm\infty$;

(b) The source at the reflecting boundary at which the condition $\partial\psi/\partial z|_{z=z_0-0} = 0$ is satisfied and radiation condition for $z \to \infty$;

(c) The source at the boundary of homogeneous half-space and radiation condition for $z \to \pm\infty$.

For x, $|\mathbf{R}| \to \infty$, from Eqs. (12.180) follow asymptotic formulas

$$\begin{aligned} G^{(2)}(x, z; z_0) &\approx \frac{1}{2ik} e^{ik|x|}\psi(|x|, z; z_0), \\ G^{(3)}(z, \mathbf{R}; z_0) &\approx \frac{1}{4\pi\sqrt{R}}\sqrt{\frac{2\pi i}{kR}} e^{ikR}\psi(R, z; z_0), \end{aligned} \tag{12.182}$$

valid under the condition that function $\psi(t, z; z_0)$ shows no exponential behavior with respect to variable t. Formulas (12.182) correspond to the small-angle scattering (the approximation of parabolic equation). Consideration of scattering at great angles requires the use of exact representations (12.180) for Green's function.

Using the Fourier transform, we can represent function $\psi(t, z; z_0)$ in the form

$$\psi_\omega(z; z_0) = \int\limits_{-\infty}^{\infty} dt \psi(t, z; z_0) e^{-i\omega t},$$

where function $\psi_\omega(z; z_0)$ satisfies the equation

$$\left[\frac{d^2}{dz^2} - 2k\omega + k^2\varepsilon(z)\right] \psi_\omega(z; z_0) = 2ik\delta(z - z_0). \tag{12.183}$$

The solution $\psi_{\omega<0}(z; z_0)$ of Eq. (12.183) corresponds to waves propagating for $\omega < 0$ and decaying for $\omega > 0$.

Our interest is in the asymptotic behavior of Green's functions for x, $|\mathbf{R}| \to \infty$. These asymptotics are as follows

$$\begin{aligned}
G^{(1)}(z; z_0) &= \frac{1}{2ik}\psi_{-k/2-i0}(z; z_0), \\
G^{(2)}(x, z; z_0) &= \frac{1}{2i\pi k}\int\limits_{-\infty}^{\infty} \frac{d\omega}{\sqrt{1 + 2\omega/k}} e^{ikx\sqrt{1+2\omega/k}} \psi_{\omega-i0}(z; z_0), \\
G^{(3)}(z, \mathbf{R}; z_0) &= -\frac{1}{8\pi^2}\sqrt{\frac{2\pi i}{kR}} \int\limits_{0}^{\infty} \frac{d\omega}{(1 + 2\omega/k)^{1/4}} e^{ikR\sqrt{1+2\omega/k}} \psi_{\omega-i0}(z; z_0),
\end{aligned} \tag{12.184}$$

and we see that formulas (12.182) follow from Eqs. (12.184) under the condition $2\omega/k \ll 1$.

Parabolic equation

Consider first statistics of Green's functions in the approximation of parabolic equation. Function $\psi(t, z; z_0)$ can be written in the form

$$\psi(t, z; z_0) = \frac{1}{2\pi}\int\limits_{0}^{\infty} d\omega \sqrt{\frac{k}{2\omega}} \widetilde{\psi}_\omega(z; z_0) e^{-i\omega t} + \frac{1}{2\pi}\int\limits_{0}^{\infty} d\omega \psi_\omega(z; z_0) e^{i\omega t}, \tag{12.185}$$

where function $\widetilde{\psi}_\omega(z; z_0)$ satisfies the equation

$$\left[\frac{d^2}{dz^2} + 2k\omega + k^2\varepsilon(z)\right] \widetilde{\psi}_\omega(z; z_0) = 2ik\sqrt{2k\omega}\,\delta(z - z_0). \tag{12.186}$$

The effect of fluctuations $\varepsilon(z)$ is insignificant for decaying waves, and function

$$\psi_{\omega>0}(z; z_0) = -i\sqrt{\frac{k}{2\omega}} e^{-\sqrt{2k\omega}|z-z_0|}$$

corresponds to the field in free space. Unknown function $\widetilde{\psi}_\omega(z; z_0)$ is the solution to Eq. (12.186) whose statistics was analyzed earlier in the section dealing with stationary one-dimensional problems.

As we have seen earlier, the principal feature of one-dimensional problems on point-source field consists in necessity of allowance for finite (even arbitrarily small) absorption in the medium $\gamma \ll 1$. Here, we introduce it as the imaginary part of function $\varepsilon(z) = \varepsilon_1(z) + i\gamma$, where $\varepsilon_1(z)$ is the random function. Our interest is in average intensity

$$\langle I(t, z; z_0)\rangle = \langle \psi(t, z; z_0)\psi^*(t, z; z_0)\rangle$$

for sufficiently great values of parameter t.

The average intensity can be represented in the form

$$\langle I(t, z; z_0)\rangle = I_{\text{fluc}}(t, z; z_0) + I_1(t, z; z_0) + I_2(t, z; z_0), \tag{12.187}$$

where functions $I_{\text{fluc}}(t, z; z_0)$ and $I_2(t, z; z_0)$ correspond to the contributions of the first and second terms of Eq. (12.185) and function $I_1(t, z; z_0)$ takes into account the cross-term $\left\langle \widetilde{\psi}_{\omega_1}(z; z_0)\psi^*_{\omega_2}(z; z_0)\right\rangle$.

Consider first the quantity

$$I_{\text{fluc}}(t, z; z_0) = \frac{k}{2\,(2\pi)^2}\int\limits_0^\infty d\Omega \int\limits_{-2\Omega}^{2\Omega} \frac{d\omega}{\sqrt{\Omega^2 - \frac{\omega^2}{4}}} e^{-i\omega t}\left\langle I_{\Omega,\omega}(z; z_0)\right\rangle,$$

where

$$\langle I_{\Omega,\omega}(z; z_0)\rangle = \left\langle \widetilde{\psi}_{\Omega+\omega/2}(z; z_0)\psi^*_{\Omega-\omega/2}(z; z_0)\right\rangle$$

is the two-frequency correlator of the solution to the corresponding boundary-value problem. For $t \to \infty$, this integral is mainly contributed by the vicinity of point $\omega \to 0$.

Statistical characteristics of the solution to Eq. (12.186) follow from the statistics of the reflection coefficient $R_{z_0}(\omega)$ of the plane wave incident on the half-space $z_0 > z$ from the homogeneous half-space. As may be easily seen, function $R_{z_0}(\omega)$ satisfies the Riccati equation

$$\frac{d}{dz_0}R_{z_0}(\omega) = \left[2i\sqrt{2k\omega} - k\sqrt{\frac{k}{2\omega}}\gamma\right] R_{z_0}(\omega) + i\frac{k}{2}\sqrt{\frac{k}{2\omega}}\varepsilon_1(z_0)\left[1 + R_{z_0}(\omega)\right]^2.$$

The one-frequency characteristics of the reflection coefficient are functions of single dimensionless parameter

$$\beta = \frac{k\gamma}{D}\sqrt{\frac{k}{2\Omega}},$$

where

$$D = D_0\frac{k}{2\Omega}, \quad D_0 = \frac{k^2\sigma_e^2 l_0}{2},$$

the wavefield functional behavior with parameter β being dependent on the particular boundary-value problem. In this case, under the assumption that function $\varepsilon_1(z)$ is the Gaussian delta-correlated random process, there exists probability density $P(u)$ of quantity $u = (1 + W)/(1 - W)$ (12.37), page 288

$$P(u) = \beta e^{-\beta(u-1)}. \tag{12.188}$$

which is independent of z_0 (half-space).

For the two-frequency function $W_{z_0}(\Omega,\omega) = R_{z_0}(\Omega+\omega/2)R^*_{z_0}(\Omega-\omega/2)$, we obtain for $\omega \to 0$ the equation

$$\frac{d}{dz_0}W_{z_0}(\Omega,\omega) = -2\sqrt{\frac{k}{2\Omega}}(k\gamma - i\omega)W_{z_0}(\Omega,\omega)$$
$$-i\frac{k}{2}\sqrt{\frac{k}{2\Omega}}\varepsilon_1(z_0)\left[R_{z_0}(\Omega) - R^*_{z_0}(\Omega)\right]\left[1 - W_{z_0}(\Omega,\omega)\right].$$

Here, we neglected terms proportional to $\omega\gamma$ and $\omega\varepsilon_1(z_0)$. Determination of the two-frequency function $W_{z_0}(\Omega,\omega)$ for small ω reduces to the analytic continuation of the corresponding one-frequency characteristics to the complex region of parameter β

$$\beta \to \beta(\Omega,\omega) = \sqrt{\frac{k}{2\Omega}}\frac{k\gamma - i\omega}{D} = -i\sqrt{\frac{2\Omega}{k}}\frac{\omega + ik\gamma}{D_0}. \tag{12.189}$$

The further analysis depends on the boundary-value problem under consideration.

(a) Source in infinite space.

Consider the average intensity at the point of source location $z_0 = z$. In this case, the one-frequency quantity is

$$\langle I_{\Omega,0}(z_0;z_0)\rangle = \left\langle \widetilde{\psi}_\Omega(z_0;z_0)\psi^*_\Omega(z_0;z_0)\right\rangle = 1 + 1/\beta.$$

After analytic continuation with respect to β, we obtain

$$I_{\text{fluc}}(t,z;z_0) = \frac{kD_0}{2i\,(2\pi)^2}\int\limits_0^\infty d\Omega \int\limits_{-2\Omega}^{2\Omega}\frac{d\omega}{\sqrt{\Omega^2 - \frac{\omega^2}{4}}}e^{-i\omega t}\sqrt{\frac{k}{2\Omega}}\frac{1}{\omega + i\gamma}. \tag{12.190}$$

In the case of the point source, Eq. (12.190) for $I_{\text{fluc}}(t,z;z_0)$ is meaningful only if wave absorption in medium γ is finite; for $\gamma \to 0$, we have

$$I_{\text{fluc}}(t,z;z_0) \sim \frac{kD_0}{\sqrt{\gamma}}.$$

Consequently, small, but finite absorption in the medium is essential for wavefield statistics (this result is similar to that of the one-dimensional problem).

(b) Source at reflecting boundary.

In this case, all conclusions made for the source in infinite space remain obviously valid.

(c) Source located at the boundary of homogeneous half-space.

If source is located at medium boundary $z = z_0$, the one-point average has the form

$$\langle I_{\Omega,0}(z_0;z_0)\rangle = \left\langle \widetilde{\psi}_\Omega(z_0;z_0)\psi^*_\Omega(z_0;z_0)\right\rangle = 1 + \left\langle |R_{z_0}(\Omega)|^2\right\rangle,$$

where averaging is performed with the use of probability density (12.188). Consequently, we have for the average intensity at medium boundary

$$\begin{aligned}\langle I(t,z;z_0)\rangle &= I_{\text{free}}(t,z;z_0) + I_{\text{fluc}}(t,z;z_0),\\ I_{\text{fluc}}(t,z;z_0) &= \frac{k}{2\,(2\pi)^2}\int\limits_0^\infty d\Omega\int\limits_{-2\Omega}^{2\Omega}\frac{d\omega}{\sqrt{\Omega^2-\frac{\omega^2}{4}}}e^{-i\omega t}\beta(\Omega,\omega)\int\limits_0^\infty\frac{u\,du}{u+2}e^{-\beta(\Omega,\omega)u},\end{aligned}$$

where parameter $\beta(\Omega, \omega)$ is given by Eq. (12.189). Integration over ω and u gives the expression ($\gamma \to 0$)

$$I_{\text{fluc}}(t, z; z_0) = \frac{k^{3/2} D_0}{2\sqrt{2\pi}} \int_0^\infty d\Omega \frac{1}{\Omega^{3/2} \left[2 + D_0 \sqrt{\frac{k}{2\Omega}} t\right]^2}.$$

Consequently, we have

$$I_{\text{fluc}}(t, z; z_0) = I_{\text{frec}}(t, z; z_0) = \frac{k}{2\pi t}$$

for $t \to \infty$, i.e., the average wavefield intensity is doubled. This result is similar to that of the one-dimensional problem.

General case

Consider the exact description of the problem on point source in infinite space by the example of the two-dimensional case; namely, we consider Green's function $G^{(2)}(x, z; z_0)$ (12.184).

Divide the integration interval into three regions: $(-\infty, -k/2)$, $(-k/2, 0)$, and $(0, +\infty)$. The contribution of the first region to Green's function is approximately $\psi_{-k/2}(z_0, z_0)/kx$ because of the exponent decaying for $x \to \infty$. This contribution corresponds to the term

$$\left\langle I_1^{(2)}(x) \right\rangle \sim \frac{D_0}{(kx)^2 \, k\gamma}$$

in the expression for average intensity.

The contribution of the second region can be estimated using the method similar to that used for analyzing the parabolic equation. The corresponding contribution to the average intensity measures

$$\left\langle I_2^{(2)}(x) \right\rangle \sim \frac{D_0}{k\sqrt{\gamma}} \tag{12.191}$$

for x sufficiently great, but such that $k\gamma x \ll 1$.

In the third region, wavefield $\psi_\omega(z_0, z_0)$ coincides with the wave propagating in free space. Its contribution to the average intensity measures

$$\left\langle I_3^{(2)}(x) \right\rangle \sim \frac{1}{kx}.$$

Note that products of integrals over different regions give no power dependence in γ in the denominators of the corresponding asymptotic expressions.

Combining all obtained terms, we see that, under the condition

$$\gamma^{3/4} \ll k\xi \ll 1,$$

term $\left\langle I_2^{(2)}(x) \right\rangle$ predominates in the expression for average intensity.

In the three-dimensional case, we obtain the similar result:

$$\left\langle I_1^{(3)}(x) \right\rangle \sim \frac{D_0 k}{(kR)^4 \gamma}, \quad \left\langle I_2^{(3)}(x) \right\rangle \sim \frac{D_0}{R\sqrt{\gamma}}, \quad \left\langle I_3^{(3)}(x) \right\rangle \sim \frac{1}{(kR)^2}$$

for $R \to \infty$, but $\gamma k R \ll 1$. Under the condition $\gamma^{5/6} \ll k\gamma R$, function $\left\langle I_2^{(3)}(x) \right\rangle$ predominates in the average intensity at the point of source location $z = z_0$ (in this case, the source generates the cylindrical wave).

A similar result holds also for the point source located at reflecting boundary.

If the source is located at the boundary of random half-space, average intensity is given by the integral

$$\langle I(x, z; z_0) \rangle = I_{\text{free}}(x, z; z_0) \left(1 + 2D_0 x \int\limits_0^1 \frac{ds}{\left[D_0 x + 2s\sqrt{1 - s^2} \right]} \right).$$

Our interest is in two asymptotic regimes, namely, $D_0 x \ll 1$, but $kx \gg 1$ and $D_0 x \gg 1$. In the first regime, we have

$$\langle I(x, z; z_0) \rangle = 2I_{\text{free}}(x, z; z_0),$$

which is similar to the result obtained for the parabolic equation. This fact shows that scattering at great angles only slightly affects the statistics in this regime.

In the second regime $D_0 x \gg 1$, average intensity is given by the expression

$$\langle I(x, z; z_0) \rangle = I_{\text{free}}(x, z; z_0) \left(1 + \frac{2}{D_0 x} \right),$$

and scattering at great angles appears significant for the formation of statistics. The effect of this scattering appears as an additional decreasing factor of the intensity in free space. A similar result holds for the tree-dimensional case.

We have seen earlier that the principal feature of one-dimensional problems on plane wave in randomly layered media consists in necessity of allowance for finite (even arbitrarily small) absorption in the medium (parameter γ). Wavefield statistics is formed by the interference of waves multiply re-reflected in random medium, which results in singular behavior of average intensity $\langle I \rangle$ as a function of parameter γ; for example, $\langle I \rangle \sim 1/\gamma$ in the case of the point source in infinite space.

In multidimensional problems on layered media, the effect of diffraction is similar to the effect of attenuation, and this fact offers a possibility of calculating wavefield statistical characteristics using analytical continuation to the complex plane of parameter γ. It could be hoped that the effect of diffraction will eliminate the singular behavior of statistical characteristics in parameter γ and will made possible the limit process $\gamma \to 0$ in multidimensional problems. Unfortunately, these hopes were not justified. Diffraction effects only reduce the degree of singularity, but not eliminate it. Thus, wave absorption in medium serves the regularizing factor in multidimensional problems on waves in random media.

In conclusion, we cite the literature on numerical simulations of statistical characteristics of the point-source field in the three-dimensional randomly layered media [110, 111, 262, 263].

12.6 Two-layer model of the medium

12.6.1 Formulation of boundary-value problems

The simplest model of wave propagation in the two-layer medium was mentioned in Chapter 1, where it was formulated as the system of wave equations (1.38), page 18

$$\frac{d^2}{dx^2}\psi_1(x) + k^2\psi_1(x) - \alpha_1 F\left(\psi_1(x) - \psi_2(x)\right) = 0,$$
$$\frac{d^2}{dx^2}\psi_2(x) + k^2\left[1+\varepsilon(x)\right]\psi_2(x) + \alpha_2 F\left(\psi_1(x) - \psi_2(x)\right) = 0, \qquad (12.192)$$

where parameters $\alpha_1 = 1/H_1$, $\alpha_2 = 1/H_2$ (H_1, H_2 are the thicknesses of the upper and lower layers), factor F characterizes wave interaction, and function $\varepsilon(x)$ describes the medium inhomogeneities in the lower layer. As earlier, we assume that function $\varepsilon(x)$ is different from zero only in region (L_0, L) and is the random function. The boundary conditions for system of equations (12.192) are formulated as the radiation condition at infinity and the conditions of continuity of wavefields and wavefield derivatives at boundaries L_0 and L. We consider the statistical description of this problem abiding by work [90].

Consider the system of equations for Green's function

$$\frac{d^2}{dx^2}\psi_1(x;x_0) + k^2\psi_1(x;x_0) - \alpha_1 F\left(\psi_1(x;x_0) - \psi_2(x;x_0)\right) = -v_1\delta(x-x_0),$$
$$\frac{d^2}{dx^2}\psi_2(x;x_0) + k^2\left[1+\varepsilon(x)\right]\psi_2(x;x_0) + \alpha_2 F\left(\psi_1(x;x_0) - \psi_2(x;x_0)\right) = -v_2\delta(x-x_0) \qquad (12.193)$$

corresponding to wave excitation in the upper and lower layers, respectively. Using the vector notation

$$\boldsymbol{\psi}(x;x_0) = \{\psi_1(x;x_0), \psi_2(x;x_0)\}, \quad \mathbf{v} = \{v_1, v_2\},$$

we can rewrite system (12.193) in the vector form

$$\left[\frac{d^2}{dx^2} + A^2 + k^2\varepsilon(x)\Gamma\right]\boldsymbol{\psi}(x;x_0) = -\mathbf{v}\delta(x-x_0), \qquad (12.194)$$

where matrixes A^2 and Γ are given by the formulas

$$A^2 = \begin{pmatrix} k^2 - \alpha_1 F & \alpha_1 F \\ \alpha_2 F & k^2 - \alpha_2 F \end{pmatrix}, \quad \Gamma = \begin{pmatrix} 0 & 0 \\ 0 & 1 \end{pmatrix}. \qquad (12.195)$$

We introduce additionally parameter $\lambda^2 = \left[1 - (\alpha_1+\alpha_2)\frac{F}{k^2}\right]$ (for $\lambda^2 > 0$, this parameter describes the mode that we will call the λ-wave) and relative layer thicknesses

$$\tilde{\alpha}_1 = \frac{\alpha_1}{\alpha_1+\alpha_2} = \frac{H_2}{H_0}, \quad \tilde{\alpha}_2 = \frac{\alpha_2}{\alpha_1+\alpha_2} = \frac{H_1}{H_0}, \quad \tilde{\alpha}_1 + \tilde{\alpha}_2 = 1.$$

In this representation, Eq. (12.194) is similar to the Helmholtz equation (12.1), page 278, where matrix A describes the constant value of the refraction coefficient and product $\varepsilon(x)\Gamma$ describes the inhomogeneities of the medium.

Consider matrix Ψ that satisfies the equation

$$\left[\frac{d^2}{dx^2} + A^2 + k^2\varepsilon(x)\Gamma\right]\Psi(x;x_0) = -E\delta(x-x_0). \tag{12.196}$$

The desired vector-function $\boldsymbol{\psi}(x;x_0)$ can be determined in terms of this matrix by the equality

$$\boldsymbol{\psi}(x;x_0) = \Psi(x;x_0)\mathbf{v} = \begin{bmatrix} \psi_{11} & \psi_{12} \\ \psi_{21} & \psi_{22} \end{bmatrix}\begin{pmatrix} v_1 \\ v_2 \end{pmatrix} = \begin{pmatrix} v_1\psi_{11} + v_2\psi_{12} \\ v_1\psi_{21} + v_2\psi_{22} \end{pmatrix}. \tag{12.197}$$

This means that vector-columns $\{\psi_{11}, \psi_{21}\}$ and $\{\psi_{12}, \psi_{22}\}$ of this matrix describe the waves generated by sources $\{v_1, 0\}$ and $\{0, v_2\}$ located in the upper and lower layer, respectively. The boundary conditions for Eq. (12.196) are formulated as follows

$$\left(\frac{d}{dx} - iA\right)\Psi(x;x_0)\bigg|_{x=L} = 0, \left(\frac{d}{dx} + iA\right)\Psi(x;x_0)\bigg|_{x=L_0} = 0, \tag{12.198}$$

where matrix A has the following form

$$A = k\begin{pmatrix} \tilde{\alpha}_2 + \lambda\tilde{\alpha}_1 & (1-\lambda)\tilde{\alpha}_1 \\ (1-\lambda)\tilde{\alpha}_2 & \tilde{\alpha}_1 + \lambda\tilde{\alpha}_2 \end{pmatrix}.$$

In further consideration, we additionally simplify the problem; namely, we will assume that the source of plane waves is located at the boundary $x_0 = L$ of the layer with inhomogeneities. In this case, using the condition of wavefield discontinuity at the point of source location x_0, we arrive at the boundary-value problem

$$\left[\frac{d^2}{dx^2} + A^2 + k^2\varepsilon(x)\Gamma\right]\Psi(x;L) = 0,$$

$$\left(\frac{d}{dx} - iA\right)\Psi(x;L)\bigg|_{x=L} = E, \qquad \left(\frac{d}{dx} + iA\right)\Psi(x;L)\bigg|_{x=L_0} = 0. \tag{12.199}$$

We can simplify this equation by diagonalizing matrix A (12.195) with the use of the matrix

$$K = \begin{bmatrix} 1 & -1 \\ \tilde{\alpha}_2 & \tilde{\alpha}_1 \end{bmatrix}, \quad K^{-1} = \begin{bmatrix} \tilde{\alpha}_1 & 1 \\ -\tilde{\alpha}_2 & 1 \end{bmatrix}.$$

After this transformation, matrixes A and Γ move to

$$B = \tilde{A} = k\begin{bmatrix} \lambda & 0 \\ 0 & 1 \end{bmatrix}; \quad \tilde{\Gamma} = K\Gamma K^{-1} = \begin{bmatrix} \tilde{\alpha}_2 & -1 \\ -\tilde{\alpha}_1\tilde{\alpha}_2 & \tilde{\alpha}_1 \end{bmatrix},$$

and we obtain new system for the transformed matrix Ψ,

$$\Psi \to U(x;L) = -2iK\,\Psi(x;L)\,K^{-1}B, \tag{12.200}$$

in the form

$$\left[\frac{d^2}{dx^2} + B^2 + k^2\varepsilon(x)\tilde{\Gamma}\right]U(x;L) = 0,$$

$$\left(\frac{d}{dx} - iB\right)U(x;L)\,|_{x=L} = -2iB, \quad \left(\frac{d}{dx} + iB\right)U(x;L)\,|_{x=L_0} = 0. \tag{12.201}$$

Boundary-value problem (12.201) describes the incidence of k- and λ-waves of unit amplitudes on the medium. In this process, incident λ-wave U_{11} generates k-wave U_{21}, and incident k-wave U_{22} generates λ-wave U_{12}.

From system (12.201) follows that amplitude U_{21} of the generated k-wave is proportional to the parameter

$$\delta = \lambda \tilde{\alpha}_1 \tilde{\alpha}_2 = \frac{\lambda H_1 H_2}{H_0^2}.$$

This parameter satisfies the condition $\delta < \lambda/4$ in the general case. However, the models of actual media usually satisfy the condition $\tilde{\alpha}_1 \tilde{\alpha}_2 \ll 1$ (for example, it is commonly assumed that $H_2 \ll H_1$, or $\tilde{\alpha}_1 \ll 1$ and $\tilde{\alpha}_2 \cong 1$ in Earth's atmosphere and $H_1 \ll H_2$, or $\tilde{\alpha}_1 \cong 1$ and $\tilde{\alpha}_2 \ll 1$ in ocean), so that parameter δ appears usually small in the problem under consideration. For medium models with $H_2/H_1 \cong 1$, parameter δ is small ($\delta \ll 1$) for $\lambda \ll 2$.

Now, we introduce matrixes $R(L) = U(L;L) - E$ and $T(L) = U(L_0;L)$ whose matrix elements R_{ij} and T_{ij} have the meaning of complex reflection and transmission coefficients of incident (for $i = j$) and generated (for $i \neq j$) λ- and k-waves, respectively.

From system (12.201) follows the existence of two integrals of motion

$$\begin{aligned}
&\tilde{\alpha}_1 \tilde{\alpha}_2 \left[U_{11}^*(x) \frac{d}{dx} U_{11}(x) - U_{11}(x) \frac{d}{dx} U_{11}^*(x) \right] \\
&+ U_{21}^*(x) \frac{d}{dx} U_{21}(x) - U_{21}(x) \frac{d}{dx} U_{21}^*(x) = \text{const}, \\
&\tilde{\alpha}_1 \tilde{\alpha}_2 \left[U_{12}^*(x) \frac{d}{dx} U_{12}(x) - U_{12}(x) \frac{d}{dx} U_{12}^*(x) \right] \\
&+ U_{22}^*(x) \frac{d}{dx} U_{22}(x) - U_{22}(x) \frac{d}{dx} U_{22}^*(x) = \text{const}
\end{aligned}$$

that correspond to conservation of energy flux densities of λ- and k-waves. In terms of reflection and transmission coefficients, these integrals can be represented in the form

$$\begin{aligned}
\delta \left[1 - |R_{11}|^2 - |T_{11}|^2 \right] &= |R_{21}|^2 + |T_{21}|^2, \\
1 - |R_{22}|^2 - |T_{22}|^2 &= \delta \left[|R_{12}|^2 + |T_{12}|^2 \right].
\end{aligned} \tag{12.202}$$

In the case of complete wave localization in the inhomogeneous layer (L_0, L), all transmission coefficients T_{ij} must tend to zero with increasing the layer thickness.

Equalities (12.202) relate the transmission coefficients to the reflection coefficients. Using the imbedding method, we can derive a closed system for the reflection coefficients. The imbedding method offers a possibility of passing from the boundary-value problem for matrix function $U(x;L)$ to the system of equations for matrix functions $U(x;L)$ and $U(L;L)$ with the initial values in parameter L (in this case, variable x is considered as a parameter)

$$\begin{aligned}
&\frac{\partial}{\partial L} U(x;L) = iU(x;L)B + \frac{i}{2} k^2 \varepsilon(L) U(x;L) B^{-1} \tilde{\Gamma} U(L;L), \\
&\qquad U(x;L)\big|_{L=x} = U(x;x); \\
&\frac{d}{dL} U(L;L) = -2iB + i \left[U(L;L)B + BU(L;L) \right] \\
&\qquad + \frac{i}{2} k^2 \varepsilon(L) U(L;L) B^{-1} \tilde{\Gamma} U(L;L), \quad U(L;L)\big|_{L=L_0} = E.
\end{aligned} \tag{12.203}$$

The last equation can be rewritten in the form of the matrix Riccati equation for matrix $R(L) = U(L;L) - E$

$$\frac{d}{dL}R(L) = i\left[R(L)B + BR(L)\right] + \frac{i}{2}k^2\varepsilon(L)\left[E + R(L)\right]B^{-1}\tilde{\Gamma}\left[E + R(L)\right],$$
$$R(L)|_{L=L_0} = 0. \qquad (12.204)$$

Rewriting this equation as the system of equations in components R_{ij}, one can easily see that the problem has an additional integral

$$R_{21} = \delta R_{12},$$

and we can consider the system of three equations for R_{11}, R_{12}, and R_{22}.

12.6.2 Statistical description

Now, we turn to the statistical description of the problem.

We introduce intensities of all reflected waves $W_{ij}(L) = |R_{ij}(L)|^2$ and indicator function

$$\varphi_L\left(W_{11}, W_{22}, W_{12}\right) = \delta\left(W_{11}(L) - W_{11}\right)\delta\left(W_{22}(L) - W_{22}\right)\delta\left(W_{12}(L) - W_{12}\right)$$

satisfying the corresponding Liouville equation. We will assume that function $\varepsilon(x)$ is the homogeneous Gaussian random process with the zero-valued mean and the following correlation and spectral functions

$$B_{\tilde{\varepsilon}}(\xi) = \langle\tilde{\varepsilon}(x)\tilde{\varepsilon}(x')\rangle, \quad \Phi_{\tilde{\varepsilon}}(q) = \int_{-\infty}^{\infty} d\xi B_{\tilde{\varepsilon}}(\xi)e^{iq\xi}, \quad \xi = x - x', \qquad (12.205)$$

where

$$\tilde{\varepsilon}(L) = \frac{k}{2\lambda}\varepsilon(L).$$

Averaging the Liouville equation for function $\varphi_L\left(W_{11}, W_{22}, W_{12}\right)$ over an ensemble of realizations of random process $\varepsilon(L)$, one can obtain that probability density

$$P\left(L; W_{11}, W_{22}, W_{12}\right) = \langle\varphi_L\left(W_{11}, W_{22}, W_{12}\right)\rangle$$

satisfies, in the diffusion approximation, the Fokker–Planck equation

$$
\begin{aligned}
&\frac{\partial}{\partial L}P(L;W_{11},W_{22},W_{12}) \\
=\; &\frac{\partial}{\partial W_{11}}[-D_1(1-W_{11})^2-4\delta^2D_4W_{12}+2\delta(D_3+D_4)W_{11} \\
&\qquad -\delta^2D_2W_{12}^2-4\delta^2D_3W_{11}W_{12}] \\
&+\frac{\partial}{\partial W_{22}}[-D_2(1-W_{22})^2-4\delta^2D_4W_{12}+2\delta(D_3+D_4)W_{22}-\delta^2D_1W_{12}^2 \\
&\qquad -4\delta^2D_3W_{22}W_{12}] \\
&+\frac{\partial}{\partial W_{12}}[\Big(D_1(1-W_{11})+D_2(1-W_{22})+2\delta(D_3+2D_4)-\delta^2D_3W_{12}\Big)W_{12} \\
&\qquad -D_3(1+W_{11}W_{22})-D_4(W_{11}+W_{22})] \\
&+\frac{\partial^2}{\partial W_{11}^2}W_{11}\left[D_1(1-W_{11})^2+4\delta^2D_4W_{12}+4\delta^2D_3W_{11}W_{12}+\delta^2D_2W_{12}^2\right] \\
&+\frac{\partial^2}{\partial W_{22}^2}W_{22}\left[D_2(1-W_{22})^2+4\delta^2D_4W_{12}+4\delta^2D_3W_{22}W_{12}+\delta^2D_1W_{12}^2\right] \\
&+\frac{\partial^2}{\partial W_{12}^2}W_{12}[W_{12}(D_1W_{11}+D_2W_{22}+\delta^2D_3W_{12}-2\delta D_3)+D_3(1+W_{11}W_{22}) \\
&\qquad +D_4(W_{11}+W_{22})] \\
&+8\delta^2D_3\frac{\partial^2}{\partial W_{11}\partial W_{22}}W_{22}W_{11}W_{12} \\
&-2\frac{\partial^2}{\partial W_{11}\partial W_{12}}W_{11}W_{12}\left[D_1(1-W_{11})+2\delta(D_3+D_4)-2\delta^2D_2W_{12}\right] \\
&-2\frac{\partial^2}{\partial W_{22}\partial W_{12}}W_{22}W_{12}[D_2(1-W_{22})+2\delta(D_3+D_4) \\
&\qquad -2\delta^2D_2W_{12}]\}P(L;W_{11},W_{22},W_{12}),
\end{aligned}
\tag{12.206}
$$

where

$$
\begin{aligned}
D_1 &= 2\tilde{\alpha}_2^2\int\limits_0^\infty d\xi B_{\tilde{\varepsilon}}(\xi)\cos(2\lambda k\xi), \quad D_2=2(\lambda\tilde{\alpha}_1)^2\int\limits_0^\infty d\xi B_{\tilde{\varepsilon}}(\xi)\cos(2k\xi), \\
D_3 &= 2\int\limits_0^\infty d\xi B_{\tilde{\varepsilon}}(\xi)\cos\left[k(1+\lambda)\xi\right], \quad D_4=2\int\limits_0^\infty d\xi B_{\tilde{\varepsilon}}(\xi)\cos\left[k(1-\lambda)\xi\right]
\end{aligned}
$$

are the diffusion coefficients that can be represented, according to Eq. (12.205), in terms of the spectral function of random process $\tilde{\varepsilon}(x)$

$$
\begin{aligned}
D_1 &= \left(\frac{k}{2\lambda}\frac{H_1}{H_0}\right)^2\Phi_{\tilde{\varepsilon}}(2\lambda k), \qquad D_2=\left(\frac{k}{2}\frac{H_2}{H_0}\right)^2\Phi_{\tilde{\varepsilon}}(2k), \\
D_3 &= \left(\frac{k}{2\lambda}\right)^2\Phi_{\tilde{\varepsilon}}(k(1+\lambda)), \qquad D_4=\left(\frac{k}{2\lambda}\right)^2\Phi_{\tilde{\varepsilon}}(k(1-\lambda)).
\end{aligned}
\tag{12.207}
$$

Deriving Eq. (12.206), we used additional averaging over fast functions, which is admissible for $k\lambda \gg D_i$. The diffusion approximation holds for $Dl_0 \ll 1$.

In the case of small-scale medium inhomogeneities ($kl_0 \ll 1$), all diffusion coefficients can be expressed in terms of the sole parameter D

$$D_1 = \left(\frac{1}{\lambda}\frac{H_1}{H_0}\right)^2 D; \quad D_2 = \left(\frac{H_2}{H_0}\right)^2 D; \quad D_3 = D_4 = \frac{1}{\lambda^2} D, \tag{12.208}$$

where

$$D = \frac{k^2}{4} \Phi_{\tilde{\varepsilon}}(0). \tag{12.209}$$

Note that, in the case of the one-layer medium model, reflection coefficient R_L satisfies the Riccati equation to which corresponds the Fokker–Planck equation (under the neglect of absorption)

$$\frac{\partial}{\partial L} P(L; W) = D \left\{ -\frac{\partial}{\partial W}(1 - W)^2 + \frac{\partial^2}{\partial W^2} W(1 - W)^2 \right\} P(L; W) \tag{12.210}$$

with reflection coefficient (12.209) in the limit of small-scale medium inhomogeneities.

As we mentioned earlier, the two-layer problem that we consider here has parameter δ whose smallness can be used to simplify the analysis. Neglect the terms of the second order in δ in Eq. (12.206) for the probability density, i.e., neglect of the effect of secondary wave re-radiation. In this approximation, quantities W_{11} and W_{22} appear statistically independent, so that probability densities $P(L, W_{11})$ and $P(L, W_{22})$ satisfy the equations

$$\begin{aligned}
\frac{\partial}{\partial L} P(L, W_{11}) &= \left\{ \frac{\partial}{\partial W_{11}} \left[-D_1 (1 - W_{11})^2 + 2\delta(D_3 + D_4) W_{11} \right] \right. \\
&\quad \left. + D_1 \frac{\partial^2}{\partial W_{11}^2} (1 - W_{11})^2 W_{11} \right\} P(L, W_{11}), \\
\frac{\partial}{\partial L} P(L, W_{22}) &= \left\{ \frac{\partial}{\partial W_{22}} \left[-D_2 (1 - W_{22})^2 + 2\delta(D_3 + D_4) W_{22} \right] \right. \\
&\quad \left. + D_2 \frac{\partial^2}{\partial W_{22}^2} (1 - W_{22})^2 W_{22} \right\} P(L, W_{22})
\end{aligned} \tag{12.211}$$

that differ from Eq. (12.210) for the one-layer model by the term

$$2\delta(D_3 + D_4) \frac{\partial}{\partial W} \left[W P(L, W) \right].$$

This means that process of λ-wave (k-wave) generation by incident k-wave (λ-wave) is statistically equivalent to the inclusion of attenuation in the initial value problem on incident waves U_{11} and U_{22} (i.e., to substitution $\varepsilon(x) \to \varepsilon(x) + i\delta(D_3 + D_4)$ in the equations for these waves). In this case, steady-state (independent of L) solutions of form (12.37), page 288 exist for Eqs. (12.211) in the limit of the half-space ($L_0 \to -\infty$)

$$P(W_{11}) = \frac{2\gamma_1}{(1 - W_{11})^2} e^{-\frac{2\gamma_1 W_{11}}{1 - W_{11}}}, \quad P(W_{22}) = \frac{2\gamma_2}{(1 - W_{22})^2} e^{-\frac{2\gamma_2 W_{22}}{1 - W_{22}}}, \tag{12.212}$$

where parameters

$$\gamma_1 = \delta \frac{D_3 + D_4}{D_1}, \quad \gamma_2 = \delta \frac{D_3 + D_4}{D_2} \tag{12.213}$$

determine the relative part of this attenuation (i.e., secondary wave generation) in comparison with the proper diffusion of these waves (i.e., multiple re-reflections of these waves by medium inhomogeneities). In the limit of small-scale inhomogeneities, attenuation parameters

$$\gamma_1 = 2\lambda \frac{H_2}{H_1}, \quad \gamma_2 = \frac{2}{\lambda} \frac{H_1}{H_2} \tag{12.214}$$

depend only on relative layer thicknesses (for a fixed wavelength of λ-wave) and are independent of inhomogeneity statistics. In this limit, attenuation parameters satisfy the identity $\gamma_1\gamma_2 = 4$, which means that smallness of one parameter implies large value of the other parameter.

Using probability distributions (12.212), we can calculate statistical characteristics of incident wave reflection coefficients. In particular, we have

$$\langle W_{11} \rangle \approx 1 - 2\gamma_1 \ln(1/\gamma_1), \quad \langle W_{22} \rangle \approx 1 - 2\gamma_2 \ln(1/\gamma_2) \tag{12.215}$$

for $\gamma_i \ll 1$. In the opposite limiting cases $\gamma_i \gg 1$, we obtain

$$\langle W_{11} \rangle \approx \frac{1}{2\gamma_1}, \quad \langle W_{22} \rangle \approx \frac{1}{2\gamma_2}. \tag{12.216}$$

It becomes clear from the above material that, in the case of sufficiently thick layer (L_0, L) (or in the limiting case of the half-space $L_0 \to -\infty$), quantities $|T_{11}|^2$ and $|T_{22}|^2$ vanish with a probability of unity, which means that incident λ- and k-waves are localized, and their localization lengths are determined by either diffusion coefficients if diffusion prevails attenuation, or attenuation coefficients if the opposite situation occurs. Indeed, if $\gamma_1 \ll 1$ $(\gamma_2 \gg 1)$, then

$$l_{\text{loc}}^{(1)} = \frac{1}{D_1} = \left(\frac{\lambda H_0}{H_1}\right)^2 l_{\text{loc}}, l_{\text{loc}}^{(2)} = \frac{1}{2\delta(D_3 + D_4)} = \frac{\lambda H_0}{4H_1H_2} l_{\text{loc}},$$

where $l_{\text{loc}} = 1/D$ is the localization length in the one-layer problem. In the opposite case $\gamma_2 \ll 1$ $(\gamma_1 \gg 1)$, we have

$$l_{\text{loc}}^{(1)} = \frac{1}{2\delta(D_3 + D_4)} = \frac{\lambda H_0}{4H_1H_2} l_{\text{loc}}, l_{\text{loc}}^{(2)} = \frac{1}{D_2} = \left(\frac{H_0}{H_2}\right)^2 l_{\text{loc}}.$$

Determination of statistics of W_{12} appears significantly more difficult, because this problem concerns correlations of W_{12} with W_{11}, W_{22}.

To estimate average transmission coefficients of generated waves, we make use of Eqs. (12.202) that we rewrite in the form

$$\begin{aligned} 1 - \langle W_{11} \rangle - \delta \langle W_{12} \rangle &= \delta \left\langle |\widetilde{T}_{21}|^2 \right\rangle, \\ 1 - \langle W_{22} \rangle - \delta \langle W_{12} \rangle &= \delta \left\langle |T_{21}|^2 \right\rangle. \end{aligned} \tag{12.217}$$

From the Fokker–Planck equation (12.206) follows that, unlike the case of the one-layer medium, the limiting case of the half-space is characterized by the absence of steady-state solutions of form $P(T_i) = \delta(T_i)$ for quantities $T_1 = 1 - W_{11} - \delta W_{12}$ and $T_2 = 1 - W_{22} - \delta W_{12}$ describing transmission coefficients of generated waves. This means that generated waves are not localized [90].

Because Eq. (12.206) is symmetric with respect to indexes 1 and 2, average quantity $\langle W_{12} \rangle$ also must be symmetric in these indexes; consequently, the order of magnitude of quantities $\langle |T_{ij}|^2 \rangle$ can be estimated, to the symmetric portion contribution, as the order of magnitude of nonsymmetric portions of Eqs. (12.215).

For example, in the asymptotic case $\gamma_1 \ll 1$ ($\gamma_2 \gg 1$), Eqs. (12.217) assume, in view of Eqs. (12.215) and (12.216), the form

$$\begin{aligned} 2\gamma_1 \ln(1/\gamma_1) &= \delta \langle W_{12} \rangle + \delta \left\langle \left| \tilde{T}_{21} \right|^2 \right\rangle, \\ 1 - \frac{1}{2\gamma_2} &= \delta \langle W_{12} \rangle + \delta \left\langle |T_{12}|^2 \right\rangle, \end{aligned}$$

i.e.,

$$\left\langle \left| \tilde{T}_{21} \right|^2 \right\rangle \sim \frac{2}{\delta} \gamma_1 \ln(1/\gamma_1), \quad \left\langle |T_{12}|^2 \right\rangle \sim \frac{1}{\delta}. \tag{12.218}$$

In the opposite asymptotic case $\gamma_2 \ll 1$ ($\gamma_1 \gg 1$), we obtain similarly

$$\left\langle \left| \tilde{T}_{21} \right|^2 \right\rangle \sim \frac{1}{\delta}, \quad \left\langle |T_{12}|^2 \right\rangle \sim \frac{2}{\delta} \gamma_2 \ln(1/\gamma_2). \tag{12.219}$$

Turning back to the initial value problem on sources located in the upper and lower medium layers (or at the boundary $x_0 = L$ of inhomogeneous layer), we can see that transmission coefficients of waves generated in both upper and lower layers are different from zero in the whole of the medium, i.e., no wave localization occurs. The concrete values of these coefficients depend on both ratio of layer thicknesses and parameter λ.

Remark 12 ***Localization of the Rossby waves under the effect of random cylindrical topography of underlying surface.***

Like the β-effect, inhomogeneities of bottom surface play important role in propagation of large-scale low-frequency oscillations in Earth's atmosphere and ocean (the Rossby waves). The effect of topography on the propagation of such waves depends mainly on the ratio of wavelength λ and the horizontal scale of topographic inhomogeneities l_h [266]. In the case $\lambda \gg l_h$ important for practice, such topographic inhomogeneities can support the propagation of large-scale waves even in the absence of the β-effect, which can be used to model generation and propagation of the Rossby waves in the laboratory conditions [57, 113].

Many investigations considered the topographic inhomogeneities as periodic and quasi-periodic functions, or represented them as superpositions of the Fourier harmonics (see, e.g., [267] for the two-layer model of medium). In actuality, the topographic inhomogeneities are highly irregular and can be considered, in essence, as specific realizations from a great ensemble of random fields with specified statistics. This fact enables us to analyze such motions (and, in particular, propagation of the Rossby waves in the absence of zonal flow) using techniques of the theory of random processes and fields [124, 274, 302], which significantly simplifies the analysis. However, in view of the fact that no ensemble exists in actuality and researchers deal with separate realizations, the final results must be formulated in the form appropriate for analyzing actual situations.

Within the framework of the quasi-geostrophic model, large-scale low-frequency motions in the two-layer medium (atmosphere, ocean) of variable depth are described by linearized equations (1.102), page 35. For functions $\psi_1(x, y)$ and $\psi_2(x, y)$,

$$\psi_1(x, y) = \psi_1(y) e^{-i(\omega t + \kappa x)}, \quad \psi_2(x, y) = \psi_2(y) e^{-i(\omega t + \kappa x)},$$

which correspond to the wave propagating to the west, these equations assume for $\kappa > 0$, $\omega > 0$ the form of the system of equations (12.192) with the parameters

$$k^2 = \kappa \left(\frac{\beta}{\omega} - \kappa \right), \quad \varepsilon(y) = \frac{\kappa f_0}{H_2 \omega k^2} \frac{d}{dy} h(y).$$

Quantity k^2 (under the condition that $k^2 > 0$) is the squared y-component of the wave vector of propagating barotropic mode of the Rossby wave with fixed κ and ω. The feature of this problem consists in the fact that system depends not on the topography, but on its spatial derivative.

Consequently, the results of the above analysis of waves in the two-layered medium are sufficient for studying the problem on localization of the Rossby waves under the effect of random cylindrical topography of underlying surface [91, 145, 175]. ♦

Chapter 13

Wave propagation in random media

Fluctuations of wavefield propagating in a medium with random large-scale (in comparison with wavelength) inhomogeneities rapidly grow with distance because of multiple forward scattering. Perturbation theory in any version fails beginning from certain distance (the boundary of the strong fluctuation region). Strong fluctuations of intensity can appear in radiowaves propagating through the ionosphere, solar corona, or interstellar medium, in experiments on transilluminating planet's atmospheres when planets shadow natural or artificial radiation sources, and in a number of other cases.

The current state of the theory of wave propagation in random media can be found in monographs and reviews [27, 70, 120, 134, 135, 205, 237, 268, 294]. Below, we will follow works [134, 135, 150, 170, 295] to describe wave propagation in random media within the framework of the parabolic equation of quasi-optics and delta-correlated approximation of medium parameter fluctuations and discuss the applicability of such an approach.

It appears worthwhile to divide this material into two parts. The first part deals with studying the statistical properties of the initial stochastic partial differential equation describing wave propagation, while the second part studies statistical properties of the solution to this stochastic equation in the explicit form of the continual integral.

13.1 Method of stochastic equation

13.1.1 Input stochastic equations and their implications

We will describe the propagation of a monochromatic wave in the medium with large-scale inhomogeneities in terms of the complex scalar parabolic equation (1.91), page 31

$$\frac{\partial}{\partial x}u(x,\mathbf{R}) = \frac{i}{2k}\Delta_{\mathbf{R}}u(x,\mathbf{R}) + i\frac{k}{2}\varepsilon(x,\mathbf{R})u(x,\mathbf{R}), \qquad (13.1)$$

where function $\varepsilon(x,\mathbf{R})$ is the fluctuating portion (deviation from unity) of dielectric permittivity, x-axis is directed along the initial direction of wave propagation, and vector $\mathbf{R}$ denotes the coordinates in the transverse plane. The initial condition to Eq. (13.1) is the condition

$$u(0,\mathbf{R}) = u_0(\mathbf{R}). \qquad (13.2)$$

Because Eq. (13.1) is the first-order equation in x and satisfies initial condition (13.2) at $x = 0$, it possesses the causality property with respect to x-coordinate (it plays here the role of time), i.e., its solution satisfies the relationship

$$\frac{\delta u(x,\mathbf{R})}{\delta\varepsilon(x',\mathbf{R}')}=0 \quad \text{for} \quad x'<0,\, x'>x. \tag{13.3}$$

The variational derivative at $x=x'$ can be obtained according to the standard procedure

$$\frac{\delta u(x,\mathbf{R})}{\delta\varepsilon(x-0,\mathbf{R}')}=\frac{ik}{2}\delta\left(\mathbf{R}-\mathbf{R}'\right)u(x,\mathbf{R}). \tag{13.4}$$

In the general case, quantity $\delta u(x,\mathbf{R})/\delta\varepsilon(x',\mathbf{R}')$ for $0\le x'<x$ can be expressed in terms of Green's function of Eq. (13.1) that relates field $u(x,\mathbf{R})$ to field $u(x',\mathbf{R}')$ for $0\le x'<x$

$$u(x,\mathbf{R})=\int d\mathbf{R}'G\left(x,\mathbf{R};x',\mathbf{R}'\right)u(x',\mathbf{R}') \tag{13.5}$$

and, in particular,

$$u(x,\mathbf{R})=\int d\mathbf{R}'G\left(x,\mathbf{R};0,\mathbf{R}'\right)u_0(\mathbf{R}').$$

The corresponding expression has the form

$$\frac{\delta u(x,\mathbf{R})}{\delta\varepsilon(x',\mathbf{R}')}=\frac{ik}{2}G\left(x,\mathbf{R};x',\mathbf{R}'\right)u(x',\mathbf{R}').$$

Here, Green's function $G\left(x,\mathbf{R};x',\mathbf{R}'\right)$ satisfies the integral equation

$$G\left(x,\mathbf{R};x',\mathbf{R}'\right)=g\left(x,\mathbf{R};x',\mathbf{R}'\right)$$
$$+\frac{ik}{2}\int\limits_{x'}^{x}dx''\int d\mathbf{R}''g\left(x,\mathbf{R};x'',\mathbf{R}''\right)\varepsilon(x'',\mathbf{R}'')G\left(x'',\mathbf{R}'';x',\mathbf{R}'\right), \tag{13.6}$$

where function

$$g\left(x,\mathbf{R};x',\mathbf{R}'\right)=e^{\frac{i(x-x')}{2k}\Delta_{\mathbf{R}}}\delta\left(\mathbf{R}-\mathbf{R}'\right)=\frac{k}{2\pi i(x-x')}e^{\frac{ik(\mathbf{R}-\mathbf{R}')^2}{2(x-x')}} \tag{13.7}$$

for $x>x'$ is Green's function of Eq. (13.1) in the absence of inhomogeneities. For $x\to x'$ Eq. (13.6) grades into the formula

$$G\left(x,\mathbf{R};x',\mathbf{R}'\right)\big|_{x\to x'}=g\left(x,\mathbf{R};x',\mathbf{R}'\right)\big|_{x\to x'}=\delta\left(\mathbf{R}-\mathbf{R}'\right).$$

Recall that Green's function $G\left(x,\mathbf{R};x',\mathbf{R}'\right)$ describes the field of spherical wave originated from point $(x',\mathbf{R}')$.

Integral equation (13.6) can be rewritten in the form of the equivalent variational differential equation

$$\frac{\delta G\left(x,\mathbf{R};x',\mathbf{R}'\right)}{\delta\varepsilon(\xi,\mathbf{R}_1)}=i\frac{k}{2}G\left(x,\mathbf{R};\xi,\mathbf{R}_1\right)G\left(\xi,\mathbf{R}_1;x',\mathbf{R}'\right) \tag{13.8}$$

with the functional initial value

$$G\left(x,\mathbf{R};x',\mathbf{R}'\right)\big|_{\varepsilon=0}=g\left(x,\mathbf{R};x',\mathbf{R}'\right).$$

In addition to Eq. (13.6), Green's function $G(x, \mathbf{R}; x', \mathbf{R}')$ satisfies the equation

$$G(x, \mathbf{R}; x', \mathbf{R}') = g(x, \mathbf{R}; x', \mathbf{R}') \\ +\frac{ik}{2}\int\limits_{x'}^{x} dx'' \int d\mathbf{R}'' G(x, \mathbf{R}; x'', \mathbf{R}'')\, \varepsilon(x'', \mathbf{R}'') g(x'', \mathbf{R}''; x', \mathbf{R}'). \qquad (13.9)$$

One can easily check this fact by comparing iterative series in $\varepsilon(x, \mathbf{R})$ for Eqs. (13.6) and (13.9).

Perform complex conjugation in Eq. (13.9) and interchange points $(x, \mathbf{R})$ and $(x', \mathbf{R}')$ (bearing in mind that $x > x'$ as before). In view of the identity

$$g^*(x', \mathbf{R}'; x, \mathbf{R}) = g(x, \mathbf{R}; x', \mathbf{R}'),$$

we obtain the equation

$$G^*(x', \mathbf{R}'; x, \mathbf{R}) = g(x, \mathbf{R}; x', \mathbf{R}') \\ +\frac{ik}{2}\int\limits_{x'}^{x} dx'' \int d\mathbf{R}'' g(x, \mathbf{R}; x'', \mathbf{R}'')\, \varepsilon(x'', \mathbf{R}'') G^*(x', \mathbf{R}'; x'', \mathbf{R}'').$$

Comparing this equation with Eq. (13.6), we obtain the equality

$$G(x, \mathbf{R}; x', \mathbf{R}') = G^*(x', \mathbf{R}'; x, \mathbf{R}) \quad (x > x'), \qquad (13.10)$$

which constitutes the reciprocity theorem in the approximation of parabolic equation. Here, function $G^*(x, \mathbf{R}; x', \mathbf{R}')$ is the spherical wave propagating from the source point $(x', \mathbf{R}')$ in the negative direction along the x-axis.

It is obvious that we can represent Eqs. (13.6) and (13.9) in the form of differential equations

$$\left(\frac{\partial}{\partial x} - \frac{i}{2k}\Delta_{\mathbf{R}}\right) G(x, \mathbf{R}; x', \mathbf{R}') = \frac{ik}{2}\varepsilon(x, \mathbf{R}) G(x, \mathbf{R}; x', \mathbf{R}'), \\ \left(\frac{\partial}{\partial x'} + \frac{i}{2k}\Delta_{\mathbf{R}'}\right) G(x, \mathbf{R}; x', \mathbf{R}') = -\frac{ik}{2}\varepsilon(x', \mathbf{R}') G(x, \mathbf{R}; x', \mathbf{R}') \qquad (13.11)$$

with the initial value

$$G(x, \mathbf{R}; x', \mathbf{R}')\big|_{x \longleftrightarrow x'} = \delta(\mathbf{R} - \mathbf{R}').$$

One can easily see that Green's function satisfies the orthogonality conditions

$$\int d\mathbf{R} G(x, \mathbf{R}; x', \mathbf{R}')\, G^*(x, \mathbf{R}; x'', \mathbf{R}'') = \delta(\mathbf{R}' - \mathbf{R}''), \\ \int d\mathbf{R}' G(x_1, \mathbf{R}_1; x', \mathbf{R}')\, G^*(x_2, \mathbf{R}_2; x', \mathbf{R}') = \delta(\mathbf{R}_1 - \mathbf{R}_2). \qquad (13.12)$$

A consequence of these conditions is the equality

$$\int d\mathbf{R} u_1(x, \mathbf{R})\, u_2^*(x, \mathbf{R}) = \int d\mathbf{R} u_1^0(\mathbf{R})\, u_2^{0*}(\mathbf{R}), \qquad (13.13)$$

where $u_1(x, \mathbf{R})$ and $u_2(x, \mathbf{R})$ are the solutions to Eq. (13.1) with initial values $u_1^0(\mathbf{R})$ and $u_2^0(\mathbf{R})$, respectively. In the special case of $u_1^0(\mathbf{R}) = u_2^0(\mathbf{R}) = u_0(\mathbf{R})$, Eq. (13.13) formulates energy conservation

$$\int d\mathbf{R} I(x, \mathbf{R}) = \int d\mathbf{R} I_0(\mathbf{R}) = \text{const} \quad \left(I(x, \mathbf{R}) = |u(x, \mathbf{R})|^2\right). \qquad (13.14)$$

13.1.2 Delta-correlated approximation for medium parameters

Consider now the statistical description of the wavefield. We will assume that random field $\varepsilon(x,\mathbf{R})$ is the homogeneous and isotropic Gaussian field with the parameters

$$\langle\varepsilon(x,\mathbf{R})\rangle = 0, \quad B_\varepsilon(x-x',\mathbf{R}-\mathbf{R}') = \langle\varepsilon(x,\mathbf{R})\varepsilon(x',\mathbf{R}')\rangle .$$

As was noted, field $u(x,\mathbf{R})$ depends functionally only on preceding values of field $\varepsilon(x,\mathbf{R})$. Nevertheless, statistically, field $u(x,\mathbf{R})$ can depend on subsequent values $\varepsilon(x_1,\mathbf{R})$ for $x_1 > x$ due to nonzero correlation between values $\varepsilon(x',\mathbf{R}')$ for $x' < x$ and values $\varepsilon(\xi,\mathbf{R})$ for $\xi > x$. It is clear that correlation of field $u(x,\mathbf{R})$ with subsequent values $\varepsilon(x',\mathbf{R}')$ is appreciable only if $x' - x \sim l_\parallel$, where $l_\parallel$ is the longitudinal correlation radius of field $\varepsilon(x,\mathbf{R})$. At the same time, the characteristic correlation radius of field $u(x,\mathbf{R})$ in the longitudinal direction is estimated in rough way as x (see, e.g., [268, 294]). Therefore, the problem under consideration has small parameter $l_\parallel/x$, and we can use it to construct an approximate solution.

In the first approximation, we can set $l_\parallel/x \to 0$. In this case, field values $u(\xi_i,\mathbf{R})$ for $\xi_i < x$ will be independent of field values $\varepsilon(\eta_j,\mathbf{R})$ for $\eta_j > x$ not only functionally, but also statistically. This is equivalent to approximating the correlation function of field $\varepsilon(x,\mathbf{R})$ by the delta function of longitudinal coordinate, i.e., to the replacement of the correlation function with the effective function

$$B_\varepsilon(x,\mathbf{R}) = B_\varepsilon^{\text{eff}}(x,\mathbf{R}) = \delta(x)A(\mathbf{R}), \quad A(\mathbf{R}) = \int\limits_{-\infty}^{\infty} dx B_\varepsilon(x,\mathbf{R}). \tag{13.15}$$

Using this approximation, we derive the equations for moment functions

$$M_{mn}(x;\mathbf{R}_1,\ldots,\mathbf{R}_m;\mathbf{R}'_1,\ldots,\mathbf{R}'_n) = \left\langle \prod_{p=1}^{m}\prod_{q=1}^{n} u(x;\mathbf{R}_p)\, u^*\left(x;\mathbf{R}'_q\right)\right\rangle . \tag{13.16}$$

In the case of $m = n$, these functions are usually called the *coherence functions* of order $2n$.

Differentiating function (13.16) with respect to x and using Eq. (13.1) and its complex conjugated version, we obtain the equation

$$\begin{aligned} &\frac{\partial}{\partial x} M_{mn}(x;\mathbf{R}_1,\ldots,\mathbf{R}_m;\mathbf{R}'_1,\ldots,\mathbf{R}'_n) \\ &= \frac{i}{2k}\left(\sum_{p=1}^{m}\Delta_{\mathbf{R}_p} - \sum_{q=1}^{n}\Delta_{\mathbf{R}'_q}\right) M_{mn}(x;\mathbf{R}_1,\ldots,\mathbf{R}_m;\mathbf{R}'_1,\ldots,\mathbf{R}'_n) \\ &+ i\frac{k}{2}\left\langle\left(\sum_{p=1}^{m}\varepsilon(x,\mathbf{R}_p) - \sum_{q=1}^{n}\varepsilon(x,\mathbf{R}'_q)\right)\left[\prod_{p=1}^{m}\prod_{q=1}^{n} u(x;\mathbf{R}_p)\, u^*(x;\mathbf{R}_q)\right]\right\rangle . \end{aligned} \tag{13.17}$$

To split the correlator in the right-hand side of Eq. (13.17), we use the Furutsu–Novikov formula that assumes here the following forms

$$\begin{aligned} \langle\varepsilon(x,\mathbf{R})u(x;\mathbf{R}_p)\rangle &= \int\limits_0^x dx' \int d\mathbf{R}' B_\varepsilon(x-x',\mathbf{R}-\mathbf{R}') \left\langle \frac{\delta u(x;\mathbf{R}_p)}{\delta\varepsilon(x',\mathbf{R}')}\right\rangle , \\ \left\langle\varepsilon(x,\mathbf{R})u^*\left(x;\mathbf{R}'_q\right)\right\rangle &= \int\limits_0^x dx' \int d\mathbf{R}' B_\varepsilon(x-x',\mathbf{R}-\mathbf{R}') \left\langle \frac{\delta u^*\left(x;\mathbf{R}'_q\right)}{\delta\varepsilon(x',\mathbf{R}')}\right\rangle . \end{aligned}$$

The delta-correlated approximation of medium parameter fluctuations with effective correlation function (13.15) simplifies these equalities; if we additionally take into account Eq. (13.4) and its complex conjugated version, then we arrive at the closed equation for the wavefield moment function

$$\begin{aligned}&\frac{\partial}{\partial x}M_{mn}(x;\mathbf{R}_1,\ldots,\mathbf{R}_m;\mathbf{R}'_1,\ldots,\mathbf{R}'_n)\\&=\frac{i}{2k}\left(\sum_{p=1}^{m}\Delta_{\mathbf{R}_p}-\sum_{q=1}^{n}\Delta_{\mathbf{R}'_q}\right)M_{mn}(x;\mathbf{R}_1,\ldots,\mathbf{R}_m;\mathbf{R}'_1,\ldots,\mathbf{R}'_n)\\&-\frac{k^2}{8}Q(\mathbf{R}_1,\ldots,\mathbf{R}_m;\mathbf{R}'_1,\ldots,\mathbf{R}'_n)M_{mn}(x;\mathbf{R}_1,\ldots,\mathbf{R}_m;\mathbf{R}'_1,\ldots,\mathbf{R}'_n),\end{aligned}\tag{13.18}$$

where

$$\begin{aligned}&Q(\mathbf{R}_1,\ldots,\mathbf{R}_m;\mathbf{R}'_1,\ldots,\mathbf{R}'_n)\\&=\sum_{i=1}^{m}\sum_{j=1}^{m}A(\mathbf{R}_i-\mathbf{R}_j)-2\sum_{i=1}^{m}\sum_{j=1}^{n}A\left(\mathbf{R}_i-\mathbf{R}'_j\right)+\sum_{i=1}^{n}\sum_{j=1}^{n}A\left(\mathbf{R}'_i-\mathbf{R}'_j\right).\end{aligned}\tag{13.19}$$

An equation for the characteristic functional of random field $u(x,\mathbf{R})$ also can be obtained (Eq. 5.190, page 144); however it will be the linear variational derivative equation.

Draw explicitly equations for average field $\langle u(x,\mathbf{R})\rangle$, second-order coherence function

$$\Gamma_2(x,\mathbf{R},\mathbf{R}')=\langle\gamma_2(x,\mathbf{R},\mathbf{R}')\rangle,\quad \gamma_2(x,\mathbf{R},\mathbf{R}')=u(x,\mathbf{R})u^*(x,\mathbf{R}'),$$

and fourth-order coherence function

$$\Gamma_4(x,\mathbf{R}_1,\mathbf{R}_2,\mathbf{R}'_1\mathbf{R}'_2)=\langle u(x,\mathbf{R}_1)u\,(x,\mathbf{R}_2)\,u^*(x,\mathbf{R}'_1)u^*(x,\mathbf{R}'_2)\rangle,$$

which follow from Eqs. (13.18) and (13.19) for $m=1$, $n=0$; $m=n=1$; and $m=n=2$. They have the forms

$$\frac{\partial}{\partial x}\langle u(x,\mathbf{R})\rangle=\frac{i}{2k}\Delta_{\mathbf{R}}\langle u(x,\mathbf{R})\rangle-\frac{k^2}{8}A(0)\langle u(x,\mathbf{R})\rangle,\quad \langle u(0,\mathbf{R})\rangle=u_0(\mathbf{R}),\tag{13.20}$$

$$\begin{aligned}&\frac{\partial}{\partial x}\Gamma_2(x,\mathbf{R},\mathbf{R}')=\frac{i}{2k}\left(\Delta_{\mathbf{R}}-\Delta_{\mathbf{R}'}\right)\Gamma_2(x,\mathbf{R},\mathbf{R}')\\&-\frac{k^2}{4}D(\mathbf{R}-\mathbf{R}_1)\Gamma_2(x,\mathbf{R},\mathbf{R}'),\quad \Gamma_2(0,\mathbf{R},\mathbf{R}')=u_0(\mathbf{R})u_0^*(\mathbf{R}'),\end{aligned}\tag{13.21}$$

$$\begin{aligned}&\frac{\partial}{\partial x}\Gamma_4(x,\mathbf{R}_1,\mathbf{R}_2,\mathbf{R}'_1,\mathbf{R}'_2)\\&=\frac{i}{2k}\left(\Delta_{\mathbf{R}_1}+\Delta_{\mathbf{R}_2}-\Delta_{\mathbf{R}'_1}-\Delta_{\mathbf{R}'_2}\right)\Gamma_4(x,\mathbf{R}_1,\mathbf{R}_2,\mathbf{R}'_1,\mathbf{R}'_2)\\&-\frac{k^2}{8}Q(\mathbf{R}_1,\mathbf{R}_2,\mathbf{R}'_1,\mathbf{R}'_2)\Gamma_4(x,\mathbf{R}_1,\mathbf{R}_2,\mathbf{R}'_1,\mathbf{R}'_2),\\&\Gamma_4(0,\mathbf{R}_1,\mathbf{R}_2,\mathbf{R}'_1,\mathbf{R}'_2)=u_0(\mathbf{R}_1)u_0\,(\mathbf{R}_2)\,u_0^*(\mathbf{R}'_1)u_0^*(\mathbf{R}'_2),\end{aligned}\tag{13.22}$$

where we introduced new functions

$$\begin{aligned}&D(\mathbf{R})=A(0)-A(\mathbf{R}),\\&Q(\mathbf{R}_1,\mathbf{R}_2,\mathbf{R}'_1,\mathbf{R}'_2)=D(\mathbf{R}_1-\mathbf{R}'_1)+D(\mathbf{R}_2-\mathbf{R}'_2)+D(\mathbf{R}_1-\mathbf{R}'_2)\\&\quad+D(\mathbf{R}_2-\mathbf{R}'_1)-D(\mathbf{R}_2-\mathbf{R}_1)-D(\mathbf{R}'_2-\mathbf{R}'_1)\end{aligned}$$

related to the structure function of random field $\varepsilon(x, \mathbf{R})$.

Introducing new variables

$$\mathbf{R} \to \mathbf{R} + \frac{1}{2}\boldsymbol{\rho}, \quad \mathbf{R}' \to \mathbf{R} - \frac{1}{2}\boldsymbol{\rho},$$

Eq. (13.21) can be rewritten in the form

$$\begin{aligned} &\left(\frac{\partial}{\partial x} - \frac{i}{k}\boldsymbol{\nabla}_{\mathbf{R}}\boldsymbol{\nabla}_{\boldsymbol{\rho}}\right)\Gamma_2(x, \mathbf{R}, \boldsymbol{\rho}) = -\frac{k^2}{4}D(\boldsymbol{\rho})\Gamma_2(x, \mathbf{R}, \boldsymbol{\rho}), \\ &\Gamma_2(0, \mathbf{R}, \boldsymbol{\rho}) = u_0\left(\mathbf{R} + \frac{1}{2}\boldsymbol{\rho}\right)u_0^*\left(\mathbf{R} - \frac{1}{2}\boldsymbol{\rho}\right). \end{aligned} \tag{13.23}$$

Remark 13 ***Small-angle approximation of the theory of radiative transfer.***

Note that Eq. (13.23) corresponds to the so-called small-angle approximation of the phenomenological theory of radiative transfer. Indeed, if we introduce function

$$J(x, \mathbf{R}, \mathbf{q}) = \frac{1}{(2\pi)^2}\int d\boldsymbol{\rho}\Gamma_2(x, \mathbf{R}, \boldsymbol{\rho})e^{-i\mathbf{q}\boldsymbol{\rho}},$$

then we obtain that it satisfies the integro-differential equation

$$\begin{aligned} &\left(\frac{\partial}{\partial x} + \frac{1}{k}\mathbf{q}\boldsymbol{\nabla}_{\mathbf{R}}\right)J(x, \mathbf{R}, \mathbf{q}) = -\gamma J(x, \mathbf{R}, \mathbf{q}) + \int d\mathbf{q}' f(\mathbf{q} - \mathbf{q}')J(x, \mathbf{R}, \mathbf{q}'), \\ &J(0, \mathbf{R}, \mathbf{q}) = \frac{1}{(2\pi)^2}\int d\boldsymbol{\rho}\Gamma_2(0, \mathbf{R}, \boldsymbol{\rho})e^{-i\mathbf{q}\boldsymbol{\rho}}. \end{aligned} \tag{13.24}$$

Here,

$$\gamma = \frac{1}{4}k^2 A(0) = \int d\mathbf{q} f(\mathbf{q}) \tag{13.25}$$

is the *extinction coefficient*, $f(\mathbf{q}) = \frac{1}{2}\pi k^2 \Phi_\varepsilon(0, q)$ is the scattering indicatrix, and

$$\Phi_\varepsilon(q_1, \mathbf{q}) = \frac{1}{(2\pi)^3}\int\limits_{-\infty}^{\infty} dx \int d\mathbf{R} B_\varepsilon(x, \mathbf{R})e^{-iq_1 x - i\mathbf{q}\mathbf{R}} \tag{13.26}$$

is the three-dimensional spectral density of field $\varepsilon(x, \mathbf{R})$. Note additionally that the above function $J(x, \mathbf{R}, \mathbf{q})$ is the average of the *Wigner function*

$$W(x, \mathbf{R}, \mathbf{q}) = \frac{1}{(2\pi)^2}\int d\boldsymbol{\rho}\gamma_2(x, \mathbf{R}, \boldsymbol{\rho})e^{-i\mathbf{q}\boldsymbol{\rho}},$$

where

$$\gamma_2(x; \mathbf{R}, \boldsymbol{\rho}) = u\left(x, \mathbf{R} + \frac{1}{2}\boldsymbol{\rho}\right)u^*\left(x, \mathbf{R} - \frac{1}{2}\boldsymbol{\rho}\right). \quad \blacklozenge$$

Equations (13.20) and (13.23) can be easily solved for arbitrary function $D(\boldsymbol{\rho})$ and arbitrary initial conditions. Indeed, the average field is given by the expression

$$\langle u(x, \mathbf{R})\rangle = u_0(x, \mathbf{R})e^{-\frac{\gamma}{2}x}, \tag{13.27}$$

where $u_0(x, \mathbf{R})$ is the solution to the problem with absent fluctuations of medium parameters,

$$u_0(x, \mathbf{R}) = \int d\mathbf{R}' g(x, \mathbf{R} - \mathbf{R}')u_0(\mathbf{R}'),$$

and function $g(x, \mathbf{R})$ is free space Green's function (13.7).

Correspondingly, the second-order coherence function is given by the expression

$$\Gamma_2(x, \mathbf{R}, \boldsymbol{\rho}) = \int d\mathbf{q}\gamma_0\left(\mathbf{q}, \boldsymbol{\rho} - \mathbf{q}\frac{x}{k}\right) \exp\left\{i\mathbf{q}\mathbf{R} - \frac{k^2}{4}\int\limits_0^x d\xi D\left(\boldsymbol{\rho} - \mathbf{q}\frac{\xi}{k}\right)\right\}, \qquad (13.28)$$

where

$$\gamma_0\left(\mathbf{q}, \boldsymbol{\rho}\right) = \frac{1}{(2\pi)^2}\int d\mathbf{R}\gamma_0(\mathbf{R}, \boldsymbol{\rho})e^{-i\mathbf{q}\mathbf{R}}.$$

The further analysis depends on the initial conditions to Eq. (13.1) and the fluctuation nature of field $\varepsilon(x, \mathbf{R})$. Three types of initial data are usually used in practice. They are

- Plane incident wave, in which case $u_0(\mathbf{R}) = u_0$;
- Spherical divergent wave, in which case $u_0(\mathbf{R}) = \delta(\mathbf{R})$; and
- Incident wave beam with the initial field distribution

  $$u_0(\mathbf{R}) = u_0 \exp\left\{-\frac{\mathbf{R}^2}{2a^2} + i\frac{k\mathbf{R}^2}{2F}\right\}, \qquad (13.29)$$

  where a is the effective beam width, F is the distance to the radiation center (in the case of free space, value $F = \infty$ corresponds to the collimated beam and value $F < 0$ corresponds to the beam focused at distance $x = |F|$).

In the case of the plane incident wave, we have

$$u_0(\mathbf{R}) = u_0 = \text{const}, \quad \gamma_0(\mathbf{R}, \boldsymbol{\rho}) = |u_0|^2, \quad \gamma_0\left(\mathbf{q}, \boldsymbol{\rho}\right) = |u_0|^2\delta(\mathbf{q}),$$

and Eqs. (13.27) and (13.28) become significantly simpler

$$\langle u(x, \mathbf{R})\rangle = u_0 e^{-\frac{1}{2}\gamma x}, \quad \Gamma_2(x, \mathbf{R}, \boldsymbol{\rho}) = |u_0|^2 e^{-\frac{1}{4}k^2 x D(\boldsymbol{\rho})} \qquad (13.30)$$

and appear independent of plane wave diffraction in random medium. Moreover, the expression for the coherence function shows the appearance of new statistical scale $\boldsymbol{\rho}_{\text{coh}}$ defined by the condition

$$\frac{1}{4}k^2 x D\left(\boldsymbol{\rho}_{\text{coh}}\right) = 1. \qquad (13.31)$$

This scale is called the *coherence radius* of field $u(x, \mathbf{R})$. Its value depends on the wavelength, distance the wave travels in the medium, and medium statistical parameters.

In the case of the wave beam (13.29), we easily obtain that

$$\gamma_0(\mathbf{q}, \boldsymbol{\rho}) = \frac{|u_0|^2 a^2}{4\pi}\exp\left\{-\frac{1}{4}\left[\frac{\boldsymbol{\rho}^2}{a^2} + \left(\frac{k\boldsymbol{\rho}}{F} - \mathbf{q}\right)^2 a^2\right]\right\}.$$

Using this expression and considering the turbulent atmosphere as an example of random medium for which the structure function $D(\mathbf{R})$ is described by the *Kolmogorov–Obukhov law* (see, e.g., [134, 135, 268, 294])

$$D(\mathbf{R}) = NC_\varepsilon^2 R^{5/3} \quad (R_{\min} \ll R \ll R_{\max}),$$

where $N = 1.46$, and C_ε^2 is the structure characteristic of dielectric permittivity fluctuations, we obtain that average intensity in the beam

$$\langle I(x, \mathbf{R})\rangle = \Gamma_2(x, \mathbf{R}, 0)$$

is given by the expression

$$\langle I(x, \mathbf{R})\rangle = \frac{2|u_0|^2 k^2 a^4}{x^2 g^2(x)} \int\limits_0^\infty dt J_0\left(\frac{2kaRt}{xg(x)}\right) \exp\left\{-t^2 - \frac{3\pi N}{32} C_\varepsilon^2 k^2 x \left(\frac{2a}{g(x)}\right)^{5/3} t^{5/3}\right\},$$

where

$$g(x) = \sqrt{1 + k^2 a^4 \left(\frac{1}{x} + \frac{1}{F}\right)^2}$$

and $J_0(t)$ is the Bessel function. Many full-scale experiments testified this formula in turbulent atmosphere and showed a good agreement between measured data and theory.

Equation (13.22) for the fourth-order coherence function cannot be solved in analitic form; the analysis of this function requires either numerical, or approximate techniques. It describes intensity fluctuations and reduces to the intensity variance for equal transverse coordinates.

In the case of the plane incident wave, Eq. (13.22) can be simplified by introducing new transverse coordinates

$$\widetilde{\mathbf{R}}_1 = \mathbf{R}_1' - \mathbf{R}_1 = \mathbf{R}_2 - \mathbf{R}_2', \quad \widetilde{\mathbf{R}}_2 = \mathbf{R}_2' - \mathbf{R}_1 = \mathbf{R}_2 - \mathbf{R}_1'.$$

In this case, Eq. (13.22) assumes the form (we omit here the tilde sign)

$$\frac{\partial}{\partial x}\Gamma_4(x, \mathbf{R}_1, \mathbf{R}_2) = \frac{i}{k}\frac{\partial^2}{\partial \mathbf{R}_1 \partial \mathbf{R}_2}\Gamma_4(x, \mathbf{R}_1, \mathbf{R}_2) - \frac{k^2}{4}F(\mathbf{R}_1, \mathbf{R}_2)\Gamma_4(x, \mathbf{R}_1, \mathbf{R}_2), \qquad (13.32)$$

where

$$F(\mathbf{R}_1, \mathbf{R}_2) = 2D(\mathbf{R}_1) + 2D(\mathbf{R}_2) - D(\mathbf{R}_1 + \mathbf{R}_2) - D(\mathbf{R}_1 - \mathbf{R}_2).$$

We give the asymptotic solution to this equation in Subsection 13.3, page 403.

Now, we note that the above approach makes it possible to analyze the problem on wave beam propagation in random media described by field $\varepsilon(x, \mathbf{R})$ showing layered (on average) structure. In this case, we deal with the dynamic equation

$$\frac{\partial}{\partial x}u(x, \mathbf{R}) = \frac{i}{2k}\Delta_{\mathbf{R}} u(x, \mathbf{R}) + \frac{ik}{2}\left[\varepsilon_0(\mathbf{R}) + \varepsilon(x, \mathbf{R})\right] u(x, \mathbf{R}). \qquad (13.33)$$

Considering propagation of beam (13.29) in a parabolic waveguide with

$$\varepsilon_0(\mathbf{R}) = -\alpha^2 \mathbf{R}^2,$$

we can easily obtain that the second-order coherence function is given by the expression [134, 135]

$$\Gamma_2(x, \mathbf{R}, \boldsymbol{\rho}) = \frac{e^{-i\alpha \mathbf{R}\boldsymbol{\rho}\tan(\alpha x)}}{\cos^2(\alpha x)} \int d\mathbf{q} \gamma_0 \left(\mathbf{q}, \frac{1}{\cos(\alpha x)}\boldsymbol{\rho} - \frac{\mathbf{q}}{\alpha k}\tan(\alpha x)\right)$$
$$\times \exp\left\{i\frac{1}{\cos(\alpha x)}\mathbf{q}\mathbf{R} - \frac{k^2}{4}\int\limits_0^x d\xi D\left(\frac{\cos(\alpha\xi)}{\cos(\alpha x)}\boldsymbol{\rho} - \frac{1}{\alpha k}\frac{\sin\left[\alpha(x-\xi)\right]}{\cos(\alpha x)}\mathbf{q}\right)\right\}. \qquad (13.34)$$

At $\alpha = 0$, this formula grades naturally into Eq. (13.28). Setting now $\boldsymbol{\rho} = 0$ in Eq. (13.34), we obtain the expression for average intensity

$$\begin{aligned}\langle I(x, \mathbf{R})\rangle &= \frac{1}{\cos^2(\alpha x)} \int d\mathbf{q}\gamma_0\left(\mathbf{q}, -\frac{\mathbf{q}}{\alpha k}\tan(\alpha x)\right) \\ &\times \exp\left\{i\frac{1}{\cos(\alpha x)}\mathbf{qR} - \frac{k^2}{4}\int\limits_0^x d\xi D\left(\frac{1}{\alpha k}\frac{\sin(\alpha\xi)}{\cos(\alpha x)}\mathbf{q}\right)\right\}, \qquad (13.35)\end{aligned}$$

and setting then $\mathbf{R} = 0$ in (13.35), we arrive at the expression for average intensity along the waveguide axis

$$\begin{aligned}\langle I(x, 0)\rangle &= \frac{1}{\cos^2(\alpha x)} \int d\mathbf{q}\gamma_0\left(\mathbf{q}, -\frac{\mathbf{q}}{\alpha k}\tan(\alpha x)\right) \\ &\times \exp\left\{-\frac{k^2}{4}\int\limits_0^x d\xi D\left(\frac{1}{\alpha k}\frac{\sin(\alpha\xi)}{\cos(\alpha x)}\mathbf{q}\right)\right\}. \qquad (13.36)\end{aligned}$$

Remark 14 ***Wave localization in a stochastic parabolic waveguide.***

Consider the wave beam

$$u_0(\mathbf{R}) = u_0 \exp\left\{-\tfrac{\mathbf{R}^2}{2a^2}\right\}, \qquad (13.37)$$

where parameter a is the beam width, propagating in a random parabolic waveguide [134, 135, 142].

We will assume that random field $\varepsilon(x, \mathbf{R})$ is described by the formula

$$\varepsilon(x, \mathbf{R}) = -\left[\alpha^2 - z(x)\right]\mathbf{R}^2, \qquad (13.38)$$

where α is the deterministic parameter and $z(x)$ is the random function.

In the absence of medium parameter fluctuations, the wavefield satisfies the equation

$$\frac{\partial}{\partial x}u_0(x, \mathbf{R}) = \frac{i}{2k}\left[\Delta_{\mathbf{R}} - \alpha^2k^2\mathbf{R}^2\right]u_0(x, \mathbf{R}). \qquad (13.39)$$

The solution to Eq. (13.39) is representable in the form

$$u_0(x, \mathbf{R}) = f(x, \mathbf{R})\tilde{u}(x, \mathbf{R}), \qquad (13.40)$$

where

$$\begin{aligned}f(x, \mathbf{R}) &= \frac{1}{\cos(\alpha x)}\exp\left\{-i\frac{\alpha k^2}{2}\mathbf{R}^2\tan(\alpha x)\right\}, \\ \tilde{u}(x, \mathbf{R}) &= \int d\mathbf{q}u_0(\mathbf{q})\exp\left\{-i\frac{\mathbf{q}^2}{2k\alpha}\tan(\alpha x) + i\frac{\mathbf{qR}}{\cos(\alpha x)}\right\}, \\ u_0(\mathbf{R}) &= \int d\mathbf{q}\, u_0(\mathbf{q})e^{i\mathbf{qR}}, \quad u_0(\mathbf{q}) = \frac{1}{(2\pi)^2}\int d\mathbf{R}u_0(\mathbf{R})e^{-i\mathbf{qR}}. \qquad (13.41)\end{aligned}$$

We note that function $f(x, \mathbf{R})$ describes the wavefield of a plane wave and is a periodic function with period $L = 2\pi/\alpha$. In addition, function $f(x, \mathbf{R})$ becomes infinite at points

$x_n = (2n+1)\frac{\pi}{2\alpha}$, which corresponds to plane wave focusing. At the same time, wavefield $u_0(x, \mathbf{R})$ assumes generally no infinite values.

In the case of wave beam (13.37), the wavefield has the following structure

$$u_0(x, \mathbf{R}) = \frac{u_0}{\cos(\alpha x)\left(1 + \frac{i}{a^2 k\alpha}\tan(\alpha x)\right)}$$
$$\times \exp\left\{i\alpha k \tan(\alpha x) + \frac{1}{a^2\cos^2(\alpha x)\left(1 + \frac{i}{a^2 k\alpha}\tan(\alpha x)\right)}\right\}, \tag{13.42}$$

so that the wavefield intensity assumes the form

$$I_0(x, \mathbf{R}) = |u_0(x, \mathbf{R})|^2 = \frac{|u_0|^2}{g_\alpha^2(x)} \exp\left\{-\frac{\mathbf{R}^2}{g_\alpha^2(x)}\right\}, \tag{13.43}$$

where

$$g_\alpha^2(x) = \cos^2(\alpha x) + \frac{1}{k^2\alpha^2 a^4}\sin^2(\alpha x).$$

If wave beam (13.37) matches the inhomogeneous waveguide, i.e., if

$$k\alpha a^2 = 1, \tag{13.44}$$

then wavefield $u_0(x, \mathbf{R})$ assumes the form

$$u_0(x, \mathbf{R}) = u_0 \exp\left\{-\frac{\mathbf{R}^2}{2a^2} - i\alpha x\right\}.$$

In this case, the amplitude of propagating wavefield remains intact, which means that this field is the eigenmode of the problem under consideration.

In the presence of dielectric permittivity fluctuations described by function $\varepsilon(x, \mathbf{R})$ of the form (13.38), the solution to Eq. (13.1) can be represented in the form

$$u(x, \mathbf{R}) = u_0 \exp\left\{-\frac{\mathbf{R}^2}{2a^2} A(x) + B(x)\right\},$$

where complex functions $A(x)$ and $B(x)$ satisfy the system of equations

$$\begin{aligned} \frac{d}{dx} A(x) &= -\frac{i}{ka^2}\left[A^2(x) - \alpha^2 k^2 a^4\right] - ika^2 z(x), \quad \grave{A}(0) = 1, \\ \frac{d}{dx} B(x) &= -\frac{i}{ka^2} A(x), \quad B(0) = 0. \end{aligned} \tag{13.45}$$

We easily obtain that

$$B(x) = -\frac{i}{ka^2}\int_0^x d\xi A(\xi),$$

and the wavefield intensity is given by the expression

$$I(x, \mathbf{R}) = \exp\left\{-\frac{\mathbf{R}^2}{2a^2}\left[A(x) + A^*(x)\right] - \frac{i}{ka^2}\int_0^x d\xi \left[A(\xi) - A^*(\xi)\right]\right\}. \tag{13.46}$$

We can exclude the imaginary part of function $A(x)$ in (13.46) by using the first equation of Eqs. (13.45); namely, we have

$$-\frac{i}{ka^2}\left[A(x) - A^*(x)\right] = \frac{d}{dx}\ln\left[A(x) + A^*(x)\right].$$

As a result, we obtain the following expression for the wavefield intensity (we assume $|u_0|^2 = 1$ for simplicity)

$$I(x, \mathbf{R}) = I(x, 0) \exp\left\{-\tfrac{\mathbf{R}^2}{a^2} I(x, 0)\right\}, \tag{13.47}$$

where

$$I(x, 0) = \frac{1}{2}\left[A(x) + A^*(x)\right] \tag{13.48}$$

is the wavefield intensity at the axis of the inhomogeneous waveguide. A consequence of these expressions is the fact that statistical characteristics of wavefield intensity are described by statistical characteristics of the solution to the sole equation (13.45) in function $A(x)$. Moreover, this equation is similar to the equation for the reflection coefficient of a wave in the one-dimensional medium, which we discussed in Sect. 12.2.

We represent function $A(x)$ in the form

$$A(x) = k\alpha a^2 \frac{1 + \psi(x)e^{-2i\alpha x}}{1 - \psi(x)e^{-2i\alpha x}}.$$

Then, function $\psi(x)$ satisfies the equation following from Eq. (13.45)

$$\frac{d}{dx}\psi(x) = -\frac{i}{2\alpha k}\left(e^{i\alpha x} - \psi(x)e^{-i\alpha x}\right)^2 z(x), \quad \psi(0) = \frac{1 - k\alpha a^2}{1 + k\alpha a^2}.$$

Now, we introduce amplitude–phase representation of function $\psi(x)$ by the formula

$$\psi(x) = \sqrt{\frac{w(x) - 1}{w(x) + 1}} e^{i\{\phi(x) - 2\alpha x\}}, \, w \geq 1.$$

Then, functions $w(x)$ and $\phi(x)$ will satisfy the system of equations

$$\begin{aligned}
\frac{d}{dx}w(x) &= -\frac{1}{\alpha k} z(x)\sqrt{w^2(x) - 1}\sin\left(\phi(x) - 2\alpha x\right), \\
w(0) &= \frac{1}{2\alpha k a^2}\left(1 + k^2\alpha^2 a^4\right); \\
\frac{d}{dx}\phi(x) &= \frac{1}{\alpha k} z(x)\left(1 - \frac{w}{\sqrt{w^2(x) - 1}}\cos\left(\phi(x) - 2\alpha x\right)\right), \\
\phi(0) &= 0.
\end{aligned} \tag{13.49}$$

Consequently, the wavefield intensity at the waveguide axis (13.48) assumes the form

$$I(x, 0) = \frac{\alpha k a^2}{w(x) + \sqrt{w^2(x) - 1}\cos\left(\phi(x) - 2\alpha x\right)}. \tag{13.50}$$

As earlier, we assume that quantity $z(x)$ is the Gaussian delta-correlated function with the parameters

$$\langle z(x) \rangle = 0; \quad \langle z(x)z(x') \rangle = 2\sigma^2 l_0\, \delta(x - x').$$

Additionally we assume sufficiently small the variance of fluctuations of function $z(x)$ ($\sigma^2 \ll 1$). In this case, statistical characteristics of functions $w(x)$ and $\phi(x)$ only slowly vary over scales about $1/\alpha$, and we can evaluate statistical characteristics of wave intensity (13.50) using additional averaging over fast oscillations, which yields statistical independence of functions $w(x)$ and $\phi(x)$ and uniform probability distribution of phase $\phi(x)$. As a result, we obtain that probability distribution of function $w(x)$

$$P(x,w) = \langle \delta\left(w(x) - w\right)\rangle$$

satisfies the Fokker–Planck equation

$$\frac{\partial}{\partial x}P(x,w) = D\frac{\partial}{\partial w}\left(w^2-1\right)\frac{\partial}{\partial w}P(x,w), \quad P(0,w) = \delta\left(w(0) - w\right) \tag{13.51}$$

with the diffusion coefficient $D = \sigma^2 l_0/2\alpha^2 k^2$.

Calculate the moments $\langle I^n(x,0)\rangle$ of the intensity at the waveguide axis in the framework of the above assumptions. We perform averaging in two steps. At the first step, we average over fast phase oscillations to obtain the expression

$$\left\langle\left(\frac{I}{\alpha k a^2}\right)^n\right\rangle_\phi = P_{n-1}(w). \tag{13.52}$$

Here, $P_n(w)$ is the Legendre polynomial of order n. At the second step, we average Eq. (13.52) using probability distribution of function w (13.51). To do this, we multiply Eq. (13.51) by $P_{n-1}(w)$ and integrate the result over all $w \geq 1$. Integrating by parts and using the equality

$$\frac{d}{dw}\left(w^2-1\right)\frac{d}{dw}P_{n-1}(w) = n(n-1)P_{n-1}(w),$$

we obtain the equation

$$\frac{\partial}{\partial x}\langle P_{n-1}(w)\rangle = Dn(n-1)\langle P_{n-1}(w)\rangle$$

whose solution has the form

$$\langle P_n(w)\rangle = P_{n-1}(w_0)e^{Dn(n-1)x},$$

so that

$$\left\langle\left(\frac{I}{\alpha k a^2}\right)^n\right\rangle = P_{n-1}(w_0)e^{Dn(n-1)x}. \tag{13.53}$$

For the wave beam (13.44) matched with the waveguide, $w_0 = 1$ and Eq. (13.53) grades into

$$\langle I^n(x,0)\rangle = e^{Dn(n-1)x}, \tag{13.54}$$

which means that quantity $I(x,0)$ is distributed according to the lognormal law. The mean intensity at the waveguide axis remains intact, and all higher moments are exponentially increasing functions of the distance the wave travels in the medium. Nevertheless, as we have seen earlier, the typical realization curve of process $I(x,0)$ exponentially decreases with this distance

$$I^*(x,0) = e^{-Dx},$$

which means that radiation must leave waveguide axis in the transverse direction in actual realizations. This is the manifestation of the dynamic localization in the x-direction. According to Eq. (13.47), the typical realization curve of the intensity in the transverse direction has the form

$$I^*(x,\mathbf{R}) = I^*(x,0)\exp\left\{-\tfrac{\mathbf{R}^2}{a^2}I^*(x,0)\right\}.$$

Thus, in the stochastic parabolic waveguide with

$$\varepsilon(x, \mathbf{R}) = -\alpha \mathbf{R}^2 + z(x)\mathbf{R}^2,$$

average intensity remains intact and its higher moments show the exponential increase with distance x, as distinct from the solution (13.36) that was obtained for homogeneous and isotropic fluctuations of field $\varepsilon(x, \mathbf{R})$ and is a decreasing function of distance x. The comparison of these results shows that the parabolic fluctuations of field $\varepsilon(x, \mathbf{R})$ have a greater impact on propagating wave beam than the homogeneous isotropic fluctuations. ♦

13.1.3 Applicability of the delta-correlated approximation for medium fluctuations and the diffusion approximation for wavefield

Perturbation method

Here, we dwell on the conditions of applicability of the delta-correlated approximation for fluctuations if field $\varepsilon(x, \mathbf{R})$. We construct a perturbation theory that improves the representation of wave statistical characteristics in terms of a functional of field $\varepsilon(x, \mathbf{R})$. The above delta-correlated approximation is the first step of this theory; the higher approximations allow for finite longitudinal correlation radius of field $\varepsilon(x, \mathbf{R})$ and yield a system of closed integro-differential equations for wavefield moments.

This perturbation theory is constructed as follows. We draw first the infinite system of connected equations for arbitrary moment function. Deriving this system, we assume that field $\varepsilon(x, \mathbf{R})$ is the Gaussian random field and use the Furutsu–Novikov formula, but we make no assumptions about delta-correlated property of field $\varepsilon(x, \mathbf{R})$. Every of thus obtained equations explicitly depends on correlation function $B_\varepsilon(x, \mathbf{R})$. If we substitute the delta-like approximation of correlation function (13.15) in the first of these equations, then we arrive at the above approximation of the delta-correlated fluctuations of field $\varepsilon(x, \mathbf{R})$ and all remaining equations appear superfluous. However, if we hold the exact function $B_\varepsilon(x, \mathbf{R})$ in the first $(n-1)$ equations and use approximation (13.15) only in the n-th equation, then we obtain the closed system of n equations for the moment function at hand. We illustrate this theory by the example of the equation for average field.

Averaging Eq. (13.1) over an ensemble of field realizations and calculating the correlator by the Furutsu–Novikov formula, we obtain that average field satisfies the equation

$$\left(\frac{\partial}{\partial x} - \frac{i}{2k}\Delta_{\mathbf{R}}\right) \langle u(x, \mathbf{R})\rangle = i\frac{k}{2}\int_0^x dx' \int d\mathbf{R}' B_\varepsilon(x - x', \mathbf{R} - \mathbf{R}') \left\langle \frac{\delta u\,(x; \mathbf{R})}{\delta\varepsilon(x', \mathbf{R}')}\right\rangle. \tag{13.55}$$

Equation (13.55) is unclosed because of new unknown function $\langle \delta u\,(x; \mathbf{R})\,\delta\varepsilon(x', \mathbf{R}')\rangle$. To derive the equation for this function, we vary Eq. (13.1) with respect to field $\varepsilon(x', \mathbf{R}')$ for $x' < x$ and average the result. We obtain the equation with initial condition

$$\begin{aligned} \left(\frac{\partial}{\partial x} - \frac{i}{2k}\Delta_{\mathbf{R}}\right) \left\langle \frac{\delta u(x, \mathbf{R})}{\delta\varepsilon(x', \mathbf{R}')}\right\rangle &= i\frac{k}{2}\left\langle \varepsilon(x, \mathbf{R})\frac{\delta u(x, \mathbf{R})}{\delta\varepsilon(x', \mathbf{R}')}\right\rangle, \\ \left\langle \frac{\delta u(x, \mathbf{R})}{\delta\varepsilon(x', \mathbf{R}')}\right\rangle_{x=x'+0} &= i\frac{k}{2}\delta(\mathbf{R} - \mathbf{R}')\,\langle u(x', \mathbf{R})\rangle. \end{aligned} \tag{13.56}$$

We again use the Furutsu–Novikov formula to calculate correlator $\left\langle \varepsilon(x,\mathbf{R})\frac{\delta u(x,\mathbf{R})}{\delta\varepsilon(x',\mathbf{R}')}\right\rangle$. In this way, we obtain the equation

$$\left(\frac{\partial}{\partial x}-\frac{i}{2k}\Delta_{\mathbf{R}}\right)\left\langle\frac{\delta u(x,\mathbf{R})}{\delta\varepsilon(x',\mathbf{R}')}\right\rangle$$
$$=i\frac{k}{2}\int_0^x dx''\int d\mathbf{R}''B_\varepsilon(x-x'',\mathbf{R}-\mathbf{R}'')\left\langle\frac{\delta^2 u\,(x;\mathbf{R})}{\delta\varepsilon(x',\mathbf{R}')\delta\varepsilon(x'',\mathbf{R}'')}\right\rangle,$$
$$\left\langle\frac{\delta u(x,\mathbf{R})}{\delta\varepsilon(x',\mathbf{R}')}\right\rangle_{x=x'+0}=i\frac{k}{2}\delta\left(\mathbf{R}-\mathbf{R}'\right)\left\langle u(x',\mathbf{R})\right\rangle. \tag{13.57}$$

Equation (13.57) is again unclosed because it includes the second variational derivative of field $u(x,\mathbf{R})$. We can draw the equations for the second variational derivative and so forth. Thus, Eqs. (13.55), (13.57), and others form an infinite system of connected equations. The initial conditions of every new equation depend on functions appearing in the equation of the previous step. As was mentioned, the closed equation is obtained by replacing the correlation function of field $\varepsilon(x,\mathbf{R})$ in Eqs. (13.55) with the delta-like effective correlation function, because the variational derivative at $x=x'$ is expressed in this case through average field $\langle u(x,\mathbf{R})\rangle$, which just corresponds to the approximation of the delta-correlated fluctuations of field $\varepsilon(x,\mathbf{R})$.

We can replace the correlation function with the effective one not in the first equation (13.55), but in one of subsequent equations. For example, if we perform such a replacement in Eq. (13.57), we obtain the equation

$$\left(\frac{\partial}{\partial x}-\frac{i}{2k}\Delta_{\mathbf{R}}\right)\left\langle\frac{\delta u(x,\mathbf{R})}{\delta\varepsilon(x',\mathbf{R}')}\right\rangle = -i\frac{k^2}{4}A(0)\left\langle\frac{\delta u(x,\mathbf{R})}{\delta\varepsilon(x',\mathbf{R}')}\right\rangle,$$
$$\left\langle\frac{\delta u(x,\mathbf{R})}{\delta\varepsilon(x',\mathbf{R}')}\right\rangle_{x=x'+0} = i\frac{k}{2}\delta\left(\mathbf{R}-\mathbf{R}'\right)\left\langle u(x',\mathbf{R})\right\rangle. \tag{13.58}$$

Equations (13.55) and (13.57) form the closed system of equations of the second approximation.

We can similarly derive the closed systems of equations for higher approximations, and the systems of equations for other moment functions of field $u(x,\mathbf{R})$ as well.

The solution to Eq. (13.58) has the form

$$\left\langle\frac{\delta u(x,\mathbf{R})}{\delta\varepsilon(x',\mathbf{R}')}\right\rangle = i\frac{k}{2}e^{-\frac{k^2}{8}A(0)(x-x')}g\left(x-x',\mathbf{R}-\mathbf{R}'\right)\left\langle u(x',\mathbf{R})\right\rangle, \tag{13.59}$$

where $g\left(x,\mathbf{R}\right)$ is Green's function of free space (13.7). Substituting Eq. (13.59) in Eq. (13.55), we obtain the integro-differential equation

$$\left(\frac{\partial}{\partial x}-\frac{i}{2k}\Delta_{\mathbf{R}}\right)\left\langle u(x,\mathbf{R})\right\rangle$$
$$= -\frac{k^2}{4}\int_0^x dx'\int d\mathbf{R}'e^{-\frac{k^2}{8}A(0)(x-x')}B_\varepsilon(x-x',\mathbf{R}-\mathbf{R}')g\left(x-x',\mathbf{R}-\mathbf{R}'\right)\left\langle u(x',\mathbf{R}')\right\rangle. \tag{13.60}$$

Equation (13.60) can be solved using the Laplace transform with respect to x and Fourier transform with respect to $\mathbf{R}$. However, we will not solve this equation here; instead, we elucidate the conditions under which the solution to Eq. (13.60) grades into the solution to the equation corresponding to the approximation of the delta-correlated random field $\varepsilon(x, \mathbf{R})$.

Green's function $g(x - x', \mathbf{R} - \mathbf{R}')$ in (13.60) appears as the delta-function of variable $\mathbf{R} - \mathbf{R}'$ smeared over scale $a = \sqrt{(x - x')/k}$. In turn, difference $x - x'$ is limited by the longitudinal scale of inhomogeneities $l_{\parallel}$ in view of the factor $B_{\varepsilon}(x - x', \mathbf{R} - \mathbf{R}')$. As a result, we obtain that $a \sim \left(l_{\parallel}/k\right)^{1/2}$. If scale a is small in comparison with the scale $l_{\perp}$of function $B_{\varepsilon}(x - x', \mathbf{R} - \mathbf{R}')$ with respect to variable $\mathbf{R} - \mathbf{R}'$, i.e., if $l_{\parallel} \ll kl_{\perp}^2$, then we can replace Green's function with the delta function. Thus, if $l_{\parallel} \ll kl_{\perp}^2$, we can rewrite Eq. (13.60) in the form

$$\left(\frac{\partial}{\partial x} - \frac{i}{2k}\Delta_{\mathbf{R}}\right)\langle u(x, \mathbf{R})\rangle = -\frac{k^2}{4}\int\limits_0^x dx' e^{-\frac{k^2}{8}A(0)x'} B_{\varepsilon}(x', 0)\langle u(x - x', \mathbf{R})\rangle. \tag{13.61}$$

If the condition

$$\frac{k^2}{8}A(0)l_{\parallel} \ll 1$$

holds additionally, i.e., if attenuation of average field is small over scales about $l_{\parallel}$, then we can replace the exponential factor with unity and neglect the shift of argument of function $\langle u(x - x', \mathbf{R})\rangle$ by setting $\langle u(x - x', \mathbf{R})\rangle \approx \langle u(x, \mathbf{R})\rangle$. As a result, the equation assumes the form

$$\left(\frac{\partial}{\partial x} - \frac{i}{2k}\Delta_{\mathbf{R}}\right)\langle u(x, \mathbf{R})\rangle = -\frac{k^2}{4}\int\limits_0^x dx' B_{\varepsilon}(x', 0)\langle u(x, \mathbf{R})\rangle.$$

Finally, if $x \gg l_{\parallel}$, the upper limit of the integral can be replaced with infinity, and we arrive at Eq. (13.20).

Thus, in the context of average field $\langle u(x, \mathbf{R})\rangle$, the delta-correlated approximation of field $\varepsilon(x, \mathbf{R})$ holds under the following three conditions

$$l_{\parallel} \ll kl_{\perp}^2, \quad \sigma_{\varepsilon}^2 k^2 l_{\parallel}^2 \ll 1, \quad x \gg l_{\parallel} \quad (A(0) \sim \sigma_{\varepsilon}^2 l_{\parallel}). \tag{13.62}$$

In a similar way, one can obtain and analyze equations of the second approximation for the coherence function $\Gamma_2(x, \mathbf{R}, \boldsymbol{\rho})$. In the context of coherence function $\Gamma_2(x, \mathbf{R}, \boldsymbol{\rho})$, applicability range of the approximation of delta-correlated fluctuations of field $\varepsilon(x, \mathbf{R})$ is described (in the case of the plane incident wave) by the inequalities

$$\rho \ll x, \quad kx|\boldsymbol{\nabla} A(\boldsymbol{\rho})| \ll 1. \tag{13.63}$$

It should be emphasized that conditions (13.62) and (13.63) are virtually independent, because they restrict different parameters. In particular, conditions (13.63) may hold even if condition $\sigma_{\varepsilon}^2 k^2 l_{\parallel}^2 \ll 1$ fails. Note additionally that conditions (13.63) restrict only local characteristics of fluctuations of field $\varepsilon(x, \mathbf{R})$; for this reason, they can be formulated even for turbulent medium. On the contrary, extinction coefficient $\gamma = k^2 A(0)/4$ (see Eq. (13.25), page 360) is governed by the most large-scale fluctuations of field $\varepsilon(x, \mathbf{R})$.

Diffusion approximation for the wavefield

Consider now the diffusion approximation as applied to describing statistical properties of the solution to parabolic equation (13.1). Note that this approximation is very close by implication to the *Chernov local method* [42]; it is more physical than the formal approximation of delta-correlated random field $\varepsilon(x,\mathbf{R})$, allows for finite longitudinal correlation radius of field $\varepsilon(x,\mathbf{R})$, and adequately describes wave propagation in media with inhomogeneities elongated in the propagation direction [270, 305].

As earlier, we will assume that $\varepsilon(x,\mathbf{R})$ is the homogeneous Gaussian random field with correlation function $B_\varepsilon(x,\mathbf{R})$.

Consider first the equation for average field. Averaging Eq. (13.1) over an ensemble of realizations of field $\varepsilon(x,\mathbf{R})$ and using the Furutsu–Novikov formula, we obtain the exact equation (13.55).

The diffusion approximation assumes that the variational derivative appeared in the exact equation satisfies the deterministic equation

$$\left(\frac{\partial}{\partial x}-\frac{i}{2k}\Delta_{\mathbf{R}}\right)\frac{\delta u(x,\mathbf{R})}{\delta\varepsilon(x',\mathbf{R}')}=0 \tag{13.64}$$

with the stochastic initial condition

$$\left.\frac{\delta u(x,\mathbf{R})}{\delta\varepsilon(x',\mathbf{R}')}\right|_{x=x'+0}=i\frac{k}{2}\delta(\mathbf{R}-\mathbf{R}')u(x',\mathbf{R}), \tag{13.65}$$

so that

$$\frac{\delta u(x,\mathbf{R})}{\delta\varepsilon(x',\mathbf{R}')}=i\frac{k}{2}e^{\frac{i(x-x')}{2k}\Delta_{\mathbf{R}}}\left[\delta(\mathbf{R}-\mathbf{R}')u(x',\mathbf{R})\right], \tag{13.66}$$

We remind that, being applied to the delta function, the operator in the right-hand side of Eq. (13.66) produces Green's function of Eq. (13.1) with $\varepsilon(x,\mathbf{R})=0$ (the point source field in free space).

Within the framework of the diffusion approximation, wavefield $u(x',\mathbf{R})$ is also related to field $u(x,\mathbf{R})$ by the relationship

$$u(x',\mathbf{R})=e^{-\frac{i(x-x')}{2k}\Delta_{\mathbf{R}}}u(x,\mathbf{R}),$$

which is a consequence of the solution to problem (13.1) for absent fluctuations. Consequently, we have

$$\left\langle\frac{\delta u(x,\mathbf{R})}{\delta\varepsilon(x',\mathbf{R}')}\right\rangle=i\frac{k}{2}e^{\frac{i(x-x')}{2k}\Delta_{\mathbf{R}}}\left[\delta(\mathbf{R}-\mathbf{R}')e^{-\frac{i(x-x')}{2k}\Delta_{\mathbf{R}}}\langle u(x,\mathbf{R})\rangle\right],$$

Substituting this expression in the right-hand side of Eq. (13.55), we obtain the closed operator equation

$$\left(\frac{\partial}{\partial x}-\frac{i}{2k}\Delta_{\mathbf{R}}\right)\langle u(x,\mathbf{R})\rangle$$
$$=-\frac{k^2}{4}\int\limits_0^x dx'\int d\mathbf{R}'B_\varepsilon(x',\mathbf{R}-\mathbf{R}')e^{\frac{ix'}{2k}\Delta_{\mathbf{R}}}\left[\delta(\mathbf{R}-\mathbf{R}')e^{-\frac{ix'}{2k}\Delta_{\mathbf{R}}}\langle u(x,\mathbf{R})\rangle\right], \tag{13.67}$$

with the initial condition $\langle u(0,\mathbf{R})\rangle=u_0(\mathbf{R})$.

Now, we introduce the two-dimensional spectral density of inhomogeneities with respect to transverse coordinates

$$B_\varepsilon(x,\mathbf{R})=\int d\mathbf{q}\Phi_\varepsilon^{(2)}(x,\mathbf{q})e^{i\mathbf{q}\mathbf{R}},\quad \Phi_\varepsilon^{(2)}(x,\mathbf{q})=\frac{1}{(2\pi)^2}\int d\mathbf{R}B_\varepsilon(x,\mathbf{R})e^{-i\mathbf{q}\mathbf{R}}$$

and the Fourier transform of wavefield $u(x, \mathbf{R})$ with respect to transverse coordinates

$$u(x, \mathbf{R}) = \int d\mathbf{q}\tilde{u}(x, \mathbf{q})e^{i\mathbf{q}\mathbf{R}}, \quad \tilde{u}(x, \mathbf{q}) = \frac{1}{(2\pi)^2}\int d\mathbf{R}u(x, \mathbf{R})e^{-i\mathbf{q}\mathbf{R}}.$$

Then, from Eq. (13.67) follows that function $\tilde{u}(x, \mathbf{q})$ satisfies the equation

$$\left(\frac{\partial}{\partial x} + \frac{i\mathbf{q}^2}{2k}\right)\langle\tilde{u}(x, \mathbf{q})\rangle = -\frac{k^2}{2}D(x, \mathbf{q})\langle\tilde{u}(x, \mathbf{q})\rangle, \quad \langle\tilde{u}(0, \mathbf{q})\rangle = \tilde{u}_0(\mathbf{q}),$$

where

$$\tilde{u}_0(\mathbf{q}) = \frac{1}{(2\pi)^2}\int d\mathbf{R}u_0(\mathbf{R})e^{-i\mathbf{q}\mathbf{R}}$$

and

$$D(x, \mathbf{q}) = \int_0^x d\xi \int d\mathbf{q}'\Phi_\varepsilon^{(2)}(\xi, \mathbf{q}')\exp\left\{-\frac{i\xi}{2k}\left(\mathbf{q}'^2 - 2\mathbf{q}'\mathbf{q}\right)\right\}.$$

Consequently, we have

$$\langle u(x, \mathbf{R})\rangle = \frac{1}{(2\pi)^2}\int d\mathbf{q}\int d\mathbf{R}'u_0(\mathbf{R}')\exp\left\{i\mathbf{q}(\mathbf{R} - \mathbf{R}') - i\frac{\mathbf{q}^2 x}{2} - \frac{k^2}{2}\int_0^x dx' D(x', \mathbf{q})\right\}. \tag{13.68}$$

For distances $x \gg l_\parallel$, where $l_\parallel$ is the longitudinal correlation radius of field $\varepsilon(x, \mathbf{R})$, Eq. (13.68) can be reduced to the form

$$\langle u(x, \mathbf{R})\rangle = \frac{1}{(2\pi)^2}\int d\mathbf{q}\int d\mathbf{R}'u_0(\mathbf{R}')\exp\left\{i\mathbf{q}(\mathbf{R} - \mathbf{R}') - i\frac{\mathbf{q}^2 x}{2} - \frac{k^2}{2}xD(\mathbf{q})\right\}, \tag{13.69}$$

where

$$D(\mathbf{q}) = \int_0^\infty d\xi \int d\mathbf{q}'\Phi_\varepsilon^{(2)}(\xi, \mathbf{q}')\exp\left\{-\frac{i\xi}{2k}\left(\mathbf{q}'^2 - 2\mathbf{q}'\mathbf{q}\right)\right\}. \tag{13.70}$$

If we introduce now the three-dimensional spectral function $\Phi_\varepsilon(q_1, \mathbf{q})$ (13.26) of field $\varepsilon(x, \mathbf{R})$, then the expression for coefficient $D(\mathbf{q})$ reduces to the form

$$D(\mathbf{q}) = \pi\int d\mathbf{q}'\Phi_\varepsilon\left(\frac{1}{2k}\left(\mathbf{q}'^2 - 2\mathbf{q}'\mathbf{q}\right), \mathbf{q}'\right). \tag{13.71}$$

Recall that the delta-correlated approximation assumes that coefficient $D(\mathbf{q})$ has the form

$$D(\mathbf{q}) = \pi\int d\mathbf{q}'\Phi_\varepsilon\left(0, \mathbf{q}'\right).$$

In the case of the plane incident wave, we have $u_0(\mathbf{R}) = 1$, and Eq. (13.69) yields the expression independent of $\mathbf{R}$

$$\langle u(x, \mathbf{R})\rangle = e^{-\frac{1}{2}k^2 x D(0)}, \quad D(0) = \pi\int d\mathbf{q}'\Phi_\varepsilon\left(\frac{\mathbf{q}'^2}{2k}, \mathbf{q}'\right). \tag{13.72}$$

It is obvious that this expression will be valid under the condition

$$\frac{k^2}{2}D(0)l_\parallel \ll 1.$$

Equations for higher moment functions of field $u(x, \mathbf{R})$ can be derived similarly. Consider the dynamic equation

$$\begin{aligned}\left(\frac{\partial}{\partial x}-\frac{i}{k}\nabla_{\mathbf{R}}\nabla_{\boldsymbol{\rho}}\right)\gamma_2(x,\mathbf{R},\boldsymbol{\rho}) &= i\frac{k}{2}\left[\varepsilon\left(x,\mathbf{R}+\frac{1}{2}\boldsymbol{\rho}\right)-\varepsilon\left(x,\mathbf{R}-\frac{1}{2}\boldsymbol{\rho}\right)\right]\gamma_2(x,\mathbf{R},\boldsymbol{\rho}),\\ \gamma_2(0,\mathbf{R},\boldsymbol{\rho}) &= u_0\left(\mathbf{R}+\frac{1}{2}\boldsymbol{\rho}\right)u_0^*\left(\mathbf{R}'-\frac{1}{2}\boldsymbol{\rho}\right)\end{aligned} \tag{13.73}$$

that follows from the initial parabolic equation (13.1) for the function

$$\gamma_2(x;\mathbf{R},\boldsymbol{\rho}) = u\left(x,\mathbf{R}+\frac{1}{2}\boldsymbol{\rho}\right)u^*\left(x,\mathbf{R}-\frac{1}{2}\boldsymbol{\rho}\right).$$

Averaging Eq. (13.73) over an ensemble of realizations of field $\varepsilon(x, \mathbf{R})$ and splitting the correlator by the Furutsu–Novikov formula, we obtain the equation

$$\begin{aligned}&\left(\frac{\partial}{\partial x}-\frac{i}{k}\nabla_{\mathbf{R}}\nabla_{\boldsymbol{\rho}}\right)\Gamma_2(x,\mathbf{R},\boldsymbol{\rho})\\ &= i\frac{k}{2}\int\limits_0^x dx_1\int d\mathbf{R}_1\left[B_\varepsilon\left(x-x_1,\mathbf{R}-\mathbf{R}_1+\frac{1}{2}\boldsymbol{\rho}\right)-B_\varepsilon\left(x-x_1,\mathbf{R}-\mathbf{R}_1-\frac{1}{2}\boldsymbol{\rho}\right)\right]\\ &\quad\times\left\langle\frac{\delta}{\delta\varepsilon(x_1,\mathbf{R}_1)}\gamma_2(x,\mathbf{R},\boldsymbol{\rho})\right\rangle.\end{aligned} \tag{13.74}$$

The diffusion approximation assumes that the variational derivative in the right-hand side of Eq. (13.74) can be represented the form

$$\begin{aligned}&\left\langle\frac{\delta}{\delta\varepsilon(x_1,\mathbf{R}_1)}\gamma_2(x,\mathbf{R},\boldsymbol{\rho})\right\rangle\\ &= i\frac{k}{2}e^{\frac{i}{k}(x-x_1)\nabla_{\mathbf{R}}\nabla_{\boldsymbol{\rho}}}\left\{\left[\delta\left(\mathbf{R}-\mathbf{R}_1+\frac{1}{2}\boldsymbol{\rho}\right)-\delta\left(\mathbf{R}-\mathbf{R}_1-\frac{1}{2}\boldsymbol{\rho}\right)\right]\right.\\ &\quad\left.\times\, e^{-\frac{i}{k}(x-x_1)\nabla_{\mathbf{R}}\nabla_{\boldsymbol{\rho}}}\Gamma_2(x,\mathbf{R},\boldsymbol{\rho})\right\}.\end{aligned} \tag{13.75}$$

Substituting this expression in Eq. (13.74), we arrive at the closed operator equation

$$\begin{aligned}&\left(\frac{\partial}{\partial x}-\frac{i}{k}\nabla_{\mathbf{R}}\nabla_{\boldsymbol{\rho}}\right)\Gamma_2(x,\mathbf{R},\boldsymbol{\rho})\\ &= -\frac{k^2}{4}\int\limits_0^x dx_1\int \mathbf{R}_1\left[B_\varepsilon\left(x_1,\mathbf{R}-\mathbf{R}_1+\frac{1}{2}\boldsymbol{\rho}\right)-B_\varepsilon\left(x_1,\mathbf{R}-\mathbf{R}_1-\frac{1}{2}\boldsymbol{\rho}\right)\right]\\ &\times e^{\frac{i}{k}x_1\nabla_{\mathbf{R}}\nabla_{\boldsymbol{\rho}}}\left[\delta\left(\mathbf{R}-\mathbf{R}_1+\frac{1}{2}\boldsymbol{\rho}\right)-\delta\left(\mathbf{R}-\mathbf{R}_1-\frac{1}{2}\boldsymbol{\rho}\right)\right]e^{-\frac{i}{k}x_1\nabla_{\mathbf{R}}\nabla_{\boldsymbol{\rho}}}\Gamma_2(x-x_1,\mathbf{R},\boldsymbol{\rho}).\end{aligned} \tag{13.76}$$

The further derivation will be similar to the derivation of the equation for average field. We express the correlation function of field $\varepsilon(x, \mathbf{R})$ in terms of its spectral density with respect to transverse coordinates to obtain

$$\left(\frac{\partial}{\partial x} - \frac{i}{k}\nabla_{\mathbf{R}}\nabla_{\boldsymbol{\rho}}\right)\Gamma_2(x,\mathbf{R},\boldsymbol{\rho})$$

$$= -\frac{k^2}{4}\int\limits_0^x dx_1 \int \mathbf{R}_1 \int d\mathbf{q}\Phi_{\varepsilon}^{(2)}(x_1,\mathbf{q})\, e^{i\mathbf{q}(\mathbf{R}-\mathbf{R}_1)}\left[e^{\frac{i}{2}\mathbf{q}\boldsymbol{\rho}} - e^{-\frac{i}{2}\mathbf{q}\boldsymbol{\rho}}\right]$$

$$\times e^{\frac{i}{k}x_1\nabla_{\mathbf{R}}\nabla_{\boldsymbol{\rho}}}\left[\delta\left(\mathbf{R}-\mathbf{R}_1+\frac{1}{2}\boldsymbol{\rho}\right) - \delta\left(\mathbf{R}-\mathbf{R}_1-\frac{1}{2}\boldsymbol{\rho}\right)\right] e^{-\frac{i}{k}x_1\nabla_{\mathbf{R}}\nabla_{\boldsymbol{\rho}}}\Gamma_2(x-x_1,\mathbf{R},\boldsymbol{\rho}). \tag{13.77}$$

Then, we introduce the Fourier transform of the coherence function with respect to variable $\mathbf{R}$

$$\Gamma_2(x,\mathbf{R},\boldsymbol{\rho}) = \int d\mathbf{q}\tilde{\Gamma}_2(x,\mathbf{q},\boldsymbol{\rho})e^{i\mathbf{q}\mathbf{R}}, \quad \tilde{\Gamma}_2(x,\mathbf{q},\boldsymbol{\rho}) = \frac{1}{(2\pi)^2}\int d\mathbf{R}\Gamma_2(x,\mathbf{R},\boldsymbol{\rho})e^{-i\mathbf{q}\mathbf{R}}.$$

As a result, we obtain that function $\tilde{\Gamma}_2(x,\mathbf{q},\boldsymbol{\rho})$ satisfies the equation

$$\left(\frac{\partial}{\partial x} + \frac{1}{k}\mathbf{q}\nabla_{\boldsymbol{\rho}}\right)\tilde{\Gamma}_2(x,\mathbf{q},\boldsymbol{\rho}) = -\frac{k^2}{4}\int\limits_0^x dx_1 \int d\mathbf{q}_1\Phi_{\varepsilon}^{(2)}(x_1,\mathbf{q}_1)$$

$$\times\left\{\cos\left[\frac{x_1}{2k}\mathbf{q}_1(\mathbf{q}_1-\mathbf{q})\right] - \cos\left[\mathbf{q}_1\boldsymbol{\rho} - \frac{x_1}{2k}\mathbf{q}_1(\mathbf{q}_1-\mathbf{q})\right]\right\}\tilde{\Gamma}_2(x,\mathbf{q}_1,\boldsymbol{\rho}) \tag{13.78}$$

with the initial condition

$$\tilde{\Gamma}_2(0,\mathbf{q},\boldsymbol{\rho}) = \tilde{\gamma}_2(0,\mathbf{q},\boldsymbol{\rho}).$$

In contrast to the equation for average field, this is the integro-differential equation.

In the case of the delta-correlated fluctuations, Eq. (13.78) grades into the differential equation

$$\left(\frac{\partial}{\partial x} + \frac{1}{k}\mathbf{q}\nabla_{\boldsymbol{\rho}}\right)\tilde{\Gamma}_2(x,\mathbf{q},\boldsymbol{\rho}) = -\frac{k^2}{4}\int\limits_0^x dx_1 \int d\mathbf{q}_1\Phi_{\varepsilon}^{(2)}(x_1,\mathbf{q}_1)\left\{1-\cos\left[\mathbf{q}_1\boldsymbol{\rho}\right]\right\}\tilde{\Gamma}_2(x,\mathbf{q}_1,\boldsymbol{\rho}),$$

equivalent to Eq. (13.23).

Note that both approximation of the delta-correlated (in x) field $\varepsilon(x,\mathbf{R})$ and diffusion approximation fail if field $\varepsilon(x,\mathbf{R})$ is independent of x as it is the case for cylindrical medium with $\varepsilon(x,\mathbf{R}) = \varepsilon(\mathbf{R})$ or layered medium with $\varepsilon(x,\mathbf{R}) = \varepsilon(z)$. Formally, random field $\varepsilon(x,\mathbf{R})$ is characterized in these cases by the infinite correlation radius along the x-axis.

13.1.4 Wavefield amplitude–phase fluctuations. Rytov's smooth perturbation method

Here, we consider the statistical description of wave amplitude–phase fluctuations.

We introduce the amplitude and phase (and the complex phase) of the wavefield by the formula

$$u(x,\mathbf{R}) = A(x,\mathbf{R})e^{iS(x,\mathbf{R})} = e^{i\varphi(x,\mathbf{R})},$$

where

$$\varphi(x,\mathbf{R}) = \chi(x,\mathbf{R}) + iS(x,\mathbf{R}),$$

$\chi(x, \mathbf{R}) = \ln A(x, \mathbf{R})$ being the level of the wave and $S(x, \mathbf{R})$ being the wave random phase addition to the phase of the incident wave kx. Starting from parabolic equation (13.1), we can obtain that the complex phase satisfies the nonlinear equation of Rytov's *smooth perturbation method* (*SPM*)

$$\frac{\partial}{\partial x}\varphi(x, \mathbf{R}) = \frac{i}{2k}\Delta_{\mathbf{R}}\varphi(x, \mathbf{R}) + \frac{i}{2k}\left[\nabla_{\mathbf{R}}\varphi(x, \mathbf{R})\right]^2 + i\frac{k}{2}\varepsilon(x, \mathbf{R}). \tag{13.79}$$

For the plane incident wave (in what follows, we will deal just with this case), we can set $u_0(\mathbf{R}) = 1$ and $\varphi(0, \mathbf{R}) = 0$ without loss of generality.

Separating the real and imaginary parts in Eq. (13.79), we obtain

$$\frac{\partial}{\partial x}\chi(x, \mathbf{R}) + \frac{1}{2k}\Delta_{\mathbf{R}}S(x, \mathbf{R}) + \frac{1}{k}\left[\boldsymbol{\nabla}_{\mathbf{R}}\chi(x, \mathbf{R})\right]\left[\boldsymbol{\nabla}_{\mathbf{R}}S(x, \mathbf{R})\right] = 0, \tag{13.80}$$

$$\frac{\partial}{\partial x}S(x, \mathbf{R}) - \frac{1}{2k}\Delta_{\mathbf{R}}\chi(x, \mathbf{R}) - \frac{1}{2k}\left[\boldsymbol{\nabla}_{\mathbf{R}}\chi(x, \mathbf{R})\right]^2 + \frac{1}{2k}\left[\boldsymbol{\nabla}_{\mathbf{R}}S(x, \mathbf{R})\right]^2 = \frac{k}{2}\varepsilon(x, \mathbf{R}). \tag{13.81}$$

Using Eq. (13.80), we can derive the equation for wave intensity $I(x, \mathbf{R}) = e^{2\chi(x, \mathbf{R})}$ in the form

$$\frac{\partial}{\partial x}I(x, \mathbf{R}) + \frac{1}{k}\boldsymbol{\nabla}_{\mathbf{R}}\left[I(x, \mathbf{R})\boldsymbol{\nabla}_{\mathbf{R}}S(x, \mathbf{R})\right] = 0. \tag{13.82}$$

If function $\varepsilon(x, \mathbf{R})$ is sufficiently small, then we can solve Eqs. (13.80) and (13.81) by constructing iterative series in field $\varepsilon(x, \mathbf{R})$. The first approximation of Rytov's SPM deals with the Gaussian fields $\chi(x, \mathbf{R})$ and $S(x, \mathbf{R})$, whose statistical characteristics are determined by statistical averaging of the corresponding iterative series. For example, the second moments (including variances) of these fields are determined from the linearized system of equations (13.80) and (13.81), i.e., from the system

$$\begin{aligned}\frac{\partial}{\partial x}\chi_0(x, \mathbf{R}) &= -\frac{1}{2k}\Delta_{\mathbf{R}}S_0(x, \mathbf{R}),\\ \frac{\partial}{\partial x}S_0(x, \mathbf{R}) &= \frac{1}{2k}\Delta_{\mathbf{R}}\chi_0(x, \mathbf{R}) + \frac{k}{2}\varepsilon(x, \mathbf{R}),\end{aligned} \tag{13.83}$$

while average values are determined immediately from Eqs. (13.80) and (13.81). Such amplitude–phase description of the wave filed in random medium was first used by A. M. Obukhov more than fifty years ago in paper [256] (see also [257]) where he pioneered to consider diffraction phenomena accompanying wave propagation in random media using the perturbation theory. Before this work, similar investigations were based on the geometric optics (acoustics) approximation. The technique suggested by Obukhov is topical till now. Basically, it forms the mathematical apparatus of different engineering applications. However, as it was experimentally shown later in papers [85, 86], wavefield fluctuations rapidly grow with distance due to the effect of multiple forward scattering, and perturbation theory fails beginning from certain distance (region of strong fluctuations).

The liner system of equations (13.83) can be solved using the Fourier transformation with respect to the transverse coordinate. Introducing the Fourier transforms of level, phase, and random field $\varepsilon(x, \mathbf{R})$

$$\begin{aligned}
\chi_0(x, \mathbf{R}) &= \int d\mathbf{q} \chi^0_{\mathbf{q}}(x) e^{i\mathbf{q}\mathbf{R}}, \quad \chi^0_{\mathbf{q}}(x) = \frac{1}{(2\pi)^2} \int d\mathbf{R} \chi_0(x, \mathbf{R}) e^{-i\mathbf{q}\mathbf{R}}; \\
S_0(x, \mathbf{R}) &= \int d\mathbf{q} S^0_{\mathbf{q}}(x) e^{i\mathbf{q}\mathbf{R}}, \quad S^0_{\mathbf{q}}(x) = \frac{1}{(2\pi)^2} \int d\mathbf{R} S_0(x, \mathbf{R}) e^{-i\mathbf{q}\mathbf{R}}; \\
\varepsilon(x, \mathbf{R}) &= \int d\mathbf{q} \varepsilon_{\mathbf{q}}(x) e^{i\mathbf{q}\mathbf{R}}, \quad \varepsilon_{\mathbf{q}}(x) = \frac{1}{(2\pi)^2} \int d\mathbf{R} \varepsilon(x, \mathbf{R}) e^{-i\mathbf{q}\mathbf{R}},
\end{aligned} \tag{13.84}$$

we obtain the solution to system (13.83) in the form

$$\begin{aligned}
\chi^0_{\mathbf{q}}(x) &= \frac{k}{2} \int_0^x d\xi \varepsilon_{\mathbf{q}}(\xi) \sin \frac{q^2}{2k}(x - \xi), \\
S^0_{\mathbf{q}}(x) &= \frac{k}{2} \int_0^x d\xi \varepsilon_{\mathbf{q}}(\xi) \cos \frac{q^2}{2k}(x - \xi).
\end{aligned} \tag{13.85}$$

For random field $\varepsilon(x, \mathbf{R})$ described by the correlation and spectral functions (13.15) and (13.26), the correlation function of random Gaussian field $\varepsilon_{\mathbf{q}}(x)$ can be easily obtained by calculating the corresponding integrals.

Indeed, in the case of the delta-correlated approximation for random field $\varepsilon(x, \mathbf{R})$, the relationship between the correlation and spectral functions has the form

$$B_\varepsilon(x_1 - x_2, \mathbf{R}_1 - \mathbf{R}_2) = 2\pi\delta(x_1 - x_2) \int d\mathbf{q} \Phi_\varepsilon(0, \mathbf{q}) e^{i\mathbf{q}(\mathbf{R}_1 - \mathbf{R}_2)}. \tag{13.86}$$

Multiplying Eq. (13.86) by $\frac{1}{(2\pi)^4} e^{-i(\mathbf{q}_1\mathbf{R}_1 + \mathbf{q}_2\mathbf{R}_2)}$, integrating the result over all $\mathbf{R}_1$ and $\mathbf{R}_2$ and taking into account the definition (13.84), we obtain the desired equality

$$\langle \varepsilon_{\mathbf{q}_1}(x_1) \varepsilon_{\mathbf{q}_2}(x_2) \rangle = 2\pi\delta(x_1 - x_2)\delta(\mathbf{q}_1 + \mathbf{q}_2)\Phi_\varepsilon(0, \mathbf{q}_1). \tag{13.87}$$

If field $\varepsilon(x, \mathbf{R})$ is different from zero only in layer $(0, \Delta x)$ and $\varepsilon(x, \mathbf{R}) = 0$ for $x > \Delta x$, then Eq. (13.87) is replaced with the expression

$$\langle \varepsilon_{\mathbf{q}_1}(x_1) \varepsilon_{\mathbf{q}_2}(x_2) \rangle = 2\pi\delta(x_1 - x_2)\theta(\Delta x - x)\delta(\mathbf{q}_1 + \mathbf{q}_2)\Phi_\varepsilon(0, \mathbf{q}_1). \tag{13.88}$$

If we deal with field $\varepsilon(x, \mathbf{R})$ whose fluctuations are caused by temperature turbulent pulsations in Earth's atmosphere, then the three-dimensional spectral density can be represented for a wide range of wave numbers in the form

$$\Phi_\varepsilon(\mathbf{q}) = AC^2_\varepsilon q^{-11/3} \quad (q_{\min} \ll q \ll q_{\max}), \tag{13.89}$$

where $A = 0.033$ is a constant, C^2_ε is the structure characteristic of dielectric permittivity fluctuations that depends on medium parameters. The use of spectral density (13.89) sometimes gives rise to the divergence of the integrals describing statistical characteristics of amplitude–phase fluctuations of the wavefield. In these cases, we can use the phenomenological spectral function

$$\Phi_\varepsilon(\mathbf{q}) = \Phi_\varepsilon(q) = AC^2_\varepsilon q^{-11/3} e^{-\mathbf{q}^2/\kappa^2_m}, \tag{13.90}$$

where κ_m is the wave number corresponding to the *turbulence microscale*.

Within the framework of the first approximation of Rytov's SPM, statistical characteristics of amplitude fluctuations are described by the variance of amplitude level, i.e., by the parameter

$$\sigma_0^2(x) = \left\langle \chi_0^2(x, \mathbf{R}) \right\rangle .$$

In the case of the medium occupying a layer of finite thickness Δx, this parameter can be represented by virtue of Eqs. (13.85) and (13.87) as

$$\begin{aligned} \sigma_0^2(x) &= \int\int d\mathbf{q}_1 d\mathbf{q}_2 \left\langle \chi^0_{\mathbf{q}_1}(x) \chi^0_{\mathbf{q}_2}(x) \right\rangle e^{i(\mathbf{q}_1+\mathbf{q}_2)\mathbf{R}} \\ &= \frac{\pi^2 k^2 \Delta x}{2} \int\limits_0^\infty dq q \Phi_\varepsilon(q) \left\{ 1 - \frac{k}{q^2 \Delta x} \left[\sin \frac{q^2 x}{k} - \sin \frac{q^2 (x - \Delta x)}{k} \right] \right\}. \end{aligned} \quad (13.91)$$

As for the average amplitude level, we determine it from Eq. (13.82). Assuming that incident wave is the plane wave and averaging this equation over an ensemble of realizations of field $\varepsilon(x, \mathbf{R})$, we obtain the equality

$$\langle I(x, \mathbf{R}) \rangle = 1.$$

Rewriting this equality in the form

$$\langle I(x, \mathbf{R}) \rangle = \left\langle e^{2\chi_0(x,\mathbf{R})} \right\rangle = e^{2\langle \chi_0(x,\mathbf{R}) \rangle + 2\sigma_0^2(x)} = 1,$$

we see that

$$\langle \chi_0(x, \mathbf{R}) \rangle = -\sigma_0^2(x)$$

in the first approximation of Rytov's SPM.

Applicability range of the first approximation of Rytov's SPM is restricted by the obvious condition

$$\sigma_0^2(x) \ll 1.$$

As for the wave intensity variance called also *flicker rate*, the first approximation of Rytov's SPM yields the following expression

$$\beta_0^2(x) = \left\langle I^2(x, \mathbf{R}) \right\rangle - 1 = \left\langle e^{4\chi_0(x,\mathbf{R})} \right\rangle - 1 \approx 4\sigma_0^2(x). \quad (13.92)$$

Therefore, the one-point probability density of field $\chi(x, \mathbf{R})$ has in this approximation the form

$$P(x; \chi) = \sqrt{\frac{2}{\pi \beta_0(x)}} \exp\left\{ -\frac{2}{\beta_0(x)} \left(\chi + \frac{1}{4}\beta_0(x) \right)^2 \right\}.$$

Thus, the wavefield intensity is the logarithmic-normal random field, and its one-point probability density is given by the expression

$$P(x; I) = \frac{1}{I\sqrt{2\pi\beta_0(x)}} \exp\left\{ -\frac{1}{2\beta_0(x)} \ln^2 \left(I e^{\frac{1}{2}\beta_0(x)} \right) \right\}. \quad (13.93)$$

The statistical analysis considers commonly two limiting asymptotic cases.

The first case corresponds to the assumption $\Delta x \ll x$ and is called the *random phase screen*. In this case, the wave first traverses a thin layer of fluctuating medium and then propagates in free space. The thin medium layer adds to the wavefield only phase fluctuations. In view of nonlinearity of Eqs. (13.80) and (13.81), the further propagation in free space transforms these phase fluctuations into amplitude fluctuations.

The second case corresponds to the *continuous medium*, i.e., to the condition $\Delta x = x$.

Consider these limiting cases in more details assuming that wavefield fluctuations are weak .

Random phase screen ($\Delta x \ll x$) In this case, the variance of amplitude level is given by the expression following from Eq. (13.91)

$$\sigma_0^2(x) = \frac{\pi^2 k^2 \Delta x}{2} \int\limits_0^\infty dq\, q \Phi_\varepsilon(q) \left\{1 - \cos\frac{q^2 x}{k}\right\}. \tag{13.94}$$

If fluctuations of field $\varepsilon(x, \mathbf{R})$ are caused by turbulent pulsations of medium, then spectrum $\Phi_\varepsilon(q)$ is described by Eq. (13.89), and integral (13.91) can be easily calculated. The resulting expression is

$$\sigma_0^2(x) = 0.144 C_\varepsilon^2 k^{7/6} x^{5/6} \Delta x, \tag{13.95}$$

and, consequently, the flicker rate is given by the expression

$$\beta_0^2(x) = 0.563 C_\varepsilon^2 k^{7/6} x^{5/6} \Delta x. \tag{13.96}$$

As regards phase fluctuations, the quantity of immediate physical interest is the angle of wave arrival at point $(x, \mathbf{R})$,

$$\alpha(x, \mathbf{R}) = \frac{1}{k} |\boldsymbol{\nabla}_{\mathbf{R}} S(x, \mathbf{R})|.$$

The derivation of the formula for its variance is similar to the derivation of Eq. (13.94); the result is as follows

$$\left\langle \alpha^2(x, \mathbf{R}) \right\rangle = \frac{\pi^2 \Delta x}{2} \int\limits_0^\infty dq\, q \Phi_\varepsilon(q) \left\{1 + \cos\frac{q^2 x}{k}\right\}.$$

Continuous medium ($\Delta x = x$) In this case, the variance of amplitude level is given by the formula

$$\sigma_0^2(x) = \frac{\pi^2 k^2 x}{2} \int\limits_0^\infty dq\, q \Phi_\varepsilon(q) \left\{1 - \frac{k}{q^2 x} \sin\frac{q^2 x}{k}\right\}, \tag{13.97}$$

so that parameters $\sigma_0^2(x)$ and $\beta_0^2(x)$ for turbulent medium pulsations assume the forms

$$\sigma_0^2(x) = 0.077 C_\varepsilon^2 k^{7/6} x^{11/6}, \quad \beta_0^2(x) = 0.307 C_\varepsilon^2 k^{7/6} x^{11/6}. \tag{13.98}$$

The variance of the angle of wave arrival at point $(x, \mathbf{R})$ is given by the formula

$$\left\langle \alpha^2(x, \mathbf{R}) \right\rangle = \frac{\pi^2 x}{2} \int\limits_0^\infty dq\, q \Phi_\varepsilon(q) \left\{1 + \frac{k}{q^2 x} \sin\frac{q^2 x}{k}\right\}. \tag{13.99}$$

We can similarly investigate the variance of the amplitude level gradient. In this case, we are forced to use the spectral function $\Phi_\varepsilon(q)$ in form (13.89). Assuming that turbulent medium occupies the whole of the space and the so-called *wave parameter* $D(x) = \kappa_m^2 x / k$ (see, e.g., [268]) is large, $D(x) \gg 1$, we can obtain for parameter $\sigma_{\mathbf{q}}^2(x) = \left\langle [\boldsymbol{\nabla}_{\mathbf{R}} \chi(x, \mathbf{R})]^2 \right\rangle$ the expression

$$\sigma_{\mathbf{q}}^2(x) = \frac{k^2 \pi^2 x}{2} \int\limits_0^\infty dq\, q^3 \Phi_\varepsilon(q) \left\{1 - \frac{k}{q^2 x} \sin\frac{q^2 x}{k}\right\} = \frac{1.476}{L_f^2(x)} D^{1/6}(x) \beta_0(x), \tag{13.100}$$

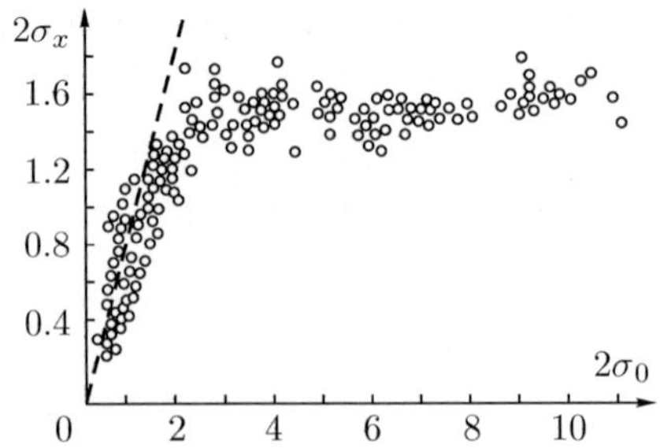

Figure 13.1: Measured variances of the amplitude level versus parameter σ_0 (the dashed line corresponds to the calculation in the first approximation of the Rytov method).

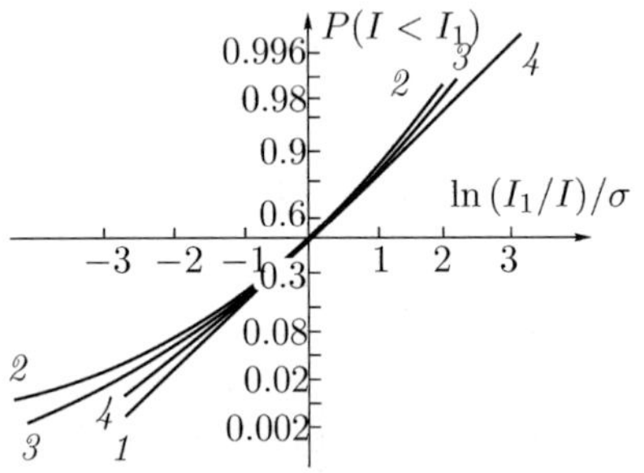

Figure 13.2: Probability distribution of light intensity in turbulent medium. Lines *1* to *4* correspond to $\sigma_0^2 < 1$, $\sigma_0^2 = 1 \div 4$, $\sigma_0^2 > 4$, and $\sigma_0^2 \geq 25$, respectively.

where we introduced the natural scale of length $L_f(x) = \sqrt{x/k}$ in plane $x = \text{const}$; this scale is independent of medium parameters and is equal in size to the first Fresnel zone that determines the size of the light–shade region in the problem on wave diffraction on the edge of an opaque screen (see, e.g., [268]).

The first approximation of Rytov's SPM for amplitude fluctuations is valid in the general case under the condition

$$\sigma_0^2(x) \ll 1.$$

The region, where this inequality is satisfied, is called the *weak fluctuations region*. In the region, where $\sigma_0^2(x) \geq 1$ (this region is called the *strong fluctuations region*), the linearization fails, and we must study the nonlinear system of equations (13.80), (13.81).

Figure 13.1 shows the measured variance of amplitude level $\beta(x) = 2\sigma_\chi(x)$ of light propagating in turbulent atmosphere as a function of parameter $\beta_0(x) = 2\sigma_0(x)$ [104]. The solution in the first approximation of Rytov's SPM is shown here as the dashed line. As may be seen, the weak fluctuation region is limited by values $\beta_0(x) \leq 1$. Moreover, we see that quantity $\beta(x) = 2\sigma_\chi(x)$ tends to a constant value for large parameters $\beta_0(x) > 2\sigma_0(x)$. Figure 13.2 shows that probability distribution of amplitude level is nearly the Gaussian distribution for both weak fluctuation region and very strong fluctuation region; deviations from the Gaussian law occur only in region $\sigma_0^2(x) \sim 1$.

As concerns the fluctuations of angle of wave arrival at the observation point $\alpha(x,\mathbf{R}) = \frac{1}{k}|\nabla_{\mathbf{R}} S(x,\mathbf{R})|$, they are adequately described by the first approximation of Rytov's SPM even for large values of parameter $\sigma_0(x)$.

Note that the approximation of the delta-correlated random field $\varepsilon(x,\mathbf{R})$ in the context of Eq. (13.1) only slightly restricts amplitude fluctuations; as a consequence, the above equations for moments of field $u(x,\mathbf{R})$ appear valid even in the region of strong amplitude fluctuations. The analysis of statistical characteristics in this case will be given later.

13.2 Geometrical optics approximation in randomly inhomogeneous media

13.2.1 Ray diffusion in random media (the Lagrangian description)

In the geometrical optics approximation, characteristic curves (rays) satisfy the system of equations (1.94), page 33

$$\frac{d}{dx}\mathbf{R}(x) = \mathbf{p}(x), \quad \frac{d}{dx}\mathbf{p}(x) = \frac{1}{2}\nabla_{\mathbf{R}}\varepsilon(x,\mathbf{R}). \tag{13.101}$$

In addition, wavefield intensity and matrix of second derivatives vary along these rays in accordance with the equations (1.95), page 33

$$\begin{aligned} &\frac{d}{dx}I(x) = -I(x)u_{ii}(x), \\ &\frac{d}{dx}u_{ij}(x) + u_{ik}(x)u_{kj}(x) = \frac{1}{2}\frac{\partial^2}{\partial R_i \partial R_j}\varepsilon(x,\mathbf{R}). \end{aligned} \tag{13.102}$$

Equations (13.101) and (13.102) form the point of departure for describing wavefields in the framework of small-angle approximation of geometrical optics. We note that Eqs. (13.101) coincide in appearance with the Hamiltonian equations describing the motion of a particle in random field of external forces.

It is clear that if $\varepsilon(x,\mathbf{R})$ is the homogeneous isotropic Gaussian delta-correlated field with the parameters

$$\langle\varepsilon(x,\mathbf{R})\rangle = 0, \quad B_\varepsilon(x - x', \mathbf{R} - \mathbf{R}') = A(\mathbf{R} - \mathbf{R}')\delta(x - x'),$$

then the one-point joint probability density

$$P(x;\mathbf{R},\mathbf{p}) = \langle\delta(\mathbf{R}(x) - \mathbf{R})\delta(\mathbf{p}(x) - \mathbf{p})\rangle$$

satisfies the Fokker–Planck equation

$$\left(\frac{\partial}{\partial x} + \mathbf{p}\frac{\partial}{\partial \mathbf{R}}\right) P(x;\mathbf{R},\mathbf{p}) = D\Delta_{\mathbf{R}} P(x;\mathbf{R},\mathbf{p}), \tag{13.103}$$

where

$$D = -\frac{1}{8}\,\Delta_{\mathbf{R}} A(\mathbf{R})|_{\mathbf{R}=0} = \pi^2 \int\limits_0^\infty d\kappa\, \kappa^3 \Phi_\varepsilon(0,\kappa)$$

is the diffusion coefficient,

$$\Phi_\varepsilon(q,\boldsymbol{\kappa}) = \frac{1}{(2\pi)^3} \int\limits_{-\infty}^{\infty} dx \int d\mathbf{R} B_\varepsilon(x,\mathbf{R}) e^{-i(qx+\boldsymbol{\kappa}\mathbf{R})}$$

is the three-dimensional spectral density of random field $\varepsilon(x, \mathbf{R})$, and

$$A(\mathbf{R}) = 2\pi \int d\boldsymbol{\kappa} \Phi_\varepsilon(0, \boldsymbol{\kappa}) e^{i\boldsymbol{\kappa}\mathbf{R}}.$$

Equation (13.103) can be easily solved, and its solution corresponding to the initial condition $P(0; \mathbf{R}, \mathbf{p}) = \delta(\mathbf{R})\delta(\mathbf{p})$ is the Gaussian probability density with the parameters

$$\langle R_j(x) R_k(x)\rangle = \frac{2}{3} D\delta_{jk} x^3, \quad \langle R_j(x) p_k(x)\rangle = D\delta_{jk} x^2, \quad \langle p_j(x) p_k(x)\rangle = 2D\delta_{jk} x. \quad (13.104)$$

The longitudinal correlation function of ray displacements also can be easily determined from Eqs. (13.101). We multiply Eqs. (13.101) by $\mathbf{R}(x')$ with $x' < x$ and average the result over an ensemble of realizations of field $\varepsilon(x, \mathbf{R})$. As a result, we obtain the system of equations

$$\begin{aligned} \frac{d}{dx} \langle \mathbf{R}(x) \mathbf{R}(x')\rangle &= \langle \mathbf{p}(x) \mathbf{R}(x')\rangle, \\ \frac{d}{dx} \langle \mathbf{p}(x) \mathbf{R}(x')\rangle &= \frac{1}{2} \langle \mathbf{R}(x') \nabla_{\mathbf{R}} \varepsilon(x, \mathbf{R})\rangle \end{aligned} \quad (13.105)$$

with predetermined initial conditions at $x = x'$

$$\langle \mathbf{R}(x) \mathbf{R}(x')\rangle_{x=x'} = \left\langle \mathbf{R}^2(x')\right\rangle, \quad \langle \mathbf{p}(x) \mathbf{R}(x')\rangle_{x=x'} = \langle \mathbf{p}(x') \mathbf{R}(x')\rangle. \quad (13.106)$$

In the framework of the model of delta-correlated (in x) inhomogeneities, quantity $\mathbf{R}(x')$ is not correlated with field $\nabla_{\mathbf{R}} \varepsilon(x, \mathbf{R})$ for consequent values of argument x,

$$\langle \mathbf{R}(x') \nabla_{\mathbf{R}} \varepsilon(x, \mathbf{R})\rangle = 0 \quad \text{for} \quad x' < x,$$

so that

$$\langle \mathbf{p}(x) \mathbf{R}(x')\rangle = \langle \mathbf{p}(x') \mathbf{R}(x')\rangle = 2D(x')^2.$$

Substituting this result in the first equation of system (13.105) and solving it, we obtain

$$\langle \mathbf{R}(x) \mathbf{R}(x')\rangle = 2D(x')^2 \left(x - \frac{1}{3} x'\right).$$

Consider now the problem on the cooperative diffusion of two rays. This problem is described by the system of equations

$$\frac{d}{dx} \mathbf{R}_\nu(x) = \mathbf{p}_\nu(x), \quad \frac{d}{dx} \mathbf{p}_\nu(x) = \frac{1}{2} \nabla_{\mathbf{R}_\nu} \varepsilon(x, \mathbf{R}_\nu), \quad (13.107)$$

where index $\nu = 1, 2$ marks the number of the corresponding ray. We obtain in the regular way that the joint probability density

$$\begin{aligned} &P(x; \mathbf{R}_1, \mathbf{p}_1, \mathbf{R}_2, \mathbf{p}_2) \\ &= \langle \delta(\mathbf{R}_1(x) - \mathbf{R}_1) \delta(\mathbf{p}_1(x) - \mathbf{p}_1) \delta(\mathbf{R}_2(x) - \mathbf{R}_2) \delta(\mathbf{p}_2(x) - \mathbf{p}_2)\rangle \end{aligned}$$

satisfies the Fokker–Planck equation

$$\begin{aligned} &\left(\frac{\partial}{\partial x} + \mathbf{p}_1 \frac{\partial}{\partial \mathbf{R}_1} + \mathbf{p}_2 \frac{\partial}{\partial \mathbf{R}_2}\right) P(x; \mathbf{R}_1, \mathbf{p}_1, \mathbf{R}_2, \mathbf{p}) \\ &= \widehat{L} \left(\frac{\partial}{\partial \mathbf{p}_1}, \frac{\partial}{\partial \mathbf{p}_2}\right) P(x; \mathbf{R}_1, \mathbf{p}_1, \mathbf{R}_2, \mathbf{p}), \end{aligned} \quad (13.108)$$

where operator $\widehat{L}$ is given by the formula

$$\widehat{L}\left(\frac{\partial}{\partial\mathbf{p}_1},\frac{\partial}{\partial\mathbf{p}_2}\right)=\frac{\pi}{4}\int d\boldsymbol{\kappa}\Phi_\varepsilon(0,\boldsymbol{\kappa})\left[\left(\boldsymbol{\kappa}\frac{\partial}{\partial\mathbf{p}_1}\right)^2+\left(\boldsymbol{\kappa}\frac{\partial}{\partial\mathbf{p}_2}\right)^2\right.$$
$$\left.+2\cos\left[\boldsymbol{\kappa}(\mathbf{R}_1-\mathbf{R}_2)\right]\left(\boldsymbol{\kappa}\frac{\partial}{\partial\mathbf{p}_1}\right)\left(\boldsymbol{\kappa}\frac{\partial}{\partial\mathbf{p}_1}\right)\right].$$

We derive the equation for probability density of relative diffusion of two rays, i.e., for function

$$P(x;\mathbf{R},\mathbf{p})=\langle\delta(\mathbf{R}_1(x)-\mathbf{R}_2(x)-\mathbf{R})\delta(\mathbf{p}_1(x)-\mathbf{p}_2(x)-\mathbf{p})\rangle,$$

by multiplying Eq. (13.103) by

$$\delta(\mathbf{R}_1(x)-\mathbf{R}_2(x)-\mathbf{R})\delta(\mathbf{p}_1(x)-\mathbf{p}_2(x)-\mathbf{p})$$

and integrating over $\mathbf{R}_1$, $\mathbf{R}_2$, $\mathbf{p}_1$, and $\mathbf{p}_2$. As a result, we arrive at the Fokker–Planck equation

$$\left(\frac{\partial}{\partial x}+\mathbf{p}\frac{\partial}{\partial\mathbf{R}}\right)P(x;\mathbf{R},\mathbf{p})=D_{\alpha\beta}(\mathbf{R})\frac{\partial^2}{\partial p_\alpha\partial p_\beta}P(x;\mathbf{R},\mathbf{p}).\qquad(13.109)$$

Here, $D_{\alpha\beta}(\mathbf{R})$ is the following matrix

$$D_{\alpha\beta}(\mathbf{R})=2\pi\int d\boldsymbol{\kappa}\left[1-\cos\left(\boldsymbol{\kappa}\mathbf{R}\right)\right]\kappa_\alpha\kappa_\beta\Phi_\varepsilon(0,\boldsymbol{\kappa}).$$

If we denote the correlation radius of random field $\varepsilon(x,\mathbf{R})$ as l_0 and assume that $R\gg l_0$, then we obtain that

$$D_{\alpha\beta}(\mathbf{R})=2D\delta_{\alpha\beta}.$$

This means that relative diffusion of two rays is characterized by the diffusion coefficient exceeding the diffusion coefficient of a separate ray by a factor of two, which in turn means that these rays are statistically independent. In this case, the joint probability density of relative diffusion is the Gaussian probability density.

In the general case, Eq. (13.109) cannot be solved in analytic form. The only clear point is that the solution to this equation is not the Gaussian distribution if the diffusion coefficient is not a constant.

Asymptotic case $R\ll l_0$ can be analyzed in sufficient details. In this case, we can expand function $\{1-\cos\left(\boldsymbol{\kappa}\mathbf{R}\right)\}$ in the Taylor series to reduce the diffusion matrix to the form

$$D_{\alpha\beta}(\mathbf{R})=\pi R_iR_j\int d\boldsymbol{\kappa}\kappa_i\kappa_j\kappa_\alpha\kappa_\beta\Phi_\varepsilon(0,\boldsymbol{\kappa}).$$

It is clear that

$$\int d\boldsymbol{\kappa}\,\kappa_i\kappa_j\kappa_\alpha\kappa_\beta\Phi_\varepsilon(0,\boldsymbol{\kappa})=B\left(\delta_{ij}\delta_{\alpha\beta}+\delta_{i\alpha}\delta_{j\beta}+\delta_{i\beta}\delta_{j\alpha}\right)$$

in the case of statistically isotropic fluctuations. Contracting this equality over index pairs i,j and α,β, we find that

$$B=\frac{\pi}{4}\int\limits_0^\infty d\kappa\,\kappa^5\Phi_\varepsilon(0,\kappa)$$

and, consequently,

$$D_{\alpha\beta}(\mathbf{R})=\pi B\left(\mathbf{R}^2\delta_{a\beta}+2R_\alpha R_\beta\right).\qquad(13.110)$$

Note that quantity B characterizes amplitude fluctuations in the geometrical optics approximation. This fact is quite expectable because amplitude fluctuations are related to variations of ray tube section, i.e., to relative ray displacements.

Diffusion coefficients $D_{\alpha\beta}(\mathbf{R})$ in form (13.110) can be used only if average square of distance between the rays is small in comparison with l_0^2. Equation (13.109) with coefficients (13.110), i.e., the equation

$$\left(\frac{\partial}{\partial x}+\mathbf{p}\frac{\partial}{\partial \mathbf{R}}\right)P(x;\mathbf{R},\mathbf{p})=\pi B\left(\mathbf{R}^2\delta_{a\beta}+2R_\alpha R_\beta\right)\frac{\partial^2}{\partial p_\alpha \partial p_\beta}P(x;\mathbf{R},\mathbf{p})$$

yields the following equations for moment functions

$$\frac{d}{dx}\left\langle \mathbf{p}^2(x)\right\rangle = 8\pi B\left\langle \mathbf{R}^2(x)\right\rangle, \quad \frac{d}{dx}\left\langle \mathbf{R}^2(x)\right\rangle = 2\left\langle \mathbf{R}(x)\mathbf{p}(x)\right\rangle,$$
$$\frac{d}{dx}\left\langle \mathbf{R}(x)\mathbf{p}(x)\right\rangle = \left\langle \mathbf{p}^2(x)\right\rangle, \tag{13.111}$$

which can be easily solved. From this solution follows that quantities $\langle \mathbf{R}^2(x)\rangle$, $\langle \mathbf{R}(x)\mathbf{p}(x)\rangle$, and $\langle \mathbf{p}^2(x)\rangle$ are exponentially increasing functions in the interval of parameter x such that $\alpha x \gg 1$ $\left(\alpha=(16\pi B)^{1/3}\right)$, but $R_0^2 e^{\alpha x} \ll l_0^2$ (such an interval always exists for sufficiently small initial distances between rays R_0). Note that this region of exponential increase begins at distances $\alpha x \sim 1$, which coincides with the onset of strong intensity fluctuations because $\alpha x \sim \sigma_I^{2/3} = \left\langle [\ln(I/I_0)]^2\right\rangle^{1/3}$.

Outline now the applicability range of the Fokker–Planck equation. The Fokker–Planck equation for ray diffusion was derived in the small-angle approximation. This implies that its applicability range is restricted by the condition (see Eqs. (13.104))

$$\left\langle \mathbf{p}^2(x)\right\rangle \ll 1, \quad \text{or} \quad Dx \ll 1. \tag{13.112}$$

As regards the corrections caused by the finite value of the longitudinal correlation radius, the requirement of their smallness obviously reduces to the condition $x \gg l_0$ and to condition (13.112).

13.2.2 Formation of caustics in randomly inhomogeneous media

As we have seen earlier, the parabolic equation of quasi-optics predicts exponentially increasing behavior of statistical characteristics related to relative ray diffusion with distance; in other words, it predicts statistical ray dispersion. At the same time, it is well known that caustics are formed for finite distances in random medium with a probability equal to unity [141, 215, 309, 321]. The problem on caustic formation is similar to the problem on statistical description of transfer phenomenon; it can be formulated in terms of statistical characteristics of phase front curvature and wave intensity in random medium, which are governed by stochastic equations (13.102).

In the two-dimensional case, the problem becomes simpler, and phase front curvature in plane (x,y) satisfies the equation

$$\frac{d}{dx}u(x) = -u^2(x)+f(x), \quad u(0)=u_0, \tag{13.113}$$

where $f(x)=\frac{1}{2}\frac{\partial^2}{\partial y^2}\varepsilon(x,y(x))$ and transverse ray displacement $y(x)$ is governed by system of equations (13.101).

In the case of the homogeneous isotropic Gaussian delta-correlated field $\varepsilon(x,y)$ with the parameters

$$\langle\varepsilon(x,y)\rangle = 0, \quad \langle\varepsilon(x,y)\varepsilon(x',y')\rangle = \delta(x-x')A(y-y'),$$

the one-point probability density of curvature is statistically independent of ray displacements and satisfies the Fokker–Planck equation

$$\frac{\partial}{\partial x}P(x;u) = \frac{\partial}{\partial u}u^2(x;u)P(x;u) + \frac{D}{2}\frac{\partial^2}{\partial u^2}P(x;u), \quad P(0;u) = \delta(u-u_0) \tag{13.114}$$

with the diffusion coefficient

$$D = \frac{1}{4}\frac{\partial^4}{\partial y^4}A(0) = \pi\int\limits_0^\infty \kappa^4 d\kappa \Phi_\varepsilon(0,\kappa),$$

where $\Phi_\varepsilon(0,\kappa)$ is the two-dimensional spectral function of random field $\varepsilon(x,y)$.

We have already considered Eq. (13.114) in Chapter 8, page 212. We showed that random process $u(x)$ is the discontinuous process; it tends to $-\infty$ at a finite distance $x(u_0)$ whose value depends of initial value u_0, which corresponds to wave focusing in random medium. In this case, the average of this distance $\langle x(u_0)\rangle$ is given by the expression

$$\langle x(u_0)\rangle = \frac{2}{D}\int\limits_{-\infty}^{u_0} d\xi \int\limits_{\xi}^{\infty} d\eta \exp\left\{\frac{2}{3D}\left(\xi^3 - \eta^3\right)\right\}, \tag{13.115}$$

from which follows that

$$D^{1/3}\langle x(\infty)\rangle \approx 6.27, \quad D^{1/3}\langle x(0)\rangle = \frac{2}{3}D^{1/3}\langle x(\infty)\rangle \approx 4.18.$$

Distance $\langle x(0)\rangle$ is the average distance to focal points formed in random medium by the plane incident wave and distance $\langle x(\infty)\rangle$ is the average distance between two successive focal points. We remind that Rytov's smooth perturbation method predicts the variance of amplitude level in the form $\sigma^2(x) \cong Dx^3$, from which immediately follows that random focusing occurs in the region of strong intensity fluctuations where $\sigma^2(x) \geq 1$.

Further analysis of Eq. (13.114) essentially depends on boundary conditions with respect to variable u. For example, if we consider function $u(x)$ as the discontinuous function determined for all x in such a way that its value of $-\infty$ at point $x \to x_0 - 0$ is immediately followed by a value of ∞ at point $x \to x_0 + 0$, then Eq. (13.114) must be supplemented with the boundary condition

$$J(x;u)|_{u\to\infty} = J(x;u)|_{u\to-\infty},$$

where

$$J(x;u) = u^2 P(x;u) + \frac{D}{2}\frac{\partial}{\partial u}P(x;u)$$

is the probability flux density. We considered this case in Sect. 8.4.4, page 214 and showed that Eq. (13.114) has the steady-state (independent of x) probability density in the limit of large x and this probability density

$$P(u) = J\int\limits_{-\infty}^{u} d\xi \exp\left\{\frac{2}{3D}\left(\xi^3 - u^3\right)\right\} \tag{13.116}$$

corresponds to the constant probability flux density

$$J = \frac{1}{\langle x(\infty\rangle}.$$

For large u, Eq. (13.114) yields the asymptotic formula

$$P(u) \approx \frac{1}{\langle x(\infty)\rangle\, u^2},$$

which means that steady-state statistics is formed by the behavior of function $u(x)$ in the vicinity of discontinuities

$$u(x) = \frac{1}{x - x_k}.$$

Near the discontinuities, the wavefield intensity behaves as

$$I(x) = \frac{x_k}{|x - x_k|}$$

(this expression follows from Eq. (13.102)). In this case, the probability density of quantity $z(x) = I^2(x)$ for sufficiently large x and z is given by the asymptotic expression

$$\begin{aligned} P(x, z) &= \sum_{k=0}^{\infty} \left\langle \delta\left(\frac{x_k^2}{(x - x_k)^2} - z\right)\right\rangle = \frac{1}{z^2}\sum_{k=0}^{\infty} \left\langle \delta\left(\frac{(x - x_k)^2}{x_k^2} - \frac{1}{z}\right)\right\rangle \\ &= \frac{x}{z\sqrt{z}} \sum_{k=0}^{\infty} \langle \delta(x - x_k)\rangle = \frac{x}{z\sqrt{z}}\frac{1}{2\pi} \int\limits_{-\infty}^{\infty} dk e^{-ikx} \frac{\Phi_0(k)}{1 + \Phi(k)}, \end{aligned}$$

where $\Phi_0(k)$ is the characteristic function of the distance to the first caustic and $\Phi(k)$ is the characteristic function of the distance between two adjacent caustics. Consequently, probability density of quantity z for $x \gg \langle x(\infty)\rangle$ has the form

$$P(x, z) = \frac{x}{\langle x(\infty)\rangle\, z\sqrt{z}},$$

and probability density of large values of wave intensity I can be represented by the asymptotic formula

$$P(x, I) = \frac{2x}{\langle x(\infty)\rangle\, I^2}.$$

This probability density depends on the distance the wave travels in the medium and decays according to the power law for large intensities I.

As we mentioned earlier, the other type of boundary conditions corresponds to the assumption that curve $u(x)$ is cut immediately after it achieves the value of $-\infty$ at point $x = x_0$. In this case, boundary conditions have the form

$$J(x, u) \to 0 \quad \text{for} \quad u \to \pm\infty,$$

and probability of focus formation at distance x is given by the expression

$$P(x > x_0) = 1 - \int\limits_{-\infty}^{\infty} du P(x, u).$$

The corresponding probability density is related to the probability flux density by the expression [215, 309, 321]

$$p(x) = \frac{\partial}{\partial x} P(x > x_0) = -\frac{\partial}{\partial x} \int\limits_{-\infty}^{\infty} du P(x, u) = \lim_{u \to -\infty} J(x, u).$$

We obtain the asymptotic behavior of probability density $p(x)$ versus small parameter $D \to 0$ by using the standard technique of analyzing the parabolic equation having a small parameter as the factor of the highest derivative.

We represent the solution to Eq. (13.114) in the form

$$P(x, u) = C(D) \exp\left\{-\frac{1}{D} A(x, u) - B(x, u)\right\}. \qquad (13.117)$$

Substituting Eq. (13.117) in Eq. (13.114) and isolating terms proportional to D^0 and D^{-1}, we obtain partial differential equations in functions $A(x, u)$ and $B(x, u)$. Constant $C(D)$ can be determined from the condition that probability density for $x \to 0$ is known; for example, in the case of the plane incident wave, it must have the form

$$P(x, u) = \frac{1}{\sqrt{2\pi D x}} \exp\left\{-\frac{u^2}{2Dx}\right\}.$$

This leads to the estimate $C(D) \cong 1/\sqrt{D}$. Substituting Eq. (13.117) in the expression for the probability density of focus formation, we obtain

$$p(x) = \lim_{u \to -\infty} P(x, u) \left[u^2 - \frac{1}{2} \frac{\partial}{\partial u} A(x, u)\right]. \qquad (13.118)$$

Note that representation of density $P(x, u)$ in form (13.117) makes it possible to immediately obtain the structure of function $p(x)$ from dimensional considerations [141]. Indeed, the respective dimensions of quantities u, D, and $P(x, u)$ are

$$[u] = x^{-1}, \quad [D] = x^{-3}, \quad [P(x, u)] = x.$$

As a consequence, from Eqs. (13.117) and (13.118), we obtain

$$p(x) = C_1 D^{-1/2} x^{-5/2} e^{-C_2/D x^3},$$

so that the task is reduced to calculating constants C_1 and C_2. These constants were determined in paper [215]; the final formula has the form

$$p(x) = 3\alpha^2 (2\pi D)^{-1/2} x^{-5/2} e^{-\alpha^4/6Dx^3}, \qquad (13.119)$$

where $\alpha = 2.85$.

Applicability range of Eq. (13.119) is restricted by the condition $Dx^3 \ll 1$. Nevertheless, simulations carried out in paper [215] showed that Eq. (13.119) adequately describes probability density even if $Dx^3 \sim 1$.

Discuss now the three-dimensional problem. In this case, matrix of wave front curvature

$$u_{ij}(x) = \frac{1}{k} \frac{\partial^2}{\partial R_i \partial R_j} S(x, \mathbf{R})$$

satisfies the stochastic equation

$$\frac{d}{dx}u(x)+u^2(x)=F\left(x,\mathbf{R}(x)\right), \tag{13.120}$$

where matrix function $F_{ij}\left(x,\mathbf{R}(x)\right)$ is given by the expression

$$F_{ij}\left(x,\mathbf{R}(x)\right)=\frac{1}{2}\frac{\partial^2}{\partial R_i\partial R_j}\varepsilon(x,\mathbf{R}),\quad i,j=1,2.$$

Matrix $u_{ij}(x)$ is symmetric, so that it can be reduced to the diagonal form by the rotation transform

$$D^T(x)u(x)D(x)=\Lambda(x), \tag{13.121}$$

where matrixes $\Lambda(x)$ and $D(x)$ ($D^T(x)$ is the transpose of matrix $D(x)$) have the forms

$$\Lambda(x)=\left\|\begin{matrix}\lambda_1(x) & 0\\ 0 & \lambda_2(x)\end{matrix}\right\|,\quad D(x)=\left\|\begin{matrix}\cos\theta(x) & -\sin\theta(x)\\ \sin\theta(x) & \cos\theta(x)\end{matrix}\right\|,$$

and quantities $\lambda_1(x)$ and $\lambda_2(x)$ are the principal curvatures of phase front $S(x,\mathbf{R})=\text{const}$.

Differentiating Eq. (13.121) with respect to x and using dynamic equation (13.120), we obtain the stochastic matrix equation in $\Lambda(x)$

$$\frac{d}{dx}\Lambda(x)=-\Lambda^2(x)+\frac{dD^T(x)}{dx}D(x)\Lambda(x)+\Lambda(x)D^T(x)\frac{dD(x)}{dx}+D^T(x)F(x)D(x). \tag{13.122}$$

This equation is equivalent to the system of three equations

$$\begin{aligned}\frac{d}{dx}\lambda_1(x) &= -\lambda_1^2(x)+F_{11}(x)\cos^2\theta(x)+F_{22}(x)\sin^2\theta(x)+F_{12}(x)\sin 2\theta(x),\\ \frac{d}{dx}\lambda_2(x) &= -\lambda_2^2(x)+F_{22}(x)\cos^2\theta(x)+F_{11}(x)\sin^2\theta(x)-F_{12}(x)\sin 2\theta(x),\\ \frac{d}{dx}\theta(x) &= \frac{1}{2}\frac{F_{22}(x)-F_{11}(x)}{\lambda_1(x)-\lambda_2(x)}\sin 2\theta(x)+\frac{F_{12}(x)}{\lambda_1(x)-\lambda_2(x)}\cos 2\theta(x).\end{aligned} \tag{13.123}$$

As a result, we arrive at the following Fokker–Planck equation for the joint probability density of quantities $\lambda_1(x)$ and $\lambda_2(x)$ [141]

$$\begin{aligned}\frac{\partial}{\partial x}P(x;\lambda_1,\lambda_2) &= \left(\frac{\partial}{\partial\lambda_1}\left[\lambda_1^2-\frac{2D}{\lambda_1-\lambda_2}\right]+\frac{\partial}{\partial\lambda_2}\left[\lambda_2^2-\frac{2D}{\lambda_1-\lambda_2}\right]\right)P(x;\lambda_1,\lambda_2)\\ &\quad +D\left(3\frac{\partial^2}{\partial\lambda_1^2}+\frac{\partial^2}{\partial\lambda_1\partial\lambda_2}+3\frac{\partial^2}{\partial\lambda_2^2}\right)P(x;\lambda_1,\lambda_2),\\ P(0;\lambda_1,\lambda_2) &= \delta(\lambda_1)\delta(\lambda_2),\end{aligned} \tag{13.124}$$

where

$$D=\frac{1}{64}\,\Delta^2_{\mathbf{R}}A(\mathbf{R})\Big|_{\mathbf{R}=0}=\frac{\pi^2}{16}\int\limits_0^\infty d\kappa\,\kappa^5\Phi_\varepsilon(0,\kappa).$$

We note that the stochastic dynamic system

$$\begin{aligned}\frac{d}{dx}\lambda_1(x) &= -\lambda_1^2+\frac{2D}{\lambda_1-\lambda_2}+F(x)+\widetilde{F}(x),\\ \frac{d}{dx}\lambda_2(x) &= -\lambda_2^2-\frac{2D}{\lambda_1-\lambda_2}+F(x)-\widetilde{F}(x)\end{aligned} \tag{13.125}$$

with random delta-correlated functions $F(x)$ and $\widetilde{F}(x)$ is equivalent to this Fokker–Planck equation; as a consequence, we obtain that the joint probability density of quantities $\lambda_1(x)$ and $\lambda_2(x)$ has for small x, namely, for $Dx^3 \ll 1$ the form

$$P(x;\lambda_1,\lambda_2) = \frac{|\lambda_1-\lambda_2|}{32\sqrt{2\pi\,(Dx)^3}} \exp\left\{-\frac{1}{32Dx}\left(3\lambda_1^2 - 2\lambda_1\lambda_2 + 3\lambda_2^2\right)\right\}. \tag{13.126}$$

It is quite natural that the probability of caustic appearance in the region where $Dx^3 \ll 1$ is negligibly small. The corresponding probability density is given by the expression similar to that derived in the two-dimensional case [309],

$$p(x) = -\int\limits_{-\infty}^{\infty} d\lambda_1 \int\limits_{-\infty}^{\infty} d\lambda_2 \boldsymbol{\nabla}_{\boldsymbol{\lambda}} \mathbf{J}(x;\boldsymbol{\lambda}), \quad \boldsymbol{\lambda} = (\lambda_1,\lambda_2), \tag{13.127}$$

where $\mathbf{J}(x;\boldsymbol{\lambda})$ is the vector of probability flux density determined from (13.124),

$$\mathbf{J}(x;\boldsymbol{\lambda}) = \begin{pmatrix} \lambda_1^2 - \frac{2D}{\lambda_1-\lambda_2} + 3D\frac{\partial}{\partial\lambda_1} + \frac{D}{2}\frac{\partial}{\partial\lambda_2} \\ \\ \lambda_2^2 + \frac{2D}{\lambda_1-\lambda_2} + 3D\frac{\partial}{\partial\lambda_2} + \frac{D}{2}\frac{\partial}{\partial\lambda_1} \end{pmatrix} P(x;\lambda_1,\lambda_2).$$

Expression (13.127) can be represented as the contour integral

$$p(x) = \oint\limits_C ds \mathbf{J}(x;\boldsymbol{\lambda})\mathbf{n} \tag{13.128}$$

in the limit of infinite diameter of contour C. Here, $\mathbf{n}$ is the vector of the exterior normal to the boundary of contour C.

As was mentioned earlier, we can obtain the asymptotic solution to Eq. (13.124) by representing the solution in the form

$$P(x;\lambda_1,\lambda_2) = D^{-3/2}\exp\left\{-\frac{1}{D}A(x;\lambda_1,\lambda_2) - B(x;\lambda_1,\lambda_2)\right\}$$

and constructing the perturbation series in parameter D.

We note that function $P(x;\lambda_1,\lambda_2)$ must have the stationary point on contour C, at which function $A(x;\lambda_1,\lambda_2)$ is minimum. This yields an additional factor $D^{1/2}$ for $Dx^3 \to 0$, so that dimensional considerations will lead in this case to the following expression for the probability density of caustic formation [141]

$$p(x) = \frac{\alpha}{Dx^4}\exp\left\{-\frac{\beta}{Dx^3}\right\},$$

where α and β are constants. This result with $\alpha = 2.74$ and $\beta = 0.66$ was derived in paper [309].

13.2.3 Wavefield amplitude–phase fluctuations (the Eulerian description)

Consideration of amplitude–phase fluctuations in the Eulerian description becomes significantly simpler if the geometrical optics approximation holds ($k \to \infty$).

Consider the quantity

$$\Theta(x,\mathbf{R}) = \frac{1}{k}S(x,\mathbf{R}),$$

which is usually called the *eikonal*, and perform limit process $k \to \infty$. The equation for quantity $\chi(x,\mathbf{R})$ (13.80), page 374 remains in this case intact; as a consequence, the equation for wavefield intensity (13.82) also remains intact

$$\frac{\partial}{\partial x}I(x,\mathbf{R}) + \nabla_{\mathbf{R}}\left[I(x,\mathbf{R})\nabla_{\mathbf{R}}\Theta(x,\mathbf{R})\right] = 0 \tag{13.129}$$

and Eq. (13.81), page 374 for the wavefield phase assumes the form of the Hamilton–Jacobi equation

$$\frac{\partial}{\partial x}\Theta(x,\mathbf{R}) + \frac{1}{2}\left[\nabla_{\mathbf{R}}\Theta(x,\mathbf{R})\right]^2 = \frac{1}{2}\varepsilon(x,\mathbf{R}). \tag{13.130}$$

Moreover, one can obtain that the transverse gradient of the wave phase satisfies the closed quasilinear equation

$$\left\{\frac{\partial}{\partial x} + \frac{\partial\Theta(x,\mathbf{R})}{\partial\mathbf{R}}\nabla_{\mathbf{R}}\right\}\nabla_{\mathbf{R}}\Theta(x,\mathbf{R}) = \frac{1}{2}\nabla_{\mathbf{R}}\varepsilon(x,\mathbf{R}). \tag{13.131}$$

Equations (13.129)–(13.131) form the starting point for analyzing amplitude–phase fluctuations in the geometrical optics approximation. In addition, the equation in the wave phase appears independent of amplitude. This equation is the first-order partial differential equation and its characteristics are the rays whose statistical description was considered earlier. Here, we consider the corollary facts that can be derived immediately from partial differential equations (13.129)–(13.131), i.e., from the Eulerian description.

Assuming spatial homogeneity of all fields in plane $x = \text{const}$, we can easily obtain from Eqs. (13.129), (13.130) the expression

$$\frac{\partial}{\partial x}\left\langle I(x,\mathbf{R})\Theta(x,\mathbf{R})\right\rangle = \left\langle I(x,\mathbf{R})\left[\nabla_{\mathbf{R}}\Theta(x,\mathbf{R})\right]^2\right\rangle + \frac{1}{2}\left\langle \varepsilon(x,\mathbf{R})I(x,\mathbf{R})\right\rangle.$$

On the other hand, the geometrical optics approximation yields the following relationship

$$\left.\frac{1}{k^2}\nabla_{\mathbf{R}_1}\nabla_{\mathbf{R}_2}\Gamma_2(x;\mathbf{R}_1,\mathbf{R}_2)\right|_{\mathbf{R}_1=\mathbf{R}_2=\mathbf{R}} = \left\langle I(x,\mathbf{R})\left[\nabla_{\mathbf{R}}\Theta(x,\mathbf{R})\right]^2\right\rangle. \tag{13.132}$$

We can calculate the left-hand side of Eq. (13.132) using the delta-correlated random field approximation. In the case of the plane incident wave ($u_0 = 1$), function $\Gamma_2(x;\mathbf{R}_1,\mathbf{R}_2)$ is given in this approximation by Eq. (13.30), and $\left\langle\varepsilon(x,\mathbf{R})I(x,\mathbf{R})\right\rangle = 0$. Consequently, we have

$$\begin{aligned}\left\langle I(x,\mathbf{R})\left[\nabla_{\mathbf{R}}\Theta(x,\mathbf{R})\right]^2\right\rangle &= -\frac{1}{4}\Delta_{\mathbf{R}}A(0)x = \gamma(x),\\ \left\langle I(x,\mathbf{R})\Theta(x,\mathbf{R})\right\rangle &= -\frac{1}{16}\Delta_{\mathbf{R}}A(0)x,\end{aligned} \tag{13.133}$$

where

$$\gamma(x) = \pi^2 x\int\limits_0^\infty dq\, q^3\Phi_\varepsilon(q) = \left\langle\left[\nabla_{\mathbf{R}}\Theta_0(x,\mathbf{R})\right]^2\right\rangle$$

is the variance of angle of wave arrival at the observation point in the first approximation of Rytov's SPM in the limit of geometrical optics.

The geometrical optics approximation combined with the approximation of delta-correlated random field $\varepsilon(x, \mathbf{R})$ allow additionally deriving the closed equation for function

$$G(x, \mathbf{R}; \Theta, \mathbf{p}) = \langle I(x, \mathbf{R})\delta\left(\Theta(x, \mathbf{R}) - \Theta\right)\delta\left(\boldsymbol{\nabla}_{\mathbf{R}}\Theta(x, \mathbf{R}) - \mathbf{p}\right)\rangle$$

(this function characterizes correlators of the intensity with wave phase and wave phase gradient). Indeed, differentiating this function with respect to x, using dynamic equations (13.129)–(13.131), splitting correlators by the Furutsu–Novikov formula, and performing some rearrangements, we obtain the equation

$$\begin{aligned}\left(\frac{\partial}{\partial x} + \mathbf{p}\boldsymbol{\nabla}_{\mathbf{R}} + \frac{1}{2}\mathbf{p}^2\frac{\partial}{\partial \Theta}\right) G(x, \mathbf{R}; \Theta, \mathbf{p})\\ = \frac{1}{4}\left(A(0)\frac{\partial^2}{\partial \Theta^2} - \frac{1}{2}\Delta_{\mathbf{R}}A(0)\frac{\partial^2}{\partial \mathbf{p}^2}\right) G(x, \mathbf{R}; \Theta, \mathbf{p}).\end{aligned} \qquad (13.134)$$

In the case of the plane incident wave, $\boldsymbol{\nabla}_{\mathbf{R}}G(x, \mathbf{R}; \Theta, \mathbf{p}) = 0$ in view of assumed statistical homogeneity of all fields in plane $x = \text{const}$. This means that function $G(x, \mathbf{R}; \Theta, \mathbf{p}) = G(x; \Theta, \mathbf{p})$ is independent of $\mathbf{R}$ and Eq. (13.134) assumes the simpler form

$$\left(\frac{\partial}{\partial x} + \frac{1}{2}\mathbf{p}^2\frac{\partial}{\partial \Theta}\right) G(x, \mathbf{R}; \Theta, \mathbf{p}) = \frac{1}{4}\left(A(0)\frac{\partial^2}{\partial \Theta^2} - \frac{1}{2}\Delta_{\mathbf{R}}A(0)\frac{\partial^2}{\partial \mathbf{p}^2}\right) G(x, \mathbf{R}; \Theta, \mathbf{p}). \qquad (13.135)$$

A distinction of this equation consists in the fact that it allows deriving the closed finite-dimensional system of the first-order equations for quantities

$$\langle I(x, \mathbf{R})\Theta^m(x, \mathbf{R})\left|\boldsymbol{\nabla}_{\mathbf{R}}\Theta(x, \mathbf{R})\right|^n\rangle\,.$$

The solution of such a system presents usually no difficulties. The above expressions (13.133) are the special case of the solution to this system of equations.

We note that, if we integrate Eq. (13.134) over Θ, i.e., if we exclude wave phase from consideration, then we arrive at the equation

$$\left(\frac{\partial}{\partial x} + \mathbf{p}\boldsymbol{\nabla}_{\mathbf{R}}\right) G(x, \mathbf{R}; \mathbf{p}) = -\frac{1}{8}\Delta_{\mathbf{R}}A(0)\frac{\partial^2}{\partial \mathbf{p}^2}G(x, \mathbf{R}; \mathbf{p}), \qquad (13.136)$$

which coincides with the equation for the probability density describing the diffusion of a separate ray. This is quite natural, because the conversion from the Lagrangian coordinates to the Eulerian ones has the Jacobian $j(x) = 1/I(x)$.

Remark 15 ***Wigner function and geometrical optics approximation.***

Earlier, we introduced the Wigner function by the formula

$$W(x, \mathbf{R}, \mathbf{q}) = \frac{1}{(2\pi)^2}\int d\rho u\left(x, \mathbf{R} + \frac{1}{2}\rho\right) u^*\left(x, \mathbf{R} - \frac{1}{2}\rho\right) e^{-i\mathbf{q}\rho}$$

whose average coincides with the Fourier transform of the second-order coherence function. Using the amplitude–phase representation of the wavefield and performing limit process

$k \to \infty$ (in this limit, one should expand all functions in $\boldsymbol{\rho}$), we obtain the expression

$$W(x, \mathbf{R}, \mathbf{p}) = \frac{1}{(2\pi)^2} \int d\boldsymbol{\rho} A\left(x, \mathbf{R} + \frac{1}{2}\boldsymbol{\rho}\right) A\left(x, \mathbf{R} - \frac{1}{2}\boldsymbol{\rho}\right)$$

$$\times \exp\left\{i\left[S\left(x, \mathbf{R} + \frac{1}{2}\boldsymbol{\rho}\right) - S\left(x, \mathbf{R} - \frac{1}{2}\boldsymbol{\rho}\right) - \mathbf{p}\boldsymbol{\rho}\right]\right\}$$

$$= \frac{1}{(2\pi)^2} \int d\boldsymbol{\rho}\, I(x, \mathbf{R})\, e^{i\left[\boldsymbol{\rho}\frac{\partial}{\partial \mathbf{R}} S(x,\mathbf{R}) - \mathbf{p}\boldsymbol{\rho}\right]} = I(x, \mathbf{R})\, \delta\left(\frac{\partial}{\partial \mathbf{R}} S(x, \mathbf{R}) - \mathbf{p}\right).$$

Consequently, the geometrical optics approximation of the second-order coherence function coincides with function $G(x, \mathbf{R}; \mathbf{p})$. If we now define function $F(x, \mathbf{R}; \boldsymbol{\rho})$ as the Fourier transform of function $G(x, \mathbf{R}; \mathbf{p})$,

$$F(x, \mathbf{R}; \boldsymbol{\rho}) = \int d\mathbf{p} G(x, \mathbf{R}; \mathbf{p}) e^{i\mathbf{p}\boldsymbol{\rho}},$$

then we obtain from Eq. (13.136) that it satisfies the equation

$$\left(\frac{\partial}{\partial x} - i\frac{\partial^2}{\partial \mathbf{R}\partial \boldsymbol{\rho}}\right) G(x, \mathbf{R}; \boldsymbol{\rho}) = \frac{1}{8}\Delta_{\mathbf{R}} A(0) \boldsymbol{\rho}^2 G(x, \mathbf{R}; \boldsymbol{\rho}),$$

which coincides with the equation for the second-order coherence function (13.23) with function $D(\boldsymbol{\rho})$ expanded in the Taylor series in argument $\boldsymbol{\rho}$ [18, 134, 135, 252]. ♦

If we attempt to seek an equation for probability density

$$P(x, \mathbf{R}; I, \Theta, \mathbf{p}) = \langle \delta\left(I(x, \mathbf{R}) - I\right) \delta\left(\Theta(x, \mathbf{R}) - \Theta\right) \delta\left(\boldsymbol{\nabla}_{\mathbf{R}} \Theta(x, \mathbf{R}) - \mathbf{p}\right)\rangle$$

parametrically dependent on spatial point $(x, \mathbf{R})$, then we quickly arrive at the fact that no closed equation can be derived in this case. Nevertheless, we can close the equation for probability density if we supplement its variables I, Θ, and $\mathbf{p}$ with the symmetrical matrix of phase front curvatures

$$u_{ij}(x, \mathbf{R}) = \frac{\partial^2}{\partial R_i \partial R_j} \Theta(x, \mathbf{R})$$

whose components satisfy the equations

$$\left\{\frac{\partial}{\partial x} + \left(\boldsymbol{\nabla}_{\mathbf{R}} \Theta(x, \mathbf{R})\right) \boldsymbol{\nabla}_{\mathbf{R}}\right\} u_{ij}(x, \mathbf{R}) + u_{il}(x, \mathbf{R}) u_{lj}(x, \mathbf{R}) = \frac{1}{2}\frac{\partial^2}{\partial R_i \partial R_j} \varepsilon(x, \mathbf{R}). \quad (13.137)$$

The possibility of closing the equation follows from the fact that namely fluctuations of phase front curvature are responsible for wave intensity fluctuations in the geometrical optics approximation.

Introduce now the indicator function

$$W(x, \mathbf{R}; I, \Theta, \mathbf{p}, u_{ij})$$

$$= \delta\left(I(x, \mathbf{R}) - I\right) \delta\left(\Theta(x, \mathbf{R}) - \Theta\right) \delta\left(\frac{\partial \Theta(x, \mathbf{R})}{\partial \mathbf{R}} - \mathbf{p}\right) \delta\left(\frac{\partial^2 \Theta(x, \mathbf{R})}{\partial R_i \partial R_j} - u_{ij}\right)$$

governed by the stochastic Liouville equation of the form

$$\left(\frac{\partial}{\partial x} + \mathbf{p}\frac{\partial}{\partial \mathbf{R}} + \frac{\mathbf{p}^2}{2}\frac{\partial}{\partial \Theta} - u_{ii}\frac{\partial}{\partial I} I - \frac{\partial}{\partial u_{ik}} u_{il} u_{lk} - u_{ii}\right) W(x, \mathbf{R}; I, \Theta, \mathbf{p}, u_{ij})$$

$$= -\frac{1}{2}\left(\varepsilon(x, \mathbf{R})\frac{\partial}{\partial \Theta} + \frac{\partial \varepsilon(x, \mathbf{R})}{\partial \mathbf{R}}\frac{\partial}{\partial \mathbf{p}} + \frac{\partial^2 \varepsilon(x, \mathbf{R})}{\partial R_i \partial R_k}\frac{\partial}{\partial u_{ik}}\right) W(x, \mathbf{R}; I, \Theta, \mathbf{p}, u_{ij}).$$

(13.138)

Average Eq. (13.138) over an ensemble of realizations of field $\varepsilon(x, \mathbf{R})$. Using the Furutsu–Novikov formula to split the correlators, we obtain that the joint probability density of all quantities

$$P(x, \mathbf{R}; I, \Theta, \mathbf{p}, u_{ij}) = \langle W(x, \mathbf{R}; I, \Theta, \mathbf{p}, u_{ij}) \rangle$$

satisfies the equation

$$\begin{aligned}
&\left(\frac{\partial}{\partial x} + \mathbf{p}\frac{\partial}{\partial \mathbf{R}} + \frac{\mathbf{p}^2}{2}\frac{\partial}{\partial \Theta} - u_{ii}\frac{\partial}{\partial I}I - \frac{\partial}{\partial u_{ik}}u_{il}u_{lk} - u_{ii} \right) P(x, \mathbf{R}; I, \Theta, \mathbf{p}, u_{ij}) \\
&= \frac{1}{4}\left[A(0)\frac{\partial^2}{\partial \Theta^2} + \Delta_{\mathbf{R}} A(0) \left(\frac{\partial^2}{\partial \Theta \partial u_{ii}} - \frac{1}{2}\frac{\partial^2}{\partial \mathbf{p}^2} \right) \right. \\
&\qquad \left. - \frac{1}{8}\Delta^2_{\mathbf{R}} A(0) \left(2\frac{\partial^2}{\partial u^2_{kl}} + \frac{\partial^2}{\partial u^2_{ii}} \right) \right] P(x, \mathbf{R}; I, \Theta, \mathbf{p}, u_{ij}). \qquad (13.139)
\end{aligned}$$

In the case of the plane incident wave, $\frac{\partial}{\partial \mathbf{R}} P(x, \mathbf{R}; I, \Theta, \mathbf{p}, u_{ij}) = 0$ in view of the assumed spatial homogeneity, so that $P(x, \mathbf{R}; I, \Theta, \mathbf{p}, u_{ij}) = P(x; I, \Theta, \mathbf{p}, u_{ij})$.

Integrating Eq. (13.139) over Θ and I, we obtain a simpler equation

$$\begin{aligned}
&\left(\frac{\partial}{\partial x} - \frac{\partial}{\partial u_{ik}}u_{il}u_{lk} - u_{ii} \right) P(x; \mathbf{p}, u_{ij}) \\
&= -\frac{1}{8}\left[\Delta_{\mathbf{R}} A(0)\frac{\partial^2}{\partial \mathbf{p}^2} + \frac{1}{4}\Delta^2_{\mathbf{R}} A(0) \left(2\frac{\partial^2}{\partial u^2_{kl}} + \frac{\partial^2}{\partial u^2_{ii}} \right) \right] P(x; \mathbf{p}, u_{ij}) \qquad (13.140)
\end{aligned}$$

that governs the probability density of phase gradient fluctuations

$$P(x; \mathbf{p}, u_{ij}) = \left\langle \delta\left(\frac{\partial \Theta(x, \mathbf{R})}{\partial \mathbf{R}} - \mathbf{p} \right) \delta\left(\frac{\partial^2 \Theta(x, \mathbf{R})}{\partial R_i \partial R_j} - u_{ij} \right) \right\rangle.$$

Similarly, integrating Eq. (13.139) over Θ and $\mathbf{p}$, we obtain the equation

$$\begin{aligned}
&\left(\frac{\partial}{\partial x} - u_{ii}\frac{\partial}{\partial I}I - \frac{\partial}{\partial u_{ik}}u_{il}u_{lk} - u_{ii} \right) P(x; I, u_{ij}) \\
&= -\frac{1}{32}\Delta^2_{\mathbf{R}} A(0) \left(2\frac{\partial^2}{\partial u^2_{kl}} + \frac{\partial^2}{\partial u^2_{ii}} \right) P(x; I, u_{ij}) \qquad (13.141)
\end{aligned}$$

governing the probability density

$$P(x; I, u_{ij}) = \left\langle \delta\left(I(x, \mathbf{R}) - I \right) \delta\left(\frac{\partial^2 \Theta(x, \mathbf{R})}{\partial R_i \partial R_j} - u_{ij} \right) \right\rangle$$

and describing wavefield intensity fluctuations.

Integrating Eq. (13.141) over I, we obtain the equation

$$\left(\frac{\partial}{\partial x} - \frac{\partial}{\partial u_{ik}}u_{il}u_{lk} - u_{ii} \right) P(x; u_{ij}) = -\frac{1}{32}\Delta^2_{\mathbf{R}} A(0) \left(2\frac{\partial^2}{\partial u^2_{kl}} + \frac{\partial^2}{\partial u^2_{ii}} \right) P(x; u_{ij}) \qquad (13.142)$$

that governs probability density of the second derivatives of phase, i.e., probability density of phase front curvatures.

Comparing Eq. (13.142) with Eq. (13.140), we see that the first and second derivatives of the wave phase are statistically independent; in addition, probability density of phase gradient satisfies the equation

$$\frac{\partial}{\partial x}P(x;\mathbf{p}) = -\frac{1}{8}\Delta_{\mathbf{R}}A(0)\frac{\partial^2}{\partial \mathbf{p}^2}P(x;\mathbf{p}). \tag{13.143}$$

From Eq. (13.143) follows that the one-point distribution of quantity $\nabla_{\mathbf{R}}\Theta(x,\mathbf{R})$ is the Gaussian distribution with the variance

$$\left\langle [\nabla_{\mathbf{R}}\Theta(x,\mathbf{R})]^2 \right\rangle = -\frac{1}{8}\Delta_{\mathbf{R}}A(0)x,$$

which coincides with the known result obtained for small amplitude fluctuations and extends it to arbitrary amplitude fluctuations. At the same time, Eq. (13.141) shows a strong statistical relationship between intensity fluctuations and phase front curvature.

Equations (13.139)–(13.142) become significantly simpler in the two-dimensional case. For example, Eq. (13.141) assumes in this case the form

$$\left(\frac{\partial}{\partial x} - u\frac{\partial}{\partial I}I - \frac{\partial}{\partial u}u^2 - u\right)P(x;I,u) = -\frac{3}{32}\frac{\partial^2 A(0)}{\partial y^2}\frac{\partial^2}{\partial u^2}P(x;I,u). \tag{13.144}$$

Nevertheless, the above equations are very complicated and little-studied.

13.3 Method of path integral

Here, we consider statistical description of characteristics of the wavefield in random medium on the basis of problem solution in the functional form (i.e., in the form of the path integral) [40, 54, 55, 134, 135, 168], [297] – [299], [318].

As earlier, we will describe wave propagation in random medium by the parabolic equation (13.1), page 355 whose solution can be represented in the operator form, or in the form of the path integral.

To obtain such a representation, we replace Eq. (13.1) with a more complicated one with an arbitrary deterministic vector function $\mathbf{v}(x)$; namely, we consider the equation

$$\begin{aligned}\frac{\partial}{\partial x}\Phi(x,\mathbf{R}) &= \frac{i}{2k}\Delta_{\mathbf{R}}\Phi(x,\mathbf{R}) + i\frac{k}{2}\varepsilon(x,\mathbf{R})\Phi(x,\mathbf{R}) + \mathbf{v}(x)\nabla_{\mathbf{R}}\Phi(x,\mathbf{R}),\\ \Phi(0,\mathbf{R}) &= u_0(\mathbf{R}).\end{aligned} \tag{13.145}$$

The solution to the original parabolic equation (13.1) is then obtained by the formula

$$u(x,\mathbf{R}) = \Phi(x,\mathbf{R})|_{\mathbf{v}(x)=0}. \tag{13.146}$$

In the standard way, we obtain the expression for the variational derivative $\frac{\delta\Phi(x,\mathbf{R})}{\delta\mathbf{v}(x-0)}$

$$\frac{\delta\Phi(x,\mathbf{R})}{\delta\mathbf{v}(x-0)} = \nabla_{\mathbf{R}}\Phi(x,\mathbf{R}), \tag{13.147}$$

and rewrite Eq. (13.145) in the form

$$\frac{\partial}{\partial x}\Phi(x,\mathbf{R}) = \frac{i}{2k}\frac{\delta^2\Phi(x,\mathbf{R})}{\delta\mathbf{v}^2(x-0)} + i\frac{k}{2}\varepsilon(x,\mathbf{R})\Phi(x,\mathbf{R}) + \mathbf{v}(x)\nabla_{\mathbf{R}}\Phi(x,\mathbf{R}). \tag{13.148}$$

We will seek the solution to Eq. (13.148) in the form

$$\Phi(x,\mathbf{R}) = e^{\frac{i}{2k}\int\limits_0^{x-0} d\xi \frac{\delta^2}{\delta \mathbf{v}^2(\xi)}} \varphi(x,\mathbf{R}). \tag{13.149}$$

Because the operator in the exponent of Eq. (13.149) commutes with function $\mathbf{v}(x)$, we obtain that function $\varphi(x,\mathbf{R})$ satisfies the first-order equation

$$\frac{\partial}{\partial x}\varphi(x,\mathbf{R}) = i\frac{k}{2}\varepsilon(x,\mathbf{R})\varphi(x,\mathbf{R}) + \mathbf{v}(x)\boldsymbol{\nabla}_{\mathbf{R}}\varphi(x,\mathbf{R}), \quad \varphi(0,\mathbf{R}) = u_0(\mathbf{R}), \tag{13.150}$$

whose solution as a functional of $\mathbf{v}(\xi)$ has the following form

$$\begin{aligned}\varphi(x,\mathbf{R}) &= \varphi\left[x,\mathbf{R};\mathbf{v}(\xi)\right] \\ &= u_0\left(\mathbf{R} + \int\limits_0^x d\xi \mathbf{v}(\xi)\right)\exp\left\{i\frac{k}{2}\int\limits_0^x d\xi\varepsilon\left(\xi,\mathbf{R} + \int\limits_\xi^x d\eta \mathbf{v}(\eta)\right)\right\}.\end{aligned} \tag{13.151}$$

As a consequence, taking into account Eqs. (13.149) and (13.146), we obtain the solution to the parabolic equation (13.1) in the operator form

$$\begin{aligned}u(x,\mathbf{R}) &= \exp\left\{\frac{i}{2k}\int\limits_0^x d\xi \frac{\delta^2}{\delta \mathbf{v}^2(\xi)}\right\} \\ &\times u_0\left(\mathbf{R} + \int\limits_0^x d\xi \mathbf{v}(\xi)\right)\exp\left\{i\frac{k}{2}\int\limits_0^x d\xi\varepsilon\left(\xi,\mathbf{R} + \int\limits_\xi^x d\eta \mathbf{v}(\eta)\right)\right\}\Bigg|_{\mathbf{v}(x)=0}.\end{aligned} \tag{13.152}$$

In the case of the plane incident wave, we have $u_0(\mathbf{R}) = u_0$, and Eq. (13.152) is simplified

$$u(x,\mathbf{R}) = u_0 e^{\frac{i}{2k}\int\limits_0^x d\xi \frac{\delta^2}{\delta \mathbf{v}^2(\xi)}} \exp\left\{i\frac{k}{2}\int\limits_0^x d\xi\varepsilon\left(\xi,\mathbf{R} + \int\limits_\xi^x d\eta \mathbf{v}(\eta)\right)\right\}\Bigg|_{\mathbf{v}(x)=0}. \tag{13.153}$$

Now, we formally consider Eq. (13.150) as the stochastic equation in which function $\mathbf{v}(x)$ is assumed the 'Gaussian' random vector function with the zero-valued mean and the imaginary 'correlation' function

$$\langle v_i(x)v_j(x')\rangle = \frac{i}{k}\delta_{ij}\delta(x-x'). \tag{13.154}$$

One can easily check that all formulas valid for the Gaussian random processes hold in this case, too.

Averaging Eq. (13.150) over an ensemble of realizations of random process $\mathbf{v}(x)$, we obtain that average function $\langle\varphi(x,\mathbf{R})\rangle_{\mathbf{v}}$ satisfies the equation that coincides with Eq. (13.1). Thus the solution to parabolic equation (13.1) can be treated in the probabilistic sense; namely, we can formally represent this solution as the following average

$$u(x,\mathbf{R}) = \langle\varphi\left[x,\mathbf{R};\mathbf{v}(\xi)\right]\rangle_{\mathbf{v}}. \tag{13.155}$$

This expression can be represented in the form of the *Feynman path integral*

$$u(x,\mathbf{R}) = \int D\mathbf{v}(x) u_0\left(\mathbf{R} + \int_0^x d\xi \mathbf{v}(\xi)\right)$$
$$\times \exp\left\{i\frac{k}{2}\int_0^x d\xi\left[\mathbf{v}^2(\xi) + \varepsilon\left(\xi, \mathbf{R} + \int_\xi^x d\eta \mathbf{v}(\eta)\right)\right]\right\}, \tag{13.156}$$

where the integral measure $D\mathbf{v}(x)$ is defined as follows

$$D\mathbf{v}(x) = \frac{\prod\limits_{\xi=0}^{x} d\mathbf{v}(\xi)}{\int \ldots \int \prod\limits_{\xi=0}^{x} d\mathbf{v}(\xi) \exp\left\{i\frac{k}{2}\int_0^x d\xi \mathbf{v}^2(\xi)\right\}}.$$

Representations (13.152) and (13.155) are equivalent. Indeed, considering the solution to Eq. (13.150) as a functional of random process $\mathbf{v}(\xi)$, we can reduce Eq. (13.155) to the following chain of equalities

$$u(x,\mathbf{R}) = \left\langle \varphi\left[x, \mathbf{R}; \mathbf{v}(\xi) + \mathbf{y}(\xi)\right]\right\rangle_{\mathbf{v}}\Big|_{\mathbf{y}=0}$$
$$= \left\langle e^{\int_0^x d\xi \mathbf{v}(\xi)\frac{\delta}{\delta \mathbf{y}(\xi)}}\right\rangle_{\mathbf{v}} \varphi\left[x, \mathbf{R}; \mathbf{y}(\xi)\right]\Bigg|_{\mathbf{y}=0} = e^{\frac{i}{2k}\int_0^x d\xi \frac{\delta^2}{\delta \mathbf{y}^2(\xi)}} \varphi\left[x, \mathbf{R}; \mathbf{y}(\xi)\right]\Bigg|_{\mathbf{y}=0}$$

and, consequently, to the operator form (13.152).

We can rewrite Eq. (13.155) in a more convenient form using probabilistic similarity. First, we represent Eq. in the form

$$u(x,\mathbf{R}) = \left\langle \varphi\left[x, \mathbf{R}; \mathbf{v}(\xi)\right]\right\rangle_{\mathbf{v}}$$
$$= \int d\mathbf{q}\tilde{u}_0(\mathbf{q}) e^{i\mathbf{q}\mathbf{R}} \left\langle \exp\left\{i\mathbf{q}\int_0^x d\xi \mathbf{v}(\xi) + i\frac{k}{2}\int_0^x d\xi \varepsilon\left(\xi, \mathbf{R} + \int_\xi^x d\eta \mathbf{v}(\eta)\right)\right\}\right\rangle_{\mathbf{v}},$$

where

$$\tilde{u}_0(\mathbf{q}) = \frac{1}{(2\pi)^2}\int d\mathbf{R} u_0(\mathbf{R}) e^{-i\mathbf{q}\mathbf{R}}.$$

Then, we can take the exponent outside averaging brackets (see Eq. (4.18), page 81). As a result, we obtain the expression

$$u(x,\mathbf{R}) = \int d\mathbf{q}\tilde{u}_0(\mathbf{q}) e^{i\left(\mathbf{q}\mathbf{R} - \frac{\mathbf{q}^2}{2k}x\right)} \psi(x, \mathbf{R}, \mathbf{q}), \tag{13.157}$$

where function

$$\psi(x, \mathbf{R}, \mathbf{q}) = \left\langle \exp\left\{i\frac{k}{2}\int_0^x d\xi \varepsilon\left(\xi, \mathbf{R} + \int_\xi^x d\eta \left[\mathbf{v}(\eta) - \frac{\mathbf{q}}{k}\right]\right)\right\}\right\rangle_{\mathbf{v}} \tag{13.158}$$

also can be represented in the operator form

$$\psi(x,\mathbf{R},\mathbf{q}) = e^{\frac{i}{2k}\int\limits_0^x d\xi \frac{\delta^2}{\delta \mathbf{v}^2(\xi)}} \exp\left\{i\frac{k}{2}\int\limits_0^x d\xi\varepsilon\left(\xi,\mathbf{R}+\int\limits_\xi^x d\eta\left[\mathbf{v}(\eta)-\frac{\mathbf{q}}{k}\right]\right)\right\}_{\mathbf{v}=0}. \tag{13.159}$$

Expressions (13.158) and (13.159) form the solution to the differential equation

$$\begin{aligned}\left(\frac{\partial}{\partial x}-\frac{i}{2k}\Delta_{\mathbf{R}}\right)\psi(x,\mathbf{R},\mathbf{q}) &= i\frac{k}{2}\varepsilon(x,\mathbf{R})\psi(x,\mathbf{R},\mathbf{q})-\frac{\mathbf{q}}{k}\nabla_{\mathbf{R}}\psi(x,\mathbf{R},\mathbf{q}),\\ \psi(0,\mathbf{R},\mathbf{q}) &= 1,\end{aligned} \tag{13.160}$$

which could be derived immediately from parabolic equation (13.1). In such a derivation, Eqs. (13.157)–(13.159) represent the decomposition of the solution in plane waves. The integrand of the right-hand side in Eq. (13.157) describes the plane wave diffraction on the inhomogeneities of field $\varepsilon(x,\mathbf{R})$, factor $\tilde{u}_0(\mathbf{q})\exp\left\{i\mathbf{qR}-\frac{i\mathbf{q}^2}{2k}x\right\}$ being responsible for the diffraction in free space (for $\varepsilon(x,\mathbf{R})=0$) and factor $\psi(x,\mathbf{R},\mathbf{q})$ being responsible for the effect of inhomogeneities on the wave diffracted in free space.

In closing, we give the expressions for Green's function of Eq. (13.1), i.e., for the field of the spherical wave corresponding to the initial condition $u(x',\mathbf{R})=\delta(\mathbf{R}-\mathbf{R}')$ at point $x=x'$

$$G(x,\mathbf{R};x',\mathbf{R}') = e^{\frac{i}{2k}\int\limits_{x'}^x d\xi \frac{\delta^2}{\delta \mathbf{v}^2(\xi)}} \times\left[\delta\left(\mathbf{R}-\mathbf{R}'+\int\limits_{x'}^x d\xi\mathbf{v}(\xi)\right)\exp\left\{i\frac{k}{2}\int\limits_{x'}^x d\xi\varepsilon\left(\xi,\mathbf{R}+\int\limits_\xi^x d\eta\mathbf{v}(\eta)\right)\right\}\right]_{\mathbf{v}=0}, \tag{13.161}$$

$$G(x,\mathbf{R};x',\mathbf{R}') = \int D\mathbf{v}(x)\delta\left(\mathbf{R}-\mathbf{R}'+\int\limits_{x'}^x d\xi\mathbf{v}(\xi)\right) \times\exp\left\{i\frac{k}{2}\int\limits_{x'}^x d\xi\left[\mathbf{v}^2(\xi)+\varepsilon\left(\xi,\mathbf{R}+\int\limits_\xi^x d\eta\mathbf{v}(\eta)\right)\right]\right\}. \tag{13.162}$$

The complex conjugated formulas specify Green's function in the form of the spherical wave propagating in the negative direction of the x-axis.

13.3.1 Statistical description of wavefield

Consider now the statistical description of the wavefield propagating in a medium with random inhomogeneities. We will assume that random field $\varepsilon(x,\mathbf{R})$ is the homogeneous and isotropic Gaussian field with the correlation function

$$B_\varepsilon(x,\mathbf{R};x',\mathbf{R}') = B_\varepsilon(x-x',\mathbf{R}-\mathbf{R}') = \langle\varepsilon(x,\mathbf{R})\varepsilon(x',\mathbf{R}')\rangle. \tag{13.163}$$

Averaging Eq. (13.157) over an ensemble of realizations of field $\varepsilon(x,\mathbf{R})$, we obtain average field in the form

$$\langle u(x,\mathbf{R})\rangle = \int d\mathbf{q}\tilde{u}_0(\mathbf{q})e^{i\left(\mathbf{qR}-\frac{\mathbf{q}^2}{2k}x\right)}\langle\psi(x,\mathbf{R},\mathbf{q})\rangle, \tag{13.164}$$

where function

$$
\begin{aligned}
&\langle\psi(x,\mathbf{R},\mathbf{q})\rangle \\
&= \left\langle \exp\left\{-\frac{k^2}{8}\int_0^x d\xi_1 \int_0^x d\xi_2 B_\varepsilon\left(\xi_1-\xi_2, \int_{\xi_1}^{\xi_2} d\eta\left[\mathbf{v}(\eta)-\frac{\mathbf{q}}{k}\right]\right)\right\}\right\rangle_{\mathbf{v}}
\end{aligned} \tag{13.165}
$$

can be represented in the operator form as follows

$$
\begin{aligned}
\langle\psi(x,\mathbf{R},\mathbf{q})\rangle &= \exp\left\{\frac{i}{2k}\int_0^x d\xi \frac{\delta^2}{\delta\mathbf{v}^2(\xi)}\right\} \\
&\times \exp\left\{-\frac{k^2}{8}\int_0^x d\xi_1 \int_0^x d\xi_2 B_\varepsilon\left(\xi_1-\xi_2, \int_{\xi_1}^{\xi_2} d\eta\left[\mathbf{v}(\eta)-\frac{\mathbf{q}}{k}\right]\right)\right\}_{\mathbf{v}=0}.
\end{aligned} \tag{13.166}
$$

The integral representation of the second-order coherence function can be obtained similarly; it has the form

$$
\begin{aligned}
\Gamma_2(x;\mathbf{R}_1,\mathbf{R}_2) &= \int d\mathbf{q}_1 \int d\mathbf{q}_2 \tilde{u}_0(\mathbf{q}_1)\tilde{u}_0(\mathbf{q}_2) \\
&\times \exp\left\{i\left(\mathbf{q}_1\mathbf{R}_1-\mathbf{q}_2\mathbf{R}_2\right) - \frac{i\left(\mathbf{q}_1^2-\mathbf{q}_2^2\right)}{2k}x\right\}\langle\psi(x,\mathbf{R}_1,\mathbf{q}_1)\psi^*(x,\mathbf{R}_2,\mathbf{q}_2)\rangle,
\end{aligned} \tag{13.167}
$$

where

$$
\begin{aligned}
&\langle\psi(x,\mathbf{R}_1,\mathbf{q}_1)\psi^*(x,\mathbf{R}_2,\mathbf{q}_2)\rangle \\
&= \left\langle \exp\left\{-\frac{k^2}{8}\int_0^x d\xi_1 \int_0^x d\xi_2 \left[B_\varepsilon\left(\xi_1-\xi_2, \int_{\xi_1}^{\xi_2} d\eta\left[\mathbf{v}_1(\eta)-\frac{\mathbf{q}_1}{k}\right]\right)\right.\right.\right. \\
&\quad + B_\varepsilon\left(\xi_1-\xi_2, \int_{\xi_1}^{\xi_2} d\eta\left[\mathbf{v}_2(\eta)-\frac{\mathbf{q}_2}{k}\right]\right) \\
&\quad \left.\left.\left. - 2B_\varepsilon\left(\xi_1-\xi_2, \mathbf{R}_1-\mathbf{R}_2+\int_{\xi_1}^{x} d\eta\left[\mathbf{v}_1(\eta)-\frac{\mathbf{q}_1}{k}\right]-\int_{\xi_2}^{x} d\eta\left[\mathbf{v}_2(\eta)-\frac{\mathbf{q}_2}{k}\right]\right)\right]\right\}\right\rangle_{\mathbf{v}_i},
\end{aligned} \tag{13.168}
$$

or, in the operator form,

$$\begin{aligned}&\langle\psi(x,\mathbf{R}_1,\mathbf{q}_1)\psi^*(x,\mathbf{R}_2,\mathbf{q}_2)\rangle\\&=\exp\left\{\frac{i}{2k}\int\limits_0^x d\xi\left[\frac{\delta^2}{\delta\mathbf{v}_1^2(\xi)}-\frac{\delta^2}{\delta\mathbf{v}_2^2(\xi)}\right]\right\}\\&\times\exp\left\{-\frac{k^2}{8}\int\limits_0^x d\xi_1\int\limits_0^x d\xi_2\left[B_\varepsilon\left(\xi_1-\xi_2,\int\limits_{\xi_1}^{\xi_2}d\eta\left[\mathbf{v}_1(\eta)-\frac{\mathbf{q}_1}{k}\right]\right)\right.\right.\\&+B_\varepsilon\left(\xi_1-\xi_2,\int\limits_{\xi_1}^{\xi_2}d\eta\left[\mathbf{v}_2(\eta)-\frac{\mathbf{q}_2}{k}\right]\right)\\&\left.\left.-\,2B_\varepsilon\left(\xi_1-\xi_2,\mathbf{R}_1-\mathbf{R}_2+\int\limits_{\xi_1}^{x}d\eta\left[\mathbf{v}_1(\eta)-\frac{\mathbf{q}_1}{k}\right]-\int\limits_{\xi_2}^{x}d\eta\left[\mathbf{v}_2(\eta)-\frac{\mathbf{q}_2}{k}\right]\right)\right]\right\}_{\mathbf{v}_i=0}.\end{aligned}\tag{13.169}$$

Unfortunately, the way of calculating path integrals (13.165) and (13.168) or corresponding operator expressions (13.166) and (13.169) is not known at the moment, and we are forced to resort to simplifying assumptions. For example, these integrals can be calculated if we assume that approximation (13.15)

$$B_\varepsilon(x,\mathbf{R})=\delta(x)A(\mathbf{R}),\quad A(\mathbf{R})=\int\limits_{-\infty}^{\infty}dxB_\varepsilon(x,\mathbf{R})$$

holds for the correlation function of field $\varepsilon(x,\mathbf{R})$, i.e., if we assume that field $\varepsilon(x,\mathbf{R})$ is delta-correlated in x. The operator form of these expressions appears more convenient for corresponding calculations.

In this case, we easily obtain the expression for function $\langle\psi(x,\mathbf{R},\mathbf{q})\rangle$,

$$\langle\psi(x,\mathbf{R},\mathbf{q})\rangle=e^{\frac{i}{2k}\int\limits_0^x d\xi\frac{\delta^2}{\delta\mathbf{v}^2(\xi)}}e^{-\frac{k^2}{8}A(0)x}\Big|_{v=0}=e^{-\frac{k^2}{8}A(0)x},$$

and Eq. (13.164) coincides with Eq. (13.27), page 360 obtained immediately by averaging stochastic parabolic equation (13.1), page 355, which is quite natural.

In the context of Eq. (13.169), we obtain similarly

$$\begin{aligned}&\langle\psi(x,\mathbf{R}_1,\mathbf{q}_1)\psi^*(x,\mathbf{R}_2,\mathbf{q}_2)\rangle=e^{\frac{i}{2k}\int\limits_0^x d\xi\left[\frac{\delta^2}{\delta\mathbf{v}_1^2(\xi)}-\frac{\delta^2}{\delta\mathbf{v}_2^2(\xi)}\right]}\\&\times\exp\left\{-\frac{k^2}{4}\int\limits_0^x d\xi D\left(\mathbf{R}_1-\mathbf{R}_2+\int\limits_\xi^x d\eta\left[\mathbf{v}_1(\eta)-\mathbf{v}_2(\eta)-\frac{\mathbf{q}_1-\mathbf{q}_2}{k}\right]\right)\right\}_{\mathbf{v}_i=0},\end{aligned}$$

where $D(\mathbf{R})=A(0)-A(\mathbf{R})$ as earlier.

Changing functional variables

$$\mathbf{v}_1(x)-\mathbf{v}_2(x)=\mathbf{v}(x),\quad\mathbf{v}_1(x)+\mathbf{v}_2(x)=2\mathbf{V}(x)$$

and introducing new variables

$$\mathbf{R}_1 - \mathbf{R}_2 = \boldsymbol{\rho}, \quad \mathbf{R}_1 + \mathbf{R}_2 = 2\mathbf{R}, \quad \mathbf{q}_1 - \mathbf{q}_2 = \mathbf{q},$$

we can rewrite last expression in the form

$$\langle \psi(x, \mathbf{R}_1, \mathbf{q}_1)\psi^*(x, \mathbf{R}_2, \mathbf{q}_2)\rangle = e^{\frac{i}{k}\int\limits_0^x d\xi \frac{\delta^2}{\delta \mathbf{v}(\xi)\delta \mathbf{V}(\xi)}}$$

$$\times \exp\left\{-\frac{k^2}{4}\int\limits_0^x d\xi D\left(\boldsymbol{\rho} + \int\limits_\xi^x d\eta \left[\mathbf{v}(\eta) - \frac{\mathbf{q}}{k}\right]\right)\right\}_{\mathbf{v}_i=0} = e^{-\frac{k^2}{4}\int\limits_0^x d\xi D\left(\boldsymbol{\rho} - \frac{\mathbf{q}}{k}(x-\xi)\right)}, \tag{13.170}$$

from which follows that the second-order coherence function is given by the expression coinciding with Eq. (13.28), page 361.

In terms of average field and second-order coherence function, the method of path integral (or the operator method) is equivalent to direct averaging of stochastic equations. It seems however essential that the operator method (or the method of path integral) offers a possibility of obtaining expressions for quantities that cannot be described in terms of closed equations (among which are, for example, the expressions related to wave intensity fluctuations). Indeed, we can derive the closed equation for the fourth-order coherence function

$$\Gamma_4(x; \mathbf{R}_1, \mathbf{R}_2, \mathbf{R}_3, \mathbf{R}_4) = \langle u(x, \mathbf{R}_1)u(x, \mathbf{R}_2)u^*(x, \mathbf{R}_3)u^*(x, \mathbf{R}_4)\rangle$$

and then determine quantity $\langle I^2(x, \mathbf{R})\rangle$ by setting $\mathbf{R}_1 = \mathbf{R}_2 = \mathbf{R}_3 = \mathbf{R}_4 = \mathbf{R}$ in the solution. However, this equation cannot be solved in the analytic form; moreover, it includes many parameters unnecessary for determining $\langle I^2(x, \mathbf{R})\rangle$, whereas the path integral representation of quantity $\langle I^2(x, \mathbf{R})\rangle$ includes no such parameters. Therefore, the path integral representation of problem solution can be useful for studying asymptotic characteristics of arbitrary moments and—as a consequence—probability distribution of wavefield intensity. In addition, the operator representation of the field sometimes makes it possible to simplify the determination of the desired average characteristics as compared with the analysis of the corresponding equations. For example, if we would desire to calculate the quantity

$$\langle \varepsilon(y, \mathbf{R}_1)I(x, \mathbf{R})\rangle \quad (y < x),$$

then, starting from Eq. (13.1), we should first derive the differential equation for quantity $\varepsilon(y, \mathbf{R}_1)u(x, \mathbf{R}_2)u^*(x, \mathbf{R}_3)$ for $y < x$, average it over an ensemble of realizations of field $\varepsilon(x, \mathbf{R})$, specify boundary condition for quantity $\langle \varepsilon(y, \mathbf{R}_1)u(x, \mathbf{R}_2)u^*(x, \mathbf{R}_3)\rangle$ at $x = y$, solve the obtained equation with this boundary condition, and only then set $\mathbf{R}_2 = \mathbf{R}_3 = \mathbf{R}$. At the same time, the calculation of this quantity in terms of the operator representation only slightly differs from the above calculation of quantity $\langle \psi\psi^*\rangle$.

Now, we turn to the analysis of asymptotic behavior of plane wave intensity fluctuations in random medium in the region of strong fluctuations. In this analysis, we will adhere to works [134, 135, 318].

13.3.2 Asymptotic analysis of plane wave intensity fluctuations

Consider statistical moment of field $u(x, \mathbf{R})$

$$M_{nn}(x, \mathbf{R}_1, \ldots, \mathbf{R}_{2n}) = \left\langle \prod_{k=1}^{n} u(x, \mathbf{R}_{2k-1})u^*(x, \mathbf{R}_{2k})\right\rangle. \tag{13.171}$$

In the approximation of the delta-correlated field $\varepsilon(x,\mathbf{R})$, function $M_{nn}(x,\mathbf{R}_1,\ldots,\mathbf{R}_{2n})$ satisfies Eq. (13.18), page 359 for $n=m$. In the case of the plane incident wave, this is the equation with the initial condition, which assumes the form (in variables $\mathbf{R}_k$)

$$\left(\frac{\partial}{\partial x}-\frac{i}{2k}\sum_{l=1}^{2n}(-1)^{l+1}\Delta_{\mathbf{R}_l}\right)M_{nn}(x,\mathbf{R}_1,\ldots,\mathbf{R}_{2n})$$
$$=\frac{k^2}{8}\sum_{l,j=1}^{2n}(-1)^{l+j}D(\mathbf{R}_l-\mathbf{R}_j)M_{nn}(x,\mathbf{R}_1,\ldots,\mathbf{R}_{2n}),\qquad(13.172)$$

where

$$D(\mathbf{R})=A(0)-A(\mathbf{R})=2\pi\int d\mathbf{q}\Phi_\varepsilon(0,\mathbf{q})\left[1-\cos(\mathbf{qR})\right]\qquad(13.173)$$

and $\Phi_\varepsilon(0,\mathbf{q})$ is the three-dimensional spectrum of field $\varepsilon(x,\mathbf{R})$ whose argument is the two-dimensional vector $\mathbf{q}$.

Using the path integral representation of field $u(x,\mathbf{R})$ (13.156), page 394 and averaging it over field $\varepsilon(x,\mathbf{R})$, we obtain the expression for $M_{nn}(x,\mathbf{R}_1,\ldots,\mathbf{R}_{2n})$ in the form

$$M_{nn}(x,\mathbf{R}_1,\ldots,\mathbf{R}_{2n})=\int\ldots\int D\mathbf{v}_1(\xi)\ldots D\mathbf{v}_{2n}(\xi)$$
$$\times\exp\left\{\frac{ik}{2x}\sum_{j=1}^{2n}(-1)^{j+1}\int\limits_0^x d\xi\mathbf{v}_j^2(\xi)\right.$$
$$\left.-\frac{k^2}{8}\sum_{j,l=1}^{2n}(-1)^{j+l+1}\int\limits_0^x d\xi D\left(\mathbf{R}_j-\mathbf{R}_l+\int\limits_\xi^x dx'\left[\mathbf{v}_j(x')-\mathbf{v}_l(x')\right]\right)\right\}.\qquad(13.174)$$

Another way of obtaining Eq. (13.174) consists in solving Eq. (13.172) immediately, by the method described earlier. We can rewrite Eq. (13.174) in the operator form

$$M_{nn}(x,\mathbf{R}_1,\ldots,\mathbf{R}_{2n})=\prod_{l=1}^{2n}\exp\left\{\frac{i}{2k}(-1)^{l+1}\int\limits_0^x d\xi\frac{\delta^2}{\delta\mathbf{v}^2(\xi)}\right\}$$
$$\times\exp\left\{-\frac{k^2}{8}\sum_{j,l=1}^{2n}(-1)^{j+l+1}\int\limits_0^x dx'D\left(\mathbf{R}_j-\mathbf{R}_l+\int\limits_{x'}^x d\xi\left[\mathbf{v}_j(\xi)-\mathbf{v}_l(\xi)\right]\right)\right\}_{\mathbf{v}=0}.\qquad(13.175)$$

If we now superpose points $\mathbf{R}_{2k-1}$ and $\mathbf{R}_{2k}$ (i.e., if we set $\mathbf{R}_{2k-1}=\mathbf{R}_{2k}$), then function $M_{nn}(x,\mathbf{R}_1,\ldots,\mathbf{R}_{2n})$ will grade into the function $\left\langle\prod\limits_{k=1}^{n}I(x,\mathbf{R}_{2k-1})\right\rangle$ that describes correlation characteristics of wave intensity. Further, if we set all $\mathbf{R}_l$ equal ($\mathbf{R}_l=\mathbf{R}$), then function

$$M_{nn}(x,\mathbf{R},\ldots,\mathbf{R})=\Gamma_{2n}(x,\mathbf{R})=\langle I^n(x,\mathbf{R})\rangle$$

will describe the n-th moment of wavefield intensity.

Prior to discuss the asymptotics of functions $\Gamma_{2n}(x,\mathbf{R})$ for the continuous random medium, we consider a simpler problem on wavefield fluctuations behind the random phase screen.

Random phase screen

Suppose that we deal with the inhomogeneous medium layer whose thickness is so small that the wave traversed this layer acquires only random phase incursion

$$S(\mathbf{R}) = \frac{k}{2}\int\limits_{0}^{\Delta x} d\xi\varepsilon(\xi,\mathbf{R}), \tag{13.176}$$

the amplitude remaining intact. As earlier, we will assume that random field $\varepsilon(x,\mathbf{R})$ is the Gaussian field delta correlated in x. After traversing the inhomogeneous layer, the wave is propagating in homogeneous medium and its propagation is governed by the equation (13.1) with $\varepsilon(x,\mathbf{R}) = 0$. The solution to this problem is given by the formulas

$$u(x,\mathbf{R}) = e^{i\frac{x}{2k}\Delta_{\mathbf{R}}}e^{iS(\mathbf{R})} = \frac{k}{2\pi i x}\int d\mathbf{v}\exp\left\{\frac{ik}{2x}\mathbf{v}^2 + iS(\mathbf{R}+\mathbf{v})\right\}, \tag{13.177}$$

which are the finite-dimensional analogs of Eqs. (13.153) and (13.156).

Consider function $M_{nn}(x,\mathbf{R}_1,\ldots,\mathbf{R}_{2n})$. Substituting Eq. (13.177) in Eq. (13.171) and averaging the result, we easily obtain the formula

$$\begin{aligned}M_{nn}(x,\mathbf{R}_1,\ldots,\mathbf{R}_{2n}) &= \left(\frac{k}{2\pi x}\right)^{2n}\int\ldots\int d\mathbf{v}_1\ldots d\mathbf{v}_{2n}\\ &\times\exp\left\{\frac{ik}{2x}\sum_{j=1}^{2n}(-1)^{j+1}\mathbf{v}_j^2 - \frac{k^2\Delta x}{8}\sum_{j,l=1}^{2n}(-1)^{j+l+1}D\left(\mathbf{R}_j-\mathbf{R}_l+\mathbf{v}_j-\mathbf{v}_l\right)\right\},\end{aligned} \tag{13.178}$$

which is analogous to Eq. (13.174).

First of all, we consider in greater detail the case with $n = 2$ for superimposed pairs of observation points

$$\mathbf{R}_1 = \mathbf{R}_2 = \mathbf{R}', \quad \mathbf{R}_3 = \mathbf{R}_4 = \mathbf{R}'', \quad \mathbf{R}'-\mathbf{R}'' = \boldsymbol{\rho}.$$

In this case, function

$$\Gamma_4(x;\mathbf{R}',\mathbf{R}',\mathbf{R}'',\mathbf{R}'') = \langle I(x,\mathbf{R}')I(x,\mathbf{R}'')\rangle$$

is the covariation of intensities $I(x,\mathbf{R}) = |u(x,\mathbf{R})|^2$. If we consider Eq. (13.178) for $n = 2$ and introduce new integration variables

$$\mathbf{v}_1-\mathbf{v}_2 = \mathbf{R}_1, \quad \mathbf{v}_1-\mathbf{v}_4 = \mathbf{R}_2, \quad \mathbf{v}_1-\mathbf{v}_3 = \mathbf{R}_3, \quad \frac{1}{2}(\mathbf{v}_1+\mathbf{v}_2) = \mathbf{R},$$

then we can perform integrations over $\mathbf{R}$ and $\mathbf{R}_3$ to obtain the simpler formula

$$\begin{aligned}&\langle I(x,\mathbf{R}')I(x,\mathbf{R}'')\rangle\\ &= \left(\frac{k}{2\pi x}\right)^2\int\int d\mathbf{R}_1 d\mathbf{R}_2\exp\left\{\frac{ik}{x}\mathbf{R}_1\left(\mathbf{R}_2-\boldsymbol{\rho}\right) - \frac{k^2\Delta x}{4}F\left(\mathbf{R}_1,\mathbf{R}_2\right)\right\},\end{aligned} \tag{13.179}$$

where $\boldsymbol{\rho} = \mathbf{R}'-\mathbf{R}''$ and function $F(\mathbf{R}_1,\mathbf{R}_2)$ is determined from Eq. (13.32),

$$\begin{aligned}F\left(\mathbf{R}_1,\mathbf{R}_2\right) &= 2D\left(\mathbf{R}_1\right)+2D\left(\mathbf{R}_2\right)-D\left(\mathbf{R}_1+\mathbf{R}_2\right)-D\left(\mathbf{R}_1-\mathbf{R}_2\right),\\ D\left(\mathbf{R}\right) &= A(0)-A\left(\mathbf{R}\right).\end{aligned}$$

The integral in Eq. (13.179) was studied in detail (including numerical methods) in many works. Its asymptotics for $x \to \infty$ has the form

$$\langle I(x,\mathbf{R}')I(x,\mathbf{R}'')\rangle = 1 + \exp\left\{-\frac{k^2\Delta x}{2}D(\boldsymbol{\rho})\right\}$$
$$+\pi k^2\Delta x\int d\mathbf{q}\Phi_\varepsilon(\mathbf{q})\left[1-\cos\frac{\mathbf{q}^2 x}{k}\right]\exp\left\{i\mathbf{q}\boldsymbol{\rho}-\frac{k^2\Delta x}{2}D\left(\frac{\mathbf{q}x}{k}\right)\right\}$$
$$+\pi k^2\Delta x\int d\mathbf{q}\Phi_\varepsilon(\mathbf{q})\left[1-\cos\left(\mathbf{q}\boldsymbol{\rho}-\frac{\mathbf{q}^2 x}{k}\right)\right]\exp\left\{-\frac{k^2\Delta x}{2}D\left(\boldsymbol{\rho}-\frac{\mathbf{q}x}{k}\right)\right\}+\ldots. \tag{13.180}$$

Note that, in addition to spatial scale ρ_{cog}, the second characteristic spatial scale

$$r_0 = \frac{x}{k\rho_{\text{cog}}} \tag{13.181}$$

appears in the problem.

Setting $\boldsymbol{\rho} = 0$ in Eq. (13.180), we can obtain the expression for the intensity variance

$$\beta^2(x) = \left\langle I^2(x,\mathbf{R})\right\rangle - 1$$
$$= 1 + \pi\Delta x\int d\mathbf{q}\, q^4\Phi_\varepsilon(\mathbf{q})\exp\left\{-\frac{k^2\Delta x}{2}D\left(\frac{\mathbf{q}x}{k}\right)\right\}+\ldots \tag{13.182}$$

If the turbulence is responsible for fluctuations of field $\varepsilon(x,\mathbf{R})$ in the inhomogeneous layer, so that spectrum $\Phi_\varepsilon(\mathbf{q})$ is given by Eq. (13.89), then Eq. (13.182) yields

$$\beta^2(x) = 1 + 0.429\beta_0^{-4/5}(x), \tag{13.183}$$

where $\beta_0^2(x)$ is the intensity variance calculated in the first approximation of Rytov's smooth perturbation method applied to the phase screen (13.96).

The above considerations can be easily extended to higher moment functions of field $u(x,\mathbf{R})$ and, in particular, to functions $\Gamma_{2n}(x,\mathbf{R}) = \langle I^n(x,\mathbf{R})\rangle$. In this case, Eq. (13.178) assumes the form

$$\begin{aligned}\langle I^n(x,\mathbf{R})\rangle &= \left(\frac{k}{2\pi x}\right)^{2n}\int\ldots\int d\mathbf{v}_1\ldots d\mathbf{v}_{2n}\\ &\times\exp\left\{\frac{ik}{2x}\sum_{j=1}^{2n}(-1)^{j+1}\mathbf{v}_j^2 - F(\mathbf{v}_1,\ldots,\mathbf{v}_{2n})\right\},\end{aligned} \tag{13.184}$$

where

$$F(\mathbf{v}_1,\ldots,\mathbf{v}_{2n}) = \frac{k^2\Delta x}{8}\sum_{j,l=1}^{2n}(-1)^{j+l+1}D(\mathbf{v}_j-\mathbf{v}_l). \tag{13.185}$$

Function $F(\mathbf{v}_1,\ldots,\mathbf{v}_{2n})$ can be expressed in terms of random phase incursions $S(\mathbf{v}_i)$ (13.176) by the formula

$$F(\mathbf{v}_1,\ldots,\mathbf{v}_{2n}) = \frac{1}{2}\left\langle\left[\sum_{j=1}^{2n}(-1)^{j+1}S(\mathbf{v}_j)\right]^2\right\rangle \geq 0.$$

This formula clearly shows that function $F(\mathbf{v}_1, \ldots, \mathbf{v}_{2n})$ vanishes if each odd point $\mathbf{v}_{2l+1}$ coincides with certain pair point among even points, because the positive and negative phase incursions cancel in this case. It becomes clear that namely the regions in which this cancellation occur will mainly contribute to moments $\langle I^n(x,\mathbf{R})\rangle$ for $\sqrt{x/k} \gg \rho_{\text{cog}}$. It is not difficult to calculate that the number of these regions is equal to $n!$. Then, replacing the integral in (13.184) with $n!$ multiplied by the integral over one of these regions A_1 in which

$$|\mathbf{v}_1 - \mathbf{v}_2| \sim |\mathbf{v}_3 - \mathbf{v}_4| \sim \ldots \sim |\mathbf{v}_{2n-1} - \mathbf{v}_{2n}| < \rho_{\text{cog}},$$

we obtain

$$\begin{aligned}\langle I^n(x,\mathbf{R})\rangle \;&\approx\; n!\left(\frac{k}{2\pi x}\right)^{2n} \int\limits_{A_1}\cdots\int d\mathbf{v}_1 \ldots d\mathbf{v}_{2n} \\ &\times \exp\left\{\frac{ik}{2x}\sum_{j=1}^{2n}(-1)^{j+1}\mathbf{v}_j^2 - F(\mathbf{v}_1, \ldots, \mathbf{v}_{2n})\right\}. \qquad (13.186)\end{aligned}$$

Terms of sum (13.185)

$$\frac{k^2\Delta x}{8}D(\mathbf{v}_1 - \mathbf{v}_2), \quad \frac{k^2\Delta x}{8}D(\mathbf{v}_3 - \mathbf{v}_4), \quad \text{and so on}$$

ensure the decreasing behavior of the integrand with respect to every of variables $\mathbf{v}_1 - \mathbf{v}_2$, $\mathbf{v}_3 - \mathbf{v}_4$, and so on. We keep these terms in the exponent and expand the exponential function of other terms in the series to obtain the following approximate expression

$$\begin{aligned}\langle I^n(x,\mathbf{R})\rangle \;&\approx\; n!\left(\frac{k}{2\pi x}\right)^{2n} \int\limits_{A_1}\cdots\int d\mathbf{v}_1 \ldots d\mathbf{v}_{2n} \\ &\times \exp\left\{\frac{ik}{2x}\sum_{j=1}^{2n}(-1)^{j+1}\mathbf{v}_j^2 - \frac{k^2\Delta x}{4}\sum_{l=1}^{n} D(\mathbf{v}_{2l-1} - \mathbf{v}_{2l})\right\} \\ &\times \left\{1 + \frac{k^2\Delta x}{8}\sum_{j,l=1}^{2n}{}'(-1)^{j+l+1}D(\mathbf{v}_j - \mathbf{v}_l) + \ldots\right\}. \qquad (13.187)\end{aligned}$$

Here, the prime of the sum sign means that this sum excludes the terms kept in the exponent. Because the integrand is negligible outside region A_1, we can extend the region of integration in Eq. (13.187) to the whole space. Then the multiple integral in Eq. (13.187) can be calculated in the analytic form, and we obtain for $\langle I^n(x,\mathbf{R})\rangle$ the formula

$$\langle I^n(x,\mathbf{R})\rangle = n!\left[1 + n(n-1)\frac{\beta^2(x) - 1}{4} + \ldots\right], \qquad (13.188)$$

where quantity $\beta^2(x)$ is given by Eq. (13.183). We discuss this formula a little later, after considering wave propagation in continuous random medium, which yields a very similar result.

Continuous medium

Consider now the asymptotic behavior of higher moment functions $M_{nn}(x, \mathbf{R}_1, \ldots, \mathbf{R}_{2n})$ of the wavefield propagating in random medium. The formal solution to this problem is given by Eqs. (13.174) and (13.175). They differ from the phase screen formulas only by the fact that ordinary integrals are replaced with path integrals. We consider first quantity $\langle I(x,\mathbf{R}')I(x,\mathbf{R}'')\rangle$ that can be obtained from moment $M_{22}(x,\mathbf{R}_1,\ldots,\mathbf{R}_4)$ by setting $\mathbf{R}_1=\mathbf{R}_2=\mathbf{R}'$, $\mathbf{R}_3=\mathbf{R}_4=\mathbf{R}''$. In the case of the plane wave ($u_0(\mathbf{R})=1$), we can use (13.175) and introduce new variables similar to those for the phase screen to obtain

$$\begin{aligned}
\langle I(x,\mathbf{R}')I(x,\mathbf{R}'')\rangle &= \exp\left\{\frac{i}{k}\int\limits_0^x d\xi\frac{\delta^2}{\delta\mathbf{v}_1(\xi)\delta\mathbf{v}_2(\xi)}\right\}\\
\times\exp\Biggl\{-\frac{k^2}{4}\int\limits_0^x dx'\Biggl[&2D\left(\boldsymbol{\rho}+\int\limits_{x'}^x d\xi\mathbf{v}_1(\xi)\right)+2D\left(\int\limits_{x'}^x d\xi\mathbf{v}_2(\xi)\right)\\
-D\left(\boldsymbol{\rho}+\int\limits_{x'}^x d\xi\left[\mathbf{v}_1(\xi)+\mathbf{v}_2(\xi)\right]\right)&-D\left(\boldsymbol{\rho}+\int\limits_{x'}^x d\xi\left[\mathbf{v}_1(\xi)-\mathbf{v}_2(\xi)\right]\right)\Biggr]\Biggr\}_{\mathbf{v}=0},
\end{aligned} \tag{13.189}$$

where $\boldsymbol{\rho}=\mathbf{R}'-\mathbf{R}''$.

Using Eq. (13.174), formula (13.189) can be represented in the form of the path integral; however, we will use here the operator representation. As in the case of the phase screen, we can represent $\langle I(x,\mathbf{R}')I(x,\mathbf{R}'')\rangle$ for $x\to\infty$ in the form

$$B_I(x,\boldsymbol{\rho})=\langle I(x,\mathbf{R}')I(x,\mathbf{R}'')\rangle-1=B_I^{(1)}(x,\boldsymbol{\rho})+B_I^{(2)}(x,\boldsymbol{\rho})+B_I^{(3)}(x,\boldsymbol{\rho}), \tag{13.190}$$

where

$$\begin{aligned}
B_I^{(1)}(x,\boldsymbol{\rho}) &= \exp\left\{-\frac{k^2x}{2}D(\boldsymbol{\rho})\right\},\\
B_I^{(2)}(x,\boldsymbol{\rho}) &= \pi k^2\int\limits_0^x dx'\int d\mathbf{q}\Phi_\varepsilon(\mathbf{q})\left[1-\cos\frac{\mathbf{q}^2}{k}(x-x')\right]\\
&\times\exp\left\{i\mathbf{q}\boldsymbol{\rho}-\frac{k^2x'}{2}D\left(\frac{\mathbf{q}}{k}(x-x')\right)-\frac{k^2}{2}\int\limits_{x'}^x dx'D\left(\frac{\mathbf{q}}{k}(x-x'')\right)\right\},\\
B_I^{(3)}(x,\boldsymbol{\rho}) &= \pi k^2\int\limits_0^x dx'\int d\mathbf{q}\Phi_\varepsilon(\mathbf{q})\left[1-\cos\left(\mathbf{q}\boldsymbol{\rho}-\frac{\mathbf{q}^2}{k}(x-x')\right)\right]\\
&\times\exp\left\{-\frac{k^2x'}{2}D\left(\boldsymbol{\rho}-\frac{\mathbf{q}}{k}(x-x')\right)-\frac{k^2}{2}\int\limits_{x'}^x dx'D\left(\boldsymbol{\rho}-\frac{\mathbf{q}}{k}(x-x'')\right)\right\}.
\end{aligned}$$

Setting $\boldsymbol{\rho}=0$ and taking into account only the first term of the expansion of function

$$1-\cos\frac{\mathbf{q}^2}{k}(x-x'),$$

we obtain that intensity variance

$$\beta^2(x)=\left\langle I^2(x,\mathbf{R})\right\rangle-1=B_I(x,0)-1$$

is given by the formula similar to Eq. (13.182)

$$\beta^2(x) = 1 + \pi \int\limits_0^x dx'(x - x') \int d\mathbf{q}\, q^4 \Phi_\varepsilon(\mathbf{q})$$
$$\times \exp\left\{ -\frac{k^2 x'}{2} D\left(\frac{\mathbf{q}}{k}(x - x')\right) - \frac{k^2}{2} \int\limits_{x'}^x dx' D\left(\frac{\mathbf{q}}{k}(x - x'')\right) \right\} + \dots \quad (13.191)$$

If we deal with the turbulent medium, then Eq. (13.191) yields

$$\beta^2(x) = 1 + 0.861 \left(\beta_0^2(x)\right)^{-2/5}, \quad (13.192)$$

where $\beta_0^2(x)$ is the wavefield intensity variance calculated in the first approximation of Rytov's smooth perturbation method (13.98).

Expression (13.191) remains valid also in the case when functions $\Phi_\varepsilon(\mathbf{q})$ and $D(\boldsymbol{\rho})$ slowly vary with x. In this case, we can easily reduce Eq. (13.191) to Eq. (13.182) by setting $\Phi_\varepsilon(\mathbf{q}) = 0$ outside layer $0 \le x' \le \Delta x \ll x$.

Concerning correlation function $B_I(x, \boldsymbol{\rho})$, we note that the main term $B_I^{(1)}(x, \boldsymbol{\rho})$ in Eq. (13.190) is the squared modulus of the second-order coherence function (see, e.g., [318], as well as [134, 135]).

Now, we turn to the higher moment functions $\langle I^n(x, \mathbf{R})\rangle = \Gamma_{2n}(x, 0)$. Similarly to the case of the phase screen, one can easily obtain that this moment of the wavefield in continuous medium is represented in the form of the expansion

$$\langle I^n(x, \mathbf{R})\rangle = n! \left[1 + n(n-1)\frac{\beta^2(x) - 1}{4} + \dots\right], \quad (13.193)$$

which coincides with expansion (13.188) for the phase screen, excluding the fact that parameter $\beta^2(x)$ is given by different formulas in these cases.

Formula (13.193) specifies two first terms of the asymptotic expansion of function $\langle I^n(x, \mathbf{R})\rangle$ for $\beta_0^2(x) \to \infty$. Because $\beta^2(x) \to 1$ for $\beta_0^2(x) \to \infty$, the second term in Eq. (13.193) is small in comparison with the first one for sufficiently great $\beta_0^2(x)$. Expression (13.193) makes sense only if

$$n(n-1)\frac{\beta^2(x) - 1}{4} \ll 1. \quad (13.194)$$

However, we can always select numbers n for which condition (13.194) will be violated for a fixed $\beta_0^2(x)$. This means that Eq. (13.193) holds only for moderate n. It should be noted additionally that the moment can approach the asymptotic behavior (13.193) for $\beta_0^2(x) \to \infty$ fairly slowly.

Formula (13.193) yields the singular probability density of intensity. To avoid the singularities, we can approximate this formula by the expression (see, e.g., [55])

$$\langle I^n(x, \mathbf{R})\rangle = n! \exp\left\{n(n-1)\frac{\beta^2(x) - 1}{4}\right\}, \quad (13.195)$$

which yields the probability density (see, e.g., [43, 55])

$$P(x, I) = \frac{1}{\sqrt{\pi\left(\beta(x) - 1\right)}} \int\limits_0^\infty dz \exp\left\{ -zI - \frac{\left[\ln z - \frac{\beta(x) - 1}{4}\right]^2}{\beta(x) - 1} \right\}. \quad (13.196)$$

Note that probability distribution (13.196) is not applicable in a narrow region $I \sim 0$ (the width of this region the narrower the greater parameter $\beta_0^2(x)$). This follows from the fact that Eq. (13.196) yields infinite values for the moments of inverse intensity $1/I(x,\mathbf{R})$. Nevertheless, moments $\langle 1/I^n(x,\mathbf{R})\rangle$ are finite for any finite-valued parameter $\beta_0^2(x)$ (arbitrarily great), and the equality $P(x,0) = 0$ must hold. It is clear that the existence of this narrow region around the point $I \sim 0$ does not affect the behavior of moments (13.195) for larger $\beta_0^2(x)$.

Asymptotic formulas (13.195) and (13.196) describe the transition to the region of *saturated* intensity fluctuations, where $\beta(x) \to 1$ for $\beta_0^2(x) \to \infty$. In this region, we have correspondingly

$$\langle I^n(x,\mathbf{R})\rangle = n!, \quad P(x,I) = e^{-I}. \tag{13.197}$$

The exponential probability distribution (13.197) means that complex field $u(x,\mathbf{R})$ is the Gaussian random field. Recall that

$$u(x,\mathbf{R}) = A(x,\mathbf{R})e^{iS(x,\mathbf{R})} = u_1(x,\mathbf{R}) + iu_2(x,\mathbf{R}), \tag{13.198}$$

where $u_1(x,\mathbf{R})$ and $u_2(x,\mathbf{R})$ are the real and imaginary parts, respectively. As a result, the wavefield intensity is

$$I(x,\mathbf{R}) = A^2(x,\mathbf{R}) = u_1^2(x,\mathbf{R}) + u_2^2(x,\mathbf{R}).$$

From the Gaussian property of complex field $u(x,\mathbf{R})$ follows that random fields $u_1(x,\mathbf{R})$ and $u_2(x,\mathbf{R})$ are also the Gaussian statistically independent fields with variances

$$\left\langle u_1^2(x,\mathbf{R})\right\rangle = \left\langle u_2^2(x,\mathbf{R})\right\rangle = \frac{1}{2}. \tag{13.199}$$

It is quite natural to assume that their gradients $\mathbf{p}_1(x,\mathbf{R}) = \nabla_{\mathbf{R}} u_1(x,\mathbf{R})$ and $\mathbf{p}_2(x,\mathbf{R}) = \nabla_{\mathbf{R}} u_2(x,\mathbf{R})$ are also statistically independent of fields $u_1(x,\mathbf{R})$ and $u_2(x,\mathbf{R})$ and are the Gaussian homogeneous and isotropic (in plane $\mathbf{R}$) fields with variances

$$\sigma_{\mathbf{p}}^2(x) = \left\langle \mathbf{p}_1^2(x,\mathbf{R})\right\rangle = \left\langle \mathbf{p}_2^2(x,\mathbf{R})\right\rangle. \tag{13.200}$$

With these assumptions, the joint probability density of fields $u_1(x,\mathbf{R})$, $u_2(x,\mathbf{R})$ and gradients $\mathbf{p}_1(x,\mathbf{R})$ and $\mathbf{p}_2(x,\mathbf{R})$ has the form

$$P(x;u_1,u_2,\mathbf{p}_1,\mathbf{p}_2) = \frac{1}{\pi^3\sigma_{\mathbf{p}}^4(x)} \exp\left\{-u_1^2 - u_2^2 - \frac{\mathbf{p}_1^2+\mathbf{p}_2^2}{\sigma_{\mathbf{p}}^2(x)}\right\}. \tag{13.201}$$

Consider now the joint probability density of wavefield intensity $I(x,\mathbf{R})$ and amplitude gradient

$$\boldsymbol{\kappa}(x,\mathbf{R}) = \nabla_{\mathbf{R}} A(x,\mathbf{R}) = \frac{u_1(x,\mathbf{R})\mathbf{p}_1(x,\mathbf{R}) + u_2(x,\mathbf{R})\mathbf{p}_2(x,\mathbf{R})}{\sqrt{u_1^2(x,\mathbf{R}) + u_2^2(x,\mathbf{R})}}.$$

We have for this probability density the expression

$$\begin{aligned} P(x;I,\boldsymbol{\kappa}) &= \left\langle \delta\left(I(x,\mathbf{R}) - I\right)\delta\left(\boldsymbol{\kappa}(x,\mathbf{R}) - \boldsymbol{\kappa}\right)\right\rangle_{u_i,\mathbf{p}_i} \\ &= \frac{1}{\pi^3\sigma_{\mathbf{p}}^4(x)} \int\limits_{-\infty}^{\infty} du_1 \int\limits_{-\infty}^{\infty} du_2 \int d\mathbf{p}_1 \int d\mathbf{p}_2 \exp\left\{-u_1^2 - u_2^2 - \frac{\mathbf{p}_1^2+\mathbf{p}_2^2}{\sigma_{\mathbf{p}}^2(x)}\right\} \\ &\times \delta\left(u_1^2 + u_2^2 - I\right)\delta\left(\frac{u_1\mathbf{p}_1 + u_2\mathbf{p}_2}{\sqrt{u_1^2+u_2^2}} - \boldsymbol{\kappa}\right) = \frac{1}{2\pi\sigma_{\mathbf{p}}^2(x)}\exp\left\{-I - \frac{\boldsymbol{\kappa}^2}{2\sigma_{\mathbf{p}}^2(x)}\right\}. \end{aligned} \tag{13.202}$$

Consequently, the transverse gradient of amplitude is statistically independent of wavefield intensity and is the Gaussian random field with the variance

$$\left\langle \boldsymbol{\kappa}^2(x,\mathbf{R})\right\rangle = 2\sigma_{\mathbf{p}}^2(x). \qquad (13.203)$$

We note that the transverse gradient of amplitude is also independent of the second derivatives of wavefield intensity with respect to transverse coordinates.

In the region of strong intensity fluctuations, the second-order coherence function is independent of diffraction phenomena and is given by the expression

$$\begin{aligned} &\Gamma_2(x,\mathbf{R}-\mathbf{R}') = \langle u(x,\mathbf{R})u^*(x,\mathbf{R}')\rangle \\ = &\langle u_1(x,\mathbf{R})u_1^*(x,\mathbf{R}')\rangle + \langle u_2(x,\mathbf{R})u_2^*(x,\mathbf{R}')\rangle = e^{-\frac{1}{4}k^2xD(\mathbf{R}-\mathbf{R}')}, \qquad (13.204)\end{aligned}$$

where $D(\mathbf{R}) = A(0) - A(\mathbf{R})$. Consequently, variance $\sigma_{\mathbf{p}}^2(x)$ appeared in Eq. (13.199) is given by the expression

$$\sigma_{\mathbf{p}}^2(x) = \frac{k^2x}{8}\Delta_{\mathbf{R}}D(\mathbf{R})\,|_{\mathbf{R}=0} = -\frac{k^2x}{8}\Delta_{\mathbf{R}}A(\mathbf{R})\,|_{\mathbf{R}=0}.$$

In the case of turbulent fluctuations of field $\varepsilon(x,\mathbf{R})$, this expression certainly coincides with Eq. (13.100), page 377

$$\sigma_{\mathbf{p}}^2(x) = \frac{1.476}{L_f^2(x)}D^{1/6}(x)\beta_0(x), \qquad (13.205)$$

where $L_f(x) = \sqrt{x/k}$ is the size of the first Fresnel zone, $D(x) = \kappa_m^2 x/k \gg 1$ is the wave parameter, and κ_m is the wave number corresponding to the turbulence microscale.

At the end of this section, we note that the path integral representation of field $u(x,\mathbf{R})$ makes it possible to investigate the applicability range of the approximation of the delta-correlated random field $\varepsilon(x,\mathbf{R})$ in the context of wave intensity fluctuations. It turns out that all conditions restricting applicability of the delta-correlated random field $\varepsilon(x,\mathbf{R})$ for calculating quantity $\langle I^n(x,\mathbf{R})\rangle$ coincide with those obtained for quantity $\langle I^2(x,\mathbf{R})\rangle$. In other words, the approximation of the delta-correlated random field $\varepsilon(x,\mathbf{R})$ does not affect the shape of the probability distribution of wavefield intensity.

In the case of turbulent temperature pulsations, the approximation of the delta-correlated random field $\varepsilon(x,\mathbf{R})$ holds in the region of weak fluctuations under the conditions

$$\lambda \ll \sqrt{\lambda x} \ll x,$$

where $\lambda = 2\pi/k$ is the wavelength.

As to the region of strong fluctuations, the applicability range of the approximation of the delta-correlated random field $\varepsilon(x,\mathbf{R})$ is restricted by the conditions

$$\lambda \ll \rho_{\text{cog}} \ll r_0 \ll x,$$

where ρ_{cog} and r_0 are given by Eqs. (13.31), page 361 and (13.181), page 401. The physical meaning of all these inequalities is simple. The delta-correlated approximation remains valid as long as the correlation radius of field $\varepsilon(x,\mathbf{R})$ (its value is given by the size of the first Fresnel zone in the case of turbulent temperature pulsations) is the smallest longitudinal scale of the problem. As the wave approaches at the region of strong intensity fluctuations, a new longitudinal scale appears; its value $\sim \rho_{\text{cog}}\sqrt{kx}$ gradually decreases and, for sufficiently large values of parameter $\beta_0^2(x)$, can become smaller than the correlation radius of field $\varepsilon(x,\mathbf{R})$. In this situation, the delta-correlated approximation fails.

We can consider the above inequalities as restrictions from above and from below on the scale of the intensity correlation function. In these terms, the delta-correlated approximation holds only if any scales appeared in the problem are small in comparison with the length of the wave path.

13.3.3 Caustic structure of wavefield in random media

The above statistical characteristics of wavefield $u(x,\mathbf{R})$, for example, the intensity correlation function in the region of strong fluctuations, have nothing in common with the actual behavior of the wavefield propagating in particular realizations of the medium (see Figs. 1.11 and 1.12, page 32 in Introduction). In order to analyze the detailed structure of wavefield, one can use methods of statistical topography; they provide an insight into the formation of the wavefield caustic structure and make it possible to ascertain the statistical parameters that describe this structure. We note that Bunkin and Gochelashvili [35, 36] (see also [76]) seemingly pioneered in analyzing wave propagation in turbulent medium using the theory of large deviations of random intensity fields.

Elements of statistical topography of random intensity field

If we deal with the plane incident wave, all one-point statistical characteristics, including probability densities, are independent of variable $\mathbf{R}$ in view of spatial homogeneity. In this case, a number of physical quantities that characterize cluster structure of wavefield intensity can be adequately described in terms of specific (per unit area) values. In addition, the role the natural medium-independent length scale in plane $x = \text{const}$ plays the size of the *first Fresnel zone* $L_f(x) = \sqrt{x/k}$, which determines the size of the transient light–shadow zone appeared in the problem on diffraction by the edge of an opaque screen. Among these quantities are

- Specific average total area of regions in plane $\{\mathbf{R}\}$, which are bounded by level lines inside which $I(x,\mathbf{R}) > I$,

$$\langle s(x,I)\rangle = \int_I^\infty dI' P(x;I'),$$

where $P(x;I)$ is the probability density of wavefield intensity $I(x,\mathbf{R})$;

- Specific average field power within these regions

$$\langle e(x,I)\rangle = \int_I^\infty I' dI' P(x;I');$$

- Specific average length of these contours

$$\langle l(x,I)\rangle = L_f(x)\,\langle |\mathbf{p}(x,\mathbf{R})|\delta\left(I(x,\mathbf{R}) - I\right)\rangle\,,$$

where $\mathbf{p}(x,\mathbf{R}) = \nabla_{\mathbf{R}} I(x,\mathbf{R})$ is the transverse gradient of wavefield intensity;

- Estimate of average difference between the numbers of contours with opposite normal orientation per first Fresnel zone

$$\langle n(x,I)\rangle = \frac{1}{2\pi}L_f^2(x)\left\langle \kappa(x,\mathbf{R};I)|\mathbf{p}(x,\mathbf{R})|\delta\left(I(x,\mathbf{R})-I\right)\right\rangle,$$

where $\kappa(x,\mathbf{R};I)$ is the curvature of the level line,

$$\kappa(x,\mathbf{R};I) = \frac{-p_y^2(x,\mathbf{R})\frac{\partial^2 I(x,\mathbf{R})}{\partial z^2} - p_z^2(x,\mathbf{R})\frac{\partial^2 I(x,\mathbf{R})}{\partial y^2} + 2p_y(x,\mathbf{R})p_z(x,\mathbf{R})\frac{\partial^2 I(x,\mathbf{R})}{\partial y\partial z}}{p^3(x,\mathbf{R})}.$$

Consider now the behavior of these quantities with the distance x (parameter $\beta_0(x)$).

Weak intensity fluctuations

The region of weak intensity fluctuations is limited by inequality $\beta_0(x) \le 1$. In this region, wavefield intensity is the lognormal process described by probability distribution (13.93).

The typical realization curve of this logarithmic-normal process is the exponentially decaying curve

$$I^*(x) = e^{-\frac{1}{2}\beta_0(x)},$$

and statistical characteristics (moment functions $\langle I^n(x,\mathbf{R})\rangle$, for example) are formed by large spikes of process $I(x,\mathbf{R})$ relative this curve.

In addition, various majorant estimates are available for lognormal process realizations. For example, separate realizations of wavefield intensity satisfy the inequality

$$I(x) < 4e^{-\frac{1}{4}\beta_0(x)}$$

for all distances $x \in (0,\infty)$ with probability $p = 1/2$. All these facts are indicative of the onset of cluster structure formations in wavefield intensity.

As we have seen earlier, the knowledge of probability density (13.93) is sufficient for obtaining certain quantitative characteristics of these cluster formations. For example, the average area of regions within which $I(x,\mathbf{R}) > I$ is

$$\langle s(x,I)\rangle = \Phi\left(\frac{1}{\sqrt{2\beta_0(x)}}\ln\left(\frac{e^{-\frac{1}{2}\beta_0(x)}}{I}\right)\right), \qquad (13.206)$$

and specific average power confined in these regions is given by the expression

$$\langle e(x,I)\rangle = \Phi\left(\frac{1}{\sqrt{2\beta_0(x)}}\ln\left(\frac{e^{\frac{1}{2}\beta_0(x)}}{I}\right)\right), \qquad (13.207)$$

where $\Phi(z) = \frac{1}{\sqrt{\pi}}\int_{-\infty}^{z} dy e^{-y^2}$ is the well-known error function.

The character of cluster structure spatial evolution versus parameter $\beta_0(x)$ essentially depends on the desired level I. In the most interesting case of $I > 1$, the values of these functions in the initial plane are $\langle s(0,I)\rangle = 0$ and $\langle e(0,I)\rangle = 0$. As $\beta_0(x)$ increases, small cluster regions appear in which $I(x,\mathbf{R}) > I$; for certain distances, these regions remain almost intact and actively absorb a considerable portion of total energy. With

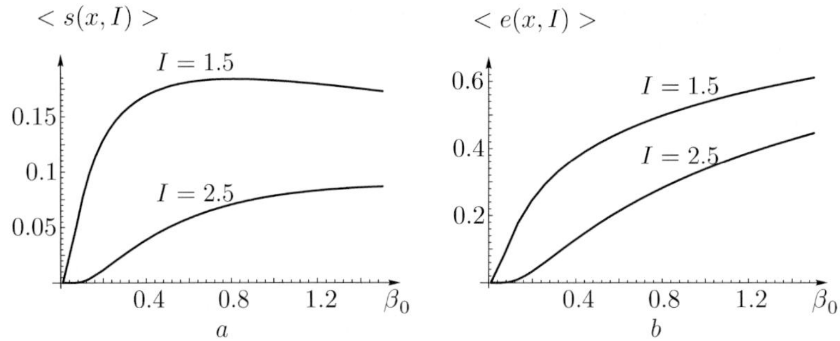

Figure 13.3: (*a*) Average specific area and (*b*) average power versus parameter $\beta_0(x)$.

further increasing $\beta_0(x)$, the areas of these regions decrease and the power within them increases, which corresponds to an increase of regions's average brightness. The cause of these processes lies in radiation focusing by separate portions of medium. Figures 13.3*a* and 13.3*b* show functions $\langle s(x,I)\rangle$ and $\langle e(x,I)\rangle$ for different parameters $\beta_0(x)$ from the given range. The specific average area is maximum at $\beta_0(x) = 2\ln(I)$, and

$$\langle s(x,I\rangle_{\max} = \Phi\left(-\frac{1}{\sqrt{\ln(I)}}\right).$$

The average power at this value of $\beta_0(x)$ is $\langle e(x,I)\rangle = 1/2$.

In the region of weak intensity fluctuations, the spatial gradient of amplitude level $\boldsymbol{\nabla}_{\mathbf{R}}\chi(x,\mathbf{R})$ is statistically independent of $\chi(x,\mathbf{R})$. This fact makes it possible both to calculate specific average length of contours $I(x,\mathbf{R}) = I$ and to estimate specific average number of these contours. Indeed, in the region of weak fluctuations, probability density of amplitude level gradient $\mathbf{q}(x,\mathbf{R}) = \boldsymbol{\nabla}_{\mathbf{R}}\chi(x,\mathbf{R})$ is the Gaussian density

$$P(x;\mathbf{q}) = \langle\delta\left(\boldsymbol{\nabla}_{\mathbf{R}}\chi(x,\mathbf{R}) - \mathbf{q}\right)\rangle = \frac{1}{\sigma_{\mathbf{q}}^2(x)}\exp\left\{-\frac{\mathbf{q}^2}{\sigma_{\mathbf{q}}^2(x)}\right\}, \tag{13.208}$$

where $\sigma_{\mathbf{q}}^2(x) = \langle\mathbf{q}^2(x,\mathbf{R})\rangle$ is the variance of amplitude level gradient given by Eq. (13.100).

As a consequence, we obtain that the specific average length of contours is described by the expression

$$\langle l(x,I)\rangle = 2L_f(x)\left\langle|q(x,\mathbf{R})|\right\rangle IP(x;I) = L_f(x)\sqrt{\pi\sigma_q^2(x)}IP(x;I). \tag{13.209}$$

For the specific average number of contours, we have similarly

$$\begin{aligned}\langle n(x,I)\rangle &= \frac{1}{2\pi}L_f^2(x)\left\langle\kappa(x,\mathbf{R},I)|\mathbf{p}(x,\mathbf{R})\delta\left(I(x,\mathbf{R}) - I\right)\right\rangle \\ &= -\frac{1}{2\pi}L_f^2(x)I\left\langle\Delta_{\mathbf{R}}\chi(x,\mathbf{R})\delta\left(I(x,\mathbf{R}) - I\right)\right\rangle \\ &= -\frac{1}{\pi}L_f^2(x)\left\langle\mathbf{q}^2(x,\mathbf{R})\right\rangle I\frac{\partial}{\partial I}IP(x;I) = \frac{L_f^2(x)\sigma_{\mathbf{q}}^2(x)}{\pi\beta_0(x)}\ln\left(Ie^{\frac{1}{2}\beta_0(x)}\right)IP(x;I).\end{aligned} \tag{13.210}$$

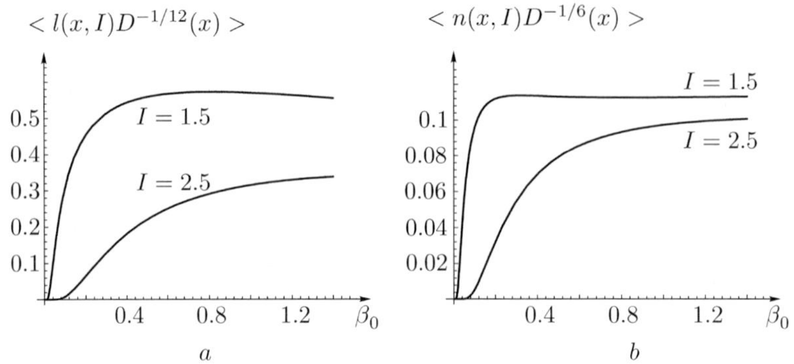

Figure 13.4: (*a*) Average specific contour length and (*b*) average contour number versus parameter $\beta_0(x)$.

We notice that Eq. (13.210) vanishes at $I = I_0(x) = \exp\left\{-\frac{1}{2}\beta_0(x)\right\}$. This means that this intensity level corresponds to the situation in which the specific average number of contours bounding region $I(x, \mathbf{R}) > I_0$ coincides with the specific average number of contours bounding region $I(x, \mathbf{R}) < I_0$. Figures 13.4*a* and 13.4*b* show functions $\langle l(x, I)\rangle$ and $\langle n(x, I)\rangle$ versus parameter $\beta_0(x)$.

Dependence of specific average length of level lines and specific average number of contours on turbulence microscale reveals the existence of small-scale ripples imposed upon large-scale random relief. These ripples do not affect the redistribution of areas and power, but increases the irregularity of level lines and causes the appearance of small contours.

As we mentioned earlier, this description holds for $\beta_0(x) \leq 1$. With increasing parameter $\beta_0(x)$ to this point, Rytov's smooth perturbation method fails, and we must consider the nonlinear equation in wavefield complex phase. This region of fluctuations called the *strong focusing region* is very difficult for analytical treatment. With further increasing parameter $\beta_0(x)$ ($\beta_0^2(x) \geq 10$) statistical characteristics of intensity approach the saturation regime, and this region of parameter $\beta_0(x)$ is called the *region of strong intensity fluctuations*.

Strong intensity fluctuations

From Eq. (13.196) for probability density follows that specific average area of regions within which $I(x, \mathbf{R}) > I$ is

$$\langle s(x, I)\rangle = \frac{1}{\sqrt{\pi\left(\beta(x) - 1\right)}} \int\limits_0^\infty \frac{dz}{z} \exp\left\{-zI - \frac{\left[\ln z - \frac{\beta(x)-1}{4}\right]^2}{\beta(x) - 1}\right\}, \tag{13.211}$$

and specific average power concentrated in these regions is given by the expression

$$\langle e(x, I)\rangle = \frac{1}{\sqrt{\pi\left(\beta(x) - 1\right)}} \int\limits_0^\infty \frac{dz}{z}\left(I + \frac{1}{z}\right) \exp\left\{-zI - \frac{\left[\ln z - \frac{\beta(x)-1}{4}\right]^2}{\beta(x) - 1}\right\}. \tag{13.212}$$

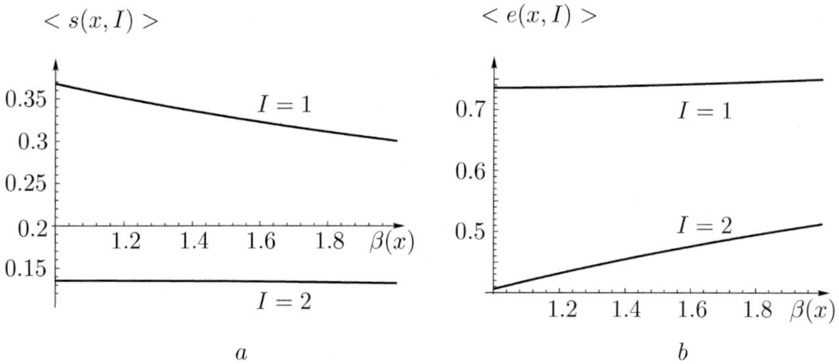

Figure 13.5: (*a*) Average specific area and (*b*) average contour number in the region of strong intensity fluctuations versus parameter $\beta_0(x)$.

Figures 13.5*a* and 13.5*b* shows functions (13.211) and (13.212) versus parameter $\beta(x)$. We note that parameter $\beta(x)$ is a very slow function of $\beta_0(x)$. Indeed, limit process $\beta_0(x) \to \infty$ corresponds to $\beta(x) = 1$ and value $\beta_0(x) = 1$ corresponds to $\beta(x) = 1.861$.

Asymptotic formulas (13.211) and (13.212) adequately describe the transition to the region of saturated intensity fluctuations ($\beta(x) \to 1$). In this region, we have

$$P(I) = e^{-I}, \quad \langle s(I) \rangle = e^{-I}, \quad \langle e(I) \rangle = (I+1)\, e^{-I}. \tag{13.213}$$

Moreover, we obtain the expression for specific average contour length

$$\begin{aligned} \langle l(x, I) \rangle &= L_f(x) \left\langle |\mathbf{p}(x, \mathbf{R})| \delta\left(I(x, \mathbf{R}) - I\right) \right\rangle \\ &= 2L_f(x)\sqrt{I} \left\langle |\boldsymbol{\kappa}(x, \mathbf{R})| \delta\left(I(x, \mathbf{R}) - I\right) \right\rangle \\ &= 2L_f(x)\sqrt{I} \left\langle |\mathbf{q}(x, \mathbf{R})| \right\rangle P(x; I) L_f(x) \sqrt{2\pi\sigma_{\mathbf{q}}^2(x) I}\, P(x; I), \end{aligned} \tag{13.214}$$

where the variance of amplitude level gradient in the region of saturated fluctuations coincides with the variance calculated in the first approximation of Rytov's smooth perturbation method. Specific average contour length (13.214) is maximum at $I = 1/\sqrt{2}$.

In the region of saturated intensity fluctuations, the estimator of specific average number of contours is given by the following chain of equalities

$$\begin{aligned} \langle n(x, I) \rangle &= \frac{L_f^2(x)}{2\pi} \left\langle \kappa(x, \mathbf{R}, I) |\mathbf{p}(x, \mathbf{R})| \delta\left(I(x, \mathbf{R}) - I\right) \right\rangle \\ &= -\frac{L_f^2(x)}{2\pi} \sqrt{I} \left\langle \Delta_{\mathbf{R}} A(x, \mathbf{R}) \delta\left(I(x, \mathbf{R}) - I\right) \right\rangle \\ &= -\frac{L_f^2(x)}{\pi} \left\langle \boldsymbol{\kappa}^2(x, \mathbf{R}) \right\rangle \sqrt{I} \frac{\partial}{\partial I} \sqrt{I} P(x; I) \\ &= -\frac{2L_f^2(x)\sigma_{\mathbf{q}}^2(x)}{\pi} \sqrt{I} \frac{\partial}{\partial I} \sqrt{I} e^{-I} = \frac{2L_f^2(x)\sigma_{\mathbf{q}}^2(x)}{\pi} \left(I - \frac{1}{2}\right) e^{-I}. \end{aligned} \tag{13.215}$$

Expression (13.215) is maximum at $I = 3/2$, and the level at which specific average number of contours bounding region $I(x, \mathbf{R}) > I_0$ is equal to specific average number of contours corresponding to $I(x, \mathbf{R}) < I_0$ is $I_0 = 1/2$.

We note that Eq. (13.215) fails in the narrow region around $I \sim 0$. The correct formula must vanish for $I = 0$ ($\langle n(x, 0)\rangle = 0$).

As may be seen from Eqs. (13.214) and (13.215), average length of level lines and average number of contours keep increasing with parameter $\beta_0(x)$ in the region of saturated fluctuations, although the corresponding average areas and powers remain fixed. The reason of this behavior consists in the fact that the leading (and defining) role in this regime is played by interference of partial waves arriving from different directions .

Behavioral pattern of level lines depends on the relationship between processes of focusing and defocusing by separate portions of turbulent medium. Focusing by large-scale inhomogeneities becomes apparent in random intensity relief as high peaks. In the regime of maximal focusing ($\beta_0(x) \sim 1$), the narrow high peaks concentrate about a half of total wave power. With increasing parameter $\beta_0(x)$, radiation defocusing begins to prevail,which results in spreading the high peaks and forming highly idented (interference) relief characterized by multiple vertices at level $I \sim 1$.

In addition to parameter $\beta_0(x)$, average length of level lines and average number of contours depend on wave parameter $D(x)$; namely, they increase with decreasing microscale of inhomogeneities. This follows from the fact that the large-scale relief is distorted by fine ripples appeared due to scattering by small-scale inhomogeneities.

Thus, in this section, we attempted to qualitatively explain the cluster (caustic) structure of the wavefield generated in turbulent medium by the plane light wave and to quantitatively estimate the parameters of such a structure in the plane transverse to the propagation direction. In the general case, this problem is multiparametric. However, if we limit ourselves to the problem on the plane incident wave and consider it for a fixed wave parameter in a fixed plane, then the solution is expressed in terms of the sole parameter, namely, the variance of intensity in the region of weak fluctuations $\beta_0(x)$. We have analyzed two asymptotic cases corresponding to weak and saturated intensity fluctuations. It should be noted that applicability range of these asymptotic formulas most likely depends on intensity level I. It is expected naturally that this applicability range will grow with decreasing the level.

As regards the analysis of the intermediate case corresponding to the developed caustic structure (this case is the most interesting from the standpoint of applications), it requires the knowledge of the probability density of intensity and its transverse gradient for arbitrary distances in the medium. Such an analysis can be carried out either by using probability density approximations for all parameters [43], or on the basis of numerical simulations (see, e.g., [64, 65, 238, 239]).

Chapter 14

Some problems of statistical hydrodynamics

In Part 2, we analyzed statistics of solutions to the nonlinear equations of hydrodynamics using the rigorous approach based on deriving and investigating the exact variational differential equations for characteristic functionals of nonlinear random fields. However, this approach encounters severe difficulties caused by the lack of development of the theory of variational differential equations. For this reason, many researchers prefer to proceed from more habitual partial differential equations for different moment functions of fields of interest. The nonlinearity of the input dynamic equations governing random fields results in the appearance of higher moment functions of fields of interest in the equations governing any moment function. As a result, even determination of average field or correlation functions requires, in the strict sense, solving an infinite system of linked equations.

Thus, the main problem of this approach consists in cutting the mentioned system of equations by means one or another physical hypothesis. The most known example of such a hypothesis is the *Millionshchikov hypothesis* according to which the higher moment functions of even orders are expressed in terms of the lower ones by the laws of the Gaussian statistics. The disadvantage of such approaches consists in the fact that the validity of hypothesis usually cannot be proved; moreover, cutting the system of equations may often yield physically contradictory results, namely, energy spectra of turbulence can appear negative for certain wave numbers. Nevertheless, these approximate approaches provide a deeper insight into physical mechanisms of forming the statistics of strongly nonlinear random fields and make it possible to derive quantitative expressions for fields' correlation functions and spectra. It seems that the Millionshchikov hypothesis provides correct spectra of the developed turbulence in viscous interval [251].

We emphasize additionally that the mentioned approximate equations reveal many nontrivial effects inherent in nonlinear random fields and having no analogs in the behavior of deterministic fields and waves. Here, we illustrate these methods of analysis by the example of an interesting physical effect, which consists in the fact that average flows of noncompressible liquids superimposed on the background of developed turbulent pulsations acquire quasi-elastic wave properties. This effect was first mentioned by Moffatt [250] who studied the reaction of turbulence on variations of the transverse gradient of average velocity. Moffatt noted that turbulent medium behaves in some sense like an elastic medium; namely, variations of average velocity profile of a plane-parallel flow satisfy the wave equation.

14.1 Quasi-elastic properties of isotropic and stationary non-compressible turbulent media

Let $\mathbf{u}^{\mathrm{T}}(\mathbf{r},t)$ is the random velocity field of developed turbulence stationary in time and homogeneous and isotropic in space (we assume that $\left\langle \mathbf{u}^{\mathrm{T}}(\mathbf{r},t)\right\rangle = 0$). In actuality, regular average flows are imposed on turbulent pulsations of liquid. We denote $\mathbf{U}(\mathbf{r},t)$ the velocity field of these regular flows. Because the turbulent pulsations and average flows interact nonlinearly, we can represent the total velocity field in the form

$$\mathbf{u}(\mathbf{r},t) = \mathbf{U}(\mathbf{r},t) + \mathbf{u}^{\mathrm{T}}(\mathbf{r},t) + \mathbf{u}'(\mathbf{r},t), \tag{14.1}$$

Here, $\mathbf{u}^{\mathrm{T}}(\mathbf{r},t)$ is the unperturbed turbulent field in the absence of average flows and $\mathbf{u}'(\mathbf{r},t)$ is the turbulent field disturbance caused by the interaction with the regular flow (we assume that $\langle \mathbf{u}'(\mathbf{r},t)\rangle = 0$).

Investigation of nonlinear interactions between regular flows and turbulent pulsations (and the analysis of turbulent pulsations $\mathbf{u}^{\mathrm{T}}(\mathbf{r},t)$ by themselves, though) is very difficult in the general case. However, in the case of weak average flows, i.e., under the condition that

$$\mathbf{U}^2(\mathbf{r},t) \ll 2T(t),$$

where $T(t) = \frac{1}{2}\left\langle \left[\mathbf{u}^{\mathrm{T}}(\mathbf{r},t)\right]^2\right\rangle$ is the average density of turbulent energy pulsation, we can discuss the effect of turbulence on the average flow evolution in sufficient detail by considering the linear approximation in small disturbances of fields $\mathbf{U}(\mathbf{r},t)$ and $\mathbf{u}'(\mathbf{r},t)$ and assuming known the spectrum of turbulent pulsations $\mathbf{u}^{\mathrm{T}}(\mathbf{r},t)$.

The total velocity field (14.1) and its turbulent component $\mathbf{u}^{\mathrm{T}}(\mathbf{r},t)$ satisfy the Navier–Stokes equation (1.97), page 33. Hence, substituting Eq. (14.1) in Eq. (1.97) and linearizing the equations in average field $\mathbf{U}(\mathbf{r},t)$ and fluctuation component $\mathbf{u}'(\mathbf{r},t)$, we arrive at the approximate system of equations

$$\frac{\partial}{\partial t}U_i(\mathbf{r},t) + \frac{\partial}{\partial r_k}T_{ik}(\mathbf{r},t) = -\frac{\partial}{\partial r_i}P(\mathbf{r},t), \tag{14.2}$$

$$\begin{aligned}&\frac{\partial}{\partial t}u_i'(\mathbf{r},t) + \frac{\partial}{\partial r_k}\left[u_i(\mathbf{r},t)u_k(\mathbf{r},t) - \langle u_i(\mathbf{r},t)u_k(\mathbf{r},t)\rangle\right]\\ &+u_l^{\mathrm{T}}(\mathbf{r},t)\frac{\partial U_i(\mathbf{r},t)}{\partial r_l} + U_l(\mathbf{r},t)\frac{\partial u_i^{\mathrm{T}}(\mathbf{r},t)}{\partial r_l} = -\frac{\partial}{\partial r_i}p'(\mathbf{r},t),\end{aligned} \tag{14.3}$$

where $T_{ik}(\mathbf{r},t) = \langle u_i(\mathbf{r},t)u_k(\mathbf{r},t)\rangle$ is the *Reynolds stress tensor* and $P(\mathbf{r},t)$ and $\mathbf{u}'(\mathbf{r},t)$ are the average and perturbed turbulent pressure components, respectively. Here and below, we neglect the effect of viscosity on dynamics and statistics of perturbed fields $\mathbf{U}(\mathbf{r},t)$ and $\mathbf{u}'(\mathbf{r},t)$. In addition, we will bear in mind that fields $\mathbf{U}(\mathbf{r},t)$ and $\mathbf{u}'(\mathbf{r},t)$ satisfy the incompressibility condition (1.97)

$$\boldsymbol{\nabla}\mathbf{U}(\mathbf{r},t) = 0, \quad \boldsymbol{\nabla}\mathbf{u}'(\mathbf{r},t) = 0. \tag{14.4}$$

Being combined with Eq. (1.97) for turbulent pulsations $\mathbf{u}^{\mathrm{T}}(\mathbf{r},t)$, Eq. (14.3) yields the following equation for the Reynolds stress tensor $T_{ik}(\mathbf{r},t)$

$$\begin{aligned}
&\frac{\partial}{\partial t}T_{ik}(\mathbf{r},t)+\frac{\partial}{\partial r_l}\left\langle u_i(\mathbf{r},t)u_k(\mathbf{r},t)u_l(\mathbf{r},t)\right\rangle\\
&+\left\langle u_k^{\mathrm{T}}(\mathbf{r},t)u_l^{\mathrm{T}}(\mathbf{r},t)\right\rangle\frac{\partial U_i(\mathbf{r},t)}{\partial r_l}+\left\langle u_i^{\mathrm{T}}(\mathbf{r},t)u_l^{\mathrm{T}}(\mathbf{r},t)\right\rangle\frac{\partial U_k(\mathbf{r},t)}{\partial r_l}\\
&+U_l(\mathbf{r},t)\frac{\partial\left\langle u_i^{\mathrm{T}}(\mathbf{r},t)u_k^{\mathrm{T}}(\mathbf{r},t)\right\rangle}{\partial r_l}\\
&=-\left\langle\left[u_k^{\mathrm{T}}(\mathbf{r},t)\frac{\partial}{\partial r_i}p'(\mathbf{r},t)+u_i^{\mathrm{T}}(\mathbf{r},t)\frac{\partial}{\partial r_k}p'(\mathbf{r},t)\right]\right\rangle.
\end{aligned}\tag{14.5}$$

Here,

$$u_i(\mathbf{r},t)u_k(\mathbf{r},t)u_l(\mathbf{r},t)=u_i^{\mathrm{T}}(\mathbf{r},t)u_k^{\mathrm{T}}(\mathbf{r},t)u_l'(\mathbf{r},t)+\ldots.$$

Pressure pulsations p' in this equations can be expressed in terms of perturbed velocities $\mathbf{U}(\mathbf{r},t)$ and $\mathbf{u}'(\mathbf{r},t)$,

$$\begin{aligned}
&-p'(\mathbf{r},t)=\Delta^{-1}(\mathbf{r},\mathbf{r}')\\
&\times\left\{\frac{\partial^2}{\partial r_m'\partial r_n'}\left[u_m(\mathbf{r}',t)u_n(\mathbf{r}',t)-\left\langle u_m(\mathbf{r}',t)u_n(\mathbf{r}',t)\right\rangle\right]+2\frac{\partial u_l^{\mathrm{T}}(\mathbf{r}',t)}{\partial r_m'}\frac{\partial U_m(\mathbf{r}',t)}{\partial r_l'}\right\},
\end{aligned}$$

where $\Delta^{-1}(\mathbf{r},\mathbf{r}')$ is the integral operator inverse to the Laplace operator. Equations (14.2), (14.5), and the expression for pressure $p'(\mathbf{r},t)$ form the system of equations in average field $\mathbf{U}(\mathbf{r},t)$ and stress tensor $T_{ik}(\mathbf{r},t)$. These equations are not closed because they depend on higher-order velocity correlators like the triple correlator $\left\langle u_i^{\mathrm{T}}(\mathbf{r},t)u_k^{\mathrm{T}}(\mathbf{r},t)u_l'(\mathbf{r},t)\right\rangle$. Assuming that such correlators only slightly affect the dynamics of perturbations and taking into account conditions (14.4) and statistical homogeneity of field $\mathbf{u}^{\mathrm{T}}(\mathbf{r},t)$, we reduce the system of equations (14.2), (14.5) to the form $\left(\tau_i(\mathbf{r},t)=\frac{\partial}{\partial r_k}T_{ik}(\mathbf{r},t)\right)$

$$\frac{\partial}{\partial t}U_i(\mathbf{r},t)+\tau_i(\mathbf{r},t)=-\frac{\partial}{\partial r_i}P(\mathbf{r},t),\tag{14.6}$$

$$\begin{aligned}
&\frac{\partial}{\partial t}\tau_i(\mathbf{r},t)+\left\langle u_k^{\mathrm{T}}(\mathbf{r},t)u_l^{\mathrm{T}}(\mathbf{r},t)\right\rangle\frac{\partial^2U_i(\mathbf{r},t)}{\partial r_l\partial r_k}\\
&=2\frac{\partial}{\partial r_k}\left[\left\langle u_k^{\mathrm{T}}(\mathbf{r},t)\frac{\partial}{\partial r_i}\Delta^{-1}(\mathbf{r},\mathbf{r}')\left(\frac{\partial u_l^{\mathrm{T}}(\mathbf{r}',t)}{\partial r_m'}\frac{\partial U_m(\mathbf{r}',t)}{\partial r_l'}\right)\right\rangle\right.\\
&\left.+\left\langle u_i^{\mathrm{T}}(\mathbf{r},t)\frac{\partial}{\partial r_k}\Delta^{-1}(\mathbf{r},\mathbf{r}')\left(\frac{\partial u_l^{\mathrm{T}}(\mathbf{r}',t)}{\partial r_m'}\frac{\partial U_m(\mathbf{r}',t)}{\partial r_l'}\right)\right\rangle\right].
\end{aligned}\tag{14.7}$$

This system of equations is closed in $\mathbf{U}(\mathbf{r},t)$. The coefficients in the left-hand side of Eq. (14.7) and the integral operator in the right-hand side can be expressed in terms of the correlation tensor of vortex field of unperturbed turbulence $\mathbf{u}^{\mathrm{T}}(\mathbf{r},t)$

$$\left\langle u_i^{\mathrm{T}}(\mathbf{r},t)u_j^{\mathrm{T}}(\mathbf{r}',t)\right\rangle=\int d\mathbf{q}\Phi_{ij}(\mathbf{q})e^{i\mathbf{q}(\mathbf{r}-\mathbf{r}')}.\tag{14.8}$$

In view of the fact that field $\mathbf{u}^{\mathrm{T}}(\mathbf{r},t)$ is the solenoidal field, we have

$$\Phi_{ij}(\mathbf{q})=\Delta_{ij}(\mathbf{q})F(q).$$

where

$$\Delta_{ij}(\mathbf{q}) = \delta_{ij} - \frac{q_i q_j}{\mathbf{q}^2}.$$

We will assume the energy spectrum of turbulent pulsations $F(q)$ known. From Eq. (14.8) follows in particular that the energy of turbulent pulsations is expressed through the energy spectrum by the relationship

$$T = \frac{1}{2}\left\langle \left[\mathbf{u}^{\mathrm{T}}(\mathbf{r}, t)\right]^2 \right\rangle = 4\pi \int\limits_0^\infty dq\, q^2 F(q),$$

and coefficients in the left-hand side of Eq. (14.7) are given by the formula

$$\left\langle u_k^{\mathrm{T}}(\mathbf{r}, t) u_l^{\mathrm{T}}(\mathbf{r}, t) \right\rangle = \frac{2}{3} T \delta_{kl}. \tag{14.9}$$

Finally, using Eq. (14.8), the right-hand side of Eq. (14.7) can be represented in the form

$$-\int d\mathbf{q} e^{i\mathbf{q}\mathbf{r}} U_m(\mathbf{q}, t) g_{im}(\mathbf{q}), \tag{14.10}$$

where $\mathbf{U}(\mathbf{q}, t)$ is the spatial Fourier transform of average velocity field $\mathbf{U}(\mathbf{r}, t)$ and tensor $g_{im}(\mathbf{q})$ is given by the formula

$$\begin{aligned} g_{im}(\mathbf{p}) &= 2 \int d\mathbf{q} \frac{q_m p_l p_k}{(\mathbf{q}+\mathbf{p})^2} \\ &\times \left[(q_i + p_i)\, \Delta_{kl}(\mathbf{q}) + (q_k + p_k)\, \Delta_{il}(\mathbf{q}) \right] F(q). \end{aligned} \tag{14.11}$$

From the obvious fact that tensor $g_{im}(\mathbf{p})$ is invariant relative rotations in space follows that it must have the form

$$g_{im}(\mathbf{p}) = A(p)\delta_{im} + B(p)\frac{p_i p_m}{\mathbf{p}^2}. \tag{14.12}$$

Moreover, the term proportional to $B(p)$ disappears in Eq. (14.12) in view of Eq. (14.4) (it expresses the property of incompressibility of average field $\mathbf{U}(\mathbf{r}, t)$) and the identity $\mathbf{p}\mathbf{U}(\mathbf{p}, t) = 0$ following from Eq. (14.4), so that the right-hand side (14.10) of Eq. (14.7) assumes the form

$$-\int d\mathbf{q} e^{i\mathbf{q}\mathbf{r}} U_i(\mathbf{q}, t) A(q).$$

Substituting this expression in the right-hand side of Eq. (14.7) and taking into account Eq. (14.9), we rewrite Eqs. (14.6), (14.7) in the form

$$\begin{aligned} &\frac{\partial}{\partial t} U_i(\mathbf{r}, t) + \tau_i(\mathbf{r}, t) = -\frac{\partial}{\partial r_i} P(\mathbf{r}, t), \\ &\frac{\partial}{\partial t} \tau_i(\mathbf{r}, t) + \frac{2}{3} T \Delta U_i(\mathbf{r}, t) = -A(-i\boldsymbol{\nabla}) U_i(\mathbf{r}, t). \end{aligned} \tag{14.13}$$

The kernel of the integral operator appeared in the second equation can be obtained from comparison of Eq. (14.11) with (14.12). The result is as follows

$$A(p) = \int d\mathbf{q} \frac{F(q)}{(q^2 + p^2)} \left[p^2 - \frac{(\mathbf{q}\mathbf{p})^2}{q^2} \right] \left[q^2 - (\mathbf{q}\mathbf{p}) - 2\frac{(\mathbf{q}\mathbf{p})^2}{p^2} \right]. \tag{14.14}$$

We can perform integration in Eq. (14.14) over angular coordinate to obtain the expression

$$A(p) = 2\pi p^2 \int_0^\infty dq\, F(q) \left\{ \frac{2q^2}{3p^2} - \frac{q^4 - p^4}{4p^4} + \frac{(q^2 - p^2)^3}{8qp^5} \ln \left| \frac{q+p}{q-p} \right| \right\}. \tag{14.15}$$

We note additionally that Eqs. (14.13) with allowance for Eq. (14.4) yield the identity $\Delta P(\mathbf{r}, t) = 0$, so that $P(\mathbf{r}, t) = P_0 = \text{const}$. Then, eliminating quantity $\tau_i(\mathbf{r}, t)$ from system of equations (14.11) and (14.4), we arrive at a single equation in the vector field of average velocity $\mathbf{U}(\mathbf{r}, t)$ of liquid

$$\frac{\partial^2}{\partial t^2} \mathbf{U}(\mathbf{r}, t) - \left[\frac{2}{3} T\Delta + A(-i\boldsymbol{\nabla}) \right] \mathbf{U}(\mathbf{r}, t) = 0. \tag{14.16}$$

The corresponding dispersion equation is as follows

$$\omega^2(p) = \frac{2}{3} T p^2 - A(p).$$

It can be rewritten in the form

$$\omega^2(p) = 2\pi p^2 \int_0^\infty dq\, F(q) f\left(\frac{q}{p}\right),$$

where

$$f(x) = \frac{2 - x^2}{3} + \frac{x^4 - 1}{8} - \frac{(x^2 - 1)^3}{16x} \ln \left| \frac{1+x}{1-x} \right|.$$

Function $f(x)$ has the following asymptotics

$$f(x) = \begin{cases} 2/3, & x \ll 1, \\ 4/15 & x \gg 1. \end{cases}$$

Moreover, the equalities

$$\frac{2}{3} \geq f(x) \geq \frac{4}{15}$$

hold for arbitrary x.

Thus, the time-dependent development of disturbances in average flow is governed by the hyperbolic equation (14.16). As a result, the turbulent medium possesses certain quasi-elastic properties; namely, disturbances diffuse in the turbulent medium as transverse waves showing dispersion property. The phase and group velocities of these waves vary in the limits

$$\sqrt{\frac{2}{3} T} \geq c \geq \sqrt{\frac{4}{15} T}.$$

14.2 Sound radiation by vortex motions

In the previous section, we considered the physical effect immediately related to statistical averaging of the nonlinear system of hydrodynamical equations in the case of noncompressible liquid. Here, we consider an effect related to weakly compressible media; namely, we consider the problem on sound radiation by a weakly compressible medium.

This problem corresponds to the inclusion of random fields in the linearized equations of hydrodynamics. Note that, within the framework of linear equations, parametric action of turbulent medium yields the equations of acoustics with random refractive index. We dealt with such problems in Chapter 13.

Turbulent motion of a liquid in certain finite spatial region excites acoustic waves outside this region. In the case of a weakly compressible liquid, this sound field is such as if it were generated by the static distribution of acoustic quadrupoles whose instantaneous power per unit volume is given by the relationship

$$T_{ij}(\mathbf{r},t) = \rho_0 v_i^{\mathrm{T}}(\mathbf{r},t) v_j^{\mathrm{T}}(\mathbf{r},t),$$

where ρ_0 is the average density and $v_i^{\mathrm{T}}(\mathbf{r},t)$ are the components of the velocity of liquid in turbulent region V inside which the turbulence is assumed homogeneous and isotropic in space and stationary in time (we use the coordinate system in which the whole of the liquid is quiescent). Turbulent motions cause the fluctuating waves of density $\rho(\mathbf{r},t)$ that satisfy the wave equation

$$\left(\frac{\partial^2}{\partial t^2} - c_0^2 \Delta\right) \rho(\mathbf{r},t) = \frac{\partial^2}{\partial r_i \partial r_j} T_{ij}(\mathbf{r},t), \qquad (14.17)$$

where c_0 is the sound velocity in the homogeneous portion of the medium, i.e., outside the region of turbulent motions.

The solution to this equation has the form of the retarded solution

$$\rho(\mathbf{r},t) = \frac{1}{4\pi c_0^2} \frac{\partial^2}{\partial r_i \partial r_j} \int\limits_V d\mathbf{y} T_{ij}\left(\mathbf{y}, t - \frac{|\mathbf{r}-\mathbf{y}|}{c_0}\right).$$

For distances significantly exceeding linear sizes of turbulent region V, this solution can be reduced to the asymptotic expression

$$\rho(\mathbf{r},t) = \frac{1}{4\pi c_0^4} \frac{r_i r_j}{x^3} \int\limits_V d\mathbf{y} \ddot{T}_{ij}\left(\mathbf{y}, t - \frac{|\mathbf{r}-\mathbf{y}|}{c_0}\right), \qquad (14.18)$$

where

$$\ddot{T}_{ij}(\mathbf{y},t) = \frac{\partial^2}{\partial t^2} T_{ij}(\mathbf{y},t).$$

Correspondingly, average energy flux density of acoustic wave excited by turbulent motions

$$q(\mathbf{r},t) = \frac{c_0^3 \left\langle \rho^2(\mathbf{r},t)\right\rangle}{\rho_0}$$

is given by the expression

$$\begin{aligned} q(\mathbf{r},t) &= \frac{1}{16\pi^2 c_0^5} \frac{r_i r_j r_k r_l}{r^6} \\ &\times \int\limits_V d\mathbf{y} \int\limits_V d\mathbf{z} \left\langle \ddot{T}_{ij}\left(\mathbf{y}, t - \frac{|\mathbf{r}-\mathbf{y}|}{c_0}\right) \ddot{T}_{kl}\left(\mathbf{z}, t - \frac{|\mathbf{r}-\mathbf{z}|}{c_0}\right)\right\rangle. \end{aligned} \qquad (14.19)$$

Further calculations require the knowledge of the space–time correlators of the velocity field. The current theory of strong turbulence is incapable of yielding the corresponding relationships. For this reason, investigators limit themselves to various plausible hypotheses that allow complete calculations to be performed (see, e.g., [224]–[226], [251]). In particular, it was shown that, with allowance for incompressibility of liquid in volume V and the use of the Kolmogorov–Obukhov hypothesis and a number of simplifying hypotheses for splitting space-time correlators, from Eq. (14.19) follows that both average energy flux density and acoustic power are proportional to $\sim M^5$, where

$$M = \frac{\sqrt{\langle \mathbf{v}^2(\mathbf{r},t)\rangle}}{c_0}$$

is the *Mach number* (significantly smaller than unity).

We note that this result can be explained purely hydrodynamically, by analyzing vortex interactions in weakly compressible medium [128]. The simplest sound-radiating vortex systems are the pair of vortex lines (radiating cylindrical waves) and the pair of vortex rings (radiating spherical waves).

14.2.1 Sound radiation by vortex lines

Consider two parallel vortex lines separated by distance $2h$ and characterized by equal intensities

$$\kappa = \frac{1}{2}\pi\xi\sigma,$$

where ξ is the vorticity (the size of the vortex uniformly distributed over the area of infinitely small section σ), so that the circulation about each of vortex line is

$$\Gamma = 2\pi\kappa.$$

We will call these vortex lines simply vortices. In a noncompressible liquid, these vortices revolve with angular velocity

$$\omega = \frac{\kappa}{2h^2}.$$

around the center of the line connecting these vortices (see, e.g., [248]).

Select the coordinate system with the origin at a fixed point and z-axis along the vortex line. In this coordinate system, velocity potential $\varphi_0(\mathbf{r},t)$ ($\mathbf{v}_0(\mathbf{r},t) = -\operatorname{Re}\boldsymbol{\nabla}\varphi_0(\mathbf{r},t)$) and squared velocity $\mathbf{v}_0^2(\mathbf{r},t)$ assume the forms

$$\begin{aligned} \varphi_0(\mathbf{r},t) &= ik\ln\left[r^2e^{2i\theta}-h^2e^{2i\omega t}\right], \\ \mathbf{v}_0^2(\mathbf{r},t) &= \frac{4\kappa^2r^2}{r^4+h^4-2r^2h^2\cos 2(\omega t-\theta)}. \end{aligned} \tag{14.20}$$

Here $re^{i\theta}$ is the radius-vector of the observation point.

According to the Bernoulli equation, local velocity pulsations described by Eq. (14.20) must produce the corresponding pressure pulsations; in the case of weakly compressible medium, namely, under the condition that $M \ll 1$ ($M = \frac{\kappa}{2hc_0}$ is the Mach number and c_0 is the sound velocity) these pressure pulsations will propagate at large distances as sound waves.

Encircle the origin with a circle of radius R such that $h \ll R \ll \lambda$, where λ is the sound wavelength. This is possible because

$$\frac{\lambda}{h} = \frac{\pi}{M} \gg 1 \quad \text{for} \quad M \ll 1.$$

In region $r < R$, dynamics of liquid approximately coincides with the dynamics of noncompressible liquid. In other words, the dynamics of liquid is described by Eqs. (14.20) in this region.

In region $r > R$, equations of motion coincide with the standard equations of acoustics (see, e.g., [217])

$$\left(\frac{\partial^2}{\partial t^2} - c_0^2\Delta\right)\varphi(\mathbf{r},t) = 0, \quad p'(\mathbf{r},t) = \rho_0 \frac{\partial}{\partial t}\varphi(\mathbf{r},t). \tag{14.21}$$

Here, $\varphi(\mathbf{r},t)$ is the potential of velocity in the acoustic wave, $p'(\mathbf{r},t)$ is the pressure in the wave, and ρ_0 is the density of the liquid.

Taking into account the fact that velocity $\mathbf{v}_0(\mathbf{r},t)$ depends on t and θ only in combination $2(\omega t - \theta)$, we will seek the solution to Eq. (14.21) in the form

$$\varphi(\mathbf{r},t) = f(r)e^{2i(\omega t-\theta)}, \tag{14.22}$$

Substituting Eq. (14.22) in Eq. (14.21) and solving the resulting equation in function $f(r)$ with allowance for the radiation condition, we obtain

$$\varphi(\mathbf{r},t) = AH_2^2\left(2\omega r/c_0\right)e^{2i(\omega t-\theta)}, \tag{14.23}$$

where $H_2^2(z)$ is the Hankel function of the second kind and A is a constant.

For $r \gg c_0/2\omega$ we have the standard divergent cylindrical wave with wavelength $\lambda = \pi c_0/\omega$

$$\varphi(\mathbf{r},t) = A\sqrt{\frac{c}{\pi\omega r}}\exp\{2i(\omega t - \omega r/c_0 - \theta - 3\pi/8)\}.$$

In the opposite case $r \ll \lambda$, we obtain

$$\varphi(\mathbf{r},t) = i\frac{A}{\pi}\left(\frac{c}{\omega r}\right)^2 e^{2i(\omega t-\theta)}.$$

Potential $\varphi(\mathbf{r},t)$ in region $h \ll r \ll \lambda$ must coincide with the oscillating portion of potential $\varphi_0(\mathbf{r},t)$ (see, e.g., [217]), i.e., with

$$\varphi_0^{(1)}(\mathbf{r},t) = -i\kappa\frac{h^2}{r^2}e^{2i(\omega t-\theta)}.$$

This condition yields the following expression for constant A

$$A = -\pi\kappa M^2 = -\frac{\pi\kappa^3}{4h^2c_0^2}.$$

Consequently, potential $\varphi(\mathbf{r},t)$ in the wave zone assumes the form

$$\varphi(\mathbf{r},t) = -\kappa M^{3/2}\sqrt{\frac{\pi h}{r}}\exp\{2i(\omega t - \omega r/c_0 - \theta - 3\pi/8)\}. \tag{14.24}$$

The pressure in sound wave can be determined from potential (14.24) using Eq. (14.21). The result is as follows

$$p'(\mathbf{r},t) = -2\kappa\omega\rho_0 M^{3/2}\sqrt{\frac{\pi h}{r}}\exp\{2i(\omega t - \omega r/c_0 - \theta - 3\pi/8)\}.$$

The sound intensity (energy radiated per unit time) can be obtained by integrating along the circle of radius $R \gg \lambda$

$$I = \frac{c_0}{\rho_0} \oint dl \left\langle [p'(r,t)]^2 \right\rangle = 2\pi^2 \rho_0 M^4 \frac{\kappa^3}{h^2}. \tag{14.25}$$

The radiated energy must coincide with the interaction energy of vortices located in region $r < R$. Total energy in region $r < R$ is

$$E = \frac{\rho_0}{2} \int dS \mathbf{v}_0^2(\mathbf{r}, t). \tag{14.26}$$

Substituting Eq. (14.20) in (14.26) and discarding infinite terms corresponding to the energy of motion of vortices themselves (we assume vortices the point vortices), we obtain the interaction energy in the form

$$E_1 = 4\pi\kappa^2 \rho_0 \ln(R/h). \tag{14.27}$$

The interaction energy can vary only at the expense of varying the distance between vortices ($h = h(t)$), because the circulation remains intact due to the fact that we consider nonviscous medium.

Differentiating Eq. (14.27) with respect to time, we obtain energy variation rate

$$I(t) = -4\pi\rho_0 \frac{\kappa^2}{h(t)} \frac{d}{dt} h(t), \tag{14.28}$$

which is just transferred into the energy of acoustic waves. Using Eqs. (14.25) and (14.28), we obtain that the distance between vortices satisfies the equation

$$\frac{d}{dt} h(t) = \frac{\pi\kappa M^4}{2h(t)}. \tag{14.29}$$

Integrating Eq. (14.29) with allowance for the fact that $M = M(t) = \kappa/(2h(t)c_0)$, we obtain

$$h(t) = h_0 \left[1 + 6\pi M_0^4 \omega_0 t\right]^{1/6}.$$

Thus, the intensity of radiated sound is proportional to M^4, $I \sim M^4$. It is obvious that, in the case of statistically distributed system of vortex line pairs, this estimate remains valid for certain portion of the plane.

14.2.2 Sound radiation by vortex rings

In a noncompressible liquid, a vortex ring of intensity κ causes the liquid to move with a velocity equal by the *Biot–Savart law* (see, e.g. [248]) to,

$$\mathbf{v}(\mathbf{r}, t) = \frac{\kappa}{2} a \int_0^{2\pi} d\phi \frac{\mathbf{s} \times \mathbf{r}}{r^3}.$$

Here, $\mathbf{s}$ is the unit vector tangent to the vortex ring (it is directed along the vortex vector), a is the radius of the ring, and $\mathbf{r}$ is the vector specifying observation point position relative to points lying on the ring. In the cylindrical coordinate system with origin at the center of ring and z-axis directed along ring axis, we have

$$v_R = \frac{\kappa}{2} a \int_0^{2\pi} d\phi \frac{\cos\phi}{r^3}, \quad v_\theta = 0, \quad v_z = \frac{\kappa}{2} a \int_0^{2\pi} d\phi \frac{a - R\cos\phi}{r^3}, \tag{14.30}$$

where

$$r = \left(R^2 + z^2 + a^2 - 2Ra\cos\phi\right)^{1/2}.$$

Here (R, θ, z) are the coordinates of the radius-vector of the observation point.

Let now we have two vortex rings of equal intensities and equal radii a_0 at distance $2h_0$. In this case, the front ring will increase in size, while the rear ring will decrease and pursue the front one. At certain instant, it will penetrate through the front ring, and the rings switch places. This phenomenon is called the *game of vortex rings*. In a weakly compressible liquid, these motions of rings produce local regions of compression and rarefaction; these regions propagate in the medium and, for large distances assume the form of spherical acoustic waves. To determine the structure of radiated sound, we must know relative motions of rings in a noncompressible liquid.

Let rings have radii $a_1(t)$ and $a_2(t)$ and are separated by distance $2h(t)$ at certain instant t. The rates of variations of ring radii are equal to radial velocities that rings induce at each other, and the rate of variation of the distance between the rings is equal to the difference of the z-components of velocities induced by the rings. Consequently, we have

$$\begin{aligned}
\frac{d}{dt} a_1(t) &= 2\kappa a_2(t) h(t) \int_0^{\pi} d\phi \frac{\cos\phi}{|a_1(t) - a_2(t)|^3}, \\
\frac{d}{dt} a_2(t) &= -2\kappa a_1(t) h(t) \int_0^{\pi} d\phi \frac{\cos\phi}{|a_1(t) - a_2(t)|^3}, \\
\frac{d}{dt} h(t) &= -\frac{\kappa}{2}\left[a_1^2(t) - a_2^2(t)\right] \int_0^{\pi} d\phi \frac{1}{|a_1(t) - a_2(t)|^3},
\end{aligned} \tag{14.31}$$

where

$$|a_1(t) - a_2(t)| = \left[a_1^2(t) + a_2^2(t) + 4h^2(t) - 2a_1(t)a_2(t)\cos\phi\right]^{1/2}.$$

Equations (14.31) should be solved with the initial conditions at $t = 0$

$$a_1(0) = a_2(0) = a_0, \quad h(0) = h_0.$$

The first two equations immediately yield the relationship between $a_1(t)$ and $a_2(t)$; namely,

$$a_1^2(t) + a_2^2(t) = 2a_0^2. \tag{14.32}$$

This relationship shows that the moment of inertia of rings relative the z-axis is conserved.

The integrals in the right-hand side of Eq. (14.31) can be expressed in terms of elliptic functions. If rings are far from each other ($\gamma = h_0/a_0 \gg 1$), they interact only slightly, and we can assume that, in the first approximation, they move independently with the velocities determined by the ring areas. In the other limiting case $\gamma \ll 1$ (namely this case will be considered in what follows), the rings interact actively. The integrals in the right-hand side of Eq. (14.31) are mainly contributed by the neighborhood of point $\phi = 0$. For this reason, we can replace $\cos\phi$ in the numerators of two first equations by unity. Thus we obtain the second integral of motion

$$4h^2(t) + [a_1(t) - a_2(t)]^2 = 4h_0^2. \tag{14.33}$$

Integral (14.33) means that the distance between points on different rings at the same polar angle is the conserved quantity.

In view of existence of integrals (14.32) and (14.33), we can reduce system (14.31) to a single equation in variable $\theta(t)$ determined by the equalities

$$\begin{aligned} a_1(t) &= \sqrt{2}a_0 \cos\left(\frac{\pi}{4} - \gamma \sin\theta(t)\right), \quad a_2(t) = \sqrt{2}a_0 \sin\left(\frac{\pi}{4} - \gamma \sin\theta(t)\right), \\ h(t) &= h_0 \cos\theta(t). \end{aligned} \tag{14.34}$$

With this definition, conservation laws (14.32) and (14.33) are satisfied automatically (in the first order with respect to γ). Substituting Eq. (14.34) in Eq. (14.31), expanding the result in the series, and calculating the integral, we obtain

$$\theta(t) = \frac{\kappa}{2h_0^2}t.$$

Consequently, the ring radii and the distance between rings are given by the expression

$$\begin{aligned} a_1(t) &= a_0\left(1 + \gamma \sin(\omega t)\right), \quad a_2(t) = a_0\left(1 - \gamma \sin(\omega t)\right), \\ h(t) &= h_0 \cos(\omega t), \end{aligned} \tag{14.35}$$

where

$$\omega = \frac{\kappa}{2h_0^2}, \tag{14.36}$$

as in the above case of vortex lines.

We note that the infinitely thin rings move with an infinite velocity. However, actual vortex rings move with a finite velocity significantly smaller than the sound velocity, but the dynamics of relative motion of rings only slightly differs from that we just obtained. The fact that angular velocity (14.36) coincides with the corresponding angular velocity in the case of vortex lines shows that the points lying on rings at equal polar angles revolve relative the center point of the line connecting them at the rotational speed coinciding with the rotational speed of vortex lines located at the same distance and having the same intensity as the rings.

To study the structure of the sound radiated by the system of rings, we need to know the velocity field for large distances from the system. We associate the coordinate system with the point located at the center of the common axis segment connecting the centers of the rings. The velocity of liquid outside rings is given by the formula

$$\mathbf{v}(\mathbf{r}, t) = \frac{\kappa}{2}a_1(t)\int_0^{2\pi} d\phi \frac{\mathbf{s}_1 \times \mathbf{r}_1}{r_1^3} + \frac{\kappa}{2}a_2(t)\int_0^{2\pi} d\phi \frac{\mathbf{s}_2 \times \mathbf{r}_2}{r_2^3}. \tag{14.37}$$

Here, $\mathbf{s}_1$ and $\mathbf{s}_2$ are the unit vectors tangent to the vortex rings and $\mathbf{r}_1$ and $\mathbf{r}_2$ are the vectors specifying observation point position relative to points lying on the rings. For large distances from the rings, the oscillating parts of the velocity have the form (in the cylindrical coordinates)

$$\begin{aligned} v_R^{(1)}(t) &= \frac{3\kappa}{4}\frac{R\left(4z^2-R^2\right)}{\left(R^2+z^2\right)^{3/2}}h\left(a_1^2-a_2^2\right), \quad v_\theta^{(1)}(t)=0, \\ v_z^{(1)}(t) &= \frac{3\kappa}{4}\frac{z\left(2z^2-3R^2\right)}{\left(R^2+z^2\right)^{3/2}}h\left(a_1^2-a_2^2\right). \end{aligned} \tag{14.38}$$

Introducing potential by the formula $\mathbf{v}^{(1)} = -\boldsymbol{\nabla}\varphi^{(1)}$, we obtain

$$\varphi^{(1)}(t) = -\frac{\kappa}{4}\frac{R^2-2z^2}{\left(R^2+z^2\right)^{5/2}}h\left(a_1^2-a_2^2\right). \tag{14.39}$$

Substituting Eq. (14.35) in Eq. (14.39) and changing to spherical coordinates, we can rewrite Eq. (14.39) in the complex form

$$\varphi^{(1)}(t) = i\frac{\kappa}{2}a_0^3\gamma^2\frac{1-3\cos^2\theta}{r^3}e^{2i\omega t}. \tag{14.40}$$

Now, we will proceed similar to the case of two vortex lines. We encircle the origin by a sphere of radius L such that $a_0 \ll L \ll \lambda$, where λ is the wavelength of radiated sound waves. We will use the fact that the motion of liquid inside the sphere approximately coincides with the motion of noncompressible liquid. Outside the sphere, the equations of motion will have the form of Eq. (14.21) with the only difference that now Δ is the spatial Laplace operator.

Represent potential φ in the form

$$\varphi(r,\theta) = f(r,\theta)e^{2i\omega t}.$$

Substituting this expression in Eq. (14.21) and taking into account that we must obtain divergent spherical waves for $r \gg \lambda$, we obtain $f(r,\theta)$ in the form

$$f(r,\theta) = \sum_{n=0}^{\infty}\frac{A_n}{\sqrt{r}}H_{n+1/2}^{(2)}\left(\frac{2\omega r}{c_0}\right)P_n(\cos\theta),$$

where $H_{n+1/2}^{(2)}(z)$ is the Hankel function of the second kind and $P_n(z)$ is the Legendre polynomial. Comparing potential $\varphi(r,\theta)$ for $r \ll \lambda$ with $\varphi^{(1)}(r,\theta)$, we obtain that $n=2$ and

$$A_2 = \frac{2\sqrt{\pi}}{3}\frac{\kappa\gamma^2 a_0^3\omega^{5/2}}{c_0^{5/2}}.$$

Consequently, potential and pressure in the wave zone have the forms

$$\begin{aligned} \varphi(r,\theta) &= \frac{2\kappa}{3}\left(\frac{\kappa}{2h_0c_0}\right)^2\frac{a_0}{r}\left(1-3\cos^2\theta\right)\exp\left\{2i\left(\omega t-\frac{\omega r}{c_0}-\frac{3\pi}{4}\right)\right\}, \\ p'(r,\theta) &= \frac{4i\kappa}{3}\left(\frac{\kappa}{2h_0c_0}\right)^2\frac{a_0\omega}{r}\rho_0\left(1-3\cos^2\theta\right)\exp\left\{2i\left(\omega t-\frac{\omega r}{c_0}-\frac{3\pi}{4}\right)\right\}, \end{aligned}$$

so that angular distribution of the energy radiated per unit time is

$$I(\theta) = \frac{8\pi}{9}\frac{\kappa^3a_0^2}{h_0^3}\left(\frac{\kappa}{2h_0c_0}\right)^5\rho_0\left(1-3\cos^2\theta\right)^2.$$

Vortex rings radiate energy as a quadrupole; the main portion of energy is radiated in a cone about z-axis with corner angle $106°$ in both positive and negative z-directions. Integrating over θ, we obtain that total energy radiated per unit time is given by the expression

$$I = \frac{64\pi}{45} \frac{\kappa^3 a_0^2}{h_0^3} \left(\frac{\kappa}{2h_0 c_0} \right)^5 \rho_0.$$

Thus, intensity of sound radiated by a pair of vortex rings is proportional to M^5. In the case of statistically distributed system of pairs of vortex rings, this proportionality will remain valid for certain portion of space, which agrees with estimates obtained in [224] – [226].

Part V

Appendixes

Appendix A

Variation (functional) derivatives

Recall first the general definition of a functional. One says that a functional is given if a rule is fixed that associates a number to every function from certain function family. Below, we give some examples of functionals.

$$\text{(a)} \qquad F[\varphi(\tau)] = \int_{t_1}^{t_2} d\tau a(\tau)\varphi(\tau),$$

where $a(t)$ is the given (fixed) function and limits t_1 and t_2 can be both finite and infinite. This is the linear functional.

$$\text{(b)} \qquad F[\varphi(\tau)] = \int_{t_1}^{t_2}\int_{t_1}^{t_2} d\tau_1 d\tau_2 B(\tau_1,\tau_2)\varphi(\tau_1)\varphi(\tau_2),$$

where $B(t_1,t_2)$ is the given (fixed) function. This is the quadratic functional.

$$\text{(c)} \qquad F[\varphi(\tau)] = f\left(\Phi[\varphi(\tau)]\right),$$

where $f(x)$ is the given function and quantity $\Phi[\varphi(\tau)]$ is by itself the functional.

Estimate the difference between the values of a functional calculated for functions $\varphi(\tau)$ and $\varphi(\tau)+\delta\varphi(\tau)$, where $\delta\varphi(\tau)\neq 0$ for $t-\frac{1}{2}\Delta t<\tau<t+\frac{1}{2}\Delta t$ (see Fig. A.1).

The variation of a functional is defined as the linear (in $\delta\varphi(\tau)$) portion of the difference

$$\delta F[\varphi(\tau)] = \left\{F\left[\varphi(\tau)+\delta\varphi(\tau)\right] - F[\varphi(\tau)]\right\}.$$

The limit

$$\frac{\delta F[\varphi(\tau)]}{\delta\varphi(t)dt} = \lim_{\Delta t\to 0}\frac{\delta F[\varphi(\tau)]}{\int_{\Delta t} d\tau\delta\varphi(\tau)} \tag{A.1}$$

is called the *variational* (or *functional*) *derivative* (see, e.g., [251]).

For short, we will use notation $\delta F[\varphi(\tau)]/\delta\varphi(t)$ instead of $\delta F[\varphi(\tau)]/\delta\varphi(t)dt$.

Note that, if we use function $\delta\varphi(\tau)=\alpha\delta(\tau)$, where $\delta(\tau)$ is the Dirac delta function in Eq. (A.1), then Eq. (A.1) can be represented in the form of the ordinary derivative

$$\frac{\delta F[\varphi(\tau)]}{\delta\varphi(t)} = \lim_{\alpha\to 0}\frac{d}{d\alpha}F[\varphi(\tau)+\alpha\delta(\tau-t)].$$

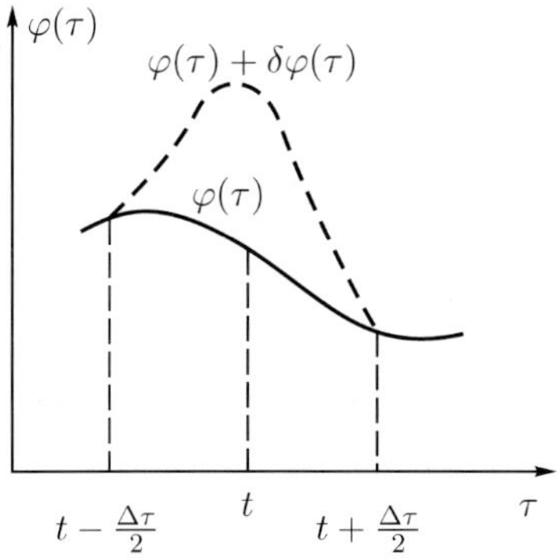

Figure A.1: To definition of variational derivative.

The variational derivative of functional $F[\varphi(\tau)]$ is again the functional of $\varphi(\tau)$, which depends additionally on point t as a parameter. As a result, this variational derivative will have two types of derivatives; one can differentiate it in the ordinary sense with respect to parameter t and in the functional sense with respect to $\varphi(\tau)$ at point $\tau = t'$, thus obtaining the second variational derivative of the initial functional

$$\frac{\delta^2 F[\varphi(\tau)]}{\delta\varphi(t')\delta\varphi(t)} = \frac{\delta}{\delta\varphi(t')}\left[\frac{\delta F[\varphi(\tau)]}{\delta\varphi(t)}\right].$$

The second variational derivative will now be the functional of $\varphi(\tau)$ dependent on two points t and t', and so forth.

Determine the variational derivatives of functionals (**a**), (**b**), and (**c**).

In the case (**a**), we have

$$\delta F[\varphi(\tau)] = F[\varphi(\tau) + \delta\varphi(\tau)] - F[\varphi(\tau)] = \int\limits_{t-\frac{1}{2}\Delta t}^{t+\frac{1}{2}\Delta t} d\tau a(\tau)\delta\varphi(\tau).$$

If function $a(t)$ is continuous on segment Δt, then, by the average theorem,

$$\delta F[\varphi(\tau)] = a(t')\int\limits_{\Delta t} d\tau\delta\varphi(\tau),$$

where point t' belongs to segment $\left[t - \frac{1}{2}\Delta t, t + \frac{1}{2}\Delta t\right]$. Consequently,

$$\frac{\delta F[\varphi(\tau)]}{\delta\varphi(t)} = \lim_{\Delta t\to 0} a(t') = a(t). \qquad \text{(A.2)}$$

In the case (**b**), we obtain similarly

$$\frac{\delta F[\varphi(\tau)]}{\delta\varphi(t)} = \int\limits_{t_1}^{t_2} d\tau\,[B(\tau,t)+B(t,\tau)]\,\varphi(\tau) \quad (t_1 < t < t_2).$$

Note that function $B(\tau_1, \tau_2)$ can always be assumed a symmetric function of its arguments here.

In the case (**c**), we have

$$F[\varphi(\tau) + \delta\varphi(\tau)] = f\left(\Phi[\varphi(\tau)]\right) + \frac{\partial f(\Phi[\varphi(\tau)])}{\partial \Phi}\delta\Phi[\varphi(\tau)] + ...$$
$$= F[\varphi(\tau)] + \frac{\partial f(\Phi[\varphi(\tau)])}{\partial \Phi}\delta\Phi[\varphi(\tau)] + ...$$

and, consequently,

$$\frac{\delta}{\delta\varphi(t)} f\left(\Phi[\varphi(\tau)]\right) = \frac{\partial f(\Phi[\varphi(\tau)])}{\partial \Phi}\frac{\delta}{\delta\varphi(t)}\Phi[\varphi(\tau)]. \tag{A.3}$$

Consider now functional $\Phi[\varphi(\tau)] = F_1[\varphi(\tau)]F_2[\varphi(\tau)]$. We have

$$\begin{aligned}\delta\Phi[\varphi(\tau)] &= \{F_1[\varphi(\tau) + \delta\varphi(\tau)]F_2[\varphi(\tau) + \delta\varphi(\tau)] - F_1[\varphi(\tau)]F_2[\varphi(\tau)]\} \\ &= F_1[\varphi(\tau)]\delta F_2[\varphi(\tau)] + F_2[\varphi(\tau)]\delta F_1[\varphi(\tau)]\end{aligned}$$

and, consequently,

$$\frac{\delta}{\delta\varphi(t)}F_1[\varphi(\tau)]F_2[\varphi(\tau)] = F_1[\varphi(\tau)]\frac{\delta}{\delta\varphi(t)}F_2[\varphi(\tau)] + F_2[\varphi(\tau)]\frac{\delta}{\delta\varphi(t)}F_1[\varphi(\tau)]. \tag{A.4}$$

We can define the expression for the variational derivative of functional $\varphi(\tau_0)$ with respect to function $\varphi(t)$ by the formal relationship

$$\frac{\delta\varphi(\tau_0)}{\delta\varphi(t)} = \delta(\tau_0 - t). \tag{A.5}$$

Formula (A.5) can be proved, for example, by considering the linear functional of the form

$$F[\varphi(\tau)] = \frac{1}{\sqrt{2\pi}\sigma}\int_{-\infty}^{\infty} d\tau\varphi(\tau)\exp\left\{-\frac{(\tau - \tau_0)^2}{2\sigma^2}\right\}. \tag{A.6}$$

According to Eq. (A.2), the variational derivative of this functional has the form

$$\frac{\delta}{\delta\varphi(t)}F[\varphi(\tau)] = \frac{1}{\sqrt{2\pi}\sigma}\exp\left\{-\frac{(t - \tau_0)^2}{2\sigma^2}\right\}. \tag{A.7}$$

Performing now formal limit process $\sigma \to 0$ in Eqs. (A.6) and (A.7), we obtain the desired formula (A.5).

Formula (A.5) is very convenient for functional differentiation of functionals explicitly dependent on $\varphi(\tau)$. Indeed, for the quadratic functional (**b**), we have

$$\frac{\delta}{\delta\varphi(t)}\int_{t_1}^{t_2}\int_{t_1}^{t_2} d\tau_1 d\tau_2 B(\tau_1, \tau_2)\varphi(\tau_1)\varphi(\tau_2)$$
$$\overset{(A.4)}{=} \int_{t_1}^{t_2}\int_{t_1}^{t_2} d\tau_1 d\tau_2 B(\tau_1, \tau_2)\left[\frac{\delta\varphi(\tau_1)}{\delta\varphi(t)}\varphi(\tau_2) + \varphi(\tau_1)\frac{\delta\varphi(\tau_2)}{\delta\varphi(t)}\right]$$
$$\overset{(A.5)}{=} \int_{t_1}^{t_2} d\tau\left[B(t, \tau) + B(\tau, t)\right]\varphi(\tau) \quad (t_1 < t < t_2).$$

Consider the functional

$$F[\varphi(\tau)] = \int\limits_{t_1}^{t_2} d\tau L\left(\tau, \varphi(\tau), \frac{d\varphi(\tau)}{d\tau}\right)$$

as another example. In this case,

$$\begin{aligned}
&\frac{\delta}{\delta\varphi(t)}F[\varphi(\tau)] \\
&\overset{(A.3)}{=} \int\limits_{t_1}^{t_2} d\tau \left[\frac{\delta L\left(\tau, \varphi(\tau), \frac{d\varphi(\tau)}{d\tau}\right)}{\delta\varphi(\tau)} + \frac{\delta L\left(\tau, \varphi(\tau), \frac{d\varphi(\tau)}{d\tau}\right)}{\delta\dot{\varphi}(\tau)} \frac{d}{d\tau}\right] \frac{\delta\varphi(\tau)}{\delta\varphi(t)} \\
&\overset{(A.5)}{=} \left(-\frac{d}{dt}\frac{\partial}{\partial\dot{\varphi}(t)} + \frac{\partial}{\partial\varphi(t)}\right) L\left(t, \varphi(t), \frac{d\varphi(t)}{dt}\right),
\end{aligned}$$

where $\dot{\varphi}(t) = \frac{d}{dt}\varphi(t)$ if point t belongs to interval (t_1, t_2).

Just as a function can be expanded in the Taylor series, a functional $F[\varphi(\tau) + \eta(\tau)]$ can be expanded in the functional Taylor series in function $\eta(\tau)$ for $\eta(\tau) \sim 0$

$$\begin{aligned}
F[\varphi(\tau) + \eta(\tau)] &= F[\varphi(\tau)] + \int\limits_{-\infty}^{\infty} dt \frac{\delta F[\varphi(\tau)]}{\delta\varphi(t)} \eta(t) \\
&+ \frac{1}{2!} \int\limits_{-\infty}^{\infty}\int\limits_{-\infty}^{\infty} dt_1 dt_2 \frac{\delta^2 F[\varphi(\tau)]}{\delta\varphi(t_1)\delta\varphi(t_2)} \eta(t_1)\eta(t_2) + \ldots.
\end{aligned} \tag{A.8}$$

Note that the operator expression

$$\begin{aligned}
&1 + \int\limits_{-\infty}^{\infty} dt\eta(t)\frac{\delta}{\delta\varphi(t)} + \frac{1}{2!}\int\limits_{-\infty}^{\infty}\int\limits_{-\infty}^{\infty} dt_1 dt_2 \eta(t_1)\eta(t_2)\frac{\delta^2}{\delta\varphi(t_1)\delta\varphi(t_2)} + \ldots \\
&= 1 + \int\limits_{-\infty}^{\infty} dt\eta(t)\frac{\delta}{\delta\varphi(t)} + \frac{1}{2!}\left[\int\limits_{-\infty}^{\infty} dt\eta(t)\frac{\delta}{\delta\varphi(t)}\right]^2 + \ldots
\end{aligned} \tag{A.9}$$

can be written shortly as the operator

$$\exp\left\{\int\limits_{-\infty}^{\infty} dt\eta(t)\frac{\delta}{\delta\varphi(t)}\right\}, \tag{A.10}$$

whose action should be treated precisely in the sense of expansion (A.9). Using this operator, we can rewrite Eq. (A.8) in the form

$$F[\varphi(\tau) + \eta(\tau)] = e^{\int_{-\infty}^{\infty} dt\eta(t)\frac{\delta}{\delta\varphi(t)}} F[\varphi(\tau)], \tag{A.11}$$

which enables us to interpret operator (A.10) as the functional shift operator.

Consider now functional $F[t; \varphi(\tau)]$ dependent on parameter t. We can differentiate this functional with respect to t and determine its variational derivative with respect to $\varphi(t')$, as well. One can easily see that these operations commute, i.e., the equality

$$\frac{\partial}{\partial t}\frac{\delta F[t; \varphi(\tau)]}{\delta\varphi(t')} = \frac{\delta}{\delta\varphi(t')}\frac{\partial F[t; \varphi(\tau)]}{\partial t} \tag{A.12}$$

holds. If the domain of τ is independent of t, the validity of Eq. (A.12) is obvious. Otherwise, for example, for functionals $F[t;\varphi(\tau)]$ with $0 \le \tau \le t$, the validity of Eq. (A.12) can be checked on by expanding functional $F[t;\varphi(\tau)]$ in the functional Taylor series.

Consider now the law of functional derivative transformation under the change of functional variables.

Let function $\varphi(t)$ is replaced by the new function $\psi(t)$ according to the formula

$$\varphi(t) = \Psi\left[\psi(\tau);t\right], \tag{A.13}$$

where $\Psi\left[\psi(\tau);t\right]$ is a functional of function $\psi(\tau)$ dependent additionally on point t. Then functional $F\left[\varphi(\tau)\right]$ is certain complex functional of $\psi(\tau)$

$$F\left[\varphi(\tau)\right] = F\left[\Psi\left[\psi(\eta);\tau\right]\right] \equiv F_1\left[\psi(\tau)\right].$$

It is obvious that we have in this case

$$\frac{\delta F_1\left[\psi(\tau)\right]}{\delta\varphi(t)} = \int dt' \frac{\delta F\left[\varphi(\tau)\right]}{\delta\varphi(t')}\frac{\delta\Psi\left[\psi(\eta);t'\right]}{\delta\psi(t)}. \tag{A.14}$$

Appendix B

Fundamental solutions of wave problems in free space and layered media

In this Appendix, we discuss several properties of fundamental solutions (Green's functions) of wave equations in free space and layered media following monograph [136] and papers [107, 142].

B.1 Free space

First of all, we consider Green's function of the one-dimensional Helmholtz equation

$$\frac{d^2}{dx^2}g(x;x_0) + k^2 g(x;x_0) = \delta(x - x_0). \tag{B.1}$$

The solution of Eq. (B.1) satisfying radiation condition for $x \to \pm\infty$ has the form

$$g(x;x_0) = g(x - x_0) = \frac{1}{2ik}e^{ik|x-x_0|}. \tag{B.2}$$

The modulus $|x - x_0|$ appeared in the right-hand side of Eq. (B.2) by virtue of the fact that Eq. (B.1) is the equation of the second order in variable x. However, if we fix mutual order of observation points and source, then Green's function will satisfy the equality (for definiteness, we assume that $x_0 > x$)

$$\frac{\partial}{\partial x_0}g(x - x_0) = ikg(x - x_0)$$

that, being supplemented with the initial condition

$$g(x - x_0)|_{x_0=x} = g(0) = \frac{1}{2ik},$$

can be considered the first-order differential equation.

Thus, the order of the equation for Green's function decreases if source and observation points obey certain order. This property is generic of wave problems (*factorization property of wave equations*) and follows from the fact that the wave radiated in direction $x < x_0$ (or $x > x_0$) travels in free space without changing the direction.

In the general case, Green's function satisfies the second-order operator equation

$$\left\{\frac{\partial^2}{\partial x^2}+\hat{M}^2(\boldsymbol{\eta})\right\} g(x-x_0,\boldsymbol{\eta}-\boldsymbol{\eta}_0)=\delta(x-x_0)g(\boldsymbol{\eta}-\boldsymbol{\eta}_0), \tag{B.3}$$

where operator $\hat{M}(\boldsymbol{\eta})$ acts on the temporal and other spatial variables denoted by $\boldsymbol{\eta}$. For example, operator $\hat{M}^2(\boldsymbol{\eta})$ in Eq. (B.1) is the number $\hat{M}^2(\boldsymbol{\eta})=k^2$.

Structurally, Green's function is similar to Eq. (B.2),

$$g(x-x_0,\boldsymbol{\eta}-\boldsymbol{\eta}_0)=e^{i|x-x_0|\hat{M}(\boldsymbol{\eta})}g(0,\boldsymbol{\eta}-\boldsymbol{\eta}_0)=e^{i|x-x_0|\hat{M}(-\boldsymbol{\eta}_0)}g(0,\boldsymbol{\eta}-\boldsymbol{\eta}_0). \tag{B.4}$$

As a consequence, it can be described for $x<x_0$ by the operator equation of the first order in variable x (or x_0)

$$\begin{aligned}\frac{\partial}{\partial x_0}g(x-x_0,\boldsymbol{\eta}-\boldsymbol{\eta}_0)&=-\frac{\partial}{\partial x}g(x-x_0,\boldsymbol{\eta}-\boldsymbol{\eta}_0)\\&=i\hat{M}(\boldsymbol{\eta})g(x-x_0,\boldsymbol{\eta}-\boldsymbol{\eta}_0)=i\hat{M}(-\boldsymbol{\eta}_0)g(x-x_0,\boldsymbol{\eta}-\boldsymbol{\eta}_0)\end{aligned}$$

with the initial condition

$$g(x-x_0,\boldsymbol{\eta}-\boldsymbol{\eta}_0)|_{x_0=x}=g(0,\boldsymbol{\eta}-\boldsymbol{\eta}_0)\equiv g(\boldsymbol{\eta}-\boldsymbol{\eta}_0).$$

For $x>x_0$, the equations are similar.

The solution of Eq. (B.3) is continuous in x, but its derivative with respect to x is discontinuous at the point of source location $x=x_0$

$$\left.\frac{\partial}{\partial x}g(x-x_0,\boldsymbol{\eta}-\boldsymbol{\eta}_0)\right|_{x=x_0+0}-\left.\frac{\partial}{\partial x}g(x-x_0,\boldsymbol{\eta}-\boldsymbol{\eta}_0)\right|_{x=x_0-0}=\delta(\boldsymbol{\eta}-\boldsymbol{\eta}_0). \tag{B.5}$$

Substituting Eq. (B.4) in Eq. (B.5), we obtain the expression

$$2i\hat{M}(\boldsymbol{\eta})g(0,\boldsymbol{\eta}-\boldsymbol{\eta}_0)=\delta(\boldsymbol{\eta}-\boldsymbol{\eta}_0). \tag{B.6}$$

In the general case, operator $\hat{M}(\boldsymbol{\eta})$ can be considered as an integral operator. Indeed, action of operator $\hat{M}(\boldsymbol{\eta})$ on arbitrary function $f(\boldsymbol{\eta})$ is representable in the form

$$\hat{M}(\boldsymbol{\eta})f(\boldsymbol{\eta})=\int\limits_{-\infty}^{\infty}d\boldsymbol{\xi}\,\hat{M}(\boldsymbol{\eta})\delta(\boldsymbol{\eta}-\boldsymbol{\xi})f(\boldsymbol{\xi})=\int\limits_{-\infty}^{\infty}d\boldsymbol{\xi}\,M(\boldsymbol{\eta}-\boldsymbol{\xi})f(\boldsymbol{\xi}),$$

where the kernel of the integral operator is defined by the equality

$$M(\boldsymbol{\eta}-\boldsymbol{\xi})=\hat{M}(\boldsymbol{\eta})\delta(\boldsymbol{\eta}-\boldsymbol{\xi}). \tag{B.7}$$

The inverse operator $\hat{M}^{-1}(\boldsymbol{\eta})$ also can be introduced by the corresponding choice of kernel $M^{-1}(\boldsymbol{\eta}-\boldsymbol{\xi})$.

Applying operator $\hat{M}(\boldsymbol{\eta})$ to Eq. (B.6), we obtain, according to Eq. (B.7), the kernel of the integral operator in the form

$$M(\boldsymbol{\eta}-\boldsymbol{\eta}_0)=2i\hat{M}^2(\boldsymbol{\eta})g(0,\boldsymbol{\eta}-\boldsymbol{\eta}_0). \tag{B.8}$$

The kernel of the inverse integral operator

$$M^{-1}(\boldsymbol{\eta}-\boldsymbol{\eta}_0)=\hat{M}^{-1}(\boldsymbol{\eta})\delta(\boldsymbol{\eta}-\boldsymbol{\eta}_0)=2ig(0,\boldsymbol{\eta}-\boldsymbol{\eta}_0). \tag{B.9}$$

is obtained by applying the inverse operator $\hat{M}^{-1}(\boldsymbol{\eta})$ to Eq. (B.6).

Thus, kernels of integral operators $\hat{M}(\boldsymbol{\eta})$ and $\hat{M}^{-1}(\boldsymbol{\eta})$ are expressed in terms of the wave equation fundamental solution.

Consider now several specific wave problems.

1. We represent the Helmholtz equation in the form

$$\left(\frac{\partial^2}{\partial x^2}+\Delta_{\mathbf{R}}+k^2\right) g(x-x_0,\mathbf{R}-\mathbf{R}_0)=\delta(x-x_0)\delta(\mathbf{R}-\mathbf{R}_0), \tag{B.10}$$

where vector $\mathbf{R}$ denotes the coordinates in the plane perpendicular to the x-axis.

The solution of Eq. (B.10) satisfying radiation conditions at infinity has the form

$$g(\mathbf{r}-\mathbf{r}_0)=-\frac{1}{4\pi|\mathbf{r}-\mathbf{r}_0|}e^{ik|\mathbf{r}-\mathbf{r}_0|},\quad \mathbf{r}=\{x,\mathbf{R}\}.$$

Function $g(\mathbf{r})$ can be represented in the integral form

$$g(x,\mathbf{R})=\frac{1}{8i\pi^2}\int\frac{d\mathbf{q}}{\sqrt{k^2-\mathbf{q}^2}}\exp\left\{i\sqrt{k^2-\mathbf{q}^2}|x|+i\mathbf{q}\mathbf{R}\right\},$$

from which follows that operator $\hat{M}(\mathbf{R})$ has in this case the form

$$\hat{M}(\mathbf{R})=\sqrt{k^2+\Delta_{\mathbf{R}}},\quad \hat{M}(\mathbf{R}_0)=\sqrt{k^2+\Delta_{\mathbf{R}_0}},$$

and the corresponding kernels of the integral operators are given, according to Eqs. (B.8) and (B.9), by the expressions

$$\begin{aligned} M(\mathbf{R})&=2i\left(k^2+\Delta_{\mathbf{R}}\right)g(\mathbf{R})=-\frac{i}{2\pi R^2}\left(\frac{1}{R}-ik\right)e^{ikR},\\ M^{-1}(\mathbf{R})&=-\frac{i}{2\pi R}e^{ikR}. \end{aligned} \tag{B.11}$$

In the two-dimensional case, we have

$$g(\mathbf{r}-\mathbf{r}_0)=-\frac{i}{4}H_0^{(1)}\left(k|\mathbf{r}-\mathbf{r}_0|\right)\quad(\mathbf{r}=\{x,y\}),$$

where $H_0^{(1)}(k|\mathbf{r}|)$ is the Hankel function. As a consequence, kernels of the corresponding integral operators

$$\hat{M}(y)=\sqrt{k^2+\frac{\partial^2}{\partial y^2}},\quad \hat{M}^{-1}(y)=\frac{1}{\sqrt{k^2+\frac{\partial^2}{\partial y^2}}}$$

are given by the expressions

$$M(y)=\frac{k}{2|y|}H_1^{(1)}(k|y|),\quad M^{-1}(y)=\frac{1}{2}H_0^{(1)}(k|y|). \tag{B.12}$$

As we mentioned earlier, in the one-dimensional case, operators $\hat{M}$ and $\hat{M}^{-1}$ are simply the numbers.

2. We represent the nonstationary wave equation in the form of Eq. (B.3)

$$\left(\frac{\partial^2}{\partial x^2}+\Delta_{\mathbf{R}}-\frac{1}{c^2}\frac{\partial^2}{\partial t^2}\right) g(x-x_0,\mathbf{R}-\mathbf{R}_0,t-t_0)=\delta(x-x_0)\delta(\mathbf{R}-\mathbf{R}_0)\delta(t-t_0). \tag{B.13}$$

In this case, operator $\hat{M}^2(\boldsymbol{\eta})$ is the differential operator

$$\hat{M}^2(\mathbf{R}, t) = \Delta_{\mathbf{R}} - \frac{1}{c^2}\frac{\partial^2}{\partial t^2}.$$

In the three-dimensional case, the solution of Eq. (B.13) satisfying radiation conditions (the retarded solution) has the form

$$g(x, \mathbf{R}, t) = -\frac{c}{2\pi}\theta(t)\delta\left(c^2t^2 - x^2 - \mathbf{R}^2\right),$$

where $\theta(t)$ is the Heaviside step function. As a consequence, kernels of the corresponding integral operators are given by the formulas

$$M(\mathbf{R}, t) = \frac{i}{\pi ct}\theta(t)\frac{\partial}{\partial t}\delta\left(c^2t^2 - \mathbf{R}^2\right), \quad M^{-1}(\mathbf{R}, t) = -\frac{ic}{\pi}\theta(t)\delta\left(c^2t^2 - \mathbf{R}^2\right). \tag{B.14}$$

In the two-dimensional case,

$$g(x, y, t) = -\frac{c}{2\pi}\frac{\theta\left(ct - \sqrt{x^2 + y^2}\right)}{\sqrt{c^2t^2 - x^2 - y^2}} = -\frac{c}{2\pi}\frac{\theta\left(c^2t^2 - x^2 - y^2\right)}{\sqrt{c^2t^2 - x^2 - y^2}}$$

and, consequently,

$$M(y, t) = \frac{i}{\pi ct}\theta(t)\frac{\partial}{\partial t}\frac{\theta\left(c^2t^2 - y^2\right)}{\sqrt{c^2t^2 - y^2}}, \quad M^{-1}(y, t) = -\frac{ic}{\pi}\frac{\theta\left(ct - y\right)}{\sqrt{c^2t^2 - y^2}}. \tag{B.15}$$

In the one-dimensional case,

$$g(x, t) = -\frac{c}{2}\theta(ct - |x|)$$

and, consequently,

$$M(t) = \frac{i}{c}\delta'(t), \quad M^{-1}(t) = -ic\theta(t). \tag{B.16}$$

Here, we considered certain properties of fundamental solutions (Green's functions) of wave equations describing the field of the point source in unbounded free space. Note that the similar analysis for problems on the point source field in a finite layer of free or layered space differs from the above analysis only in insignificant details.

B.2 Layered space

For a layered medium in which $\varepsilon(x, y, z) = \varepsilon(z)$, wave equations can be factorized because waves spread in plane (x, y) and do not scatter in the backward direction.

We denote $G^{(1)}(z; z_0)$ the point source field in the one-dimensional space. This function satisfies the equation

$$\left[\frac{d^2}{dz^2} + k^2(z)\right] G^{(1)}(z; z_0) = \delta(z - z_0),$$

whose solution can be represented in the operator form,

$$G^{(1)}(z; z_0) = \hat{L}^{-2}(z)\delta(z - z_0),$$

where

$$\hat{L}^2(z) = \frac{d^2}{dz^2} + k^2(z).$$

In the two-dimensional space, the wave field of the point source is described by Green's function $G^{(2)}(x, z; z_0)$ satisfying the equation

$$\left[\frac{\partial^2}{\partial x^2} + \hat{L}^2(z)\right] G^{(2)}(x, z; z_0) = \delta(x)\delta(z - z_0).$$

The solution of this equation has the form

$$G^{(2)}(x, z; z_0) = e^{i|x|\hat{L}(z)} G^{(2)}(0, z; z_0), \tag{B.17}$$

where function $G^{(2)}(0, z; z_0)$ describes the wave field on axis $x = 0$. The discontinuity of derivative $\frac{\partial}{\partial x} G^{(2)}(x, z; z_0)$ at $x = 0$ is given by the expression

$$\left.\frac{\partial}{\partial x} G^{(2)}(x, z; z_0)\right|_{x=+0} - \left.\frac{\partial}{\partial x} G^{(2)}(x, z; z_0)\right|_{x=-0} = \delta(z - z_0).$$

Being combined with Eq. (B.17), this discontinuity yields the equality

$$2i\hat{L}(z) G^{(2)}(0, z; z_0) = \delta(z - z_0), \tag{B.18}$$

from which follows that

$$G^{(2)}(0, z; z_0) = \frac{1}{2i} \hat{L}^{-1}(z)\delta(z - z_0). \tag{B.19}$$

Applying now operator $\hat{L}^2(z)$ to Eq. (B.18), we obtain the equality

$$\hat{L}^2(z) G^{(2)}(0, z; z_0) = \frac{1}{2i} \hat{L}(z)\delta(z - z_0). \tag{B.20}$$

We can consider operators $\hat{L}(z)$ and $\hat{L}^{-1}(z)$ as the integral operators; in this case, Eqs. (B.20), (B.19) define the kernels of these operators. With this fact in mind, we see that Eq. (B.19) is the nonlinear integral equation for function $G^{(2)}(0, z; z_0)$ describing the wave field on axis $x = 0$,

$$\int\limits_{-\infty}^{\infty} d\xi G^{(2)}(0, z; \xi) G^{(2)}(0, \xi; z_0) = -\frac{1}{4} G^{(1)}(z; z_0),$$

where $G^{(1)}(z; z_0)$ is Green's function of the one-dimensional problem.

In the three-dimensional case, Green's function of layered medium satisfies the equation

$$\left[\frac{\partial^2}{\partial x^2} + \frac{\partial^2}{\partial y^2} + \hat{L}^2(z)\right] G^{(3)}(x, y, z; z_0) = \delta(x)\delta(y)\delta(z - z_0).$$

We represent the solution to this equation in the form

$$G^{(3)}(x, y, z; z_0) = e^{i|x|\hat{L}(y,z)} G^{(3)}(0, y, z; z_0), \tag{B.21}$$

where $\hat{L}(y,z) = \frac{\partial^2}{\partial y^2} + \hat{L}^2(z)$ and function $G^{(3)}(0,y,z;z_0)$ describes the wave field in plane (y,z). The condition of discontinuity of derivative $\frac{\partial}{\partial x}G^{(3)}(x,y,z;z_0)$ in plane $x=0$ yields the operator equality

$$G^{(3)}(0,y,z;z_0) = \frac{1}{2i}\hat{L}^{-1}(y,z)\delta(y)\delta(z-z_0),$$

which can be rewritten in terms of the Hankel function of the first kind

$$G^{(3)}(0,y,z;z_0) = -\frac{i}{4}H_0^{(1)}\left[y\hat{L}(z)\right]\delta(z-z_0).$$

Using the Hankel function integral representation

$$H_0^{(1)}(\beta\mu) = \frac{1}{i\pi}\int\limits_0^\infty \frac{dx}{x}\exp\left\{i\frac{\mu}{2}\left(x+\frac{\beta^2}{x}\right)\right\},$$

we obtain that function $G^{(3)}(0,y,z;z_0)$ is related to the solution of the parabolic equation

$$\frac{\partial}{\partial t}u(t,z;z_0) = \frac{i}{2k}\hat{L}^2(z)u(t,z;z_0), \quad u(0,z;z_0) = \delta(z-z_0)$$

with respect to auxiliary parameter t by the quadrature

$$G^{(3)}(0,y,z;z_0) = -\frac{1}{4\pi}\int\limits_0^\infty \frac{dt}{t}\exp\left[i\frac{k}{2t}y^2\right]u(t,z;z_0),$$

or by the expression

$$G^{(3)}(0,y,z;z_0) = -\frac{1}{4\pi}\int\limits_0^\infty \frac{dt}{t}\exp\left\{i\frac{k}{2t}\left(y^2+t^2\right)\right\}\psi(t,z;z_0),$$

where function $\psi(t,z;z_0)$ is the solution to the parabolic equation

$$\frac{\partial}{\partial t}\psi(t,z;z_0) = \frac{i}{2k}\left[\frac{\partial^2}{\partial z^2} + k^2(z) - k^2\right]\psi(t,z;z_0), \quad \psi(0,z;z_0) = \delta(z-z_0). \tag{B.22}$$

In view of arbitrary direction of the x-axis, we obtain that, for $y>0$, function

$$G^{(3)}(x,y,z;z_0) = G^{(3)}(\rho,z;z_0),$$

where $\rho^2 = x^2+y^2$, defines Green's function in the whole of the space,

$$G^{(3)}(x,y,z;z_0) = -\frac{1}{4\pi}\int\limits_0^\infty \frac{dt}{t}\exp\left\{i\frac{k}{2t}\left(x^2+y^2+t^2\right)\right\}\psi(t,z;z_0). \tag{B.23}$$

Integrating Eq. (B.23) first over y and x, we obtain the corresponding integral representations of two- and one-dimensional Green's functions

$$G^{(2)}(x,z;z_0) = \frac{1}{2ik}\left(\frac{k}{2\pi i}\right)^{1/2}\int\limits_0^\infty \frac{dt}{\sqrt{t}}\exp\left\{i\frac{k}{2t}\left(x^2+t^2\right)\right\}\psi(t,z;z_0), \tag{B.24}$$

$$G^{(1)}(z;z_0) = \frac{1}{2ik}\int\limits_0^\infty dt\exp\left\{i\frac{kt}{2}\right\}\psi(t,z;z_0). \tag{B.25}$$

Appendix C

Imbedding method in boundary-value wave problems

Different statistical methods are used to statistically describe dynamic systems; however, these statistical methods are applicable only to the problems of special types, namely, the problems that possess the dynamic causality property, in which case the solution to the problem depends only on preceding (in time or in space) parameter values and is independent of consequent ones. Boundary-value problems are not among these problems. In such cases, it is desirable to transform the problem at hand into the equivalent evolution-type initial value problem. Such a conversion is necessary if we deal with statistical problems and can appear practicable in the context of numerical procedures for solving deterministic problems.

The *imbedding method* (or *invariant imbedding method*, as it is usually called in mathematical literature) offers a possibility of reducing boundary-value problems at hand to the evolution-type initial value problems possessing the property of dynamic causality with respect to an auxiliary parameter.

The idea of this method was first suggested by V.A. Ambartsumyan (the so-called *Ambartsumyan invariance principle*) [4]–[6] for solving the equations of linear theory of radiative transfer. Further, mathematicians grasped this idea and used it to convert boundary-value (nonlinear, in the general case) problems into evolution-type initial value problems that are more convenient for simulations. Several monographs (see, e.g., [26, 39, 123]) deal with this method and consider both physical and computational aspects.

In the context of boundary-value wave problems, the imbedding method was developed in papers and books [14, 16, 135, 136]. A noteworthy feature of wave problems consists in the fact that the imbedding parameter, i.e., the parameter used to construct evolution-type equations, has a clear geometric meaning — it is the coordinate of the boundary interfacing the media. It seems that the imbedding method is the simplest among the methods capable of correct formulation of statistical wave problems in the general case.

The imbedding equations were obtained for many stationary and nonstationary, linear and nonlinear boundary-value wave problems in spaces of different dimensions. They are nonlinear integro-differential equations in finite- and often in infinite-dimension space (in the latter case, they are variational differential equations). These equations are very complicated and only little investigated in the general case. Stationary problems on plane waves in layered media form an exception, because they can be reduced to the one-dimensional problems that allow sufficiently complete systematic analysis [15], [134]–[139] and [142].

Note that natural media such as Earth's atmosphere and oceans can be considered layered media in the first approximation. In turn, the problems on plane waves in layered media can serve a primitive base for analyzing more complicated problems.

It is interesting to note that the imbedding method developed primarily for solving certain simplest equations of the theory of radiative transfer seems now to appear the instrument that can vindicate the linear theory of radiative transfer and indicate how this theory can be modified to extend its applicability range.

The imbedding method is convenient to numerically solve deterministic problems dealing with naturally stratified media, because it is capable of using the medium parameters measured by immediate sounding. In the context of statistical problems possessing ergodicity property with respect to the imbedding parameter, the obtained equations appear very convenient for determining and analyzing wavefield statistical characteristics from numerical simulations. This is especially important because full-scale experiments deal with only one realization (or a few realizations) of medium parameters, so that there is no possibility of averaging over an ensemble of realizations. With the ergodicity property available, a sole realization of medium parameters is sufficient to perform the ensemble averaging.

Different imbedding parameters and different procedures can be used to derive the imbedding equations. However, all such equations are equivalent to the input boundary-value problem, despite they may have different forms and structures. Both imbedding parameter and derivation procedure are usually governed by the convenience of the corresponding equations in the context of the problem under investigation.

In this Appendix, we consider different boundary-value wave problems, derive and analyze the corresponding imbedding equations with initial values.

C.1 Boundary-value problems formulated in terms of ordinary differential equations

Consider the dynamic system described in terms of the system of ordinary differential equations

$$\frac{d}{dt}\mathbf{x}(t) = \mathbf{F}\left(t, \mathbf{x}(t)\right), \tag{C.1}$$

defined on segment $t \in [0, T]$ with the boundary conditions

$$g\mathbf{x}(0) + h\mathbf{x}(T) = \mathbf{v}, \tag{C.2}$$

where g and h are the constant matrixes.

Dynamic problem (C.1), (C.2) possesses no dynamic causality property, which means that the solution to this problem $\mathbf{x}(t)$ at instant t functionally depends on external forces $\mathbf{F}\left(\tau, \mathbf{x}(\tau)\right)$ for all $0 \le \tau \le T$. Moreover, even boundary values $\mathbf{x}(0)$ and $\mathbf{x}(T)$ are functionals of field $\mathbf{F}\left(\tau, \mathbf{x}(\tau)\right)$. The absence of dynamic causality in problem (C.1), (C.2) prevents us from using the known statistical methods of analyzing statistical characteristics of the solution to Eq. (C.1) if external force functional $\mathbf{F}\left(t, \mathbf{x}\right)$ is the random space- and time-domain field. Introducing the one-time probability density $P(t; \mathbf{x})$ of the solution to Eq. (C.1), we can easily see that condition (C.2) is insufficient for determining the value of this probability at any point. The boundary condition imposes only certain functional restriction.

Note that the solution to problem (C.1), (C.2) parametrically depends on T and $\mathbf{v}$, i.e., $\mathbf{x}(t) = \mathbf{x}(t; T, \mathbf{v})$. Adhering to paper [83], we introduce functions

$$\mathbf{R}(T, \mathbf{v}) = \mathbf{x}(T; T, \mathbf{v}), \quad \mathbf{S}(T, \mathbf{v}) = \mathbf{x}(0; T, \mathbf{v})$$

that describe the boundary values of the solution to Eq. (C.1).

Differentiate Eq. (C.1) with respect to T and $\mathbf{v}$. We obtain two linear equations in the corresponding derivatives

$$\begin{aligned} \frac{d}{dt}\frac{\partial x_i(t; T, \mathbf{v})}{\partial T} &= \frac{\partial F_i(t, \mathbf{x})}{\partial x_l}\frac{\partial x_l(t; T, \mathbf{v})}{\partial T}, \\ \frac{d}{dt}\frac{\partial x_i(t; T, \mathbf{v})}{\partial v_k} &= \frac{\partial F_i(t, \mathbf{x})}{\partial x_l}\frac{\partial x_l(t; T, \mathbf{v})}{\partial v_k}. \end{aligned} \tag{C.3}$$

These equations are identical in form; consequently, we can expect that their solutions are related by the linear expression

$$\frac{\partial x_i(t; T, \mathbf{v})}{\partial T} = \lambda_k(T, \mathbf{v})\frac{\partial x_i(t; T, \mathbf{v})}{\partial v_k} \tag{C.4}$$

if vector quantity $\boldsymbol{\lambda}(T, \mathbf{v})$ is such that boundary conditions (C.2) are satisfied and the solution is unique. To determine vector quantity $\boldsymbol{\lambda}(T, \mathbf{v})$, we first set $t = 0$ in Eq. (C.4) and multiply the result by matrix g; then, we set $t = T$ and multiply the result by matrix h; and, finally, we combine the obtained expressions. Taking into account Eq. (C.2), we obtain

$$g\frac{\partial \mathbf{x}(0; T, \mathbf{v})}{\partial T} + h\left.\frac{\partial \mathbf{x}(t; T, \mathbf{v})}{\partial T}\right|_{t=T} = \boldsymbol{\lambda}(T, \mathbf{v}).$$

In view of the fact that

$$\left.\frac{\partial \mathbf{x}(t; T, \mathbf{v})}{\partial T}\right|_{t=T} = \frac{\partial \mathbf{x}(T; T, \mathbf{v})}{\partial T} - \left.\frac{\partial \mathbf{x}(t; T, \mathbf{v})}{\partial t}\right|_{t=T} = \frac{\partial \mathbf{R}(T, \mathbf{v})}{\partial T} - \mathbf{F}\left(T, \mathbf{R}(T, \mathbf{v})\right)$$

(with allowance for Eq. (C.1)), we obtain the desired expression for quantity $\boldsymbol{\lambda}(T, \mathbf{v})$,

$$\boldsymbol{\lambda}(T, \mathbf{v}) = -h\mathbf{F}\left(T, \mathbf{R}(T, \mathbf{v})\right). \tag{C.5}$$

Expression (C.4) with parameter $\boldsymbol{\lambda}(T, \mathbf{v})$ defined by Eq. (C.5), i.e., the expression

$$\frac{\partial x_i(t; T, \mathbf{v})}{\partial T} = -h_{kl}F_l\left(T, \mathbf{R}(T, \mathbf{v})\right)\frac{\partial x_i(t; T, \mathbf{v})}{\partial v_k}, \tag{C.6}$$

can be considered as the linear differential equation; one needs only to supplement it with the corresponding initial condition

$$\mathbf{x}(t; T, \mathbf{v})|_{T=t} = \mathbf{R}(t, \mathbf{v})$$

assuming that function $\mathbf{R}(T, \mathbf{v})$ is known.

The equation for this function can be obtained from the equality

$$\frac{\partial \mathbf{R}(T, \mathbf{v})}{\partial T} = \left.\frac{\partial \mathbf{x}(t; T, \mathbf{v})}{\partial t}\right|_{t=T} + \left.\frac{\partial \mathbf{x}(t; T, \mathbf{v})}{\partial T}\right|_{t=T}. \tag{C.7}$$

The right-hand side of Eq. (C.7) is the sum of the right-hand sides of Eqs. (C.1) and (C.4) at $t = T$. As a result, we obtain the closed nonlinear (quasilinear) equation

$$\frac{\partial \mathbf{R}(T,\mathbf{v})}{\partial T} = -h_{kl}F_l\left(T,\mathbf{R}(T,\mathbf{v})\right)\frac{\partial \mathbf{R}(T,\mathbf{v})}{\partial v_k} + \mathbf{F}\left(T,\mathbf{R}(T,\mathbf{v})\right). \tag{C.8}$$

The initial condition for Eq. (C.8) follows from Eq. (C.2) for $T \to 0$

$$\mathbf{R}(T,\mathbf{v})|_{T=0} = (g+h)^{-1}\mathbf{v}. \tag{C.9}$$

Setting now $t = 0$ in Eq. (C.3), we obtain for the secondary boundary quantity $\mathbf{S}(T,\mathbf{v}) = \mathbf{x}(0;T,\mathbf{v})$ the equation

$$\frac{\partial \mathbf{S}(T,\mathbf{v})}{\partial T} = -h_{kl}F_l\left(T,\mathbf{R}(T,\mathbf{v})\right)\frac{\partial \mathbf{S}(T,\mathbf{v})}{\partial v_k} \tag{C.10}$$

with the initial condition

$$\mathbf{S}(T,\mathbf{v})|_{T=0} = (g+h)^{-1}\mathbf{v}$$

following from Eq. (C.9).

Thus, the problem reduces to the closed quasilinear equation (C.8) with initial value (C.9) and linear equation (C.4) whose coefficients and initial value are determined by the solution of Eq. (C.8).

In the problem under consideration, input 0 and output T are symmetric. For this reason, one can solve it not only from T to 0, but also from 0 to T. In the latter case, functions $\mathbf{R}(T,\mathbf{v})$ and $\mathbf{S}(T,\mathbf{v})$ switch the places.

An important point consists in the fact that, despite the initial problem (C.1) is nonlinear, Eq. (C.4) is the linear equation, because it is essentially the equation in variations. It is Eq. (C.8) that is responsible for nonlinearity.

Note that the above technique of deriving imbedding equations for Eq. (C.1) can be easily extended to the boundary condition of the form [83]

$$\mathbf{g}\left(\mathbf{x}(0)\right) + \mathbf{h}\left(\mathbf{x}(T)\right) + \int_0^T d\tau \mathbf{K}\left(\tau,\mathbf{x}(\tau)\right) = \mathbf{v},$$

where $\mathbf{g}\left(\mathbf{x}\right)$, $\mathbf{h}\left(\mathbf{x}\right)$ and $\mathbf{K}\left(T,\mathbf{x}\right)$ are arbitrary given vector functions.

If function $\mathbf{F}\left(t,\mathbf{x}\right)$ is linear in $\mathbf{x}$, $F_i\left(t,\mathbf{x}\right) = A_{ij}(t)x_j(t)$, then boundary-value problem (C.1), (C.2) assumes the simpler form

$$\frac{d}{dt}\mathbf{x}(t) = A\left(t\right)\mathbf{x}(t), \quad g\mathbf{x}(0) + h\mathbf{x}(T) = \mathbf{v},$$

and the solution of Eqs. (C.4), (C.8) and (C.10) will be the function linear in $\mathbf{v}$

$$\mathbf{x}(t;T,\mathbf{v}) = X(t;T)\mathbf{v}. \tag{C.11}$$

As a result, we arrive at the closed matrix Riccati equation for matrix $R(T) = X(T;T)$

$$\frac{d}{dT}R(T) = A(T)R(T) - R(T)hA(T)R(T), \quad R(0) = (g+h)^{-1}. \tag{C.12}$$

As regards matrix $X(t,T)$, it satisfies the linear matrix equation with the initial condition

$$\frac{\partial}{\partial T}X(t;T) = -X(t;T)hA(T)R(T), \quad X(t;T)_{T=t} = R(t). \tag{C.13}$$

Consider now how the above formalism can be used in the context of different stationary and nonstationary, linear and nonlinear boundary-value wave problems of different dimension. In practice, it appears more convenient to derive the imbedding equations immediately from the concrete problem statement.

C.2 Stationary boundary-value wave problems

Linear wave problems describing propagation of acoustic and electromagnetic waves in layered media and their extensions to both multidimensional and nonlinear cases are of immediate interest of many physical applications. In the simplest statement, the input equation is either the one-dimensional Helmholtz equation (the problem on the plane wave incident on the medium layer), or the equation for Green's function (the problem on plane wave generation by a point source).

C.2.1 One-dimensional stationary boundary-value wave problems

Helmholtz equation with unmatched boundary

Consider the one-dimensional stationary boundary-value problem

$$\left(\frac{d^2}{dx^2}+k^2(x)\right)u(x)=0,$$
$$\left(\frac{d}{dx}+ik_1\right)u(x)\bigg|_{x=L_0}=0,\quad \left(\frac{d}{dx}-ik_0\right)u(x)\bigg|_{x=L}=-2ik_0. \tag{C.14}$$

It describes the incidence of plane wave $u(x)=e^{-ik_0(x-L)}$ from the homogeneous half-space $x>L$ characterized by wave parameter $k_0\neq k(L)$ on inhomogeneous medium layer $L_0<x<L$. The half-space $x<L_0$ is assumed homogeneous and is described by wave parameter k_1. Note that boundary-value problem (C.14) describes the spatial structure of a monochromatic wave (proportional to $e^{-i\omega t}$) in inhomogeneous medium characterized by wave parameter $k(x)=\omega/c(x)$, where $c(x)$ is the velocity of wave propagation in the medium layer.

We represent function $k^2(x)$ in the form

$$k^2(x)=k_0^2\left[1+\varepsilon(x)\right],$$

where function $\varepsilon(x)$ describes the inhomogeneities of the medium (such as inhomogeneities of the velocity of wave propagation in the medium and inhomogeneities of the refractive index, or dielectric permittivity). In the general case, function $\varepsilon(x)$ is the complex function $\varepsilon(x)=\varepsilon_1(x)+i\gamma$, where parameter γ describes absorption of the wave in the medium. With this substitution, boundary-value problem (C.14) assumes the form

$$\left(\frac{d^2}{dx^2}+k_0^2\left[1+\varepsilon(x)\right]\right)u(x)=0,$$
$$\left(\frac{d}{dx}+ik_1\right)u(x)\bigg|_{x=L_0}=0,\quad \left(\frac{d}{dx}-ik_0\right)u(x)\bigg|_{x=L}=-2ik_0. \tag{C.15}$$

In this case under consideration, the reason of wave reflection at boundary $x=L$ lies not only in medium inhomogeneities inside the layer, but also in discontinuity of function $k(x)$ at this boundary. Therefore, we will call boundary problem (C.14) the problem with *unmatched boundary* $x=L$.

The values of the wavefield at layer boundaries determine the reflection and transmission coefficients of the layer $R_L=u(L)-1$ and $T_L=u(L_0)$.

Remark 16 ***Structure of problem solution in the case of a homogeneous medium.***

Consider the structure of the solution to boundary-value problem (C.14) in the case of a homogeneous medium with $k(x) \equiv k = \text{const}$. We consider the problems with two ($k_1 = k$) and tree ($k_1 \neq k$) layers. It is obvious that the solution to the two-layer boundary-value problem (C.14) has the form $u(x) = (1 + R_0)\, e^{-ik(x-L)}$, where the reflection coefficient R_0 is given by the equality

$$R_0 = \frac{k_0 - k}{k_0 + k}. \tag{C.16}$$

In the three-layer case, the solution to boundary-value problem (C.14) has the form

$$u(x) = (1 + R_0)\, \frac{e^{ik(L-x)} + R_1 e^{-ik(L-x)+2ik(L-L_0)}}{1 + R_0 R_1 e^{2ik(L-L_0)}}, \tag{C.17}$$

where

$$R_1 = \frac{k - k_1}{k + k_1}. \tag{C.18}$$

As a result, the wave field at layer boundary and transmission coefficient are given by the expressions

$$\begin{aligned} u(L) &= 1 + R_L = (1 + R_0)\, \frac{1 + R_1 e^{2ik(L-L_0)}}{1 + R_0 R_1 e^{2ik(L-L_0)}}, \\ T_L &= (1 + R_0)\,(1 + R_1)\, \frac{e^{ik(L-L_0)}}{1 + R_0 R_1 e^{2ik(L-L_0)}}. \quad \blacklozenge \end{aligned} \tag{C.19}$$

Reformulate now boundary-value problem (C.15) in terms of the boundary-value system of equations in functions $u(x) = u(x; L)$, $v(x; L) = \frac{\partial}{\partial x} u(x; L)$

$$\begin{aligned} &\frac{d}{dx} u(x; L) = v(x; L), \quad \frac{d}{dx} v(x; L) = -k_0^2\, [1 + \varepsilon(x)]\, u(x; L), \\ &v(L_0; L) + ik_1 u(L_0; L) = 0, \quad v(L; L) - ik_0 u(L; L) = -2ik_0, \end{aligned} \tag{C.20}$$

where new variable L is added to follow the spirit of the imbedding method.

For clarity, we repeat in a few words the derivation of imbedding equations for problem (C.20). Considering the solution to this problem as a function of parameter L, we obtain the boundary-value problem in the derivatives with respect to this parameter

$$\begin{aligned} &\frac{d}{dx} \frac{\partial u(x; L)}{\partial L} = \frac{\partial v(x; L)}{\partial L}, \quad \frac{d}{dx} \frac{\partial v(x; L)}{\partial L} = -k_0^2\, [1 + \varepsilon(x)]\, \frac{\partial u(x; L)}{\partial L}, \\ &\frac{\partial v(L_0; L)}{\partial L} + ik_1 \frac{\partial u(L_0; L)}{\partial L} = 0, \\ &\left.\frac{\partial v(x; L)}{\partial L}\right|_{x=L} - ik_0 \left.\frac{\partial u(x; L)}{\partial L}\right|_{x=L} = -\left.\frac{\partial v(x; L)}{\partial x}\right|_{x=L} + ik_0 \left.\frac{\partial u(x; L)}{\partial x}\right|_{x=L} \\ &= 2k_0^2 + k_0^2 \varepsilon(L) u(L; L). \end{aligned} \tag{C.21}$$

Then, correlating boundary-value problem (C.21) with boundary-value problem (C.20), we obtain the imbedding equations in the form

$$\begin{aligned} &\frac{\partial}{\partial L} u(x; L) = ik_0 \left\{ 1 + \frac{1}{2} \varepsilon(L) u(L; L) \right\} u(x; L), \\ &\frac{\partial}{\partial L} v(x; L) = ik_0 \left\{ 1 + \frac{1}{2} \varepsilon(L) u(L; L) \right\} v(x; L), \\ &u(x; L)|_{L=x} = u(x; x), \quad v(x; L)_{L=x} = -ik_0\, [2 - u(x; x)]\,, \end{aligned} \tag{C.22}$$

from which follows that

$$v(x;L) = \frac{\partial}{\partial x}u(x;L) = -ik_0\frac{2-u(x;x)}{u(x;x)}u(x;L). \tag{C.23}$$

Function $u(L;L)$ satisfies the equality

$$\frac{d}{dL}u(L;L) = \left.\frac{\partial u(x;L)}{\partial x}\right|_{x=L} + \left.\frac{\partial u(x;L)}{\partial L}\right|_{x=L}.$$

Using this equality and taking into account Eqs. (C.20) and (C.22), we obtain the Riccati equation

$$\frac{d}{dL}u(L;L) = 2ik_0\left[u(L;L)-1\right] + i\frac{k_0}{2}\varepsilon(L)u^2(L;L), \quad u(L;L)|_{L=L_0} = \frac{2k_0}{k_0+k_1}. \tag{C.24}$$

Introducing reflection coefficient $R_L = u(L;L)-1$, we can rewrite Eqs. (C.22) and (C.24) in the form

$$\begin{aligned}
&\frac{\partial}{\partial L}u(x;L) = ik_0\left\{1+\frac{1}{2}\varepsilon(L)\left(1+R_L\right)\right\}u(x;L), \quad u(x;L)|_{L=x} = 1+R_x, \\
&\frac{d}{dL}R_L = 2ik_0R_L + i\frac{k_0}{2}\varepsilon(L)\left(1+R_L\right)^2, \quad R_{L_0} = \frac{k_0-k_1}{k_0+k_1}.
\end{aligned} \tag{C.25}$$

In addition, Eq. (C.23) grades into the formula

$$\left(\frac{\partial}{\partial x} + ik_0\frac{1-R_x}{1+R_x}\right)u(x;L) = 0$$

that extends boundary conditions given in the second row of Eq. (C.15) to arbitrary point x inside the layer.

If parameter $\gamma = 0$, i.e., if is k_0 is the real-valued parameter, then we have the following equality for the intensity of wavefield

$$I(x;L) = |u(x;L)|^2 = |1+R_x|^2\frac{|1-R_L|^2}{|1-R_x|^2}. \tag{C.26}$$

Note that the initial value of the reflection coefficient in Eq. (C.25) coincides with to the solution of the two-layer problem (see Remark 16).

In boundary-value problem (C.15) and, consequently, in Eqs. (C.25), wave parameter k_1 describes reflecting properties of half-space $x < L_0$. If $k_1 = k_0$, then the initial condition of the Riccati equation (C.24) assumes the form $R_{L_0} = 0$; we will call boundary $x = L_0$ of such type the free-transmission boundary. Reflecting boundaries can be described using limit processes with respect to k_1. For example, limit process $k_1 \to 0$ corresponds to reflecting boundary $x = L_0$ at which $\left.\frac{d}{dx}u(x)\right|_{x=L_0} = 0$; in this case, $R_{L_0} = 1$. Another limit process $k_1 \to \infty$ corresponds to reflecting boundary $x = L_0$ at which $u(x)|_{x=L_0} = 0$; in this case, $R_{L_0} = -1$.

We note that the knowledge of reflection coefficient R_L in the form of a functional of function $\varepsilon(x)$ offers a possibility of determining the wavefield structure. Indeed, variational derivative $\delta R_L/\delta\varepsilon(x)$ satisfies the linear equation

$$\begin{aligned}
&\frac{d}{dL}\frac{\delta R_L}{\delta\varepsilon(x)} = 2ik_0\frac{\delta R_L}{\delta\varepsilon(x)} + ik_0\varepsilon(L)\left(1+R_L\right)\frac{\delta R_L}{\delta\varepsilon(x)}, \\
&\left.\frac{\delta R_L}{\delta\varepsilon(x)}\right|_{L=x} = i\frac{k_0}{2}\left(1+R_x\right)^2,
\end{aligned}$$

so that,

$$\frac{\delta R_L}{\delta\varepsilon(x)} = i\frac{k_0}{2}u^2(x;L).$$

Consider now the problem on the field generated by a point source located at point x_0 inside the medium layer. This problem also is the boundary-value problem; the corresponding equation and boundary conditions are

$$\begin{aligned}&\left(\frac{d^2}{dx^2}+k_0^2\left[1+\varepsilon(x)\right]\right)G(x;x_0) = 2ik_0\delta(x-x_0),\\ &\left(\frac{d}{dx}+ik_1\right)G(x;x_0)\Big|_{x=L_0} = 0, \quad \left(\frac{d}{dx}-ik_0\right)G(x;x_0)\Big|_{x=L} = 0.\end{aligned} \tag{C.27}$$

Factor $2ik_0$ of the delta function in the right-hand side of Eq. (C.27) ensures the problem solution $G(x;L)$ in the case of the source located at boundary $x_0 = L$ to coincide with the solution $u(x;L)$ to problem (C.15), i.e. $G(x;L) = u(x;L)$. Indeed, function $G(x;x_0)$ is continuous at each point x and its derivative with respect to x is discontinuous at the point of source location

$$\frac{d}{dx}G(x;x_0)\Big|_{x=x_0+0} - \frac{d}{dx}G(x;x_0)\Big|_{x=x_0-0} = 2ik_0.$$

Setting now $x_0 = L$ in Eq. (C.27) and using the above condition of derivative discontinuity, we arrive at boundary-value problem (C.15).

Rewrite the boundary-value problem (C.27) in the form of the system of equations similar to Eqs. (C.20),

$$\begin{aligned}&\frac{d}{dx}G(x;x_0;L) = V(x;x_0;L),\\ &\frac{d}{dx}V(x;x_0;L) = -k_0^2\left[1+\varepsilon(x)\right]G(x;x_0;L) + 2ik_0\delta(x-x_0),\\ &V(L_0;x_0;L) + ik_1G(L_0;x_0;L) = 0, \quad V(L;x_0;L) - ik_0G(L;x_0;L) = 0.\end{aligned} \tag{C.28}$$

In Eq. (C.28), we again included parameter L to explicitly show the dependence of the solution on this parameter.

Differentiating system (C.28) with respect to L, we obtain the boundary-value problem in derivatives

$$\begin{aligned}&\frac{d}{dx}\frac{\partial G(x;x_0;L)}{\partial L} = \frac{\partial V(x;x_0;L)}{\partial L},\\ &\frac{d}{dx}\frac{\partial V(x;x_0;L)}{\partial L} = -k_0^2\left[1+\varepsilon(x)\right]\frac{\partial G(x;x_0;L)}{\partial L},\\ &\frac{\partial V(L_0;x_0;L)}{\partial L} + ik_1\frac{\partial G(L_0;x_0;L)}{\partial L} = 0,\\ &\frac{\partial V(x;x_0;L)}{\partial L}\Big|_{x=L} - ik_0\frac{\partial G(x;x_0;L)}{\partial L}\Big|_{x=L}\\ &= -\frac{\partial V(x;x_0;L)}{\partial x}\Big|_{x=L} + ik_0\frac{\partial G(x;x_0;L)}{\partial x}\Big|_{x=L} = k_0^2\varepsilon(L)G(L;x_0;L).\end{aligned}$$

Correlation of this system with boundary-value (C.20) yields the equality

$$\frac{\partial}{\partial L}G(x;x_0;L) = i\frac{k_0}{2}\varepsilon(L)G(L;x_0;L)u(x;L) \tag{C.29}$$

that, being supplemented with the initial condition of continuity

$$G(x;x_0;L)|_{L=\max\{x,x_0\}} = \begin{cases} G(x;x_0;x), & x \geq x_0, \\ G(x;x_0;x_0), & x \leq x_0, \end{cases} \tag{C.30}$$

can be considered as the imbedding equation with respect to variable L.

Equation (C.29) and initial condition (C.30) depend on new unknown function $G(L;x_0;L)$. It satisfies the obvious equality

$$\frac{\partial}{\partial L}G(L;x_0;L) = \frac{\partial}{\partial x}G(x;x_0;L)\bigg|_{x=L} + \frac{\partial}{\partial L}G(x;x_0;L)\bigg|_{x=L},$$

that can be reduced in view of Eqs. (C.28) and (C.29) to the imbedding equation

$$\frac{\partial}{\partial L}G(L;x_0;L) = ik_0\left\{1+\frac{1}{2}\varepsilon(L)u(L;L)\right\}G(L;x_0;L). \tag{C.31}$$

It is obvious that the initial condition for this equation is

$$G(L;x_0;L)|_{L=x_0} = G(x_0;x_0;x_0) = u(x_0;x_0).$$

Correlating now Eq. (C.31) with the first equation of Eqs. (C.24), we see that

$$G(L;x_0;L) = G(x_0;L;L) = u(x_0;L). \tag{C.32}$$

Equality (C.32) expresses the *reciprocity theorem* in the context our problem.

Thus, the system of imbedding equations for the field $G_\alpha(x;x_0)$ of a point source located in the layer of inhomogeneous medium, which is described by boundary-value problem (C.27)

$$\begin{aligned} \left(\frac{d^2}{dx^2}+k_0^2\left[1+\varepsilon(x)\right]\right)G_\alpha(x;x_0) &= 2ik_0\delta(x-x_0), \\ \left(\frac{d}{dx}+i\alpha k_0\right)G_\alpha(x;x_0)\bigg|_{x=L_0} &= 0, \quad \left(\frac{d}{dx}-ik_0\right)G_\alpha(x;x_0)\bigg|_{x=L} = 0, \end{aligned} \tag{C.33}$$

where $\alpha = k_1/k_0$, has the form

$$\begin{aligned} &\frac{\partial}{\partial L}G_\alpha(x;x_0;L) = i\frac{k_0}{2}\varepsilon(L)u_\alpha(x_0;L)u_\alpha(x;L), \\ &\qquad G_\alpha(x;x_0;L)|_{L=\max\{x,x_0\}} = \begin{cases} u_\alpha(x_0;x), & x \geq x_0, \\ G(x;x_0), & x \leq x_0, \end{cases} \\ &\frac{\partial}{\partial L}u_\alpha(x;L) = ik_0\left\{1+\frac{1}{2}\varepsilon(L)u_\alpha(L;L)\right\}u_\alpha(x;L), \quad u_\alpha(x;L)|_{L=x} = u_\alpha(x;x), \\ &\frac{d}{dL}u_\alpha(L;L) = 2ik_0\left[u_\alpha(L;L)-1\right]+i\frac{k_0}{2}\varepsilon(L)u_\alpha^2(L;L), \quad u_\alpha(L;L)|_{L=L_0} = \frac{2}{1+\alpha}. \end{aligned} \tag{C.34}$$

Here, index α is introduced to reveal the wavefield dependence on the boundary condition at $x = L_0$.

Remark 17 ***Consideration of different boundary conditions at boundary*** $x = L$.

With the solution of boundary problem (C.33) (or imbedding equations (C.34)) at hand, we can easily obtain solutions to boundary-value wave problems that differ from problem (C.33) in the value of the wave parameter in free half-space $x > L$. Consider the boundary-value problem

$$\left(\frac{d^2}{dx^2}+k_0^2\left[1+\varepsilon(x)\right]\right)G(x;x_0)=2ik_0\delta(x-x_0),$$
$$\left(\frac{d}{dx}+i\alpha k_0\right)G(x;x_0)\bigg|_{x=L_0}=0,\quad \left(\frac{d}{dx}-ik_2\right)G(x;x_0)\bigg|_{x=L}=0. \quad \text{(C.35)}$$

Represent the solution to problem (C.35) in the form

$$G(x;x_0)=G_\alpha(x;x_0)+A(x_0;L)u_\alpha(x;L), \quad \text{(C.36)}$$

where $G_\alpha(x;x_0)$ and $u_\alpha(x;x_0)$ are the solutions to boundary-value problems (C.33), (C.14), respectively (these solutions satisfy imbedding equations (C.34)) and quantity $A(x_0;L)$ is independent of variable x. It is obvious that function (C.36) satisfies differential equation and boundary condition at $x = L_0$ of Eq. (C.35). Function (C.36) will satisfy the boundary condition at $x = L$ if we represent quantity $A(x_0;L)$ in the form

$$A(x_0;L)=\frac{1}{G-u_\alpha(L;L)}u_\alpha(x_0;L),$$

where we introduced constant G,

$$G=\frac{2}{1-k_2/k_0}\quad \text{or}\quad k_2=\frac{G-2}{G}k_0. \quad \text{(C.37)}$$

Thus, the solution to boundary problem (C.35) is given by the expression

$$G(x;x_0)=G_\alpha(x;x_0)+\tilde{G}(x;x_0), \quad \text{(C.38)}$$

where

$$\tilde{G}(x;x_0)=\frac{1}{G-u_\alpha(L;L)}u_\alpha(x_0;L)u_\alpha(x;L). \quad \text{(C.39)}$$

Note that, dealing with the problem with unmatched boundary $x = L$ (C.35), the imbedding method uses the problem solution assuming that region $x > L$ is characterized by wave number k_2 independently of boundary position L (see Fig. C.1a).

If the source is located at boundary $x = L$, i.e., if. $x_0 = L$, then we obtain from Eqs. (C.38), (C.39) that

$$G(x;L)=\frac{G}{G-u_\alpha(L;L)}u_\alpha(x;L)\quad \text{and}\quad G(L;L)=\frac{Gu_\alpha(L;L)}{G-u_\alpha(L;L)}.$$

Function $G(x;L)$ describes the incidence of wave $u(x) = \frac{k_0}{k_2}e^{-ik_2(x-L)}$ on the medium layer from the half-space $x > L$ characterized by wave number k_2. As a consequence, the reflection coefficient is given in this problem by the equality

$$\tilde{R}_L=\frac{k_2}{k_0}G(L;L)-1=\frac{(k_2-k_0)+(k_2+k_0)R_L}{(k_2+k_0)+(k_2-k_0)R_L},$$

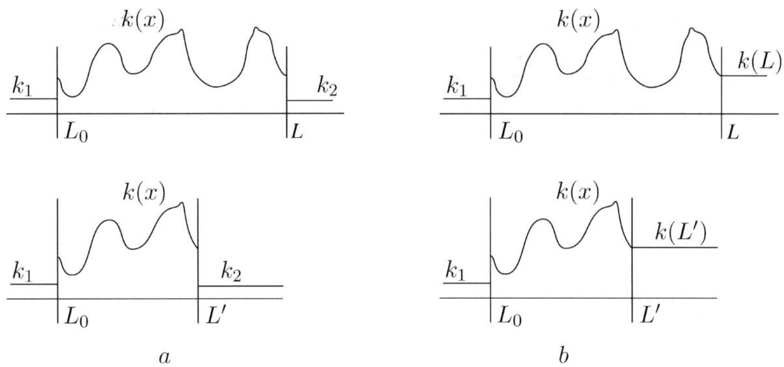

Figure C.1: Stationary wave boundary problems on a wave incident on medium layer in the cases of (a) unmatched boundary at $x = L$ and (b) matched boundary at $x = L$.

where R_L is the reflection coefficient in problem (C.14). The effects of boundaries $x = L_0$ and $x = L$ appear different in problem (C.35). The effect of boundary $x = L_0$ concerns the initial condition to the equation in function $u_\alpha(L; L)$, while the effect of boundary $x = L$ concerns the immediate structure of function $\tilde{G}(x; x_0)$.

Limit processes with respect to k_2 offer a possibility of considering boundary-value wave problems in which boundary $x = L$ is characterized by specific reflecting properties. For example, in the case of free-transmission boundary $x = L$, wave number $k_2 = k_0$ and constant $G = \infty$, so that function $\tilde{G}(x; x_0) = 0$. Limiting case $k_2 \to 0$ corresponds to reflecting boundary $x = L$ at which the boundary condition is $\frac{\partial}{\partial x} G(x; x_0)\Big|_{x=L} = 0$. In this case, constant $G = 2$ and Eq. (C.38) assumes the form

$$G(x; x_0) = G_\alpha(x; x_0) + \frac{1}{2 - u_\alpha(L; L)} u_\alpha(x_0; L) u_\alpha(x; L).$$

If the source is located at this boundary, i.e., if $x_0 = L$, then

$$G(x; L) = \frac{2}{2 - u_\alpha(L; L)} u_\alpha(x; L) = \frac{2}{1 - R_L} u_\alpha(x; L). \tag{C.40}$$

The limiting case $k_2 \to \infty$ also corresponds to reflecting boundary $x = L$, but the boundary condition has in this case the form $G(L; x_0) = 0$. In this case, constant $G = 0$ and Eq. (C.38) assumes the form

$$G(x; x_0) = G_\alpha(x; x_0) - \frac{1}{u_\alpha(L; L)} u_\alpha(x_0; L) u_\alpha(x; L).$$

In physical problems on propagation of acoustic (electromagnetic) waves in inhomogeneous media, great attention is focused on the effect of boundary impedance on the acoustic (electromagnetic) field in the medium. The obtained representation appears very convenient and 'economically' efficient for analyzing problems of namely this type. Indeed, as we mentioned earlier, the solution of every boundary-value problem taken separately requires solving the Riccati equation and calculating two quadratures of fast oscillating functions.

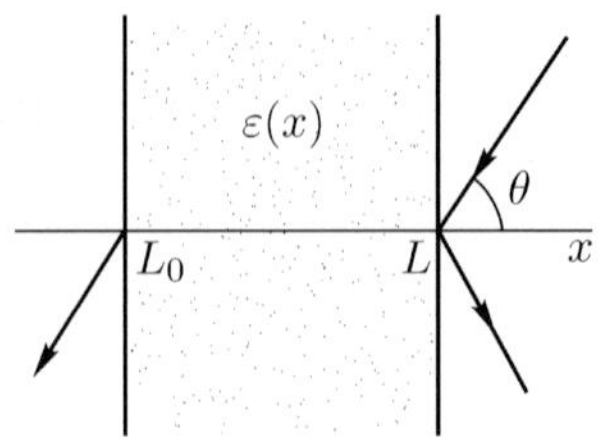

Figure C.2: Plane wave obliquely incident at angle θ.

The simultaneous consideration of two such problems offers a possibility of doing with solving two Riccati equations and calculating one quadrature. All other wave characteristics of both problems can be then derived from the obtained solutions algebraically. If we have a third problem in addition to the two considered problems, then Green's formula gives the solution of this problem immediately

$$G_{\gamma}(x;x_0) = G_{\alpha}(x;x_0) - \frac{(\alpha-\gamma)}{1+(\alpha-\gamma)G_{\alpha}(L_0;L_0)} G_{\alpha}(L_0;x_0)G_{\alpha}(x;L_0). \qquad \blacklozenge$$

Remark 18 ***Oblique wave incidence.***

The above consideration dealt with the wave incident on the inhomogeneous medium layer along the normal. The case of the wave incident on boundary $x = L$ obliquely can be considered similarly. In this case, the problem is formulated in terms of the three-dimensional Helmholtz equation. We represent this equation in the form

$$\left\{\frac{\partial^2}{\partial x^2} + \Delta_{\mathbf{R}} + k_0^2\,[1+\varepsilon(x)]\right\} U(x,\mathbf{R}) = 0, \tag{C.41}$$

where $\mathbf{R} = \{y, z\}$ denotes the coordinates in the plane perpendicular to the x-axis. We assume that inhomogeneous medium occupies, as earlier, the portion of space $L_0 < x < L$. For simplicity, we will additionally assume that function $\varepsilon(x) = 0$ outside the medium, i.e., we will assume that wave numbers in free half-spaces $x > L$ and $x < L_0$ are equal to k_0. Let now the unit-amplitude wave is incident on the inhomogeneous layer from the homogeneous half-space $x > L$ at angle θ (Fig. C.2)

$$U_0(x,\mathbf{R}) = e^{i\sqrt{k_0^2-q^2}(L-x)+i\mathbf{q}\mathbf{R}} = e^{ip(L-x)+i\mathbf{q}\mathbf{R}},$$

where $q = k_0 \sin\theta$, and $p = \sqrt{k_0^2 - q^2} = k\cos\theta$. The case of normal incidence corresponds to $\theta = 0$.

Medium inhomogeneities cause the appearance of the reflected wave in the half-space $x > L$; this means that wavefield for $x > L$ has the following structure

$$U(x,\mathbf{R}) = e^{ip(L-x)+i\mathbf{q}\mathbf{R}} + R_L e^{ip(x-L)+i\mathbf{q}\mathbf{R}}.$$

In the half-space $x < L_0$, we have only the transmitted wave of the form

$$U(x,\mathbf{R}) = T_L e^{-ip(x-L_0)+i\mathbf{q}\mathbf{R}}.$$

Boundary conditions for Eq. (C.41) are the continuity conditions of the field and field's normal derivative (with respect to x in this case) at layer boundaries. Inside the layer, the wavefield structure is $U(x, \mathbf{R}) = u(x)e^{i\mathbf{qR}}$, where function $u(x)$ is the solution to the boundary-value problem for the one-dimensional Helmholtz equation

$$\begin{aligned}\left(\frac{d^2}{dx^2} + P^2(x)\right) u(x) &= 0,\\ \left(\frac{d}{dx} + ip\right) u(x)\bigg|_{x=L_0} &= 0, \quad \left(\frac{d}{dx} - ip\right) u(x)\bigg|_{x=L} = -2ip,\end{aligned} \tag{C.42}$$

where

$$P^2(x) = p^2\left[1 + \frac{k^2}{p^2}\varepsilon(x)\right] = p^2\left[1 + \frac{1}{\cos^2\theta}\varepsilon(x)\right].$$

Boundary-value problem (C.42) coincides with boundary-value problem (C.14) to notation. Consequently, considering the solution to this problem as a function of parameter L, we obtain the imbedding equations of type (C.35); in the case at hand, these imbedding equations have the form

$$\begin{aligned}&\frac{\partial}{\partial L}u(x; L) = ik\left\{\cos\theta + \frac{1}{2\cos\theta}\varepsilon(L)\,(1+R_L)\right\}u(x; L), \quad u(x; L)|_{L=x} = 1 + R_x;\\ &\frac{d}{dL}R_L = 2ik(\cos\theta)R_L + \frac{ik}{2\cos\theta}\varepsilon(L)\,(1+R_L)^2, \quad R_{L_0} = 0. \quad \blacklozenge\end{aligned} \tag{C.43}$$

Remark 19 ***Method of integral equation.***

Deriving imbedding equations, we dealt with boundary-value problems in differential formulation. However, representation of the input boundary-value problem in the form of the corresponding integral equation may sometimes significantly simplify the derivation. In this case, we have no need in differentiating the boundary conditions with respect to the imbedding parameter. For example, boundary-value problem (C.31) at $\alpha = 1$ corresponds to the integral equation

$$G(x; x_0) = e^{ik_0|x-x_0|} + i\frac{k_0}{2}\int_{L_0}^{L} d\xi e^{ik_0|x-\xi|}\varepsilon(\xi)G(\xi; x_0), \tag{C.44}$$

and boundary-value problem (C.14) at $k_1 = k_0$ corresponds to the integral equation

$$u(x; L) = e^{ik_0(L-x)} + i\frac{k_0}{2}\int_{L_0}^{L} d\xi e^{ik_0|x-\xi|}\varepsilon(\xi)u(\xi; L) \tag{C.45}$$

coinciding with Eq. (C.44) at $x_0 = L$ (i.e., $u(x; L) = G(x; L)$).

Equation (C.44) can be represented in the form

$$G(x; x_0) = e^{ik_0|x-x_0|} + i\frac{k_0}{2}\int_{L_0}^{L} d\xi G(x; \xi)\varepsilon(\xi)e^{ik_0|\xi-x_0|}. \tag{C.46}$$

Interchange points x and x_0 in Eq. (C.46). Then, we obtain the equation

$$G(x_0; x) = e^{ik_0|x-x_0|} + i\frac{k_0}{2}\int_{L_0}^{L} d\xi e^{ik_0|x-\xi|}\varepsilon(\xi)G(x_0; \xi)$$

whose correlation with Eq. (C.44) yields the reciprocity theorem

$$G(x;x_0) = G(x_0;x). \tag{C.47}$$

Differentiate Eq. (C.44) with respect to parameter L. In view of dependence of function $G(x;x_0) \equiv G(x;x_0;L)$ of parameter L, we arrive at the integral equation

$$\frac{\partial}{\partial L}G(x;x_0;L) = i\frac{k}{2}e^{ik_0(L-x)}\varepsilon(L)G(L;x_0;L) + i\frac{k_0}{2}\int\limits_{L_0}^{L} d\xi e^{ik_0|x-\xi|}\varepsilon(\xi)\frac{\partial}{\partial L}G(\xi;x_0;L)$$

whose solution can obviously be expressed in terms of function $u(x;L)$ by the equality

$$\frac{\partial}{\partial L}G(x;x_0;L) = i\frac{k_0}{2}\varepsilon(L)G(L;x_0;L)u(x;L).$$

By the reciprocity theorem, this equality can be represented in the form

$$\frac{\partial}{\partial L}G(x;x_0;L) = i\frac{k_0}{2}\varepsilon(L)u(x_0;L)u(x;L)$$

coinciding with the first equation of the imbedding method (C.32).

Differentiate now Eq. (C.45) with respect to parameter L. We obtain the integral equation in derivative $\frac{\partial}{\partial L}u(x;L)$

$$\frac{\partial}{\partial L}u(x;L) = a(L)e^{ik_0(L-x)|} + i\frac{k_0}{2}\int\limits_{L_0}^{L} d\xi e^{ik_0|x-\xi|}\varepsilon(\xi)\frac{\partial}{\partial L}u(\xi;L),$$

where $a(L) = ik_0\left\{1+\frac{1}{2}\varepsilon(L)u(L;L)\right\}$. This equation is equivalent to the equality

$$\frac{\partial}{\partial L}u(x;L) = ik_0\left\{1+\frac{1}{2}\varepsilon(L)u(L;L)\right\}u(x;L),$$

which coincides with the second equation of the imbedding method (C.32).

For function $u(L;L)$, we have

$$u(L;L) = 1 + i\frac{k_0}{2}\int\limits_{L_0}^{L} d\xi e^{ik_0(L-\xi)}\varepsilon(\xi)u(\xi;L).$$

As a consequence, we obtain the chain of equalities for derivative $\frac{d}{dL}u(L;L)$

$$\begin{aligned}
&\frac{d}{dL}u(L;L) - i\frac{k_0}{2}\varepsilon(L)u(L;L) \\
&= ik_0\left\{2+\frac{1}{2}\varepsilon(L)u(L;L)\right\}[u(L;L)-1] \\
&= -i\frac{k_0}{2}\varepsilon(L)u(L;L) + 2ik_0[u(L;L)-1] + i\frac{k_0}{2}\varepsilon(L)u^2(L;L),
\end{aligned}$$

which yields the third equation—the Riccati equation—in Eq. (C.32). ♦

Remark 20 *Matrix Helmholtz equation.*

In physics, many wave problems are formulated in terms of the boundary-value problems concerning not only the second-order linear differential equations, but generally the systems of the second-order linear differential equations with boundary conditions of the form

$$\left(\frac{d^2}{dx^2} + \gamma(x)\frac{d}{dx} + K(x)\right) U(x) = 0,$$
$$\left(\frac{d}{dx} + B\right) U(x)\Big|_{x=L} = D, \quad \left(\frac{d}{dx}+C\right) U(x)\Big|_{x=0} = 0, \tag{C.48}$$

where $\gamma(x)$, $K(x)$, and $U(x)$ are the matrix variable and B, C and D are matrix constants. Using the procedure similar to the above procedure of deriving imbedding equations, we can reformulate boundary-value problem (C.48) in terms of the initial value problem immediately, i.e., without representing problem (C.48) in the form of the system of the first-order differential equations [90, 91, 145, 175]. Indeed, the solution to boundary-value problem (C.48) depends on parameter L (i.e.,. $U(x) = U(x; L)$) an we can rewrite problem (C.48) in the form

$$\left(\frac{d^2}{dx^2} + \gamma(x)\frac{d}{dx} + K(x)\right) U(x; L) = 0,$$
$$\left(\frac{d}{dx} + B\right) U(x; L)\Big|_{x=L} = D, \quad \left(\frac{d}{dx}+C\right) U(x; L)\Big|_{x=0} = 0. \tag{C.49}$$

Differentiating equation in matrix $U(x; L)$ of problem (C.49) with respect to parameter L, we obtain the equation

$$\left(\frac{d^2}{dx^2} + \gamma(x)\frac{d}{dx} + K(x)\right) \frac{\partial}{\partial L} U(x; L) = 0 \tag{C.50}$$

coinciding with the input equation (C.49). As a consequence, we can draw the equality

$$\frac{\partial}{\partial L} U(x; L) = U(x; L)\Lambda(L). \tag{C.51}$$

Being supplemented with the initial condition

$$U(x; L)|_{L=x} = U(x; x),$$

this equality can be considered the differential equation with respect to parameter L.

The expression for matrix $\Lambda(L)$ can be derived from boundary conditions at $x = L$. Applying operator $\left(\frac{d}{dx}+B\right)$ to Eq. (C.51), we obtain the equality

$$\left(\frac{d}{dx}+B\right) \frac{\partial}{\partial L} U(x; L) = \frac{\partial}{\partial L}\left(\frac{d}{dx}+B\right) U(x; L) = \left(\frac{d}{dx}+B\right) U(x; L)\Lambda(L). \tag{C.52}$$

Set now $x = L$. The right-hand side of Eq. (C.52) grades into $D\Lambda(L)$ in view of boundary condition (C.49). For the right-hand side of Eq. (C.52), we have

$$\left[\frac{\partial}{\partial L}\left(\frac{d}{dx} + B\right) U(x; L)\right]_{x=L}$$
$$= \frac{\partial}{\partial L}\left[\left(\frac{d}{dx}+B\right) U(x; L)\Big|_{x=L}\right] - \frac{d}{dx}\left(\frac{d}{dx} + B\right) U(x; L)\Big|_{x=L}$$
$$= [\gamma(L) - B]\, D + \left[K(L) + B^2 - \gamma(L)B\right] U(L; L).$$

As a consequence, we have

$$\Lambda(L) = D^{-1}\left[\gamma(L) - B\right]D + D^{-1}\left[K(L) + B^2 - \gamma(L)B\right]U(L;L). \tag{C.53}$$

Matrix $U(L;L)$ satisfies the obvious equality

$$\frac{dU(L;L)}{dL} = \left.\frac{\partial U(x;L)}{\partial x}\right|_{x=L} + \left.\frac{\partial U(x;L)}{\partial L}\right|_{x=L} = D - BU(L;L) + U(L;L)\Lambda(L).$$

We can consider this equality as the matrix Riccati equation

$$\frac{d}{dL}U(L;L) = D - \left\{BU(L;L) + U(L;L)D^{-1}BD\right\}$$
$$+U(L;L)D^{-1}\gamma(L)D + U(L;L)D^{-1}\left\{K(L) - \gamma(L)B + B^2\right\}U(L;L), \tag{C.54}$$

with the initial condition at $L = 0$

$$U(0;0) = (B - C)^{-1}D$$

following from boundary condition(C.49). The boundary-value problem (C.14) considered earlier corresponds to problem (C.48) with parameters $\gamma = 0$, $B = -ik_0$, $C = ik_1$, and $D = -2ik_0$. ♦

Helmholtz equation with matched boundary

Similar equations can be derived in the case when boundary conditions themselves explicitly depend on parameter L. As an example, we consider the boundary-value problem

$$\left(\frac{d^2}{dx^2} + k^2(x)\right)u(x;L) = 0,$$
$$\left.\left(\frac{d}{dx} + ik_1\right)u(x;L)\right|_{x=L_0} = 0, \quad \left.\left(\frac{d}{dx} - ik(L)\right)u(x;L)\right|_{x=L} = -2ik(L) \tag{C.55}$$

that describes the incidence of plane wave $u(x) = e^{-ik(L)(x-L)}$ from the homogeneous half-space $x > L$ characterized by wave parameter $k = k(L)$ on the layer of inhomogeneous medium $L_0 < x < L$. In this case, function $k(x)$ has no discontinuity at layer boundary $x = L$ for arbitrary boundary position (Fig. C.1*b*), and we will call this problem the problem with the *matched boundary*.

Rewrite boundary-value problem (C.55) in the form of the boundary-value system of equations

$$\frac{d}{dx}u(x;L) = v(x;L), \quad \frac{d}{dx}v(x;L) = -k^2(x)u(x;L),$$
$$v(L_0;L) + ik_1u(L_0;L) = 0, \quad v(L;L) - ik(L)u(L;L) = -2ik(L). \tag{C.56}$$

Considering the solution to this system as a function of parameter L, we obtain the boundary-value problem for the derivatives with respect to parameter L,

$$\frac{d}{dx}\frac{\partial u(x;L)}{\partial L} = \frac{\partial v(x;L)}{\partial L}, \quad \frac{d}{dx}\frac{\partial v(x;L)}{\partial L} = -k^2(x)\frac{\partial u(x;L)}{\partial L},$$
$$\frac{\partial v(L_0;L)}{\partial L} + ik_1\frac{\partial u(L_0;L)}{\partial L} = 0,$$
$$\left.\frac{\partial v(x;L)}{\partial L}\right|_{x=L} - ik(L)\left.\frac{\partial u(x;L)}{\partial L}\right|_{x=L}$$
$$= -\left.\frac{\partial v(x;L)}{\partial x}\right|_{x=L} + ik(L)\left.\frac{\partial u(x;L)}{\partial x}\right|_{x=L} + ik'(L)u(L;L) - 2ik'(L)$$
$$= 2k^2(L) + ik'(L)\left[u(L;L) - 2\right], \tag{C.57}$$

where $k'(L) = \frac{dk(L)}{dL}$. Correlating now boundary-value problem (C.57) with boundary-value problem (C.56), we obtain the imbedding equations in the form

$$\frac{\partial}{\partial L}u(x;L) = \left\{ik(L) + \frac{1}{2}\frac{k\prime(L)}{k(L)}\left[2 - u(L;L)\right]\right\}u(x;L),$$
$$u(x;L)|_{L=x} = u(x;x),$$
$$\frac{\partial}{\partial L}v(x;L) = \left\{ik(L) + \frac{1}{2}\frac{k\prime(L)}{k(L)}\left[2 - u(L;L)\right]\right\}v(x;L),$$
$$v(x;L)_{L=x} = v(x;x) = -ik(L)\left[2 - u(x;x)\right]. \tag{C.58}$$

As distinct from Eqs. (C.17), these equations depend on the derivative of function $k(L)$. Consequently, we have in this case

$$v(x;L) = \frac{\partial}{\partial x}u(x;L) = -ik(L)\frac{2 - u(x;x)}{u(x;x)}u(x;L). \tag{C.59}$$

Function $u(L;L)$ satisfies the equality

$$\frac{d}{dL}u(L;L) = \left.\frac{\partial u(x;L)}{\partial x}\right|_{x=L} + \left.\frac{\partial u(x;L)}{\partial L}\right|_{x=L},$$

which yields, in view of Eqs. (C.57) and (C.58), the Riccati equation

$$\frac{d}{dL}u(L;L) = 2ik(L)\left[u(L;L) - 1\right] + \frac{1}{2}\frac{k'(L)}{k(L)}\left[2 - u(L;L)\right]u(L;L),$$
$$u(L;L)|_{L=L_0} = \frac{2k(L_0)}{k(L_0) + k_1}. \tag{C.60}$$

In terms of reflection coefficient $R_L = u(L;L) - 1$, Eqs. (C.58) and (C.60) assume the form

$$\frac{\partial}{\partial L}u(x;L) = \left\{ik(L) + \frac{1}{2}\frac{k'(L)}{k(L)}(1 - R_L)\right\}u(x;L), \quad u(x;L)|_{L=x} = 1 + R_x,$$
$$\frac{d}{dL}R_L = 2ik(L)R_L + \frac{1}{2}\frac{k'(L)}{k(L)}\left(1 - R_L^2\right), \quad R_{L_0} = \frac{k(L_0) - k_1}{k(L_0) + k_1}. \tag{C.61}$$

If we introduce function $\varepsilon(x)$ by the equality $k^2(x) = k^2\left[1+\varepsilon(x)\right]$ and assume that $|\varepsilon(x)| \ll 1$, then Eqs. (C.61) become simpler

$$\frac{\partial}{\partial L}u(x;L) = \left\{ik(L) + \frac{1}{4}\varepsilon'(L)(1 - R_L)\right\}u(x;L), \quad u(x;L)|_{L=x} = 1 + R_x,$$
$$\frac{d}{dL}R_L = 2ik(L)R_L + \frac{1}{4}\varepsilon'(L)\left(1 - R_L^2\right), \quad R_{L_0} = \frac{k(L_0) - k_1}{k(L_0) + k_1}, \tag{C.62}$$

where $k(L) = k\left[1 + \frac{1}{2}\varepsilon(L)\right]$.

In the case of oblique wave incidence, we can represent the wavefield in the form $U(x, \mathbf{R}) = u(x)e^{i\mathbf{qR}}$, as it was done in Remark C.3. Here, function $u(x)$ satisfies the boundary-value problem for the one-dimensional Helmholtz equation (for simplicity, we assume that boundary $x = L_0$ is also the matched boundary, i.e., $k_1 = k(L_0)$)

$$\begin{aligned}
&\left(\frac{d^2}{dx^2} + k^2(x) - q^2\right) u(x; L) = 0, \\
&\left(\frac{d}{dx} + i\sqrt{k^2(L_0) - q^2}\right) u(x)\bigg|_{x=L_0} = 0, \\
&\left(\frac{d}{dx} - i\sqrt{k^2(L) - q^2}\right) u(x)\bigg|_{x=L} = -2i\sqrt{k^2(L) - q^2}.
\end{aligned} \tag{C.63}$$

Consequently, the imbedding equations with respect to parameter L will have the form

$$\begin{aligned}
&\frac{\partial}{\partial L} u(x; L) = \left\{i\sqrt{k^2(L) - q^2} + \frac{1}{2}\frac{k(L)k'(L)}{k^2(L) - q^2}\left[2 - u(L; L)\right]\right\} u(x; L), \\
&u(x; L)|_{L=x} = u(x; x), \\
&\frac{d}{dL} u(L; L) = 2ik\sqrt{k^2(L) - q^2}\left[u(L; L) - 1\right] \\
&\qquad + \frac{1}{2}\frac{k(L)k'(L)}{k^2(L) - q^2}\left[2 - u(L; L)\right] u(L; L), \\
&u(L; L)|_{L=L_0} = 1.
\end{aligned} \tag{C.64}$$

As distinct from Eqs. (C.43), these equations depend on the derivative of function $k(L)$.

If we introduce now function $\varepsilon(x)$ by the equality

$$k^2(L) - q^2 = k_0^2 \cos^2\theta + k_0^2 \varepsilon(L),$$

where θ is the angle of wave incidence (Fig. C.2) and assume that $|\varepsilon(L)| \ll 1$ and attenuation is absent, then we obtain the imbedding equations in the form

$$\begin{aligned}
&\frac{\partial}{\partial L} u(x; L) = \left\{ik_0 \cos\theta + \frac{1}{4\cos^2\theta}\varepsilon'(L)\left(1 - R_L\right)\right\} u(x; L), \\
&u(x; L)|_{L=x} = 1 + R_x, \\
&\frac{d}{dL} R_L = 2ik_0 \left(\cos\theta\right) R_L + \frac{1}{4\cos^2\theta}\varepsilon'(L)\left(1 - R_L^2\right), \quad R_{L_0} = 0.
\end{aligned} \tag{C.65}$$

These equations fail in the narrow region of angles of incidence $\pi/2 - \theta \sim |\varepsilon(L)|$.

Acoustic waves in variable-density media and electromagnetic waves in layered inhomogeneous media

The above boundary-value wave problems describe different physical processes such, for example, as acoustic waves in media with uniform density and certain types of electromagnetic waves. In this case, function $\varepsilon(x)$ in Eq. (C.20) describes inhomogeneity of the velocity of wave propagation (refractive index or dielectric permittivity). If the medium (for example, in the acoustic case) is such that not only $\varepsilon(x)$, but also its density $\rho(x)$ varies with x, then the wave equation assumes the form $\left(\rho'(x) = \frac{d\rho(x)}{dx}\right)$

$$\left\{\frac{\partial^2}{\partial x^2} + \Delta_{\mathbf{R}} - \frac{\rho'(x)}{\rho(x)}\frac{\partial}{\partial x} + k_0^2\left[1 + \varepsilon(x)\right]\right\} U(x, \mathbf{R}) = 0, \tag{C.66}$$

where $\mathbf{R} = \{y, z\}$ denotes the coordinates in the plane perpendicular to the x-axis. As earlier, we assume that inhomogeneities of the medium occupy only the layer $L_0 < x < L$. For simplicity, we will assume additionally that function $\varepsilon(x) = 0$ outside the medium layer; namely, we assume that wave numbers are equal to k_0 and medium density is uniform and equal to unity (the density is normalized by the characteristic value in the medium layer and is, consequently, the dimensionless quantity) in free half-spaces $x > L$ and $x < L_0$.

Now, let the oblique plane wave

$$U_0(x, \mathbf{R}) = e^{-ip(x-L)+i\mathbf{qR}} \qquad \left(p = \sqrt{k_0^2 - q^2}\right)$$

is incident on the layer of inhomogeneous medium from the homogeneous half-space. The case of normal incidence on boundary $x = L$ corresponds to $\mathbf{q} = 0$.

Medium inhomogeneities cause the appearance of the reflected wave in the half-space $x > L$; this means that wavefield for $x > L$ has the following structure

$$U(x, \mathbf{R}) = e^{-ip(x-L)+i\mathbf{qR}} + R_L e^{ip(x-L)+i\mathbf{qR}}.$$

In the half-space $x < L_0$, we have only the transmitted wave of the form

$$U(x, \mathbf{R}) = T_L e^{-ip(x-L)+i\mathbf{qR}}.$$

Boundary conditions for Eq. (C.66) are the continuity conditions of the field and quantity $\frac{1}{\rho(x)} \frac{\partial}{\partial x} U(x, \mathbf{R})$ at layer boundaries.

Inside the layer, the wavefield structure is

$$U(x, \mathbf{R}) = u(x) e^{i\mathbf{qR}},$$

where function $u(x)$ is the solution to the boundary-value problem for the one-dimensional wave equation

$$\begin{gathered}\left\{\frac{d^2}{dx^2} - \frac{\rho'(x)}{\rho(x)} \frac{d}{dx} + p^2 \left[1 + \frac{k_0^2}{p^2} \varepsilon(x)\right]\right\} u(x) = 0, \\ \left(\frac{1}{\rho(x)} \frac{d}{dx} + ip\right) u(x)\Big|_{x=L_0} = 0, \quad \left(\frac{1}{\rho(x)} \frac{d}{dx} - ip\right) u(x)\Big|_{x=L} = -2ip. \end{gathered} \tag{C.67}$$

Remark 21 ***Conversion to the Helmholtz equation.***

Using the functional change $\tilde{u}(x) = u(x)/\sqrt{\rho(x)}$, we can convert Eq. (C.67) into the Helmholtz equation with the effective wave number $\tilde{k}(x)$ dependent on the first and second derivatives of density. However, the appearance of derivatives of density in the wave equation gives rise to a number of restrictions concerning the smoothness of function $\rho(x)$. This fact appears especially inconvenient when function $\rho(x)$ is an experimentally measured function. Below, we show that this difficulty is imaginary and is completely caused by the replacement of function $u(x)$ with function $\tilde{u}(x)$. ♦

Remark 22 ***Conversion to the integral equation.***

We note that boundary-value problem (C.67) is equivalent to the integral equation

$$u(x) = g(x; L) + \int\limits_{L_0}^{L} d\xi g(x; \xi) \varphi(\xi) u(\xi), \tag{C.68}$$

where Green's function in free space ($\varepsilon(x) = 0$) with a given density distribution $\rho(x)$

$$g(x; x_0) = \exp\left\{ ip \operatorname{sgn}(x - x_0) \int\limits_{x_0}^{x} d\eta \rho(\eta) \right\} \tag{C.69}$$

(function $\operatorname{sgn}(x)$ is equal to 1 if $x > 0$, and -1 if $x < 0$) satisfies the boundary-value problem

$$\begin{aligned}
\left\{ \frac{d}{dx} \frac{1}{\rho(x)} \frac{d}{dx} + p^2 \rho(x) \right\} g(x; x_0) &= 2ip\delta(x - x_0), \\
\left(\frac{1}{\rho(x)} \frac{d}{dx} + ip \right) g(x; x_0) \bigg|_{x=L_0} &= 0, \quad \left(\frac{1}{\rho(x)} \frac{d}{dx} - ip \right) g(x; x_0) \bigg|_{x=L} = 0,
\end{aligned}$$

and function $\varphi(x)$ is determined by the equality

$$\varphi(x) = \frac{ip}{2\rho(x)} \left[1 + \frac{k_0^2}{p^2} \varepsilon(x) - \rho^2(x) \right]. \quad \blacklozenge \tag{C.70}$$

Now, we pass on to deriving the imbedding equations.

We rewrite boundary-value problem (C.67) in the form of the system of equations

$$\begin{aligned}
&\frac{d}{dx} u(x; L) = -\rho(x) v(x; L), \\
&\frac{d}{dx} v(x; L) = \frac{p^2}{\rho(x)} \left[1 + \frac{k_0^2}{p^2} \varepsilon(x) \right] u(x; L), \\
&v(L_0; L) - ipu(L_0; L) = 0, \quad v(L; L) + ipu(L; L) = 2ip,
\end{aligned} \tag{C.71}$$

where parameter L is included as new variable.

Then, we proceed as in the foregoing sections. We differentiate system of equations (C.71) with respect to parameter L to obtain the boundary-value system of equations in derivatives $\frac{\partial u(x;L)}{\partial L}$ and $\frac{\partial v(x;L)}{\partial L}$

$$\begin{aligned}
&\frac{d}{dx} \frac{\partial u(x; L)}{\partial L} = -\rho(x) \frac{\partial v(x; L)}{\partial L}, \\
&\frac{d}{dx} \frac{\partial v(x; L)}{\partial L} = \frac{p^2}{\rho(x)} \left[1 + \frac{k_0^2}{p^2} \varepsilon(x) \right] \frac{\partial u(x; L)}{\partial L}, \\
&\frac{\partial v(L_0; L)}{\partial L} - ip \frac{\partial u(L_0; L)}{\partial L} = 0, \\
&\left\{ \frac{\partial v(x; L)}{\partial L} + ip \frac{\partial u(x; L)}{\partial L} \right\} \bigg|_{x=L} = 2ip \left\{ ip\rho(L) + \varphi(L) u(L; L) \right\},
\end{aligned} \tag{C.72}$$

where function $\varphi(L)$ is given by Eq. (C.70).

Correlating now systems of equations (C.71) and (C.72), we obtain the imbedding equations (i.e., the equations with respect to parameter L) for the field inside the medium

$$\begin{aligned}
&\frac{\partial}{\partial L} u(x; L) = \left\{ ip\rho(L) + \varphi(L) u(L; L) \right\} u(x; L), \\
&\qquad u(x; L)|_{L=x} = u(x; x); \\
&\frac{\partial}{\partial L} v(x; L) = \left\{ ip\rho(L) + \varphi(L) u(L; L) \right\} v(x; L), \\
&\qquad v(x; L)|_{L=x} = v(x; x) = ip[2 - u(L; L)].
\end{aligned} \tag{C.73}$$

Function $u(L;L)$ satisfies the equality

$$\frac{d}{dL}u(L;L) = \frac{\partial}{\partial x}u(x;L)\Big|_{x=L} + \frac{\partial}{\partial L}u(x;L)\Big|_{x=L}$$

from which we obtain in view of Eqs. (C.71) and (C.73) the Riccati equation

$$\frac{d}{dL}u(L;L) = 2ip\rho(L)\left[u(L;L)-1\right] + \varphi(L)u^2(L;L),$$
$$u(L;L)|_{L=L_0} = 1 \qquad \text{(C.74)}$$

Equations (C.73) and (C.74) are the equations of the imbedding method for boundary-value problem (C.67) [162]. Of course, we could derive them from the integral equation (C.68). The feature of these equations is that they have no terms explicitly dependent on the derivatives of density.

We can similarly consider the point-source field that satisfies the boundary-value problem

$$\left\{\frac{d^2}{dx^2} - \frac{\rho'(x)}{\rho(x)}\frac{d}{dx} + p^2\left[1 + \frac{k_0^2}{p^2}\varepsilon(x)\right]\right\} G(x;x_0)$$
$$= 2ip\rho(x_0)\delta(x-x_0),$$
$$\left(\frac{1}{\rho(x)}\frac{d}{dx} + ip\right) G(x;x_0)\Big|_{x=L_0} = 0, \quad \left(\frac{1}{\rho(x)}\frac{d}{dx} - ip\right) G(x;x_0)\Big|_{x=L} = 0, \qquad \text{(C.75)}$$

or equivalent integral equation

$$G(x;x_0) = g(x;x_0) + \int\limits_{L_0}^{L} d\xi g(x;\xi)\varphi(\xi)G(\xi;x_0), \qquad \text{(C.76)}$$

where function $g(x;x_0)$ is given by Eq. (C.69).

We can easily obtain the equality similar to Eq. (C.38). In this case, it describes the solution of the problem on the point source field inside the layer of inhomogeneous medium under the assumption that wave parameters and densities of the outside half-spaces $x > L$ and $x < L_0$ are equal to k_2, ρ_2 and k_1, ρ_1, respectively [31, 135, 136, 162].

Equation (C.67) describes also the propagation of electromagnetic waves. As is well known, considering a linear layered media, we can content themselves with analyzing the incident fields of only two polarization directions, namely, the fields with electric field E perpendicular and parallel to the plane of incidence. The first case is equivalent to density $\rho(x) = 1$, while the second, to density $\rho(x) = 1 + \varepsilon(x)$ (we assume here that magnetic permeability is equal to unity). Thus, imbedding equations (C.73) and (C.74) are suitable for analyzing boundary-value problems appeared in the theory of electromagnetic waves. This approach was used in papers [33, 34, 81], [189]–[197], [199, 200] to study the propagation of short and ultra-short radio waves in tropospheric layered waveguide over the ocean surface.

Acoustic-gravity waves in layered ocean

In the foregoing sections, we considered problems with free-transmission boundaries as the reference problem. However, it often appears in physics that input (reference) problems are the problems with reflecting boundaries. It is clear that these problems can be converted into imbedding equations using representation (C.38), but it appears simpler to consider them as reference problems. We consider the problem on excitation of acoustic-gravity waves in layered ocean as an example of such a problem. More details can be found in papers [92]– [98], [162]–[164], where the depth distribution of low-frequency acoustic noise in ocean was considered for different models of surface noise, medium stratification, and impedance of the sea bottom.

The input equations are the equations of hydrodynamics in the adiabatic approximation

$$
\begin{aligned}
&\rho(\mathbf{r},t)\frac{d}{dt}\mathbf{v}(\mathbf{r},t) = -\boldsymbol{\nabla}p(\mathbf{r},t) + \rho(\mathbf{r},t)\mathbf{g} + \mathbf{F}(\mathbf{r},t),\\
&\frac{\partial}{\partial t}\rho(\mathbf{r},t) + \operatorname{div}\left(\rho(\mathbf{r},t)\mathbf{v}(\mathbf{r},t)\right) = Q(\mathbf{r},t),\\
&\frac{d}{dt}p\left(\mathbf{r},t\right) = c^2(\mathbf{r},t)\frac{d}{dt}\rho\left(\mathbf{r},t\right).
\end{aligned}
\tag{C.77}
$$

Here, $\rho(\mathbf{r},t)$ is the medium density; $p(\mathbf{r},t)$ is the pressure; $\mathbf{v}(\mathbf{r},t)$ is the medium velocity; $c(\mathbf{r},t)$ is the sound velocity in the medium; $\mathbf{g} = \{0,0,-g\}$ is the gravity acceleration (the z-axis is directed upright, against the gravity force); $\mathbf{F}(\mathbf{r},t)$ and $Q(\mathbf{r},t)$ are the force and mass sources, respectively; and

$$
\frac{d}{dt} = \frac{\partial}{\partial t} + \mathbf{v}(\mathbf{r},t)\boldsymbol{\nabla}.
$$

Equations (C.77) describe small-scale motions slightly affected by the rotation of the Earth, in which case the Coriolis force can be neglected.

Let the unperturbed state of the medium is described by parameters $\mathbf{v}_0(z) = \{\mathbf{U}_0(z), 0\}$, $\rho_0(z)$, $c(z)$, $p_0(z)$, where $\mathbf{U}_0(z)$ is the velocity horizontal component, and functions $p_0(z)$ and $\rho_0(z)$ are related by the equation of hydrostatics

$$
\frac{d}{dz}p_0\left(z\right) = -g\rho_0\left(z\right).
$$

Consider small oscillations generated by the force and mass sources. We set

$$
\mathbf{v}(\mathbf{r},t) = \mathbf{v}_0(z) + \tilde{\mathbf{v}}(\mathbf{r},t), \quad \rho(\mathbf{r},t) = \rho_0(z) + \tilde{\rho}(\mathbf{r},t), \quad p(\mathbf{r},t) = p_0(z) + \tilde{p}(\mathbf{r},t). \tag{C.78}
$$

Then, substituting Eqs. (C.78) in Eqs. (C.77) and linearizing the system, we obtain the system of equations for oscillating quantities (they are marked by the tilde sign in Eqs. (C.78); below, we omit this sign)

$$
\begin{aligned}
&\rho_0(z)\left[\frac{D}{Dt}\mathbf{u}(\mathbf{r},t) + w(\mathbf{r},t)\frac{d}{dz}\mathbf{U}_0(z)\right] = -\boldsymbol{\nabla}_\perp p(\mathbf{r},t) + \mathbf{F}_\perp(\mathbf{r},t),\\
&\rho_0(z)\frac{D}{Dt}w(\mathbf{r},t) = -\frac{\partial}{\partial z}p(\mathbf{r},t) - g\rho(\mathbf{r},t) + F_z(\mathbf{r},t),\\
&\frac{D}{Dt}\rho(\mathbf{r},t) + w(\mathbf{r},t)\frac{d}{dz}\rho_0(z)\\
&\qquad + \rho_0(z)\left(\frac{d}{dz}w(\mathbf{r},t) + \nabla_\perp\mathbf{u}(\mathbf{r},t)\right) = Q(\mathbf{r},t),\\
&\frac{D}{Dt}\rho(\mathbf{r},t) = \frac{1}{c^2(z)}\frac{D}{Dt}p(\mathbf{r},t) + \frac{1}{g}N^2(z)\rho_0(z)w(\mathbf{r},t),
\end{aligned}
\tag{C.79}
$$

where $\mathbf{u}(\mathbf{r},t)$ and $w(\mathbf{r},t)$ are the horizontal and vertical oscillating components of velocity, respectively;

$$
\frac{D}{Dt} = \frac{\partial}{\partial t} + \left(\mathbf{U}_0(z)\boldsymbol{\nabla}_\perp\right); \quad \boldsymbol{\nabla}_\perp = \left\{\frac{\partial}{\partial x}, \frac{\partial}{\partial y}\right\};
$$

$\mathbf{F}_\perp(\mathbf{r},t)$ and $F_z(\mathbf{r},t)$ are the projections of force $\mathbf{F}(\mathbf{r},t)$ on the horizontal plane and z-axis, respectively; and

$$N^2(z) = -g\left\{\frac{1}{\rho_0(z)}\frac{d\rho_0(z)}{dz} + \frac{g}{c^2(z)}\right\}$$

is the square of the *Brunt–Väisälä frequency*, which is the fundamental characteristics of the internal gravity waves.

If we introduce spectral densities of all quantities $\varphi = \{\mathbf{u}, w, p, \rho, \mathbf{F}, Q\}$

$$\begin{aligned}
\varphi(t,\mathbf{R};z) &= \int d\omega \int d\mathbf{q}\varphi(\omega,\mathbf{q};z)e^{-i\omega t+i\mathbf{qR}}, \\
\varphi(\omega,\mathbf{q};z) &= \frac{1}{(2\pi)^3}\int dt \int d\mathbf{R}\varphi(t,\mathbf{R};z)e^{i\omega t-i\mathbf{qR}},
\end{aligned}$$

where $\mathbf{R} = \{x, y\}$, and eliminate density and velocity horizontal component from system (C.79), then we arrive at the closed system of equations in pressure and vertical velocity

$$\begin{aligned}
&\left(\frac{d}{dz}+\frac{g}{c^2(z)}\right)p(\omega,\mathbf{q};z) \\
&\quad -iA(\omega,\mathbf{q};z)\rho_0(z)\left(1-\frac{N^2(z)}{c^2(z)}\right)w(\omega,\mathbf{q};z) = F_z(\omega,\mathbf{q};z), \\
&\left(\frac{d}{dz}-\frac{g}{c^2(z)}-\frac{1}{A(\omega,\mathbf{q};z)}\frac{dA(\omega,\mathbf{q};z)}{dz}\right)w(\omega,\mathbf{q};z) \\
&\quad -i\frac{A(\omega,\mathbf{q};z)}{\rho_0(z)}\left(\frac{1}{c^2(z)}-\frac{q^2}{A^2(\omega,\mathbf{q};z)}\right)p(\omega,\mathbf{q};z) \\
&\quad = \frac{1}{\rho_0(z)}\left\{Q(\omega,\mathbf{q};z)+\frac{1}{A(\omega,\mathbf{q};z)}\mathbf{q}F_\perp(\omega,\mathbf{q};z)\right\},
\end{aligned} \tag{C.80}$$

where

$$A(\omega,\mathbf{q};z) = \omega - \mathbf{q}\mathbf{U}_0(z).$$

The density and the velocity horizontal component are expressed in terms of the solution of system (C.80) by the equalities

$$\begin{aligned}
&\rho(\omega,\mathbf{q};z) = \frac{1}{c^2(z)}p(\omega,\mathbf{q};z) + \frac{iN^2(z)\rho_0(z)}{gA(\omega,\mathbf{q};z)}w(\omega,\mathbf{q};z), \\
&\mathbf{u}(\omega,\mathbf{q};z) \\
&\quad = \frac{1}{\rho_0(z)A(\omega,\mathbf{q};z)}\left\{-i\rho_0(z)\frac{d}{dz}w(\omega,\mathbf{q};z)+\mathbf{q}p(\omega,\mathbf{q};z)+i\mathbf{F}_\perp(\omega,\mathbf{q};z)\right\}.
\end{aligned} \tag{C.81}$$

We should supplement system (C.80) with the linearized boundary conditions. These conditions (one of them is formulated as vanishing of oscillating vertical velocity at the bottom $z = L_0$ and the other is the condition on free surface at $z = L$) have the forms

$$\begin{aligned}
&w(\omega,\mathbf{q};L_0) = 0, \\
&iA(\omega,\mathbf{q};L)p(\omega,\mathbf{q};L) + g\rho_0(L)w(\omega,\mathbf{q};L) = iA(\omega,\mathbf{q};L)p_a(\omega,\mathbf{q}),
\end{aligned}$$

where $p_a(\omega,\mathbf{q})$ is the spectral component of atmospheric pressure disturbances above the ocean surface. The vertical displacement of the free surface is described by the equality

$$\xi(\omega,\mathbf{q};L) = \frac{i}{A(\omega,\mathbf{q};L)}w(\omega,\mathbf{q};L).$$

Introducing new variables

$$\begin{aligned}\xi(\omega,\mathbf{q};z) &= \frac{i}{A(\omega,\mathbf{q};L)}w(\omega,\mathbf{q};z),\\ P(\omega,\mathbf{q};z) &= p(\omega,\mathbf{q};z)-g\rho_0(z)\xi(\omega,\mathbf{q};z),\end{aligned}$$

($\xi(\omega,\mathbf{q};z)$ is the displacement of liquid particles and $P(\omega,\mathbf{q};z)$ is the pressure in such particles), we can rewrite the boundary-value problem (C.80) in the form

$$\begin{aligned}&\left(\frac{d}{dz}-\frac{gq^2}{A^2(\omega,\mathbf{q};z)}\right)\xi(\omega,\mathbf{q};z)\\ &\quad+\frac{1}{\rho_0(z)}\left(\frac{1}{c^2(z)}-\frac{q^2}{A^2(\omega,\mathbf{q};z)}\right)P(\omega,\mathbf{q};z)\\ &\quad=\frac{1}{\rho_0(z)A(\omega,\mathbf{q};z)}\left(Q+\frac{\mathbf{q}}{A(\omega,\mathbf{q};z)}\mathbf{F}_\perp(\omega,\mathbf{q};z)\right),\\ &\left(\frac{d}{dz}+\frac{gq^2}{A^2(\omega,\mathbf{q};z)}\right)P(\omega,\mathbf{q};z)\\ &\quad-\rho_0(z)\left(A^2(\omega,\mathbf{q};z)-\frac{g^2q^2}{A^2(\omega,\mathbf{q};z)}\right)\xi(\omega,\mathbf{q};z)\\ &\quad=F_z(\omega,\mathbf{q};z)-\frac{ig}{A(\omega,\mathbf{q};z)}\left(Q+\frac{\mathbf{q}}{A(\omega,\mathbf{q};z)}\mathbf{F}_\perp(\omega,\mathbf{q};z)\right),\\ &\xi(\omega,\mathbf{q};L_0)=0,\quad P(\omega,\mathbf{q};L)=p_a(\omega,\mathbf{q}).\end{aligned}\tag{C.82}$$

An advantage of Eqs. (C.82) against Eqs. (C.80) consists in the fact that Eqs. (C.82) has no terms dependent on the derivatives of medium stratification parameters.

The solution to boundary-value problem (C.82) is the sum of two partial solutions of which the first corresponds to the impact of sources in the right-hand side of system (C.82) under the condition that $p_a(\omega,\mathbf{q})=0$ and the second corresponds to the absence of sources $\mathbf{F}(\omega,\mathbf{q};z)$ and $Q(\omega,\mathbf{q};z)$. We content themselves with the consideration of the second boundary-value problem. Normalizing the solution of system (C.82) by $p_a(\omega,\mathbf{q})$, we can rewrite it in the form

$$\begin{aligned}\left(\frac{d}{dz}-\frac{gq^2}{A^2(\omega,\mathbf{q};z)}\right)\xi(\omega,\mathbf{q};z)+K(\omega,\mathbf{q};z)P(\omega,\mathbf{q};z) &= 0,\\ \xi(\omega,\mathbf{q};L_0) &= 0,\\ \left(\frac{d}{dz}+\frac{gq^2}{A^2(\omega,\mathbf{q};z)}\right)P(\omega,\mathbf{q};z)-L(\omega,\mathbf{q};z)\xi(\omega,\mathbf{q};z) &= 0,\\ P(\omega,\mathbf{q};L) &= 1,\end{aligned}\tag{C.83}$$

where

$$\begin{aligned}K(\omega,\mathbf{q};z) &= \frac{1}{\rho_0(z)}\left(\frac{1}{c^2(z)}-\frac{q^2}{A^2(\omega,\mathbf{q};z)}\right),\\ L(\omega,\mathbf{q};z) &= \rho_0(z)\left(A^2(\omega,\mathbf{q};z)-\frac{g^2q^2}{A^2(\omega,\mathbf{q};z)}\right).\end{aligned}$$

Then, we proceed as in the foregoing sections. Considering the problem solution as a function of imbedding parameter L, we obtain the imbedding equations in the form

$$\begin{aligned}
&\frac{\partial}{\partial L}\xi(\omega,\mathbf{q};z;L)\\
&\quad=\left\{\frac{gq^2}{A^2(\omega,\mathbf{q};z)}-L(\omega,\mathbf{q};L)\xi_L(\omega,\mathbf{q})\right\}\xi(\omega,\mathbf{q};z;L),\\
&\frac{\partial}{\partial L}P(\omega,\mathbf{q};z;L)\\
&\quad\left\{\frac{gq^2}{A^2(\omega,\mathbf{q};z)}-L(\omega,\mathbf{q};L)\xi_L(\omega,\mathbf{q})\right\}P(\omega,\mathbf{q};z;L),\\
&\quad\xi(\omega,\mathbf{q};z;L)|_{L\to z}=\xi_z(\omega,\mathbf{q}),\\
&\quad P(\omega,\mathbf{q};z;L)|_{L\to z}=P(\omega,\mathbf{q};z;z)=1,
\end{aligned}\tag{C.84}$$

where $\xi_L(\omega,\mathbf{q})=\xi(\omega,\mathbf{q};L;L)$.

The solution to Eqs. (C.84) has the form

$$\begin{aligned}
\xi(\omega,\mathbf{q};z;L)&=\xi_z(\omega,\mathbf{q})P(\omega,\mathbf{q};z;L),\\
P(\omega,\mathbf{q};z;L)&=\exp\left\{\int\limits_z^L d\eta\left[\frac{gq^2}{A^2(\omega,\mathbf{q};z)}-L(\omega,\mathbf{q};\eta)\xi_\eta(\omega,\mathbf{q})\right]\right\}.
\end{aligned}\tag{C.85}$$

Function $\xi_L(\omega,\mathbf{q})$ satisfies the equality

$$\frac{d}{dL}\xi_L(\omega,\mathbf{q})=\left.\frac{\partial}{\partial z}\xi(\omega,\mathbf{q};z;L)\right|_{z=L}+\left.\frac{\partial}{\partial L}\xi(\omega,\mathbf{q};z;L)\right|_{z=L}.$$

Consequently, in view of Eqs. (C.83) and (C.84), we obtain the Riccati equation

$$\begin{aligned}
&\frac{d}{dL}\xi_L(\omega,\mathbf{q})=-K(\omega,\mathbf{q};L)\\
&\quad+\frac{2gq^2}{A^2(\omega,\mathbf{q};z)}\xi_L(\omega,\mathbf{q})-L(\omega,\mathbf{q};L)\xi_L^2(\omega,\mathbf{q}).
\end{aligned}\tag{C.86}$$

Using Eq. (C.86), we can rewrite Eqs. (C.85) in the form

$$\begin{aligned}
\xi(\omega,\mathbf{q};z;L)&=\xi_L(\omega,\mathbf{q})\exp\left\{\int\limits_z^L d\eta\left[\frac{K(\omega,\mathbf{q};\eta)}{\xi_\eta(\omega,\mathbf{q})}-\frac{gq^2}{A^2(\omega,\mathbf{q};z)}\right]\right\},\\
P(\omega,\mathbf{q};z;L)&=\frac{\xi_L(\omega,\mathbf{q})}{\xi_z(\omega,\mathbf{q})}\exp\left\{\int\limits_z^L d\eta\left[\frac{K(\omega,\mathbf{q};\eta)}{\xi_\eta(\omega,\mathbf{q})}-\frac{gq^2}{A^2(\omega,\mathbf{q};z)}\right]\right\}.
\end{aligned}\tag{C.87}$$

It should be emphasized that both Eq. (C.86) and quadratures for different hydrophysical fields include only stratification parameters, but not their derivatives. This fact offers a possibility of using numerical procedures to solve Eq. (C.86) and calculating the corresponding quadratures not only for sufficiently smooth model medium parameter profiles, but also for actual profiles obtained from ocean sounding. Derivatives of stratification parameters appear in Eqs. (C.81) that describe other hydrophysical parameters.

Introduce now the function

$$f_L(\omega, \mathbf{q}) = 1/\xi_L(\omega, \mathbf{q})$$

satisfying the Riccati equation

$$\begin{aligned}\frac{d}{dL} f_L(\omega, \mathbf{q}) &= L(\omega, \mathbf{q}; L) - \frac{2gq^2}{A^2(\omega, \mathbf{q}; z)} f_L(\omega, \mathbf{q}) \\ &\quad + K(\omega, \mathbf{q}; L) f_L^2(\omega, \mathbf{q})\end{aligned} \tag{C.88}$$

that follows from Eq. (C.86),

The solution of problem (C.83) has resonance structure. This means that poles of function $\xi_L(\omega, \mathbf{q})$ (or zeros of function $f_L(\omega, \mathbf{q})$) describe eigenvalues (dispersion curves) and eigenfunctions of the homogeneous boundary-value problem (C.83). Namely, eigenvalues (dispersion curves) of our problem are described by the equation

$$f_L(\omega_n(\mathbf{q}; L), \mathbf{q}) = 0,$$

and quadratures (C.87) describe unnormalized eigenfunctions

$$\begin{aligned}&\xi_n\left(\omega_n(\mathbf{q}; L), \mathbf{q}; z; L\right) \\ &\quad = \exp\left\{\int_z^L d\eta \left[\frac{K\left(\omega_n(\mathbf{q}; L), \mathbf{q}; \eta\right)}{\xi_\eta\left(\omega_n(\mathbf{q}; L), \mathbf{q}\right)} - \frac{gq^2}{A^2\left(\omega_n(\mathbf{q}; L), \mathbf{q}; \eta\right)}\right]\right\}, \\ &P_n\left(\omega_n(\mathbf{q}; L), \mathbf{q}; z; L\right) \\ &\quad = \frac{1}{\xi_z\left(\omega_n(\mathbf{q}; L), \mathbf{q}\right)} \exp\left\{\int_z^L d\eta \left[\frac{K\left(\omega_n(\mathbf{q}; L), \mathbf{q}; \eta\right)}{\xi_\eta\left(\omega_n(\mathbf{q}; L), \mathbf{q}\right)} - \frac{gq^2}{A^2\left(\omega_n(\mathbf{q}; L), \mathbf{q}; \eta\right)}\right]\right\}.\end{aligned} \tag{C.89}$$

This feature can be immediately used for determining the spectral characteristics of the boundary-value problem. In particular, one can immediately derive dynamic equations for these characteristics (the initial value problem), and these equations appear practicable for analyzing both deterministic and statistical problems [77]–[80], [269, 272].

The analysis of eigenvalues is based on the analysis of zeros of the solution to the Riccati equation the general form of which is

$$\frac{d}{dL} f_L = a_L(\lambda) + b_L(\lambda) f_L + c_L(\lambda) f_L^2, \tag{C.90}$$

where λ is the spectral parameter. Eigenvalues are determined as the solution of the equation

$$f_L(\lambda_L) = 0, \tag{C.91}$$

where we introduced the dependence of the spectral parameter on parameter L. Because eigenvalues are functions of parameter L, they satisfy the equation

$$a_L(\lambda_L) + A_L(\lambda_L) \frac{d}{dL} \lambda_L = 0, \tag{C.92}$$

where

$$A_L(\lambda) = \frac{\partial}{\partial \lambda} f_L(\lambda).$$

The initial condition for $L \to 0$ (we consider here $L_0 = 0$) must be determined from the asymptotic behavior of every particular eigenvalue.

C.2.2 Waves in periodically inhomogeneous media

In the foregoing sections of this Appendix, we derived the imbedding equations for a wide class of boundary-value problems related to wave propagation in layered inhomogeneous media. The current methods of analyzing such problems are based on the use of approximate methods, and the correlation of the results with an exact solution can be of certain interest. The above imbedding equations appear convenient for obtaining exact solutions. To illustrate practicability of the imbedding method, we consider the simplest problem on wave propagation in the layer of periodically inhomogeneous medium.

The problem on waves in periodic media attracts attention of physicists by tradition, because of its importance for almost all fields of physics. The current state of the theory is given in review [60]. Commonly, investigators content themselves with the analysis of dispersion relations (determination of transparency and opaqueness zones), i.e., with the determination of the relationship between the frequency and wave number of a monochromatic wave, which allows the wave to propagate. However, the problem on propagation of a given wave (with a given frequency and wave number) in periodically inhomogeneous media is also of great interest. The problem on radio wave propagation in Earth's ionosphere, where inhomogeneities are created by a powerful pump wave, is an example of such a problem. The analysis of such problems is based on different approximate methods, the main of which is the method of averaging over fast oscillations (conversion to abridged equations). In the strict sense, this method is not asymptotic, and its main advantage consists in the simplicity and physical clarity of the results. It is interesting to compare results of this approximate method with an exact solution of the problem [158]. Note that numerical simulation of time-domain impulses in periodically inhomogeneous media was performed for the first time in papers [37, 99, 100] (see also [316]).

Wave incident on the layer of periodically inhomogeneous medium Let the inhomogeneous medium, as earlier, occupy the layer $L_0 < x < L$ and let the unit-amplitude plane wave $e^{-ik(x-L)}$ is incident on this layer from the right-hand homogeneous half-space $x > L$. Then, the wavefield inside the layer is described by the boundary-value problem for the Helmholtz equation (C.14)

$$\left(\frac{d^2}{dx^2} + k_0^2\left[1+\varepsilon(x)\right]\right) u(x) = 0,$$

$$\left(\frac{d}{dx} + ik_0\right) u(x)\bigg|_{x=L_0} = 0, \qquad \left(\frac{d}{dx} - ik_0\right) u(x)\bigg|_{x=L} = -2ik_0. \tag{C.93}$$

We assume that $\varepsilon(x) = 0$ outside the layer. Inside the layer, we specify function $\varepsilon(x)$ by the formula

$$\varepsilon(x) = -4\mu\cos(2Kx) + 2i\gamma, \tag{C.94}$$

where 2γ is the attenuation coefficient.

In this case, complex reflection and transmission coefficients are determined through the solution to boundary-value problem (C.93) by the equalities

$$R_L = u(L) - 1, \quad T_L = u(L_0).$$

Using dimensionless distances (i.e., setting $k_0 = 1$), we rewrite boundary-value problem

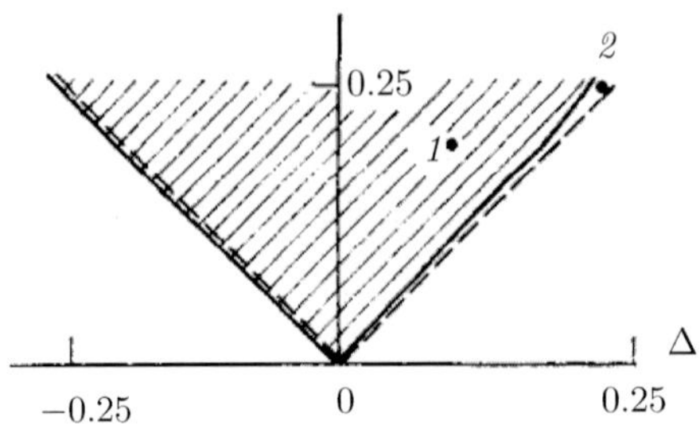

Figure C.3: Zones of parametric instability of the solution to Eq. (C.95) in parameter plane (μ, Δ) at $\gamma = 0$.

(C.93) in the form $(\Delta = (K - k_0)/k_0)$:

$$\left(\frac{d^2}{dx^2} + [1 - 4\mu\cos(2(1+\Delta)x) + 2i\gamma]\right) u(x) = 0,$$

$$\left(i\frac{d}{dx} - 1\right) u(x)\Big|_{x=L_0} = 0, \quad \left(i\frac{d}{dx} + 1\right) u(x)\Big|_{x=L} = 2. \tag{C.95}$$

Without boundary conditions, Eq. (C.95) is the well investigated *Mathieu equation* (see, e.g., [2]). At $\gamma = 0$, plane (μ, Δ) has regions corresponding to parametric instability (parametric resonance), and Fig. C.3 shows the first such region (crosshatched region). For $\mu \to 0$, these regions correspond to $\Delta_n = 1/n - 1$, $n = 1, 2, ...$ $(K = k_0/n)$. In the context of our boundary-value problem, these regions correspond to increased reflectivity of the layer. Outside these regions, the wave relatively freely transverses the medium layer.

The solution to boundary-value problem (C.95) can be represented in terms of *Mathieu functions* and their derivatives. Nevertheless, despite these functions are well investigated and adequately tabulated, construction of the wavefield pattern inside the medium layer (and, consequently, reflection and transmission coefficients) appears far form being an easy task in view of high variability of the wavefield. It appears much simpler to obtain the solution to boundary-value problem (C.95) using numerical methods. Imbedding equations (C.24) (recall, that they consider the solution to boundary-value problem (C.95) as a function of parameter L) appear very convenient here; in our case, these equations have the form

$$\frac{\partial}{\partial L} u(x; L) = i\left\{1 + \frac{1}{2}\tilde{\varepsilon}(L)(1 + R_L)\right\} u(x; L), \quad u(x; L)|_{L=x} = 1 + R_x,$$

$$\frac{d}{dL} R_L = 2iR_L + \frac{i}{2}\tilde{\varepsilon}(L)(1 + R_L)^2, \quad R_{L_0} = 0, \tag{C.96}$$

where

$$\tilde{\varepsilon}(L) = -4\mu\cos(2(1+\Delta)L) + 2i\gamma.$$

The first equation in Eqs. (C.96) can be integrated in the analytic form. Accordingly, solving boundary-value problem (C.95) reduces to solving the Riccati equation and calculating the quadrature. Moreover, in the absence of attenuation $(\gamma = 0)$, the quadrature expressing wavefield intensity $I(x;L) = |u(x;L)|^2$ can be calculated in the analytic form (C.26)

$$I(x;L) = |1 + R_x|^2 \frac{1 - |R_L|^2}{1 - |R_x|^2}, \tag{C.97}$$

so that the solution of our problem reduces to solving the sole Riccati equation.

Now, we dwell on the approximate procedure of solving the Riccati equation with the use of solution averaging over fast oscillations.

We represent reflection coefficient R_L in the form

$$R_L = -i\rho_L e^{2i(1+\Delta)L}.$$

As follows from Eqs. (C.96), function ρ_L must satisfy the equation

$$\frac{d}{dL}\rho_L = -2(\gamma + i\Delta)\rho_L + \mu\left(1 - \rho_L^2\right) + \{...\}, \quad \rho_{L_0} = 0, \tag{C.98}$$

where $\{...\}$ stands for oscillating terms proportional to functions $e^{\pm 2i(1+\Delta)L}$ and $e^{\pm 4i(1+\Delta)L}$. Assuming that function ρ_L only slightly varies on the oscillation period, we can average Eq. (C.98) over these fast oscillations to obtain the approximate equation

$$\frac{d}{dL}\rho_L = -2(\gamma + i\Delta)\rho_L + \mu\left(1 - \rho_L^2\right), \quad \rho_{L_0} = 0$$

whose solution has the form $\left(\alpha = \sqrt{\mu^2 + (\gamma + i\Delta)^2}\right)$

$$\rho_L = \frac{\mu}{\alpha} \frac{\sinh \alpha(L - L_0)}{\cosh \alpha(L - L_0) + \frac{\gamma + i\Delta}{\alpha}\sinh \alpha(L - L_0)}. \tag{C.99}$$

Consider the case of absent attenuation ($\gamma = 0$) in more detail. In this case, the square of the reflection coefficient modulus $|R_L|^2$ coincides with $|\rho_L|^2$ ($|R_L|^2 = |\rho_L|^2$) and, consequently,

$$|R_L|^2 = \frac{\sinh^2 \alpha(L - L_0)}{\cosh^2 \alpha(L - L_0) - \frac{\Delta^2}{\mu^2}}, \quad \alpha = \sqrt{\mu^2 - \Delta^2}. \tag{C.100}$$

Formula (C.97) yields in this case the following expression for the wavefield intensity inside the medium layer

$$I(x;L) = \frac{\cosh 2\alpha(x - L_0) - \frac{\Delta^2}{\mu^2}}{\cosh^2 \alpha(L - L_0) - \frac{\Delta^2}{\mu^2}}. \tag{C.101}$$

In the limiting case $L_0 \to -\infty$ corresponding to the incidence of the wave on half-space $x < L$, the intensity is given by the expression

$$I(x;L) = e^{-2\alpha(L-x)}. \tag{C.102}$$

A consequence of Eqs. (C.100)–(C.102) is the fact that $|R_L|^2 \to 1$ for $\mu^2 \geqslant \Delta^2$, and the wavefield intensity exponentially decays with the distance in medium. On the contrary, for $\mu^2 < \Delta^2$, all these functions appear periodic functions with the period dependent on layer thickness. From the procedure of deriving Eqs. (C.100) – (C.102) clearly follows that these formulas must fail for $\Delta \sim -1$. Moreover, one can expect that Eqs. (C.100) – (C.102) will also fail for $\mu \sim |\Delta|$, i.e., in the region where the solution changes the type of behavior, because they were obtained from physical considerations, rather than from an asymptotic

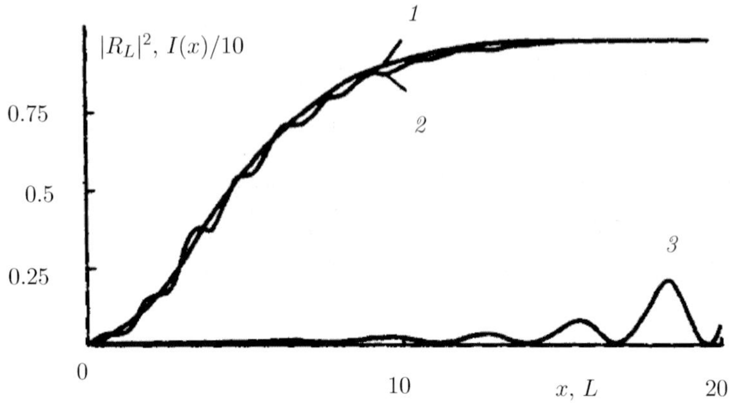

Figure C.4: Squared reflection coefficient modulus $|R_L|^2$ as a function of layer thickness. Curve *1* corresponds to Eq. C.100, curve *2* is the calculated curve, and curve *3* shows function $I(x)/10$ at $L = 20$ ($\mu = 0.2$, $\Delta = 0.1$, $\gamma = 0$).

analysis. The region of parameters $\mu^2 \geqslant \Delta^2$, in which the above theory predicts increased reflectivity of the medium layer, is shown in Fig. C.3 by dashed lines.

Numerical analysis of the problem shows first of all that the solution in the absence of absorption indeed appears periodic in transparence regions, and, in opacity regions, it shows increased reflectivity characterized by high variability. In the transparence regions (far from the boundaries), the solution obtained by the approximate method of averaging agrees with the simulated result. In the first opacity region and again far from the boundaries, the approximate solution also agrees with the simulated result.

Figure C.4 shows quantity $|R_L|^2$ as a function of layer thickness L and wave intensity $I(x; L)$ in the layer for $L = 20$ (the parameters of this curve correspond to point 1 in Fig. C.3). The situation becomes more complicated near the boundaries of these regions. Figure C.5 shows the reflection coefficient as a function of layer thickness (this curve was simulated for the parameters corresponding to point 2 in Fig. C.2) and wavefield intensity in the layer for $L = 100$. If the layer is sufficiently thin (to $L \sim 10$), it behaves as the reflecting layer and the formula (C.100) of the averaging method appears adequate. However, layer reflectivity decreases with further increasing layer thickness. For $L \approx 53$, the layer becomes perfectly transparent. Then, the described pattern is periodically repeated as far as the parameters of point 2 in Fig. C.3 correspond to the transparence region.

The above calculations assume the absence of attenuation. In the presence of attenuation, reflection coefficient shows qualitatively identical behaviors both in and outside transparence regions. For sufficiently thick layers, the modulus of reflection coefficient behaves as a periodic function even in the opacity region.

Bragg resonance in inhomogeneous media Above, we showed that the choice of $\varepsilon(x)$ in the form (C.94) results in the fact that, under the condition $\mu^2 \geqslant \Delta^2$ (it corresponds to the first zone of parametric instability of the solution to the Mathieu equation), reflection coefficient modulus $|R_L|$ tends to unity with increasing layer thickness, and wavefield in-

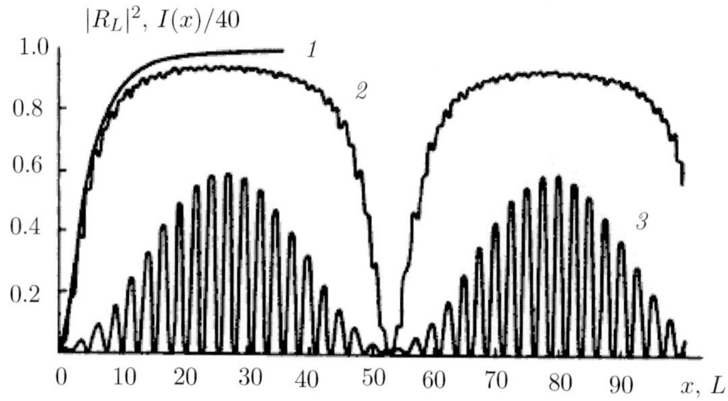

Figure C.5: Squared reflection coefficient modulus $|R_L|^2$ as a function of layer thickness. Curve *1* corresponds to Eq. C.100, curve *2* is the calculated curve, and curve *3* shows function $I(x)/40$ at $L = 100$ ($\mu = 0.25$, $\Delta = 0.24$, $\gamma = 0$).

tensity $I(x) = |u(x; L)|^2$ averaged over oscillations exponentially decreases with distance from boundary $x = L$.

If reflecting boundary $x = L_0$ with boundary condition $u(L_0; L) = 0$ is available and function $\tilde{\varepsilon}(x)$ is specified in the form

$$\tilde{\varepsilon}(x) = -4\mu \cos\left(2(1+\Delta)x+\delta\right) \quad (|\Delta| < \mu) \tag{C.103}$$

differing from Eq. (C.94) by the presence of constant phase shift δ, the wavefield intensity can exponentially increase with the distance in the medium for certain values of parameter δ, which corresponds to excitation of the mirror–grating resonator operating at the Bragg resonance. Indeed, in this case reflection coefficient modulus is equal to unity, $|R_L| = 1$, so that the reflection coefficient can be represented as

$$R_L = e^{i\phi_L},$$

and imbedding equations (C.96) assume the forms (here, we set $L_0 = 0$)

$$\begin{aligned} \frac{\partial}{\partial L} I(x; L) &= -I(x; L)\tilde{\varepsilon}(L)\sin\phi_L, \quad I(x; x) = 2\left(1 + \cos\phi_x\right), \\ \frac{d}{dL}\phi_L &= 2 + \tilde{\varepsilon}(L)\left(1 + \cos\phi_L\right)^2, \quad \phi_0 = \pi. \end{aligned} \tag{C.104}$$

Substituting Eq. (C.103) in Eq. (C.104) and averaging the result over fast oscillations ($\phi_L = \phi_0 + 2L$), we obtain the approximate system of equations

$$\begin{aligned} &\frac{\partial}{\partial L} \ln I(x; L) = 2\mu \sin\left[\phi_L - 2(1+\Delta)L - \delta\right], \\ &\frac{\partial}{\partial L}\phi_L = 2 - 2\mu \cos\left[\phi_L - 2(1+\Delta)L - \delta\right]. \end{aligned} \tag{C.105}$$

If we replace ϕ_L with the new variable $\tilde{\phi}_L$,

$$\tilde{\phi}_L = \phi_L - 2(1+\Delta)L - \delta,$$

system (C.105) assumes the form

$$\begin{aligned} \frac{\partial}{\partial L} \ln I(x;L) &= 2\mu \sin\tilde{\phi}_L, \\ \frac{\partial}{\partial L}\tilde{\phi}_L &= -2\left(\Delta + \mu\cos\tilde{\phi}_L\right), \quad \tilde{\phi}_0 = \pi - \delta. \end{aligned} \tag{C.106}$$

Now, it becomes clear that, if $\Delta + \mu\cos\tilde{\phi}_0 = 0$, i.e., if.

$$\delta = \frac{\pi}{2} - \arcsin\frac{\Delta}{\mu}, \tag{C.107}$$

then $\tilde{\phi}_L \equiv \tilde{\phi}_0$, and, consequently,

$$I(x;L) = 2\left(1+\cos\phi_x\right)\exp\left\{2\sqrt{(\mu^2-\Delta^2)}(L-x)\right\}, \tag{C.108}$$

from which follows that the intensity exponentially increases with distance in the medium and achieves the maximum near the boundary at which $I(0;L) = 0$.

The described effect is subtle, because even small variations of parameter δ result in failure of resonance excitation. Nevertheless, we derived this effect using an approximate approach (the method of averaging).

Equations (C.96) and (C.97) supplemented with the initial condition were integrated numerically [159] for different parameters μ, Δ, and δ of function $\tilde{\varepsilon}(x)$ specified in Eq. (C.103). Phenomenon of parametric excitation was observed both in and outside the first zone of parametric instability. Figure C.6 shows examples of such excitation. Curve 1 corresponds to the intensity distribution inside the medium in the first zone of parametric instability and curve 2, to the intensity distribution in the second zone. Small variations of parameter δ (± 0.05) cause a decrease of the wave intensity in the medium at least by a factor of 10.

C.2.3 Boundary-value stationary nonlinear wave problem on self-action

General equation

Consider now the problem on incidence of plane wave $U(x) = ve^{-ik_0(x-L)}$, where v is the amplitude, on the nonlinear medium occupying the layer $L_0 < x < L$, and assume that function

$$\varepsilon(x) = \varepsilon(x, J(x))$$

depends additionally on wavefield intensity $J(x) = |U(x)|^2$ inside the medium (the *nonlinear problem on wave self-action*). As earlier, we assume that $\varepsilon(x) = 0$ outside the medium. In the deterministic case, this problem was formulated and analyzed in detail in papers [13, 14, 172, 179] (see also [136]) and, in the statistical case, in papers [122, 182, 291].

The stationary nonlinear problem on wave self-action is described by the nonlinear boundary-value problem

$$\begin{gathered} \left(\frac{d^2}{dx^2} + k_0^2\left[1+\varepsilon\left(x, J(x)\right)\right]\right)U(x) = 0, \\ \left(\frac{d}{dx} + ik_0\right)U(x)\bigg|_{x=L_0} = 0, \quad \left(\frac{d}{dx} - ik_0\right)U(x)\bigg|_{x=L} = -2ik_0 v. \end{gathered} \tag{C.109}$$

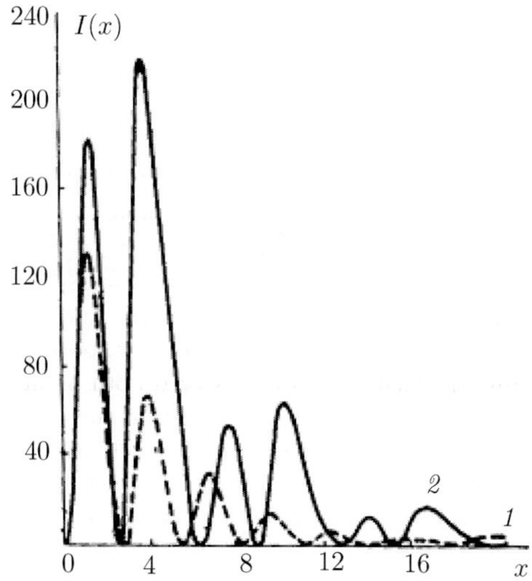

Figure C.6: Parametric excitation of mirror–lattice resonator. Curve *1* corresponds to $\mu = 0.2, \Delta = 0.15, \delta = \pi/2 - \arcsin(\Delta/\mu) + 0.1$ and curve *2* corresponds to $\mu = 0.25$, $\Delta = -0.5$, $\delta = \pi/2 - 0.75$.

We note additionally that problems formulated in terms of the Schrödinger equation are similar to boundary-value problem (C.109) (see [219]–[221], [242]).

We represent the solution of this problem in the form

$$U(x) = vu(x).$$

Then, function $u(x)$ satisfies the boundary-value problem

$$\left(\frac{d^2}{dx^2} + k_0^2\left[1+\varepsilon\left(x, wI(x)\right)\right]\right)u(x) = 0,$$
$$\left(\frac{d}{dx} + ik_0\right)u(x)\bigg|_{x=L_0} = 0, \quad \left(\frac{d}{dx} - ik_0\right)u(x)\bigg|_{x=L} = -2ik_0, \tag{C.110}$$

where $w = |v|^2$ is the intensity of the incident wave and $I(x) = |u(x)|^2$ is the wavefield intensity in the medium layer.

The solution to boundary-value problem (C.110) depends on parameters L and w, i.e.,.

$$u(x) = u(x; L, w).$$

Here, we derive the imbedding equations using the method of integral equation. Boundary-value problem (C.110) is equivalent to integral equation (C.45) that assumes in our case the form

$$u(x; L, w) = e^{ik_0(L-x)} + i\frac{k_0}{2}\int\limits_{L_0}^{L} d\xi e^{ik_0|x-\xi|}\varepsilon\left(\xi, wI(\xi; L, w)\right)u(\xi; L, w). \tag{C.111}$$

Differentiating Eq. (C.111) with respect to parameter L, we obtain that derivative $\frac{\partial}{\partial L}u(x; L, w)$ satisfies the integral equation

$$\frac{\partial}{\partial L}u(x; L, w) = a(L, w)e^{ik_0(L-x)|}$$
$$+i\frac{k_0}{2}\int\limits_{L_0}^{L} d\xi e^{ik_0|x-\xi|}\left\{\varepsilon\left(\xi, wI(\xi; L, w)\right)\frac{\partial u(\xi; L, w)}{\partial L}\right.$$
$$\left.+\,u(\xi; L, w)\frac{\partial\varepsilon\left(\xi, wI(\xi; L, w)\right)}{\partial I(\xi; L, w)}\frac{\partial I(\xi; L, w)}{\partial L}\right\}, \tag{C.112}$$

where

$$a(L, w) = ik_0\left\{1 + \frac{1}{2}\varepsilon\left(L, wI_L(w)\right)u_L(w)\right\} \tag{C.113}$$

and

$$u_L(w) = u(L; L, w), \quad I_L(w) = I(L; L, w).$$

If we set

$$\frac{\partial}{\partial L}u(x; L, w) = a(L, w)u(x; L, w) + \psi(x; L, w)$$

then function $\psi(x; L, w)$ will satisfy the integral equation

$$\psi(x; L, w) = i\frac{k_0}{2}\int\limits_{L_0}^{L} d\xi e^{ik_0|x-\xi|}\varepsilon\left(\xi, wI(\xi; L, w)\right)\psi(\xi; L, w)$$
$$+i\frac{k_0}{2}\int\limits_{L_0}^{L} d\xi e^{ik_0|x-\xi|}u(\xi; L, w)\frac{\partial\varepsilon\left(\xi, wI(\xi; L, w)\right)}{\partial I(\xi; L, w)}\frac{\partial I(\xi; L, w)}{\partial L}. \tag{C.114}$$

Differentiating now Eq. (C.111) with respect to parameter w, we obtain the integral equation in derivative $\frac{\partial}{\partial w}u(x; L, w)$

$$\frac{\partial}{\partial w}u(x; L, w) = i\frac{k_0}{2}\int\limits_{L_0}^{L} d\xi e^{ik_0|x-\xi|}\varepsilon\left(\xi, wI(\xi; L, w)\right)\frac{\partial}{\partial w}u(\xi; L, w)$$
$$+i\frac{k_0}{2}\int\limits_{L_0}^{L} d\xi e^{ik_0|x-\xi|}u(\xi; L, w)\frac{\partial\varepsilon\left(\xi, wI(\xi; L, w)\right)}{w\partial I(\xi; L, w)}\left[1 + w\frac{\partial}{\partial w}\right]I(\xi; L, w). \tag{C.115}$$

From definition of quantity $I(x; L, w) = u(x; L, w)u^*(x; L, w)$, where $u^*(x; L, w)$ is the complex conjugated wavefield, we can derive, in view of Eq. (C.113), the relationships

$$\frac{\partial}{\partial L}I(x; L, w) = \left[a(L, w) + a^*(L, w)\right]I(x; L, w)$$
$$+u(x; L, w)\psi^*(x; L, w) + u^*(x; L, w)\psi(x; L, w),$$
$$\frac{1}{w}\left(I(x; L, w) + w\frac{\partial I(x; L, w)}{\partial w}\right) = \frac{I(x; L, w)}{w}$$
$$+u(x; L, w)\frac{\partial u^*(x; L, w)}{\partial w} + u^*(x; L, w)\frac{\partial u(x; L, w)}{\partial w}.$$

Consequently, Eqs. (C.114) and (C.115) coincide under the condition that

$$\psi(x; L, w) = w\left[a(L, w) + a^*(L, w)\right]\frac{\partial}{\partial w}u(x; L, w).$$

Thus, assuming the uniqueness of problem solution, we obtain the equality

$$\frac{\partial}{\partial L}u(x; L, w) = \left[a(L, w) + wb(L, w)\frac{\partial}{\partial w}\right]u(x; L, w) \quad (x < L), \tag{C.116}$$

where

$$b(L, w) = a(L, w) + a^*(L, w). \tag{C.117}$$

Supplementing this equality with the initial condition for $L \to x$

$$u(x; L, w)|_{L=x} = u_x(w) \tag{C.118}$$

we can consider it as the differential equation.

It is obvious that function $u_L(w)$ satisfies the relationship

$$\frac{\partial}{\partial L}u_L(w) = \left.\frac{\partial}{\partial L}u(x; L, w)\right|_{x=L} + \left.\frac{\partial}{\partial x}u(x; L, w)\right|_{x=L}. \tag{C.119}$$

The first term in the right-hand side of Eq. (C.119) can be determined from Eq. (C.116) by setting $x = L$, and the second term can be determined from the boundary condition in Eq. (C.110). As a result, we obtain the closed nonlinear equation

$$\frac{\partial}{\partial L}u_L(w) = 2ik_0\left[u_L(w) - 1\right] + i\frac{k_0}{2}\varepsilon\left(L, wI_L(w)\right)u_L^2(w)$$
$$+wb(L, w)\frac{\partial}{\partial w}u_L(w) \quad \left(I_L(w) = |u_L(w)|^2\right) \tag{C.120}$$

with the initial condition

$$u_{L_0}(w) = 1$$

following from Eq. (C.110). Equations (C.116) and (C.120) are equivalent to both integral equation (C.111) and input boundary-value problem (C.110); consequently, the problem is reduced now to the initial value problem, and Eqs. (C.116) and (C.120) are the equations of the imbedding method in the context of the problem under consideration.

If we set $x = L_0$ in Eq. (C.116), then we obtain the equation for the transmission coefficient $T_L(w) = u(L_0; L, w)$

$$\frac{\partial}{\partial L} T_L(w) = \left[a(L, w) + w b(L, w) \frac{\partial}{\partial w} \right] T_L(w), \quad T_L(w) = 1. \tag{C.121}$$

The reflection coefficient given by the formula $\rho_L(w) = u_L(w) - 1$ satisfies the closed equation following from Eq. (C.120),

$$\begin{aligned} \frac{\partial}{\partial L} \rho_L(w) &= 2ik_0 \rho_L(w) + i\frac{k_0}{2}\varepsilon\left(L, w|1 + \rho_L(w)|^2\right)(1 + \rho_L(w))^2 \\ &+ w b(L, w) \frac{\partial}{\partial w} \rho_L(w), \quad \rho_{L_0}(w) = 0. \end{aligned} \tag{C.122}$$

If the medium is linear, then dependence on w disappears and all equations grade into the corresponding equations of the linear problem. Note that the equation for wavefield intensity $J(x; L, w) = w|u(x; L, w)|^2$ can be derived as a consequence of Eq. (C.116)

$$\begin{aligned} \frac{\partial}{\partial L} J(x, L, w) &= w b(w) \frac{\partial}{\partial w} J(x, L, w), \\ J(x; x, w) &= J_x(w) = w|u_x(w)|^2. \end{aligned} \tag{C.123}$$

As is well known, first-order partial differential equations are equivalent to systems of ordinary differential equations. If we introduce characteristic curves

$$w_L = w(L, w_0)$$

by the equality

$$\frac{d}{dL} w_L = -b(L, w_L) w_L, \quad w_{L_0} = w_0, \tag{C.124}$$

then the field at layer boundary $u_L(w)$ will be described along the characteristics by the equation ($I_L = |u_L|^2$)

$$\frac{d}{dL} u_L = 2ik_0 \left[u_L - 1\right] + i\frac{k_0}{2}\varepsilon\left(L, w_L I_L\right) u_L^2, \quad u_{L_0} = 1, \tag{C.125}$$

which coincides in appearance with the equation of the linear problem, and Eq. (C.123) will grade into the equality

$$\frac{d}{dL} J(x, L) = 0, \quad J(x; x) = J_x = w_x|u_x|^2. \tag{C.126}$$

As a consequence, wavefield intensity inside the medium remains intact along a characteristic curve, i.e.,

$$J(x; L) = J_x = w_x|u_x|^2. \tag{C.127}$$

Thus, the solution to problem (C.124), (C.125), i.e., the field at the layer boundary completely determines the wavefield intensity inside the medium. Moreover, if we know the behavior of characteristics w_L as functions of L and wave intensity distribution inside the layer of some fixed thickness $J(x;L)$, then the behavior of the intensity will remain valid for any other layer thickness $L_1 \leqslant L$, but will correspond to the incident wave intensity w_{L_1}, i.e.,

$$J(x;L_1) = J(x;L).$$

Consequently, Eq. (C.127) reflects the property of invariance of wavefield intensity inside the medium layer with respect to layer thickness and intensity of the wave incident on the layer. This is a general property that can be extended to the three-dimensional problems.

In view of Eq. (C.127), we have at $x = L_0$

$$J(0;L) = w_0.$$

Taking into account that the field at layer boundary $x = L_0$ coincides with the complex transmission coefficient $T_L = u(L_0;L)$, we obtain that the squared modulus of the transmission coefficient is given by the expression

$$|T_L|^2 = \frac{1}{w_L} J(L_0;L) = \frac{w_0}{w_L}.$$

This expression reveals physical meaning of characteristics $w_L = w(L, w_0)$, and the quantity $|T_L|^2$ by itself satisfies the equation

$$\frac{d}{dL}|T_L|^2 = b(L, w_L)|T_L|^2, \quad |T_{L_0}|^2 = 1.$$

In the presence of attenuation in the medium, wave intensity at boundary $x = L_0$ (and, consequently, quantity $|T_L|^2$) must decrease with increasing layer thickness. It becomes clear therefore that quantity w_L must increase with increasing L for sufficiently large L.

We divide wavefield along the characteristic into real and imaginary parts

$$u_L = R(L) + iS(L).$$

Then, Eqs. (C.124) and (C.125) assume the form

$$\begin{aligned}
\frac{d}{dL} w_L &= [\gamma(L, J_L) R(L) + \varepsilon_1(L, J_L) S(L)] w_L, \quad w_{L_0} = w_0, \\
\frac{d}{dL} R(L) &= -2S(L) - \varepsilon_1(L, J_L) R(L) S(L) \\
&\quad -\frac{1}{2}\gamma(L, J_L)\left[R^2(L) - S^2(L)\right], \\
\frac{d}{dL} S(L) &= 2[R(L) - 1] + \frac{1}{2}\varepsilon_1(L, J_L)\left[R^2(L) - S^2(L)\right] \\
&\quad -\gamma(L, J_L) R(L) S(L),
\end{aligned} \tag{C.128}$$

where

$$J_L = w_L\left[R^2(L) + S^2(L)\right].$$

The squared modulus of the reflection coefficient from the medium layer is defined by the expression

$$|\rho_L|^2 = [R(L) - 1]^2 + S^2(L).$$

From Eq. (C.128) at $\gamma = 0$ follows the equality

$$|\rho_L|^2 + |T_L|^2 = 1$$

that corresponds to conservation of the energy flux density. Note that if we specify

$$\varepsilon(L, J_L) \equiv \varepsilon(J_L) = \varepsilon_1(J_L) + i\gamma(J_L),$$

where $\varepsilon_1^*(J) = \varepsilon_1(J)$ and quantity $\gamma(J)$ describes wave absorption, and eliminate variable L from Eq. (C.128), then we arrive at the system of equations whose solution determines field $u_L = u(w_L)$:

$$\begin{aligned} & w_L \left[\gamma\left(J_L\right) R(L) + \varepsilon_1\left(J_L\right) S(L)\right] \frac{dR(L)}{dw_L} \\ & \quad = -2S(L) - \varepsilon_1\left(J_L\right) R(L) S(L) - \frac{1}{2}\gamma\left(J_L\right)\left[R^2(L) - S^2(L)\right], \\ & w_L \left[\gamma\left(J_L\right) R(L) + \varepsilon_1\left(J_L\right) S(L)\right] \frac{dS(L)}{dw_L} \\ & \quad = 2\left[R(L) - 1\right] + \frac{1}{2}\varepsilon_1\left(J_L\right)\left[R^2(L) - S^2(L)\right] - \gamma\left(J_L\right) R(L) S(L). \end{aligned} \tag{C.129}$$

Thus, in this case, the behavior of quantity u_L as a function of layer thickness L is governed only by the dependence of w_L on L.

If characteristic curves do not cross, then the continuous increase of w_L at a fixed L corresponds to the continuous increase of the corresponding values w_0. The region of values w_0 contracts with increasing L at a fixed w_L to value $w_0 = 0$. Taking into account that this value is associated with characteristic curve $w_L = 0$ (the case of the linear problem), we can obviously take the solution of the linear problem as the initial condition of Eq. (C.129) for $L \to \infty$. If quantity $b(w_L)$ increases with increasing w_L for sufficiently large w_L, then, for arbitrary w_0, there exists a finite layer thickness $L(w_0)$ such that $w_L = \infty$. And vice versa, for any finite thickness L, there exists a limiting value $\tilde{w}_0$ such that the corresponding value $w_L = \infty$. Variation of quantity w_0 in region $0 \leqslant w_0 \leqslant \tilde{w}_0$ corresponds to the continuous variation of quantity w_L in region $0 \leqslant w_L < \infty$. With increasing layer thickness L, quantity $\tilde{w}_0 \to 0$. Below, we consider some special examples to make sure that this situation really takes place in a number of cases.

In this section, we considered the problem on wave incidence on medium layer. One could consider also the problem on the source located inside the medium layer. We will not dwell on this problem because it is of little physical interest. In addition, we note that the problem on the plane wave oblique incidence reduces, for simplest types of nonlinearity, to the considered one by simple variable renaming.

Wave incidence on a half-space of nonlinear medium

If function $\varepsilon(x, wI(x))$ has no explicit dependence on x, i.e., if $\varepsilon(x, wI(x)) = \varepsilon(wI(x))$, then we can perform limit process $L_0 \to -\infty$ in Eq. (C.122), which corresponds to the wave incident on half-space $x < L$. In this case, we obtain that the field at the medium boundary satisfies the first-order nonlinear differential equation $(I(w) = |u(w)|^2)$

$$wb(w)\frac{d}{dw}u(w) = -2i\left[u(w) - 1\right] - i\frac{1}{2}\varepsilon\left(wI(w)\right)u^2(w). \tag{C.130}$$

The initial condition to this equation at $w = 0$ is defined by the solution to the linear problem and has the form

$$u(0) = \frac{2}{1+\alpha}, \quad \alpha = \sqrt{1+\varepsilon(0)}, \quad \operatorname{Im}\alpha > 0, \quad \operatorname{Re}\alpha > 0.$$

The field inside the medium $u(x, w)$ satisfies the linear equation ($k(L - x)= \xi$)

$$\frac{\partial}{\partial \xi} u(\xi, w) = \left[a(w) + wb(w)\frac{\partial}{\partial w}\right] u(\xi, w) \quad (\xi > 0) \tag{C.131}$$

with the initial condition $u(0, w) = u(w)$. Here,

$$a(w) = i\left\{1 + \frac{1}{2}\varepsilon\left(wI(w)\right)u(w)\right\}, \quad b(w) = a(w) + a^*(w).$$

The equation for wavefield intensity inside the medium $J(\xi, w) = w|u(\xi, w)|^2$ follows from Eq. (C.131) and has the form

$$\frac{\partial}{\partial \xi} J(\xi, w) = wb(w)\frac{\partial}{\partial w} J(\xi, w), \quad J(0, w) = wI(w) = w|u(w)|^2. \tag{C.132}$$

The parametric representation of the solution to Eq. (C.132) can be easily constructed by the method of characteristics with characteristic parameter $\bar{w}$

$$\xi = -\int\limits_{\bar{w}}^{w} \frac{dw}{wb(w)}, \quad J(\xi, w) = \bar{w}I(\bar{w}). \tag{C.133}$$

Eliminating parameter $\bar{w}$, we arrive at the intensity $J(\xi, w)$ in the explicit form.

Thus, we reduced the solution of the problem to the determination of either the field at medium boundary $u(w)$, or the reflection coefficient $\rho(w) = u(w) - 1$.

Note that if we set $\varepsilon(0) = 0$, $u(0) = 1$ and assume attenuation absent, so that function $u(w)$ is the real function, the partial solution can be easily found. In these conditions $b(w) \equiv 0$, and Eq. (C.130) yields the transcendental equation in $u(w)$

$$4\left[u(w) - 1\right] = -\varepsilon(wI(w))u^2(w).$$

Then, from Eq. (C.131) follows the solution in the form of a plane wave propagating in the nonlinear medium

$$u(\xi, w) = u(w)\exp\left\{i\xi\frac{2 - u(w)}{u(w)}\right\}. \tag{C.134}$$

Consider in more detail the structure of the obtained equations and their solutions. We set

$$u(w) = R(w) + iS(w)$$

and separate the real and imaginary parts in Eq. (C.130)

$$\begin{aligned}
wb(w)\frac{d}{dw}R(w) &= 2S(w) + \varepsilon_1\left(wI(w)\right)R(w)S(w) \\
&\quad + \frac{1}{2}\gamma\left(wI(w)\right)\left[R^2(w) - S^2(w)\right], \\
wb(w)\frac{d}{dw}S(w) &= 2\left[1 - R(w)\right] + \gamma\left(wI(w)\right)R(w)S(w) \\
&\quad - \frac{1}{2}\varepsilon_1\left(wI(w)\right)\left[R^2(w) - S^2(w)\right],
\end{aligned} \tag{C.135}$$

where

$$\begin{aligned} I(w) &= R^2(w) + S^2(w), \\ b(w) &= -\left[\gamma\left(wI(w)\right)R(w) + \varepsilon_1\left(wI(w)\right)S(w)\right]. \end{aligned}$$

Note that system of equations (C.135) formally coincides with system of equations (C.129). The initial conditions to system (C.135) follow from Eq. (C.130). Condition $|\rho(w)|^2 \leqslant 1$ yields the following restrictions

$$0 \leqslant R(w) \leqslant 2, \quad |S(w)| \leqslant 1,$$

the equality being realized only at $\gamma = 0$.

The equalities

$$\begin{aligned} b(w)\frac{d}{dw}wR(w) &= 2S(w) - \frac{1}{2}\gamma\left(wI(w)\right)I(w), \\ b(w)\frac{d}{dw}wI(w) &= 4S(w) \end{aligned} \tag{C.136}$$

can be obtained as consequences of Eqs. (C.135).

We consider first the case of absent attenuation, i.e., we set. $\gamma\left(wI(w)\right) = 0$. Then system of equations (C.135) is simplified,

$$\begin{aligned} &\varepsilon_1\left(wI(w)\right)wS(w)\frac{d}{dw}R(w) = -S(w)\left[2 + \varepsilon_1\left(wI(w)\right)R(w)\right], \\ &\varepsilon_1\left(wI(w)\right)wS(w)\frac{d}{dw}S(w) = 2\left[R(w) - 1\right] \\ &+\frac{1}{2}\varepsilon_1\left(wI(w)\right)\left[R^2(w) - S^2(w)\right] \end{aligned} \tag{C.137}$$

and Eqs. (C.136) assume the form

$$\begin{aligned} \varepsilon_1\left(wI(w)\right)wS(w)\frac{d}{dw}wR(w) &= -2S(w), \\ \varepsilon_1\left(wI(w)\right)wS(w)\frac{d}{dw}wI(w) &= -4S(w). \end{aligned} \tag{C.138}$$

Considering Eqs. (C.137) as the system of ordinary differential equations (without taking initial conditions into account), we see that any solution to this system belongs to one of two types corresponding to $S(w) = 0$ and $S(w) \neq 0$, respectively.

In the first case $S(w) = 0$. Then, the first equation of Eqs. (C.137) is satisfied identically and the second equation yields the transcendental equation in $R(w)$

$$4\left[1 - R(w)\right] = R^2(w)\varepsilon_1\left(wR^2(w)\right). \tag{C.139}$$

In this case $b(w) \equiv 0$ and the solution to Eq. (C.133) has the form

$$J(\xi, w) = wR^2(w). \tag{C.140}$$

This type of solutions corresponds to the regime of a plane wave propagating in the nonlinear medium. As it follows from Eq. (C.131), the wavefield has in this case the form

$$u(\xi, w) = R(w)\exp\left\{i\xi\frac{2 - R(w)}{R(w)}\right\}. \tag{C.141}$$

The second case corresponds to $S(w) \neq 0$. We set $R(w_0) = R_0$, $S(w_0) = S_0$ at $w = w_0$. Then, we can cancel $S(w)$ in Eqs. (C.137), (C.138) to obtain the system of equations

$$\begin{aligned} &\varepsilon_1 (wI(w))\, w\frac{d}{dw}R(w) = -\left[2 + \varepsilon_1 (wI(w))\, R(w)\right], \\ &\varepsilon_1 (wI(w))\, wS(w)\frac{d}{dw}S(w) = 2\left[R(w) - 1\right] \\ &+\frac{1}{2}\varepsilon_1 (wI(w)) \left[R^2(w) - S^2(w)\right], \end{aligned} \tag{C.142}$$

and equalities

$$\begin{aligned} \varepsilon_1 (wI(w)) \frac{d}{dw}wR(w) &= -2, \\ \varepsilon_1 (wI(w)) \frac{d}{dw}wI(w) &= -4. \end{aligned} \tag{C.143}$$

Integrating Eqs. (C.143), we obtain

$$\int\limits_{w_0 I_0}^{wI(w)} dt\varepsilon_1(t) = -4(w - w_0), \quad I_0 = R_0^2 + S_0^2, \tag{C.144}$$

$$wI(w) - 2wR(w) = w_0\left[I(w_0) - 2R(w_0)\right]. \tag{C.145}$$

Equality (C.144) defines $I(w)$ as a function of w and Eq. (C.145) defines function $R(w)$. Function $S(w)$ is defined by the equality

$$S(w) = \pm\sqrt{I(w) - R^2(w)}, \tag{C.146}$$

where the choice of sign of the root depends on the initial value; if the initial value $S(w_0) = 0$, this choice of the sign follows from the requirement of wavefield finiteness for $\xi > 0$. From Eq. (C.144) follows the expression for the reflection coefficient squared modulus

$$|\rho(w)|^2 = [R(w) - 1]^2 + S^2(w) = 1 - \frac{w_0}{w}(2R_0 - I_0). \tag{C.147}$$

As a consequence, we have $2R_0 > I_0$, so that this type of solutions yields the increase of the reflection coefficient modulus with increasing the incident wave intensity. This type of solutions can exist only if the radicand in Eq. (C.146) is positive.

At points where

$$I(w_1) = R^2(w_1),$$

the regime of the solution can change. For the solutions of this type, one can use Eq. (C.133) to obtain the wavefield intensity inside the medium in the implicit form. Of course, all these formulas can be obtained by the immediate integration of Eq. (C.109), page 470 (for $\gamma = 0$, $\varepsilon = \varepsilon(wI)$) using two integrals

$$\begin{aligned} U(x)\frac{d}{dx}U^*(x) - U^*(x)\frac{d}{dx}U(x) &= \text{const}, \\ \frac{dU(x)}{dx}\frac{dU^*(x)}{dx} + k^2 \int\limits_{J_0}^{J(x)} dt\left[1 + \varepsilon(t)\right] &= \text{const}. \end{aligned} \tag{C.148}$$

However, the formulas obtained in this way explicitly express all quantities in terms of the incident wave intensity that hardly can be derived from integrals (C.148). A common practice consists in the use of integrals (C.148) only for analyzing possible types of the solutions; then, these possible solutions are sewed together with the incident wave at the layer boundary. In a number of cases, this process is accompanied by ambiguities; namely, several values of reflection coefficient can correspond to the same field inside the medium. Even allowance for attenuation cannot sometimes kill this ambiguity. Our approach rests on other grounds. For small incident wave intensities w, we deal with the linear problem. The further evolution of the field with increasing w is described by the nonlinear system of equations (C.135) with given initial conditions. It is reasonable to suppose that this evolution must select among the possible types of the solutions that can really occur, so that we automatically obtain the solution of the type (first or second) that corresponds to the initial data and, moreover, possible changes from one type to the other. It is assumed that the incident wave intensity is varied adiabatically. In the presence of attenuation, there is no way of solving Eqs. (C.135) in the analytic form. The analysis of system (C.135) in the absence of attenuation reveals singularities in the solutions, and the solution behavior in the vicinity of these singularities can be established by numerical simulations.

Examples of wavefield calculations in nonlinear medium

Consider specifically two simplest examples of nonlinearity $\varepsilon_1(t) = \pm\beta t, \quad \beta > 0$. Here, our concern is in the case of small attenuation only. For other types of nonlinearity, see, e.g., [136].

Example 1

Let $\varepsilon_1(t) = \beta t, \quad \beta > 0, \quad \gamma = 0$. In this case $\varepsilon_1(0) = 0$ and the initial conditions of system (C.135) have the form

$$R(0) = 1, \quad S(0) = 0.$$

In view of the fact that parameter β appears only as a factor in the product $\beta w > 0$, we can unrestrictedly set it equal to unity. Thus, we have the system

$$\begin{aligned}
&\left[R^2(w) + S^2(w)\right] w^2 S(w) \frac{d}{dw} R(w) \\
&\quad = S(w) \left\{2 + wR(w) \left[R^2(w) + S^2(w)\right]\right\}, \quad R(0) = 1, \\
&\left[R^2(w) + S^2(w)\right] w^2 S(w) \frac{d}{dw} S(w) \\
&\quad = 2\left[R(w) - 1\right] + \frac{1}{2} w \left[R^4(w) - S^4(w)\right], \quad S(0) = 0.
\end{aligned} \tag{C.149}$$

We assume that function $S(w)$ is not identically equal to zero in the vicinity of the origin. Dividing out Eq. (C.149) by $S(w)$ and linearizing the result, we obtain the equation

$$w^2 \frac{d}{dw} R(w) = -2 - wR(w)$$

whose integral has the form

$$wR(w) = -2 \ln w + \text{const}\,.$$

There is no value of the constant in this integral that satisfies the initial condition $R(0) = 1$. Consequently, $S(w) = 0$ about the origin, and the first equation in Eqs. (C.149) is satisfied identically. Thus, we have the solution of the first type, and function $R(w)$ is to be determined from the algebraic equation (C.139) whose form in this case is

$$4\left[1 - R(w)\right] = wR^4(w).$$

This equation always has two real roots of different signs. According to the Ferrary formula, the branch satisfying the initial condition $R(0) = 1$ is defined by the relationship

$$R(w) = \sqrt{\frac{2}{w\sqrt{2y}} - \frac{y}{2}} - \frac{1}{2}\sqrt{2y}, \tag{C.150}$$

where

$$y = \frac{4}{\sqrt{3w}}\sinh\frac{\varphi}{3}, \qquad \sinh\varphi = \frac{9}{8\sqrt{3w}}.$$

For small w, we have

$$R(w) \approx 1 - w/4, \quad \rho(w) \approx -w/4 \quad (w \to 0).$$

For large arguments w

$$R(w) \approx \sqrt{2}w^{-1/4}, \quad \rho(w) \approx -1 + \sqrt{2}w^{-1/4} \quad (w \to \infty),$$

so that function $R(w)$ monotonically decreases to zero with increasing w, and the reflection coefficient tends to -1. Such a solution corresponds to the plane wave regime in the nonlinear medium, and the wavefield intensity inside the layer is described by Eq. (C.140).

Consider now the layer of a finite thickness and trace how the solution for the finite layer grades into the above solution for the half-space. In the case of nonlinearity of type $\varepsilon_1(J) = J$, all characteristic curves w_L are smooth functions of layer thickness L, and these curves cross or touch each other nowhere. Therefore, there is a unique solution for any incident wave intensity w and any given layer thickness. Figure C.7*a* shows examples of the wavefield intensity inside the thin medium layer $L = 10$ for different incident wave intensities w and $\gamma = 0.05$. These curves show the oscillating behavior caused by the interference of the direct and reflected waves, the oscillation amplitude being the greater the greater is parameter w. As layer thickness increases, the oscillation amplitude decreases (see Fig. C.7*b*), and the curves become monotonically decaying in the limiting case of the wave incident on the half-space (Fig. C.7*c*). For the layer of thickness $L = 100$ and distances $\xi = L - x \approx 60$ from the layer boundary on which the wave is incident, the solution coincides with the solution to the linear problem. As regards the field intensity on the layer boundary and the reflection coefficient squared modulus, they strongly oscillate as functions of w for sufficiently thin layers; however, this oscillations disappear on going to the half-space (Fig. C.8). ♦

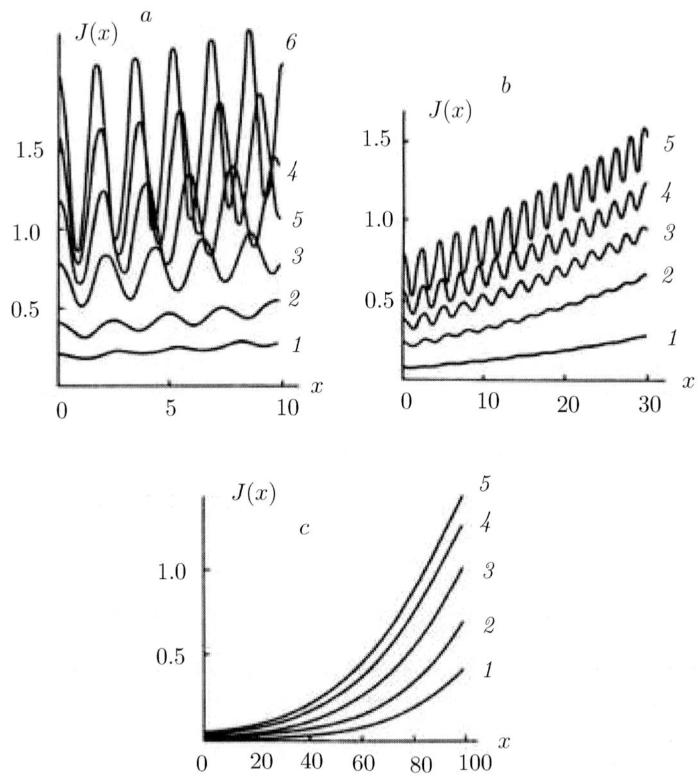

Figure C.7: Wave intensity $J(x)$ in the medium layer for $\varepsilon_1(J) = J$ and $\gamma = 0.05$ at (a) $L = 10$ (curves *1* to *6* correspond to $w = 0.32$, 0.61, 1.23, 1.76, 2.58, and 2.95, respectively), (b) $L = 30$ (curves *1* to *5* correspond to $w = 0.32$, 0.87, 1.35, 1.79, and 2.45, respectively), and (c) $L = 100$ (curves *1* to *5* correspond to $w = 0.49$, 0.86, 1.46, 1.95, and 2.39, respectively).

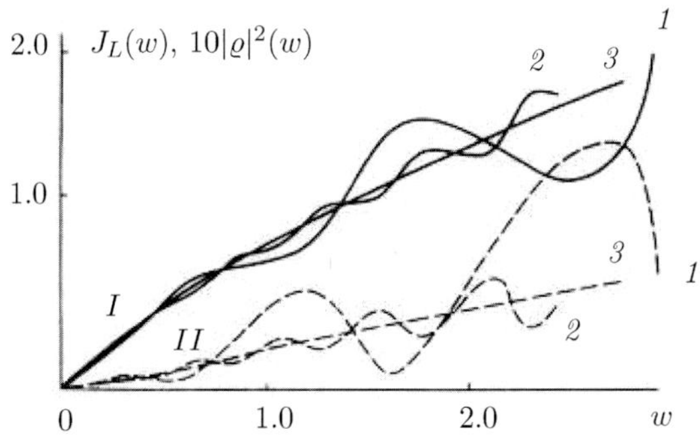

Figure C.8: Functions $J_L(w)$ (the solid lines (I)) and $10|\rho|^2(w)$ (the dashed lines (II)) for $\varepsilon_1(J) = J$ and $\gamma = 0.05$. Curves *1* to *3* correspond to $L = 10$, 30, and 100, respectively.

Example 2

Let now $\varepsilon_1(t) = -\beta t, \quad \beta > 0, \quad \gamma = 0$. In this case, we can again set $\beta = 1$, so that the problem is described by the system of equations

$$\begin{aligned}&\left[R^2(w) + S^2(w)\right] w^2 S(w) \frac{d}{dw} R(w)\\&= S(w)\left\{2+wR(w)\left[R^2(w) + S^2(w)\right]\right\}, \quad R(0) = 1,\\&\left[R^2(w) + S^2(w)\right] w^2 S(w) \frac{d}{dw} S(w)\\&= 2\left[R(w) - 1\right] + \frac{1}{2} w \left[R^4(w) - S^4(w)\right], \quad S(0) = 0.\end{aligned} \tag{C.151}$$

One can easily see that, as in Example 1, function $S(w) = 0$ about point $w = 0$ and function $R(w)$ is to be derived from the algebraic equation

$$4\left[R(w) - 1\right] = wR^4(w). \tag{C.152}$$

A simple analysis shows that this equation has two real roots for

$$0 < w < w_{\rm cr} = (3/4)^3.$$

The desired branch satisfying the condition $R(0) = 1$ can vary in the limits $0 < R < R_{\rm cr} = 3/4$. The solution can again be obtained by the Ferrary formula,

$$R(w) = \frac{1}{2}\sqrt{2y} - \sqrt{\frac{2}{w\sqrt{2y}} - \frac{y}{2}}, \tag{C.153}$$

where now

$$y = \frac{4}{\sqrt{3w}} \cosh\frac{\varphi}{3}, \quad \cosh\varphi = \frac{9}{8\sqrt{3w}} = \left(\frac{w_0}{w}\right)^{1/2}.$$

For small w, we have

$$R(w) \approx 1 + w/4, \quad \rho(w) \approx w/4 \quad (w \to 0).$$

For $w \to w_{\rm cr} - 0$, one can easily obtain the asymptotic expression

$$R(w) \approx \frac{4}{3} - \frac{2}{3}\sqrt{1 - \frac{w}{w_{\rm cr}}},$$

from which follows in particular that $dR(w)/dw \to \infty$ for $w \to w_{\rm cr} - 0$. This type of solution also corresponds to the plane wave regime, and wavefield intensity inside the layer is given by the formula $J(\xi, w) = wR^2(w)$. At the critical point, we have $J(\xi, w_{\rm cr}) = 3/4$, so that $1 + \varepsilon(J) = 1/4$ in this case. .

For $w > w_{\rm cr}$, Eq. (C.152) has no real roots, and this means that $S(w) \neq 0$ and we arrive at the solution of the second type. For $w > w_{\rm cr}$, this problem formally has a continuum of solutions which are the solutions to system of equations (C.142) with arbitrary initial conditions at $w = w_{\rm cr}$. It is reasonable to suppose that the solution to our problem must be a continuous function of w. Then, we can set $R(w_{\rm cr}) = 4/3$, $S(w_{\rm cr}) = 0$ at $w = w_{\rm cr}$ to obtain from Eqs. (C.144)–(C.146) the solution of the form ($w \geqslant w_{\rm cr}$)

$$\begin{aligned} I(w) &= \frac{1}{4w}Q(w), \quad R(w) = \frac{1}{8w}\left[Q(w)+\frac{3}{2}\right], \\ S(w) &= \frac{1}{16\sqrt{2w}}\left[Q(w) - 3\right]\sqrt{Q(w) - 2}, \end{aligned} \tag{C.154}$$

where

$$Q(w) = \sqrt{128w - 45}.$$

According to Eq. (C.147), the reflection coefficient squared modulus is described by the formula

$$|\rho(w)|^2 = 1 - \frac{3}{8w}, \quad |\rho(w_{\rm cr})|^2 = \frac{1}{9}.$$

Taking into account that

$$wb(w) = \frac{1}{64\sqrt{2}}Q(w)\left[Q(w) - 3\right]\sqrt{Q(w) - 2}$$

in this case, we can easily calculate the characteristic curve (integral in Eq. (C.133)) to obtain the final expression for the field intensity inside the medium

$$J(\xi, w) = \frac{1}{2}\left\{1 + \frac{1}{2}\left[\frac{q(w)e^{\xi/\sqrt{2}}+1}{q(w)e^{\xi/\sqrt{2}}-1}\right]^2\right\}, \tag{C.155}$$

where

$$q(w) = \frac{\sqrt{Q(w) - 2} + 1}{\sqrt{Q(w) - 2} - 1}.$$

In view of the fact that $\varepsilon(J) = -J$ in our problem, Eq. (C.155) describes the dielectric permittivity formed by the incident wave as a function of w and ξ. We can see that the change of field behavior from the plane wave regime to the more complicated regime (C.155) starts earlier than quantity $\tilde{\varepsilon}(J) = 1+\varepsilon(J)$ vanishes. For $w_{\rm cr} < w < w_1 = 61/128$, quantity $\tilde{\varepsilon}(J)$ does not at all vanish. For $w \geqslant w_1$, we always have the point

$$\xi_0(w) = \sqrt{2}\ln\left[\frac{\sqrt{2}+1}{\sqrt{2}-1}\right]\frac{\sqrt{Q(w) - 2}-1}{\sqrt{Q(w) - 2}+1}$$

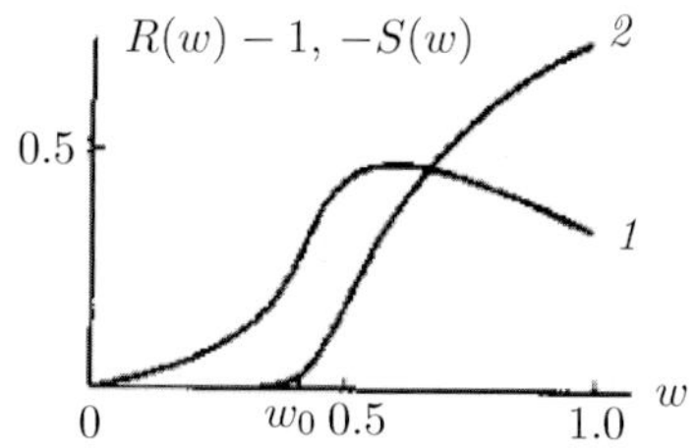

Figure C.9: Field at the boundary as a function of w. Curve *1* shows quantity $(R(w) - 1)$ and curve *2* shows quantity $-S(w)$ $(\gamma = 0.01)$.

at which $\tilde{\varepsilon}(J) = 0$. In addition, quantity $\tilde{\varepsilon}(J) \leqslant 0$ for $0 \leqslant \xi \leqslant \xi_0$. We have $\xi_0 = 0$ at $w = w_1$ and

$$\xi_0(w) = \sqrt{2} \ln \left[\frac{\sqrt{2}+1}{\sqrt{2}-1} \right] \approx 2.5.$$

for $w \gg w_1$. In the remainder of the space $(\xi > \xi_0)$, quantity $\tilde{\varepsilon}(J) > 0$.

Thus, a narrow layer (with thickness about the wavelength) in which $\tilde{\varepsilon}(J) < 0$ appears near the medium boundary, and it is this layer that allows the field to penetrate far in the medium with increasing the incident wave intensity $(J(\xi, w) \approx 3/4$ for $\xi \gg 1)$.

The above consideration assumed the continuous prolongation of the solution through critical point w_{cr}. As we have seen, in this case the derivatives of all considered quantities appear discontinuous at the critical point. We can check whether this fact occurs in actuality by studying the solution of the problem with allowance for a finite (even arbitrarily small) attenuation. We integrated Eqs. (C.135) with $\varepsilon_1(t) = -\beta t$ numerically for different small and constant attenuation coefficients γ. With decreasing parameter γ, the continuous solution tends to the solution obtained here, i.e., to the solution continuous in w, but having a discontinuous derivative (Fig. C.9).

Consider now the layer of a finite thickness. As in Example 1, our interest is in when and how the solution of the problem on finite layer grades into the solution to the problem on wave incidence on the half-space. Here, we must consider the absorptive medium, i.e., we must assume that quantity γ is different from zero, though it can be arbitrarily small.

For nonlinearity described by the relationship $\varepsilon_1(J) = -J$, the pattern essentially depends on parameter γ. For example, if $\gamma > 0.05$, the characteristic curves cross nowhere as in Example1, and the solution grades into the solution of the problem on the wave incident on the half-space practically at $L \sim 70$. For smaller parameters γ, characteristic curves begin to cross, and values of J_L and $|\rho|^2$ at crossing points appear different on different characteristics. For example, at $\gamma = 0.01$, we have a bundle of characteristic curves corresponding to the initial values from the interval $0.25 < w_0 < 0.33$, and the curves cross in this bundle for $7.4 < L < 33$. The values of w_L at crossing points of characteristics vary in the interval $0.36 < w_L < 0.41$. Remind that the problem on wave incidence on the half-space with $\gamma = 0$ is characterized by a critical incident wave intensity $w_{\text{cr}} = (3/4)^2 \approx 0.42$ at which the structure of the field is drastically changed. The existence of crossing points, in turn, is indicative of the fact that the layer of a finite thickness is characterized by the field many-valuedness both at layer boundary and inside the medium. Figure C.10b shows the solution of the problem for the layer of thickness $L = 14.15$ as an example. Curves 3 and 4 in Fig. C.10a correspond to the characteristics that cross each other at a given layer thickness and bound the other crossing characteristics. Curve 5 corresponds to one of the nearest characteristics that crosses no other characteristics. We see that this characteristics extends practically at infinity for the given layer thickness.

The curves showing the field at layer boundary versus w were obtained by successively joining the end points of solutions to system (C.128) in the order coinciding with the order of points w_0 (we used the ascending order of initial values w_0). Figure C.10b shows an ambiguous behavior of the field about point $w \sim 0.40$, which is indicative of the discontinuous behavior of functions such as the reflection coefficient modulus and the existence of hysteretic behavior with both increasing and decreasing parameter w.

For the nonlinearity type considered here, any characteristic curve extends at infinity for a finite layer thickness. For layer thickness $L \approx 33$, the lowermost curve of the family of crossing characteristics extends at infinity and problem solution becomes unique and smooth for arbitrary layer thickness and incident wave intensity. Figure C.10c shows the problem solution for $L = 155.23$, which is practically equivalent to the solution of the problem on wave incidence on the half-space. The problem at hand is characterized by dielectric permittivity $\tilde{\varepsilon}(x) = 1 - J(x)$, and Fig. C.10$c$ shows that a thin layer with $\tilde{\varepsilon}(x) < 0$ is formed near the boundary, in agreement with the above results. Outside this layer, dielectric permittivity $\tilde{\varepsilon}(x) > 0$ and the solution rapidly tends to the solution of the linear problem. ♦

C.2.4 Stationary multidimensional boundary-value problem

Let the inhomogeneous medium occupies the layer $L_0 < x < L$ and let the point source is located at point $(x_0, \mathbf{R}_0)$, where $\mathbf{R}$ stands for the coordinates in the plane perpendicular to the x-axis. Then, the wavefield inside the layer $G(x, \mathbf{R}; x_0, \mathbf{R}_0)$ is described by the boundary-value problem for Green's function of the Helmholtz equation

$$\begin{gathered}\left\{\frac{\partial^2}{\partial x^2} + \Delta_{\mathbf{R}} + k_0^2\left[1 + \varepsilon(x, \mathbf{R})\right]\right\} G(x, \mathbf{R}; x_0, \mathbf{R}_0) = \delta(x - x_0)\delta(\mathbf{R} - \mathbf{R}_0), \\ \left(\frac{\partial}{\partial x} + i\sqrt{k_0^2 + \Delta_{\mathbf{R}}}\right) G(x, \mathbf{R}; x_0, \mathbf{R}_0)\Big|_{x=L_0} = 0, \\ \left(\frac{\partial}{\partial x} - i\sqrt{k_0^2 + \Delta_{\mathbf{R}}}\right) G(x, \mathbf{R}; x_0, \mathbf{R}_0)\Big|_{x=L} = 0,\end{gathered} \tag{C.156}$$

where k_0 is the wave number and $\varepsilon(x, \mathbf{R})$ is the deviation of the refractive index (or dielectric permittivity) from unity. We assume that $\varepsilon(x, \mathbf{R}) = 0$ outside the layer. Operator $i\sqrt{k_0^2 + \Delta_{\mathbf{R}}}$ appeared in Eqs. (C.156), can be considered as the linear integral operator whose kernel coincides with Green's function of free space (see Appendix B). Its action on arbitrary function $F(\mathbf{R})$ is representable in the form of the integral operator

$$\sqrt{k_0^2 + \Delta_{\mathbf{R}}} F(\mathbf{R}) = \int d\mathbf{R}' K(\mathbf{R} - \mathbf{R}') F(\mathbf{R}') \tag{C.157}$$

whose kernel is defined by the equality

$$K(\mathbf{R} - \mathbf{R}') = \sqrt{k^2 + \Delta_{\mathbf{R}}}\delta(\mathbf{R} - \mathbf{R}') = 2i\left(k_0^2 + \Delta_{\mathbf{R}}\right) g_0(0, \mathbf{R} - \mathbf{R}'), \tag{C.158}$$

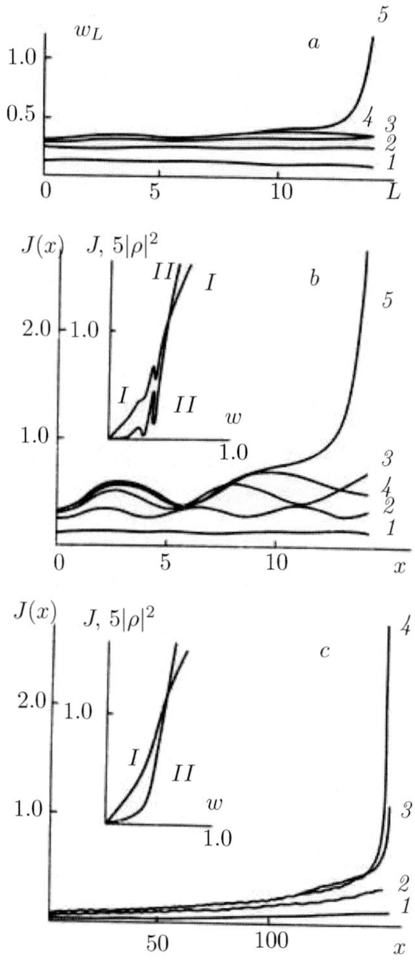

Figure C.10: Problem solution for $\varepsilon_1(J) = -J$ and $\gamma = 0.01$. (a) Quantity w_L as a function of layer thickness, (b) wave intensity $J(x)$ in the layer at $L = 14.15$ ($J_L = 3.75$, curves *1* to *5* correspond to $w = 0.14$, 0.29, 0.40, 1.40, and 2.13, respectively; setting-in shows functions (I) $J(w)$, and (II) $5|\rho|^2(w)$), and (c) wave intensity $J(x)$ in the layer at $L = 155.23$ ($J_L = 3.93$, curves *1* to *4* correspond to $w = 0.13$, 0.30, 0.55, and 2.33, respectively; setting-in shows functions (I) $J(w)$, and (II) $5|\rho|^2(w)$).

where $g_0(x, \mathbf{R})$ is Green's function of free space. For example, Green's function is given by the formula

$$g_0(x, \mathbf{R}) = -\frac{1}{4\pi r} e^{ik_0 r} \quad (\mathbf{r} = \{x, \mathbf{R}\}),$$

in the three-dimensional case; the integral representation of this function has the form

$$g_0(x, \mathbf{R}) = \int d\mathbf{q} g_0(\mathbf{q}) e^{i\sqrt{k_0^2 - q^2}|x| + i\mathbf{q}\mathbf{R}}, \quad g_0(\mathbf{q}) = \frac{1}{8i\pi^2\sqrt{k_0^2 - q^2}}. \tag{C.159}$$

The corresponding kernel of the inverse operator is defined by the equality

$$L(\mathbf{R} - \mathbf{R}') = \left(k_0^2 + \Delta_{\mathbf{R}}\right)^{-1/2} \delta(\mathbf{R} - \mathbf{R}') = 2ig_0(0, \mathbf{R} - \mathbf{R}'). \tag{C.160}$$

Boundary-value problem (C.156) is equivalent to the integral equation

$$\begin{aligned} G(x, \mathbf{R}; x_0, \mathbf{R}_0) &= g_0(x - x_0, \mathbf{R} - \mathbf{R}_0) \\ &- k_0^2 \int\limits_{L_0}^{L} dx_1 \int d\mathbf{R}_1 g_0(x - x_1, \mathbf{R} - \mathbf{R}_1) \varepsilon(x_1, \mathbf{R}_1) G(x_1, \mathbf{R}_1; x_0, \mathbf{R}_0). \end{aligned} \tag{C.161}$$

Note that Eq. (C.161) can be rewritten in the form

$$\begin{aligned} G(x, \mathbf{R}; x_0, \mathbf{R}_0) &= g_0(x - x_0, \mathbf{R} - \mathbf{R}_0) \\ &- k_0^2 \int\limits_{L_0}^{L} dx_1 \int d\mathbf{R}_1 G(x, \mathbf{R}; x_1, \mathbf{R}_1) \varepsilon(x_1, \mathbf{R}_1) g_0(x_1 - x_0, \mathbf{R}_1 - \mathbf{R}_0). \end{aligned} \tag{C.162}$$

Function $G(x, \mathbf{R}; x_0, \mathbf{R}_0)$ is continuous everywhere inside the layer. As regards quantity $\frac{\partial}{\partial x} G(x, \mathbf{R}; x_0, \mathbf{R}_0)$, it has a discontinuity at the point of source location $x = x_0$

$$\frac{\partial}{\partial x} G(x, \mathbf{R}; x_0, \mathbf{R}_0)\bigg|_{x = x_0 + 0} - \frac{\partial}{\partial x} G(x, \mathbf{R}; x_0, \mathbf{R}_0)\bigg|_{x = x_0 - 0} = \delta(\mathbf{R} - \mathbf{R}').$$

If the point source is located at layer boundary $x_0 = L$, then the wavefield inside the layer (i.e., for $L_0 < x < L$) is described by the boundary-value problem

$$\begin{aligned} &\left\{ \frac{\partial^2}{\partial x^2} + \Delta_{\mathbf{R}} + k_0^2 \left[1 + \varepsilon(x, \mathbf{R})\right] \right\} G(x, \mathbf{R}; L, \mathbf{R}_0) = 0, \\ &\left(\frac{\partial}{\partial x} + i\sqrt{k_0^2 + \Delta_{\mathbf{R}}} \right) G(x, \mathbf{R}; L, \mathbf{R}_0)\bigg|_{x = L_0} = 0, \\ &\left(\frac{\partial}{\partial x} - i\sqrt{k_0^2 + \Delta_{\mathbf{R}}} \right) G(x, \mathbf{R}; L, \mathbf{R}_0)\bigg|_{x = L} = -\delta(\mathbf{R} - \mathbf{R}'), \end{aligned} \tag{C.163}$$

Boundary-value problem (C.163) is equivalent to the integral equation

$$\begin{aligned} G(x, \mathbf{R}; L, \mathbf{R}_0) &= g_0(x - L, \mathbf{R} - \mathbf{R}_0) \\ &- k_0^2 \int\limits_{L_0}^{L} dx_1 \int d\mathbf{R}_1 g_0(x - x_1, \mathbf{R} - \mathbf{R}_1) \varepsilon(x_1, \mathbf{R}_1) G(x_1, \mathbf{R}_1; L, \mathbf{R}_0), \end{aligned} \tag{C.164}$$

which corresponds to setting $x_0 = L$ in Eq.(C.161). Setting $x = L$ in Eq. (C.162) and comparing the result with Eq. (C.164), we see that the equality

$$G(L, \mathbf{R}; x_0, \mathbf{R}_0) = G(x_0, \mathbf{R}_0; L, \mathbf{R}) \tag{C.165}$$

holds, which expresses the reciprocity theorem.

Remark 23 *Wave incidence on medium layer.*

We note that boundary-value problem (C.163) describes the wave incidence from half-space $x > L$ on the inhomogeneous medium layer. Indeed, if a wave $u_0(x-L,\mathbf{R})$ is incident on the medium layer from region $x > L$ (in the negative direction of the x-axis), then it creates the distribution of sources $f(\mathbf{R}_0)$ at boundary $x = L$ such that

$$f(\mathbf{R}_0) = 2i\sqrt{k_0^2 + \Delta_{\mathbf{R}}}\,u_0(0,\mathbf{R}_0). \tag{C.166}$$

In this case, wavefield $U(x,\mathbf{R})$ inside the layer is related to the solution to Eq. (C.163) by the equality

$$U(x,\mathbf{R}) = \int d\mathbf{R}_0 G(x,\mathbf{R};L,\mathbf{R}_0)f(\mathbf{R}_0), \tag{C.167}$$

and is described by the boundary-value problem

$$\begin{aligned}
&\left\{\frac{\partial^2}{\partial x^2} + \Delta_{\mathbf{R}} + k_0^2\left[1+\varepsilon(x,\mathbf{R})\right]\right\}U(x,\mathbf{R}) = 0,\\
&\left(\frac{\partial}{\partial x} + i\sqrt{k_0^2+\Delta_{\mathbf{R}}}\right)U(x,\mathbf{R})\Big|_{x=L_0} = 0,\\
&\left(\frac{\partial}{\partial x} - i\sqrt{k_0^2+\Delta_{\mathbf{R}}}\right)U(x,\mathbf{R})\Big|_{x=L} = -2i\sqrt{k_0^2+\Delta_{\mathbf{R}}}\,u_0(0,\mathbf{R}),
\end{aligned} \tag{C.168}$$

or by the equivalent integral equation

$$U(x,\mathbf{R}) = u_0(x,\mathbf{R}) - k_0^2\int\limits_{L_0}^{L} dx_1 \int d\mathbf{R}_1 g_0(x-x_1,\mathbf{R}-\mathbf{R}_1)\varepsilon(x_1,\mathbf{R}_1)U(x_1,\mathbf{R}_1). \quad \blacklozenge \tag{C.169}$$

Derive the imbedding equations for boundary-value problem (C.163). Differentiating Eq. (C.164) with respect to parameter L, we obtain the integral equation in $\frac{\partial}{\partial L}G(x,\mathbf{R};L,\mathbf{R}_0)$

$$\begin{aligned}
\frac{\partial}{\partial L}G(x,\mathbf{R};L,\mathbf{R}_0) &= \frac{\partial}{\partial L}g_0(x-L,\mathbf{R}-\mathbf{R}_0)\\
&-k_0^2\int d\mathbf{R}_1 g_0(x-L,\mathbf{R}-\mathbf{R}_1)\varepsilon(L,\mathbf{R}_1)H_L(\mathbf{R}_1;\mathbf{R}_0)\\
&-k_0^2\int\limits_{L_0}^{L} dx_1\int d\mathbf{R}_1 g_0(x-x_1,\mathbf{R}-\mathbf{R}_1)\varepsilon(x_1,\mathbf{R}_1)\frac{\partial}{\partial L}G(x_1,\mathbf{R}_1;L,\mathbf{R}_0).
\end{aligned} \tag{C.170}$$

In Eq. (C.170), function

$$H_L(\mathbf{R};\mathbf{R}_0) = G(L,\mathbf{R};L,\mathbf{R}_0) \tag{C.171}$$

describes the wavefield in the source plane $x_0 = L$.

In view of the fact that free space Green's function can be factorized (see Appendix B), it satisfies the first-order equation

$$\frac{\partial}{\partial L}g_0(x-L,\mathbf{R}) = i\sqrt{k_0^2+\Delta_{\mathbf{R}}}\,g_0(x-L,\mathbf{R}).$$

As a result, we can rewrite integral equation (C.170) in the form

$$\begin{aligned}
\frac{\partial}{\partial L}G(x,\mathbf{R};L,\mathbf{R}_0) &= \widehat{A}(L,\mathbf{R}_0)g(x-L,\mathbf{R}-\mathbf{R}_0)\\
&-k_0^2\int\limits_{L_0}^{L} dx_1\int d\mathbf{R}_1 g_0(x-x_1,\mathbf{R}-\mathbf{R}_1)\varepsilon(x_1,\mathbf{R}_1)\frac{\partial}{\partial L}G(x_1,\mathbf{R}_1;L,\mathbf{R}_0),
\end{aligned} \tag{C.172}$$

where operator $\hat{A}(L, \mathbf{R}_0)$ acts on arbitrary function $f(\mathbf{R}_0)$ of variable $\mathbf{R}_0$ in accordance with the formula

$$\hat{A}(L, \mathbf{R}_0) f(\mathbf{R}_0) = i\sqrt{k_0^2 + \Delta_{\mathbf{R}_0}} f(\mathbf{R}_0) - k_0^2 \int d\mathbf{R}_1 \varepsilon(L, \mathbf{R}_1) H_L(\mathbf{R}_1; \mathbf{R}_0) f(\mathbf{R}_1).$$

Operator $\hat{A}(L, \mathbf{R}_0)$ extends function $a(L)$ appeared in the corresponding one-dimensional problem to the multidimensional case.

Correlating now integral equations (C.172) and (C.164), we see that they are identical in structure; consequently, their solutions are related by the integral equality

$$\left[\frac{\partial}{\partial L} - i\sqrt{k_0^2 + \Delta_{\mathbf{R}_0}}\right] G(x, \mathbf{R}; L, \mathbf{R}_0) = -k_0^2 \int d\mathbf{R}_1 G(x, \mathbf{R}; L, \mathbf{R}_1) \varepsilon(L, \mathbf{R}_1) H_L(\mathbf{R}_1; \mathbf{R}_0), \tag{C.173}$$

which, being supplemented with the initial condition (continuity condition at $x = L$)

$$G(x, \mathbf{R}; x, \mathbf{R}_0) = H_x(\mathbf{R}; \mathbf{R}_0),$$

can be considered as the equation of the imbedding method.

We can rewrite Eq. (C.173) in the form of the integral equation

$$G(x, \mathbf{R}; L, \mathbf{R}_0) = g(x, \mathbf{R}; L, \mathbf{R}_0) - k_0^2 \int_x^L dx_1 \int d\mathbf{R}_1 G(x, \mathbf{R}; L, \mathbf{R}_1) \varepsilon(x_1, \mathbf{R}_1) g(x_1, \mathbf{R}_1; L, \mathbf{R}_0),$$

where

$$g(x, \mathbf{R}; L, \mathbf{R}_0) = e^{i\sqrt{k_0^2 + \Delta_{\mathbf{R}_0}}(L - x)} H_x(\mathbf{R}; \mathbf{R}_0). \tag{C.174}$$

In the case of a wave $u_0(x, \mathbf{R})$ incident on the medium layer in the negative direction of the x-axis, the source distribution $f(\mathbf{R}_0)$ created by the incident wave at boundary $x = L$ is given by Eq. (C.166); in this case, wavefield $U(x, \mathbf{R})$ (C.167) is described by the integral equation

$$U(x, \mathbf{R}) = u_0(x, \mathbf{R}) - k_0^2 \int_x^L dx_1 \int d\mathbf{R}_1 g(x, \mathbf{R}; x_1, \mathbf{R}_1) \varepsilon(x_1, \mathbf{R}_1) U(x_1, \mathbf{R}_1). \tag{C.175}$$

An essential difference of Eq. (C.175) with a given function $g(x, \mathbf{R}; x_1, \mathbf{R}_1)$ from Eq. (C.169) consists in the fact that wavefield $U(x, \mathbf{R})$ at point $(x, \mathbf{R})$ is governed by field $\varepsilon(x_1, \mathbf{R}_1)$ in region $x \leqslant x_1 \leqslant L$, which means that the wavefield is quasi-causal. For $L_0 \leqslant x_1 \leqslant x$, the functional dependence of field $U(x, \mathbf{R})$ on $\varepsilon(x_1, \mathbf{R}_1)$ is realized implicitly, in terms of function $g(x, \mathbf{R}; L, \mathbf{R}_0)$.

Function $H_L(\mathbf{R}; \mathbf{R}_0)$ satisfies the equality

$$\frac{\partial}{\partial L} H_L(\mathbf{R}; \mathbf{R}_0) = \frac{\partial}{\partial L} G(x, \mathbf{R}; L, \mathbf{R}_0)\Big|_{x=L} + \frac{\partial}{\partial x} G(x, \mathbf{R}; L, \mathbf{R}_0)\Big|_{x=L}. \tag{C.176}$$

The first term in the right-hand side of Eq. (C.176) can be obtained from Eq. (C.173) at $x = L$, and the second term, from the boundary condition in Eq. (C.163). As a result, we obtain the closed integro-differential equation

$$\left[\frac{\partial}{\partial L} - i\sqrt{k_0^2 + \Delta_{\mathbf{R}}} - i\sqrt{k_0^2 + \Delta_{\mathbf{R}_0}}\right] H_L(\mathbf{R}; \mathbf{R}_0) = -\delta(\mathbf{R} - \mathbf{R}_0) - k_0^2 \int d\mathbf{R}_1 H_L(\mathbf{R}; \mathbf{R}_1) \varepsilon(L, \mathbf{R}_1) H_L(\mathbf{R}_1; \mathbf{R}_0) \tag{C.177}$$

with the initial condition

$$H_{L_0}(\mathbf{R};\mathbf{R}_0) = g_0(0,\mathbf{R}-\mathbf{R}_0) \tag{C.178}$$

following from Eq. (C.163).

Thus, the input boundary-value problem (C.163) is equivalent to Eqs. (C.173) and (C.177). These equations are the equations of the imbedding method for the problem under consideration. An essential difference between these equations and the input problem (C.163) consists in the fact that they form an initial value problem with respect to parameter L.

We notice that function $H_L(\mathbf{R};\mathbf{R}_0)$ (it describes the wavefield in the source plane and is the sum of the incident and backscattered fields) satisfies the closed nonlinear equation (C.177). As regards Eq. (C.173), it is the linear equation.

Having the solution of Eqs. (C.177) and (C.173), we can easily write the solution to the problem in regions $x > L$ (the reflected wave) and $x < L_0$ (the transmitted wave). Moreover, function $G(L_0,\mathbf{R};L,\mathbf{R}_0)$ is also described by Eq. (C.173) with the initial condition

$$G(L_0,\mathbf{R};L_0,\mathbf{R}_0) = g(0,\mathbf{R}-\mathbf{R}_0).$$

In the context of statistical problems in the general statement, the literature on the backscattered field is practically lacking. Papers [22] and [282]–[288] form an exception. For the qualitative and quantitative results on the backscattering effects obtained with the use of different approximate methods, see reviews [23, 206] and [207].

Remark 24 ***Conversion to the parabolic equation of quasi-optics.***

Now, we trace the conversion to the approximation of parabolic equation. Equation (C.177) describes the backscattered field. The effect of backscattering is an essentially nonlinear effect and is described by the last term in Eq. (C.177). If we neglect this term, then the solution of the resulting equation will have the form

$$H_L(\mathbf{R};\mathbf{R}_0) = g_0(0,\mathbf{R}-\mathbf{R}_0),$$

which corresponds to the assumption that only the incident wave is present in plane $x = L$. In this case, function $g(x,\mathbf{R};L,\mathbf{R}_0)$ (C.174) grades into Green's function of free space

$$g(x,\mathbf{R};L,\mathbf{R}_0) \equiv g_0(x-L,\mathbf{R}-\mathbf{R}_0)$$

and Eq. (C.175) assumes the form of the causal integral equation

$$U(x,\mathbf{R}) = u_0(x,\mathbf{R}) - k_0^2\int\limits_x^L dx_1 \int d\mathbf{R}_1 g_0(x-x_1,\mathbf{R}-\mathbf{R}_1)\varepsilon(x_1,\mathbf{R}_1)U(x_1,\mathbf{R}_1), \tag{C.179}$$

which describes the propagation of a wave in the approximation generally valid for moderate (not exceeding $\pi/2$) scattering angles.

Equation (C.179) can be rewritten in the form of an operator equation. Indeed, differentiating Eq. (C.179) with respect to x, using Eq. (C.160) in the form

$$g_0(0,\mathbf{R}-\mathbf{R}_0) = \frac{1}{2i\sqrt{k_0^2+\Delta_{\mathbf{R}}}}\delta(\mathbf{R}-\mathbf{R}_0),$$

and the factorization property of the incident field $u_0(x, \mathbf{R})$

$$\frac{\partial}{\partial x} u_0(x, \mathbf{R}) = -i\sqrt{k_0^2 + \Delta_{\mathbf{R}}} u_0(x, \mathbf{R}),$$

we obtain the equation

$$\left(\frac{\partial}{\partial x} + i\sqrt{k_0^2 + \Delta_{\mathbf{R}}}\right) U(x, \mathbf{R}) = -i \frac{k_0^2}{2\sqrt{k_0^2 + \Delta_{\mathbf{R}}}} \{\varepsilon(x, \mathbf{R}) U(x, \mathbf{R})\},$$

$$U(L, \mathbf{R}) = u_0(\mathbf{R}).$$

The parabolic equation is the result of the small-angle approximation corresponding to the Fresnel expansion of Green's function, which, in turn, corresponds to the condition $\Delta_{\mathbf{R}} \ll k_0^2$. ♦

The problem on the field of a point source located inside the layer of inhomogeneous medium can be considered similarly. Indeed, let the inhomogeneous medium occupies, as earlier, layer $L_0 < x < L$. Then, the point source field (Green's function) satisfies integral equation (C.161) (in this case, x_0 and $\mathbf{R}_0$ are the coordinates of the source)

$$G(x, \mathbf{R}; x_0, \mathbf{R}_0; L) = g_0(x - x_0, \mathbf{R} - \mathbf{R}_0)$$
$$-k_0^2 \int\limits_{L_0}^{L} dx_1 \int d\mathbf{R}_1 g_0(x - x_1, \mathbf{R} - \mathbf{R}_1) \varepsilon(x_1, \mathbf{R}_1) G(x_1, \mathbf{R}_1; x_0, \mathbf{R}_0; L), \quad \text{(C.180)}$$

where we explicitly included imbedding parameter L as the argument of function G,

$$G(x, \mathbf{R}; x_0, \mathbf{R}_0) \equiv G(x, \mathbf{R}; x_0, \mathbf{R}_0; L).$$

Differentiating Eq. (C.180) with respect to parameter L, we obtain the integral equation in function $\frac{\partial}{\partial L} G(x, \mathbf{R}; x_0, \mathbf{R}_0; L)$

$$\frac{\partial}{\partial L} G(x, \mathbf{R}; x_0, \mathbf{R}_0; L)$$
$$= -k_0^2 \int d\mathbf{R}_1 g_0(x - L, \mathbf{R} - \mathbf{R}_1) \varepsilon(L, \mathbf{R}_1) G(L, \mathbf{R}_1; x_0, \mathbf{R}_0; L)$$
$$-k_0^2 \int\limits_{L_0}^{L} dx_1 \int d\mathbf{R}_1 g_0(x - x_1, \mathbf{R} - \mathbf{R}_1) \varepsilon(x_1, \mathbf{R}_1) \frac{\partial}{\partial L} G(x_1, \mathbf{R}_1; x_0, \mathbf{R}_0; L) \quad \text{(C.181)}$$

Correlating now Eq. (C.181) with Eq. (C.180), we see that the equality

$$\frac{\partial}{\partial L} G(x, \mathbf{R}; x_0, \mathbf{R}_0; L) = -k_0^2 \int d\mathbf{R}_1 G(x, \mathbf{R}; L, \mathbf{R}_1) G(x_0, \mathbf{R}_0; L, \mathbf{R}_1) \varepsilon(L, \mathbf{R}_1), \quad \text{(C.182)}$$

holds. If we supplement this equality with the initial condition

$$G(x, \mathbf{R}; x_0, \mathbf{R}_0; L)|_{L=\max\{x, x_0\}} = \begin{cases} G(x, \mathbf{R}; x_0, \mathbf{R}_0) & (x_0 \geq x), \\ G(x_0, \mathbf{R}; x, \mathbf{R}_0) & (x \geq x_0), \end{cases} \quad \text{(C.183)}$$

(which is the condition of solution continuity with respect to parameter L) we can consider it as the integro-differential equation in function $G(x, \mathbf{R}; x_0, \mathbf{R}_0; L)$. Deriving Eq. (C.182), we used additionally Eq. (C.165) (the reciprocity theorem).

Thus, Eqs. (C.183), (C.173), and (C.177) form the closed system of imbedding equations in the context of the problem under consideration. The limit process $L_0 \to -\infty$, $L \to \infty$ corresponds to the problem on a point source located in the inhomogeneous medium that occupies the whole of space.

Equation (C.182) with condition (C.183) can be integrated in the analytic form. Thus, the field of a point source located inside the medium layer appears simply related (through a quadrature) to the field in the problem on the wave incident on the layer (i.e., the problem on the point source located at the layer boundary).

Remark 25 ***Layered medium.***

Consider in more detail the case of layered medium with $\varepsilon(L, R) \equiv \varepsilon(L)$. In this case all functions G are functions of the difference $(\mathbf{R} - \mathbf{R}_0)$ and we can use the Fourier transform,

$$G(x, x_0, \mathbf{R}) = \int d\mathbf{q} G(x; x_0; \mathbf{q}) e^{i\mathbf{qR}}, \quad G(x; x_0; \mathbf{q}) = \frac{1}{4\pi^2} \int d\mathbf{R} G(x, x_0, \mathbf{R}) e^{-i\mathbf{qR}},$$

to convert the system of integro-differential equations into the system of ordinary differential equations

$$\frac{\partial}{\partial L} G(x; x_0; L, \mathbf{q}) = -(2\pi k_0)^2 \varepsilon(L) G(x; L, \mathbf{q}) G(x_0; L, \mathbf{q}),$$

$$G(x; x_0; L, \mathbf{q})|_{L=\max\{x, x_0\}} \left\{ \begin{array}{ll} G(x, ; x_0, \mathbf{q}) & (x_0 \geq x), \\ G(x_0; x, \mathbf{q}) & (x \geq x_0), \end{array} \right. \tag{C.184}$$

$$\left[\frac{\partial}{\partial L} - i\sqrt{k_0^2 - q^2}\right] G(x; L, \mathbf{q}) = -(2\pi k_0)^2 \varepsilon(L) G(x; L, \mathbf{q}) H_L(\mathbf{q}),$$

$$G(x; x, \mathbf{q}) = H_x(\mathbf{q}), \tag{C.185}$$

$$\left[\frac{\partial}{\partial L} - 2i\sqrt{k_0^2 - q^2}\right] = H_L(\mathbf{q}) - \frac{1}{4\pi^2} - (2\pi k_0)^2 \varepsilon(L) H_L^2(\mathbf{q}),$$

$$H_{L_0}(\mathbf{q}) = g_0(\mathbf{q}) = \frac{1}{8\pi^2\sqrt{k_0^2 - q^2}}. \tag{C.186}$$

Equations (C.185) and (C.186) describe the propagation of the plane wave of amplitude $g_0(q)$ obliquely incident on boundary $x = L$. Being correspondingly renormalized to the unit amplitude, these equations grade into the equations for the plane incident wave $u_0(x, R) = e^{i\{k_0(L-x)+\mathbf{qR}\}}$

$$\frac{\partial}{\partial L} G(x; x_0; L, \mathbf{q}) = \frac{ik_0^2}{2\sqrt{k_0^2 - q^2}} \varepsilon(L) G(x; L, \mathbf{q}) G(x_0; L, \mathbf{q}),$$

$$G(x; x_0; L, \mathbf{q})|_{L=\max\{x, x_0\}} = \left\{ \begin{array}{ll} G(x, ; x_0, \mathbf{q}) & (x_0 \geq x), \\ G(x_0; x, \mathbf{q}) & (x \geq x_0), \end{array} \right.$$

$$\left[\frac{\partial}{\partial L} - i\sqrt{k_0^2 - q^2}\right] G(x; L, \mathbf{q}) = \frac{ik_0^2}{2\sqrt{k_0^2 - q^2}} \varepsilon(L) G(x; L, \mathbf{q}) H_L(\mathbf{q}),$$

$$G(x; x, \mathbf{q}) = H_x(\mathbf{q}),$$

$$\left[\frac{\partial}{\partial L} - 2i\sqrt{k_0^2 - q^2}\right] H_L(\mathbf{q}) = -2i\sqrt{k_0^2 - q^2} + \frac{ik_0^2}{2\sqrt{k_0^2 - q^2}} \varepsilon(L) H_L^2(\mathbf{q}),$$

$$H_{L_0}(\mathbf{q}) = 1,$$

which were discussed in detail in Sect. C.2.1 of this Appendix. The case of the normal wave incidence corresponds to setting $q = 0$. ♦

Thus, we reduced the three-dimensional boundary-value problem on wave propagation to the causal equations with respect to parameter L.

Stationary nonlinear multidimensional boundary-value problem

Consider now the problem on a wave $u_0(x, \mathbf{R})$ incident from free half-space $x > L$ on the layer of medium $L_0 < x < L$ under the assumption that medium inhomogeneities are formed by the wavefield intensity. This problem extends the one-dimensional problem on wave self-action to the multidimensional case.

Thus, Eq. (C.169) is replaced with the integral equation

$$
\begin{aligned}
U(x, \mathbf{R}) &= u_0(x, \mathbf{R}) \\
&- k_0^2 \int\limits_{L_0}^{L} dx_1 \int d\mathbf{R}_1 g_0(x - x_1, \mathbf{R} - \mathbf{R}_1) \varepsilon\left(x_1, \mathbf{R}_1; I(x_1, \mathbf{R}_1)\right) U(x_1, \mathbf{R}_1),
\end{aligned}
\tag{C.187}
$$

where $I(x, \mathbf{R}) = |U(x, \mathbf{R})|^2$.

Consider the equation for function $G(x, \mathbf{R}; L, \mathbf{R}_0)$ (this function is similar to Green's function of the linear problem with the source at point $(L, \mathbf{R}_0)$) for $x < L$

$$
\begin{aligned}
G(x, \mathbf{R}; L, \mathbf{R}_0) &= g_0(x - L, \mathbf{R} - \mathbf{R}_0) \\
&- k_0^2 \int\limits_{L_0}^{L} dx_1 \int d\mathbf{R}_1 g_0(x - x_1, \mathbf{R} - \mathbf{R}_1) \varepsilon\left(x_1, \mathbf{R}_1; I(x_1, \mathbf{R}_1)\right) G(x_1, \mathbf{R}_1; L, \mathbf{R}_0).
\end{aligned}
\tag{C.188}
$$

This equation is equivalent to the boundary-value problem

$$
\begin{aligned}
&\left\{ \frac{\partial^2}{\partial x^2} + \Delta_{\mathbf{R}} + k_0^2 \left[1 + \varepsilon\left(x, \mathbf{R}; I(x, \mathbf{R})\right)\right] \right\} G(x, \mathbf{R}; L, \mathbf{R}_0) = 0, \\
&\left(\frac{\partial}{\partial x} + i\sqrt{k_0^2 + \Delta_{\mathbf{R}}} \right) G(x, \mathbf{R}; L, \mathbf{R}_0) \Big|_{x=L_0} = 0, \\
&\left(\frac{\partial}{\partial x} - i\sqrt{k_0^2 + \Delta_{\mathbf{R}}} \right) G(x, \mathbf{R}; L, \mathbf{R}_0) \Big|_{x=L} = -\delta(\mathbf{R} - \mathbf{R}').
\end{aligned}
\tag{C.189}
$$

Then, wavefield $U(x, \mathbf{R})$ is given by Eq. (C.167)

$$
U(x, \mathbf{R}) = \int d\mathbf{R}_0 G(x, \mathbf{R}; L, \mathbf{R}_0) f(\mathbf{R}_0),
$$

where function $f(\mathbf{R}_0)$ is the source distribution in plane $x = L$ given by Eq. (C.166)

$$
f(\mathbf{R}_0) = 2i\sqrt{k_0^2 + \Delta_{\mathbf{R}}} u_0(0, \mathbf{R}_0).
$$

Consequently,

$$
I(x, \mathbf{R}) = \int\int d\mathbf{R}_1 d\mathbf{R}_2 G(x, \mathbf{R}; L, \mathbf{R}_1) G^*(x, \mathbf{R}; L, \mathbf{R}_2) W(\mathbf{R}_1, \mathbf{R}_2),
$$

where

$$W(\mathbf{R}_1, \mathbf{R}_2) = f(\mathbf{R}_1) f^*(\mathbf{R}_2).$$

Now, we introduce function

$$H_L(\mathbf{R}; \mathbf{R}_0) = G(L, \mathbf{R}; L, \mathbf{R}_0)$$

that describes the wavefield in the source plane. Equation (C.187) is similar to the one-dimensional equation (C.111), page 472, excluding the fact that wave number k is replaced with operator $\sqrt{k_0^2 + \Delta_{\mathbf{R}}}$ and parameter w is replaced with function $W(\mathbf{R}_1; \mathbf{R}_2)$. Therefore, we can simply replicate the derivation of the equations of the imbedding method. In this replica, quantity $a(L, w)$ will be replaced with an integro-differential operator and partial derivative $\partial/\partial w$ will be replaced with variational derivative $\delta/\delta W(\mathbf{R}_1, \mathbf{R}_2)$. As a result, we obtain the relationship

$$\begin{aligned}&\left(\frac{\partial}{\partial L} - \hat{A}(L, \mathbf{R}_0)\right) G(x, \mathbf{R}; L, \mathbf{R}_0)\\ &\quad = \int d\mathbf{R}_1 \int d\mathbf{R}_2 W(\mathbf{R}_1, \mathbf{R}_2) \hat{B}(L, \mathbf{R}_1, \mathbf{R}_2) \frac{\delta G(x, \mathbf{R}; L, \mathbf{R}_0)}{\delta W(\mathbf{R}_1, \mathbf{R}_2)}. \end{aligned} \tag{C.190}$$

Being supplemented with the initial condition

$$G(x, \mathbf{R}; L, \mathbf{R}_0)|_{L=x} = H_x(\mathbf{R}; \mathbf{R}_0), \tag{C.191}$$

this relationship can be considered as the equation in quantity $G(x, \mathbf{R}; L, \mathbf{R}_0)$.

Operators $\hat{A}(L, \mathbf{R})$ and $\hat{B}(L, \mathbf{R}_1, \mathbf{R}_2)$ act on a function $G(x, \mathbf{R}; L, \mathbf{R}_0)$ according to the equalities

$$\begin{aligned}&\hat{A}(L, \mathbf{R}_0) G(x, \mathbf{R}; L, \mathbf{R}_0) = i\sqrt{k_0^2 + \Delta_{\mathbf{R}_0}} G(x, \mathbf{R}; L, \mathbf{R}_0)\\ &-k_0^2 \int d\mathbf{R}_1 G(x, \mathbf{R}; L, \mathbf{R}_1) \varepsilon\left(L, \mathbf{R}_1; I(L, \mathbf{R}_1)\right) H_L(\mathbf{R}_1; \mathbf{R}_0),\\ &\hat{B}(L, \mathbf{R}_1, \mathbf{R}_2) = \hat{A}(L, \mathbf{R}_1) + \hat{A}^*(L, \mathbf{R}_2). \end{aligned} \tag{C.192}$$

Function $H_L(\mathbf{R}; \mathbf{R}_0)$ satisfies the relationship

$$\frac{\partial}{\partial L} H_L(\mathbf{R}; \mathbf{R}_0) = \frac{\partial}{\partial L} G(x, \mathbf{R}; L, \mathbf{R}_0)\bigg|_{x=L} + \frac{\partial}{\partial x} G(x, \mathbf{R}; L, \mathbf{R}_0)\bigg|_{x=L}. \tag{C.193}$$

The first term in the right-hand side of Eq. (C.193) can be obtained from Eq. (C.190) at $x = L$ and the second term, from the boundary condition in Eq. (C.189). As a result, we obtain the closed integro-differential equation

$$\begin{aligned}&\left[\frac{\partial}{\partial L} - i\sqrt{k_0^2 + \Delta_{\mathbf{R}}} - i\sqrt{k_0^2 + \Delta_{\mathbf{R}_0}}\right] H_L(\mathbf{R}; \mathbf{R}_0) = -\delta(\mathbf{R} - \mathbf{R}_0)\\ &-k_0^2 \int d\mathbf{R}_1 H_L(\mathbf{R}; \mathbf{R}_1) \varepsilon\left(L, \mathbf{R}_1; I(L, \mathbf{R}_1)\right) H_L(\mathbf{R}_1; \mathbf{R}_0)\\ &+\int d\mathbf{R}_1 \int d\mathbf{R}_2 W(\mathbf{R}_1, \mathbf{R}_2) \hat{B}(L, \mathbf{R}_1, \mathbf{R}_2) \frac{\delta H_L(\mathbf{R}; \mathbf{R}_0)}{\delta W(\mathbf{R}_1, \mathbf{R}_2)} \end{aligned} \tag{C.194}$$

with the initial condition

$$H_{L_0}(\mathbf{R}; \mathbf{R}_0) = g_0(0, \mathbf{R} - \mathbf{R}_0). \tag{C.195}$$

following from Eq. (C.189).

Equations (C.190) and (C.194) with initial conditions (C.191) and (C.195) (and relationships (C.192) as well) are the imbedding equations of the input three-dimensional nonlinear boundary-value problem. In the case of the linear medium, solution dependence on W disappears. A consequence of Eq. (C.190) is the equation for the wavefield intensity inside the medium $I(x, \mathbf{R}; L)$

$$\frac{\partial}{\partial L} I(x, \mathbf{R}; L) = \int d\mathbf{R}_1 \int d\mathbf{R}_2 W(\mathbf{R}_1, \mathbf{R}_2) \hat{b}(\mathbf{R}_1, \mathbf{R}_2) \frac{\delta I(x, \mathbf{R}; L)}{\delta W(\mathbf{R}_1, \mathbf{R}_2)}. \qquad \text{(C.196)}$$

Now, we will proceed as in the case of the one-dimensional problem. Variational differential equations (C.190), (C.194), and (C.196) are equivalent to the system of integro-differential equations. If we introduce the characteristic surface by the equality

$$\begin{aligned} \frac{\partial}{\partial L} W_L(\mathbf{R}_1, \mathbf{R}_2) &= -\hat{B}(L, \mathbf{R}_1, \mathbf{R}_2) W_L(\mathbf{R}_1, \mathbf{R}_2), \\ W_{L_0}(\mathbf{R}_1, \mathbf{R}_2) &= W_0(\mathbf{R}_1, \mathbf{R}_2), \end{aligned} \qquad \text{(C.197)}$$

then the field at layer boundary will be described by the equation

$$\begin{aligned} &\left[\frac{\partial}{\partial L} - i\sqrt{k_0^2 + \Delta_{\mathbf{R}}} - i\sqrt{k_0^2 + \Delta_{\mathbf{R}_0}}\right] H_L(\mathbf{R}; \mathbf{R}_0) = -\delta(\mathbf{R} - \mathbf{R}_0) \\ &-k_0^2 \int d\mathbf{R}_1 H_L(\mathbf{R}; \mathbf{R}_1) \varepsilon\left(L, \mathbf{R}_1; I_L(\mathbf{R}_1)\right) H_L(\mathbf{R}_1; \mathbf{R}_0), \\ &H_{L_0}(\mathbf{R}; \mathbf{R}_0) = g_0(0, \mathbf{R} - \mathbf{R}_0), \end{aligned} \qquad \text{(C.198)}$$

which coincides in appearance with the equation of the linear problem. In Eq. (C.198), we introduced quantity

$$I_L(\mathbf{R}) = \int d\mathbf{R}_1 \int d\mathbf{R}_2 H_L(\mathbf{R}; \mathbf{R}_1) H_L^*(\mathbf{R}; \mathbf{R}_2) W_L(\mathbf{R}_1, \mathbf{R}_2).$$

Thus, we reduced the variational differential equation (C.194) for the field at layer boundary to the system of integro-differential equations (C.197), (C.198). In addition, Eq. (C.196) assumes now the form

$$\frac{\partial}{\partial L} I(x, \mathbf{R}; L) = 0, \quad I(x, \mathbf{R}; x) = I_x(\mathbf{R}), \qquad \text{(C.199)}$$

i.e.,

$$\begin{aligned} I(x, \mathbf{R}; L; W_L) &= I_x(\mathbf{R}; W_L) \\ &= \int\int d\mathbf{R}_1 d\mathbf{R}_2 H_x(\mathbf{R}; \mathbf{R}_1) H_x^*(\mathbf{R}; \mathbf{R}_2) W_x(\mathbf{R}_1, \mathbf{R}_2). \end{aligned} \qquad \text{(C.200)}$$

Equality (C.200) reflects the property of invariance of the wavefield intensity distribution inside the medium. This property is similar to that appeared in the one-dimensional problem; namely, we have

$$I(x, \mathbf{R}; L_1; W_{L_1}) \equiv I(x, \mathbf{R}; L; W_L) \quad (L_1 > L)$$

with decreasing layer thickness, i.e., intensity distribution remains intact, but it refers now to the source distribution at layer boundary $W_{L_1}(\mathbf{R}_1, \mathbf{R}_2)$, which is the result of evolution of characteristic surface $W_L(\mathbf{R}_1, \mathbf{R}_2)$ from L to L_1.

If we neglect backscattering, then function

$$H_L(\mathbf{R},\mathbf{R}_0) \equiv g_0(0,\mathbf{R}-\mathbf{R}_0),$$

which means that intensity distribution inside the medium is governed only by the characteristic surface dynamics, i.e., by Eq. (C.197). If function $\varepsilon\left(x,\mathbf{R};I(x,\mathbf{R})\right)$ has no explicit dependence on coordinates, we can as earlier consider the case of the medium occupying half-space $x < L$. This can be done by limit process $L_0 \to -\infty$. In particular, we obtain in this way that function

$$H(\mathbf{R};\mathbf{R}_0) = H_L(\mathbf{R};\mathbf{R}_0)|_{L_0\to-\infty}$$

(it describes the backscattered field) satisfies the equation

$$\begin{aligned}\int\int d\mathbf{R}_1 d\mathbf{R}_2 W(\mathbf{R}_1,\mathbf{R}_2)\hat{B}(L,\mathbf{R}_1,\mathbf{R}_2)\frac{\delta H(\mathbf{R};\mathbf{R}_0)}{\delta W(\mathbf{R}_1,\mathbf{R}_2)} &= \delta(\mathbf{R}-\mathbf{R}_0)\\ -i\left[\sqrt{k_0^2+\Delta_{\mathbf{R}}}+\sqrt{k_0^2+\Delta_{\mathbf{R}_0}}\right]H(\mathbf{R};\mathbf{R}_0)&\\ +k^2\int d\mathbf{R}_1 H(\mathbf{R};\mathbf{R}_1)\varepsilon\left(I(\mathbf{R}_1)\right)H(\mathbf{R}_1;\mathbf{R}_0).&\end{aligned} \tag{C.201}$$

Remark 26 ***Another nonlinearity type.***

Note that if we consider the problem on a wave $u_0(x-L,R)$ incident from free half-space $x > L$ on the layer of medium $L_0 < x < L$ and assume that medium inhomogeneities are formed by the wavefield itself, then the wavefield inside the layer will satisfy the integral equation

$$\begin{aligned}U(x,\mathbf{R}) &= u_0(x-L,\mathbf{R})\\ &-k_0^2\int\limits_{L_0}^{L}dx_1\int d\mathbf{R}_1 g_0(x-x_1,\mathbf{R}-\mathbf{R}_1)\varepsilon\left(x_1,\mathbf{R}_1;U(x_1,\mathbf{R}_1)\right)U(x_1,\mathbf{R}_1).\end{aligned} \tag{C.202}$$

Correspondingly, function $G(x,R;L,R_0)$ for $x < L$ will satisfy the equation

$$\begin{aligned}G(x,\mathbf{R};L,\mathbf{R}_0) &= g_0(x-L,\mathbf{R}-\mathbf{R}_0)\\ &-k_0^2\int\limits_{L_0}^{L}dx_1\int d\mathbf{R}_1 g_0(x-x_1,\mathbf{R}-\mathbf{R}_1)\varepsilon\left(x_1,\mathbf{R}_1;U(x_1,\mathbf{R}_1)\right)G(x_1,\mathbf{R}_1;L,\mathbf{R}_0),\end{aligned} \tag{C.203}$$

where

$$U(x,\mathbf{R}) = \int d\mathbf{R}_0 G(x,\mathbf{R};L,\mathbf{R}_0)f(\mathbf{R}_0)$$

and function $f(R_0)$ (source distribution in plane $x = L$) is given by the formula

$$f(\mathbf{R}_0) = 2i\sqrt{k_0^2+\Delta_{\mathbf{R}}}u_0(0,\mathbf{R}_0).$$

Equation (C.203) is equivalent now to the boundary-value problem

$$\left\{\frac{\partial^2}{\partial x^2} + \Delta_{\mathbf{R}} + k_0^2\left[1 + \varepsilon\left(x, \mathbf{R}; U(x, \mathbf{R})\right)\right]\right\} G(x, \mathbf{R}; L, \mathbf{R}_0) = 0,$$
$$\left(\frac{\partial}{\partial x} + i\sqrt{k_0^2 + \Delta_{\mathbf{R}}}\right) G(x, \mathbf{R}; L, \mathbf{R}_0)\bigg|_{x=L_0} = 0,$$
$$\left(\frac{\partial}{\partial x} - i\sqrt{k_0^2 + \Delta_{\mathbf{R}}}\right) G(x, \mathbf{R}; L, \mathbf{R}_0)\bigg|_{x=L} = -\delta(\mathbf{R} - \mathbf{R}'). \tag{C.204}$$

Proceeding as earlier to derive the imbedding equations, we obtain the linear variational differential equation for the field inside the medium

$$\left(\frac{\partial}{\partial L} - \hat{A}(L, \mathbf{R}_0)\right) G(x, \mathbf{R}; L, \mathbf{R}_0)$$
$$= \int d\mathbf{R}_1 f(\mathbf{R}_1) \hat{A}(L, \mathbf{R}_1) \frac{\delta}{\delta f(\mathbf{R}_1)} G(x, \mathbf{R}; L, \mathbf{R}_0),$$
$$G(x, \mathbf{R}; L, \mathbf{R}_0)|_{L=x} = H_x(\mathbf{R}; \mathbf{R}_0), \tag{C.205}$$

where operator $\hat{A}(L, R)$ acts on a function $G(x, R; L, R_0)$ according to the equality

$$\hat{A}(L, \mathbf{R}_0) G(x, \mathbf{R}; L, \mathbf{R}_0) = i\sqrt{k_0^2 + \Delta_{\mathbf{R}_0}} G(x, \mathbf{R}; L, \mathbf{R}_0)$$
$$-k_0^2 \int d\mathbf{R}_1 G(x, \mathbf{R}; L, \mathbf{R}_1) \varepsilon\left(L, \mathbf{R}_1; U(L, \mathbf{R}_1)\right) H_L(\mathbf{R}_1; \mathbf{R}_0), \tag{C.206}$$

and function

$$H_L(\mathbf{R}; \mathbf{R}_0) = G(L, \mathbf{R}; L, \mathbf{R}_0)$$

describes the wavefield in the source plane. Quantity $H_L(R; R_0)$ satisfies the closed integro-differential equation

$$\left[\frac{\partial}{\partial L} - i\sqrt{k_0^2 + \Delta_{\mathbf{R}}} - i\sqrt{k_0^2 + \Delta_{\mathbf{R}_0}}\right] H_L(\mathbf{R}; \mathbf{R}_0) = -\delta(\mathbf{R} - \mathbf{R}_0)$$
$$-k_0^2 \int d\mathbf{R}_1 H_L(\mathbf{R}; \mathbf{R}_1) \varepsilon\left(L, \mathbf{R}_1; U(L, \mathbf{R}_1)\right) H_L(\mathbf{R}_1; \mathbf{R}_0)$$
$$+ \int d\mathbf{R}_1 f(\mathbf{R}_1) \hat{A}(L, \mathbf{R}_1) \frac{\delta}{\delta f(\mathbf{R}_1)} H_L(\mathbf{R}; \mathbf{R}_0) \tag{C.207}$$

with the initial condition.

$$H_{L_0}(\mathbf{R}; \mathbf{R}_0) = g_0(0, \mathbf{R} - \mathbf{R}_0). \quad \blacklozenge$$

C.3 One-dimensional nonstationary boundary-value wave problem

In the foregoing sections, we considered in detail the linear stationary boundary-value wave problems. Here, consider the conversion of the boundary-value problem for the scalar wave equation into the initial value problem. Such problems are characteristic of the time-domain analysis of impulses propagating in stationary and nonstationary media; they appear also in the consideration of scattering of waves of one type by the waves of the other type (for example, light scattering by ultrasound and sound scattering by internal waves). Consider the simplest one-dimensional problem with unmatched boundary.

C.3.1 Nonsteady medium

Let the inhomogeneous medium occupies, as earlier, layer $L_0 < x < L$ and let the point source is located at the space-time point (x_0, t_0). We define Green's function of the wave equation (the point source wave field) as the solution to the equation

$$\left(\frac{\partial^2}{\partial x^2} - \frac{\partial^2}{\partial t^2}\frac{1}{c^2(x,t)}\right) G(x,t;x_0,t_0) = -\frac{2}{c_0}\delta(x-x_0)\delta(t-t_0), \tag{C.208}$$

where function $c(x,t)$ describes the space-time inhomogeneities of the velocity of wave propagation in the medium. In this case, function $G(x,t;x_0,t_0)$ will be the dimensionless function. We assume that the space outside the layer is homogeneous and characterized by the velocity of wave propagation c_0. If $c(L,t) \neq c_0$, then the velocity of wave is discontinuous at boundary $x = L$. As in the case of the stationary problem, we will call such a boundary the *unmatched boundary*. Conversely, if $c(L,t) = c_0$, then the discontinuity of wave velocity disappears, and we will call such a boundary the *matched boundary*. Consider the case of the unmatched boundary. In the case of the matched boundary, equations of the imbedding method are derived in paper [142].

Introducing function

$$\varepsilon(x,t) = \frac{c_0^2}{c^2(x,t)} - 1, \tag{C.209}$$

we can rewrite wave equation (C.208) in the form

$$\begin{aligned}\left(\frac{\partial^2}{\partial x^2} - \frac{\partial^2}{c_0^2\partial t^2}\right) G(x,t;x_0,t_0) - \frac{\partial^2}{c_0^2\partial t^2}\left[\varepsilon(x,t)G(x,t;x_0,t_0)\right]&\\ = -\frac{2}{c_0}\delta(x-x_0)\delta(t-t_0).&\end{aligned} \tag{C.210}$$

Outside the medium layer, the solution has the form of outgoing waves

$$\begin{aligned} G(x,t;x_0,t_0) &= T_1(x - L - c_0 t) \quad (x \geqslant L), \\ G(x,t;x_0,t_0) &= T_2(x - L_0 + c_0 t) \quad (x \leqslant L_0), \end{aligned}$$

and boundary conditions for this problem are, as earlier, the continuity of field $u(x,t)$ and derivative $\partial u(x,t)/\partial x$ at layer boundaries. These conditions can be represented as

$$\begin{aligned} \left(\frac{\partial}{\partial x} + \frac{\partial}{c_0\partial t}\right) G(x,t;x_0,t_0)\Big|_{x=L} &= 0, \\ \left(\frac{\partial}{\partial x} - \frac{\partial}{c_0\partial t}\right) G(x,t;x_0,t_0)\Big|_{x=L_0} &= 0. \end{aligned} \tag{C.211}$$

Function $G(x,t;x_0,t_0)$ is continuous everywhere and its spatial derivative with respect to x is discontinuous at the point of source location

$$\frac{\partial}{\partial x}G(x,t;x_0,t_0)\Big|_{x=x_0+0} - \frac{\partial}{\partial x}G(x,t;x_0,t_0)\Big|_{x=x_0-0} = -\frac{2}{c_0}\delta(t-t_0). \tag{C.212}$$

The absence of inhomogeneities of the velocity of wave propagation ($\varepsilon(x,t) = 0$) corresponds to free space Green's function

$$g_0(x,t;x_0,t_0) = g_0(x - x_0; t - t_0)$$

given by the expression

$$g_0(x;t) = \theta\left(c_0 t - |x|\right) = -\frac{1}{2i\pi}\int_{-\infty}^{\infty}\frac{d\omega}{\omega + i0}e^{-i\omega(c_0 t - |x|)}. \tag{C.213}$$

Under the condition that the source and observation points are fixed (for $x < x_0$, for example), it satisfies the equalities

$$\begin{aligned}\frac{\partial}{\partial x_0}g(x - x_0; t - t_0) &= -\frac{\partial}{\partial x}g(x - x_0; t - t_0)\\ &= -\frac{\partial}{c_0\partial t}g(x - x_0; t - t_0) = \frac{\partial}{c_0\partial t_0}g(x - x_0; t - t_0)\end{aligned} \tag{C.214}$$

expressing the factorization property of wave equation (see Appendix B).

Boundary-value problem (C.210), (C.211) is equivalent to the integral equation

$$\begin{aligned}G(x,t;x_0,t_0;L) &= g_0(x - x_0; t - t_0)\\ &-\frac{1}{2c_0}\int_{L_0}^{L}dx_1\int_{-\infty}^{\infty}dt_1 g_0(x - x_1; t - t_1)\frac{\partial^2}{\partial t_1^2}\left[\varepsilon(x_1,t_1)G(x_1,t_1;x_0,t_0;L)\right].\end{aligned} \tag{C.215}$$

Let now the source is located at layer boundary $x_0 = L$. Then, boundary-value problem (C.210), (C.211) assumes the form (with allowance for Eq. (C.212))

$$\begin{aligned}\left(\frac{\partial^2}{\partial x^2} - \frac{\partial^2}{c_0^2\partial t^2}\right)G(x,t;L,t_0) &= \frac{\partial^2}{c_0^2\partial t^2}\left[\varepsilon(x,t)G(x,t;L,t_0)\right],\\ \left(\frac{\partial}{\partial x} + \frac{\partial}{c_0\partial t}\right)G(x,t;L,t_0)\Big|_{x=L} &= \frac{2}{c_0}\delta(t - t_0),\\ \left(\frac{\partial}{\partial x} - \frac{\partial}{c_0\partial t}\right)G(x,t;L,t_0)\Big|_{x=L_0} &= 0.\end{aligned} \tag{C.216}$$

This problem is equivalent to the integral equation

$$\begin{aligned}G(x,t;L,t_0) &= g_0(x - L; t - t_0)\\ &-\frac{1}{2c_0}\int_{L_0}^{L}dx_1\int_{-\infty}^{\infty}dt_1 g_0(x - x_1; t - t_1)\frac{\partial^2}{\partial t_1^2}\left[\varepsilon(x_1,t_1)G(x_1,t_1;L,t_0)\right].\end{aligned} \tag{C.217}$$

Remark 27 ***Problem on a wave incident on medium layer.***

Note that integral equation (C.217) (or the corresponding boundary-value problem (C.216)) describes the problem on a wave incident on the layer of inhomogeneous medium. Let the wave $u_0(x - L + c_0 t)$ (c_0 is the velocity of wave propagation in free space) is incident on this layer from the right, i.e., from region $x > L$. Then, the wavefield in region $x > L$ is given by the equality

$$u(x,t) = u_0(x - L + c_0 t) + R(x - L - c_0 t) \quad (x \geqslant L),$$

where $R(x-L-c_0t)$ is the reflected wave. In region $x < L_0$, we have only the transmitted wave

$$u(x,t) = T(x - L_0 + c_0 t) \quad (x \leqslant L_0),$$

and the wavefield in region $L_0 < x < L$ satisfies the wave equation

$$\left(\frac{\partial^2}{\partial x^2} - \frac{\partial^2}{c_0^2 \partial t^2}\right) u(x,t) = \frac{\partial^2}{c_0^2 \partial t^2}\left[\varepsilon(x,t)u(x,t)\right] \tag{C.218}$$

with the boundary conditions

$$\begin{aligned} \left(\frac{\partial}{\partial x} + \frac{\partial}{c_0 \partial t}\right) u(x,t)\bigg|_{x=L} &= 2\frac{\partial}{c_0 \partial t} u_0(c_0 t), \\ \left(\frac{\partial}{\partial x} - \frac{\partial}{c_0 \partial t}\right) u(x,t)\bigg|_{x=L_0} &= 0. \end{aligned} \tag{C.219}$$

In addition, at layer boundary $x = L$, incident field $u_0(x - L + c_0 t)$ creates the source distribution $f(t_0)$ such that

$$u_0(c_0 t) = \int\limits_{-\infty}^{\infty} dt_0 \theta(t - t_0) f(t_0), \quad f(t_0) = \frac{\partial}{\partial t_0} u_0(c_0 t_0),$$

so that the wavefield inside the layer $u(x,t)$ can be represented in the form

$$u(x,t) = \int\limits_{-\infty}^{\infty} dt_0 G(x,t;L,t_0) f(t_0).$$

Note that the incident wave in the form of the Heaviside step function (this incident wave corresponds to Eq. (C.213))

$$g_0(x - L, t) = \theta(x - L + c_0 t)$$

creates the source distribution

$$f(t_0) = \delta(t_0),$$

and we obtain that the wavefield inside the medium in this case is

$$u(x,t) = G(x,t;L,0). \quad ♦$$

Derive now the equations of the imbedding method for the boundary-value problem (C.216). Differentiating Eq. (C.217) with respect to parameter L and taking into account Eq. (C.214), we obtain the integral equation in quantity $\partial G(x,t;L,t_0)/\partial L$

$$\begin{aligned} \frac{\partial}{\partial L} G(x,t;L,t_0) &= \hat{A}(L,t_0) g_0(x - L; t - t_0) \\ &- \frac{1}{2c_0} \int\limits_{L_0}^{L} dx_1 \int\limits_{-\infty}^{\infty} dt_1 g_0(x - x_1; t - t_1) \frac{\partial^2}{\partial t_1^2}\left[\varepsilon(x_1,t_1) \frac{\partial}{\partial L} G(x_1,t_1;L,t_0)\right], \end{aligned} \tag{C.220}$$

where operator $\hat{A}(L,t_0)$ acts on arbitrary function $F(t_0)$ of variable t_0 in accordance with the formula

$$\hat{A}(L,t_0)F(t_0) = \frac{\partial}{c_0\partial t_0}F(t_0) - \frac{1}{2c_0}\int\limits_{-\infty}^{\infty} dt_1 F(t_1)\frac{\partial^2}{\partial t_1^2}\left[\varepsilon(L,t_1)G_L(t_1;t_0)\right], \qquad \text{(C.221)}$$

and function $G_L(t;t_0) = G(L,t;L,t_0)$ describes the wavefield in the source plane $x = L$.

The solution of integral equation (C.220) can be related to the wavefield $G(x,t;L,t_0)$ either by the operator relationship

$$\frac{\partial}{\partial L}G(x,t;L,t_0) = \hat{A}(L,t_0)G(x,t;L,t_0),$$

or in the form

$$\left(\frac{\partial}{\partial L} - \frac{\partial}{c_0\partial t_0}\right)G(x,t;L,t_0) = -\frac{1}{2c_0}\int\limits_{-\infty}^{\infty} dt_1 G(x,t;L,t_1)\frac{\partial^2}{\partial t_1^2}\left[\varepsilon(L,t_1)G_L(t_1;t_0)\right]. \quad \text{(C.222)}$$

We can consider the latter relationship as the integro-differential equation by supplementing it with the initial condition

$$G(x,t;x,t_0) = G_x(t;t_0). \qquad \text{(C.223)}$$

Function $G_L(t;t_0) = G(L,t;L,t_0)$ satisfies the relationship

$$\frac{\partial}{\partial L}G_L(t;t_0) = \frac{\partial}{\partial L}G(x,t;L,t_0)\bigg|_{x=L} + \frac{\partial}{\partial x}G(x,t;L,t_0)\bigg|_{x=L}. \qquad \text{(C.224)}$$

The first term in Eq. (C.224) can be determined from Eq. (C.222) at $x = L$ and the second, from the boundary condition in Eq. (C.216). As a result, we obtain the closed integro-differential equation with the initial condition following from Eq. (C.216)

$$\begin{aligned}\left(\frac{\partial}{\partial L} - \frac{\partial}{c_0\partial t_0} + \frac{\partial}{c_0\partial t}\right)G_L(t;t_0) &= \frac{2}{c_0}\delta(t-t_0)\\ &-\frac{1}{2c_0}\int\limits_{-\infty}^{\infty} dt_1 G_L(t;t_1)\frac{\partial^2}{\partial t_1^2}\left[\varepsilon(L,t_1)G_L(t_1;t_0)\right],\\ G_{L_0}(t;t_0) = g_0(0,t-t_0) &= \theta(t-t_0). \end{aligned} \qquad \text{(C.225)}$$

Equations (C.222), (C.223), and (C.225) form the equations of the imbedding method in the context of the problem with unmatched boundary [17, 136].

Remark 28 ***Consideration of boundary condition at*** $x = L_0$.

Above, we assumed that half-space $x < L_0$ is free and characterizes by the free space velocity of wave propagation c_0. If this velocity differs from the velocity in half-space $x > L$ and is equal to c_1, then all above equations remain obviously valid. In this case, only the boundary condition at boundary $x = L_0$ is replaced in Eq. (C.216) with the condition

$$\begin{aligned}\left(\frac{\partial}{\partial x} + \frac{\partial}{c_0\partial t}\right)G(x,t;L,t_0)\bigg|_{x=L} &= \frac{2}{c_0}\delta(t-t_0),\\ \left(\frac{\partial}{\partial x} - \frac{\partial}{c_1\partial t}\right)G(x,t;L,t_0)\bigg|_{x=L_0} &= 0.\end{aligned}$$

As a consequence, the initial condition of function $G_L(t;t_0)$ is also replaced with the condition

$$G_L(t;t_0) = \frac{2c_1}{c_0+c_1}\theta(t-t_0). \qquad \blacklozenge$$

C.3.2 Steady medium

In the steady medium, velocity of wave propagation is independent of time and function $\varepsilon(x,t) \equiv \varepsilon(x)$. As a result, all above equations are simplified because the solutions depend only on time difference $(t - t_0)$. In this case, we can set $t_0 = 0$ and rewrite, for example, Eqs. (C.222) and (C.225) in the form

$$\left(\frac{\partial}{\partial L} + \frac{\partial}{c_0 \partial t}\right) G(x,t;L) = \frac{1}{2c_0}\left(1 - \frac{c_0^2}{c^2(L)}\right) \int_{-\infty}^{\infty} dt_1 \frac{\partial G(x,t-t_1;L)}{\partial t} \frac{\partial G_L(t_1)}{\partial t_1}, \tag{C.226}$$

$$\left(\frac{\partial}{\partial L} + \frac{2}{c_0}\frac{\partial}{\partial t}\right) G_L(t) = \frac{2}{c_0}\delta(t) + \frac{1}{2c_0}\left(1 - \frac{c_0^2}{c^2(L)}\right) \int_{-\infty}^{\infty} dt_1 \frac{\partial G_L(t-t_1)}{\partial t} \frac{\partial G_L(t_1)}{\partial t_1},$$

$$G_{L_0}(t) = g_0(0,t) = \frac{2c_1}{c_0 + c_1}\theta(t), \tag{C.227}$$

where c_1 is the velocity of wave propagation in free half-space $x < L_0$.

Setting $x = L_0$ in Eq. (C.226), we obtain the equation for the wave $T_L(t) = G(L_0, t; L)$ outgoing from the layer

$$\left(\frac{\partial}{\partial L} + \frac{\partial}{c_0 \partial t}\right) T_L(t) = \frac{1}{2c_0}\left(1 - \frac{c_0^2}{c^2(L)}\right) \int_{-\infty}^{\infty} dt_1 \frac{\partial T_L(t-t_1)}{\partial t} \frac{\partial G_L(t_1)}{\partial t_1},$$

$$T_{L_0}(t) = g_0(0,t) = \frac{2c_1}{c_0 + c_1}\theta(t). \tag{C.228}$$

Remark 29 ***Structure of the solution in the layer of homogeneous medium.***

If medium parameters remain constant ($c(x) \equiv c$), we can easily draw the solution to imbedding equations (or to the corresponding boundary-value problem) using the Fourier transform. Namely, we obtain the following expression for the field at layer boundary $x = L$ (for simplicity, we assume $t_0 = 0$ in what follows)

$$\begin{aligned} G_L(t) &= -\frac{1+R_1}{2\pi i} \int \frac{d\omega}{\omega + i0} e^{-i\omega t} \frac{1 + R_2 e^{2i\omega \tau_{L_0}}}{1 + R_1 R_2 e^{2i\omega \tau_{L_0}}} \\ &= (1+R_1)\left[\theta(t) + R_2(1-R_1)\theta(t - 2\tau_{L_0}) + \ldots\right], \end{aligned} \tag{C.229}$$

where $\tau_{L_0} = (L - L_0)/c$ is the time required for the wave to traverse the medium layer and R_i are the respective reflection coefficients of the plane harmonic wave from boundaries $x = L$ and $x = L_0$,

$$R_1 = \frac{c - c_0}{c + c_0}, \quad R_2 = \frac{c_1 - c}{c_1 + c}.$$

From Eq. (C.229) follows that

$$G_{L_0}(t) = -\frac{(1+R_1)(1+R_2)}{1 + R_1 R_2}\theta(t) = \frac{2c_1}{c_0 + c_1}\theta(t) \tag{C.230}$$

for $L \to L_0$ (i.e., when layer thickness tends to zero), and we must take into consideration all multiple re-reflections from layer boundaries. On the contrary, value $G_L(t = +0)$ is governed solely by boundary $x = L$:

$$G_L(t = +0) = 1 + R_1 = \frac{2c}{c + c_0}. \tag{C.231}$$

At $t = 2\tau_{L_0} + 0$, i.e., at the instant the wave reflected from boundary $x = L_0$ arrives at boundary $x = L$, we have

$$G_L(2\tau_{L_0} + 0) = (1 + R_1)\left[1 + R_2(1 - R_1)\right]. \tag{C.232}$$

In a similar way, we obtain the expression for the wavefield inside the layer

$$G(x,t;L) = (1 + R_1)\left[\theta(t - \tau_x) + R_2\theta(t - 2\tau_{L_0} + \tau_x) + \ldots\right],$$

where $\tau_x = (L - x)/c$ is the time of arrival of the wave at point x. From this expression follows in particular that

$$G(x, \tau_x + 0; L) = \frac{2c}{c + c_0}, \qquad T_L(\tau_{L_0} = 0) = \frac{4cc_1}{(c + c_0)(c + c_1)}, \tag{C.233}$$

where $T_L(t) = G(L_0, t; L)$ is the wave transmitted through the layer. It will be shown below that Eqs. (C.230)–(C.233) can be easily extended to the case of inhomogeneous medium. ♦

Remark 30 ***Conversion to the stationary wave problem.***

Represent the solution in the form

$$G_L(t) = -\frac{1}{2i\pi}\int\limits_{-\infty}^{\infty}\frac{d\omega}{\omega + i0}G_L(\omega)e^{-i\omega t}.$$

Then, Eq. (C.227), for example, will assume the form of the ordinary differential equation

$$\frac{d}{dL}G_L(\omega) = 2i\frac{\omega}{c_0}\left[G_L(\omega) - 1\right] + i\frac{\omega}{2c_0}\varepsilon(L)G_L^2(\omega).$$

From this equation follows that the reflection coefficient at frequency ω

$$R_L(\omega) = G_L(\omega) - 1$$

satisfies the Riccati equation

$$\frac{d}{dL}R_L(\omega) = 2i\frac{\omega}{c_0}R_L(\omega) + i\frac{\omega}{2c_0}\varepsilon(L)\left[1 + R_L(\omega)\right]^2, \quad R_{L_0}(\omega) = \frac{c_1 - c_0}{c_1 + c_0}$$

corresponding to the stationary problem. ♦

As we mentioned earlier, function $G(x,t;L)$ for $t > 0$ describes the wavefield in the medium illuminated by the wave of the form

$$g_0(x - L, t) = \theta(x - L + c_0 t).$$

In addition, function $G_L(t)$ describing the wavefield in plane $x = L$ (i.e., the backscattered field) has the form

$$G_L(t) = H_L(t)\theta(t). \tag{C.234}$$

Substituting Eq. (C.234) in Eq. (C.227) and separating, in accordance with the method of singularity spreading (see, e.g., [30]), the singular ($\delta(t)$) and regular ($\theta(t)$) parts, we obtain the equality

$$H_L(+0) = \frac{2c(L)}{c(L) + c_0}. \tag{C.235}$$

It is clearly this equality that expresses the feature that, at the moment the wave arrives at boundary $x = L$, reflection is realized only due to the discontinuity of velocity $c(x)$ at layer boundary $x = L$. We could assume this equality as a basis from the outset. As regards function $H_L(t)$ for $t > 0$, it satisfies the equation

$$\begin{aligned} \left(\frac{\partial}{\partial L} + \frac{2}{c(L)}\frac{\partial}{\partial t}\right) H_L(t) &= \frac{1}{2c_0}\left(1 - \frac{c_0^2}{c^2(L)}\right)\int\limits_0^t dt_1 \frac{\partial H_L(t-t_1)}{\partial t}\frac{\partial H_L(t_1)}{\partial t_1}, \\ H_{L_0}(t) &= \frac{2c_1}{c_0 + c_1}. \end{aligned} \tag{C.236}$$

Due to the structure of Eq. (C.236), we can successively calculate the coefficients of the Taylor series of function $H_L(t)$ about point $t = 0$. Indeed, setting $t = 0$ in Eq. (C.236), we obtain

$$\left.\frac{\partial H_L(t)}{\partial t}\right|_{t=0} = -\frac{c(L)}{2}\frac{\partial}{\partial L}H_L(0) = -\frac{c_0 c(L) c'(L)}{\left(c(L) + c_0\right)^2}, \tag{C.237}$$

where $c'(L) = dc(L)/dL$. Differentiating Eq. (C.236) with respect to t and setting again $t = 0$, we obtain quantity $\partial^2 H_L(t)/\partial t^2$ that determines the second derivative $c''(L) = d^2c(L)/dL^2$, and so forth.

Wavefield inside the layer $G(x, t; L)$ satisfies Eq. (C.226) and has the following structure

$$G(x, t; L) = H(x, t; L)\theta(t - \tau_x(L)), \tag{C.238}$$

where $\tau_x(L)$ is the time the wave travels from boundary $x = L$ to point x. Substituting Eq. (C.238) in Eq. (C.226) and putting to zero the coefficient of $\theta(t - \tau_x(L)$, we obtain the corresponding equation for quantity $\tau_x(L)$, from which follows that

$$\tau_x(L) = \int\limits_x^L \frac{d\xi}{c(\xi)}. \tag{C.239}$$

For $t > \tau_x(L)$, function $H(x, t; L)$ satisfies the equation

$$\begin{aligned} &\left(\frac{\partial}{\partial L} + \frac{2}{c(L)}\frac{\partial}{\partial t}\right) H(x, t; L) \\ &= \frac{1}{2c_0}\left(1 - \frac{c_0^2}{c^2(L)}\right) H(x, \tau_x(L); L)\frac{\partial H_L(t - \tau_x(L))}{\partial t} \\ &+ \frac{1}{2c_0}\left(1 - \frac{c_0^2}{c^2(L)}\right)\int\limits_0^{t-\tau_x(L)} dt_1 \frac{\partial H(x, t - t_1; L)}{\partial t}\frac{\partial H_L(t_1)}{\partial t_1} \end{aligned} \tag{C.240}$$

with the initial condition $H(x, t; L)|_{L=x} = H_x(t)$. Equation (C.240) is unclosed in function $H(x, t; L)$ because the right-hand side depends on quantity $H(x, \tau_x(L); L)$. To specify this quantity, we set $t = \tau_x(L)$ in Eq. (C.240). Then, taking into account Eqs. (C.239) and (C.237), we obtain the equation

$$\frac{\partial}{\partial L}H(x, \tau_x(L); L) = -\frac{c'(L)}{2c(L)}\frac{c(L) - c_0}{c(L) + c_0}H(x, \tau_x(L); L) \tag{C.241}$$

whose solution satisfying the initial condition

$$H(x, \tau_x(L); L)|_{L=x} = H_x(0) = \frac{2c(x)}{c(x) + c_0}$$

has the form

$$H(x,\tau_x(L);L)=\frac{2\sqrt{c(x)c(L)}}{c(L)+c_0}. \tag{C.242}$$

The differential equation for quantity $T_L(\tau_{L_0}(L))$, which is the wave outgoing from the layer, can be obtained quite similarly, by setting $x=L_0$ in Eq. (C.241):

$$\frac{d}{dL}T_L(\tau_{L_0}(L))=-\frac{c'(L)}{2c(L)}\frac{c(L)-c_0}{c(L)+c_0}T_L(\tau_{L_0}(L)).$$

The initial condition for this equation is the equality (C.233) for $L\to L_0$

$$T_L(\tau_{L_0}=0)|_{L\to L_0}=\frac{4c(L_0)c_1}{(c(L_0)+c_0)(c(L_0)+c_1)}.$$

Consequently, we have

$$T_L(\tau_{L_0}(L))=\frac{4c_1\sqrt{c(L_0)c(L)}}{(c(L_0)+c_0)(c(L_0)+c_1)}.$$

Thus, the wavefield amplitude (step) at the instant of wave arrival is determined by the local value of quantity $c(x)$ at this point and is independent of wave propagation prehistory.

The above equations hold only for times t from the interval during which no wave reflected from boundary $x=L_0$ is present. For example, function $G_L(t)$ is governed by Eq. (C.236) only for $0\le t<2\tau_{L_0}(L)$. For $0\le t<4\tau_{L_0}(L)$, function $G_L(t)$ has the form

$$G_L(t)=H_L(t)\theta(t)+F_L(t)\theta(t-2\tau_{L_0}(L)),$$

which means that a step occurs at instant $t=2\tau_{L_0}(L)+0$ and this step is caused by the arrival of the wave reflected from boundary $x=L_0$. Substituting this expression in Eq. (C.227), one can obtain the equation for function $F_L(t)$ and the expression for $F_L(2\tau_{L_0})$ [32]

$$F_L(2\tau_{L_0})=\frac{4c_0c(L)(c_1-c(L_0))}{(c(L)+c_0)^2(c(L_0)+c_1)}. \tag{C.243}$$

Thus, the amplitude of the step of backscattered field at the time of arrival of the wave reflected by boundary $x=L_0$ is also determined by the local characteristics of quantity $c(x)$ at reflecting boundaries.

The Eq. (C.236) offers a possibility of obtaining the asymptotic behavior of function $H_L(t)$ for $t\to\infty$. To do this, we must take into account that the solution of this equation for $t\to\infty$ is independent of the initial (in time) value $H_L(0)$. Performing the Laplace transform with respect to time and neglecting the initial condition, we obtain the equation

$$\frac{d}{dL}H_L(p)=-\frac{2p}{c(L)}H_L(p)-\frac{p}{2c_0}\varepsilon(L)H_L^2(p),\quad H_{L_0}(p)=\frac{2c_1}{c_0+c_1},$$

whose solution has the form

$$H_L(p)=H_{L_0}(p)\frac{e^{-2p\tau_{L_0}(L)}}{1+\frac{p}{2c_0}H_{L_0}(p)\int\limits_{L_0}^{L}d\xi\varepsilon(\xi)e^{-2p\tau_{L_0}(\xi)}},\quad \tau_{L_0}(L)=\int\limits_{L_0}^{L}\frac{d\eta}{c(\eta)}.$$

The corresponding time-domain solution has the form

$$H_L(t) = \frac{1}{2\pi i} \int\limits_{-i\infty+\sigma}^{i\infty+\sigma} dp e^{p(t-2p\tau_{L_0}(L))} \frac{H_{L_0}(p)}{1 + \frac{p}{2c_0} H_{L_0}(p) \int\limits_{L_0}^{L} d\xi \varepsilon(\xi) e^{-2p\tau_{L_0}(\xi)}}, \tag{C.244}$$

and arrives at stationary value $H_L(t) = 1$ for $t \to \infty$.

Remark 31 ***Inverse problem solution.***

The above relationships and equations offer a possibility of solving the inverse problem on recovering velocity of wave propagation $c(x)$ from the known temporal behavior of the backscattered field [32], [208].

In the case when the time-dependent behavior of wavefield at certain point inside the medium is known, the inverse problem was analyzed in papers [176]–[178]. Indeed, the backscattered field is described by function $H_L(t)$ whose expansion in the Taylor series about $t = 0$ determines quantities $c(L)$, $c'(L)$, and so on. If we consider now Eq. (C.236) as an auxiliary equation and rewrite it in the form

$$\begin{aligned} &\left(\frac{\partial}{\partial x} + \frac{2}{c(x)} \frac{\partial}{\partial t}\right) H_x(t) \\ &= \frac{1}{2c_0} \left(1 - \frac{c_0^2}{c^2(x)}\right) \int\limits_0^t dt_1 \frac{\partial H_x(t-t_1)}{\partial t} \frac{\partial H_x(t_1)}{\partial t_1} \end{aligned} \tag{C.245}$$

with the initial condition

$$H_x(t)|_{x=L} = H(t),$$

then we can solve Eq. (C.245) to determine $H_x(t)$ for $x = L - \delta$ from the known behavior of $c(x)$ in the vicinity of point $x = L$. From the determined $H_x(t)$, we again determine $c(x)$, $c'(x)$, and so on by the formulas

$$H_x(0) = \frac{2c(x)}{c(x) + c_0}, \quad \frac{\partial}{\partial t} H_x(t)\bigg|_{t=0} = -\frac{c_0 c(x) c'(x)}{(c_0 + c(x))^2}, \ldots. \tag{C.246}$$

The above procedure of solving inverse problem allows analytic solutions in two cases corresponding to the exponential and linear functions $H_L(t)$.

Indeed, if

$$H_L(t) = \alpha e^{\beta t}, \tag{C.247}$$

then quantities α and β determine values $c(L)$ and $c'(L)$. In this case, the solution of Eq. (C.245) is also the exponential function of time; namely, we have

$$H_x(t) = \alpha(x) e^{\beta(x)t}, \quad \alpha(x) = \frac{2c(x)}{c(x) + c_0}, \quad \beta(x) = -\frac{c_0 c'(x)}{2\,(c(x) + c_0)}. \tag{C.248}$$

in accordance with Eq. (C.246). Substituting Eq. (C.248) in Eq. (C.245), we obtain that $c(x)$ satisfies the second-order differential equation

$$c''(x) - \frac{[c'(x)]^2}{2c(x)} = 0 \tag{C.249}$$

with the initial conditions

$$c'(x)|_{x=L} = c'(L), \quad c(x)|_{x=L} = c(L).$$

The solution of this equation has two branches

$$c(x) = c(L)\left(1 \pm \frac{\xi}{2}\right)^2, \quad \xi = \frac{|c'(L)|}{c(L)}(L - x), \tag{C.250}$$

where the upper sign refers to the case $c'(L) > 0$ and the lower sign, to the case $c'(L) < 0$. An interesting feature of solution (C.250) for $c'(L) > 0$ consists in the fact that the time required for the wave to arrive at point $\xi_0 = 2$ at which $c(x) = 0$ appears infinite. In this case, the incident wave is totally reflected from the layer, and the maximum depth it can reach in the layer is

$$L - x_0 = 2c(L)/c'(L).$$

For the linear time-dependent function

$$H_L(t) = \alpha + \beta t, \tag{C.251}$$

the analytic solution can be obtained similarly. In this case, we have, in accordance with Eq. (C.246),

$$H_x(t) = \alpha(x) + \beta(x)t, \quad \alpha(x) = \frac{2c(x)}{c(x) + c_0}, \quad \beta(x) = -\frac{c_0 c(x) c'(x)}{(c(x) + c_0)^2}, \tag{C.252}$$

and the substitution of Eq. (C.252) in Eq. (C.245) yields the differential equation of the form

$$c''(x) - \frac{1}{2c(x)} \frac{c(x) - c_0}{c(x) + c_0} \left[c'(x)\right]^2 = 0. \tag{C.253}$$

The solution to Eq. (C.253) can be easily obtained from the transcendental equation

$$\arctan\sqrt{\frac{c(x)}{c_0}} - \sqrt{\frac{c(x)}{c_0}} - \arctan\sqrt{\frac{c(L)}{c_0}} + \sqrt{\frac{c(L)}{c_0}}$$
$$= \pm\sqrt{\frac{c(L)}{c_0}} \frac{|c'(L)|}{c(L) + c_0}(L - x), \tag{C.254}$$

where, as earlier, the upper sign refers to the case $c'(L) > 0$ and the lower sign, to the case $c'(L) < 0$. However, this solution depends on the discontinuity of the velocity of wave propagation at boundary L. For example, for $c'(L) < 0$, the time of arrival of the wave at point x is given in this case by the expression

$$\tau_x(L) = \frac{2\left(c(L) + c_0\right)}{\sqrt{c(L)c_0}|c'(L)|}\left(\arctan\sqrt{\frac{c(x)}{c_0}} - \arctan\sqrt{\frac{c(L)}{c_0}}\right).$$

The relationship between the two analytic solutions can be easily established by considering limiting cases $c(x) \gg c_0$ and $c(x) \ll c_0$ in Eq. (C.253). For $c(x) \gg c_0$, Eq. (C.253) grades into Eq. (C.249) everywhere in the layer, while for $c(x) \ll c_0$, it grades into the equation

$$c(x) + \frac{[c'(x)]^2}{2c(x)} = 0, \tag{C.255}$$

whose solution also has two branches

$$c(x) = c(L)\left(1 \mp \frac{3}{2}\xi\right)^{2/3}, \quad \xi = \frac{|c'(L)|}{c(L)}(L - x). \tag{C.256}$$

We should emphasize in this connection that these limit processes can result in instability of the direct problem solution in the case $c'(L) < 0$. This instability is related to the fact that function $H_L(t)$ corresponding to solutions (C.249) and (C.256) increases exponentially, whereas function $H_L(t)$ corresponding to the exact solution of Eq. (C.253) increases linearly. Note that, if the inverse problem is formulated as above in terms of the field at boundary in the form of Eqs. (C.247) and (C.251), the respective field inside the medium also appears either exponential or linear function of time.

Finally, Eq. (C.243) is used with

$$G_L\left(2\tau_{L_0}(L)\right) = H_L\left(2\tau_{L_0}(L)\right) + F_L\left(2\tau_{L_0}(L)\right),$$

to obtain quantity c_1 characterizing half-space $x < L_0$.

Here, we considered the case of the unmatched boundary. In the case of the matched boundary, one can derive similar imbedding equations that also offer a possibility of solving the inverse problem, i.e., recovering function $c(x)$ from the known time-dependent field at layer boundary $H_L(t)$ [44]–[46], [209]–[214]. ♦

C.3.3 One-dimensional nonlinear wave problem

The above equations can be easily extended to the case when the right-hand side of Eq. (C.218), page 501 includes the nonlinear operator, such as

$$\Im(u) = \frac{\partial^2}{c_0^2\partial t^2}\left[\varepsilon\left(x, t; u(x,t)\right)u(x,t)\right],$$

for example. In this case, boundary-value problem (C.218), (C.219) is replaced with the nonlinear boundary-value [136]

$$\left(\frac{\partial^2}{\partial x^2} - \frac{\partial^2}{c_0^2\partial t^2}\right)u(x,t) = \frac{\partial^2}{c_0^2\partial t^2}\left[\varepsilon\left(x, t; u(x,t)\right)u(x,t)\right],$$

$$\left(\frac{\partial}{\partial x} + \frac{\partial}{c_0\partial t}\right)u(x,t)\Big|_{x=L} = 2\frac{\partial}{c_0\partial t}u_0(c_0 t),$$

$$\left(\frac{\partial}{\partial x} - \frac{\partial}{c_0\partial t}\right)u(x,t)\Big|_{x=L_0} = 0.$$

The incident field $u_0(c_0 t)$ creates at layer boundary $x = L$ the source distribution $f(t)$, such that

$$u_0(c_0 t) = \int dt_0 g_0(0, t - t_0) f(t_0),$$

where

$$g_0(x - L, t - t_0) = \theta(c_0(t - t_0) - L + x)$$

is Green's function in the free half-space $x > L$. As a consequence, we have

$$u(x, t; L) = \int dt_0 G(x, t; L, t_0) f(t_0),$$

where function $G(x,t;L,t_0)$ is described by the boundary-value problem

$$\left(\frac{\partial^2}{\partial x^2}-\frac{\partial^2}{c_0^2\partial t^2}\right)G(x,t;L,t_0)=\frac{\partial^2}{c_0^2\partial t^2}\left[\varepsilon\left(x,t;u(x,t)\right)G(x,t;L,t_0)\right],$$
$$\left(\frac{\partial}{\partial x}+\frac{\partial}{c_0\partial t}\right)G(x,t;L,t_0)\Big|_{x=L}=\frac{2}{c_0}\delta(t-t),$$
$$\left(\frac{\partial}{\partial x}-\frac{\partial}{c_0\partial t}\right)G(x,t;L,t_0)\Big|_{x=L_0}=0, \tag{C.257}$$

or the equivalent integral equation

$$G(x,t;L,t_0)=g_0(x-L;t-t_0)$$
$$-\frac{1}{2c_0}\int\limits_{L_0}^{L}dx_1\int\limits_{-\infty}^{\infty}dt_1 g_0(x-x_1;t-t_1)$$
$$\times\frac{\partial^2}{\partial t_1^2}\left[\varepsilon\left(x_1,t_1;u(x_1,t_1;L)\right)G(x_1,t_1;L,t_0)\right]. \tag{C.258}$$

Now, we can easily obtain that the solution to Eq. (C.258), i.e., function $G(x,t;L,t_0)$ satisfies the variational differential equality

$$\frac{\partial G(x,t;L,t_0)}{\partial L}=\hat{A}(t_0)G(x,t;L,t_0)+\int\limits_{-\infty}^{\infty}dt' f(t')\hat{A}(t')\frac{\delta G(x,t;L,t_0)}{\delta f(t')}, \tag{C.259}$$

which, after supplementing it with the initial condition

$$G(x,t;L,t_0)|_{L=x}=G_x(t;t_0), \tag{C.260}$$

can be considered as the functional equation.

In Eq. (C.259) operator $\hat{A}(t_0)$ acts on arbitrary function $F(t_0)$ according to the relationship

$$\hat{A}(t_0)F(t_0)=\frac{\partial}{c_0\partial t_0}F(t_0)$$
$$-\frac{1}{2c_0}\int\limits_{-\infty}^{\infty}dt_1F(t_1)\frac{\partial^2}{\partial t_1^2}\left[\varepsilon\left(L,t_1;\int dt_2 G_L(t_1;t_2)f(t_2)\right)G_L(t_1;t_0)\right],$$

where function

$$G_L(t;t_0)=G(L,t;L,t_0)$$

describes the wavefield at boundary $x=L$, i.e., the backscattered wave.

With allowance for boundary conditions (C.97), function $G_L(t;t_0)$ satisfies the equation

$$\left(\frac{\partial}{\partial L}+\frac{\partial}{c_0\partial t}\right)G_L(t;t_0)=\frac{2}{c_0}\delta(t-t_0)$$
$$+\hat{A}(t_0)G_L(t;t_0)+\int\limits_{-\infty}^{\infty}dt' f(t')\hat{A}(t')\frac{\delta G(x,t;L,t_0)}{\delta f(t')}, \tag{C.261}$$

with the initial condition

$$G_{L_0}(t;t_0)=\theta(t-t_0). \tag{C.262}$$

Equations (C.259)–(C.262) are the equations of the imbedding method in the context of the nonlinear problem under consideration. Note that, as in the linear problem [105], these equations can be used for analyzing the problem on propagation of the incident wave front.

If we omit terms containing $\varepsilon(x, t; u)$ in Eq. (C.261), then the solution of the simplified equation will assume the form

$$G_L(t; t_0) = g_0(0, t - t_0)$$

that corresponds to the neglect of the backscattering. Substituting this solution in Eq. (C.259), we can obtain the integral equation in function $G(x, t; L, t_0)$,

$$\begin{aligned} G(x, t; L, t_0) &= g_0(x - L; t - t_0) \\ &- \frac{1}{2c_0} \int\limits_x^L dx_1 \int\limits_{-\infty}^{\infty} dt_1 g_0(x - x_1; t - t_1) \\ &\times \frac{\partial^2}{\partial t_1^2} \left[\varepsilon\left(x_1, t_1; u(x_1, t_1; L)\right) G(x_1, t_1; L, t_0)\right]. \end{aligned} \tag{C.263}$$

For the wavefield, we have in this case the equation

$$\begin{aligned} u(x, t) &= u_0(x, t) \\ &- \frac{1}{2c_0} \int\limits_x^L dx_1 \int\limits_{-\infty}^{\infty} dt_1 g_0(x - x_1; t - t_1) \frac{\partial^2}{\partial t_1^2} \left[\varepsilon\left(x_1, t_1; u(x_1, t_1)\right) u(x_1, t_1)\right], \end{aligned} \tag{C.264}$$

which can be rewritten in the equivalent form

$$\begin{aligned} &\left(\frac{\partial}{\partial x} - \frac{\partial}{c_0 \partial t}\right) u(x, t) \\ &= -\frac{1}{2c_0} \int\limits_{-\infty}^{\infty} dt_1 g_0(0; t - t_1) \frac{\partial^2}{\partial t_1^2} \left[\varepsilon\left(x, t_1; u(x, t_1)\right) u(x, t_1)\right] \\ &= -\frac{1}{2c_0} \frac{\partial}{\partial t} \left[\varepsilon\left(x, t; u(x, t)\right) u(x, t)\right]. \end{aligned} \tag{C.265}$$

Bibliography

1. Abramovich, B. S. and A. I. Dyatlov, 1975, *Theory of wave propagation in a one-dimensional randomly inhomogeneous absorptive medium in the diffusion approximation*, Izv. Vuzov., Radiofizika **18**(8), 1222–1224 [Radiophys. & Quantum Electron. **18**(8), 909–910, 1975].

2. Abramowitz, M. and I. A. Stegun, 1970, *Handbook of Mathematical Functions*, Dover, New York.

3. Adler, R. J., 1991, *The Geometry of Random Fields*, Wiley, New York.

4. Ambartsumyan, V. A., 1943, *Diffuse reflection of light by a foggy medium*, Comptes Rendus (Doklady) de l'USSR, **38**(8), 229–232.

5. Ambartsumyan, V. A., 1944, *On the problem of diffuse reflection of light*, Journal of Physics USSR, **8**(1), 65.

6. Ambartsumyan, V. A., 1989, *On the principle of invariance and its some applications*, in: *Principle of Invariance and its Applications*, 9–18, eds. M. A. Mnatsakanyan, H. V. Pickichyan, Armenian SSR Acad. of Sciences, Yerevan.

7. Anderson, P. W., 1958, *Absence of diffusion in certain random lattices*, Phys. Rev. **109**, 1492–1505.

8. Anishchenko, V. S., A. B. Neiman, F. Moss and L. Schimansky-Geier, 1999, *Stochastic resonance: noise enhanced order*, Uspekhi Fiz. Nauk **169**(1), 7–38 [Physics - Uspekhi **42**(1), 1999].

9. Asch, M., G. Papanicolaou, M. Postel, P. Sheng and B. White, 1990, *Frequency content of randomly scattered signals*, Part I, Wave Motion, **12**(4), 429–450.

10. Asch, M., W. Kohler, G. Papanicolaou, M. Postel and B. White, 1991, *Frequency content of randomly scattered signals*, SIAM Review, **33**(4), 519–625.

11. Asch, M., W. Kohler, G. Papanicolaou, M. Postel and B. White, 1996, *Statistical invertion from reflections of spherical waves by a randomly layered medium*, Waves in Random Media **6**(4), 293–334.

12. Avellaneda, M. and A. J. Majda, 1990, *Mathematical models with exact renormalization for turbulent transport*, Comm. Math. Phys., **131**, 381–429.

13. Babkin, G. I. and V. I. Klyatskin, 1980, *Theory of wave propagation in nonlinear inhomogeneous media*, Zh. Éxper. Teor. Fiz. **79**(3), 817–827 [Sov. Phys. JETP **52**(3), 416–420, 1980].

14. Babkin, G. I. and V. I. Klyatskin, 1982a, *Invariant imbedding method for wave problems*, Wave Motion **4**(2), 195–207.

15. Babkin, G. I. and V. I. Klyatskin, 1982b, *Statistical theory of radiative transfer in layered media*, Wave Motion **4**(3), 327–339.

16. Babkin, G. I., V. I. Klyatskin and L. Ya. Lyubavin, 1980, *Theory of invariant imbedding and waves in statistically nonuniform media*, Dokl. Akad. Nauk **250**(5), 1112–1115 [Sov. Phys. Doklady **25**(2), 98–100, 1980].

17. Babkin, G. I., V. I. Klyatskin and L. Ya. Lyubavin, 1982, *Boundary-value problems for wave equation*, Akust. Zh. **28**(1), 1–7 [Sov. Phys.- Acoust. **28**(1), 1–4, 1982].

18. Babkin, G. I., V. I. Klyatskin and B. M. Shevtsov, 1980, *The geometrical approximation in statistical wave theory*, Akust. Zh. **26**(4), 488–493 [Sov. Phys.–Acoust. **26**(4), 269–272, 1980].

19. Babkin, G. I., V. I. Klyatskin, V. F. Kozlov and E. V. Yaroshchuk, 1981, *Wave intensity fluctuations in a one-dimensional randomly inhomogeneous medium. V. Numerical integration of radiative transfer equations*, Izv. Vuzov., Radiofizika **24**(8), 952–959 [Radiophys. & Quantum Electron. **24**(8), 649–655, 1981].

20. Balkovsky, E., G. Falkovich and A. Fouxon, 2000, *Clustering of inertial particles in turbulent flows,* // http://arxiv.org/abs/chao-dyn/9912027.

21. Balkovsky, E., G. Falkovich and A. Fouxon, 2001, *Intermittent Distribution of Inertial Particles in Turbulent Flows*, Phys. Rev. Letters **86**(13), 2790–2793.

22. Barabanenkov, Yu. N. and D. I. Kryukov, 1992, *Functional Fokker-Planck formalism for wave propagation in random media under delocalized and weak localized regimes*, Waves in Random Media **2**(1), 1–6.

23. Barabanenkov, Yu. N., Yu. A. Kravtsov, V. D. Ozrin and A. I. Saichev, 1991, *Enhanced backscattering in optics*, Progress in Optics **XXIX**, ed. E. Wolf, 67–197, North-Holland, Amsterdam.

24. Batchelor, G. K., 1959, *Small-scale variation of convected quantities like temperature in turbulent fluid. 1. General discussion and the case of small conductivity*, J. Fluid Mech. **5**(1), 113–133.

25. Batchelor, G. K., I. D. Howells and A. A. Townsend, 1959, *Small-scale variation of convected quantities like temperature in turbulent fluid. 2. The case of large conductivity*, J. Fluid Mech. **5**(1), 134–139.

26. Bellman, R. and G. M. Wing, 1992, *An Introduction to Invariant Imbedding*, Classics in Applied Mathematics **8**, SIAM, Philadelphia.

27. Beran, M. J. and J. Oz-Vogt, 1994, *Imaging through turbulence in the atmosphere*, Progress in Optics **XXXIII**, 321–388, ed. E. Wolf, North-Holland, Amsterdam.

28. Born, M. and E. Wolf, 1964, *Principles of Optics*, MacMillan, New York.

29. Bourret R. C., U. Frish and A. Pouquet, 1973, *Brownian motion of harmonical oscillator with stochastic frequency*, Physica A **65** (2), 303–320.

30. Bruckstein A. M., B. C. Levy and T. Kailath, 1985, *Differential methods in inverse scattering*, SIAM. J. Appl. Math. **45**(2), 312–335.

31. Bugrov, A. G., 1988, *Theory of sound propagation in quasistratified ocean*, Dokl. Akad. Nauk SSSR **301**(6), 1472–1474 [Transactions (doclady) USSR Acad. Sciences, Earth Science Section **301**(4), 284–286, 1989].

32. Bugrov, A. G. and V. I. Klyatskin, 1989, *An embedding method and the solution of inverse problems in layered media*, Izv. Vuzov., Radiofizika **32**(3), 321–330 [Radiophys. & Quantum Electron. **32**(3), 246–254, 1989].

33. Bugrov, A. G., V. I. Klyatskin and B. M. Shevtsov, 1984, *Theory of radiowave propagation above the ocean*, Dokl. Akad. Nauk SSSR **275**(6), 1372–1376 [Sov. Phys. Doklady **29**(4), 308–310, 1984].

34. Bugrov, A. G., V. I. Klyatskin and B. M. Shevtsov, 1985, *Theory of propagation of short radio waves in a spherically layered atmosphere*, Radiotekh. i Electr. **30**(4), 684–690 [Sov. J. Commun. Techn. & Electr. **30**(7), 60–67, 1985].

35. Bunkin, F. V. and K. S. Gochelashvili, 1968, *Bursts of a random scalar field*, Izv. Vuzov., Radiofizika **11**(12), 1864–1870 [Radiophys. & Quantum Electron. **11**(12), 1059–1063, 1968].

36. Bunkin, F. V. and K. S. Gochelashvili, 1969, *Random spatial overshoots of the field-strength when a wave travels through a turbulent*, Izv. Vuzov., Radiofizika **12**(6), 875–881 [Radiophys. & Quantum Electron. **12**(6), 697–702, 1969].

37. Burridge, R., G. Papanicolaou and B. White, 1988, *One-dimensional wave propagation in a highly discontinues medium*, Wave Motion **10**(1), 19–44.

38. Burridge, R., G. Papanicolaou, P. Sheng and B. White, 1989, *Probing a random medium with a pulse*, SIAM J. Appl. Math. **49**, 582–607.

39. Casti, J. and R. Kalaba, 1973, *Imbedding Methods in Applied Mathematics*, Addison-Wesley, Reading, MA.

40. Charnotskii, M. I., J. Gozani, V. I. Tatarskii and V. U. Zavorotny, 1993, *Wave propagation theories in random media based on the path-integral approach*, Progress in Optics **XXXII**, 203–266, ed. E. Wolf, North-Holland, Amsterdam.

41. Chen, H., S. Chen and R. H. Kraichnan, 1989, *Probability distribution of a stochastically advected scalar field*, Phys. Rev. Letters **63**(24), 2657–2660.

42. Chernov, L. A., 1975, *Wolny v cluchainykhi sredakh* (*Waves in Random Media)*, Nauka, Moscow (in Russian).

43. Churnside, J. H. and S. F. Clifford, 1987, *Log-normal Rician probability - density function of optical scintillations in the turbulent atmosphere*, JOSA **A4**(10), 1923–1930.

44. Corones, J. and Z. Sun, 1993, *Simultaneous reconstruction of material and transient source parameters using the invariant imbedding method*, J. Math. Phys. **34**(5), 1824–1825.

45. Corones, J. P., M. E. Davison and R. J. Krueger, 1983, *Direct and inverse scattering in the time domain via invariant imbedding equations*, J. Acoust. Soc. Am. **74**(5), 1535–1541.

46. Corones, J. P., G. Kristensson, P. Nelson and D. L. Seth, 1992, eds. *Invariant Imbedding and Inverse Problems*, Proceedings in Applied Mathematics **63**, SIAM, Philadelphia.

47. Cramer, H. and M. R. Leadbetter, 1967, *Stationary and Related Stochastic Processes*, Wiley, New York.

48. Cressman, J. R. and W. I. Goldburg, 2003, *Compressible flow: Turbulence at the surface*, J. Stat. Phys., **113**(5/6), 875–883.

49. Cressman, J. R., W. I. Goldburg and J. Schumacher, 2004, *Dispersion of tracer particles in a compressible flow*, Europhys. Letters, **66**(2), 219–225.

50. Cristiani, A. and A. Vulpiani, 1993, *On the effect of noise and drift on diffusion in fluids*, J. Stat. Phys. **70**, 197–211.

51. Csanady, G. T., 1980, *Turbulent Diffusion in the Environment*, Reidel Publ. Co., Dordrecht.

52. Dagan, G., 1987, *Theory of solute transport by groundwater*, Annu. Rev. Fluid Mech., **19**, 183–215.

53. Dahm, W. J .A. and K. A. Buch, 1989, *Lognormality of the scalar dissipation pdf in turbulent flows*, Phys. Fluids **A1**(7): 1290.

54. Dashen, R., 1979, *Path integrals for waves in random media*, J. Math. Phys. **20**, 894–920.

55. Dashen R., 1984, *Distribution of intensity in a multiply scattering medium*, Opt. Lett. **9**, 110–112.

56. Ditkin, V. A. and A. P. Prudnikov, 1974, *Integral'nye preobrazovaniya i operazionnoe ischislenie* (*Integral Transforms and Operational Calculus*), Nauka, Moscow (in Russian) [*Integral Transforms and Operation Calculus*, Pergamon Press, Oxford, New York, 1965].

57. Dolzhanskii, F. V., M. V. Kurganskiy and Yu. L. Chernous'ko, 1979, *Laboratory and theoretical study of barotropic Rossby waves in a rotating annular chanel*, Izvestiya AN USSR, Fiz. Atm. i okeana **15**(6), 597–608 [Izvestiya, Atmospheric and Oceanic Physics **15**(6), 408–414, 1979].

58. Dolzhanskii, F. V., V. I. Klyatskin, A. M. Obukhov and M. A. Chusov, 1974, *Nonlinear hydrodynamic type systems*, Nauka, Moscow (in Russian).

59. Donsker, M. D., 1964, *On functional space integrals*, in: Proc. Conf. Theory and Appl. of Analysis in Funct. Space, **17**, The M.I.I. Press.

60. Elachi, C., 1976, *Waves in active and passive periodic structures: a review*, Proc. IEEE **64**, 1666.

61. Er-El, J. and R. L. Peskin, 1981, *Relative diffusion of constant-level balloons in the Southern Hemisphere*, J. Atmos. Sci. **38**(10), 2264–2274.

62. Fannjiang, A., L. Ryzhin and G. Papanicolaou, 1998, *Evolution of trajectory correlations in steady random flows*, Proc. Symposia in Appl. Math. **54**, 105–130.

63. Feller, W., 1968, *An Introduction to Probability Theory and Its Applications*, vol. **I**, 3rd edition; vol. **II**, 2nd edition, Wiley, New York.

64. Flatté, S. M., G.-Y. Wang and J. Martin, 1993, *Irradiance variance of optical waves through atmospheric turbulence by numerical simulation and comparison with experiment*, JOSA **A10**, 2363–2370.

65. Flatté, S. M., C. Bracher and G.Y. Wang, 1994, *Probability - density functions of irradiance for waves in atmospheric turbulence calculated by numerical simulation*, JOSA **A11**, 2080–2092.

66. Fofonoff, N., 1954, *Steady flow in a frictionless homogeneous ocean*, J. Mar. Res. **13**, 254–262.

67. Freilikher, V. D. and S. A. Gredescul, 1990, *Randomly layered media: fluctuating waveguide*, J. Opt. Soc. Am. **A.7**(5), 868–874.

68. Freilikher, V. D. and S. A. Gredescul, 1992, *Localization of waves in media with one-dimensional disorder*, Progress in Optics **XXX**, 137–203, ed. E. Wolf, North-Holland, Amsterdam.

69. Furutsu, K., 1963, *On the statistical theory of electromagnetic waves in a fluctuating medium*, J. Res. NBS, **D-67**, 303.

70. Furutsu, K., 1993, *Random Media and Boundaries. Unified Theory, Two-Scale Method, and Applications*, Springer-Verlag, New York.

71. Gao, F., 1991, *An analytical solution for the scalar probability density function in homogeneous turbulence*, Phys. Fluids **A3**(4), 511–513.

72. Gardiner, C. W., 1985, *Handbook of Stochastic Methods for Physics, Chemistry and the Natural Sciences*, Springer - Verlag, Berlin.

73. Gazaryan, Yu. L., 1969, *The one-dimensional problem of propagation of waves in a medium with random inhomogeneties*, Zh. Éxper. Teor. Fiz. **56**(6), 1856–1871 [Sov. Phys. JETP **29**(6), 996–1003, 1969].

74. Gel'fand, I. M. and G. E.Shilov, 1959, *Obobshchennye funkzii (Generalized Functions)*, vol. **1**, *Properties and Operation*, Gos. izd-vo fiz.-mat. lit-ry, Moscow (in Russian) [1964, Academic Press, New York].

75. Gledzer, E. B., F. V. Dolzhanskii and A. M. Obukhov, 1981, *Sistemy gidrodinamicheskogo tipa i ikh primeneniya (Systems of hydrodynamical type and their applications, Nauka)*, Moscow, (in Russian).

76. Gochelashvili, K. S. and V. I. Shishov , 1981, *Volny v sluchaino neodnorodnoi srede (Waves in Randomly Inhomogeneous Media)*, VINITI, Sov. Academy of Sc., Moscow (in Russian).

77. Goland, V. I., 1987, *The imbedding method in the problem of determining dispersion curves for internal waves in a stratified ocean*, Morsk. gidrofiz. zh. No 1, 45–51.

78. Goland, V. I., 1988, *Statistical characteristics of normal modes of a sound field in a randomly inhomogeneous ocean*, Akust. Zh. **34**(6), 1020–1022 [Sov. Phys.- Acoust. **34**(6), 585–586, 1988].

79. Goland, V. I. and V. I. Klyatskin, 1988, *Statistics of the eigenvalue and eigenfunctions in a one-dimensional normal mode boundary-value problem*, Akust. Zh. **34**(5), 828–833 [Sov. Phys.–Acoust. **34**(5), 480–483, 1988].

80. Goland, V. I. and V. I. Klyatskin, 1989, *Asymptotic method for analyzing the stochastic Sturm–Liouville problem*, Akust. Zh. **35**(5), 942–944 [Sov. Phys.–Acoust. **35**(5), 548–549, 1989].

81. Goland, V. I. and K. V. Koshel', 1990, *The method of spectral parameter evolution in problems of over-the-horizon propagation of ultrashort radio waves*, Radiotekh. i Electr. **35**(9), 1805–1809 [Sov. J. Commun. Techn. & Electr. **36**(1), 28–33, 1991].

82. Goland, V. I., V. I. Klyatskin and I. O. Yaroschuk, 1991, *Some aspects of wave propagation theory in layer random media, in: Mathematics of Random Media*, eds. W. Kohler, B. S. White, Lectures in Appl. Math. 27, 477–486, AMS, Providence RI.

83. Golberg, M. A., 1975, *Invariant imbedding and Riccati transformations*, Appl. Math. and Com. **1**(1), 1–24.

84. Gollub, J., J. Crarke, M. Gharib, B. Lane and O. Mesquita, 1991, *Fluctuations and transport in a stirred fluid with a mean gradient*, Phys. Rev. Lett. **67**(25), 3507–3510.

85. Gracheva, M. E. and A. S. Gurvich, 1965, *On strong intensity fluctuation of light propagating in atmospheric layer near the Earth,* Izv. Vuzov., Radiofizika **8**(4), 717–724 [Radiophys. & Quantum Electron. **8**(4), 1965].

86. Gracheva, M. E., A. S. Gurvich, S. O. Lomadze, V. V. Pokasov and A. S. Khrupin, 1974, *Probability distribution of "strong" fluctuations of light intensity in the atmosphere,* Izv. Vuzov., Radiofizika **17**(1), 105–112 [Radiophys. & Quantum Electron. **17**(1), 83–87, 1974].

87. Gredescul, S. A. and V. D. Freilikher, 1988, *Waveguiding properties of randomly stratified media,* Izv. Vuzov., Radiofizika **31**(10), 1210–1217 [Radiophys. & Quantum Electron. **31**(10), 861–867, 1988].

88. Gredescul, S. A. and V. D. Freilikher, 1990, *Localization and wave propagation in randomly layered media,* Uspekhi Fiz. Nauk **160**(2), 239–262 [Soviet Physics Usp. **33**(1), 134–136, 1990].

89. Griffa, A., 1996, *Applications of stochastic particle models to oceanographic problems,* in: Stochastic Modeling in Oceanography, Progress in Probability, **39**, 114–140, eds. R. J. Adler, P. Müller, B. L. Razovsky, Birkhäuser, Boston.

90. Gryanik, N. V. and V. I. Klyatskin, 1997a, *Contribution to the statistical theory of wave localization in a two-layer medium,* Zh. Éxper. Teor. Fiz. **111**(6), 2030–2043 [JETP, **84**(6), 1106–1113, 1997].

91. Gryanik, N. V. and V. I. Klyatskin, 1997b, *Localization of Rossby waves under the influence of random cylindrical topography (two-layer model)*, Izvestiya AN, Fiz. Atm. i okeana **33**(6), 723–732 [Izvestiya, Atmospheric and Oceanic Physics **33**, 1997].

92. Gulin, O. E., 1985, *Theory of acoustic noise in deep layered ocean*, Akust. Zh. **31**(4), 524–527 [Sov. Phys.- Acoust. **31**(4), 311–313, 1985].

93. Gulin, O. E., 1987a, *Numerical simulation of low-frequency acoustic noise in a layered ocean*, Akust. Zh. **33**(1), 113–116 [Sov. Phys.- Acoust. **33**(1), 68–70, 1987].

94. Gulin, O. E., 1987b, *Spectra of low-frequency noise in a plane-layered ocean with impedance properties of the bottom*, Akust. Zh. **33**(4), 618–623 [Sov. Phys.–Acoust. **33**(4), 361–364, 1987].

95. Gulin, O. E. and V. I. Klyatskin, 1986a, *On the resonant structure of the spectral components of the acoustic field in the ocean when acted upon by atmospheric pressure*, Izvestiya AN SSSR, Fiz. Atm. i okeana **22**(3), 282–291 [Izvestiya, Atmospheric and Oceanic Physics **22**(3), 213–218, 1986].

96. Gulin, O. E. and V. I. Klyatskin, 1986b, *On the theory of acoustical noise in randomly inhomogeneous ocean*, Dokl. Akad. Nauk SSSR **288**(1), 226–228 [Transactions (doclady) USSR Acad. Sciences, Earth Science Section].

97. Gulin, O. E. and V. I. Klyatskin, 1989, *Atmospheric excitation of low-frequency acoustic noise in a stratified ocean for various stratification models*, in: Acoustics of the Ocean, ed. L. M. Brekhovskikh, 133–140, Nauka, Moscow (in Russian).

98. Gulin, O. E. and V. I. Klyatskin, 1993, *Generation of low-frequency acoustic noise in a stratified ocean*, in: Book Natural Physical Sources of Underwater Acoustical Noise, ed. B. Kerman, 247–253, Kluver Academic Publishers, Netherlands.

99. Gulin, O. E. and V. V. Temchenko, 1990, *Some analytical solutions of the one-dimensional problem of scattering of pulses by a layer of an inhomogeneous medium in the time domain*, Akust. Zh. **36**(4), 644–648 [Sov. Phys.- Acoust. **36**(4), 364–366, 1990].

100. Gulin, O. E. and V. V. Temchenko, 1992, *Scattering of pulses by periodically inhomogeneous media: numerical simulation results*, Akust. Zh. **38**(3), 450–455 [Sov. Phys.- Acoust. **38**(3), 247–249, 1992].

101. Gurbatov, S., A. Malakhov and A. Saichev, 1990, *Nelineinye Sluchainye Volny v Sredach bez Dispersii, Nauka, Moscow* [English Translation *Nonlinear Random waves and Turbulence in Nondispersive Media: Waves, Rays and Particles*, Manchester University Press, Manchester].

102. Gurvich, A. S. and A.M. Yaglom, 1967, *Breakdown of eddies and probability distributions for small-scale turbulence*, Phys. Fluids Suppl. **10**(9): 559.

103. Gurvich, A. S., M. A. Kallistratova and F. E. Martvel', 1977, *An investigation of strong fluctuations of light intensity in a turbulent medium at a small wave parameter*, Izv.Vuzov., Radiofizika **20**(7), 1020–1031 [Radiophys. & Quantum Electron. **20**(7), 705–714, 1977].

104. Gurvich, A. S., A. I. Kon, V. L. Mironov and all, 1976, *Rasprostranenie lazernogo izlyucheniya v turbulentnoi atmosfere* (*Propagation of Laser Radiation in Turbulent Atmosphere*), Nauka, Moscow (in Russian).

105. Guzev, M. A., 1991, *Wavefront propagation theory for a nonlinear layered medium*, Zh. Prikl. Mat. Tekh. Fiz., No.4(188), 63–67 [J. Appl. Mech. and Techn. Phys. **32**(4), 523–527, 1991].

106. Guzev, M. A. and V. I. Klyatskin, 1991a, *Plane waves in a layered weakly dissipative randomly inhomogeneous medium*, Waves in Random Media **1**(1), 7–19.

107. Guzev, M. A. and V. I. Klyatskin, 1991b, *Approximation of the parabolic equation and the wavefield of a point source in a layered random medium*, Waves in Random Media **1**(4), 275–286.

108. Guzev, M. A. and V. I. Klyatskin, 1991c, *Theory of wave propagation in one-dimensional random media containing two-scale inhomogeneties*, Izv.Vuzov., Radiofizika **34**(3), 274–279 [Radiophys. & Quantum Electron. **34**(3), 230–234, 1991].

109. Guzev, M. A. and V. I. Klyatskin, 1993, *Influence of boundary conditions on statistical characteristics of wavefield in layered randomly inhomogeneous medium*, Waves in Random Media **3**(4), 307–315.

110. Guzev, M. A. and G. V. Popov, 1995, *Statistical characteristics of a point-source field in a randomly stratified medium in the absence of regular refraction*, Izv. Vuzov., Radiofizika **38**(10), 1023–1032 [Radiophys. & Quantum Electron. **38**(10), 666–672, 1995].

111. Guzev, M. A. and G. V. Popov, 1998, *Investigation of statistical characteristics of waves in layered media with regular and random inhomogeneities based on the invariant imbedding method*, Waves in Random Media **8**(2), 193–201.

112. Guzev, M. A., V. I. Klyatskin and G. V. Popov, 1992, *Phase fluctuations and localization length in layered randomly inhomogeneous media*, Waves in Random Media **2**(2), 117–123.

113. Holton, Y. R., 1971, *An experimental study of forced barotropic Rossby waves*, Geophys. Fluid Dyn. **2**(4), 323–342.

114. Holzer, M. and A. Pumir, 1993, *Simple models on non-Gaussian statistics for a turbulently advected passive scalar*, Phys. Rev. **E47**(1), 202–219.

115. Holzer, M. and E. Siggia, 1994, *Turbulent mixing of a passive scalar*, Phys. Fluids **6**(5), 1820–1837.

116. Hopf, E., 1952, *Statistical hydromechanics and functional calculus*, J. Ration. Mech. Anal. **1**, 87–123.

117. Hopf, E., 1957, *On the application of functional calculus to the statistical theory of turbulence*, in: Proc. Symp. Appl. Math **7**, 41–50, Am. Math. Society.

118. Hopf, E., 1962, *Remarks on the functional-analytic approach to turbulence*, in: Proc. Symp. Appl. Math **13**, 157–163, Am. Math. Society.

119. Hopf, E. and E. W. Titt, 1953, *On certain special solution of the equation of statistical hydrodynamics*, J. Rat. Mech.Anal. **2**, 587.

120. Ishimaru, A., 1978, *Wave Propagation and Scattering in Random Media*, vols. **1**, **2**, Academic Press, New York.

121. Isichenko, M. B., 1992, *Percolation, statistical topography, and transport in random media*, Rev. Modern Phys. **64**(4), 961–1043.

122. Jordan, K. E., G. Papanicolaou and R. Spigler, 1986, *On the numerical solution of a nonlinear stochastic Helmholtz equation with a multigrid preprocessor*, Appl. Math. and Comp. **19**, 145–157.

123. Kagiwada, H. H. and R. Kalaba, 1974, *Integral Equations Via Imbedding Methods*, 1974, Addison-Wesley, Reading, MA.

124. Keller, J. B. and G. Veronis, 1969, *Rossby waves in presence of random currents*, J. Geoph. Res. **74**(8), 1941–1951.

125. Kerstein, A. R. and P. A. McMurtry, 1994, *Mean-field theories of random advection*, Phys. Rev. **E49**(1), 474–482.

126. Kimura, Y. and R. H. Kraichnan, 1993, *Statistics of an advected passive scalar*, Phys. of Fluids **A5**(9), 2264–2277.

127. Klimontovich, Yu., L., 1999, *What are stochastic filtration and stochastic resonance?*, Uspekhi Fiz. Nauk **169**(1), 39–47 [Physics–Uspekhi **42**(1), 1999].

128. Klyatskin, V. I., 1966, *Sound radiation from vortex system*, Izvestiya AN SSSR, Mekhanika Zhidkosti i Gaza No **6**, 87–92.

129. Klyatskin, V. I., 1969a, *On Statistical theory of two-dimensional turbulence*, Prikladnaya Mathematica i Mechanika **33**(5), 889–891 [J. Appl. Math. and Mech. **33**(5), 864–866, 1969].

130. Klyatskin, V. I., 1969b, *On the validity of the Markovian random process approximation in problems related to the propagation of light in a medium with random refractive index inhomogeneities*, Zh. Éxper. Teor. Fiz. **57**(9), 952–958 [Sov. Phys. JETP **30**(9), 1969].

131. Klyatskin, V. I., 1971, *Space-time description of stationary and homogeneous turbulence*, Izvestiya AN SSSR, Mekhanika Zhidkosti i Gaza No **4**, 120–127 [Fluid Dynamics 6(4), 655–661, 1971].

132. Klyatskin, V. I., 1975, *Statisicheskoe Opisanie Dinamicheskich Sistem s Flukyuiruyushchimi Parametrami* (*Statistical Description of Dynamical Systems with Fluctuating Parameters*), Nauka, Moscow (in Russian).

133. Klyatskin, V. I., 1977, *Dynamic systems with parameter fluctuations as telegraphic-process type*, Izv. Vuzov., Radiofizika **20**(4), 562–575 [Radiophys. & Quantum Electron. **20**(4), 382–392, 1977].

134. Klyatskin, V. I., 1980, *Stochasticheskie Uravneniya i Volny v Sluchaino-Neodnorodnoi Srede (Stochastic Equations and Waves in Randomly Inhomogeneous Media)*, Nauka, Moscow (in Russian).

135. Klyatskin, V. I., 1985, *Ondes et Équations Stochastiques dans les milieus Aléatoirement non Homogènes*, Les éditions de Physique, Besançon-Cedex (in French).

136. Klyatskin, V. I., 1986, *Metod Pogruzheniya v Teorii Rasprostraneniya Voln (The Imbedding Method in Wave Propagation Theory)*, Nauka, Moscow (in Russian).

137. Klyatskin, V. I., 1991a, *Plane waves in layered random media. The role of boundary conditions*, in: Applied and Industrial Math., ed. R. Spigler, Mathematics and Its Applications **56**, 291–299, Kluwer Academic Publishers, Dordrecht.

138. Klyatskin, V. I., 1991b, *Statistical theory of radiation transport in stratified random media*, Izvestiya AN SSSR, Fiz. Atm. i okeana **27**(1), 45–66 [Izvestiya, Atmospheric and Oceanic Physics **27**(1), 31–44, 1991].

139. Klyatskin, V. I., 1991c, *A statistical theory of radiative transfer in layered random media*, in: Mathematical and Numerical Aspects of Wave Propagation Phenomena, eds. G. Cohen, L. Halpern, P. Joly, 595–608, SIAM, Philadelphia.

140. Klyatskin, V. I., 1991d, *Approximations by delta-correlated random processes and diffusion approximation in stochastic problems*, in : Mathematics of Random Media, eds. W. Kohler, B. S. White, Lectures in Appl. Math. **27**, 447–476, AMS, Providence RI.

141. Klyatskin, V. I., 1993, *Caustics in random media,* Waves in random media **3**(2), 93–100.

142. Klyatskin, V. I., 1994a, *The imbedding method in statistical boundary-value wave problems*, Progress in Optics **XXXIII**, 1–128, ed. E. Wolf, North-Holland, Amsterdam.

143. Klyatskin, V. I., 1994b, *Statistical description of the diffusion of a passive tracer in a random velocity field*, Uspekhi Fiz. Nauk **164**(5), 531–544 [Physics - Uspekhi **37**(5), 501–513, 1994].

144. Klyatskin, V. I., 1995, *Equilibrium states for quasigeostrophic flows with random topography*, Izvestiya AN, Fiz. Atm. i okeana **31**(6), 749–754 [Izvestiya, Atmospheric and Oceanic Physics **31**(6), 717–722, 1995].

145. Klyatskin, V. I., 1996, *Localization of Rossby waves over a random cylindrical topography of the ocean bottom*, Izvestiya AN, Fiz. Atm. i okeana **32**(6), 824–833, [Atmospheric and Oceanic Physics **32**(6), 757–765, 1996].

146. Klyatskin, V. I., 2000, *Stochastic transport of passive tracer by random flows*, Izvestiya AN, Fiz. Atm. i okeana **36**(2), 177–201, [Atmospheric and Oceanic Physics **36**(2), 757–765, 2000].

147. Klyatskin, V. I., 2001, *Stochasticheskie Uravneniya Glazami Fizika (Osnovnye Idei, Tochnye Resultaty i Asimptoticheskie Pribligeniya) (Stochastic Equations through the Eye of the Physicist (Basic Ideas, Exact Results, and Asymptotic Approximations))*, Moscow, Fizmatlit (in Russian).

148. Klyatskin, V. I., 2002, *Dinamika Stochasticheskih Sistem (Dynamics of Stochastic Systems)*, Moscow, Fizmatlit (in Russian).

149. Klyatskin, V. I., 2003, *Clustering and diffusion of particles and density field of passive tracers in random hydrodynamic flows*, Uspekhi Fiz. Nauk **173**(7), 689–710 [Physics-Uspekhi **46**(7), 667–688, 2003].

150. Klyatskin, V. I., 2004a, *Propagation of electromagnetic waves in randomly inhomogeneous media as a problem of statistical mathematical physics*, Uspekhi Fiz. Nauk **174**(2) 177–195 [Physics–Uspekhi **47**(2), 169–186, 2004].

151. Klyatskin, V. I., 2004b, *Diffusion and Clustering of Sedimenting tracer in Random Hydrodynamic Flows*, Zh. Éxper. Teor. Fiz. **126**(5), 1153–1166 [JETP **99**(5), 1005-1017, 2004].

152. Klyatskin, V. I., 2005, *Statistics and reality in stochastic dynamic systems,* In: *Nonlinear waves* 2004, Nignii Novgorod, IAP RAS.

153. Klyatskin, V. I. and T. Elperin, 2002a, *Clustering of the Low-Inertia Particle Number Density Field in Random Divergence-Free Hydrodynamic Flows*, Zh. Éxper. Teor. Fiz. **122**(2), 327–340 [JETP **95**(2), 328-340, 2002].

154. Klyatskin, V. I. and T. Elperin, 2002b, *Diffusion of Low-Inertia Particles in a Field of Random Forces and the Kramers Problem*, Izvestiya AN, Fiz. Atm. i okeana **38**(6), 817–823 [Atmospheric and Oceanic Physics **38**(6), 725-731, 2002].

155. Klyatskin, V. I. and D. Gurarie, 1996a, *Random topography in geophysical models, in: Stochastic Models in Geosystems*, eds. S. A. Molchanov and W. A. Woyczynski, IMA Volumes in Mathematics and its Applications, **85**, 149–170, Springer-Verlag, New York.

156. Klyatskin, V. I. and D. Gurarie, 1996b, *Equilibrium states for quasigeostrophic flows with random topography*, Physica D **98**, 466–480.

157. Klyatskin, V. I. and D. Gurarie, 1999, *Coherent phenomena in stochastic dynamic systems*, Uspekhi Fiz. Nauk **169**(2), 171–207 [Physics–Uspekhi **42**(2), 165–198, 1999].

158. Klyatskin, V. I. and K. V. Koshel', 1983, *Numerical modeling of wave propagation in periodic media*, Zh. Éxper. Teor. Fiz. **84**(6), 2092–2098 [Sov. Phys. JETP **57**(6), 1217–1220, 1983].

159. Klyatskin, V. I. and K. V. Koshel', 1984, *Numerical simulation of Bragg resonator in nonhomogeneous media*, Izv. Vuzov., Radiofizika **27**(2), 263–265.

160. Klyatskin, V. I. and K. V. Koshel', 1986, *The field of a point source in a layered media*, Dokl. Acad. Nauk SSSR **288**(6), 1478–1481 [Transactions (doclady) USSR Acad. Sciences, Earth Science Section].

161. Klyatskin, V. I. and K. V. Koshel', 2000, *The simplest example of the development of a cluster-structured passive tracer field in random flows*, Uspekhi Fiz. Nauk **170**(7), 771–778 [Physics–Uspekhi **43**(7), 717–723, 2000].

162. Klyatskin, V. I., and L. Ya. Lyubavin, 1983a, *Theory of sound propagation in a layered variable-density ocean*, Akust. Zh. **29**(1), 64–68 [Sov. Phys.- Acoust. **29**(1), 36–39, 1983].

163. Klyatskin, V. I. and L. Ya. Lyubavin, 1983b, *On the boundary-value problems of the theory of internal wave propagation in a stratified ocean*, Dokl. Akad. Nauk SSSR **271**(6), 1496–1498 [Transactions (doclady) USSR Acad. Sciences, Earth Science Section].

164. Klyatskin, V. I. and L. Ya. Lyubavin, 1984, *Theory of the excitation and propagation of acoustic-gravity waves in a layered ocean*, Izvestiya AN SSSR, Fiz. Atm. i okeana **20**(5), 422–430 [Izvestiya, Atmospheric and Oceanic Physics **20**(5), 415–420, 1984].

165. Klyatskin, V. I. and O. G. Nalbandyan, 1997, *Diffusion of passive settling inclusions in an isotropic field of random flows*, Izvestiya AN, Fiz. Atm. i okeana **32**(3), 291–297 [Izvestiya, Atmospheric and Oceanic Physics **32**(3), 263–268, 1997].

166. Klyatskin, V. I. and A. I. Saichev, 1992, *Statistical and dynamical localization of plane waves in randomly layered media*, Uspekhi Fiz. Nauk **162**(3), 161–194 [Soviet Physics Usp. **35**(3), 231–247, 1992].

167. Klyatskin, V. I. and A. I. Saichev, 1997, *Statistical theory of the diffusion of a passive tracer in a random velocity field*, Zh. Éksp. Teor. Fiz. **111**(4), 1297–1313 [JETP **84**(4), 716–724, 1997].

168. Klyatskin, V. I. and V. I. Tatarskii, 1970, *The parabolic equation approximation for propagation of waves in a random medium with random inhomogeneities*, Zh. Éxper. Teor. Fiz. **58**(2), 624–634 [Sov. Phys. JETP **31**(2), 335–339, 1970].

169. Klyatskin, V. I. and W. A. Woyczynski, 1995, *Fluctuations of passive scalar with nonzero mean concentration gradient in random velocity fields*, Zh. Éxper. Teor. Fiz. **108**(10), 1403–1410 [JETP **81**(4), 770–773, 1995].

170. Klyatskin, V. I. and I. G. Yakushkin, 1997, *Statistical theory of the propagation of optical radiation in turbulent media*, Zh. Éxper. Teor. Fiz. **111**(6), 2044–2059 [JETP **84**(6), 1114–1121, 1997].

171. Klyatskin, V. I. and I. G. Yakushkin, 2000, *Stochastic transport in random wave fields*, Zh. Éxper. Teor. Fiz. **118**(4), 849–862 [JETP **91**(4), 736–747, 2000].

172. Klyatskin, V. I. and E. V. Yaroshchuk, 1985, *Numerical solution of the one-dimensional problem of self-interaction of a wave in a layer of nonlinear medium*, Izv. Vuzov., Radiofizika **28**(3), 320–329 [Radiophys. & Quantum Electron. **28**(3), 214–222, 1985].

173. Klyatskin, V. I. and I. O. Yaroshchuk, 1983a, *Fluctuations of wave intensity in a one-dimensional randomly inhomogeneous medium. VI. Effect of a dividing boundary influence*, Izv. Vuzov., Radiofizika **26**(9), 1092–1099 [Radiophys. & Quantum Electron. **26**(9), 802–808, 1983].

174. Klyatskin, V. I. and I. O. Yaroshchuk, 1983b, *Fluctuations of intensity of a wave in randomly inhomogeneous medium. VII. Numerical modeling wave propagation in a*

stochastic media, Izv. Vuzov., Radiofizika **26**(10), 1241–1250 [Radiophys. & Quantum Electron. **26**(10), 900–907, 1983].

175. Klyatskin, V. I., N. V. Gryanik and D. Gurarie, 1998, *Localization of Rossby waves under the influence of random topography (two-layer model)*, Wave Motion **28**(4), 333–352.

176. Klyatskin, V. I., K. V. Koshel' and B. M. Shevtsov, 1994, *Time-pulse propagation and inverse problem solution for layered medium*, in: Atmospheric Propagation and Remote Sensing III, SPIE proc., eds. W. A. Flood, W. B. Miller **2222**.

177. Klyatskin, V. I., K. V. Koshel' and B. M. Shevtsov, 1995a, *Solution of inverse problems for layered media*, Izvestiya AN, Fiz. Atm. i okeana **31**(4), 517–525, [Izvestiya, Atmospheric and Oceanic Physics **31**(4), 494–502, 1996].

178. Klyatskin, V. I., K. V. Koshel' and B. M. Shevtsov, 1995b, *The Embedding method and differential run equations for inverse scattering problem in a layered medium*, Radio Science **30**(6), 1689–1698.

179. Klyatskin, V. I., V. F. Kozlov and E. V. Yaroshchuk, 1982, *On the reflection coefficient in a one-dimensional problem of self-influence of waves*, Zh. Éxper. Teor. Fiz. **82**(2), 386–396 [Sov. Phys. JETP **55**(2), 1982].

180. Klyatskin, V. I., W. A. Woyczynski and D. Gurarie, 1996a, *Short-time correlation approximation for diffusing tracers in random velocity fields: a functional approach*, in: *Stochastic Modeling in Oceanography*, Progress in Probability, **39**, 221–269, eds. R. J. Adler, P. Müller, B. L. Razovsky , Birkhäuser, Boston.

181. Klyatskin, V. I., W. A. Woyczynski and D. Gurarie, 1996b, *Diffusing passive tracers in random incompressible velocity flows: Statistical topography aspects*, J. Stat. Phys. **84**(3/4), 797–836.

182. Knapp, R., G. Papanicolaou and B. White,1989, *Nonlinearity and localization in one dimensional random media*, in: *Nonlinearity and Disorder*, eds. B. Bishop, D. K. Campbell, S. Pnevmaticos, Springer Series in Solid-State Science, Springer - Verlag, Berlin.

183. Kohler, W. and G. Papanicolaou, 1974, *Power statistics for wave propagation in one dimension and comparison with radiative transport theory. II*, J.Math. Phys. **15**(12), 2186–2197.

184. Kohler, W. and G. Papanicolaou, 1976, *Power reflection from a lossy one - dimensional random medium*, SIAM J. Appl. Math. **30**(2), 263–267.

185. Kohler, W., G. Papanicolaou and B. White, 1991, *Reflection of waves generated by a point source over a randomly layered medium*, Wave Motion **13**(1), 53–87.

186. Kohler, W., G. Papanicolaou and B. White, 1996a, *Localization and mode conversion for elastic waves in randomly layered media I*, Wave Motion **18**(1), 1–22.

187. Kohler, W., G. Papanicolaou and B. White, 1996b, *Localization and mode conversion for elastic waves in randomly layered media II*, Wave Motion **18**, 181–201.

188. Kohler, W., G. Papanicolaou, M. Postel and B. White, 1991, *Reflection of pulsed electromagnetic waves from a randomly stratified half space*, J. Optical Soc. of Amer. **A.8**(7), 1109–1125.

189. Koshel', K. V., 1986, *Numerical solution of the problem of the propagation of short radiowaves in tropospheric waveguide*, Radiotekh. i Electr. **31**(12), 2313–2318 [Sov. J. Commun. Techn. & Electr. **32**(3), 168–174, 1987].

190. Koshel', K. V., 1987, *Numerical solution of the tropospherical propagation problem for ultrashort radio waves. An elevated receiver*, Radiotekh. i Electr. **32**(6), 1305–1308 [Sov. J. Commun. Techn. & Electr. **32**(11), 151–153, 1987].

191. Koshel', K. V., 1990a, *On the effect of raised inversion layers on over-the-horizon transmission of short radio waves*, Radiotekh. i Electr. **35**(3), 647–649 [Sov. J. Commun. Techn. & Electr. **35**(9), 123–125, 1990].

192. Koshel', K. V., 1990b, *Numerical analysis of the problem of tropospheric propagation of short radio waves for a finite surface impedance*, Radiotekh. i Electr. **35**(6), 1326–1329 [Sov. J. Commun. Techn. & Electr. **35**(8), 137–140, 1990].

193. Koshel', K. V., 1990c, *The effect radially layered fluctuations of the refractive index on the over-the-horizon propagation of ultrashort radio waves in tropospheric waveguides*, Radiotekh. i Electr. **35**(12), 2502–2507 [Sov. J. Commun. Techn. & Electr. **36**(11), 34–39, 1991].

194. Koshel', K. V., 1990d, *On the effect of horizontal nonuniformities of the atmosphere on over-the-horizon propagation of ultrashort radio waves*, Izvestiya AN SSSR, Fiz. Atm. i okeana **26**(10), 1069–1077 [Izvestiya, Atmospheric and Oceanic Physics **26**(10), 783–786, 1990].

195. Koshel', K. V., 1992a, *Effect of random stratification on over-the-horizon propagation of microwaves in the tropospheric evaporation duct above the ocean*, Izvestiya AN SSSR, Fiz. Atm. i okeana **28**(10–11), 1054–1061 [Izvestiya, Atmospheric and Oceanic Physics **28**(10–11), 788–793, 1992].

196. Koshel', K. V., 1992b, *Application of invariant imbedding method to simulate numerically beyond-the-horizon propagation of SHF over the sea*, J. Electromag. Waves and Applic. **6**(10), 1433–1453.

197. Koshel', K. V., 1993, *Influence of mode interaction on the propagation of SHF in horizontally inhomogeneous tropospheric guides*, Izvestiya AN, Fiz. Atm. i okeana **29**(7), 728–732 [Izvestiya, Atmospheric and Oceanic Physics **29**(7), 1993].

198. Koshel', K. V. and O. V. Aleksandrova, 1999, *Some resullts of numerical modeling of passive scalar diffusion in randon velocity field*, Izvestiya AN, Fiz. Atm. i okeana **35**(5), 638–648 [Izvestiya, Atmospheric and Oceanic Physics **35**(5), 578–588, 1999].

199. Koshel', K. V. and A. A. Shishkarev, 1993a, *Influence of layer and anisotropic fluctuations of the refractive index on the beyond-the-horizon SHF propagation in the troposphere over the sea when there is evaporation duct*, Waves in Random Media **3**(1), 25–38.

200. Koshel', K. V. and A. A. Shishkarev, 1993b, *Effect of anisotropic fluctuations of the index of refraction on the over-the-horizon propagation of ultrashort waves in the troposphere*, Izvestiya AN, Fiz. Atm. i okeana **29**(1), 86–91 [Izvestiya, Atmospheric and Oceanic Physics **29**(1), 78–83, 1993].

201. Kraichnan, R. H., 1967, *Inertial waves in two-dimensional turbulence*, Phys. Fluids **10**, 1417–1423.

202. Kraichnan, R. H., 1975, *Statistical dynamics of two-dimensional flows*, J. Fluid Mech. **67**, 155–175.

203. Kraichnan, R. H. and D. Montgomery, 1980, *Two-dimensional turbulence*, Rep. Prog. Phys. **43**, 547–619.

204. Kravtsov, Yu. A., 1970, *"Geometrical" depolarization of light in a turbulent atmosphere*, Izv. Vuzov., Radiofizika **13**(2), 281–284 [Radiophys. & Quantum Electron. **13**(2), 1970].

205. Kravtsov, Yu. A., 1992, *Propagation of electromagnetic waves through a turbulent atmosphere*, Rep. Prog. Phys. **55**, 39–112.

206. Kravtsov, Yu. A. and A. I. Saichev, 1982, *Effects of double passage of waves in randomly inhomogeneous media*, Uspekhi Fiz. Nauk **137**(7), 501–527 [Sov. Phys. Usp. **25**(7), 494–508, 1982].

207. Kravtsov, Yu. A. and A. I. Saichev, 1985, *Properties of coherent waves reflected in a turbulent medium*, J. Opt. Soc. Am. **A 2**(12), 2100–2105.

208. Kreider, K. L., 1989, *Time-dependent inverse scattering from gradient -type interfaced using an exact solution*, J. Math. Phys. **30**(1), 53–58.

209. Kristensson, G. and R. J. Krueger, 1986a, *Direct and inverse scattering in the time domain for a dissipative wave equation. I. Scattering operators*, J. Math. Phys. **27**(6), 1667–1682.

210. Kristensson, G. and R. J. Krueger, 1986b, *Direct and inverse scattering in the time domain for a dissipative wave equation. II. Simultaneous reconstruction of dissipation and phase velocity profiles*, J. Math. Phys. **27**(6), 1683–1693.

211. Kristensson, G. and R. J. Krueger, 1987, *Direct and inverse scattering in the time domain for a dissipative wave equation. III. Scattering operators in the presence of a phase velocity mismatch*, J. Math. Phys. **28**(2), 360–370.

212. Kristensson, G. and R. J. Krueger, 1989, *Direct and inverse scattering in the time domain for a dissipative wave equation. IV. Use of the phase velocity mismatches to simplify inversions*, Inverse Problems **5**(2), 375–388.

213. Kristensson, G. and S. Rikte, 1993, *Transient wave propagation in reciprocal bi-isotropic media at oblique incidence*, J. Math. Phys. **34**(4), 1339–1359.

214. Krueger, R. J. and R. L. Ochs, Jr., 1989, *A Green's function approach to the determination of internal fields*, Wave Motion **11**(6), 525–543.

215. Kulkarny, V. A. and B. S. White, 1982, *Focusing of rays in a turbulent inhomogeneous medium*, Phys. Fluids **25**(10), 1770–1784.

216. Kuzovlev, Yu. E. and G. N. Bochkov, 1977, *Operator methods of analysing stochastic non-gaussian processes and systems*, Izv. Vuzov., Radiofizika **20**(10), 1505–1515 [Radiophys. & Quantum Electron. **20**(10), 1036–1044, 1977].

217. Landau, L. D. and E. M. Lifshitz, 1986, *Gidrodinamika (Hydrodynamic)*, Nauka, Moscow [*Fluid Mechanics*, Second edition, Pergamon Press, London, 1987].

218. Lamb, H., 1932, *Hydrodinamics*, Sixth Edition, Dover Publications, New York.

219. LeMesurier, B., G. Papanicolaou, C. Sulem and P.L. Sulem, 1987, *The focusing singularity of the nonlinear Schrödinger equation*, in: Directions in Partial Differential Equations, eds. M. G. Crandall, P. H. Rabinowitz, R. E. L. Turner, 159–201, Academic Press.

220. LeMesurier, B., G. Papanicolaou, C. Sulem and P.L. Sulem, 1988a, *Focusing an multifocusing solutions of the nonlinear Schrödinger equation*, Physica **D 31**(5), 78–102.

221. LeMesurier, B., G. Papanicolaou, C. Sulem and P.L. Sulem, 1988b, *Local structure of the self focusing singularity of the nonlinear Schrödinger equation*, Physica **D 32**, 210–226. 573.

222. Lesieur M., 1990, *Turbulence in Fluids*, Kluwer, Boston.

223. Lifshits, I. M., S. A. Gredeskul and L. A. Pastur, 1982, *Vvedenie v Teoriyu Neuporyadochennych Sistem* (*Introduction to the Theory of Disordered Systems*), Nauka, Moscow (in Russian) [*Introduction to the Theory of Disordered Solids*, John Wiley, New York, 1988].

224. Lighthill, J., 1952, *On sound generated aerodynamically. 1. General theory*, Proc. Roy. Soc. **A.211**(1107), 564–587.

225. Lighthill, J., 1954, *On sound generated aerodynamicaly. II. Turbulence as a source of sound*, Proc. Roy. Soc. **A.222**(1148), 1–32.

226. Lighthill, J., 1962, *Sound generated aerodynamicaly*, Proc. Roy. Soc. **A.267**(1329), 147–182.

227. Longuet-Higgins, M. S., 1956, *Statistical properties of a moving wave-form*, Proc. Cambridge Philos. Soc., **52**, 234–245.

228. Longuet-Higgins, M. S., 1957a, *On the velocity of the maxima in a moving wave-form*, Proc. Cambridge Philos. Soc., **53**, 230–233.

229. Longuet-Higgins, M. S., 1957b, *The statistical analysis of a random, moving surface*, Philos. Trans. R. Soc. London, Ser. **A249**(966), 321–387.

230. Longuet-Higgins, M. S., 1957c, *Statistical properties of an isotropic random surface*, Philos. Trans. R. Soc. London, Ser. **A250**(975), 157–174.

231. Longuet-Higgins, M. S., 1962, *The statistical geometry of random surfaces*, in: Hydrodynamic Stability, Proc. 13th Symp. Appl. Math., Am. Math. Soc., 105–142.

232. Lutovinov, V. S. and Chechetkin V. R., 1989, *The Kolmogorov–Obukhov spectrum for the paired correlation function of passive tracers in a turbulent fluid*, Izvestiya AN SSSR, Fiz. atm. i okeana **25**(3), 195–198 [Izvestiya, Atmospheric and Oceanic Physics **25**(3), 195–198, 1989].

233. Majda, A. J., 1993, *The random uniform shear layer: An explicit example of turbulent diffusion with broad tail probability distribution*, Phys. Fluids **A5**(8), 1963–1970.

234. Majda, A. J., 1994, *Random shearing direction models for isotropic turbulent diffusion*, J. Stat. Phys. **75**(5/6), 1153–1165.

235. Malakhov, A. N., 1978, *Kumuliantnyi analiz ne gaussovykh sluchainykh prozessov i ikh preobrazovanii* (*Kumulant analysis of non-Gaussian processes and their transforms*), Sov. Radio, Moscow (in Russian).

236. Malakhov, A. N. and A. I. Saichev, 1979, *Representation of a wave reflected from a randomly inhomogeneous layer in the form of a series satisfying the causality condition*, Izv. Vuzov. Radiofizika **22**(11), 1324–1333 [Radiophys. & Quantum Electron. **22**(11), 922–928, 1979].

237. Manning, R. M., 1993, *Stochastic Electromagnetic Image Propagation And Adaptive Compensation*, McGraw-Hill, Inc., New York.

238. Martin, J. M. and S. M. Flatté, 1988, *Intensity images and statistics from numerical simulation of wave propagation in 3-D random media*, Appl. Opt. **27**(11), 2111–2126.

239. Martin, J. M. and S. M. Flatté, 1990, *Simulation of point-source scintillation through three-dimensional random media*, J. Opt. Soc. Am. **A7**, 838–847.

240. Maxey, M. R., 1987, *The Gravitational Settling of Aerosol Particles in Homogeneous turbulence and random flow field*, J. Fluid Mech. **174**, 441–465.

241. McComb, W., 1990, *The Physics of Fluid Turbulence*, Clarendon Press, Oxford.

242. McLaughlin, D., G. Papanicolaou, P.L. Sulem and C. Sulem, 1986, *Focusing singularity of the nonlinear Schrödinger equation*, Phys. Rev. **A34**(2), 1200–1210.

243. Mesinger, F., 1965, *Behavior of a very large number of constant-volume trajectories*, J. Atmos. Sci. **22**(5), 479–492.

244. Mesinger, F., 1971, *Numerical integration of the primitive equations with a floating set of computation points: experiments with a barotropic global model*, Monthly Weather Review **99**(1), 15–29.

245. Mesinger, F. and Y. Mintz, 1970a, *Numerical simulation of the 1970–1971 Eole experiment*, Numerical simulation of weather and climate, Technical Report No. 4, Dep. Meteorology, Univ. of California, Los Angeles.

246. Mesinger, F. and Y. Mintz, 1970b, *Numerical simulation of the clustering of the constant-volume balloons in the global domain*, Numerical simulation of weather and climate, Technical Report No. 5, Dep. Meteorology, Univ. of California, Los Angeles.

247. Miller, J., P. Weichman and M. Cross, 1992, *Statistical mechanics, Euler's equation and Jupiter's red spot*, Phys. Review **A45**(4), 2328–2359.

248. Milne-Thomson, L. M., 1996, *Theoretical Hydrodynamics*, 5th edition, Dover Publication, Inc., New York.

249. Mirabel', A. P. and A. S. Monin, 1983, *On statistical properties of admixture mixing by two-dimensional turbulence*, Izvestiya AN SSSR, Fiz. Atm. i okeana **19**(9), 902–912 [Izvestiya, Atmospheric and Oceanic Physics **25**(3), 195–198, 1989].

250. Moffatt H. K., 1967, *The interaction of turbulence with strong wind shear*, in: Atmospheric Turbulence and Radio Wave Propagation, eds. A. M. Yaglom, V. I. Tatarskii, 139–154, Nauka, Moscow.

251. Monin, A. S. and A. M. Yaglom, 1965, 1967, *Statisticheskaya Gidromekhanika* (*Statistical Hydromechanics*), v.**1**, v. **2**, Nauka, Moscow (in Russian) [*Statistical Fluid Mechanics*, MIT Press, Cambridge Massachusetts, 1980].

252. Nair, B. and B. S. White, 1991, *High frequency wave propagation in random media - a unified approach*, SIAM J. Appl. Math. **51**(2), 374–411.

253. Nalbandyan, O. G., 1977, *Possible approximations of the gaussian random process in some nonlinear stochastic problems*, Izv. Vuzov., Radiofizika **20**(5), 721–725 [Radiophys. & Quantum Electron. **20**, 1977].

254. Nicolis, G. and I. Prigogin, 1989, *Exploring Complexity, an Introduction*, W. H. Freeman and Company, New York.

255. Novikov, E. A., 1964, *Functionals and the random force method in turbulence theory*, Zh. Éksper. Teor. Phys. **47**(5), 1919–1926 [Sov. Phys. JETP **20**(5), 1290–1294, 1965].

256. Obukhov, A. M., 1953, *Sound and light propagation in a weakly inhomogeneous atmosphere*, Izv. AN SSSR, Geofiz., No 2, 155–165 [Waves in Random Media, **4**, No 1, 9–19, 1994].

257. Obukhov, A. M., 1988, *Turbulentnost' i Dinamika Atmosfery* (*Turbulence and Dynamics of Atmosphere),* Leningrad, Gidrometeoizdat (in Russian).

258. Okubo, A., 1980, *Diffusion and Ecological Problems: Mathematical Models*, Springer-Verlag, New York.

259. Papanicolaou, G., M. Postel, P. Sheng and B. White, 1990, *Frequency content of randomly scattered signals. Part II: Inversion*, Wave Motion **12**(4), 527–549.

260. Pedlosky, J., 1982, *Geophysical Fluid Dynamics*, Springer-Verlag, New York.

261. Piterbarg, L., 1995, *Short-correlation approximation in models of turbulent diffusion in: Stochastic Models in Geosystems*, IMA Volumes in Mathematics and its Applications, **85**, 313, Springer-Verlag, New York.

262. Popov, G. V. and I. O. Yaroshchuk, 1988, *On the oblique plane wave incidence on a stratified stochastic medium*, Izv.Vuzov. Radiofizika **31**(10), 1266–1267.

263. Popov, G. V. and I. O. Yaroshchuk, 1990, *Spectral field components in the radiation from a point source in a randomly inhomogeneous medium*, Izv. Vuzov. Radiofizika **33**(11), 1232–1240 [Radiophys. & Quantum Electron. **33**(11), 902–906, 1990].

264. Pumir, A., 1994, *A numerical study of the mixing of a passive scalar in three dimensions in the presence of a mean gradient*, Phys. Fluids **6**(6), 2118–2132.

265. Pumir, A., B. Shraiman and E. Siggia, 1991, *Exponential tails and random advection*, Phys. Rev. Lett. **66**(23), 2984–2987.

266. Rhines, P. and F. Bretherton, 1973, *Topographic Rossby waves in a rough-bottom ocean,* J. Fluid Mech. **61**(3), 583–608.

267. Reznik, G. M. and T. B. Tsybaneva, 1994, *On the influence of topography and stratification on planetary waves in the ocean (two-layer model)*, Izvestiya AN, Okeanologiya **34**(1), 5–14.

268. Rytov, S. M., Yu. A. Kravtsov and V. I. Tatarskii, 1977–1978, *Vvedenie v Statisticheskuyu Radiofiziku* (*Principles of Statistical Radiophysics)*, Nauka, Moscow (in Russian) [*Principles of Statistical Radiophysics*, v. **1–4**, Springer-Verlag, Berlin 1987–1989].

269. Saichev, A. I., 1980, *Eigenvalue statistics of a one-dimensional randomly inhomogeneous boundary-value problem*, Izv. Vuzov., Radiofizika **23**(2), 183–188 [Radiophys. & Quantum Electron. **23**(2), 132–136, 1980].

270. Saichev, A. I., and M. M. Slavinskii, 1985, *Equations for moment functions of waves propagating in random inhomogeneous media with elongated inhomogeneities*, Izv. Vuzov., Radiofizika **28**(1), 75–83 [Radiophys. & Quantum Electron. **28**(1), 55–61, 1985].

271. Saichev, A. I. and W. A. Woyczynski, 1996, *Probability distributions of passive tracers in randomly moving media,* in: Stochastic Models in Geosystems, eds. S. A. Molchanov and W. A. Woyczynski, IMA Volumes in Mathematics and its Applications, **85**, 359–400, Springer-Verlag, New York.

272. Scott, M. R., L. F. Shampine and G. M. Wing, 1969, *Invariant imbedding and the calculation of eigenvalues for Sturm-Liouville systems*, Computing **4**(1), 10–17.

273. *Sedimentation of Small Particles in a Viscous Fluid*, 1996, Ed. E. M. Tory, Advances in Fluid Mechanics **7**, Computational mechanics Publications Southampton Boston, UK.

274. Segupta, D., L. Piterbarg and G. Reznik, 1992, *Localization of topographic Rossby waves over random relief*, Dynamics of Atm. and Oceans **17**(1), 1–21.

275. Shandarin, S. F. and Ya. B. Zel'dovich, 1989, *Turbulence, intermittency, structures in a self-gravitating medium: the large scale structure of the Universe*, Rev. Modern Phys. **61**, 185–220.

276. Shapiro, V. E. and V. M. Loginov, 1978, *'Formulae of differentiation' and their use for solving stochastic equations*, Physica **91A**, 563–574.

277. Shapiro, V. E. and V. M. Loginov, 1983, *Dinamicheskie systemy pod vozdeistviem sluchainykh vliyanii* (*Dynamical systems under random influences*), Nauka, Novosibirsk (in Russian).

278. Sheng, P. (ed.), 1990, *Scattering and Localization of Classical Waves in Random Media*, World Scien., Singapore.

279. Sheng, P, B. White, Z. Zhang and G. Papanicolaou, 1986a, *Multiple scattering noise in one dimension: universality through localization length scales*, Phys. Rev. Lett. **57**(8), 1000–1003.

280. Sheng, P, B. White, Z. Zhang and G. Papanicolaou, 1986b, *Minimum wave localization length in a one dimensional random medium*, Phys. Rev. **B 34**(7), 4757–4761.

281. Sheng, P., B. White, Z. Zhang and G. Papanicolaou, 1990, *Wave localization and multiple scattering in randomly - layered media*, in: Scattering and Localization of Classical Waves in Random Media, ed. P. Sheng, 563–619, World Scien., Singapore.

282. Shevtsov, B. M., 1981, *Statistical theory of backscattering in randomly nonhomogeneous media*, Izv. Vuzov., Radiofizika **24**(11), 1351–1355 [Radiophys. & Quantum Electron. **24**(11), 913–917, 1981].

283. Shevtsov, B. M., 1982, *Three-dimensional problem of inverse scattering in stratified randomly inhomogeneous media*, Izv. Vuzov., Radiofizika **25**(9), 1032–1040 [Radiophys. & Quantum Electron. **25**(9), 742–749, 1982].

284. Shevtsov, B. M., 1983, *The problem of backscattering in three-dimensional, randomly inhomogeneous media*, Izv. Vuzov., Radiofizika **26**(4), 434–439 [Radiophys. & Quantum Electron. **26**(4), 326–331, 1983].

285. Shevtsov, B. M., 1985, *Statistical characteristics of backscatter field*, Izv. Vuzov., Radiofizika **28**(6), 717–724 [Radiophys. & Quantum Electron. **28**(6), 495–500, 1985].

286. Shevtsov, B. M., 1987, *Statistical characteristics for wave-packet scattering in a layered randomly inhomogeneous medium above a reflecting surface*, Izv. Vuzov., Radiofizika **30**(8), 1007–1012 [Radiophys. & Quantum Electron. **30**(8), 750–754, 1987].

287. Shevtsov, B. M., 1989, *Backscattering of a wave in layered regularly and randomly inhomogeneous media*, Izv. Vuzov., Radiofizika **32**(9), 1079–1083 [Radiophys. & Quantum Electron. **32**(9), 798–801, 1998].

288. Shevtsov, B. M., 1990, *Statistical characteristics of waves in a layered randomly and regularly inhomogeneous medium*, Izv. Vuzov., Radiofizika **33**(2), 191–195 [Radiophys. & Quantum Electron. **33**(2), 146–150, 1990].

289. Shraiman, B. I. and E. D. Siggia, 1994, *Lagrangian path integrals and fluctuations in random flow*, Phys. Rev. **E49**(4), 2912–2927.

290. Sinai, Ya. G. and V. Yakhot, 1989, *Limiting probability distributions of a passive scalar in a random velocity field*, Phys. Rev. Lett. **63**(18), 1962–1964.

291. Spigler, R. J., 1986, *Mean power reflection from a one-dimensional nonlinear random medium*, J. Math. Phys. **27**(7), 1760–1771.

292. Stokes, G. G., 1851, *On the Effect of the Internal Friction of Fluids on the Motion of Pendulums*, Trans. Camb. Phil. Soc. **9**, 8.

293. Swerling, P., 1962, *Statistical properties of the contours of random surfaces*, I.R.E. Tranc. Inf. Theory **IT-8**, 315–321.

294. Tatarskii, V. I., 1967, *Rasprostranenie Voln v Turbulentnoi Atmosfere* (*Wave Propagation in Turbulent Atmosphere*), Nauka, Moscow (in Russian) [*The Effects of the Turbulent Atmosphere on Wave Propagation*, National Technical Information Service, Springfield, Va., 1977].

295. Tatarskii, V. I., 1969, *Light propagation in a medium with random index refraction inhomogeneities in the Markov process approximation*, Zh. Éxper. Teor. Fiz. **56**(6), 2106–2117 [Sov. Phys. JETP **29**(6), 1133–1138, 1969].

296. Tatarskii, V. I., 1967, *The estimation of depolarization of light of a turbulent inhomogenetious of atmosphere,* Izv. Vuzov. Radiofizika **10** (12), 1762–1765

297. Tatarskii, V. I. and V. U. Zavorotny, 1980, *Strong fluctuations in light propagation in a randomly inhomogeneous medium*, Progress in Optics **XVIII**, 204–256, ed. E. Wolf, North-Holland, Amsterdam.

298. Tatarskii, V. I., A. Ishimaru and V. U. Zavorotny, 1993, eds, *Wave propagation in random media (scintillation)*, SPIE, Bellingham, Wa.

299. Tatarskii, V. I., M. I. Charnotskii, J. Gozani and V. U. Zavorotny, 1993, *Path integral approach to wave propagation in random media*, in: Wave propagation in random media (scintillation), 383–441, SPIE, Bellingham, Wa.

300. Taylor, G. I., 1923, *Diffusions by continuous movements*, Proc. London Math. Society, Ser.2, **20**, 196–211.

301. Taylor, G. I., 1938, *The spectrum of turbulence*, Proc. Royal Society, **A164**(919), 476–490.

302. Thomson, R. E., 1982, *The propagation of planetary waves over a random topography*, J. Fluid Mech. **70**(2), 267–285.

303. Uhlenbeck, G. E., 1971, *The fundamental problems of statistical mechanics*, Uspekhi Fiz. Nauk **103**(2), 275–318 [Soviet Phys., Uspekhi].

304. Ungarish, M., 1993, *Hydrodynamics of Suspensions: Fundamentals of Centrifugal and Gravity Separation*, Springer-Verlag, New York.

305. Virovlyanskii, A. L., A. I. Saichev and M. M. Slavinskii, 1985, *Moment functions of waves propagation in waveguides with elongated random inhomogeneities of the refractive index*, Izv. Vuzov., Radiofizika **28**(9), 1149–1159 [Radiophys. & Quantum Electron. **28**(9), 794–803, 1985].

306. Weichman, P. B. and R. E. Glazman, 1999, *Turbulent fluctuation and transport of passive scalar by random wave fields*, Phys. Rev. Letters, **83**(24), 5011–5018.

307. Weichman, P. B. and R. E. Glazman, 2000, *Passive scalar transport by travelling wave fields*, J. Fluid Mech., **420**, 147–200.

308. Weichman, P. B. and R. E. Glazman, 2002, *Spatial variations of a passive tracer in a random wave field*, J. Fluid Mech., **453**, 263–287.

309. White, B. S., 1983, *The stochastic caustic*, SIAM J. Appl. Math. **44**(1), 127–149.

310. Whitham, G. B., 1974, *Linear and Nonlinear Waves*, John Wiley & Sons Inc., New York.

311. Yaroshchuk, I. O.,1984, *On numerical modeling of one-dimensional stochastic wave problems*, Zh. Vichisl. Mat. i Mat. Fiz. **24**(11), 1748–1751 [USSR Comput. Math. and Mathemat. Phys. **24**(11/12), 97–99, 1984].

312. Yaroshchuk, I. O., 1986, *Chislennoe Modelirovanie Rasprostraneniya Ploskikh Voln v Sluchaino Sloistykh Lineinykh i Nelineinykh Sredakh* (*Numerical Modeling of*

313. Yaroshchuk, I. O., 1986a, *Chislennoe Modelirovanie Rasprostraniniya Ploskikh Voln v Sluchaino Sloistykh Lineinykh i Nelineinykh Sredakh* (*Numerical Modeling of the Plane Wave Propagation in Randomly Layered Linear and Nonlinear Media)*, Ph.D dissertation, Pacific Oceanological Institute, Vladivostok.

314. Yaroshchuk, I. O., 1988a, *Numerical modelling of one-dimensional problem of self-action of a wave in a stochastic nonlinear medium*, Izv. Vuzov., Radiofizika **31**(1), 53–60 [Radiophys. & Quantum Electron. **31**(1), 42–48, 1988].

315. Yaroshchuk, I. O., 1988b, *A method for the numerical modeling of waves in a one-dimensional nonlinear medium with random inhomogeneities*, Zh. Vichisl. Mat. i Mat. Fiz. **28**(5), 760–764 [USSR Comput. Math. and Mathemat. Phys. **28**(5/6), 92–95, 1988].

316. Yaroshchuk, I. O. and O. E. Gulin, 2002, *Metod Statisticheskogo Modelirovaniya v zadachah gidroakustiki* (*Statistical Modeling Method for Hydroacoustic Problems*), Dalnauka, Vladivostok.

317. Zambianchi, E. and A. Griffa, 1994, *Effects of finite scales of turbulence on dispersion estimates*, J. Mar. Res. **52**, 129–148.

318. Zavorotny, V. U., V. I. Klyatskin and V. I. Tatarskii, 1977, *Strong fluctuations of the intensity of electromagnetic waves in randomly inhomogeneous media*, Zh. Éksper. Teor. Phys. **73**(2), 481–497 [Sov. Phys. JETP **46**(2), 252–260, 1977].

319. Ziman, J. M., 1979, *Models of Disorder, The theoretical physics of homogeneously disordered systems*, Cambr. Univ. Press, Cambridge.

320. Zirbel, C. L. and E. Çinlar, 1996, *Mass transport by Brownian motion*, in: *Stochastic Models in Geosystems*, eds. S. A. Molchanov and W. A. Woyczynski, IMA Volumes in Mathematics and its Applications, **85**, 459–492, Springer-Verlag, New York.

321. Zwillinger, D. and B. S. White, 1985, *Propagation of initially plane waves in the region of random caustics*, Wave Motion 7(2), 207–227.

Index